U0155980

nature

The Living Record of Science

《自然》学科经典系列

总顾问：李政道（Tsung-Dao Lee）

英方总主编：Sir John Maddox
Sir Philip Campbell　中方总主编：路甬祥

《自然》百年物理经典 I

（英汉对照）

英方主编：Philip Ball　中方主编：赵忠贤

外语教学与研究出版社 · 麦克米伦教育 · 自然科研

FOREIGN LANGUAGE TEACHING AND RESEARCH PRESS · MACMILLAN EDUCATION · NATURE RESEARCH

北京 BEIJING

图书在版编目（CIP）数据

《自然》百年物理经典．Ⅰ：英汉对照 /（英）菲利普·鲍尔（Philip Ball），赵忠贤主编．-- 北京：外语教学与研究出版社，2020.9
（《自然》学科经典系列）
ISBN 978-7-5213-1945-3

Ⅰ．①自… Ⅱ．①菲… ②赵… Ⅲ．①物理学－文集－英、汉 Ⅳ．①O4-53

中国版本图书馆 CIP 数据核字（2020）第 126022 号

地图审图号：GS（2020）3943 号

出 版 人　徐建忠
项目统筹　章思英
项目负责　刘晓楠　王丽霞
责任编辑　王丽霞
责任校对　黄小斌
封面设计　孙莉明　高　蕾
版式设计　孙莉明
出版发行　外语教学与研究出版社
社　　址　北京市西三环北路 19 号（100089）
网　　址　http://www.fltrp.com
印　　刷　北京华联印刷有限公司
开　　本　787×1092　1/16
印　　张　57
版　　次　2020 年 8 月第 1 版　2020 年 8 月第 1 次印刷
书　　号　ISBN 978-7-5213-1945-3
定　　价　568.00 元

购书咨询：（010）88819926　电子邮箱：club@fltrp.com
外研书店：https://waiyants.tmall.com
凡印刷、装订质量问题，请联系我社印制部
联系电话：（010）61207896　电子邮箱：zhijian@fltrp.com
凡侵权、盗版书籍线索，请联系我社法律事务部
举报电话：（010）88817519　电子邮箱：banquan@fltrp.com
物料号：319450001

《自然》学科经典系列

（英汉对照）

《自然》百年物理经典

（英汉对照）

英方主编：Philip Ball　　中方主编：赵忠贤

审稿专家（以姓氏笔画为序）

于　贵	于　渌	马宇蒨	王　琛	王乃彦	邓祖淦	厉光烈
石锦卫	朱永生	朱道本	刘　纯	刘京国	刘朝阳	江丕栋
杜江峰	李　淼	李芝芬	李兴中	李军刚	肖伟科	何香涛
狄增如	汪长征	沈宝莲	宋心琦	张元仲	张泽渤	张焕乔
陆朝阳	陈　方	尚仁成	郑东宁	赵见高	郝　伟	夏海鸿
顾镇南	郭建栋	陶宏杰	曹　俊	曹庆宏	葛墨林	韩汝珊
鲍重光	蔡荣根	翟天瑞	熊秉衡			

翻译工作组稿人（以姓氏笔画为序）

王耀杨	刘　明	何　铭	沈乃澂	郭红锋	黄小斌	蔡　迪

翻译人员（以姓氏笔画为序）

王　锋　王　静　王耀杨　牛慧冲　邓铭瑞　史春晖　刘　霞
刘东亮　安宇森　孙惠南　李　琦　李世媛　李忠伟　吴　彦
何　钧　沈乃澂　金世超　周　杰　孟　洁　胡雪兰　姜　克
姜　薇　钱　磊　高如丽　黄　娆　崔　宁　葛聆沨　韩　然
韩少卿　曾红芳

校对人员（以姓氏笔画为序）

于平蓉　于同旭　马　荣　马晨晨　王　羽　王帅帅　王阳兰
王丽霞　王晓萌　王晓蕾　王赛儿　元旭津　牛慧冲　公　晗
甘秋玲　田胜聪　史未卿　丛　岚　冯　翀　吕秋莎　朱　玥
乔萌萌　刘　明　刘子怡　刘本琼　刘晓楠　齐文静　闫　妍
许梅梅　孙　娟　孙瑞静　杜赛赛　李　芳　李　娟　李　琦
李　景　李　渝　李世媛　李红菊　李若男　李盎然　杨　茜
杨学良　吴　茜　邱彩玉　何　钧　何　铭　何　敏　何思源
邹伯夏　沈乃澂　张　帆　张　敏　张向东　张亦卓　张美月
张竞凤　张梦璇　张琦玮　张媛媛　陈　雄　陈思原　陈露芸
范艳璇　罗小青　周玉凤　郑　琪　郑婧澜　郑期彤　宗伟凯
赵凤轩　胡海霞　侯鉴璇　顾海成　钱　磊　徐　玲　徐秋燕
郭晓博　黄小斌　黄雪嫚　曹则贤　崔天明　梁　瑜　葛　越
葛云霄　葛聆沨　董静娟　韩少卿　曾红芳　曾芃斐　蔡　迪
蔡军茹　Eric Leher（澳）

Foreword by Tsung Dao Lee

We can appreciate the significance of natural science to human life in two aspects. Materially, natural science has achieved many breakthroughs, particularly in the past hundred years or so, which have brought about revolutionary changes to human life. At the same time, the spirit of science has taken an ever-deepening root in the hearts of the people. Instead of alleging that science is omnipotent, the spirit of science emphasizes down-to-earth and scrupulous research, and critical and creative courage. More importantly, it stands for the dedication to working for the wellbeing of humankind. This is perhaps more meaningful than scientific and technological achievements themselves, which may be closely related to specific backgrounds of the times. The spirit of science, on the other hand, constitutes a most valuable and constant component of humankind's spiritual civilization.

In this sense, *Nature: The Living Record of Science* presents not only the historical paths of the various fields of natural science for almost a century and a half, but also the unremitting spirit of numerous scientists in their pursuit of truth. One of the most influential science journals in the whole world, *Nature*, reflects a general picture of different branches of science in different stages of development. It has also reported many of the most important discoveries in modern science. The collection of papers in this series includes breakthroughs such as the special theory of relativity, the maturing of quantum mechanics and the mapping of the human genome sequence. In addition, the editors have not shunned papers which were proved to be wrong after publication. Included also are the academic debates over the relevant topics. This speaks volumes of their vision and broadmindedness. Arduous is the road of science; behind any success are countless failures unknown to outsiders. But such failures have laid the foundation for success in later times and thus should not be forgotten. The comprehensive and thoughtful coverage of these volumes will enable readers to gain a better understanding of the achievements that have tremendously promoted the progress of science and technology, the evolution of key and cutting-edge issues of the relevant fields, the inspiration brought about by academic controversies, the efforts and hardships behind these achievements, and the true meaning of the spirit of science.

China now enjoys unprecedented opportunities for the development of science and technology. At the policy level, the state has created a fine environment for scientific research by formulating medium- and long-term development programs. As for science and technology, development in the past decades has built up a solid foundation of research and a rich pool of talent. Some major topics at present include how to introduce the cream of academic research from abroad, to promote Sino-foreign exchange in science and technology, to further promote the spirit of science, and to raise China's development in this respect to the advanced international level. The co-publication of *Nature: The Living Record of Science* by the Foreign Language Teaching and Research

李政道序

如何认识自然科学对人类生活的意义，可以从两个方面来分析：一是物质层面，尤其是近百年来，自然科学取得了很多跨越性的发展，给人类生活带来了许多革命性的变化；二是精神层面，科学精神日益深入人心，这种科学精神并不是认为科学万能、科学可以解决一切问题，它应该是一种老老实实、严谨缜密、又勇于批判和创造的精神，更重要的是，它具有一种坚持为人类福祉而奋斗的信念。这种科学精神可能比物质意义上的科技成就更重要，因为技术进步的影响可能与时代具体的背景有密切关系，但科学精神却永远是人类精神文明中最可宝贵的一部分。

从这个意义上，这套《〈自然〉百年科学经典》丛书的出版，不仅为读者呈现了一个多世纪以来自然科学各个领域发展的历史轨迹，更重要的是，它展现了无数科学家在追求真理的过程中艰难求索、百折不回的精神世界。《自然》作为全世界最有影响力的科学期刊之一，反映了各个学科在不同发展阶段的概貌，报道了现代科学中最重要的发现。这套丛书的可贵之处在于，它不仅汇聚了狭义相对论的提出、量子理论的成熟、人类基因组测序完成这些具有开创性和突破性的大事件、大成就，还将一些后来被证明是错误的文章囊括进来，并展现了围绕同一论题进行的学术争鸣，这是一种难得的眼光和胸怀。科学之路是艰辛的，成功背后有更多不为人知的失败，前人的失败是我们今日成功的基石，这些努力不应该被忘记。因此，《〈自然〉百年科学经典》这套丛书不但能让读者了解对人类科技进步有着巨大贡献的科学成果，以及科学中的焦点和前沿问题的演变轨迹，更能使有志于科学研究的人感受到思想激辩带来的火花和收获背后的艰苦努力，帮助他们理解科学精神的真意。

当前，中国科学技术的发展面临着历史上前所未有的机遇，国家已经制定了中长期科学和技术发展纲要，为科学研究创造了良好的制度环境，同时中国的科学技术经过多年的积累也已经具备了很好的理论和人才基础。如何进一步引进国外的学术精华，促进中外科技交流，使科学精神深入人心，使中国的科技水平迅速提升至世界前列就成为这一阶段的重要课题。因此，外语教学与研究出版社和麦克米伦出

Press, Macmillan Publishers Limited and the Nature Publishing Group will prove to be a huge contribution to the country's relevant endeavors. I sincerely wish for its success.

Science is a cause that does not have a finishing line, which is exactly the eternal charm of science and the source of inspiration for scientists to explore new frontiers. It is a cause worthy of our uttermost exertion.

T. D. Lee

Editor's note: The foreword was originally written for the ten-volume *Nature: The Living Record of Science.*

版集团合作出版这套《〈自然〉百年科学经典》丛书，对中国的科技发展可谓贡献巨大，我衷心希望这套丛书的出版获得极大成功，促进全民族的科技振兴。

科学的事业永无止境。这是科学的永恒魅力所在，也是我们砥砺自身、不断求索的动力所在。这样的事业，值得我们全力以赴。

李政道

编者注：此篇原为《〈自然〉百年科学经典》（十卷本）的序。

Foreword by Lu Yongxiang

Since the birth of modern science, and in particular throughout the 20th century, we have continuously deepened our understanding of Nature, and developed more means and methods to make use of natural resources. Technological innovation and industrial progress have become decisive factors in promoting unprecedented development of productive forces and the progress of society, and have greatly improved the mode of production and the way we live.

The 20th century witnessed many revolutions in science. The establishment and development of quantum theory and the theory of relativity have changed our concept of time and space, and have given us a unified understanding of matter and energy. They served as a theoretical foundation upon which a series of major scientific discoveries and technological inventions were made. The discovery of the structure of DNA transformed our understanding of heredity and helped to unify our vision of the biological world. As a corner-stone in biology, DNA research has exerted a far-reaching influence on modern agriculture and medicine. The development of information science has provided a theoretical basis for computer science, communication technology, intelligent manufacturing, understanding of human cognition, and even economic and social studies. The theory of continental drift and plate tectonics has had important implications for seismology, geology of ore deposits, palaeontology, and palaeoclimatology. New understandings about the cosmos have enabled us to know in general terms, and also in many details, how elementary particles and chemical elements were formed, and how this led to the formation of molecules and the appearance of life, and even the origin and evolution of the entire universe.

The 20th century also witnessed revolutions in technology. Breakthroughs in fundamental research, coupled to the stimulus of market forces, have led to unparalleled technological achievements. Energy, materials, information, aviation and aeronautics, and biological medicine have undergone dramatic changes. Specifically, new energy technologies have helped to promote social development; new materials technologies promote the growth of manufacturing and industrial prosperity; information technology has ushered in the Internet and the pervasive role of computing; aviation and aeronautical technology has broadened our vision and mobility, and has ultimately led to the exploration of the universe beyond our planet; and improvements in medical and biological technology have enabled people to live much better, healthier lives.

Outstanding achievements in science and technology made in China during its long history have contributed to the survival, development and continuation of the Chinese nation. The country remained ahead of Europe for several hundred years before the 15th century. As Joseph Needham's studies demonstrated, a great many discoveries and innovations in understanding or practical capability—from the shape of snowflakes to the art of cartography, the circulation of the blood, the invention of paper and sericulture

路甬祥序

自近代科学诞生以来，特别是 20 世纪以来，随着人类对自然的认识不断加深，随着人类利用自然资源的手段与方法不断丰富，技术创新、产业进步已成为推动生产力空前发展和人类社会进步的决定性因素，极大地改变了人类的生产与生活方式，使人类社会发生了显著的变化。

20 世纪是科学革命的世纪。量子理论和相对论的创立与发展，改变了人类的时空观和对物质与能量统一性的认识，成为了 20 世纪一系列重大科学发现和技术发明的理论基石；DNA 双螺旋结构模型的建立，标志着人类在揭示生命遗传奥秘方面迈出了具有里程碑意义的一步，奠定了生物技术的基础，对现代农业和医学的发展产生了深远影响；信息科学的发展为计算机科学、通信技术、智能制造提供了知识源泉，并为人类认知、经济学和社会学研究等提供了理论基础；大陆漂移学说和板块构造理论，对地震学、矿床学、古生物地质学、古气候学具有重要的指导作用；新的宇宙演化观念的建立为人们勾画出了基本粒子和化学元素的产生、分子的形成和生命的出现，乃至整个宇宙的起源和演化的图景。

20 世纪也是技术革命的世纪。基础研究的重大突破和市场的强劲拉动，使人类在技术领域获得了前所未有的成就，能源、材料、信息、航空航天、生物医学等领域发生了全新变化。新能源技术为人类社会发展提供了多元化的动力；新材料技术为人类生活和科技进步提供了丰富的物质材料基础，推动了制造业的发展和工业的繁荣；信息技术使人类迈入了信息和网络时代；航空航天技术拓展了人类的活动空间和视野；医学与生物技术的进展极大地提高了人类的生活质量和健康水平。

历史上，中国曾经创造出辉煌的科学技术，支撑了中华民族的生存、发展和延续。在 15 世纪之前的数百年里，中国的科技水平曾遥遥领先于欧洲。李约瑟博士曾经指出，从雪花的形状到绘图的艺术、血液循环、造纸、养蚕，包括更有名的指南针和

and, most famously, of compasses and gunpowder—were first made in China. The Four Great Inventions in ancient China have influenced the development process of the world. Ancient Chinese astronomical records are still used today by astronomers seeking to understand astrophysical phenomena. Thus Chinese as well as other long-standing civilizations in the world deserve to be credited as important sources of modern science and technology.

Scientific and technological revolutions in 17th and 18th century Europe, the First and Second Industrial Revolutions in the 18th and 19th centuries, and the spread of modern science education and knowledge sped up the modernization process of the West. During these centuries, China lagged behind.

Defeat in the Opium War (1840–1842) served as strong warning to the ancient Chinese empire. Around and after the time of the launch of *Nature* in 1869, elite intellectuals in China had come to see the importance that science and technology had towards the country's development. Many scholars went to study in Western higher education and research institutions, and some made outstanding contributions to science. Many students who had completed their studies and research in the West returned to China, and their work, together with that of home colleagues, laid the foundation for the development of modern science and technology in the country.

In the six decades since the founding of the People's Republic of China, the country has made a series of achievements in science and technology. Chinese scientists independently developed the atomic bomb, the hydrogen bomb and artificial satellite within a short period of time. The continental oil generation theory led to the discovery of the Daqing oil field in the northeast. Chinese scientists also succeeded in synthesizing bovine insulin, the first protein to be made by synthetic chemical methods. The development and popularization of hybrid rice strains have significantly increased the yields from rice cultivation, benefiting hundreds of millions of people across the world. Breakthroughs in many other fields, such as materials science, aeronautics and life science, all represent China's progress in modern science and technology.

As the Chinese economy continues to enjoy rapid growth, scientific research is also producing increasing results. Many of these important results have been published in first-class international science journals such as *Nature*. This has expanded the influence of Chinese science research, and promoted exchange and cooperation between Chinese scientists with colleagues in other countries. All these indicate that China has become a significant global force in science and technology and that greater progress is expected in the future.

Science journals, which developed alongside modern science, play an essential role in faithfully recording the path of science, as well as spreading and promoting modern science. Such journals report academic development in a timely manner, provide a platform for scientists to exchange ideas and methods, explore the future direction of science, stimulate academic debates, promote academic prosperity, and help the public

火药，都是首先由中国人发现或发明的。中国的“四大发明”影响了世界的发展进程，古代中国的天文记录至今仍为天文学家在研究天体物理现象时所使用。中华文明同其他悠久的人类文明一样，成为了近代科学技术的重要源泉。

但我们也要清醒地看到，发生在17~18世纪欧洲的科学革命、18~19世纪的第一次和第二次工业革命，以及现代科学教育与知识的传播，加快了西方现代化的进程，同时也拉大了中国与西方的差距。

鸦片战争的失败给古老的中华帝国敲响了警钟。就在《自然》创刊前后，中国的一批精英分子看到了科学技术对于国家发展的重要性，一批批中国学子到西方高校及研究机构学习，其中一些人在科学领域作出了杰出的贡献。同时，一大批留学生回国，同国内的知识分子一道，为现代科学技术在中国的发展奠定了基础。

新中国成立60年来，中国在科学技术方面取得了一系列成就。在很短的时间里，独立自主地研制出“两弹一星”；在陆相生油理论指导下，发现了大庆油田；成功合成了牛胰岛素，这是世界上第一个通过化学方法人工合成的蛋白质；杂交水稻研发及其品种的普及，显著提高了水稻产量，造福了全世界几亿人。中国人在材料科学、航天、生命科学等许多领域，也取得了一批重要成果。这些都展现了中国在现代科技领域所取得的巨大进步。

当前，中国经济持续快速增长，科研产出日益增加，中国的许多重要成果已经发表在像《自然》这样的世界一流的科技期刊上，扩大了中国科学研究的影响，推动了中国科学家和国外同行的交流与合作。现在，中国已成为世界重要的科技力量。可以预见，在未来，中国将在科学和技术方面取得更大的进步。

伴随着现代科学产生的科技期刊，忠实地记录了科学发展的轨迹，在传播和促进现代科学的发展方面发挥了重要的作用。科技期刊及时地报道学术进展，交流科学思想和方法，探讨未来发展方向，以带动学术争鸣与繁荣，促进公众对科学的理解。中国在推动科技进步的同时，应更加重视科技期刊的发展，学习包括《自然》在内

to better understand science. While promoting science and technology, China should place greater emphasis on the betterment of science journals. We should draw on the philosophies and methods of leading science journals such as *Nature*, improve the standards of digital access, and enable some of our own science journals to extend their impact beyond China in the not too distant future so that they can serve as an advanced platform for the development of science and technology in our country.

In the 20th century, *Nature* published many remarkable discoveries in disciplines such as biology, geoscience, environmental science, materials science, and physics. The selection and publication of the best of the more than 100,000 articles in *Nature* over the past 150 years or so in English-Chinese bilingual format is a highly meaningful joint undertaking by the Foreign Language Teaching and Research Press, Macmillan Publishers Limited and the Nature Publishing Group. I believe that *Nature: The Living Record of Science* will help bridge cultural differences, promote international cooperation in science and technology, prove to be high-standard readings for its intended large audience, and play a positive role in improving scientific and technological research in our country. I fully endorse and support the project.

The volumes offer a picture of the course of science for nearly 150 years, from which we can explore how science develops, draw inspiration for new ideas and wisdom, and learn from the unremitting spirit of scientists in research. Reading these articles is like vicariously experiencing the great discoveries by scientific giants in the past, which will enable us to see wider, think deeper, work better, and aim higher. I believe this collection will also help interested readers from other walks of life to gain a better understanding of and care more about science, thus increasing their respect for and confidence in science.

I should like to take this opportunity to express my appreciation for the vision and joint efforts of Foreign Language Teaching and Research Press, Macmillan Publishers Limited and the Nature Publishing Group in bringing forth this monumental work, and my thanks to all the translators, reviewers and editors for their exertions in maintaining its high quality.

President of Chinese Academy of Sciences

Editor's note: The foreword was originally written for the ten-volume *Nature: The Living Record of Science.*

的世界先进科技期刊的办刊理念和方法，提高期刊的数字化水平，使中国的一些科技期刊早日具备世界影响力，为中国科学技术的发展创建高水平的平台。

20 世纪的生物学、地球科学、环境科学、材料科学和物理学等领域的许多重大发现，都被记录在《自然》上。外语教学与研究出版社、麦克米伦出版集团和自然出版集团携手合作，从《自然》创刊近一百五十年来发表过的十万余篇论文中撷取精华，并译成中文，以双语的形式呈现，纂为《〈自然〉百年科学经典》丛书。我认为这是一项很有意义的工作，并相信本套丛书的出版将跨越不同的文化，促进国际间的科技交流，向广大中国读者提供高水平的科学技术知识文献，为提升我国科学技术研发水平发挥积极的作用。我赞成并积极支持此项工作。

丛书将带领我们回顾近一百五十年来科学的发展历程，从中探索科学发展的规律，寻求思想和智慧的启迪，感受科学家们百折不挠的钻研精神。阅读这套丛书，读者可以重温科学史上一些科学巨匠作出重大科学发现的历程，拓宽视野，拓展思路，提升科研能力，提高科学道德。我相信，这套丛书一定能成为社会各界的良师益友，增强他们对科学的了解与热情，加深他们对科学的尊重与信心。

借此机会向外语教学与研究出版社、麦克米伦出版集团、自然出版集团策划出版本丛书的眼光和魄力表示赞赏，对翻译者、审校者和编辑者为保证丛书质量付出的辛苦劳动表示感谢。

是为序。

路甬祥

中国科学院院长

编者注：此篇原为《〈自然〉百年科学经典》（十卷本）的序。

Foreword by Philip Ball

The papers in these two volumes are drawn from the selection that appeared in *Nature: The Living Record of Science*, a multi-volume compilation of the most important contributions to the international science journal *Nature* from 1869, when it began, until 2007. Progress in Physics collects together the papers specifically in the discipline of physics, ranging from the theoretical foundations of the discipline to astronomy and applications in areas such as semiconductor microelectronics. *Nature* has had a particularly strong tradition of publishing in this field; in the early part of the twentieth century especially, it was the regular journal of choice for physicists presenting their new discoveries. This selection therefore offers a picture of how physics has changed from the era of James Clerk Maxwell and Lord Kelvin to today. These papers are made readily accessible here to readers in China for the first time by simultaneous publication in English and in Mandarin Chinese, as well as being accompanied by short introductions that explain the context and implications of the work described.

When *Nature* began, physics was still relatively new as a recognized discipline. Of course, what we now regard as physics has a much longer history. Aristotle's wide-ranging treatise on the natural world in the 4th century BCE is simply given that one-word title today, and the studies of the laws of motion by Galileo and Isaac Newton in the seventeenth century sit at the heart of the so-called Scientific Revolution. In a sense, one might argue that it was the character of physics as the foundation for all natural philosophy that delayed for so long its being recognized as a distinct discipline: physics seems to be about everything. It is often noted that the laws governing the interactions of atoms and fundamental particles are ultimately responsible for all of chemistry and biology, and that physical law also defines and constrains our view of the entire cosmos and how it has evolved since the universe began 13.8 billion years ago.

But the recognition that physics needs to be both pursued as a distinct fundamental and experimental science, and that its applications are central to the development of socially transformative technologies, was reflected in the inauguration, just a few years after *Nature* began, of the Cavendish Laboratory at the University of Cambridge—which, with Maxwell as its first director, quickly became and has remained one of the most important centres of academic study in the discipline. Maxwell's kinetic theory of gases was presented and discussed in the journal in 1873.

In the late nineteenth century, some of the important contributions to physics in *Nature* concerned questions in astronomy: from where, for example, do stars like the sun get their enormous, seemingly inexhaustible energy? It was only through the discoveries in

菲利普·鲍尔序

《〈自然〉百年物理经典》（共两册）的文章皆出自十卷本论文精选集《〈自然〉百年科学经典》，后者收录了国际学术期刊《自然》自 1869 年创刊至 2007 年发表过的最具影响力的经典文献。《〈自然〉百年物理经典》以物理学为线索，收录了从基础理论物理到天文学和半导体微电子学等应用物理方向的文章。在学术出版领域，物理学是《自然》的传统优势学科，特别是在二十世纪早期，物理学家更是倾向于将自己的最新成果刊发在《自然》上。《〈自然〉百年物理经典》展现了物理学从詹姆斯·克拉克·麦克斯韦和开尔文勋爵时代发展至今的光辉历程。这些经典文献首次以中英对照的形式出版，每篇同时配有简短的导读，以便中国读者快速了解文献的内容和研究的意义。

《自然》创刊伊始，物理学作为独立的学科，尚处于起步阶段。诚然，如今人们所说的物理学有着更为悠久的历史。公元前四世纪，亚里士多德便为自己的一部论述自然界万物的著作冠以了"物理学"之名。十七世纪，伽利略和艾萨克·牛顿对于物体运动法则的研究，也成为后世所说的科学革命的核心所在。在某种意义上，人们或许可以说，物理是一切自然哲学之基础。正是这一特征，使得物理长久以来未能被人们确立为一门独立的学科：似乎世间万物之道皆归于物理。我们经常可以发现，支配原子和基本粒子相互作用的定律同样适用于化学和生物学，而物理学定律也定义和约束了我们对整个宇宙的理解，以及自 138 亿年前宇宙诞生以来它是如何演化的。

物理学的研究应该从理论基础和实验观察两个方面着手，同时，其应用对于社会的革命性技术的发展也起到了至关重要的作用。上述两点认识在剑桥大学卡文迪什实验室的创立（《自然》创刊几年之后）中得到了印证。麦克斯韦担任实验室的首位主任，之后实验室迅速成长为世界上最重要的物理学学术研究中心之一。1873 年，《自然》上就曾刊发并讨论过麦克斯韦的气体动力学理论。

十九世纪末，《自然》上刊发了多篇重要论文，所涉及的问题也是天文学关心的问题：例如，太阳那样的恒星是从哪里获得巨大的且看似无穷无尽的能量的？回答

fundamental physics—the identification of X-rays in 1895 and then, very soon after, of radioactivity—that an answer to this question began to emerge, via the discovery of the energy residing in the atomic nucleus. And the particles emitted by radioactive decay could be used to probe the internal structure of atoms themselves, leading to a rationalization of the properties of the chemical elements as well as to the invention of the first of the devices—particle accelerators—that today supply the principal tools for exploring the laws that govern the properties and composition of matter at scales far smaller than those of atoms.

That we needed to understand the atomic nucleus before we could understand the sun is a perfect illustration of the unity that physics often reveals. It is now clear that understanding the cosmos at the grandest scales requires deeper theoretical insight into its constitution at the smallest. In fact, a unification of the theories that provide the conceptual frameworks at the former and the latter scales—general relativity and quantum mechanics, both of which developed in the early twentieth century (and both being discussed in *Nature* at that time)—now represents one of the most important goals in physics.

It is sometimes said that physics was the foremost science of the first half of the twentieth century, while biology claimed that role in the second half. But much of the current understanding in the life sciences has been, and continues to be, dependent on advances in physics. X-rays become the primary tool for probing the structure of proteins and other biomolecules, most notably revealing the chemical nature of DNA in 1953. Measuring signals from magnetic atomic nuclei in the technique of nuclear magnetic resonance has also been vital for studying the structure of biological matter. The use of this technique for imaging, reported in *Nature* in 1973 and rewarded with a Nobel prize, is now essential for advances in neuroscience and for biomedical research and clinical medicine.

Nature followed closely developments in the application of physics, not least in telecommunications technologies such as the telephone and television. The equivalent today is the application of physics to microelectronics, leading first to digital solid-state integrated circuits and computers and now to the implementation of quantum-mechanical rules of information processing in quantum computing and cryptography. Information technology and artificial intelligence look sure to become one of the most transformative physics-based technologies of the twenty-first century.

Well before achieving its current global prominence, China established a solid presence in physics. Chen-Ning Yang and Tsung-Dao Lee won the 1957 Nobel prize in physics for showing theoretically how the conservation principle governing the property of parity—a kind of symmetry exhibited by fundamental particles—could be violated. They proposed an experiment for testing their ideas using nuclear physics, specifically by looking at the particles emitted in a radioactive decay process. This experiment was conducted in 1956

上述问题有赖于基础物理学的相关发现——1895 年 X 射线的发现及紧随其后放射性的发现，以及随后关于原子核内能量的研究。放射性衰变放出的粒子可用来探究原子自身的内部结构，从而使化学元素的性质得到合理化的解释，并使得粒子加速器等设备的建造得以实现。今天，粒子加速器是探索远小于原子尺度的物质的性质和构成规律的主要实验设备。

了解太阳之前需要先了解原子核，这很好地阐释了物理学中经常揭示的统一性。在极大尺度上理解宇宙需要在极小尺度上对其构成有更为深入的理论研究，这点现在已经很清楚。事实上，为极大尺度和极小尺度分别提供了概念框架的理论——广义相对论和量子力学——都是在二十世纪初被提出的（当时也在《自然》上引发了讨论），而统一上述两种理论现已成为物理学最重要的目标之一。

有时，人们会说二十世纪前半叶是物理学取得空前发展的时代，二十世纪后半叶是生物学蓬勃兴起的时代。但关于生命科学的很多理解都曾经是，也一直是基于物理学的进展。X 射线成为测定蛋白质结构和其他生物分子结构的主要技术，最著名的发现是在 1953 年揭示了 DNA 的化学本质。利用核磁共振技术测量来自磁性原子核的信号也是研究生物物质结构非常重要的方法。1973 年发表在《自然》上的关于利用该技术进行成像分析的研究工作之后也获得了诺贝尔奖，核磁共振成像现已成为神经科学、生物医学和临床医学研究必不可少的手段。

《自然》也密切跟踪物理学在实际应用领域的最新进展，其中最为著名的当属电话、电视等电信技术。当今物理学的应用领域主要集中在微电子学：从数字固态集成电路和计算机的实现，到如今在量子计算和量子密码学领域内基于量子力学基本原理对信息进行处理。植根于物理学的信息技术和人工智能似乎正在成长为引领二十一世纪技术变革的中坚力量。

中国在各行各业取得了举世瞩目的辉煌成就，在物理学领域亦打下了坚实的基础。杨振宁和李政道从理论上推翻了宇称守恒定律从而荣获 1957 年的诺贝尔物理学奖，宇称是基本粒子表现出来的一种对称性。他们提出了通过核物理的实验，具体

by Chinese-born physicist Chien-Shiung Wu, and it was her success in verifying Lee and Yang's predictions that opened the door for their Nobel prize. If it were being awarded today, Wu would surely have shared it.

Among the more recent Chinese-born Nobel laureates is Daniel Chee Tsui, whose award in 1998 recognized his work in identifying a fundamental quantum-mechanical effect in semiconductor microstructures. That blend of quantum fundamentals and device engineering, with potentially powerful applications, is found too in the work in China today on quantum information technologies, an area in which Chinese scientists have become world leaders. The Micius satellite, launched in 2016 and operated by the Chinese Academy of Sciences, is the first of its kind dedicated to the transmission of quantum-encrypted data for secure telecommunications, and is a key component of the project Quantum Experiments at Space Scale (QUESS).

This collection of physics papers from *Nature* will be valuable for physicists both in academia and industry, and for science historians. The short editorial introductions will also help to make the papers accessible to and useful for students of physics, as well as to more general audiences. It is a necessarily incomplete sample of physics in from the mid-nineteenth to the early twenty-first centuries, but we hope that the selection of papers here will contain something to satisfy and interest all tastes. It includes papers of historical interest as well as those whose conclusions remain relevant to the subject today. What emerges, we hope, is more than a record of a discipline; it is a portrait of the development of some of the most profound ideas in intellectual global culture over the past 150 years.

Former Consultant Editor for *Nature*

来说即通过观察放射性衰变过程中释放出来的粒子来验证他们的理论。这个实验于 1956 年由华裔女科学家吴健雄领导完成，她成功验证了李政道和杨振宁的设想，从而帮助两人成功斩获诺贝尔奖。现在回首这段历史，吴健雄无疑也应该分享这一荣誉。

华裔科学家崔琦因发现半导体微结构中的基本量子力学效应（分数量子霍尔效应）而荣获 1998 年诺贝尔物理学奖。今天，中国的量子信息技术研究融合了量子原理和器件工程学，具有无限的应用潜力，这也是一个中国科学研究领先世界水平的领域。由中国科学院联合研究团队研制的“墨子号”量子卫星于 2016 年成功发射，这是首颗用于传输量子加密数据以实现安全通信的卫星，也是中国空间尺度量子实验项目（QUESS）的核心部分。

这套《自然》杂志物理论文选集的出版，无论对学术界和产业界的物理学家，抑或是科学史研究人员，都极具参考价值。简短的编辑导读也会帮助物理专业的学生及大众读者更好地理解和运用这些经典文献。这套选集或许并不能完整地呈现二十世纪中期至二十一世纪初所有物理学的研究成果，但我们仍期望各位读者能从中找到自己喜欢并感兴趣的内容。为此，这套选集不仅收录了那些具有历史价值的文献，同时也收录了那些研究结论至今仍具有重要意义的文章。我们希望，呈现在诸位面前的，不仅仅是一部物理学的编年史，更是一幅荟萃过去 150 年全人类智慧文明中最为深刻思想的灿烂画卷。

菲利普 · 鲍尔

《自然》前任顾问编辑

Contents
目　　录

1928

1930

1931

1933

1934

1935

1938

1939

Volume I

The Atomic Controversy

Editor's Note

In 1869, the atomic hypothesis of the constitution of matter, as proposed 60 years earlier by the Englishman John Dalton, had yet to receive definitive confirmation. This short essay reviews arguments for and against the atomic hypothesis which, as Dalton had pointed out, lends order to many otherwise puzzling phenomena. It can explain the law of definite proportion: the fact that chemical compounds always involve elements in simple rational proportions. And the hypothesis had been shown to explain why atomic elements in the gaseous state all had similar heat capacities (dependence of temperature on heat input). Even so, the essay concludes that such arguments for atomism remained circumstantial.

IT is one of the most remarkable circumstances in the history of men, that they should in all times have sought the solution of human problems in the heavens rather than upon the earth. Sixty years ago a memorable instance of this truth occurred when Dalton borrowed from the stars an explanation of the fundamental phenomena of chemical combination. Carbon and oxygen unite in a certain proportion to form "carbonic acid"; and this proportion is found to be invariable, no matter from what source the compound may have been prepared. But carbon and oxygen form one other combination, namely, "carbonic oxide"—the gas whose delicate blue flame we often see in our fires. Carbonic oxide may be obtained from many sources; but, like carbonic acid, its composition is always exactly the same. These two bodies, then, illustrate the law of *Definite* Proportions. But Dalton went a step further. He found that, for the same weight of carbon, the amount of oxygen in "carbonic acid" was *double* that which exists in carbonic oxide. Several similar instances were found of two elements forming compounds in which, while the weight of the one remained constant, the other doubled, trebled, or quadrupled itself. Hence the law of *Multiple* Proportions. The question was—in fact, the question is—how to account for these laws. Dalton soon persuaded himself that matter was made up of very small particles or *minima naturae*, not by any possibility to be reduced to a smaller magnitude. Matter could not be divisible without limit; there must be a barrier somewhere. No doubt, as a chemist, he would have rejected the famous couplet—

Big fleas have little fleas, upon their backs, to bite 'em;
And little fleas have smaller fleas, and so *ad infinitum*.

"Let the divisions be ever so minute," he said, "the number of particles must be finite; just as in a given space of the universe, the number of stars and planets cannot be infinite. We might as well attempt to introduce a new planet into the solar system, or to annihilate one already in existence, as to create or destroy a particle of hydrogen." All substances, then, are composed of atoms; and these attract each other, but at the same time keep their distance, just as is the case with the heavenly bodies. The atoms of one compound

原子论战

编者按

1869年，关于物质组成的原子理论仍旧没有得到确定性的验证，尽管这一理论已经由英国人约翰·道尔顿提出达60年之久。这篇短文综述了人们对原子理论的各种支持意见和反对意见。正如道尔顿当初说的，原子理论揭示了许多分散的、令人困惑的现象背后的规律。原子理论可以解释定比定律（自然界所有化合物中的各种组成元素之间总是存在简单的比例），通过原子理论也可以解释为什么各种元素在气态时都具有相似的比热容（温度变化对于供热的依赖程度）。不过，本文最后还是认为，关于原子理论的这些论述都只是描述性的，尚未得到证实。

人类历史上最值得注意的情况就是，我们在任何时候都应该从天空而不是从大地去探求人类问题的解决之道。60年前发生了一件证实这个真理的令人难忘的事例，道尔顿先生藉由恒星获得灵感，解释了化合物组成的基本现象。碳和氧以一定的比例结合形成"碳酸"，不管是用什么原料来制备碳酸这种化合物，其中碳和氧的比例都是恒定不变的。但是碳和氧还能形成另一种叫作"一氧化碳"的化合物，这种气体的火焰是蓝色的，就像我们经常在火中见到的那样。尽管一氧化碳来源广泛，但和碳酸一样，不同来源的一氧化碳在组成上总是完全相同的。这两个例子说明了什么是**定比**定律。但是道尔顿的研究更进了一步，他发现对于一定质量的碳，"碳酸"中氧的量是一氧化碳中的**两倍**。还有几个类似的例子，由两种元素组成的多种化合物，当其中一种元素的质量一定时，另外一种元素的量可能相差一倍、两倍甚至三倍，于是道尔顿提出了**倍比**定律。实际上，问题在于如何解释这些定律。他很快作出猜测：物质是由很小的颗粒或者**不可再分的微粒**组成的，这些颗粒不能再被分解成更小尺度的颗粒。物质不能被无限分割，肯定存在某种界限。作为一名化学家，他应该会批判下面这首著名的小诗：

大蚤生小蚤，蝇蝇啮其背，
小蚤复微蚤，夫何有止归。

他说："即使分割得非常小，微粒的数目也一定是有限的，就像在一定的宇宙空间内行星和恒星的数目是有限的一样。我们可以试图向太阳系中引入一颗新的行星，或者毁掉其中一颗已存在的行星，同样地，我们也可以创造或剔除一个氢粒子。"所有的物质都是由原子构成的，这些原子就像天体一样相互吸引而又保持一定的距离。不同化合物中原子的质量、大小以及相互间的引力是不同的。但因为原子是不可再

do not resemble those of another in weight, or size, or mutually gravitating power. But as they are indivisible, it is between them that we must conceive all chemical action to take place; and an atom of any particular kind must always have the same weight. The atom of carbon weighs 5; the atom of oxygen weighs 7. Carbonic oxide, containing one of each must therefore be invariably constituted of 5 carbon, and 7 oxygen: carbonic acid must in like manner contain 5 carbon, and 14 oxygen. Here, then, Dalton not only states that he has accounted for the two laws we have mentioned by making a single assumption; but he evidently intends his theory to be used as a criterion or control in all future analytical results, and already views it as the birth-place of chemical enterprise.

Such, and so great, was the atomic theory of Dalton; founded, certainly, on erroneous numbers, but containing in itself the germ of their correction; aspiring to the command in innumerable conquests; and setting itself for the rise or fall of the chemical spirit.

It is hardly necessary to make any detailed review of the history of the atomic theory. Berzelius made it a starting-point for researches which, on the whole, have been unsurpassed in their practical importance, and engrafted upon it his celebrated electrical doctrine. Davy and Faraday refused to admit it; Laurent and Gerhardt accepted it doubtfully, or in a much modified form. Henry declared that it did not rest on an inductive basis. There can be no doubt, however, that the atomic theory has been accepted by the majority of chemists, as may be seen on even a cursory inspection of the current literature of their science. Our present intention is to give such a summary of the atomic question as may be serviceable to those who take an interest in the discussion at the Chemical Society on Thursday last.

The modern supporters of the atomic theory agree with Dalton in the fundamental suppositions we have given above; but assert that they have a much stronger case. The phenomena of gaseous combination and specific heat have indeed changed the numerical aspect of the theory, but not its substance. The simplicity of all the results we have accumulated with respect to combining proportions is itself a great argument for the existence of atoms. They all, for example, have the same capacity for heat; they all, when in the gaseous state, have a volume which is an even multiple of that of one part by weight of hydrogen. But bodies in the free or uncombined state—such, in fact, as we *see* them—more commonly consist of many clusters of atoms (*molecules*) than of simple atoms. These molecules are determined by the fact that when in the gaseous state they all have the same volume. Again, select a series of chemical equations, in which water is formed, and eliminate between them so as to obtain the smallest proportion of water, taking part in the transformations they represent. It will be found that the number is 18; which necessarily involves the supposition that the oxygen (16) in water (18) is an indivisible quantity. To put this last point another way: hydrochloric acid, if treated with soda, no matter in what amount, only forms one compound (common salt). Now we know that the action in this case consists in the exchange of hydrogen for sodium. But if hydrogen were infinitely divisible, we ought to be able to effect an inexhaustible number of such exchanges, and produce an interminable variety of compounds of hydrogen, sodium, and chlorine;

分的，所以我们必须设想所有的化学反应发生在原子之间，并且一种特定原子的质量是不变的。假设碳原子的质量是 5，氧原子的质量是 7。一氧化碳分子包含一个碳原子和一个氧原子，因此，它的质量肯定恒定为一个碳（5）与一个氧（7）之和。同理，碳酸分子的质量一定为一个碳（5）与两个氧（14）之和。由此，道尔顿通过一个简单的假设解释了前面提到的两个定律，同时，他显然还希望他的理论被当作将来所有分析结果的标准或参照，并且他已把这视为化学工业的起源。

这就是道尔顿伟大的原子论。虽然它建立在错误的数据之上，但其中包含了正确理论的萌芽。它立志征服无数的困难，引领化学精神的荣衰。

我们就不去回顾原子论的历史细节了。伯齐利厄斯以原子论为起点进行了研究，总体上来说，他的这些研究的应用价值还未被超越，他还在原子论的基础上衍生出了著名的电子学说。戴维和法拉第拒绝接受原子论。洛朗和凯哈德则半信半疑，确切地说是认可经过了很大修正后的原子论。亨利则声称原子论并非建立在归纳学的基础之上。然而，毫无疑问原子论已被大部分科学家接受，这只要大致浏览一下目前这方面的科学文献就知道了。这里我们简单总结一下关于原子论的探讨，以供那些对上周四化学学会举行的讨论感兴趣的读者参考。

原子论的现代支持者同意前面所述的道尔顿的基本假设，同时他们声称拥有说服力更强的证据。气体化合现象和比热方面的研究确实改变了原子论的数值形式，但并没有改变其本质。我们积累的有关化合比例的所有结果都非常简单，这本身就是原子存在的强力论据。例如，它们都具有同样的热容；处于气态时它们的体积都是单位质量氢气体积的偶数倍。但实际上，就像我们**看到**的，游离态或非化合态的物质在大多数情况下是由许多原子团（**分子**）而非单个原子组成。实验表明，处于气态时它们具有相同的体积，这就证明了分子的存在。再者，我们挑选一些有水生成的化学方程式，通过比较、消约可以推算出水的最小组分，我们发现其分子量是 18。根据这一结果，很自然地可以推测：氧（16）在水（18）中是个不可分割的量。另外一个例子也能够说明这个观点：如果向盐酸中加入苏打，无论苏打的用量是多少，反应都只能产生一种化合物（食盐）。现在我们知道这个例子中发生的反应就是氢被钠替换了。但如果假设氢是无限可分的话，那么通过控制用量应该能够产生多种多样的交换，从而产生无数种由氢、氯、钠形成的化合物，盐酸只是反应这一侧的终点，食盐（氯化钠）是另一侧的终点。可是这种现象从来都没有发生过。可以肯定，物质要么是无限可分的要么是有限可分的，既然现在已经证明不是前者，那么必然

hydrochloric acid being the limit on the one side, and common salt (sodic chloride) terminating the other. No such phenomenon occurs; and, since matter must be infinitely or finitely divisible, and has been thus proved not to be the former, it must be the latter. Atoms, therefore, really exist; and chemical combination is inconsistent with any other supposition. Those who hold the contrary opinion are bound to produce an alternative theory, which shall explain the facts in some better way.

Now let us hear the plaintiff in reply.

The atomic theory has undoubtedly been of great service to science, since the laws of definite and multiple proportions would probably not have received the attention they deserve, but for being stated in terms of that theory. Yet we must discriminate between these laws, which are the simple expression of experimental facts, and the assumption of atoms, which preceded them historically, and therefore has no necessary connection with them. For it was the Greek atomic theory which Dalton revived. Nor has any substance yet been produced by the atomists, which we cannot find means to divide. If, moreover, we have no alternative but to admit the infinite divisibility of matter, even that is consistent with the simple ratios in which bodies combine; for two or more infinites may have a finite ratio. Therefore, the observed simplicity, if used as an argument, cuts both ways. Possibly we are mistaken in connecting the ideas of matter and division at all; at any rate, the connection has never been justified by the opposite side. Again, admitting the argument based on the formation of common salt, the atomic theory does not tell us why only one third of the hydrogen in tartaric acid can be exchanged for sodium; why, indeed, only a fraction of the hydrogen in most organic substances can be so exchanged. Yet, the explanation of the one fact, when discovered, will evidently include that of the other. On the whole, it appears that the atomic theory demands from us a belief in the existence of a limit to division. No such limit has been exhibited to our senses; and the facts themselves do not raise the idea of a limit, which Dalton really borrowed from philosophy. The apparent simplicity of chemical union we do not profess to explain, but to be waiting for any experimental interpretation that may arise. The atomists, in bringing forward their theory, are bound to establish it, and with them lies the *onus probandi*.

The above are a few broad outlines of the existing aspect of the atomic controversy, and may somewhat assist in forming an estimate of it. The general theoretical tone of the discussion last Thursday must have surprised most who were present. Our own position is necessarily an impartial one; but it will probably be agreed that between the contending parties there is a gulf, deeper and wider than at first appears, and perhaps unprovided with a bridge.

(**1**, 44-45; 1869)

只能是后者了。因此，可以肯定原子是确实存在的，并且化合反应的结果与其他任何假设都是不相吻合的。那些持有相反意见的人，必须提出另一种可以更好地解释这些现象的理论。

现在让我们听听反对者的意见。

毫无疑问原子论对科学是极有帮助的，因为如果不是用原子论的形式表述出来，倍比定律和定比定律的法则恐怕就不会受到应得的重视。但是，我们必须将这两大法则与原子假设加以区分，前者是对实验事实的简单描述，而后者在历史上先于两大法则出现，因此它们之间并无必然联系。道尔顿接受的只是古希腊的原子论而已。原子论者从来都没有得到过无法分割的物质。再者，如果我们别无选择而只能承认物质的无限可分性的话，那么这与物质化合中的简单比例也是相符合的，因为两种或两种以上的无限可分物之间也可以有一个有限的确定的比例。因此，就化合过程中观察到的简单比例来说，两种理论都能解释。很可能，我们把物质和分割这两个概念联系起来本身就是完全错误的。至少从来没有证明这种联系的正确性。另外，原子论虽然能够解释食盐形成的现象，但它并未告诉我们为什么酒石酸中只有三分之一的氢可以与钠交换，还有为什么大多数有机物中只有一部分氢可以这样交换。作为一种正确的理论，不但要能解释已经发现的某种现象，同样也要能够解释其他的现象。总体来说，原子论似乎就是要求我们坚信存在分割的极限，但我们从来没有感到过这种极限的存在,事实本身也并没有使人们想到极限这一概念。可以说，道尔顿只是从哲学那里借来了分割的极限这个想法。我们并未声称已经解释了化学结合过程中明显的简单比例，这要等待可能出现的实验性的解释。原子论者自从提出他们理论的那天起就背负着提供证据的责任，他们必须证实这一理论。

上面概述了目前原子论战的各方的观点，可能会有助于人们形成自己对原子论的评价。上周四的讨论中众人的观点可能使一些与会人员受惊了。我们的立场必须保持中立，但是有一点大家达成了共识，就是在论战双方之间存在着一条鸿沟，它比刚出现时更深、更宽，或许根本无法架通。

（高如丽 翻译；汪长征 审稿）

Mathematical and Physical Science

J. C. Maxwell

Editor's Note

James Clerk Maxwell, probably the most perceptive physical scientist in the second half of the nineteenth century, was president of the physics section of the British Association for the Advancement of Science in 1870. His presidential address, he explained, was stimulated by the previous year's address by the mathematician J. J. Cayley. It dealt with the role of molecules in the explanation of physical phenomena.

The president delivered the following address:—

At several of the recent meetings of the British Association the varied and important business of the Mathematical and Physical Section has been introduced by an Address, the subject of which has been left to the selection of the president for the time being. The perplexing duty of choosing a subject has not, however, fallen to me. Professor Sylvester, the president of Section A at the Exeter meeting, gave us a noble vindication of pure mathematics by laying bare, as it were, the very working of the mathematical mind, and setting before us, not the array of symbols and brackets which form the armoury of the mathematician, or the dry results which are only the monuments of his conquests, but the mathematician himself, with all his human faculties directed by his professional sagacity to the pursuit, apprehension, and exhibition of that ideal harmony which he feels to be the root of all knowledge, the fountain of all pleasure, and the condition of all action. The mathematician has, above all things, an eye for symmetry; and Professor Sylvester has not only recognised the symmetry formed by the combination of his own subject with those of the former presidents, but has pointed out the duties of his successor in the following characteristic note:—

"Mr. Spottiswoode favoured the Section, in his opening address, with a combined history of the progress of mathematics and physics; Dr. Tyndall's address was virtually on the limits of physical philosophy; the one here in print," says Professor Sylvester, "is an attempted faint adumbration of the nature of mathematical science in the abstract. What is wanting (like a fourth sphere resting on three others in contact) to build up the ideal pyramid is a discourse on the relation of the two branches (mathematics and physics) to, and their action and reaction upon, one another—a magnificent theme, with which it is to be hoped that some future president of Section A will crown the edifice, and make the tetralogy (symbolisable by A+A', A, A', AA') complete."

The theme thus distinctly laid down for his successor by our late President is indeed a magnificent one, far too magnificent for any efforts of mine to realise. I have endeavoured

数学和物理科学

麦克斯韦

编者按

詹姆斯·克拉克·麦克斯韦可以算是19世纪后半叶最敏锐的物理学家了，他是英国科学促进会1870年会议的物理分会主席。他认为自己的主席演说受到了数学家凯莱在前一年发表的演说的启发。这篇演讲稿论述了在解释物理现象时分子所起到的作用。

主席发表了如下演讲：

在最近几次英国科学促进会的会议上，有一次演讲指出了数学和物理分会中各种重要的事务，因而这一演讲主题被纳入了现任主席的选择。不过，选择主题这一艰巨的任务并没有落在我身上。埃克塞特会议中担任第一分会主席的西尔维斯特教授为我们带来了对纯数学的庄严拥护，通过直截了当地提出数学头脑所从事的真正工作，他呈现给我们的不仅仅是从数学家的武器库中取出的符号和括号的堆砌品，也不仅仅是只有数学家自得其乐的干巴巴的结果，而是一位数学家本身，在职业睿智的指引下，将全部的个人才华都用于追求、理解和展示在他看来是一切知识之本、快乐之源与行为之因的理想的和谐一致。最重要的是，数学家有着调和的眼光；西尔维斯特教授不仅考虑到将他本人的主题与此前各位主席所演讲的主题相结合，还在下面这段典型的文字中指出了后继者的职责：

“斯波蒂斯伍德先生在他的公开演讲中称这一分会为数学与物理发展历史的结合；廷德尔博士的演讲将其本质界定为物理哲学；在这份出版物中，”西尔维斯特教授讲到，“它是一种尝试性的、对于数学科学抽象本质的模糊概括。建立这座理想的金字塔还需要的（如同置于三个彼此接触的球之上的第四个球）是一场涉及两大分支学科（数学和物理）的关系及其相互作用与反作用的讨论——这是一个宏大的主题，期望第一分会未来的某位主席能够为这座大厦剪彩，并完成这出四部曲（可以符号化地表示为A+A'，A，A'，AA'）。”

我们的前任主席为其继任者清晰规划出的这一主题的确是宏伟的，以至于尽我所有的努力都无法将其实现。我曾努力追随斯波蒂斯伍德先生，他以深远的见地将

to follow Mr. Spottiswoode, as with far-reaching vision he distinguishes the systems of science into which phenomena, our knowledge of which is still in the nebulous stage, are growing. I have been carried by the penetrating insight and forcible expression of Dr. Tyndall into that sanctuary of minuteness and of power where molecules obey the laws of their existence, clash together in fierce collision, or grapple in yet more fierce embrace, building up in secret the forms of visible things. I have been guided by Professor Sylvester towards those serene heights

> "Where never creeps a cloud, or moves a wind,
> Nor ever falls the least white star of snow,
> Nor ever lowest roll of thunder moans,
> Nor sound of human sorrow mounts, to mar
> Their sacred everlasting calm."

But who will lead me into that still more hidden and dimmer region where Thought weds Fact; where the mental operation of the mathematician and the physical action of the molecules are seen in their true relation? Does not the way to it pass through the very den of the metaphysician, strewed with the remains of former explorers, and abhorred by every man of science? It would indeed be a foolhardy adventure for me to take up the valuable time of the section by leading you into those speculations which require, as we know, thousands of years even to shape themselves intelligibly.

But we are met as cultivators of mathematics and physics. In our daily work we are led up to questions the same in kind with those of metaphysics; and we approach them, not trusting to the native penetrating power of our own minds, but trained by a long-continued adjustment of our modes of thought to the facts of external nature. As mathematicians, we perform certain mental operations on the symbols of number or of quantity, and, by proceeding step by step from more simple to more complex operations, we are enabled to express the same thing in many different forms. The equivalence of these different forms, though a necessary consequence of self-evident axioms, is not always, to our minds, self-evident; but the mathematician, who, by long practice, has acquired a familiarity with many of these forms, and has become expert in the processes which lead from one to another, can often transform a perplexing expression into another which explains its meaning in more intelligible language.

As students of physics, we observe phenomena under varied circumstances, and endeavour to deduce the laws of their relations. Every natural phenomenon is, to our minds, the result of an infinitely complex system of conditions. What we set ourselves to do is to unravel these conditions, and by viewing the phenomenon in a way which is in itself partial and imperfect, to piece out its features one by one, beginning with that which strikes us first, and thus gradually learning how to look at the whole phenomenon so as to obtain a continually greater degree of clearness and distinctness. In this process, the feature which presents itself most forcibly to the untrained inquirer may not be that which is considered most fundamental by the experienced man of science; for the success of any physical investigation depends on the judicious selection of what is to be observed as of primary importance, combined with a voluntary abstraction of the mind from those

科学系统归类成各种不断增长的现象，而我们对于这些现象的认识还处于蒙昧状态。我也曾被廷德尔博士那敏锐的洞察力和有力的陈述带入那个力量与精微的圣殿，在那里，分子遵循着自身存在的定律，猛烈地撞在一起，或者在更为热烈的拥抱中纠缠，就这样悄然形成了事物的可见形式。我曾被西尔维斯特教授引领向那些宁静的峰巅

> "那里从没有云的踪迹，或风的迹象，
> 从不曾有些微雪花的斑痕，
> 从不曾有丝毫雷电的呼啸，
> 或是人类的悲怨之声，能够破坏，
> 他们那庄严持久的宁静。"

然而，谁能引领我进入那更为隐蔽与晦暗的思想与事实交汇的地带，数学家的头脑运算与分子间的物理作用呈现出其真实关联的所在在哪里呢？难道这条路不会经过那遍布着早期探索者的遗迹并被每一位科研工作者所痛恨的形而上学家的巢穴吗？对于我来说，占用整个分会的宝贵时间将各位引入那些就我们所知需要用几千年时间才能建立成型的思考中，实在是一种莽撞的冒险。

然而，我们是别人眼中数学和物理的耕耘者。在日常工作中，我们向形而上学家遇到的同类型的问题进军；我们着手处理那些问题，但并不完全寄希望于我们与生俱来的洞察力，而是通过长期持续地调整自身思维模式使之符合客观自然现象来训练自己。作为数学家，我们对数字与数量符号进行特定的大脑运算，并且通过从易到难的运算一步一步地推导，我们可以用多种不同形式表达同一事物。尽管这些形式之间的等价性是不证自明的公理的必然推论，但是对我们的头脑来说却并不总是不言而喻的；而数学家们，经过长期的实践已经对其中诸多形式颇为熟悉，并且已经成为将一种形式转变为另一种形式这一过程的专家，他们经常能够将一种复杂的表达形式转变为另外一种能以更容易被理解的语言解释其含义的形式。

作为物理研究者，我们在各种条件下观察现象，并致力于归纳出表达这些现象之间关联的定律。对我们的头脑来说，每一种自然现象都是一个无限复杂的条件体系的结果。我们努力做的就是将这些条件分解，用本身就是局部的、不完全的方式来观察现象，然后从我们最初遇到的那些开始，将现象的特征一个接一个地拼凑起来，并由此逐渐了解到该如何看待整个现象才能使明确度与清晰度不断增加。在这一过程中，那些在未经训练的研究者看来表现得最为强烈的特征，可能在有经验的科学工作者看来并不是最根本的性质；因为任何物理研究的成功都取决于在所观察到的一切中对于首要因素的明智选择，还要结合对那些尽管看来很诱人却还未能充

features which, however attractive they appear, we are not yet sufficiently advanced in science to investigate with profit.

Intellectual processes of this kind have been going on since the first formation of language, and are going on still. No doubt the feature which strikes us first and most forcibly in any phenomenon, is the pleasure or the pain which accompanies it, and the agreeable or disagreeable results which follow after it. A theory of nature from this point of view is embodied in many of our words and phrases, and is by no means extinct even in our deliberate opinions. It was a great step in science when men became convinced that, in order to understand the nature of things, they must begin by asking, not whether a thing is good or bad, noxious or beneficial, but of what kind is it? and how much is there of it? Quality and quantity were then first recognised as the primary features to be observed in scientific inquiry. As science has been developed, the domain of quantity has everywhere encroached on that of quality, till the process of scientific inquiry seems to have become simply the measurement and registration of quantities, combined with a mathematical discussion of the numbers thus obtained. It is this scientific method of directing our attention to those features of phenomena which may be regarded as quantities which brings physical research under the influence of mathematical reasoning. In the work of the section we shall have abundant examples of the successful application of this method to the most recent conquests of science; but I wish at present to direct your attention to some of the reciprocal effects of the progress of science on those elementary conceptions which are sometimes thought to be beyond the reach of change.

If the skill of the mathematician has enabled the experimentalist to see that the quantities which he has measured are connected by necessary relations, the discoveries of physics have revealed to the mathematician new forms of quantities which he could never have imagined for himself. Of the methods by which the mathematician may make his labours most useful to the student of nature, that which I think is at present most important is the systematic classification of quantities. The quantities which we study in mathematics and physics may be classified in two different ways. The student who wishes to master any particular science must make himself familiar with the various kinds of quantities which belong to that science. When he understands all the relations between these quantities, he regards them as forming a connected system, and he classes the whole system of quantities together as belonging to that particular science. This classification is the most natural from a physical point of view, and it is generally the first in order of time. But when the student has become acquainted with several different sciences, he finds that the mathematical processes and trains of reasoning in one science resemble those in another so much that his knowledge of the one science may be made a most useful help in the study of the other. When he examines into the reason of this, he finds that in the two sciences he has been dealing with systems of quantities, in which the mathematical forms of the relations of the quantities are the same in both systems, though the physical nature of the quantities may be utterly different. He is thus led to recognise a classification of quantities on a new principle, according to which the physical nature of the quantity is subordinated to its mathematical form. This is the point of view which is characteristic

分有益于科学研究的想法的自觉提炼。

这种理性过程自从语言形成以来就一直在进行着，并且还要进行下去。无疑，在任何现象中，最先刺激我们并表现得最为强烈的特征，就是伴随该现象而来的喜悦或烦恼，以及接踵而至的一致或不一致的结果。一种基于此看法的自然理论体现在很多词汇和短语中，而且即使在经过深思熟虑的观点中也不会消逝。当人们最终深信，为了理解事物的本质，他们一开始必须询问的不是该事物是好还是坏、是有害的还是有益的，而是它属于哪一类别、具体有多少时，科学便迈出了伟大的一步。自此，定性和定量第一次被认为是科学研究中要观察的首要特征。科学一旦建立起来，定量的疆域就会从各个角落侵占定性的疆域，直到科学研究的过程逐渐变为只是数量的测量与记录再加上对由此获得的数字的数学讨论。正是这种将我们的注意力引向自然现象中那些可以被视为量的特征的科学方法，把我们带入了数学推理影响下的物理研究中。在本分会所涉及的研究工作中，我们有丰富的实例可以说明这种方法在最近的科学成就中的成功应用；但是现在，我希望可以将各位的注意力引向科学发展对那些有时被认为是亘古不变的基本概念的相反的影响。

如果说数学家的技艺使实验家看到了他已测量的量之间有着必然的联系，物理的发现则已向数学家揭示出他们自己绝对无法想象出来的量的新形式。在数学家作为自然界的研究者付出辛劳时所用到的最有益的方法中，我认为当前最重要的方法就是量的系统分类。我们在数学和物理中所研究的量可以用两种不同的方式分类。想要掌握任何一门特定科学的学生必须使自己熟悉该科学的各种量。一旦他理解了这些量之间的全部关联，就会认为它们形成了一个联结起来的体系，并把整个量的体系归在一起作为属于该特定科学的类。从物理的观点来看这种分类法是最自然的，并且在时间上一般也是最先出现的。但是当这名学生逐渐通晓了若干个不同学科时，他会发现在一门科学中的数学过程和推理训练与另外一门中的十分相似，以至于一门科学中的知识对于学习另外一门科学极有帮助。在分析其中的道理时，他会发现，在需要处理包含各自量的系统的两门科学中，尽管各自的量的物理本质可能是完全不同的，但两个系统中量之间关系的数学形式是一样的。由此，他认识到基于一种新原则的量的分类法，根据这一原则，量的物理本质服从于其数学形式。这是带有数学家特征的观点；不过在时间顺序上它位于物理观点之后，因为，为了使人类的头脑能够想象出不同种类的量，首先它们本身就必须是确实存在的。这里我并没有提及这样的事实，即所有的量本身都必须服从代数和几何规则，并因此服从于很多

of the mathematician; but it stands second to the physical aspect in order of time, because the human mind, in order to conceive of different kinds of quantities, must have them presented to it by nature. I do not here refer to the fact that all quantities, as such, are subject to the rules of arithmetic and algebra, and are therefore capable of being submitted to those dry calculations which represent, to so many minds, their only idea of mathematics. The human mind is seldom satisfied, and is certainly never exercising its highest functions, when it is doing the work of a calculating machine. What the man of science, whether he be a mathematician or a physical inquirer, aims at is, to acquire and develop clear ideas of the things he deals with. For this purpose he is willing to enter on long calculations, and to be for a season a calculating machine, if he can only at last make his ideas clearer. But if he finds that clear ideas are not to be obtained by means of processes, the steps of which he is sure to forget before he has reached the conclusion, it is much better that he should turn to another method, and try to understand the subject by means of well-chosen illustrations derived from subjects with which he is more familiar. We all know how much more popular the illustrative method of exposition is found, than that in which bare processes of reasoning and calculation form the principal subject of discourse. Now a truly scientific illustration is a method to enable the mind to grasp some conception or law in one branch of science, by placing before it a conception or a law in a different branch of science, and directing the mind to lay hold of that mathematical form which is common to the corresponding ideas in the two sciences, leaving out of account for the present the difference between the physical nature of the real phenomena. The correctness of such an illustration depends on whether the two systems of ideas which are compared together are really analogous in form, or whether, in other words, the corresponding physical quantities really belong to the same mathematical class. When this condition is fulfilled, the illustration is not only convenient for teaching science in a pleasant and easy manner, but the recognition of the mathematical analogy between the two systems of ideas leads to a knowledge of both, more profound than could be obtained by studying each system separately.

There are men who, when any relation or law, however complex, is put before them in a symbolical form, can grasp its full meaning as a relation among abstract quantities. Such men sometimes treat with indifference the further statement that quantities actually exist in nature which fulfil this relation. The mental image of the concrete reality seems rather to disturb than to assist their contemplations. But the great majority of mankind are utterly unable, without long training, to retain in their minds the unembodied symbols of the pure mathematician; so that if science is ever to become popular and yet remain scientific, it must be by a profound study and a copious application of those principles of truly scientific illustration which, as we have seen, depend on the mathematical classification of quantities. There are, as I have said, some minds which can go on contemplating with satisfaction pure quantities presented to the eye by symbols, and to the mind in a form which none but mathematicians can conceive. There are others who feel more enjoyment in following geometrical forms, which they draw on paper, or build up in the empty space before them. Others, again, are not content unless they can project their whole physical energies into the scene which they conjure up. They learn at what a rate the planets rush through space, and they experience a delightful feeling of exhilaration. They calculate the

人认为的是代表其唯一数学观念的那些干巴巴的计算。人类的头脑在从事计算机器的工作时，很少会得到满足，当然也不会用到其最高级的功能。无论是数学家还是物理学者，科学工作者所希求的，是获得和发展有关他所研究的事物的清晰观念。为了这个目的，他甘愿投身于漫长的计算，并暂时充当一台计算机器，只要最终能使自己的观念变得清晰一些。不过，要是他发现通过某种过程（在得到计算结果之前就必然已把前面的步骤忘光的过程）不可能获得清晰的观念，那么他最好还是换一种方法，并通过在他比较熟悉的学科衍生出的种种解释中仔细挑选出的解释，来理解这个学科。我们都知道，直观的解释方法比从讨论主题出发进行干巴巴的推导和计算的过程受欢迎得多。那么，一个真正的科学解释是使头脑能够掌握某一科学分支中的一些概念和定律的方法，这是通过以下方式实现的：在头脑中呈现出另一科学分支中的一个概念或一条定律，并指引头脑去把握两个科学分支的对应观念中所共有的数学形式，而不去考虑那些真实现象的物理本质之间的不同。这样一种解释的正确性取决于两个放在一起比较的观念体系是否真的在形式上相似，或者换句话说，相应的物理量是否真的属于同一数学类。一旦满足了这一条件，不仅这种解释便于以轻松愉快的方式进行科学教学，而且对两个观念体系之间数学上的相似性的认识还会促进这两个系统的知识的发展，这就比我们单独研究任何一门科学所能获得的知识更加深刻。

任何一种无论多么复杂的关系或定律以符号形式呈现在人们面前时，总有人能够把握到它作为抽象的量之间关系的全部含义。有时，这些人对那些满足这一关系的自然界真实存在的量的进一步阐述毫不关心。具体现实在头脑中的印象似乎对他们思考的干扰多于帮助。但是，若非经过长期训练，绝大多数人绝对没法在其头脑中记住纯数学家的抽象符号。因此，如果科学想要广受欢迎同时保持科学性，它就必须经历深入的研究，以及能用真正的科学进行解释的原理的大量应用。而这，如同我们已经看到的，有赖于量的数学分类。正如我在前面提到的，有些头脑满足于利用那些纯粹的量进行思考，这些量以符号的形式呈现在他们眼前，并以某种除了数学家之外无人可以理解的形式展示于他们的头脑中。而另外一些人则以研究几何形式为乐，他们将几何形式描绘在纸上，或搭建在面前的空间中。还有一些人，只满足于将其全部精力投入到他们在头脑中幻想出的景象。他们研究行星掠过太空时的速度，从中体会喜悦的满足感。他们计算天体彼此之间的牵引力，并感觉自己的

forces with which the heavenly bodies pull at one another, and they feel their own muscles straining with the effort.

To such men impetus, energy, mass, are not mere abstract expressions of the results of scientific inquiry. They are words of power which stir their souls like the memories of childhood. For the sake of persons of these different types, scientific truths should be presented in different forms, and should be regarded as equally scientific, whether it appears in the robust form and the vivid colouring of a physical illustration, or in the tenuity and paleness of a symbolical expression. Time would fail me if I were to attempt to illustrate by examples the scientific value of the classification of quantities. I shall only mention the name of that important class of magnitudes having direction in space which Hamilton has called Vectors, and which form the subject-matter of the Calculus of Quaternions—a branch of mathematics which, when it shall have been thoroughly understood by men of the illustrative type, and clothed by them with physical imagery, will become, perhaps under some new name, a most powerful method of communicating truly scientific knowledge to persons apparently devoid of the calculating spirit. The mutual action and reaction between the different departments of human thought is so interesting to the student of scientific progress, that, at the risk of still further encroaching on the valuable time of the Section, I shall say a few words on a branch of science which not very long ago would have been considered rather a branch of metaphysics: I mean the atomic theory, or, as it is now called, the molecular theory of the constitution of bodies. Not many years ago, if we had been asked in what regions of physical science the advance of discovery was least apparent, we should have pointed to the hopelessly distant fixed stars on the one hand, and to the inscrutable delicacy of the texture of material bodies on the other. Indeed, if we are to regard Comte as in any degree representing the scientific opinion of his time, the research into what takes place beyond our own solar system seemed then to be exceedingly unpromising, if not altogether illusory. The opinion that the bodies which we see and handle, which we can set in motion or leave at rest, which we can break in pieces and destroy, are composed of smaller bodies which we cannot see or handle, which are always in motion, and which can neither be stopped nor broken in pieces, nor in any way destroyed or deprived of the least of their properties, was known by the name of the Atomic Theory. It was associated with the names of Democritus and Lucretius, and was commonly supposed to admit the existence only of atoms and void, to the exclusion of any other basis of things from the universe.

In many physical reasonings and mathematical calculations we are accustomed to argue as if such substances as air, water, or metal, which appear to our senses uniform and continuous, were strictly and mathematically uniform and continuous. We know that we can divide a pint of water into many millions of portions, each of which is as fully endowed with all the properties of water as the whole pint was, and it seems only natural to conclude that we might go on subdividing the water for ever, just as we can never come to a limit in subdividing the space in which it is contained. We have heard how Faraday divided a grain of gold into an inconceivable number of separate particles, and we may see Dr. Tyndall produce from a mere suspicion of nitrite of butyle an immense cloud, the

肌肉在这种作用下的变形。

对这样的人来说，冲量、能量和质量不仅仅是科学研究结果的抽象表达，它们还是如同童年记忆一样可以触动这些人灵魂的有力文字。因为有上述不同类型的人，所以科学真理应该以不同的形式表达，而且无论它出现在充实的表格和生动多彩的物理图解中，还是单调苍白的方程中，都应该被视为具有同等的科学性。时间不允许我在这里通过举例来说明量的分类的科学意义。我只能稍微说说被哈密顿称为矢量的一类具有空间方向的重要量，这正是四元数微积分的主题。四元数微积分是这样一个数学分支：当它完全被善于作举例说明的人所理解并被他们赋予丰富的物理想象的时候，它（也可能起了新名字）就会成为向那些看似缺乏计算头脑的人们传达真正科学知识的最有力的方法。人类思维的各个不同部分之间的相互作用和反作用对科学过程的研究者来说是如此有趣，以至于即使冒着继续侵占整个分会宝贵时间的风险，我也要说一说一个在不久之前还被看作是形而上学的科学分支，即，关于物体构成的原子理论或者说分子理论（现在是这样叫的）。几年前，如果有人问物理学中哪个领域进展最缓慢，我们会一手指向令人绝望的远地星体，另一手指向神秘的物质结构。的确，如果我们认为在任何情况下孔德都可以代表他所在时代的科学观点，那么对于发生在太阳系之外的事物的研究即使不是幻想，也只是毫无希望的事情。我们可以看到和触碰的、可以发动和制动的、可以打碎和毁灭的物体都是由我们看不到也触碰不到、一直处于运动状态并且不能被我们阻止、打碎、毁灭或剥夺其任何性质的更小的物体组成的，这种观点就是所谓的原子理论。这一理论与德谟克利特和卢克莱修这两个名字相联系，并且一般假定只承认原子和空隙的存在，而排除宇宙中其他一切物质基础的存在。

在很多物理推导和数学计算中，我们习惯地认为诸如空气、水或金属这些给我们均匀、连续的感觉的物质在数学上是严格均匀和连续的。我们知道，我们可以将一品脱水分成数百万份，每一份水都具有这一品脱水作为一个整体时所具有的所有性质，很自然地我们会得出这样的结论：我们可以无限细分这些水，就如同我们可以无限细分盛有这些水的空间一样。我们已经听说法拉第是如何将一粒金子分成不计其数的分离颗粒，而且我们可能会看到廷德尔博士用少量的丁基硝酸盐生成了一个巨大的云团，这个大云团的可见的微小部分仍是云团，因此一定包含很多丁基硝

minute visible portion of which is still cloud, and therefore must contain many molecules of nitrite of butyle. But evidence from different and independent sources is now crowding in upon us which compels us to admit that if we could push the process of subdivision still further we should come to a limit, because each portion would then contain only one molecule, an individual body, one and indivisible, unalterable by any power in nature. Even in our ordinary experiments on very finely divided matter we find that the substance is beginning to lose the properties which it exhibits when in a large mass, and that effects depending on the individual action of molecules are beginning to become prominent. The study of these phenomena is at present the path which leads to the development of molecular science. That superficial tension of liquids which is called capillary attraction is one of these phenomena. Another important class of phenomena are those which are due to that motion of agitation by which the molecules of a liquid or gas are continually working their way from one place to another, and continually changing their course, like people hustled in a crowd. On this depends the rate of diffusion of gases and liquids through each other, to the study of which, as one of the keys of molecular science, that unwearied inquirer into nature's secrets, the late Prof. Graham, devoted such arduous labour.

The rate of electrolytic conduction is, according to Wiedemann's theory, influenced by the same cause; and the conduction of heat in fluids depends probably on the same kind of action. In the case of gases, a molecular theory has been developed by Clausius and others, capable of mathematical treatment, and subjected to experimental investigation; and by this theory nearly every known mechanical property of gases has been explained on dynamical principles, so that the properties of individual gaseous molecules are in the fair way to become objects of scientific research. Now Sir William Thomson has shown by several independent lines of argument, drawn from phenomena so different in themselves as the electrification of metals by contact, the tension of soap-bubbles, and the friction of air, that in ordinary solids and liquids the average distance between contiguous molecules is less than the hundred-millionth, and greater than the two-thousand-millionth of a centimetre. This of course is an exceedingly rough estimate, for it is derived from measurements, some of which are still confessedly very rough; but if, at the present time, we can form even a rough plan for arriving at a result of this kind, we may hope that as our means of experimental inquiry become more accurate and more varied, our conception of a molecule will become more definite, so that we may be able at no distant period to estimate its weight. A theory which Sir W. Thomson has founded on Helmholtz's splendid hydrodynamical theorems, seeks for the properties of molecules in the ring-vortices of a uniform, frictionless, incompressible fluid. Such whirling rings may be seen when an experienced smoker sends out a dexterous puff of smoke into the still air, but a more evanescent phenomenon it is difficult to conceive. This evanescence is owing to the viscosity of the air; but Helmholtz has shown that in a perfect fluid such a whirling ring, if once generated, would go on whirling for ever, would always consist of the very same portion of the fluid which was first set whirling, and could never be cut in two by any natural cause. The generation of a ring-vortex is of course equally beyond the power of natural causes, but once generated, it has the properties of individuality, permanence in quantity, and indestructibility. It is also the recipient of impulse and of energy, which is all

酸盐分子。然而，从各个独立的渠道所获得的证据正朝我们涌来，这迫使我们承认如果将细分过程继续推进，我们将会遇到一个极限，因为这时每一部分只包含一个分子，这是一种不能被任何自然界的力量分割或改变的独立个体。即使是在平时关于分得很细的物质的实验中，我们也会发现这些物质已经开始失去作为大块材料时所具有的一些性质，产生这一结果的原因是分子的个别作用开始变得显著。目前，研究这种现象是分子科学发展的一种途径。被称为毛细引力的液体表面张力就是这种现象中的一个。另一类重要的现象是那些由激发运动产生的现象，由于激发运动，液态或气态分子不断从一个地方运动到另一个地方，并且不断变化它们的航向，就如同人群中奔忙的人那样。这种运动决定了气体和液体相互扩散的速度，对此类现象的研究正是永不停歇地探寻自然之谜的分子科学的关键点之一，已故的格雷姆教授为此付出了非常艰辛的努力。

根据维德曼的理论，电解导电速率也受到相同原因的影响；流体中的热传导可能也取决于同类作用。气体分子理论已经由克劳修斯和其他研究者发展成熟，可以用数学方法处理，并且与实验研究结果相符。应用这个理论，气体的几乎所有已知的力学性质都可以用动力学原理解释，因此，单个气体分子的性质就理所当然地成了科学研究的目标。现在，威廉·汤姆孙爵士通过一些各自独立的现象，例如金属的接触带电、肥皂泡的张力和空气的摩擦力，从不同角度证明了在普通的固体和流体中近邻分子之间的平均距离在 $5 \times 10^{-10} \sim 1 \times 10^{-8}$ 厘米之间。当然这是一个非常粗略的估算，因为导出这个估算值的测量中包含着一些仍旧被公认为非常粗略的结果，但是如果到目前为止，可以建立一套能够获得这类结果的哪怕只是粗略的计划，我们可能就会希望随着实验研究手段的越来越精确和多样化，我们对单个分子的概念会越来越清晰，这样，在不远的将来，我们就能够估算它的质量。基于亥姆霍兹精妙的流体动力学定理，汤姆孙爵士建立了一个研究分子在均匀、无摩擦且不可压缩的液体环形漩涡中的性质的理论。当一个有经验的吸烟者向静止的空气中吐出一个烟圈时就能够看到这样的涡流环，但是这是一个非常短暂的现象，很难去想象。这种短暂性是空气的粘性造成的；但是，亥姆霍兹认为在完全流体中，这样的涡流环一旦产生就会永远涡旋下去，而且一直是由开始成为涡旋的那部分流体组成，决不会被任何自然力分成两半。当然，环形漩涡的产生也同样不是出于自然原因，但是一旦产生，它就具有个体特征、量的恒定和不可摧毁性。我们能够断言的仅仅是漩涡，也就是冲量和能量的接受者，可以认为是一种物质；这些环形漩涡具有如此多样的关系和自卷结，以至于不同的涡流结一定像不同的分子一样属于不同的类型。

we can affirm of matter; and these ring-vortices are capable of such varied connections, and knotted self-involutions, that the properties of differently knotted vortices must be as different as those of different kinds of molecules can be.

If a theory of this kind should be found, after conquering the enormous mathematical difficulties of the subject, to represent in any degree the actual properties of molecules, it will stand in a very different scientific position from those theories of molecular action which are formed by investing the molecule with an arbitrary system of central forces invented expressly to account for the observed phenomena. In the vortex theory we have nothing arbitrary, no central forces or occult properties of any other kind. We have nothing but matter and motion, and when the vortex is once started its properties are all determined from the original impetus, and no further assumptions are possible. Even in the present undeveloped state of the theory, the contemplation of the individuality and indestructibility of a ring vortex in a perfect fluid cannot fail to disturb the commonly received opinion that a molecule, in order to be permanent, must be a very hard body. In fact one of the first conditions which a molecule must fulfil is, apparently, inconsistent with its being a single hard body. We know from those spectroscopic researches which have thrown so much light on different branches of science, that a molecule can be set into a state of internal vibration, in which it gives off to the surrounding medium light of definite refrangibility—light, that is, of definite wave-length and definite period of vibration. The fact that all the molecules, say of hydrogen, which we can procure for our experiments, when agitated by heat or by the passage of an electric spark, vibrate precisely in the same periodic time, or, to speak more accurately, that their vibrations are composed of a system of simple vibrations having always the same periods, is a very remarkable fact. I must leave it to others to describe the progress of that splendid series of spectroscopic discoveries by which the chemistry of the heavenly bodies has been brought within the range of human inquiry. I wish rather to direct your attention to the fact that not only has every molecule of terrestrial hydrogen the same system of periods of free vibration, but that the spectroscope examination of the light of the sun and stars shows that in regions the distance of which we can only feebly imagine there are molecules vibrating in as exact unison with the molecules of terrestrial hydrogen as two tuning forks tuned to correct pitch, or two watches regulated to solar time. Now this absolute equality in the magnitude of quantities, occurring in all parts of the universe, is worth our consideration. The dimensions of individual natural bodies are either quite indeterminate, as in the case of planets, stones, trees, &c., or they vary within moderate limit, as in the case of seeds, eggs, &c.; but, even in these cases, small quantitive differences are met with which do not interfere with the essential properties of the body. Even crystals, which are so definite in geometrical form, are variable with respect to their absolute dimensions. Among the works of man we sometimes find a certain degree of uniformity. There is a uniformity among the different bullets which are cast in the same mould, and the different copies of a book printed from the same type. If we examine the coins, or the weights and measures, of a civilised country, we find a uniformity, which is produced by careful adjustment to standards made and provided by the State. The degree of uniformity of these national standards is a measure of that spirit of justice in the nation which has enacted laws to

如果在解决了这个问题的大量数学疑难之后，可以找到一个这种类型的理论来表示分子的所有真实性质，那么这个理论将处于与那些在研究具有主观中心力系统（这个系统是为了解释宏观现象而臆造的）的分子时形成的分子相互作用理论非常不同的科学地位。在涡旋理论中，没有什么是主观的，没有中心力或其他任何类型的超自然性质。只存在物质和运动，而且一旦涡旋产生，它的性质就完全取决于初始的动力，可以不用做任何进一步的假设。即使是在目前这种理论尚未发展成熟的情况下，关于完全流体中环形漩涡的个别性和不可毁灭性的思考也会扰乱人们的普遍观点：为了具有永久性，分子必须是非常坚硬的个体。事实上，分子必须满足的一个首要条件似乎与它是一个单一坚硬个体这一观点是不一致的。我们从已经惠及多门科学分支的光谱研究可以知道，分子能够调节到内部振动状态，在这种状态下，分子向周围介质辐射具有确定折射率的光（即，该光具有确定的波长和确定的振动周期）。所有可以为我们的实验服务的分子，比如氢分子，当它们受热激发时或在电火花经过而引起激发时，都会精确地以相同的周期振动，或者更准确地说，它们的振动是由一个具有相同振动周期的简单振动系统组成的，事实的确如此。我不得不把一系列精彩的光谱学发现留给别人去描述，尽管正是这些精彩的发现将天体化学这一领域纳入了人类探索的范围。我宁愿将你们的注意力引到这样的事实中：不仅地球上每一个氢分子都有相同的自由振动周期系统，而且对太阳和行星发出的光的光谱检测结果表明，在那些与地球之间的距离大到无法想象的行星上，也有与地球上氢分子的振动完全一致的分子振动，就好像两个被调节到相同音调的音叉，或者两块校准到太阳时的表。那么这种发生在宇宙各个部分的量在量值上的绝对等价性就值得我们思考。自然个体的尺度不是非常不确定（比如，行星、石块、树木等），就是能够在适度的范围内变化（比如，种子、卵等）；但是即使在这些例子中所遇到的量值上微小的不同也不会影响物体的本质特征。即使是具有非常确定的几何形状的晶体，其绝对尺寸也是可变的。在人类的工作中我们时常会发现一定程度的一致性。比如，同一模子中铸造出的不同子弹具有一致性，同一版次印刷出的书籍的不同副本具有一致性。如果我们观察一个文明国家的货币和度量衡，也会发现一致性，一种产生于按国家规定的标准仔细调节的一致性。这些国家标准的一致程度是对这个制定了法律来规范各种标准并且委派了官员来检测各种标准的国家的公平意识程度的衡量。作为科学个体来说，这个主题是我们很感兴趣的问题，并且大家都意识到大量科学工作已经投入并且有利地投入到为商业或科学目的提供度量衡之中。地球作为长度的永久基准已经得到了测量，同时金属的每一个性质都已经得到了研究，

regulate them and appointed officers to test them. This subject is one in which we, as a scientific body, take a warm interest, and you are all aware of the vast amount of scientific work which has been expended, and profitably expended, in providing weights and measures for commercial and scientific purposes. The earth has been measured as a basis for a permanent standard of length, and every property of metals has been investigated to guard against any alteration of the material standards when made. To weigh or measure anything with modern accuracy, requires a course of experiment and calculation in which almost every branch of physics and mathematics is brought into requisition.

Yet, after all, the dimensions of our earth and its time of rotation, though, relatively to our present means of comparison, very permanent, are not so by any physical necessity. The earth might contract by cooling, or it might be enlarged by a layer of meteorites falling on it, or its rate of revolution might slowly slacken, and yet it would continue to be as much a planet as before. But a molecule, say of hydrogen, if either its mass or its time of vibration were to be altered in the least, would no longer be a molecule of hydrogen. If, then, we wish to obtain standards of length, time, and mass which shall be absolutely permanent, we must seek them not in the dimensions, or the motion, or the mass of our planet, but in the wave-length, the period of vibration, and the absolute mass of these imperishable and unalterable and perfectly similar molecules. When we find that here, and in the starry heavens, there are innumerable multitudes of little bodies of exactly the same mass, so many, and no more, to the grain, and vibrating in exactly the same time, so many times, and no more, in a second, and when we reflect that no power in nature can now alter in the least either the mass or the period of any one of them, we seem to have advanced along the path of natural knowledge to one of those points at which we must accept the guidance of that faith by which we understand that "that which is seen was not made of things which do appear." One of the most remarkable results of the progress of molecular science is the light it has thrown on the nature of irreversible processes,—processes, that is, which always tend towards, and never away from, a certain limiting state. Thus if two gases be put into the same vessel they become mixed, and the mixture tends continually to become more uniform. If two unequally heated portions of the same gas are put into the vessel, something of the kind takes place, and the whole tends to become of the same temperature. If two unequally heated solid bodies be placed in contact, a continual approximation of both to an intermediate temperature takes place. In the case of the two gases, a separation may be effected by chemical means; but in the other two cases the former state of things cannot be restored by any natural process. In the case of the conduction or diffusion of heat the process is not only irreversible, but it involves the irreversible diminution of that part of the whole stock of thermal energy which is capable of being converted into mechanical work. This is Thomson's theory of the irreversible dissipation of energy, and it is equivalent to the doctrine of Clausius concerning the growth of what he calls Entropy. The irreversible character of this process is strikingly embodied in Fourier's theory of the conduction of heat, where the formulae themselves indicate a possible solution of all positive values of the time which continually tends to a uniform diffusion of heat. But if we attempt to ascend the stream of time by giving to its symbol continually diminishing values, we are led up to a state of things in which the

目的是为了在制造过程中避免材料标准的任何改变。以现在的精度去称量或测量任何东西，都需要经过一个实验和计算的过程，这个过程几乎会用到物理和数学的所有分支。

虽然地球的尺寸及其自转的时间相对于我们现有的比较方法来说是永恒不变的，但毕竟不是任何情况下都是这样。地球可能会因为冷却而收缩，可能会因为落在其表面的陨石层而增大，它的公转速度可能会逐渐变缓，尽管这样地球仍然和以前一样是一个星球。但是，如果一个分子（如氢分子）的质量或者振动周期发生轻微的变化，那么它将不再是氢分子。因此，如果我们想要得到绝对永恒的长度、时间和质量基准，就不能在行星的尺寸、运动和质量中寻找答案，而只能在这些不灭、不变，并且完全相同的分子的波长、振动周期和绝对质量中寻求。当我们发现在地球上和星空中存在无数质量完全相同的小物体以及无数周期完全相同的振动，并且到目前为止，没有任何自然力可以使它们之中任何一个的质量或者周期发生丝毫的改变，我们似乎是沿着自然知识的道路向那些观点中的一点前进的。从这种观点来看，我们必须接受这种信仰的指引，通过这个信仰我们明白了“我们看见的事物并不是由可见的事物组成的”。分子科学进程中最显著的成就之一就是对不可逆过程（一种一直向特定最终状态发展，从不偏离的过程）本质的解释。因此，如果将两种气体装入同一个容器中，它们会混合在一起，并且混合得越来越均匀。如果将两部分温度不同的同种气体装入这个容器中，类似的过程将会发生，容器中的气体将趋于相同的温度。如果两块温度不同的固体相互接触，它们的温度将会向着一个中间的温度转变。在有两种气体的例子中，可以通过化学方法将混合在一起的两种气体再次分开，但是在另外两种情况中，不可能通过任何自然方法将系统再恢复到初始状态。热传导或热扩散过程不仅是不可逆的，而且还伴随着可以转变为机械能的储热的不可逆减少。这就是汤姆孙的能量不可逆损耗理论，它等价于克劳修斯的熵增加理论。这一过程的不可逆性显著地体现在傅立叶的热传导理论中，其方程本身就表明了热扩散达到均匀状态所需时间的所有可能的正值解。然而，如果我们试图通过给符号赋予不断减小的值来增大时间的流动，便会得到事物的一个状态，在这个状态中，公式具有所谓的临界值。而如果我们研究这个状态之前的一个状态，就会发现这个公式变得很荒谬。因此，我们得出这样的观点：事物的状态并不能看作是前一状态的物理结果。我们还发现临界状态实际上并不存在于过去的一个时期，而是被一个有限的间隔与现在的时间分开。这个新起点的观念是最近物理研究带给我们的，是任

formula has what is called a critical value; and if we inquire into the state of things the instant before, we find that the formula becomes absurd. We thus arrive at the conception of a state of things which cannot be conceived as the physical result of a previous state of things, and we find that this critical condition actually existed at an epoch not in the utmost depths of a past eternity, but separated from the present time by a finite interval. This idea of a beginning is one which the physical researches of recent times have brought home to us, more than any observer of the course of scientific thought in former times would have had reason to expect. But the mind of man is not like Fourier's heated body, continually settling down into an ultimate state of quiet uniformity, the character of which we can already predict; it is rather like a tree shooting out branches which adapt themselves to the new aspects of the sky towards which they climb, and roots which contort themselves among the strange strata of the earth into which they delve. To us who breathe only the spirit of our own age, and know only the characteristics of contemporary thought, it is as impossible to predict the general tone of the science of the future as it is to anticipate the particular discoveries which it will make. Physical research is continually revealing to us new features of natural processes, and we are thus compelled to search for new forms of thought appropriate to these features. Hence the importance of a careful study of those relations between mathematics and physics which determine the conditions under which the ideas derived from one department of physics may be safely used in forming ideas to be employed in a new department. The figure of speech or of thought by which we transfer the language and ideas of a familiar science to one with which we are less acquainted may be called scientific metaphor. Thus the words velocity, momentum, force, &c., have acquired certain precise meanings in elementary dynamics. They are also employed in the dynamics of a connected system in a sense which, though perfectly analogous to the elementary sense, is wider and more general. These generalised forms of elementary ideas may be called metaphorical terms in the sense in which every abstract term is metaphorical. The characteristic of a truly scientific system of metaphors is that each term in its metaphorical use retains all the formal relations to the other terms of the system which it had in its original use. The method is then truly scientific, that is, not only a legitimate product of science, but capable of generating science in its turn. There are certain electrical phenomena, again, which are connected together by relations of the same form as those which connect dynamical phenomena. To apply to these the phrases of dynamics with proper distinctions and provisional reservations is an example of a metaphor of a bolder kind; but it is a legitimate metaphor if it conveys a true idea of the electrical relations to those who have been already trained in dynamics. Suppose, then, that we have successfully introduced certain ideas belonging to an elementary science by applying them metaphorically to some new class of phenomena. It becomes an important philosophical question to determine in what degree the applicability of the old ideas to the new subject may be taken as evidence that the new phenomena are physically similar to the old. The best instances for the determination of this question are those in which two different explanations have been given of the same thing. The most celebrated case of this kind is that of the corpuscular and the undulatory theories of light. Up to a certain point the phenomena of light are equally well explained by both; beyond this point one of them fails. To understand the true relation of these theories in that part of the field where they

何之前的具有科学思维的观察者所不能想象的。但是人类的思维并不像傅立叶的热体那样，一直向着完全均匀的最终状态发展，而我们已经可以预测这个状态的性质。这就好像一棵树伸展枝杈，去适应一片新的生长天空，而根则在陌生的地层中委屈自身，因为那才是它们钻研的目标。对于我们这些只呼吸同龄人的精神、只知道当代思维特征的人，是不可能预言未来科学的普遍状况和特别发现的。物理研究不断向我们揭示自然过程的新性质，这也迫使我们寻求可以适应这些性质的新的思维方式。因此，对数学和物理学之间关系的细致研究的重要性在于可以决定从物理的一个领域推导出的观点被安全地用于将在新领域中使用而形成的观点。通过演讲或者思考中的符号，我们可以将一门熟悉的科学中的语言和观点迁移到我们认识较少的科学中，这些符号可以称为科学比喻。因此，速度、动量、力这些词在基础动力学中获得了特定的精确含义。它们也被用在连通系统的动力学中，虽然与其基本含义完全类似，但却有了更加广泛和普遍的意义。这些基本概念的推广形式可以称之为比喻性的词汇，因为其中每一个抽象词汇都是比喻的。比喻这个真正科学系统的特征就是比喻时所用到的每一个词都保留着自己在最初被使用的时候与系统中其他词汇之间的形式关系。这样，这种方法是真正科学的，即，不仅是科学合法的产物，而且能够继续产生科学。一些电现象之间的相互联系和动力学现象之间的相互联系具有相同的形式。在作出恰当区分和提出临时条件后，将这些动力学词汇应用到电现象中就是一个大胆的比喻的例子，但是如果能够向受过动力学方面培训的人传达真正的电学关系，这就是一个合理的比喻。假设我们通过比喻的方式，成功地将某些属于一门基础科学的观点引入到某类新现象，那么旧观点在新课题上的应用能在多大程度上被当作新现象在物理上与旧观点相似的证据，这就成了一个重要的哲学问题。解决这一问题的最好例子就是，对于同种事物存在两种不同的解释。最著名的当属光的粒子理论和波动理论了。在某些情况下，光现象可以用两种理论很好地解释，而在这些情况之外，其中一种理论就会失效。为了理解两种理论在它们同样适用的领域中的真正关系，我们必须看到哈密顿对它们作出的解释，他发现任何瞬时问题都相应于一个自由运动问题，虽然这个自由运动问题会涉及到不同的速度和时间，但是会导致相同的几何路径。泰特教授曾经写了一篇关于这个主题的有趣的论文。根据一个在德国取得重大进步的电学理论，两个带电粒子会相隔一定距离直接相互作用，但是据韦伯所说，相互作用力取决于它们的相对速度，而根据高斯提出并由黎曼、洛伦茨和纽曼发展的理论，相互作用不是即时的，而是发生在由距离决定的一段时间之后。为了得到认同，这些杰出的研究者所支持的这一理论解释每

seem equally applicable we must look at them in the light which Hamilton has thrown upon them by his discovery that to every brachystochrone problem there corresponds a problem of free motion, involving different velocities and times, but resulting in the same geometrical path. Professor Tait has written a very interesting paper on this subject. According to a theory of electricity which is making great progress in Germany two electrical particles act on one another directly at a distance, but with a force which, according to Weber, depends on their relative velocity, and according to a theory hinted at by Gauss, and developed by Riemann, Lorenz, and Neumann, acts not instantaneously, but after a time depending on the distance. The power with which this theory, in the hands of these eminent men, explains every kind of electrical phenomena must be studied in order to be appreciated. Another theory of electricity which I prefer denies action at a distance and attributes electric action to tensions and pressures in an all-pervading medium, these stresses being the same in kind with those familiar to engineers, and the medium being identical with that in which light is supposed to be propagated. Both these theories are found to explain not only the phenomena by the aid of which they were originally constructed, but other phenomena which were not thought of, or perhaps not known at the time, and both have independently arrived at the same numerical result which gives the absolute velocity of light in terms of electrical quantities. That theories, apparently so fundamentally opposed, should have so large a field of truth common to both is a fact the philosophical importance of which we cannot fully appreciate till we have reached a scientific altitude from which the true relation between hypotheses so different can be seen.

I shall only make one more remark on the relation between mathematics and physics. In themselves, one is an operation of the mind, the other is a dance of molecules. The molecules have laws of their own, some of which we select as most intelligible to us and most amenable to our calculation. We form a theory from these partial data, and we ascribe any deviation of the actual phenomena from this theory to disturbing causes. At the same time, we confess that what we call disturbing causes are simply those parts of the true circumstances which we do not know or have neglected, and we endeavour in future to take account of them. We thus acknowledge that the so-called disturbance is a mere figment of the mind, not a fact of nature, and that in natural action there is no disturbance. But this is not the only way in which the harmony of the material with the mental operation may be disturbed. The mind of the mathematician is subject to many disturbing causes, such as fatigue, loss of memory, and hasty conclusions; and it is found that from these and other causes mathematicians make mistakes. I am not prepared to deny that, to some mind of a higher order than ours, each of these errors might be traced to the regular operation of the laws of actual thinking; in fact we ourselves often do detect, not only errors of calculation, but the causes of these errors. This, however, by no means alters our conviction that they are errors, and that one process of thought is right and another process wrong. One of the most profound mathematicians and thinkers of our time, the late George Boole, when reflecting on the precise and almost mathematical character of the laws of right thinking as compared with the exceedingly perplexing, though perhaps equally determinate, laws of actual and fallible thinking, was led to

一种电学现象的能力必须得到验证。另一种我更喜欢的电学理论，否认相隔一定距离的相互作用，它将电作用归因于扩散介质中的张力和压力，这些应力与工程师所熟悉的力属于相同的类型，而其中的介质与光传播的介质一样。这两种理论不仅能够解释那些它们最初建立时要解释的现象，还能够解释一些没有想到或者现在还不知道的现象，并且这些理论都各自独立地根据电学量得到了相同的光的绝对速度的数值结果。其实这些看上去完全相反的理论，在很大程度上都是真实的，我们只有在达到可以看清不同假说之间的真实关系的科学高度时，才能理解其中的哲学价值。

我应该再次强调数学和物理学之间的关系。就它们本身而言，一个是大脑的运算，另一个是分子的舞蹈。分子有其自身的法则，我们只是选择其中一些最容易的去理解和计算。我们根据这些片面的数据建立理论，还将任何偏离这个理论的实际现象归因于干扰因素。同时，我们也承认所谓的干扰因素只是真实环境中我们不知道或者已经忽略的那部分，并且在将来我们会努力对这些因素进行全面考虑。因此，我们认识到，所谓的干扰仅仅是大脑的虚构，而不是自然事实，在自然作用中不存在干扰。然而，这并不是大脑运算中物质受干扰的唯一途径。数学家的大脑受到很多干扰因素的支配，如疲劳、失忆和犹豫，并且人们发现数学家会因为这些或其他因素犯错。我并不准备否认，对于那些比我们高级的大脑，这些错误都会被归咎于实际思维法则的规则运算。实际上，我们自身也会经常发现，不只是计算的错误，还有导致这些错误发生的原因。然而，这决不会改变我们的判断：它们是错误，一种思维过程是正确的，而另一种则是错误的。我们这个时代最深奥的数学家和思想家——已故的乔治·布尔，当他考虑到正确思维（相对于极其复杂的，虽然可能是同等正确的，真实的且易错的思维来说），其法则的精确性和近乎数学性时，也被引向了另一种观点，这种观点中科学似乎寻求一个不属于她的区域。“我们必须承认，”他说，“确实存在（思维的）法则，即使是它们严格的数学形式也没能免于被破坏。

another of those points of view from which science seems to look out into a region beyond her own domain. "We must admit," he says, "that there exist laws" (of thought) "which even the rigour of their mathematical forms does not preserve from violation. We must ascribe to them an authority, the essence of which does not consist in power, a supremacy which the analogy of the inviolable order of the natural world in no way assists us to comprehend."

(**2**, 419-422; 1870)

我们必须把它们归因于权威，这个原因的本质并不在于力量，神圣不可破坏的自然界秩序的类比不会帮助我们理解至高法则。”

（王耀杨 翻译；江丕栋 审稿）

On Colour Vision*

J. C. Maxwell

Editor's Note

Among the lesser known scientific contributions of James Clerk Maxwell is his work on colour theory. Maxwell showed how three primary colours of light—red, green and blue—can be mixed to generate almost any colour, paving the way for colour photography and projection. In this contribution based on a talk at London's Royal Institution, he reviews this work and expands on its implications for colour vision. Thomas Young had proposed in 1801 that this involved three kinds of light receptor in the eye. Maxwell shows that these receptors have now been identified as rod- and cone-shaped cells in the retina. The existence of three distinct types of colour-sensitive cone cells had not yet been proven, but Maxwell clearly suspects it.

ALL vision is colour vision, for it is only by observing differences of colour that we distinguish the forms of objects. I include differences of brightness or shade among differences of colour.

It was in the Royal Institution, about the beginning of this century, that Thomas Young made the first distinct announcement of that doctrine of the vision of colours which I propose to illustrate. We may state it thus:—We are capable of feeling three different colour-sensations. Light of different kinds excites these sensations in different proportions, and it is by the different combinations of these three primary sensations that all the varieties of visible colour are produced. In this statement there is one word on which we must fix our attention. That word is, Sensation. It seems almost a truism to say that colour is a sensation; and yet Young, by honestly recognising this elementary truth, established the first consistent theory of colour. So far as I know, Thomas Young was the first who, starting from the well-known fact that there are three primary colours, sought for the explanation of this fact, not in the nature of light, but in the constitution of man. Even of those who have written on colour since the time of Young, some have supposed that they ought to study the properties of pigments, and others that they ought to analyse the rays of light. They have sought for a knowledge of colour by examining something in external nature—something out of ourselves.

Now, if the sensation which we call colour has any laws, it must be something in our own nature which determines the form of these laws; and I need not tell you that the only evidence we can obtain respecting ourselves is derived from consciousness.

* Lecture delivered before the Royal Institution, March. 24th.

论色觉*

麦克斯韦

编者按

詹姆斯·克拉克·麦克斯韦在颜色理论方面的研究是其不太著名的科学贡献之一。早先，麦克斯韦就向人们展示了如何通过混合光的三原色（红、绿、蓝）来得到几乎所有的颜色，这为彩色照相技术和投影技术铺平了道路。在这篇以一次在伦敦皇家研究院所作的报告为基础而形成的文稿中，麦克斯韦回顾了这项研究工作并进一步阐述了其对色觉的意义。1801 年，托马斯·杨就提出眼睛中有三种光感受体。麦克斯韦在这里指出，人们已经确认这些光感受体就是视网膜上的杆状细胞和锥形细胞。尽管当时还没有证实存在三种不同类型的对颜色敏感的锥形细胞，但麦克斯韦仍然明确地支持这种猜想。

所有的视觉都是色觉，因为我们只有通过观察颜色的差别才能区分物体的形态。我把明暗的区别也包含在了颜色的区别当中。

约在本世纪初，托马斯·杨在皇家研究院第一次明确地宣布了这个关于色觉的学说，这里，我将要对它进行阐述。我们可以这样来表述：我们能够体验到三种不同的颜色感觉。不同类型的光会以不同比例激发三种颜色感觉，所有可见的颜色就是由这三种基本感觉经过不同的组合而形成的。在这里，有一个词是值得我们注意的，那就是“感觉”。说颜色是一种感觉简直就是一个起码的常识；但杨真正确认了这个基本事实，首先建立了与之一致的关于颜色的理论。据我所知，托马斯·杨是第一个从人类的知觉而不是从光的本质来解释众所周知的三原色的人。即便是那些在杨以后撰写了有关颜色的著作的人们，不是认为应该去研究颜料的特性就是认为应该去分析光线。他们试图用人类自身之外的那些外在本质来揭示颜色的奥秘。

现在，如果说我们称之为颜色的这种感觉遵循某种规律的话，那么一定是我们自身的本质决定了这种规律的形式。无需由我来告诉你，我们所能获得的关于自身的唯一证据就来自于我们的意识。

* 这是 3 月 24 日向皇家研究院所作的演讲。

The science of colour must therefore be regarded as essentially a mental science. It differs from the greater part of what is called mental science in the large use which it makes of the physical sciences, and in particular of optics and anatomy. But it gives evidence that it is a mental science by the numerous illustrations which it furnishes of various operations of the mind.

In this place we always feel on firmer ground when we are dealing with physical science. I shall therefore begin by showing how we apply the discoveries of Newton to the manipulation of light, so as to give you an opportunity of feeling for yourselves the different sensations of colour.

Before the time of Newton, white light was supposed to be of all known things the purest. When light appears coloured, it was supposed to have become contaminated by coming into contact with gross bodies. We may still think white light the emblem of purity, though Newton has taught us that its purity does not consist in simplicity.

We now form the prismatic spectrum on the screen. These are the simple colours of which white light is always made up. We can distinguish a great many hues in passing from the one end to the other; but it is when we employ powerful spectroscopes, or avail ourselves of the labours of those who have mapped out the spectrum, that we become aware of the immense multitude of different kinds of light, every one of which has been the object of special study. Every increase of the power of our instruments increases in the same proportion the number of lines visible in the spectrum.

All light, as Newton proved, is composed of these rays taken in different proportions. Objects which we call coloured when illuminated by white light, make a selection of these rays, and our eyes receive from them only a part of the light which falls on them. But if they receive only the pure rays of a single colour of the spectrum, they can appear only of that colour. If I place a disc containing alternate quadrants of red and green paper in the red rays, it appears all red, but the red quadrants brightest. If I place it in the green rays both papers appear green, but the red paper is now the darkest. This, then, is the optical explanation of the colours of bodies when illuminated with white light. They separate the white light into its component parts, absorbing some and scattering others.

Here are two transparent solutions. One appears yellow, it contains bichromate of potash; the other appears blue, it contains sulphate of copper. If I transmit the light of the electric lamp through the two solutions at once, the spot on the screen appears green. By means of the spectrum we shall be able to explain this. The yellow solution cuts off the blue end of the spectrum, leaving only the red, orange, yellow, and green. The blue solution cuts off the red end, leaving only the green, blue, and violet. The only light which can get through both is the green light, as you see. In the same way most blue and yellow paints, when mixed, appear green. The light which falls on the mixture is so beaten about between the yellow particles and the blue, that the only light which survives is the green. But yellow and blue light when mixed do not make green, as you will see if we allow them to fall on the same part of the screen together.

因此，色彩学在本质上应该被当作是一种精神科学。它与大多数所谓的精神科学有很大的区别，它要用到物理学，特别是要用到光学和解剖学。但是，种种精神活动提供了大量的例证，可以证明色彩学是一种精神科学。

当我们利用物理学来处理这一问题的时候，总是感到有更坚实的理论基础。因此，我将从说明我们如何把牛顿的发现运用于光线的操控入手，以此给你一个机会，让你知道你对颜色的不同感觉。

在牛顿之前，白光被认为是所有已知物质中最纯粹的物质。有色光被认为是白光接触到物体而受到了污染。我们也许仍然可以认为白光象征着纯粹，但是牛顿已经告诉我们，白光的纯粹并不意味着简单。

现在，我们在屏幕上呈现棱镜光谱，得到的就是构成白光的基本颜色。当我们从一端向另一端观察的时候，可以分辨出很多不同的色彩；但是当我们使用功能更为强大的分光镜，或者利用别人已经制好的光谱时，我们就会发现大量不同种类的光线，每一种都值得专门研究。光谱中可分辨谱线数量增加的比例与仪器分辨率提高的比例是一致的。

牛顿已经证实，所有的光都是由上面所提到的光线以不同的比例组合而成的。当我们所谓的有色物体被白光照亮的时候，它会选择光线，而我们的眼睛能接受到的只是照射在其上的一部分光线。如果物体只被光谱中纯粹的单色光线所照射，那么它就只能呈现出那种颜色。如果我把红纸和绿纸交替放在一个盘子的不同象限里，用红光照射，整个盘子都会呈现出红色，但是红纸所在部分最明亮。如果把盘子放在绿光中，那么红纸和绿纸都会呈现出绿色，但这一次红纸部分是最暗的。这就是物体被白光照射时所呈现的颜色的光学解释。它们把白光拆分成不同的组成部分，然后吸收一部分，反射另外的部分。

这里有两种透明的溶液。一种是黄色的重铬酸钾溶液，另一种是蓝色的硫酸铜溶液。如果我让电灯发射的光通过这两种溶液，那么投射到屏幕上的是绿色光斑，这可以用光谱来解释。黄色溶液将光谱中的蓝色一端切断，只剩下了红色、橙色、黄色和绿色；蓝色溶液则将光谱中的红色一端切断，只剩下绿色、蓝色和紫色。正如你看到的那样，只有绿色光才能通过两种溶液。同样的道理，蓝色和黄色的颜料混合在一起通常会呈现出绿色。光照射在混合颜料上，黄色和蓝色的颜料颗粒吸收各自范围的光线，只有绿色光线可以反射出来。但是黄光和蓝光却不能混合成绿光，如果我们把它们投射到屏幕上的同一个区域，就可以看出来。

It is a striking illustration of our mental processes that many persons have not only gone on believing, on the evidence of the mixture of pigments, that blue and yellow make green, but that they have even persuaded themselves that they could detect the separate sensations of blueness and of yellowness in the sensation of green.

We have availed ourselves hitherto of the analysis of light by coloured substances. We must now return, still under the guidance of Newton, to the prismatic spectrum. Newton not only

Untwisted all the shining robe of day,

but showed how to put it together again. We have here a pure spectrum, but instead of catching it on a screen, we allow it to pass through a lens large enough to receive all the coloured rays. These rays proceed, according to well-known principles in optics, to form an image of the prism on a screen placed at the proper distance. This image is formed by rays of all colours, and you see the result is white. But if I stop any of the coloured rays, the image is no longer white, but coloured; and if I only let through rays of one colour, the image of the prism appears of that colour.

I have here an arrangement of slits by which I can select one, two, or three portions of the light of the spectrum, and allow them to form an image of the prism while all the rest are stopped. This gives me a perfect command of the colours of the spectrum, and I can produce on the screen every possible shade of colour by adjusting the breadth and the position of the slits through which the light passes. I can also, by interposing a lens in the passage of the light, show you a magnified image of the slits, by which you will see the different kinds of light which compose the mixture.

The colours are at present red, green, and blue, and the mixture of the three colours is, as you see, nearly white. Let us try the effect of mixing two of these colours. Red and blue form a fine purple or crimson, green and blue form a sea-green or sky-blue, red and green form a yellow.

Here again we have a fact not universally known. No painter, wishing to produce a fine yellow, mixes his red with his green. The result would be a very dirty drab colour. He is furnished by nature with brilliant yellow pigments, and he takes advantage of these. When he mixes red and green paint, the red light scattered by the red paint is robbed of nearly all its brightness by getting among particles of green, and the green light fares no better, for it is sure to fall in with particles of red paint. But when the pencil with which we paint is composed of the rays of light, the effect of two coats of colour is very different. The red and the green form a yellow of great splendour, which may be shown to be as intense as the purest yellow of the spectrum.

I have now arranged the slits to transmit the yellow of the spectrum. You see it is similar in colour to the yellow formed by mixing red and green. It differs from the mixture, however, in being strictly homogeneous in a physical point of view. The prism, as you see, does not

这是一个与我们的精神活动过程有关的惊人例子：根据颜料混合物的实验结果，许多人不仅相信蓝色加上黄色会呈现绿色，而且还认为自己可以从绿色的视觉感受中分离出黄色和蓝色的部分。

到目前为止，我们都在用有色的物质来分析光学问题。现在，我们仍然需要按照牛顿的理论回到棱镜光谱上来。牛顿不仅

解开了日光那耀眼的罩袍，

而且还展示了如何把它重新整合起来。我们有一束纯的分光光谱，但没有将它投射到屏幕上，而是让它通过一个足够大的棱镜以便接收各种颜色的光。依照我们熟知的光学原理，这些光线在其前方一定距离处的屏幕上会形成分光光谱的图像。这个图像由各种颜色的光线组成，而你看到的结果是白色的。但如果我挡住任何一种颜色的光，图像将不再是白色的，而是有色的；如果我只让一种颜色的光通过，那么分光光谱的图像上所呈现的就是那种颜色。

这里，我可以利用设置狭缝的方法选择光谱中的一部分、两部分或三部分光谱线，使其成像，而其他部分则被挡住。这样我就可以极好地控制光谱中的颜色了，通过调整光路中各狭缝的宽度和位置，可以使屏幕上呈现出每一种颜色的图像。我还可以在光路中插入透镜，使你看到狭缝的像，这样你就可以观察到混合在一起的不同种类的光。

现在，选取红、绿、蓝三种颜色，正如你所见，它们混合在一起后几乎是白色的。我们也可以尝试一下混合三种当中的任意两种颜色。红色和蓝色形成纯紫色或者深红色，绿色和蓝色形成海绿色或者天蓝色，红色和绿色形成黄色。

这里，我们又得到了一个并非被广泛了解的事实。没有哪一个画家会用他的红颜料和绿颜料混合在一起调出纯黄色。这样做只能得到一种很脏的灰黄色。他自己本来就有鲜亮的黄颜料，用这个就行了。当他混合红绿两种颜料的时候，红颜料颗粒所反射出来的红光因为被绿颜料颗粒吸收而几乎失去了全部亮度，绿光的情况也好不了多少，因为绿颜料颗粒反射出的绿光也会被红颜料颗粒吸收。但是如果我们作画时所用的笔是由光线组成的，那么涂覆两种颜色得到的效果就会完全不同。红光和绿光会形成非常漂亮的黄色，和光谱中最纯的黄光一样鲜艳。

我现在调整狭缝，选取光谱中的黄光。你会发现它与红光和绿光混合在一起的颜色非常相似。然而，用物理学的观点来看，它与混合物不同，因为它是严格均质的。正如你所见，棱镜并没有像对待混合光线那样把它分成两部分。让我们把这束

divide it into two portions as it did the mixture. Let us now combine this yellow with the blue of the spectrum. The result is certainly not green; we may make it pink if our yellow is of a warm hue, but if we choose a greenish yellow we can produce a good white.

You have now seen the most remarkable of the combinations of colours—the others differ from them in degree, not in kind. I must now ask you to think no more of the physical arrangements by which you were enabled to see these colours, and to concentrate your attention upon the colours you saw, that is to say, on certain sensations of which you were conscious. We are here surrounded by difficulties of a kind which we do not meet with in purely physical inquiries. We can all feel these sensations, but none of us can describe them. They are not only private property, but they are incommunicable. We have names for the external objects which excite our sensations, but not for the sensations themselves.

When we look at a broad field of uniform colour, whether it is really simple or compound, we find that the sensation of colour appears to our consciousness as one and indivisible. We cannot directly recognise the elementary sensations of which it is composed, as we can distinguish the component notes of a musical chord. A colour, therefore, must be regarded as a single thing, the quality of which is capable of variation.

To bring a quality within the grasp of exact science, we must conceive it as depending on the values of one or more variable quantities, and the first step in our scientific progress is to determine the number of these variables which are necessary and sufficient to determine the quality of a colour. We do not require any elaborate experiments to prove that the quality of colour can vary in three and only in three independent ways.

One way of expressing this is by saying, with the painters, that colour may vary in hue, tint, and shade.

The finest example of a series of colours varying in hue, is the spectrum itself. A difference in hue may be illustrated by the difference between adjoining colours in the spectrum. The series of hues in the spectrum is not complete; for, in order to get purple hues, we must blend the red and the blue.

Tint may be defined as the degree of purity of a colour. Thus, bright yellow, buff, and cream-colour, form a series of colours of nearly the same hue, but varying in tint. The tints corresponding to any given hue form a series, beginning with the most pronounced colour, and ending with a perfectly neutral tint.

Shade may be defined as the greater or less defect of illumination. If we begin with any tint of any hue, we can form a gradation from that colour to black, and this gradation is a series of shades of that colour. Thus we may say that brown is a dark shade of orange.

黄光和光谱中的蓝光混合起来。结果当然不是绿光；如果采用暖色调的黄光，我们得到的将是粉色，但如果我们选的是偏绿的黄光，就会得到很好的白色。

你已经看到了一些最显著的颜色组合，其他颜色组合与这些相比只有程度上的差别，而没有本质上的差别。现在，我请你别去考虑那些让你看到颜色的实验装置，而把你的注意力集中在你所看到的颜色上，也就是说，把注意力集中在你的感受上。我们遇到的困难是，我们无法进行纯粹的物理意义上的研究。我们都能感觉得到，但是谁也无法描述这一切。感觉不仅是个人的感受，而且难以表达出来。我们能说出那些刺激我们感受的外部物体的名字，但是无法描述那种感受本身。

当我们注视一大块均匀的颜色时，不论这种颜色是简单的还是复合的，我们发现在我们的意识中对颜色的感觉是一个不可分割的整体。我们无法像分辨和弦中的音符那样，直接把构成这种感受的元素分离出来。所以颜色应该被看作是一种单一的东西，而它的性质可以改变。

为了能用精确的科学术语描述一个量，我们必须建立这个量与一个或几个变量之间的依赖关系，我们研究工作的第一步是确定能充分必要地决定一种颜色性质的变量的数目。我们不需要任何复杂的实验就能证明颜色的性质依赖且只依赖于三个独立的变量。

画家们对此有一种说法，即颜色之间的区别是：色相、纯度和明度。

一系列色彩依色相变化的最好例子就是光谱本身。光谱中相邻颜色的差异可以用来说明色相的区别。光谱中的色相系列并不完全；因为要想得到紫色色相，我们必须混合红光和蓝光。

纯度可以被定义为一种颜色纯净的程度。这样，明黄色、浅黄色和奶黄色形成了一个具有几乎相同色相的系列，但纯度不同。对应于某个给定的色相，纯度不同的一组颜色可以形成一个系列，从最浓的颜色开始，到最淡的颜色结束。

明度可以被定义为光照程度的多少。如果我们从任意色相的任意纯度出发，就可以在这种颜色和黑色之间形成一个渐变，这个渐变就是这种颜色的一个明度序列。这样，我们就可以说，棕色是橙色明度变暗得到的结果。

The quality of a colour may vary in three different and independent ways. We cannot conceive of any others. In fact, if we adjust one colour to another, so as to agree in hue, in tint, and in shade, the two colours are absolutely indistinguishable. There are therefore three, and only three, ways in which a colour can vary.

I have purposely avoided introducing at this stage of our inquiry anything which may be called a scientific experiment, in order to show that we may determine the number of quantities upon which the variation of colour depends by means of our ordinary experience alone.

Here is a point in this room: if I wish to specify its position. I may do so by giving the measurements of three distances—namely, the height above the floor, the distance from the wall behind me, and the distance from the wall at my left hand.

This is only one of many ways of stating the position of a point, but it is one of the most convenient. Now, colour also depends on three things. If we call these the intensities of the three primary colour sensations, and if we are able in any way to measure these three intensities, we may consider the colour as specified by these three measurements. Hence the specification of a colour agrees with the specification of a point in the room in depending on three measurements.

Let us go a step farther, and suppose the colour sensations measured on some scale of intensity, and a point found for which the three distances, or co-ordinates, contain the same number of feet as the sensations contain degrees of intensity. Then we may say, by a useful geometrical convention, that the colour is represented to our mathematical imagination by the point so found in the room; and if there are several colours, represented by several points, the chromatic relations of the colours will be represented by the geometrical relations of the points. This method of expressing the relations of colours is a great help to the imagination. You will find these relations of colours stated in an exceedingly clear manner in Mr. Benson's "Manual of Colour", one of the very few books on colour in which the statements are founded on legitimate experiments.

There is a still more convenient method of representing the relations of colours, by means of Young's triangle of colours. It is impossible to represent on a plane piece of paper every conceivable colour, to do this requires space of three dimensions. If, however, we consider only colours of the same shade, that is, colours in which the sum of the intensities of the three sensations is the same, then the variations in tint and in hue of all such colours may be represented by points on a plane. For this purpose we must draw a plane cutting off equal lengths from the three lines representing the primary sensations. The part of this plane within the space in which we have been distributing our colours will be an equilateral triangle. The three primary colours will be at the three angles, white or gray will be in the middle, the tint or degree of purity of any colour will be expressed by its

一种颜色的性质可以按三种不同且独立的方式变化。我们想不出任何其他的变化方式。事实上，如果我们将一种颜色调整为另一种颜色，使得前者和后者在色相、纯度、明度上都一致，那么这两种颜色绝对没有任何差异。因此一种颜色的变化方式有且只有三种。

我有意避免在科学实验的层面上去介绍我们的研究，这样做是为了表明，我们仅凭日常经验就可以确定描述颜色变化的变量数目。

房间里有一个点，如果我想确定这个点的位置，我可以给出三个距离的测量值：即相对于地面的高度、到我身后的墙的距离以及到我左手边的墙的距离。

这只是描述一个点位置的多种方法中的一种，但它是最方便的一种。现在，颜色也同样依赖于三个变量。如果我们把这些称作三种原色的感觉强度，并且如果我们能用某种方法测量出这三种强度，那么我们就可以认为颜色能够通过对这三者的测量而被确定下来。因此，描述一种颜色和描述空间某点的位置一样，都要依赖于对三个变量的测量。

让我们再深入一步。假设用强度等级来衡量的色觉所包含的强度数值，与空间中某点所包含的用英尺数表示的三个距离，或三个坐标的数值相同，那么我们就可以说，通过实用几何学的常规做法，我们可以在数学上将色觉想象成空间中的某一点；如果有多种颜色，就用多个点来表示，那么这些颜色之间的关系也可以用点之间的几何学关系来描述。这样的描述对于我们想象不同颜色之间的关系大有帮助。在本森先生所著的《颜色手册》一书中，你会发现颜色之间的这些关系被叙述得非常清晰。基于正规实验的关于颜色方面的著作少之又少，而这本书就是其中之一。

还有一种更方便的描述颜色之间关系的方法，即杨的颜色三角形法。我们无法在一张纸的平面上描述任何一种可见的颜色，要做到这一点需要三维的空间。但是，如果我们只考虑具有相同明度的颜色，也就是说，在这些颜色中三种色觉的强度之和是一样的，那么纯度和色相的变化就可以用平面上的点来描述了。为此，我们用三条等长的、代表原色感觉的线来切割一个平面。它们所围的区域是一个等边三角形，我们将在这个区域中分配我们的颜色。三种原色将位于三个顶角，中间是白色或者灰色，颜色的纯度用这种颜色到中点的距离来表示，颜色的色相则取决于它与中点连线的角度。

distance from the middle point, and its hue will depend on the angular position of the line which joins it with the middle point.

Thus the ideas of tint and hue can be expressed geometrically on Young's triangle. To understand what is meant by shade, we have only to suppose the illumination of the whole triangle increased or diminished, so that by means of this adjustment of illumination Young's triangle may be made to exhibit every variety of colour. If we now take any two colours in the triangle and mix them in any proportions, we shall find the resultant colour in the line joining the component colours at the point corresponding to their centre of gravity.

I have said nothing about the nature of the three primary sensations, or what particular colours they most resemble. In order to lay down on paper the relations between actual colours, it is not necessary to know what the primary colours are. We may take any three colours, provisionally, as the angles of a triangle, and determine the position of any other observed colour with respect to these, so as to form a kind of chart of colours.

Of all colours which we see, those excited by the different rays of the prismatic spectrum have the greatest scientific importance. All light consists either of some one kind of these rays, or of some combination of them. The colours of all natural bodies are compounded of the colours of the spectrum. If, therefore, we can form a chromatic chart of the spectrum, expressing the relations between the colours of its different portions, then the colours of all natural bodies will be found within a certain boundary on the chart defined by the positions of the colours of the spectrum.

But the chart of the spectrum will also help us to the knowledge of the nature of the three primary sensations. Since every sensation is essentially a positive thing, every compound colour-sensation must be within the triangle of which the primary colours are the angles. In particular, the chart of the spectrum must be entirely within Young's triangle of colours, so that if any colour in the spectrum is identical with one of the colour-sensations, the chart of the spectrum must be in the form of a line having a sharp angle at the point corresponding to this colour.

I have already shown you how we can make a mixture of any three of the colours of the spectrum, and vary the colour of the mixture by altering the intensity of any of the three components. If we place a compound colour side by side with any other colour, we can alter the compound colour till it appears exactly similar to the other. This can be done with the greatest exactness when the resultant colour is nearly white. I have therefore constructed an instrument which I may call a colour-box, for the purpose of making matches between two colours. It can only be used by one observer at a time, and it requires daylight, so I have not brought it with me to-night. It is nothing but the realisation of the construction of one of Newton's propositions in his "Lectiones Opticae", where he shows how to take a beam of light, to separate it into its components, to deal with these

这样，纯度和色相就可以利用杨氏三角形得到一个几何上的描述。为了理解明度的含义，我们只需增加或者减弱整个三角的照明度就可以了。因此用调整照明度的方法，杨氏三角形可以表示所有的颜色。如果我们从杨氏三角形中选取任意两种颜色，然后把二者以任意比例混合在一起，混合后的颜色对应于这两种颜色连线上重心的位置。

我没有做任何有关三种原色感觉本质的说明，也没有说它们与哪些颜色更接近。要在本文中解释清楚实际颜色之间的关系，不需要知道三原色到底是什么。我们可以选取任意的三种颜色，暂且把它们放在杨氏三角形的三个顶点上，然后就可以确定其他可见颜色与它们的相对位置，这样就得到了一种色卡。

我们见到的所有被分光光谱中的不同光线所激发的颜色在科学上都具有极高的重要性。所有的光都是其中的一种光线，或者是其中几种光线的组合。自然界中实物的颜色都是由光谱中的颜色构成的。因此，如果我们能构造一个光谱的色卡，用颜色的不同位置来表示它们之间的关系，那么，自然界中所有物体的颜色都可以在用光谱中颜色的位置界定的色卡上找到它们的位置。

色卡还有助于我们了解三原色的本质。由于每一种感觉都是一种实实在在的东西，每一种复合的色觉都必然包含在以三原色为顶角的三角形中。特别是光谱的色卡一定完全包含在杨氏三角形内部，这样，如果光谱中任何一种颜色和某种色觉相一致，那么光谱在杨氏三角形中的形式一定是一条和这种颜色所在的点成很小角度的直线。

我已经告诉大家怎样将光谱中的任意三种颜色混合，并且用改变颜色三分量中任意一个的强度来改变这种混合的颜色。如果我们把一种混合颜色和另外一种颜色并列在一起，我们可以调整这种混合颜色，直到它和另外一种颜色完全相同为止。当最终要得到的颜色接近白色时，这个过程可以最为精确地完成。于是我构造了一种我称作色箱的装置，用来匹配两种颜色。这个装置每次实验只能允许一个人进行观察，而且需要在日光下进行，所以今晚我没有把它带来。这个装置没什么大不了，只不过实现了牛顿在《光学讲义》中谈到的一个构想而已，牛顿告诉我们如何获得一束光，并将其分离成不同组分，以及如何用狭缝来获取这些组分，然后再把它们

components as we please by means of slits, and afterwards to unite them into a beam again. The observer looks into the box through a small slit. He sees a round field of light, consisting of two semicircles divided by a vertical diameter. The semicircle on the left consists of light which has been enfeebled by two reflexions at the surface of glass. That on the right is a mixture of colours of the spectrum, the positions and intensities of which are regulated by a system of slits.

The observer forms a judgment respecting the colours of the two semicircles. Suppose he finds the one on the right hand redder than the other, he says so, and the operator, by means of screws outside the box, alters the breadth of one of the slits, so as to make the mixture less red; and so on, till the right semicircle is made exactly of the same appearance as the left, and the line of separation becomes almost invisible.

When the operator and the observer have worked together for some time they get to understand each other, and the colours are adjusted much more rapidly than at first.

When the match is pronounced perfect, the positions of the slits, as indicated by a scale, are registered, and the breadth of each slit is carefully measured by means of a gauge. The registered result of an observation is called a "colour equation". It asserts that a mixture of three colours is, in the opinion of the observer (whose name is given), identical with a neutral tint, which we shall call Standard White. Each colour is specified by the position of the slit on the scale, which indicates its position in the spectrum, and by the breadth of the slit, which is a measure of its intensity.

In order to make a survey of the spectrum we select three points for purposes of comparison, and we call these the three Standard Colours. The standard colours are selected on the same principles as those which guide the engineer in selecting stations for a survey. They must be conspicuous and invariable, and not in the same straight line.

In the chart of the spectrum you may see the relations of the various colours of the spectrum to the three standard colours, and to each other. It is manifest that the standard green which I have chosen cannot be one of the true primary colours, for the other colours do not all lie within the triangle formed by joining them. But the chart of the spectrum may be described as consisting of two straight lines meeting in a point. This point corresponds to a green about a fifth of the distance from *b* towards F. This green has a wavelength of about 510 millionths of a millimetre by Ditscheiner's measure. This green is either the true primary green, or at least it is the nearest approach to it which we can ever see. Proceeding from this green towards the red end of the spectrum, we find the different colours lying almost exactly in a straight line. This indicates that any colour is chromatically equivalent to a mixture of any two colours on opposite sides of it and in the same straight line. The extreme red is considerably beyond the standard red, but it is in the same straight line, and therefore we might, if we had no other evidence, assume the extreme red as the true primary red. We shall see, however, that the true primary red is

重新整合成一束光。观察者通过一个狭缝来观察箱子的内部。他将看到一个圆形的发光区域，一条垂直方向的直径把它分割成左右两个半圆。左边半圆是经过两次镜面反射而减弱的光；右边半圆是由光谱中的颜色混合而成的，其位置和强度都可以通过狭缝来调节。

观测者将对两边半圆的颜色进行判断。假如他认为右边的光比左边的光更红，那么他就可以让色箱的操作者通过拧紧箱外的螺丝来调节某个狭缝的宽度，使得混合光线的红色变浅，如此这般，直到左右两个半圆看起来完全相同，中间的分界线几乎看不出来为止。

操作者和观察者在一起工作过一段时间以后，他们的合作会更加默契，调整颜色的速度也会比初次合作时更快。

当颜色匹配完成之后，每一个狭缝位置的刻度都被记录下来，狭缝的宽度用刻度尺仔细测量。一次观察的记录结果被称为一个“颜色方程”。它说明观察者（他的名字将被记录下来）认为，三种颜色混合而成的颜色是一种中性色，我们称之为标准白色。每一种颜色在光谱中的位置都由狭缝的位置确定，而狭缝的宽度则表示了它的强度。

为了考察光谱的特性，我们选择三个点用以比较，我们称它们为三个标准色。标准色的选择原则与工程师选取观测点的原则相同，这些点必须既突出又稳定，且不在同一条直线上。

在光谱的色卡上，你可以看到光谱中的不同颜色和三个标准色之间的关系，以及不同颜色之间的关系。事实表明，我选择的标准绿色不可能是三原色之一，因为其他颜色并不全在三点之间的区域中。但是光谱色卡可以描述成两条相交的直线。相应的交点对应一种绿色，它到标准绿色的距离为 b 到 F 距离的 1/5。根据迪特沙纳的测量，这种绿色的波长是 510 纳米。这种绿色即便不是真正的原色，至少也是我们曾经见到过的最接近原色的颜色。从这种绿色向光谱中的红色一端连线，我们发现不同的颜色几乎都落在了这条直线上。这表明，从色度上来说，任一种颜色都等价于位于该直线两端的两种颜色的混合。极端红色应该在比标准红色更远的位置，但是它与标准红色在同一条直线上，因此，如果没有相反的证据，我们可以将极端红色视为原红。然而，我们可以看到，真正的原红并没有出现在光谱的任何部分。它在比极端红色更远的位置，但仍然在同一条直线上。

not exactly represented in colour by any part of the spectrum. It lies somewhat beyond the extreme red but in the same straight line.

On the blue side of primary green the colour equations are seldom so accurate. The colours, however, lie in a line which is nearly straight. I have not been able to detect any measurable chromatic difference between the extreme indigo and the violet. The colours of this end of the spectrum are represented by a number of points very close to each other. We may suppose that the primary blue is a sensation differing little from that excited by the parts of the spectrum near G.

Now, the first thing which occurs to most people about this result is that the division of the spectrum is by no means a fair one. Between the red and the green we have a series of colours apparently very different from either, and having such marked characteristics that two of them, orange and yellow, have received separate names. The colours between the green and the blue, on the other hand, have an obvious resemblance to one or both of the extreme colours, and no distinct names for these colours have ever become popularly recognised.

I do not profess to reconcile this discrepancy between ordinary and scientific experience. It only shows that it is impossible, by a mere act of introspection, to make a true analysis of our sensations. Consciousness is our only authority; but consciousness must be methodically examined in order to obtain any trustworthy results.

I have here, through the kindness of Professor Huxley, a picture of the structure upon which the light falls at the back of the eye. There is a minute structure of bodies like rods and cones or pegs, and it is conceivable that the mode in which we become aware of the shapes of things is by a consciousness which differs according to the particular rods on the ends of which the light falls, just as the pattern on the web formed by a Jacquard loom depends on the mode in which the perforated cards act on the system of movable rods in that machine. In the eye we have on the one hand light falling on this wonderful structure, and on the other hand we have the sensation of sight. We cannot compare these two things; they belong to opposite categories. The whole of Metaphysics lies like a great gulf between them. It is possible that discoveries in physiology may be made by tracing the course of the nervous disturbance

Up the fine fibres to the sentient brain;

but this would make us no wiser than we are about those colour-sensations which we can only know by feeling them ourselves. Still, though it is impossible to become acquainted with a sensation by the anatomical study of the organ with which it is connected, we may make use of the sensation as a means of investigating the anatomical structure.

A remarkable instance of this is the deduction of Helmholtz's theory of the structure of the retina from that of Young with respect to the sensation of colour. Yong asserts that there are three elementary sensations of colour; Helmholtz asserts that there are three systems of nerves in the retina, each of which has for its function, when acted on by light or any other disturbing agent, to excite in us one of these three sensations.

在原绿的蓝端，颜色方程就不是那么精确了。但色点近似分布于一条直线上。我现在还无法测量出极端靛青和紫色的区别。在光谱中这一端的颜色是用许多非常接近的点来表示的。我们可以假设原蓝这种色觉略微区别于由光谱中靠近 G 的部分所激发的感觉。

现在，面对这样的结果，摆在大家面前的首要问题是，光谱的划分并不公平。红绿之间的一系列颜色都有明显的区别，它们的区分非常明显，以至于黄色和橙色需要有各自不同的名字。反之，绿蓝之间的颜色却与这两种极端颜色或其中之一很相似，这些被广泛认可的颜色也没有自己的名字。

我并非是要调和这种一般经验和科学实验之间的差异和矛盾。只是事实表明，仅仅用自省的方式不可能对我们的感觉作出正确的分析。感觉是我们唯一的凭据，但是感觉必须经过系统的检验才能得到可靠的结果。

我从赫胥黎教授那里得到了一张描述光线落在眼睛后部成像的结构图。这里有很多棒状、锥状、钉状的微结构。我们很可能就是通过确定光线到底落在哪些棒状体的末端而感觉到物体的形状的，就像提花织布机织出什么样的花纹取决于打孔卡作用于机器中可移动棒的方式。在眼睛里，一方面光线照射在这种精密的结构之上，另一方面我们有视觉感受。我们无法比较这两个方面，因为它们属于不同的范畴。形而上学就是二者之间的鸿沟。跟踪

从神经纤维到大脑之间

的神经扰动可能会在生理学上得到一些发现；但是，这些对我们关于色觉的认识并没什么帮助，因为我们只能靠自己去感受颜色。虽然我们不可能通过解剖相关的器官来增加对色觉的了解，但是我们可以利用我们的感觉，把它作为研究组织结构的一种手段。

这里有一个著名的例子，就是从杨的色觉理论推出亥姆霍兹关于视网膜结构的理论。杨声称有三种基本的色觉；而亥姆霍兹则声称在视网膜里有三种神经系统，每一种系统都有自己的功能，当有光照或者其他扰动作用时，每一种系统都会激发出我们这三种感觉中的一种。

No anatomist has hitherto been able to distinguish these three systems of nerves by microscopic observation. But it is admitted in physiology that the only way in which the sensation excited by a particular nerve can vary is by degrees of intensity. The intensity of the sensation may vary from the faintest impression up to an insupportable pain; but whatever be the exciting cause, the sensation will be the same when it reaches the same intensity. If this doctrine of the function of a nerve be admitted, it is legitimate to reason from the fact that colour may vary in three different ways, to the inference that these three modes of variation arise from the independent action of three different nerves or sets of nerves.

Some very remarkable observations on the sensation of colour have been made by M. Sigmund Exner in Prof. Helmholtz's physiological laboratory at Heidelberg. While looking at an intense light of a brilliant colour, he exposed his eye to rapid alternations of light and darkness by waving his fingers before his eyes. Under these circumstances a peculiar minute structure made its appearance in the field of view, which many of us may have casually observed. M. Exner states that the character of this structure is different according to the colour of the light employed. When red light is used a veined structure is seen; when the light is green, the field appears covered with minute black dots, and when the light is blue, spots are seen, of a larger size than the dots in the green, and of a lighter colour.

Whether these appearances present themselves to all eyes, and whether they have for their physical cause any difference in the arrangement of the nerves of the three systems in Helmholtz's theory I cannot say, but I am sure that if these systems of nerves have a real existence, no method is more likely to demonstrate their existence than that which M. Exner has followed.

Colour Blindness

The most valuable evidence which we possess with respect to colour vision is furnished to us by the colour-blind. A considerable number of persons in every large community are unable to distinguish between certain pairs of colours which to ordinary people appear in glaring contrast. Dr. Dalton, the founder of the atomic theory of chemistry, has given us an account of his own case.

The true nature of this peculiarity of vision was first pointed out by Sir John Herschel in a letter written to Dalton in 1832, but not known to the world till the publication of "Dalton's Life" by Dr. Henry. The defect consists in the absence of one of the three primary sensations of colour. Colour-blind vision depends on the variable intensities of two sensations instead of three. The best description of colour-blind vision is that given by Prof. Pole in his account of his own case in the "Phil. Trans.", 1859.

In all cases which have been examined with sufficient care, the absent sensation appears to resemble that which we call red. The point P on the chart of the spectrum represents the

到目前为止，还没有任何一个解剖学家能够在微观尺度的观察中分辨出这三种神经系统。但是生理学上却认为，由某种神经激发的感觉只能在强度上有所变化。感觉强度的变化可以从最微弱的触感到难以忍受的疼痛；但不论激发感觉的原因是什么，只要激发的强度相同，那么感受也相同。如果这种神经功能的学说得到认可，那么从颜色能够以三种不同的方式变化这个事实中，就可以推断出这三种不同的色觉模式起源于三种不同的神经或者神经集合。

在亥姆霍兹教授位于海德堡的生理学实验室里，西格蒙德·埃克斯纳作出了一些非常引人注目的关于色觉的观察结果。当注视一种色彩耀眼的强光时，他不断地在眼前挥动手指，使眼睛迅速地在明亮和黑暗之间切换。在这种情况下，一种奇异的微结构出现在了视野当中，可能很多人都曾偶然发现过这个现象。埃克斯纳声称，该结构的特征随着光源颜色的不同而变化。红光照射时，见到的是叶脉结构；绿光照射时，视野中好像布满了小黑点；蓝光照射时，看到的斑点比绿光中的更大，颜色也更淡。

我不知道是否每个人都能感受到这些现象，也不知道亥姆霍兹理论中所说的三种神经系统的排布是否会由于个体之间的差异而在不同人中有所不同，但是我确信，如果这样的神经系统真的存在的话，那么埃克斯纳所用的方法就是最好的证明办法。

色 盲

色盲现象为我们提供了关于色觉的最有价值的证据。在每一个大型社区中都有相当多的人无法分辨出在正常人看来区别很明显的一些颜色。化学原子理论的奠基人道尔顿博士本人就为我们提供了例子。

1832 年，约翰·赫歇尔爵士在他写给道尔顿博士的一封信中第一次指出了这种异常色觉现象的本质，但是直到亨利博士出版《道尔顿生平》一书时，这封信的内容才公诸于世。这种缺陷是由于缺少三种原色感觉中的一种而造成的。色盲者的视觉只依赖于两种色觉的强度变化，而不是三种。波尔教授在 1859 年的《自然科学会报》上给出了他自己的亲身体验，这是迄今为止对色盲现象的最佳描述。

在所有经过精心检验的例子中，我们发现，那种缺失的色觉好像类似于我们所称的红色。光谱色卡上的 P 点代表了缺失的色觉和光谱中颜色的关系，这是根据波

relation of the absent sensation to the colours of the spectrum, deduced from observations with the colour box furnished by Prof. Pole.

If it were possible to exhibit the colour corresponding to this point on the chart, it would be invisible, absolutely black, to Prof. Pole. As it does not lie within the range of the colours of the spectrum we cannot exhibit it; and, in fact, colour-blind people can perceive the extreme end of the spectrum which we call red, though it appears to them much darker than to us, and does not excite in them the sensation which we call red. In the diagram of the intensities of the three sensations excited by different parts of the spectrum, the upper figure, marked P, is deduced from the observations of Prof. Pole; while the lower one, marked K, is founded on observations by a very accurate observer of the normal type.

The only difference between the two diagrams is that in the upper one the red curve is absent. The forms of the other two curves are nearly the same for both observers. We have great reason therefore to conclude that the colour sensations which Prof. Pole sees are what we call green and blue. This is the result of my calculations; but Prof. Pole agrees with every other colour-blind person whom I know in denying that green is one of his sensations. The colour-blind are always making mistakes about green things and confounding them with red. The colours they have no doubts about are certainly blue and yellow, and they persist in saying that yellow, and not green, is the colour which they are able to see.

To explain this discrepancy we must remember that colour-blind persons learn the names of colours by the same method as ourselves. They are told that the sky is blue, that grass is green, that gold is yellow, and that soldiers' coats are red. They observe difference in the colours of these objects, and they often suppose that they see the same colours as we do, only not so well. But if we look at the diagram we shall see that the brightest example of their second sensation in the spectrum is not in the green, but in the part which we call yellow, and which we teach them to call yellow. The figure of the spectrum below Prof. Pole's curves is intended to represent to ordinary eyes what a colour-blind person would see in the spectrum. I hardly dare to draw your attention to it, for if you were to think that any painted picture would enable you to see with other people's vision I should certainly have lectured in vain.

On the Yellow Spot

Experiments on colour indicate very considerable differences between the vision of different persons, all of whom are of the ordinary type. A colour, for instance, which one person on comparing it with white will pronounce pinkish, another person will pronounce greenish. This difference, however, does not arise from any diversity in the nature of the colour sensations in different persons. It is exactly of the same kind as would be observed if one of the persons wore yellow spectacles. In fact, most of us have near the middle of the retina a yellow spot through which the rays must pass before they reach the sensitive

尔教授提供的色箱进行观察而推出的结果。

如果可以将色卡上这一点所代表的颜色呈现在波尔教授眼前，那么他将什么也看不见，或者说眼前一片黑暗。由于这种颜色不在光谱的范围之内，所以我们无法将其呈现出来。事实上，色盲者能够看到光谱的红端，尽管他们看到的红色比我们看到的暗很多，因而无法激发出我们正常人所感受到的红色。在由光谱不同部分激发的三种基本色觉的强度图中，上图中标为P的点是从波尔教授的观察结果推导出来的，而下图中标为K的点是一个色觉正常的观察者经过精确实验得到的结果。

两张图之间的唯一区别是，上图中缺少红色的曲线。对这两个观察者来说，另外两条曲线的形状几乎是完全相同的。因此，我们可以非常肯定地说，波尔教授能感受到的颜色是我们所称的绿色和蓝色。这就是我的计算结果，但是波尔教授，还有我认识的其他色盲者都不承认他们能感受到绿色。色盲者经常会把绿色的东西看错或者把红色和绿色搞混。色盲者肯定能看到的颜色是蓝色和黄色，他们坚持认为，他们能看到的颜色是黄色而不是绿色。

要想解释这个矛盾，我们必须知道，色盲者了解颜色名字的方式和我们正常人相同。别人告诉他们，天空是蓝色的，草地是绿色的，金子是黄色的，军装是红色的。他们观察这些物体在颜色上的差别，他们以为他们看到的颜色与我们看到的一样，只是不那么清楚而已。但如果我们看一看这张图，就会发现，在他们的第二种色觉中，最明亮的部分并不是光谱中的绿色，而是我们称为黄色的部分，我们告诉他们这就是黄色。波尔教授所绘曲线下面的光谱图向色觉正常的人展示了一个色盲者的眼睛所看到的光谱。其实我不敢让大家注意它，因为要是你认为自己可以用别人的视觉来看一幅画的话，那我前面所说的就都白费了。

关于黄色斑点

关于颜色的种种实验表明，人与人之间在视觉上的差别很明显，虽然他们都是视觉正常的人。比如，把一种颜色和白色作比较时，有人认为它偏粉色，有人认为它偏绿色。然而，这种差别并不能说明每个人的色觉有本质上的不同。这就好像有人戴上了黄色眼镜观察事物。事实上，我们中的大多数人在接近视网膜中部的地方都有一个黄色的斑点，光线必须穿过这个黄斑才能到达感觉器官。这个斑点之所以

organ: this spot appears yellow because it absorbs the rays near the line F, which are of a greenish-blue colour. Some of us have this spot strongly developed. My own observations of the spectrum near the line F are of very little value on this account. I am indebted to Professor Stokes for the knowledge of a method by which any one may see whether he has this yellow spot. It consists in looking at a white object through a solution of chloride of chromium, or at a screen on which light which has passed through this solution is thrown. This light is a mixture of red light with the light which is so strongly absorbed by the yellow spot. When it falls on the ordinary surface of the retina it is of a neutral tint, but when it falls on the yellow spot only the red light reaches the optic nerve, and we see a red spot floating like a rosy cloud over the illuminated field.

Very few persons are unable to detect the yellow spot in this way. The observer K, whose colour equations have been used in preparing the chart of the spectrum, is one of the very few who do not see everything as it through yellow spectacles. As for myself, the position of white light in the chart of the spectrum is on the yellow side of true white even when I use the outer parts of the retina; but as soon as I look direct at it, it becomes much yellower, as is shown by the point W C. It is a curious fact that we do not see this yellow spot on every occasion, and that we do not think white objects yellow. But if we wear spectacles of any colour for some time, or if we live in a room lighted by windows all of one colour, we soon come to recognise white paper as white. This shows that it is only when some alteration takes place in our sensations that we are conscious of their quality.

There are several interesting facts about the colour sensation which I can only mention briefly. One is that the extreme parts of the retina are nearly insensible to red. If you hold a red flower and a blue flower in your hand as far back as you can see your hand, you will lose sight of the red flower, while you still see the blue one. Another is, that when the light is diminished red objects become darkened more in proportion than blue ones. The third is, that a kind of colour blindness in which blue is the absent sensation can be produced artificially by taking doses of santonine. This kind of colour blindness is described by Dr. Edmund Rose, of Berlin. It is only temporary, and does not appear to be followed by any more serious consequences than headaches. I must ask your pardon for not having undergone a course of this medicine, even for the sake of becoming able to give you information at first hand about colour-blindness.

(**4**, 13-16; 1871)

呈现黄色，是因为它吸收了F线附近蓝绿色的光线。有些人的这个斑点非常粗大。由于这个原因，我自己看到的光谱在F线附近的区域就很微弱。我要感谢斯托克斯教授教给我判断一个人是否长有这种黄斑的方法。方法如下：让观察者通过氯化铬溶液观察一个白色物体，或者将一束光透过氯化铬溶液投射到屏幕上，让观察者去看这个屏幕。这束光是由红光和会被黄斑强烈吸收的光混合而成。当这束光线投射到正常人的视网膜上时，看到的会是一种中性色；但当它投射到黄斑上时，只有红光能够到达视神经，于是我们在被照亮的区域将看到一团像红云一般浮动着的红色斑点。

用这种方法检测发现只有极少数人没有这个黄斑。观察者K，就是其颜色方程曾被用于制作光谱色卡的那位，是极少数不用透过黄色眼镜看世界的人之一。至于我，在我的色卡上，白光的位置在真正的白色偏黄的一侧，即便我用视网膜的外侧观察也是如此；不过，当我直视的时候，我看到的白色位置就更加偏黄了，正如W C点所示。奇怪的是，我们并不是在所有情况下都能看到这个黄斑，而且我们也不会认为白色物体是黄色的。但如果我们戴上任意颜色的眼镜后过一段时间，或者我们所住房间的窗户都是同一种颜色时，我们很快就会认出白纸是白色的。这表明，只有当我们的感觉发生了一些改变的时候，我们才能意识到它们的性质。

最后，我只能简单介绍几个关于色觉的有趣事例。一个是，视网膜最外面的部分几乎感受不到红色。如果你手里拿着一枝红花和一枝蓝花，然后把手放到你身后几乎看不见的位置，这时你可能就看不见红花了，但仍然能看见蓝花。另一个是，当光减弱的时候，红色物体将比蓝色物体更快地变暗。第三，服用大剂量的山道年能人为地造成一种不能识别蓝色的假性色盲状态。柏林的埃德蒙·罗泽博士谈到过这种色盲。这只是一种暂时的状态，除了头痛以外不会有其他更严重的后果。我必须请求大家的原谅，因为我没有服用过这种药物，尽管我知道这样做可以给你们提供关于色盲的第一手资料。

（王静 翻译；江丕栋 审稿）

The Copley Medalist of 1870

J. Tyndall

Editor's Note

James prescott Joule was born in the city of Salford in Lancashire and worked either there or in the adjacent city of Manchester. His scientific work centred on electricity and the rules by which energy of one kind—a current of electricity, for example—may be converted into energy of some other kind (such as mechanical work). Joule is now commemorated by the use of his name as the unit of energy in the MKS system. The Copley medal, which was awarded to Joule in 1870, is the Royal Society's most venerable award. The author of this article, John Tyndall, was a professor at the institution in Manchester that eventually became its first university, but had by 1871 become a professor at the Royal Institution in London.

THIRTY years ago Electro-magnetism was looked to as a motive power which might possibly compete with steam. In centres of industry, such as Manchester, attempts to investigate and apply this power were numerous, as shown by the scientific literature of the time. Among others Mr. James prescott Joule, a resident of Manchester, took up the subject, and in a series of papers published in Sturgeon's "Annals of Electricity" between 1839 and 1841, described various attempts at the construction and perfection of electro-magnetic engines. The spirit in which Mr. Joule pursued these inquiries is revealed in the following extract: "I am particularly anxious," he says, "to communicate any new arrangement in order, if possible, to forestal the monopolising designs of those who seem to regard this most interesting subject merely in the light of pecuniary speculation." He was naturally led to investigate the laws of electro-magnetic attractions, and in 1840 he announced the important principle that the attractive force exerted by two electro-magnets, or by an electro-magnet and a mass of annealed iron, is directly proportional to the square of the strength of the magnetising current; while the attraction exerted between an electro-magnet and the pole of a permanent steel magnet varies simply as the strength of the current. These investigations were conducted independently of, though a little subsequently to, the celebrated inquiries of Henry, Jacobi, and Lenz and Jacobi on the same subject.

On the 17th of December, 1840, Mr. Joule communicated to the Royal Society a paper on the production of heat by Voltaic electricity; in which he announced the law that the calorific effects of equal quantities of transmitted electricity are proportional to the resistance overcome by the current, whatever may be the length, thickness, shape, or character of the metal which closes the circuit; and also proportional to the square of the quantity of transmitted electricity. This is a law of primary importance. In another paper, presented to but declined by the Royal Society, he confirmed this law by new experiments, and materially extended it. He also executed experiments on the heat consequent on the passage of Voltaic electricity through electrolytes, and found in all cases that the heat

1870 年的科普利奖章获得者

廷德尔

编者按

詹姆斯·普雷斯科特·焦耳出生在兰开夏郡的索尔福德，他在他的家乡（或者说曼彻斯特临近的城市）工作。他的研究工作集中在电学和一种能量形式（比如电流）转化成另外的能量形式（比如动能）的规律上。为了纪念焦耳，米–千克–秒单位制系统采用他的名字作为能量单位。1870 年，焦耳被授予科普利奖章，这是皇家学会的最高荣誉。这篇文章的作者约翰·廷德尔是曼彻斯特研究所的教授，该研究所最终成为了当地的第一所大学。但在 1871 年之前作者就已经成为伦敦皇家研究院的一名教授。

30 年前，人们希望电磁能能够成为一种可以和蒸汽相媲美的动力形式。在像曼彻斯特那样的工业中心，人们对这种动力的研究和应用十分广泛，这可以从当时的科学文献中看出来。和其他人一样，曼彻斯特人詹姆斯·普雷斯科特·焦耳也研究了这个课题，并于 1839 ~ 1841 年间在斯特金的《电学年鉴》上发表了一系列文章，描述了设计和完善电磁发动机的各种尝试。焦耳先生的探索精神可以从下面的话中体现出来，他说，“我非常急切地要把所有适宜的新成果公之于众，为的是尽量抢在那些仅仅为了金钱而想把这个极为有趣的课题垄断掉的人前面。”他很自然地把研究转向了电磁吸引作用的规律，并在 1840 年宣布了一个重要的定律：两个电磁体或者电磁体和退火铁之间的吸引力正比于磁化电流强度的平方；电磁体和永磁体磁极之间的相互作用只与电流强度成正比。这些研究是焦耳独立进行的，尽管比亨利、雅各比以及楞次和雅各比通过著名实验得出的相同结果略晚一些。

1840 年 12 月 17 日，焦耳先生在皇家学会宣读了他关于伏打电流发热的论文；在论文中，他宣称，等量的传输电流产生的热效应跟电流所克服的电阻成正比，与闭合电路中金属的长度、厚度、形状和特性无关，并且与电流的平方成正比。这是一个非常重要的定律。在另一篇提交给皇家学会但遭退稿的文章中，焦耳用新的实验证实了这个定律，并将其大大扩展。他还针对伏打电流通过电解液时的热效应进行了实验研究，最终发现，在所有情况下，伏打电流产生的热量都正比于电流强度的平方与电阻的乘积。通过这个定律，他推出了一些在电化学领域极为重要的结论。

evolved by the proper action of any Voltaic current is proportional to the square of the intensity of that current multiplied by the resistance to conduction which it experiences. From this law he deduced a number of conclusions of the highest importance to electro-chemistry.

It was during these inquiries, which are marked throughout by rare sagacity and originality, that the great idea of establishing quantitative relations between Mechanical Energy and Heat arose and assumed definite form in his mind. In 1843 Mr. Joule read before the meeting of the British Association at Cork a paper "On the Calorific Effects of Magneto-Electricity and on the Mechanical Value of Heat". Even at the present day this memoir is tough reading, and at the time it was written it must have appeared hopelessly entangled. This I should think was the reason why Prof. Faraday advised Mr. Joule not to submit the paper to the Royal Society. But its drift and results are summed up in these memorable words by its author, written some time subsequently: "In that paper it was demonstrated experimentally that the mechanical power exerted in turning a magneto-electric machine is converted into the heat evolved by the passage of the currents of induction through its coils, and on the other hand, that the motive power of the electro-magnetic engine is obtained at the expense of the heat due to the chemical reaction of the battery by which it is worked."* It is needless to dwell upon the weight and importance of this statement.

Considering the imperfections incidental to a first determination, it is not surprising that the "mechanical values of heat," deduced from the different series of experiments published in 1843, varied somewhat widely from each other. The lowest limit was 587, and the highest 1,026 foot-pounds for 1°F of temperature.

One noteworthy result of his inquiries, which was pointed out at the time by Mr. Joule, had reference to the exceedingly small fraction of the heat which is actually converted into useful effect in the steam-engine. The thoughts of the celebrated Julius Robert Mayer, who was then engaged in Germany upon the same question, had moved independently in the same groove; but to his labours due reference will doubtless be made on a future occasion. In the memoir now referred to Mr. Joule also announced that he had proved heat to be evolved during the passage of water through narrow tubes; and he deduced from these experiments an equivalent of 770 foot-pounds, a figure remarkably near to the one now accepted. A detached statement regarding the origin and convertibility of animal heat strikingly illustrates the penetration of Mr. Joule and his mastery of principles at the period now referred to. A friend had mentioned to him Haller's hypothesis, that animal heat might arise from the friction of the blood in the veins and arteries. "It is unquestionable," writes Mr. Joule, "that heat is produced by such friction, but it must be understood that the mechanical force expended in the friction is a part of the force of affinity which causes the venous blood to unite with oxygen, so that the whole heat of the system must still be referred to the chemical changes. But if the animal were engaged in turning a piece of machinery, or in ascending a mountain, I apprehend that in proportion

* *Phil. Mag.* May 1845.

在从事这些充满超人智慧和独创性的研究过程中，焦耳萌生并逐渐形成了在机械能和热量之间建立定量关系的想法。1843 年焦耳先生在英国科学促进会于科克召开的会议上宣读了一篇《论磁电的热效应和热的机械值》的论文。即便在今天，这个研究报告都是很难读懂的，在当时要读懂它简直是毫无希望的。我想这就是法拉第教授建议焦耳先生不要将它提交给皇家学会的原因。但隔了一段时间以后，焦耳对文章的主旨和结论作出了如下的概括，给人留下极深的印象："那篇文章用实验证明了施加于磁电式电机上的机械能被转化成线圈中感应电流所释放的热量，另一方面，电磁发动机的动力则来源于电池化学反应所提供的热量。"* 这个结论的重要性和价值也就不必在此赘述了。

在 1843 年发表的论文中，一系列实验所得到的"热的机械值"差别较大，考虑到首次得出结论的实验所要面对的各种偶然因素，这也就不足为怪了。对应于温度上升华氏 1° 所做的功，测量得到的最低值是 587 英尺磅，最高值是 1,026 英尺磅。

当时焦耳先生指出，在他的研究中有一个著名的结论，即蒸汽机的热能转化为有效能量的效率非常低。著名的尤利乌斯·罗伯特·迈尔后来在德国也研究了这个问题，并独立地得到了同样的结论，但是他的观点还需要在未来的实验中去确证。焦耳先生在他的研究报告中还指出，他发现水流过细管子时会放出热量，通过实验他得到的数值是 770 英尺磅，这个值非常接近于今天我们承认的数值。另一份关于动物热量起源和转化的报告突出地表明了焦耳先生敏锐的洞察力和他在那个年代对规律的掌控能力。一位朋友曾经向他提到过哈勒尔的假说，该假说认为动物的热量可能起源于血液在静脉和动脉中流动时的摩擦。"这是毫无疑问的，"焦耳在回信中写到，"热量就是由这样的摩擦产生的，但是要知道，在摩擦中消耗的机械力也是使静脉血与氧结合的亲和力引发的，所以动物系统的全部热量必然来自于化学反应。但动物在搬运东西或者上山的时候，我认为，动物将消耗其通过化学反应释放出的热量，而系统热量的**减少**与动物肌肉为完成这项工作所做的功成正比。"句子中突出强调的部分是焦耳在 1843 年的原稿中就标注了的。

*《哲学杂志》，1845 年 5 月。

to the muscular effort put forth for the purpose, a *diminution* of the heat evolved in the system by a given chemical action would be experienced." The italics in this memorable passage, written it is to be remembered in 1843, are Mr. Joule's own.

The concluding paragraph of this British Association paper equally illustrates his insight and precision regarding the nature of chemical and latent heat. "I had," he writes, "endeavoured to prove that when two atoms combine together, the heat evolved is exactly that which would have been evolved by the electrical current due to the chemical action taking place, and is therefore proportional to the intensity of the chemical force causing the atoms to combine. I now venture to state more explicitly, that it is not precisely the attraction of affinity, but rather the mechanical force expended by the atoms in falling towards one another, which determines the intensity of the current, and, consequently, the quantity of heat evolved; so that we have a simple hypothesis by which we may explain why heat is evolved so freely in the combination of gases, and by which indeed we may account 'latent heat' as a mechanical power prepared for action as a watch-spring is when wound up. Suppose, for the sake of illustration, that 8 lbs. of oxygen and 1 lb. of hydrogen were presented to one another in the gaseous state, and then exploded; the heat evolved would be about 1°F in 60,000 lbs. of water, indicating a mechanical force expended in the combination equal to a weight of about 50,000,000 lbs. raised to the height of one foot. Now if the oxygen and hydrogen could be presented to each other in a liquid state, the heat of combination would be less than before, because the atoms in combining would fall through less space." No words of mine are needed to point out the commanding grasp of molecular physics, in their relation to the mechanical theory of heat, implied by this statement.

Perfectly assured of the importance of the principle which his experiments aimed at establishing, Mr. Joule did not rest content with results presenting such discrepancies as those above referred to. He resorted in 1844 to entirely new methods, and made elaborate experiments on the thermal changes produced in air during its expansion: firstly, against a pressure, and therefore performing work; secondly, against no pressure, and therefore performing no work. He thus established anew the relation between the heat consumed and the work done. From five different series of experiments he deduced five different mechanical equivalents; the agreement between them being far greater than that attained in his first experiments. The mean of them was 802 foot-pounds. From experiments with water agitated by a paddle-wheel, he deduced, in 1845, an equivalent of 890 foot-pounds. In 1847 he again operated upon water and sperm-oil, agitated them by a paddle-wheel, determined their elevation of temperature, and the mechanical power which produced it. From the one he derived an equivalent of 781.5 foot-pounds; from the other an equivalent of 782.1 foot-pounds. The mean of these two very close determinations is 781.8 foot-pounds.

At this time the labours of the previous ten years had made Mr. Joule completely master of the conditions essential to accuracy and success. Bringing his ripened experience to bear upon the subject, he executed in 1849 a series of 40 experiments on the friction of

焦耳对于化学能与潜热的精确洞察力在这篇英国科学促进会论文的结束语中也有所体现。他写到，“我曾经竭力去证明两个原子结合时所释放的热量与化学反应产生的电流所释放的热量相等，也就是与使原子结合的化学力的强度成正比。现在，我要大胆地说得更明确一点，吸引力这个说法并不确切，应该说是两个原子在相互靠近时所克服的机械力，这个力决定了电流的强度，也决定了释放的热量。这样，我们就得到了一个简单的假说，可以用它来解释为什么在气体发生化合反应的过程中热量可以自由地释放，根据这个假说我们真的可以把‘潜热’看作是一种储备的机械力，就像上好的手表发条一样。例如，假设 8 磅氧和 1 磅氢在气态下混合在一起，然后发生了爆炸。爆炸所释放的热量差不多可以使 60,000 磅水的温度上升华氏 1°，说明化合反应释放的机械能相当于把 50,000,000 磅的重物从地面提升 1 英尺所做的功。如果氧和氢可以在液态下发生反应，那么化合反应放出的热将小于气态的情况，因为原子之间的结合只需克服更小的距离。”这些陈述高屋建瓴地把握了分子物理学和热力学之间的关系，对此我无需再赘述了。

焦耳先生极好地证实了他的实验所要建立的原理的重要性，但是他并没有满足于此，因为正如上面所提到的那样，实验结果还存在较大的差异。他于 1844 年采用全新的方法，精确地测量了空气在膨胀过程中的热量变化：首先，在抵抗外部压力的膨胀中，空气对外做功；其次，空气自由膨胀时，不对外做功。他用这种方式重新建立了热量消耗和做功之间的关系。他从 5 组不同的实验中推出了 5 个不同的热功当量；这些数据的吻合程度远远好于他以前的实验结果。测量得到的平均值是 802 英尺磅。1845 年，焦耳用叶轮搅动水的实验测得的热功当量值是 890 英尺磅。1847 年，他又用叶轮搅动水和鲸油，重复了上面的实验，测定了温度升高的度数，也就得到了致使温度升高的机械能的大小。一次实验得到的结果是 781.5 英尺磅，另一次实验的结果是 782.1 英尺磅。两者非常接近，它们的平均值是 781.8 英尺磅。

到目前为止，焦耳前 10 年的工作经验已经使他能够相当熟练地掌控与实验精度和成败密切相关的各种条件。1849 年焦耳将他的成熟经验应用于热功当量的测定之中，做了 40 组水的摩擦实验，50 组水银的摩擦实验和 20 组铸铁盘的摩擦实验。他

water, 50 experiments on the friction of mercury, and 20 experiments on the friction of plates of cast-iron. He deduced from these experiments our present mechanical equivalent of heat, justly recognised all over the world as "Joule's equivalent".

There are labours so great and so pregnant in consequences, that they are most highly praised when they are most simply stated. Such are the labours of Mr. Joule. They constitute the experimental foundation of a principle of incalculable moment, not only to the practice, but still more to the philosophy of Science. Since the days of Newton, nothing more important than the theory of which Mr. Joule is the experimental demonstrator has been enunciated.

I have omitted all reference to the numerous minor papers with which Mr. Joule has enriched scientific literature. Nor have I alluded to the important investigations which he has conducted jointly with Sir William Thomson. But sufficient, I think, has been here said to show that, in conferring upon Mr. Joule the highest honour of the Royal Society, the Council paid to genius not only a well-won tribute, but one which had been fairly earned twenty years previously.*

Comparing this brief history with that of the Copley Medalist of 1871, the differentiating influence of "environment" on two minds of similar natural cast and endowment comes up in an instructive manner. Withdrawn from mechanical appliances, Mayer fell back upon reflection, selecting with marvellous sagacity from existing physical data the single result on which could be founded a calculation of the mechanical equivalent of heat. In the midst of mechanical appliances, Joule resorted to experiment, and laid the broad and firm foundation which has secured for the mechanical theory the acceptance it now enjoys. A great portion of Joule's time was occupied in actual manipulation; freed from this, Mayer had time to follow the theory into its most abstruse and impressive applications. With their places reversed, however, Joule might have become Mayer, and Mayer might have become Joule.

(5, 137-138; 1871)

* Had I found it in time, this notice should have preceded that of the Copley Medalist of 1871.

从这些实验中推出了我们现在所使用的热功当量，即全世界都认可的“焦耳当量”。

这些工作取得的成果非常重要也非常出名，每逢被人提及时总能得到高度评价。这就是焦耳先生的工作，它们构成了一个意义重大的定理的实验基础，不仅仅在实践层面上意义重大，对于科学哲学更是如此。自牛顿时代以来，还没有什么科学成果能比焦耳先生用实验所阐明的理论更重大。

我没有提及焦耳先生科学著作中很多的小文章，也没有介绍他和威廉·汤姆孙爵士一同完成的重要研究。但是，我觉得我在这里介绍的内容已经足以说明：授予焦耳先生皇家学会的最高奖项，不仅仅是委员会给天才颁发的恰如其分的奖励，也是他在 20 年以前就应当得到的嘉奖。*

将焦耳工作的简短历史和 1871 年科普利奖章获得者的相比，就可以从“环境”对两个具有相同资质的人的不同影响中得到启发。迈尔在放弃了机械设备的研究之后，经过深思熟虑颇有远见地选择了利用现有数据探索出一个计算热功当量的简便算法。而焦耳身处机械设备当中，致力于实验工作，为我们今天所接受的力学理论打下了广泛而坚实的基础。焦耳的大部分时间都花在了实验操作上；而迈尔则从繁琐的工作中解放出来，他有充足的时间将理论带入最深奥和最有价值的应用之中。如果将他们的位置交换一下，也许焦耳会成为迈尔，而迈尔则会成为焦耳。

（王静 翻译；李军刚 审稿）

---

* 如果我及时地发现这一点，那么这段说明应该早于 1871 年科普利奖。

Clerk Maxwell's Kinetic Theory of Gases

J. C. Maxwell

Editor's Note

James Clerk Maxwell's recent kinetic theory of gases gave theoretical foundation to the macroscopic properties of gases. Here Maxwell defends his theory against a recent criticism. The theory predicts that a vertical column of gas should have the same temperature at all heights. However, the action of gravity implies that any molecule moving downward on some path should pick up energy, while those going upward should lose it. So shouldn't there be a net flow of energy downward? This argument, says Maxwell, assumes that particles projected upwards tend to have the same mean energy as those projected downward. Yet this cannot be so: at equilibrium, more must be ejected downwards, recovering the agreement with observations.

YOUR correspondent, Mr. Guthrie, has pointed out an, at first sight, very obvious and very serious objection to my kinetic theory of a vertical column of gas. According to that theory, a vertical column of gas acted on by gravity would be in thermal equilibrium if it were at a uniform temperature throughout, that is to say, if the mean energy of the molecules were the same at all heights. But if this were the case the molecules in their free paths would be gaining energy if descending, and losing energy if ascending. Hence, Mr. Guthrie argues, at any horizontal section of the column a descending molecule would carry more energy down with it that an ascending molecule would bring up, and since as many molecules descend as ascend through the section, there would on the whole be a transfer of energy, that is, of heat, downwards; and this would be the case unless the energy were so distributed that a molecule in any part of its course finds itself, on an average, among molecules of the same energy as its own. An argument of the same kind, which occurred to me in 1866, nearly upset my belief in calculation, and it was some time before I discovered the weak point in it.

The argument assumes that, of the molecules which have encounters in a given stratum, those projected upwards have the same mean energy as those projected downwards. This, however, is not the case, for since the density is greater below than above, a greater *number* of molecules come from below than from above to strike those in the stratum, and therefore a greater number are projected from the stratum downwards than upwards. Hence since the total momentum of the molecules temporarily occupying the stratum remains zero (because, as a whole, it is at rest), the smaller number of molecules projected upwards must have a greater initial velocity than the larger number projected downwards. This much we may gather from general reasoning. It is not quite so easy, without calculation, to show that this difference between the molecules projected upwards and downwards from the same stratum exactly counteracts the tendency to a downward transmission of energy pointed out by Mr. Guthrie. The difficulty lies chiefly in forming

克拉克·麦克斯韦的气体动力学理论

麦克斯韦

编者按

詹姆斯·克拉克·麦克斯韦最近提出的气体动力学理论为研究气体的宏观性质奠定了理论基础。在这篇文章中麦克斯韦坚持自己的理论，反驳了最近对其理论提出的批评。麦克斯韦的理论认为一个垂直气体柱在各个高度的温度应该保持一致。然而重力作用会使沿某一路径向下运动的分子获得能量，使向上运动的粒子失去能量。所以似乎应该出现一个向下的净能流。麦克斯韦说，上述论点假设向上运动的粒子和向下运动的粒子平均能量相同。但这是不可能的：在准静态，向下运动的粒子更多，这样和观测结果就不会矛盾了。

贵刊的通讯员格思里先生指出了我的垂直气体柱理论中一个初看上去非常明显而又严重的缺陷。根据那个理论，重力场中的一个垂直气体柱，如果各处温度相同，或者说如果处于不同高度的气体分子的平均动能相同，那么它将处于热平衡状态。但是如果事实确实如此，则自由运动的分子在向下运动的过程中动能将增加，向上运动的时候动能将减少。格思里先生论证说，如此一来，在气体柱的任何一个水平截面上，由于向上和向下运动的分子数目相同，向下运动的分子带走的动能将超过向上运动的分子带来的动能，所以整体上就会出现一个动能的转移，也就是热量的下移过程。要避免格思里所说的这个矛盾，只能假定动能沿高度的分布能够保证一个分子不管运动到哪里都与周边分子的动能相同。我在 1866 年也遇到了同样的问题，当时几乎动摇了我对计算结果的信心。一段时间之后我才发现这个观点中站不住脚的地方。

上述观点假定：对于任一个给定水平层上发生碰撞的分子，向上运动的分子和向下运动的分子平均动能相同。这一点并不符合实际情况。由于水平层下的分子密度大于水平层以上的，所以水平层下与水平层中分子发生碰撞的分子**数目**要大于水平层上与水平层中分子发生碰撞的分子数目，相应地，在水平层中发生碰撞而向下运动的分子数目也要大于发生碰撞而向上运动的分子数目。因此，既然该时刻水平层中的分子总动量保持为零（因为，从总体上来看水平层是静止的），那么相对于数目较大的向下运动的分子，数目较少的向上运动的分子就必须拥有更大的初始速度。从普通的推理中我们只能了解到这么多。不经过计算，想要说明从同一个水平层中向上和向下运动的分子在数目和速度上的差别正好能够抵消格思里先生指出的

exact expressions for the state of the molecules which instantaneously occupy a given stratum in terms of their state when projected from the various strata in which they had their last encounters. In my paper in the *Philosophical Transaction*, for 1867, on the "Dynamical Theory of Gases", I have entirely avoided these difficulties by expressing everything in terms of what passes through the boundary of an element, and what exists or takes place inside it. By this method, which I have lately carefully verified and considerably simplified, Mr. Guthrie's argument is passed by without ever becoming visible. It is well, however, that he has directed attention to it, and challenged the defenders of the kinetic theory to clear up their ideas of the result of those encounters which take place in a given stratum.

(**8**, 85; 1873)

那个向下的动能转移趋势，是非常困难的。主要困难在于需要找到一个确切的表达式，将某一时刻各水平层中刚发生碰撞而运动到某一特定水平层上所有分子的状态描述出来。在我在 1867 年的《自然科学会报》上发表的一篇题为《论气体动力学》的文章中，我把一切都用穿过一个单元边界的通量和单元内部本来就存在或者生成的量来表达，这样就回避了上面提到的种种麻烦。不久前我曾仔细检验并大大地简化了这个方法，利用这个方法，格思里先生提出的问题被绕过去了，没有表现出来。但是，现在他将其提出来引起大家的注意是很有益处的，因为这对分子动力论的支持者是一个挑战，促使他们整理自己关于一个特定水平层上分子碰撞结果的构想。

（何钧 翻译；鲍重光 审稿）

Molecules*

J. C. Maxwell

Editor's Note

James Clerk Maxwell was sometimes accused of being a poor lecturer, but this talk delivered to the British Association offers a lucid, engaging picture of the current understanding of the molecular nature of matter. Maxwell made a decisive contribution himself with his kinetic theory of gases, which explained how the macroscopic properties of gases, such as the laws relating pressure, volume and temperature, could be explained from the microscopic motions of the constituent particles. Maxwell's estimate of the size of a hydrogen molecule is only slightly bigger than the modern view. And his discussion of molecular diffusion anticipates the work of Albert Einstein and Jean Perrin on Brownian motion that provided the first real evidence for molecules as physical entities.

AN atom is a body which cannot be cut in two. A molecule is the smallest possible portion of a particular substance. No one has ever seen or handled a single molecule. Molecular science, therefore, is one of those branches of study which deal with things invisible and imperceptible by our senses, and which cannot be subjected to direct experiment.

The mind of man has perplexed itself with many hard questions. Is space infinite, and if so in what sense? Is the material world infinite in extent, and are all places within that extent equally full of matter? Do atoms exist, or is matter infinitely divisible?

The discussion of questions of this kind has been going on ever since men began to reason, and to each of us, as soon as we obtain the use of our faculties, the same old questions arise as fresh as ever. They form as essential a part of the science of the nineteenth century of our era, as of that of the fifth century before it.

We do not know much about the science organisation of Thrace twenty-two centuries ago, or of the machinery then employed for diffusing an interest in physical research. There were men, however, in those days, who devoted their lives to the pursuit of knowledge with an ardour worthy of the most distinguished members of the British Association; and the lectures in which Democritus explained the atomic theory to his fellow-citizens of Abdera realised, not in golden opinions only, but in golden talents, a sum hardly equalled even in America.

---

* Lecture delivered before the British Association at Bradford, by Prof. Clerk Maxwell, F. R. S.

分　子*

麦克斯韦

编者按

有人认为詹姆斯·克拉克·麦克斯韦缺乏演讲天赋，但这篇在英国科学促进会所作的报告中，麦克斯韦对物质分子的那些得到普遍接受的性质描述得非常透彻，给人留下了深刻的印象。麦克斯韦在分子学方面作出了有决定意义的贡献，他提出的气体动力学理论能够说明如何用组成粒子的微观运动来解释气体的宏观性质，如与压力、体积和温度相关的定律。麦克斯韦对氢分子大小的估计只比现在的公认值略大一些。他对分子扩散现象的讨论促使阿尔伯特·爱因斯坦和让·佩兰开始了关于布朗运动的研究，这使人们第一次认识到分子是一种物理实体。

原子是不能被一分为二的实体。分子是组成物质的最小单位。没有人看见或者摆弄过单个分子。因此，分子科学是研究不可见也不可感觉的事物的一门学问，我们无法对它进行直接实验。

人类经常思索很多难以回答的问题。空间是无限的吗？如果是，是从什么意义上讲的？物质世界的范围是无限的吗？在这个范围内是不是每个地方都同等地充满了物质？原子存在吗？或物质是否无限可分？

自从人类开始理性思考以来，关于这类问题的讨论就一直没有停止过。对于我们每个人来说，一旦开始用心智思考，那些古老的问题就会像从前一样令人觉得新奇。不论是在我们所处的 19 世纪，还是在公元前 5 世纪，这些问题都构成了科学的基本部分。

我们对 2,200 年前位于色雷斯的科学组织所知甚少，也不知道他们用何种方式来传播对自然研究的兴趣。不过那时候确实有人毕生追求知识，热情不亚于英国科学促进会中最杰出的成员。当德谟克利特向他的阿布德拉市民开设讲座讲解自己的原子理论时，他获得的高度评价和丰厚报酬即使在今天的美国也很少有人能比得上。

* 皇家学会会员克拉克·麦克斯韦教授在布拉德福德对英国科学促进会作的报告。

To another very eminent philosopher, Anaxagoras, best known in the world as the teacher of Socrates, we are indebted for the most important service to the atomic theory, which, after its statement by Democritus, remained to be done. Anaxagoras, in fact, stated a theory which so exactly contradicts the atomic theory of Democritus that the truth or falsehood of the one theory implies the falsehood or truth of the other. The question of the existence or non-existence of atoms cannot be presented to us this evening with greater clearness than in the alternative theories of these two philosophers.

Take any portion of matter, say a drop of water, and observe its properties. Like every other portion of matter we have ever seen, it is divisible. Divide it in two, each portion appears to retain all the properties of the original drop, and among others that of being divisible. The parts are similar to the whole in every aspect except in absolute size.

Now go on repeating the process of division till the separate portions of water are so small that we can no longer perceive or handle them. Still we have no doubt that the sub-division night be carried further, if our senses were more acute and our instruments more delicate. Thus far all are agreed, but now question arises, Can this sub-division be repeated for ever?

According to Democritus and the atomic school, we must answer in the negative. After a certain number of sub-divisions, the drop would be divided into a number of parts each of which is incapable of further sub-division. We should thus, in imagination, arrive at the atom, which, as its name literally signifies, cannot be cut in two. This is the atomic doctrine of Democritus, Epicurus, and Lucretius, and, I may add, of your lecturer.

According to Anaxagoras, on the other hand, the parts into which the drop is divided, are in all respects similar to the whole drop, the mere size of a body counting for nothing as regards the nature of its substance. Hence if the whole drop is divisible, so are its parts down to the minutest sub-divisions, and that without end.

The essence of the doctrine of Anaxagoras is that the parts of a body are in all respects similar to the whole. It was therefore called the doctrine of Homoiomereia. Anaxagoras did not of course assert this of the parts of organised bodies such as men and animals, but he maintained that those inorganic substances which appear to us homogeneous are really so, and that the universal experience of mankind testifies that every material body, without exception, is divisible.

The doctrine of atoms and that of homogeneity are thus in direct contradiction.

But we must now go on to molecules. Molecule is a modern word. It does not occur in *Johnson's Dictionary*. The ideas it embodies are those belonging to modern chemistry.

另一位杰出的哲学家，即以身为苏格拉底的老师而闻名于世的阿那克萨哥拉在德谟克利特之后对原子学说作出了最重要的贡献。实际上，阿那克萨哥拉和德谟克利特两人的原子学说是如此地针锋相对，一方正确则另一方必错。今晚我们对原子到底是否存在这一问题的讨论，用这两位哲学家的对立理论来表达，是最清楚不过了。

随便取一份物质，比如一滴水，来观察它的性质。就像我们看到的其他物质一样，它是可以分割的。把它分成两份，每一份都保持原来那滴水的所有性质，其他可以分割的物质也是一样。每一个部分除了尺寸比整体小些，其他各方面都和整体相似。

就这么一直分下去，直到分出来的水滴小到我们再也看不见，也无法对它们进行操作。但是大家都明白，如果我们的感官更敏锐，我们的设备更精密，细分过程还是可以接着进行下去的。到此为止不会有什么异议，但是现在问题就来了：这样的细分过程可以永远继续下去吗？

德谟克利特和原子学派的回答是否定的。经过一定次数的分割后，水滴就被分成很多很小的部分，每一部分都不能进一步细分了。也就是我们已经细分到了想象中的原子，原子这个名字的字面含义就是不可分的意思。这就是德谟克利特、伊壁鸠鲁、卢克莱修还有我，你们的演讲者，所赞成的原子论。

另一方面，阿那克萨哥拉认为，水滴被分割成的各个部分，除了和物质性质无关的物体尺寸发生了变化以外，其他一切方面都和整个的水滴类似。因此，如果原来的水滴是可分的，那么分割得到的各个部分也应该是可分的，哪怕分到极小，永无止境。

阿那克萨哥拉学说的根本点是：一个物体的部分和它的整体在所有方面都是相似的，所以被称为同质性学说。阿那克萨哥拉当然没有把它应用到人和动物这样的有机体的身上，但是他认为那些看上去是均质的无机物的确是同质的，并且人类的普遍经验也能证实每一种物质实体毫无例外都是可分的。

原子学说和同质性学说就是这样地针锋相对。

但是现在我们必须转入对分子这个现代名词的讨论，它在《约翰逊词典》里是没有的。分子所包含的概念属于现代化学的范畴。

A drop of water, to return to our former example, may be divided into a certain number, and no more, of portions similar to each other. Each of these the modern chemist calls a molecule of water. But it is by no means an atom, for it contains two different substances, oxygen and hydrogen, and by a certain process the molecule may be actually divided into two parts, one consisting of oxygen and the other of hydrogen. According to the received doctrine, in each molecule of water there are two molecules of hydrogen and one of oxygen. Whether these are or are not ultimate atoms I shall not attempt to decide.

We now see what a molecule is, as distinguished from an atom.

A molecule of a substance is a small body such that if, on the one hand, a number of similar molecules were assembled together they would form a mass of that substance, while on the other hand, if any portion of this molecule were removed, it would no longer be able, along with an assemblage of other molecules similarly treated, to make up a mass of the original substance.

Every substance, simple or compound, has its own molecule. If this molecule be divided, its parts are molecules of a different substance or substances from that of which the whole is a molecule. An atom, if there is such a thing, must be a molecule of an elementary substance. Since, therefore, every molecule is not an atom, but every atom is a molecule, I shall use the word molecule as the more general term.

I have no intention of taking up your time by expounding the doctrines of modern chemistry with respect to the molecules of different substances. It is not the special but the universal interest of molecular science which encourages me to address you. It is not because we happen to be chemists or physicists or specialists of any kind that we are attracted towards this centre of all material existence, but because we all belong to a race endowed with faculties which urge us on to search deep and ever deeper into the nature of things.

We find that now, as in the days of the earliest physical speculations, all physical researches appear to converge towards the same point, and every inquirer, as he looks forward into the dim region towards which the path of discovery is leading him, sees, each according to his sight, the vision of the same quest.

One may see the atom as a material point, invested and surrounded by potential forces. Another sees no garment of force, but only the bare and utter hardness of mere impenetrability.

But though many a speculator, as he has seen the vision recede before him into the innermost sanctuary of the inconceivably little, has had to confess that the quest was not for him, and though philosophers in every age have been exhorting each other to direct their minds to some more useful and attainable aim, each generation, from the earliest dawn of science to the present time, has contributed a due proportion of its ablest intellects to the quest of the ultimate atom.

回到我们原来的例子，一滴水能够最大限度地被分割成一定数量的彼此相似的部分。每一个这样的部分都是现代化学家称作的水分子。水分子包含两种不同的物质，氧和氢，所以它决不是一个原子。通过一定的处理方式，确实可以把水分子分解成两部分，一部分含氢，另一部分含氧。根据公认的理论，每一个水分子都包含两个氢分子和一个氧分子。至于这些氢分子和氧分子是不是不能再分解的原子，我这里先不去确定。

现在我们明白了分子是什么，它和原子有什么不同。

一种物质的分子是很小的，一方面，如果许多同样的分子聚合在一起，就会形成大量这种物质；另一方面，如果这些分子的某个部分缺失，它们与其他经过相同处理的分子聚合在一起也不能形成原来的物质。

任何物质，无论是简单的还是复合的，都有自己的分子。如果这个分子再被分割，形成的部分就是其他物质的分子。一个原子，如果确实存在的话，应该是一种基本物质的分子。不是每个分子都是原子，但是每个原子都是分子，所以我将使用含义更广的分子这个术语。

我不想浪费时间去详细解释现代化学关于各种物质分子的理论。我作这个报告的目的是讲述分子科学中普遍的而不是具体的问题。不是因为我们恰好是化学家、物理学家或者其他某个领域的专家才对这个与所有物质息息相关的中心问题感兴趣，而是因为我们都属于人类这个物种，其所具备的资质促使我们不断地深入研究事物的本质。

现在我们发现，就像早期物理猜想时代一样，所有的物理研究似乎都汇集到了同一点上；每一个探求者，在眺望发现之途指向的茫茫区域时，虽然目力各有不同，看到的却都是同一件宝物的幻像。

有些人眼中的原子是一个物质点，被有势力场包围着。另一些人则看不到力的存在，只看到裸露而坚硬的不可穿透的实体。

很多人看到幻像在眼前消退而去，躲进那不可思议的渺小之物最隐秘的庇护所之后，不得不承认宝物非他所属；各个时代的哲人们互相劝诫对方去追求更实际、更容易达到的目标。虽然如此，自从科学的启蒙时期直到今天，每一代都不乏最富才智的人投身于对最终原子的探求。

Our business this evening is to describe some researches in molecular science, and in particular to place before you any definite information which has been obtained respecting the molecules themselves. The old atomic theory, as described by Lucretius and revived in modern times, asserts that the molecules of all bodies are in motion, even when the body itself appears to be at rest. These motions of molecules are in the case of solid bodies confined within so narrow a range that even with our best microscopes we cannot detect that they alter their places at all. In liquids and gases, however, the molecules are not confined within any definite limits, but work their way through the whole mass, even when that mass is not disturbed by any visible motion.

This process of diffusion, as it is called, which goes on in gases and liquids and even in some solids, can be subjected to experiment, and forms one of the most convincing proofs of the motion of molecules.

Now the recent progress of molecular science began with the study of the mechanical effect of the impact of these moving molecules when they strike against any solid body. Of course these flying molecules must beat against whatever is placed among them, and the constant succession of these strokes is, according to our theory, the sole cause of what is called the pressure of air and other gases.

This appears to have been first suspected by Daniel Bernoulli, but he had not the means which we now have of verifying the theory. The same theory was afterwards brought forward independently by Lesage, of Geneva, who, however, devoted most of his labour to the explanation of gravitation by the impact of atoms. Then Herapath, in his "Mathematical Physics", published in 1847, made a much more extensive application of the theory to gases, and Dr. Joule, whose absence from our meeting we must all regret, calculated the actual velocity of the molecules of hydrogen.

The further development of the theory is generally supposed to have been begun with a paper by Krönig, which does not, however, so far as I can see, contain any improvement on what had gone before. It seems, however, to have drawn the attention of Prof. Clausius to the subject, and to him we owe a very large part of what has been since accomplished.

We all know that air or any other gas placed in a vessel presses against the sides of the vessel, and against the surface of any body placed within it. On the kinetic theory this pressure is entirely due to the molecules striking against these surfaces, and thereby communicating to them a series of impulses which follow each other in such rapid succession that they produce an effect which cannot be distinguished from that of a continuous pressure.

If the velocity of the molecules is given, and the number varied, then since each molecule, on an average, strikes the side of the vessel the same number of times, and with an impulse of the same magnitude, each will contribute an equal share to the whole pressure. The pressure in a vessel of given size is therefore proportional to the number of molecules in it, that is to the quantity of gas in it.

我们今天晚上的任务是介绍分子科学中的一些研究成果，特别是向你们展示那些关于分子本身的现在已经比较确定的认识。卢克莱修所描述的旧原子论到了现代重获新生。他认为所有物体中的分子都在不停地运动，即便当物体本身处于静止状态时也不例外。在固体中，分子的运动只局限于一个很小的范围，哪怕是利用当前最好的显微镜我们也察觉不到它们的移动。但是对于液体和气体的情况，分子的运动没有受到确切的范围限制，可以在整个物质中移动，哪怕这个整体没有受到任何可见的运动的干扰。

这个过程被称为扩散。它在气体、液体甚至一些固体中持续进行着，可以由实验验证，同时它也是分子运动论最有力的证明之一。

现在分子科学的最新进展，是从研究这些运动着的分子碰撞固体表面的机械效应开始的。飞行中的分子一定会撞击所有置于其中的物质。根据我们的理论，这种持续不断的撞击，就是产生所谓的空气和其他气体压力的唯一原因。

丹尼尔·伯努利可能是第一个想到这一点的人，但是他当时没有我们今天的实验手段来验证他的理论。后来日内瓦的勒萨热也独立提出过这个理论，但是他的工作主要是用原子碰撞来解释重力现象。赫拉帕斯在他 1847 年出版的《数学物理学》一书中，将这个理论更广泛地应用于各种气体。焦耳博士计算了氢分子的实际速度，今天他没有在场实在让人感到遗憾。

大家普遍认为这一理论的进一步发展是从克勒尼希的一篇论文开始的。但在我个人看来，这篇文章本身并没有在前人工作的基础上作出什么改进。不过它引起了克劳修斯教授对这个问题的关注，而后者对以后的理论发展起了很大作用。

我们都知道放置在一个容器中的空气或其他气体会对容器壁以及放置在其中的其他物体表面产生压力。根据气体动力学理论，这个压力完全是由分子碰撞这些表面产生的。这种碰撞带给表面的一系列冲击之间的间隔非常小，产生的效应和连续的压力没有什么两样。

假定分子的速率一定，但数量不同。那么平均来说，每个分子碰撞器壁的次数相同，产生的冲击强度也相同，因此对总压力的贡献也相同。这样一来，一个固定容积的容器承受的压强和其中分子的总数量，也就是其中的气体总量，成正比。

This is the complete dynamical explanation of the fact discovered by Robert Boyle, that the pressure of air is proportional to its density. It shows also that of different portions of gas forced into a vessel, each produces its own part of the pressure independently of the rest, and this whether these portions be of the same gas or not.

Let us next suppose that the velocity of the molecules is increased. Each molecule will now strike the sides of the vessel a greater number of times in a second, but besides this, the impulse of each blow will be increased in the same proportion, so that the part of the pressure due to each molecule will vary as the *square* of the velocity. Now the increase of the square of velocity corresponds, in our theory, to a rise of temperature, and in this way we can explain the effect of warming the gas, and also the law discovered by Charles that the proportional expansion of all gases between given temperatures is the same.

The dynamical theory also tells us what will happen if molecules of different masses are allowed to knock about together. The greater masses will go slower than the smaller ones, so that, on an average, every molecule, great or small, will have the same energy of motion.

The proof of this dynamical theorem, in which I claim the priority, has recently been greatly developed and improved by Dr. Ludwig Boltzmann. The most important consequence which flows from it is that a cubic centimetre of every gas at standard temperature and pressure contains the same number of molecules. This is the dynamical explanation of Gay Lussac's law of the equivalent volumes of gases. But we must now descend to particulars, and calculate the actual velocity of a molecule of hydrogen.

A cubic centimetre of hydrogen, at the temperature of melting ice and at a pressure of one atmosphere, weighs 0.00008954 grammes. We have to find at what rate this small mass must move (whether altogether or in separate molecules makes no difference) so as to produce the observed pressure on the sides of the cubic centimetre. This is the calculation which was first made by Dr. Joule, and the result is 1,859 metres per second. This is what we are accustomed to call a great velocity. It is greater than any velocity obtained in artillery practice. The velocity of other gases is less, as you will see by the table, but in all cases it is very great as compared with that of bullets.

We have now to conceive the molecules of the air in this hall flying about in all directions, at a rate of about seventeen miles in a minute.

If all these molecules were flying in the same direction, they would constitute a wind blowing at the rate of seventeen miles a minute, and the only wind which approaches this velocity is that which proceeds from the mouth of a cannon. How, then, are you and I able to stand here? Only because the molecules happen to be flying in different directions, so that those which strike against our backs enable us to support the storm which is beating against our faces. Indeed, if this molecular bombardment were to cease, even for an instant, our veins would swell, our breath would leave us, and we should, literally,

这就是罗伯特·玻意耳发现的空气压强正比于其密度这一事实的完整动力学解释。它还表明，如果我们将不同批次的气体加入容器，则无论它们的种类是否相同，每个部分都将独立地产生自己的分压强。

下一步让我们假定分子速率增加的情形。因为每秒钟内，每个分子碰撞器壁的次数相应增加，同时每次碰撞的冲击强度也成比例增加，所以每个分子对压强的贡献和它的速度的**平方**成正比。在我们现在的理论中，分子速率平方的增加和温度的增加相对应。这样我们就能解释加热气体导致压强增加的效应，以及随温度增加各种气体体积同比增大的查理定律。

动力论还告诉我们不同质量的气体分子互相碰撞会发生什么结果。质量大的分子比质量小的分子运动速率要小一些，所以平均来说，每个分子，不论质量大小，其动能都相同。

对这一动力学定律的证明是我最先提出的，近来被路德维希·玻尔兹曼博士加以改进和发展。它的一个重要推论就是在标准温度和压强下，1 立方厘米的任何气体都含有同等数量的分子，这就是盖·吕萨克定律关于相同体积气体的动力学解释。现在我们举一个具体的例子来计算一个氢分子的实际速率。

在温度为冰的熔点，压力为一个大气压时，1 立方厘米氢气的质量是 0.00008954 克。这么小的质量的运动（是合在一起还是分散到各个分子，对结果没有影响）到底需要多大速率，才能在 1 立方厘米的容器壁上产生测量到的压强呢？焦耳博士首先对此进行了计算，他的结果是每秒 1,859 米。通常对我们来说，这是一个很大的速率，比任何炮弹的速率都要大。从附表中大家可以看到，其他气体的速率要小一些，但是无论如何都远大于子弹的速率。

现在我们要设想一下这个大厅里的空气分子以每分钟 17 英里的速率向各个方向飞行。

如果所有分子都向同一个方向飞行，它们就会形成速率为每分钟 17 英里的强风，只有从加农炮炮口出膛的风速能够接近这个速率。那么，你我怎么能够在这里保持站立？这只是因为这些分子飞行的方向各不相同，在前面和后面撞击我们的分子冲击作用互相抵消。事实上，如果这种分子碰撞哪怕停止一刻，我们也会静脉肿胀，不能呼吸，一命呜呼。这些分子不只是撞击我们和房间四周的墙壁。考虑到它

expire. But it is not only against us or against the walls of the room that the molecules are striking. Consider the immense number of them, and the fact that they are flying in every possible direction, and you will see that they cannot avoid striking each other. Every time that two molecules come into collision, the paths of both are changed, and they go off in new directions. Thus each molecule is continually getting its course altered, so that in spite of its great velocity it may be a long time before it reaches any great distance from the point at which it set out.

I have here a bottle containing ammonia. Ammonia is a gas which you can recognise by its smell. Its molecules have a velocity of six hundred metres per second, so that if their course had not been interrupted by striking against the molecules of air in the hall, everyone in the most distant gallery would have smelt ammonia before I was able to pronounce the name of the gas. But instead of this, each molecule of ammonia is so jostled about by the molecules of air, that it is sometimes going one way and sometimes another. It is like a hare which is always doubling, and though it goes a great pace, it makes very little progress. Nevertheless, the smell of ammonia is now beginning to be perceptible at some distance from the bottle. The gas does diffuse itself through the air, though the process is a slow one, and if we could close up every opening of this hall so as to make it air-tight, and leave everything to itself for some weeks, the ammonia would become uniformly mixed through every part of the air in the hall.

This property of gases, that they diffuse through each other, was first remarked by Priestley. Dalton showed that it takes place quite independently of any chemical action between the inter-diffusing gases. Graham, whose researches were especially directed towards those phenomena which seem to throw light on molecular motions, made a careful study of diffusion, and obtained the first results from which the rate of diffusion can be calculated.

Still more recently the rates of diffusion of gases into each other have been measured with great precision by Prof. Loschmidt of Vienna.

He placed the two gases in two similar vertical tubes, the lighter gas being placed above the heavier, so as to avoid the formation of currents. He then opened a sliding valve, so as to make the two tubes into one, and after leaving the gases to themselves for an hour or so, he shut the valve, and determined how much of each gas had diffused into the other.

As most gases are invisible, I shall exhibit gaseous diffusion to you by means of two gases, ammonia and hydrochloric acid, which, when they meet, form a solid product. The ammonia, being the lighter gas, is placed above the hydrochloric acid, with a stratum of air between, but you will soon see that the gases can diffuse through this stratum of air, and produce a cloud of white smoke when they meet. During the whole of this process no currents or any other visible motion can be detected. Every part of the vessel appears as calm as a jar of undisturbed air.

们数量巨大，正在四处乱飞，你会发现它们不可能不互相碰撞。每当两个分子撞到一起，其轨道就都会改变，而它们会飞向新的方向。这样每个分子都在不断地改变轨道，因而虽然其速率很快，但要从出发点移开一定距离也需要不少时间。

我这里有一个里面装着氨气的瓶子。氨气的味道大家可以闻得出来。它的分子运动速率是每秒 600 米，因而如果它们的运动轨迹没有因为和大厅里的空气分子碰撞而改变，就算坐得最远的听众都会在我说出氨气这个名称之前闻到它。但是实际上并非如此，每个氨气分子都被空气分子撞来撞去，一会儿向东一会儿向西，就像一只野兔，老是改变方向，虽然步子很快，但是跑不了多远。虽然如此，在离瓶子一定距离以内的听众还是开始闻到氨气的味道了。氨气确实在空气中不断地扩散着，只是速率较慢而已，如果我们把大厅的所有出口都封起来，不让空气流走，并保持几个星期不动它，那么氨气就会均匀地散布在大厅的每个角落。

气体的这个相互扩散的性质是由普里斯特利最先指出的。道尔顿指出这种扩散的发生与相互扩散的气体间发生的具体化学反应无关。格雷姆的研究工作集中在那些能为分子运动提供线索的现象上。他仔细研究了扩散现象，最早得到了可以用来计算扩散速率的结果。

再晚些时候，维也纳的洛施密特教授精确地测量了气体分子相互扩散的速率。

他把两种不同气体分别装入两个相似的竖直试管中，轻的气体放在重的气体上面以防止对流产生。随后打开滑动阀门让两个试管相通。将它们静置大约一个小时以后，他关上阀门，然后测量每种气体中有多少已经扩散到了另一种气体中。

因为大多数气体都是不可见的，所以我得用氨气和盐酸这两种相遇后能生成固体反应物的气体来展示气体的扩散过程。氨气比较轻，所以放在盐酸上面，中间隔着一层空气。但是你们马上就可以看到，这两种气体能通过扩散穿过空气层而相遇，并产生一股白烟。整个过程看不到气体流动或者其他任何可见的运动。容器的每个部分都像一罐未受扰动的空气一样平静。

But, according to our theory, the same kind of motion is going on in calm air as in the inter-diffusing gases, the only difference being that we can trace the molecules from one place to another more easily when they are of a different nature from those through which they are diffusing.

If we wish to form a mental representation of what is going on among the molecules in calm air, we cannot do better than observe a swarm of bees, when every individual bee is flying furiously, first in one direction, and then in another, while the swarm, as a whole, either remains at rest, or sails slowly through the air.

In certain seasons, swarms of bees are apt to fly off to a great distance, and the owners, in order to identify their property when they find them on other people's ground, sometimes throw handfulls of flour at the swarm. Now let us suppose that the flour thrown at the flying swarm has whitened those bees only which happened to be in the lower half of the swarm, leaving those in the upper half free from flour.

If the bees still go on flying hither and thither in an irregular manner, the floury bees will be found in continually increasing proportions in the upper part of the swarm, till they have become equally diffused through every part of it. But the reason of this diffusion is not because the bees were marked with flour, but because they are flying about. The only effect of the marking is to enable us to identify certain bees.

We have no means of marking a select number of molecules of air, so as to trace them after they have become diffused among others, but we may communicate to them some property by which we may obtain evidence of their diffusion.

For instance, if a horizontal stratum of air is moving horizontally, molecules diffusing out of this stratum into those above and below will carry their horizontal motion with them, and so tend to communicate motion to the neighbouring strata, while molecules diffusing out of the neighbouring strata into the moving one will tend to bring it to rest. The action between the strata is somewhat like that of two rough surfaces, one of which slides over the other, rubbing on it. Friction is the name given to this action between solid bodies; in the case of fluids it is called internal friction or viscosity.

It is in fact only another kind of diffusion—a lateral diffusion of momentum, and its amount can be calculated from data derived from observations of the first kind of diffusion, that of matter. The comparative values of the viscosity of different gases were determined by Graham in his researches on the transpiration of gases through long narrow tubes, and their absolute values have been deduced from experiments on the oscillation of discs by Oscar Meyer and myself.

Another way of tracing the diffusion of molecules through calm air is to heat the upper stratum of the air in a vessel, and so observe the rate at which this heat is communicated

但是根据我们的理论，这种气体间相互扩散的运动进程，同样在平静的空气中发生着。区别只是当扩散发生在不同气体间时，追踪分子从一处到另一处的移动要容易一些。

如果我们想要在头脑中构思出一幅表现分子在平静空气中运动的图像，最好去观察一群蜜蜂，每只蜜蜂都拼命地飞来飞去，先朝一个方向飞，然后再朝另一个方向飞，但是整个蜂群不是停着不动，就是在空中缓慢地移动。

有的季节蜂群可以飞得很远，养蜂人为了能在别人的地盘上也能认出自己的蜂群，有时会向蜂群洒一把面粉。现在我们假定面粉正好把蜂群下面一半的蜜蜂染白，而上面一半没有沾上面粉。

如果这群蜜蜂继续散乱地飞来飞去，蜂群上半部就会有越来越多的沾上面粉的蜜蜂，直到它们均匀地分布于上下两部分。但是这种扩散的原因不是因为蜜蜂沾上了面粉，而是因为它们到处乱飞。沾面粉标记的目的只是为了帮助我们识别特定的蜜蜂。

我们没有办法标记一定数目的空气分子，使得它们在扩散到其他分子中之后还能被追踪到。但是我们可以用某些参数的传递来证明它们的扩散。

比如说，如果一个水平空气层向水平方向移动，那么从该层扩散到上下两个相邻层的分子将携带水平动量，并把这个动量传递给相邻的上下空气层。而从相邻空气层中扩散进入这个水平移动的水平层的分子，会减慢水平层的移动。相邻水平层之间的作用，就像两个粗糙固体表面之间的滑动和摩擦。固体之间的这种作用叫做摩擦，在流体的情况下则称为内摩擦或者粘滞力。

实际上这只不过是另一种类型的扩散——动量的横向扩散。其大小可以通过观察第一种扩散，也就是物质扩散的数据计算出来。研究气体通过长细管的流逸过程，格雷姆确定了不同气体的相对粘滞系数。粘滞系数的绝对值由奥斯卡·迈耶和我通过圆盘振动的实验结果推导得到。

另一种跟踪分子在平静空气中扩散的方法是加热容器顶层的空气，然后观察热量向下层传递的速率。这实际上是第三种类型的扩散——能量扩散。在直接进行热

to the lower strata. This, in fact, is a third kind of diffusion—that of energy, and the rate at which it must take place was calculated from data derived from experiments on viscosity before any direct experiments on the conduction of heat had been made. Prof. Stefan, of Vienna, has recently, by a very delicate method, succeeded in determining the conductivity of air, and he finds it, as he tells us, in striking agreement with the value predicted by the theory.

All these three kinds of diffusion—the diffusion of matter, of momentum, and of energy—are carried on by the motion of the molecules. The greater the velocity of the molecules and the farther they travel before their paths are altered by collision with other molecules, the more rapid will be the diffusion. Now we know already the velocity of the molecules, and therefore by experiments on diffusion we can determine how far, on an average, a molecule travels without striking another. Prof. Clausius, of Bonn, who first gave us precise ideas about the motion of agitation of molecules, calls this distance the mean path of a molecule. I have calculated, from Prof. Loschmidt's diffusion experiments, the mean path of the molecules of four well-know gases. The average distance travelled by a molecule between one collision and another is given in the table. It is a very small distance, quite imperceptible to us even with our best microscopes. Roughly speaking, it is about the tenth part of the length of a wave of light, which you know is a very small quantity. Of course the time spent on so short a path by such swift molecules must be very small. I have calculated the number of collisions which each must undergo in a second. They are given in the table and are reckoned by thousands of millions. No wonder that the travelling power of the swiftest molecule is but small, when its course is completely changed thousands of millions of times in a second.

The three kinds of diffusion also take place in liquids, but the relation between the rates at which they take place is not so simple as in the case of gases. The dynamical theory of liquids is not so well understood as that of gases, but the principal difference between a gas and a liquid seems to be that in a gas each molecule spends the greater part of its time in describing its free path, and is for a very small portion of its time engaged in encounters with other molecules, whereas in a liquid the molecule has hardly any free path, and is always in a state of close encounter with other molecules.

Hence in a liquid the diffusion of motion from one molecule to another takes place much more rapidly than the diffusion of the molecules themselves, for the same reason that it is more expeditious in a dense crowd to pass on a letter from hand to hand than to give it to a special messenger to work his way through the crowd. I have here a jar, the lower part of which contains a solution of copper sulphate, while the upper part contains pure water. It has been standing here since Friday, and you see how little progress the blue liquid has made in diffusing itself through the water above. The rate of diffusion of a solution of sugar has been carefully observed by Voit. Comparing his results with those of Loschmidt on gases, we find that about as much diffusion takes place in a second in gases as requires a day in liquids.

传导的实验之前，这种传递的速率是由粘滞系数的实验结果计算出来的。维也纳的斯特藩教授最近通过一个极其精巧的方法，成功地确定了空气的传导系数，他发现测量值和理论预测值惊人地符合。

所有这三种扩散——物质、动量和能量的扩散，都是由分子的运动完成的。分子的运动速率越大，在它与其他分子碰撞而使其运动方向发生改变之前所走的行程就越长，扩散的速率就越快。既然分子速率已知，通过扩散速率的实验，我们就可以确定一个分子在两次碰撞之间所走的平均距离。波恩的克劳修斯教授是第一个给出分子受激运动精确构想的人，他把这个距离叫做分子的平均自由程。根据洛施密特教授的扩散实验，我计算了 4 种常见气体的分子平均自由程。附表中列出了一个分子在两次碰撞之间的平均距离，这个距离非常之小，即使用最好的显微镜也不能分辨。它大概是光波长的 1/10，这样说你们就知道它有多小了。以分子的运动速率之大，行走这么短的距离，所用的时间当然是非常之短。我曾计算过一个分子每秒钟内碰撞的次数，结果也列在表中，都是几十亿次的水平。一秒钟之内方向要被改变几十亿次，难怪分子的运动速率虽大，却走不了多远。

这三种扩散过程在液体中也会发生，但是它们之间的速率关系就不像在气体中那么简单。液体的动力学理论不像气体的理论那么完善。看上去它们之间的根本区别是：相对而言，气体分子大部分时间是自由飞行，和其他分子发生碰撞的时间不多；液体分子则正好相反，平均自由程很短，总是在和附近的分子发生密切接触。

这样一来，液体中的动量扩散就比液体分子本身的扩散要快很多，这就好比在密集的人群中递送一封信，通过众人之手相传要比找一个专门的送信人穿越人群快一样。我这里有一个罐子，下半部装的是硫酸铜溶液，上半部装的是纯水。这个罐子从星期五就一直放在这里，而现在你们几乎看不出蓝色的硫酸铜扩散到上面的纯水中。沃伊特仔细计算了一种蔗糖溶液的扩散速率。把他的结果和洛施密特得到的气体扩散速率进行对比，我们发现在气体中一秒钟就能完成的扩散过程，在液体中则需要一整天。

The rate of diffusion of momentum is also slower in liquids than in gases, but by no means in the same proportion. The same amount of motion takes about ten times as long to subside in water as in air, as you will see by what takes place when I stir these two jars, one containing water and the other air. There is still less difference between the rates at which a rise of temperature is propagated through a liquid and through a gas.

In solids the molecules are still in motion, but their motions are confined within very narrow limits. Hence the diffusion of matter does not take place in solid bodies, though that of motion and heat takes place very freely. Nevertheless, certain liquids can diffuse through colloid solids, such as jelly and gum, and hydrogen can make its way through iron and palladium.

We have no time to do more than mention that most wonderful molecular motion which is called electrolysis. Here is an electric current passing through acidulated water, and causing oxygen to appear at one electrode and hydrogen at the other. In the space between, the water is perfectly calm, and yet two opposite currents of oxygen and of hydrogen must be passing through it. The physical theory of this process has been studied by Clausius, who has given reasons for asserting that in ordinary water the molecules are not only moving, but every now and then striking each other with such violence that the oxygen and hydrogen of the molecules part company, and dance about through the crowd, seeking partners which have become dissociated in the same way. In ordinary water these exchanges produce, on the whole, no observable effect, but no sooner does the electromotive force begin to act than it exerts its guiding influence on the unattached molecules, and bends the course of each toward its proper electrode, till the moment when, meeting with an unappropriated molecule of the opposite kind, it enters again into a more or less permanent union with it till it is again dissociated by another shock. Electrolysis, therefore, is a kind of diffusion assisted by electromotive force.

Another branch of molecular science is that which relates to the exchange of molecules between a liquid and a gas. It includes the theory of evaporation and condensation, in which the gas in question is the vapour of the liquid, and also the theory of the absorption of a gas by a liquid of a different substance. The researches of Dr. Andrews on the relations between the liquid and the gaseous state have shown us that though the statements in our own elementary text-books may be so neatly expressed that they appear almost self-evident, their true interpretation may involve some principle so profound that, till the right man has laid hold of it, no one ever suspects that anything is left to be discovered.

These, then, are, some of the fields from which the data of molecular science are gathered. We may divide the ultimate results into three ranks, according to the completeness of our knowledge of them.

To the first rank belong the relative masses of the molecules of different gases, and their velocities in metres per second. These data are obtained from experiments on the pressure and density of gases, and are known to a high degree of precision.

液体中动量的扩散速率也比气体中动量的扩散速率慢，但是没有差那么多。等量的运动在水中的衰减时间大概是空气中的10倍。让我来搅动一下这两个罐子，一个装的是水，另一个只有空气，大家看看结果是什么？液体和气体在温度传递上的速率差别还要更小一些。

固体中的分子也在不停地运动，但是它们的运动被限制在很小的范围内。所以物质扩散不会发生在固体中，但动量和能量扩散可以非常自由地进行。不过某些液体可以渗过像果冻和树胶一类的胶质固体，而氢气能够透过铁和钯。

时间有限，我们只能简单提一下电解这个最奇妙的分子运动现象。这里有电流通过酸性的水时，两个电极分别产生氧气和氢气。电极之间的水是完全静止的，但是其中必然有氧和氢的相对流动。克劳修斯研究了这一过程的物理学原理，给出理由认为普通水中的分子不但是运动的，而且相互碰撞的力量还很大，造成水分子中氧和氢的分离，分离的氧和氢在水中四处游动，寻找因同样原因而游离的其他对象来结合。在普通水中，这个交换过程在整体上没有造成可观测的效应。但是一旦电动势开始起作用，就对独立分子的运动产生定向影响，驱使游离分子向相应的电极运动，直到碰上异性游离分子而形成相对稳定的结合体。这个结合体还有可能再次被冲击离解。这样来看，电解是一种电动势协助下的分子扩散过程。

分子科学的另一个分支是关于液体和气体之间的分子交换。它既包括同一种物质气液两态之间的蒸发和凝结理论，也包括液体吸收不同种物质气体分子的理论。安德鲁斯博士关于气液两态之间关系的研究表明，虽然我们基本教科书中的结论看上去如此简洁，似乎是天经地义，但其实际的意义可能包含着非常深奥的原理。在有合适的人将其搞清楚之前，人们认为一切已经尽善尽美了。

以上就是分子科学取得数据成果的一些领域。我们根据完整的相关知识，把这些最终的成果分成三个等级。

第一个等级中有不同气体分子的相对质量以及它们以米每秒为单位的速度。这些数据是根据气体压强和密度的实验结果得出的，精度很高。

In the second rank we must place the relative size of the molecules of different gases, the length of their mean paths, and the number of collisions in a second. These quantities are deduced from experiments on the three kinds of diffusion. Their received values must be regarded as rough approximations till the methods of experimenting are greatly improved.

There is another set of quantities which we must place in the third rank, because our knowledge of them is neither precise, as in the first rank, nor approximate, as in the second, but is only as yet of the nature of a probable conjecture. These are the absolute mass of a molecule, its absolute diameter, and the number of molecules in a cubic centimetre. We know the relative masses of different molecules with great accuracy, and we know their relative diameters approximately. From these we can deduce the relative densities of the molecules themselves. So far we are on firm ground.

The great resistance of liquids to compression makes it probable that their molecules must be at about the same distance from each other as that at which two molecules of the same substance in the gaseous form act on each other during an encounter. This conjecture has been put to the test by Lorenz Meyer, who has compared the densities of different liquids with the calculated relative densities of the molecules of their vapours, and has found a remarkable correspondence between them.

Now Loschmidt has deduced from the dynamical theory the following remarkable proportion:—As the volume of a gas is to the combined volume of all the molecules contained in it, so is the mean path of a molecule to one-eighth of the diameter of a molecule.

Assuming that the volume of the substance, when reduced to the liquid form, is not much greater than the combined volume of the molecules, we obtain from this proportion the diameter of a molecule. In this way Loschmidt, in 1865, made the first estimate of the diameter of a molecule. Independently of him and of each other, Mr. Stoney in 1868, and Sir W. Thomson in 1870, published results of a similar kind, those of Thomson being deduced not only in this way, but from considerations derived from the thickness of soap bubbles, and from the electric properties of metals.

According to the table, which I have calculated from Loschmidt's data, the size of the molecules of hydrogen is such that about two million of them in a row would occupy a millimetre, and a million million million million of them would weigh between four and five grammes.

In a cubic centimetre of any gas at standard pressure and temperature there are about nineteen million million million molecules. All these numbers of the third rank are, I need not tell you, to be regarded as at present conjectural. In order to warrant us in putting any confidence in numbers obtained in this way, we should have to compare together a greater number of independent data than we have as yet obtained, and to show that they lead to consistent results.

在第二个等级中，我们应该归入的是不同种气体分子的相对尺寸、平均自由程和每秒钟的碰撞次数。这些参数的量值是从三种扩散的实验结果推导出来的。在实验方法得到极大改进之前，这些公认的结果只能被看作是大概的近似。

还有一批参数只能放进第三个等级，这是因为我们在这个等级中的相关知识只是基于可能的推测，既不像第一等级中的那么精确，也不像第二等级中的那样近似精确。这里面包括分子的绝对质量、绝对直径和1立方厘米中的分子数目。我们有不同分子的相对质量的准确结果，也大概知道它们的相对直径。由此我们可以得到分子的相对密度。这些都是确实有据的结果。

液体强大的抗压缩性似乎表明：其分子之间的距离已经很近，和同一物质的两个气态分子在碰撞时发生相互作用的距离差不多。洛伦茨·迈耶想了一个办法验证这个猜想，他对各种液体的密度和相应液体蒸气的相对密度的计算值进行比较，发现它们明显相关。

现在洛施密特根据动力学原理推导出一个不寻常的比例关系：气体的体积和该气体中所有分子体积之和的比值，等于分子平均自由程与其直径的1/8之比。

假定气体液化后的体积，比相应分子的体积之和大不了多少，我们就可以通过这个比例关系得到分子的直径。1865年洛施密特由此第一次估计出了分子的直径。斯托尼先生在1868年，汤姆孙爵士于1870年也都各自独立发表了类似的结果。后者除了以上的考虑，还考虑了与肥皂泡厚度以及金属电特性有关的结果。

表中的结果是我根据洛施密特的数据计算得到的。按照这个表，200万个氢分子排成一列，只有1毫米长。100万的四次方那么多的氢分子合起来的质量只有4~5克。

在标准压强和温度下，1立方厘米的任何气体都含有1.9×10^{19}那么多的分子。这些量值都属于前面所说的第三个等级，不用我说你也知道，它们只是当前的推论。除非由独立方法得到的比现在多得多的数据经过比较后都趋向一致的结果，否则我们不能轻易接受以上的结果。

Thus far we have been considering molecular science as an inquiry into natural phenomena. But though the professed aim of all scientific work is to unravel the secrets of nature, it has another effect, not less valuable, on the mind of the worker. It leaves him in possession of methods which nothing but scientific work could have led him to invent, and it places him in a position from which many regions of nature, besides that which he has been studying, appear under a new aspect.

The study of molecules has developed a method of its own, and it has also opened up new views of nature.

When Lucretius wishes us to form a mental representation of the motion of atoms, he tells us to look at a sunbeam shining through a darkened room (the same instrument of research by which Dr. Tyndall makes visible to us the dust we breathe,) and to observe the motes which chase each other in all directions through it. This motion of the visible motes, he tells us, is but a result of the far more complicated motion of the invisible atoms which knock the motes about. In his dream of nature, as Tennyson tells us, he

"saw the flaring atom-streams
And torrents of her myriad universe,
Ruining along the illimitable inane,
Fly on to clash together again, and make
Another and another frame of things
For ever."

And it is no wonder that he should have attempted to burst the bonds of Fate by making his atoms deviate from their courses at quite uncertain times and places, thus attributing to them a kind of irrational free will, which on his materialistic theory is the only explanation of that power of voluntary action of which we ourselves are conscious.

As long as we have to deal with only two molecules, and have all the data given us, we can calculate the result of their encounter, but when we have to deal with millions of molecules, each of which has millions of encounters in a second, the complexity of he problem seems to shut out all hope of a legitimate solution.

The modern atomists have therefore adopted a method which is I believe new in the department of mathematical physics, though it has long been in use in the Section of Statistics. When the working members of Section F get hold of a Report of the Census, or any other document containing the numerical data of Economic and Social Science, they begin by distributing the whole population into groups, according to age, income-tax, education, religious belief, or criminal convictions. The number of individuals is far too great to allow of their tracing the history of each separately, so that, in order to reduce their labour within human limits, they concentrate their attention on a small number of artificial groups. The varying number of individuals in each group, and not the varying state of each individual, is the primary datum from which they work.

到现在为止，我们一直认为分子科学是对自然现象的一种探索。虽然所有科学工作的目的都是为了揭示自然的秘密，但是它还有另外一个至少同样重要的作用，就是对科学工作者心灵的影响。只有科学工作才能让他们发明并掌握新的方法，这使他们站在了一个新的高度，看到除了自己的研究领域之外，其他许多自然领域也呈现出新的面貌。

分子领域的研究发展了一套自己的方法，也打开了一扇观察自然的窗户。

为了让我们在头脑中建立一幅原子运动的图像，卢克莱修让我们观察一束射入暗室的日光（廷德尔博士用同样的设备展示了我们呼吸的灰尘的运动情况），在这束日光中我们可以看到向各个方向追逐乱窜的尘埃。他告诉我们，这些可见尘埃的运动，是更加复杂的不可见的原子运动不断将它们撞来撞去的结果。就像丁尼生告诉我们的一样，在卢克莱修梦想的自然中，他

“看见闪耀的原子流
和她在无穷宇宙中的激流，
在无尽的虚空中衰耗，
又飞来撞在一起，
创造一个又一个事物的构架，
永不止息。”

难怪他试图打破必然的枷锁，让他的原子在很不确定的时刻和地点改变轨迹，因而赋予它们一种非理性的自由意志，这就是他在物质理论中对我们所知的随机行为的产生所做的唯一解释。

如果我们只需要研究两个分子，而且知道所有的数据，我们就可以计算它们相互碰撞的结果。但实际上有极多的分子，每个分子每秒钟都要经历极多的碰撞，问题的复杂性似乎超越了任何合理解答的可能性。

现代原子论者采取了一种我认为在数学物理学科中是全新的方法，尽管它在统计部门中已经应用很久了。当F部门的工作人员得到一份人口普查报告，或者是其他包含经济和社会科学统计数据的文件时，他们先把整个人口按照年龄、所得税、教育水平、宗教信仰或犯罪记录分组。人口的数量太大，很难一一跟踪每个人的实际情况。为了在人力资源有限的情况下把工作量控制在合理范围，他们把注意力集中在少数几个人为划分出来的组群上，每个组群个体人数的变化，而非每个个体的状态变化是他们进行研究工作的最初数据。

This, of course, is not the only method of studying human nature. We may observe the conduct of individual men and compare it with that conduct which their previous character and their present circumstances, according to the best existing theory, would lead us to expect. Those who practise this method endeavour to improve their knowledge of the elements of human nature, in much the same way as an astronomer corrects the elements of a planet by comparing its actual position with that deduced from the received elements. The study of human nature by parents and schoolmasters, by historians and statesmen, is therefore to be distinguished from that carried on by registrars and tabulators, and by those statesmen who put their faith in figures. The one may be called the historical, and the other the statistical method.

The equations of dynamics completely express the laws of the historical method as applied to matter, but the application of these equations implies a perfect knowledge of all the data. But the smallest portion of matter which we can subject to experiment consists of millions of molecules, not one of which ever becomes individually sensible to us. We cannot, therefore, ascertain the actual motion of any one of these molecules, so that we are obliged to abandon the strict historical method, and to adopt the statistical method of dealing with large groups of molecules.

The data of the statistical method as applied to molecular science are the sums of large numbers of molecular quantities. In studying the relations between quantities of this kind, we meet with a new kind of regularity, the regularity of averages, which we can depend upon quite sufficiently for all practical purposes, but which can make no claim to that character of absolute precision which belongs to the laws of abstract dynamics.

Thus molecular science teaches us that our experiments can never give us anything more than statistical information, and that no law deduced from them can pretend to absolute precision. But when we pass from the contemplation of our experiments to that of the molecules themselves, we leave the world of chance and change, and enter a region where everything is certain and immutable.

The molecules are conformed to a constant type with a precision which is not to be found in the sensible properties of the bodies which they constitute. In the first place the mass of each individual molecule, and all its other properties, are absolutely unalterable. In the second place the properties of all molecules of the same kind are absolutely identical.

Let us consider the properties of two kinds of molecules, those of oxygen and those of hydrogen.

We can procure specimens of oxygen from very different sources—from the air, from water, from rocks or every geological epoch. The history of these specimens has been very different, and if, during thousands of years, difference of circumstances could produce difference of properties, these specimens of oxygen would show it.

这显然不是研究人类性质的唯一方法。我们可以观察个体的行为，也可以按照当前最好的理论，根据以往的特征预测目前情况下的行为，并把观察结果和预测的行为相比较。人们采用这种方法以增进他们对人类性质各个方面的认识，就像天文学家通过对比行星的实际位置和依据已知参数预测的位置之间的差别来修正轨道参数一样。父母、校长、历史学家、政治家对人性的研究，和登记员、制表人还有注重数据的政治家的研究是不同的。一个可以叫历史方法，另一个则是统计方法。

动力学方程完全是历史方法定律在物质研究上的应用，但应用这些方程需要知道所有的数据，然而哪怕是实验中物质最小的部分也包含着极多我们看不见的分子。我们不可能确定任何一个分子的实际运动状况，所以必须放弃严格的历史方法，采用统计方法来处理大量分子的情况。

应用在分子科学中的统计方法牵涉的数据是大量分子的参量的总和。在研究这类参量之间的关系时，我们碰上了一种新的规律，也就是平均值的规律。对于实际应用，依赖这些规律就足够了，但是它们不能给出理论动力学定律所具有的绝对精确性。

分子科学告诉我们：实验永远只能提供统计信息，从实验中总结的定律都没有绝对的精确性。但是当我们的关注点从实验转向分子本身时，我们就离开了充满偶然性和变化的世界，进入到一切都是确定不变的领域。

分子都精确地具有恒定不变的属性，这种属性是分子所组成的物体的可测宏观参数所不具备的。首先，所有分子的质量和所有其他性质都是绝对不变的。其次，同种物质所有分子的所有性质都是绝对相同的。

让我们来考虑一下氧和氢这两种分子的性质。

我们可以从空气、水和各个地质时代的岩石等不同来源制备氧气样品。这些样品形成的历史完全不同。如果在几千年的时间里，不同的环境能造就不同的性质，这些氧气样品应该就能表现出这些不同的性质。

In like manner we may procure hydrogen from water, from coal, or, as Graham did, from meteoric iron. Take two litres of any specimen of hydrogen, it will combine with exactly one litre of any specimen of oxygen, and will form exactly two litres of the vapour of water.

Now if, during the whole previous history of either specimen, whether imprisoned in the rocks, flowing in the sea, or careering through unknown regions with the meteorites, any modification of the molecules had taken place, these relations would no longer be preserved.

But we have another and an entirely different method of comparing the properties of molecules. The molecule, though indestructible, is not a hard rigid body, but is capable of internal movements, and when these are excited it emits rays, the wavelength of which is a measure of the time of vibration of the molecule.

By means of the spectroscope the wavelengths of different kinds of light may be compared to within one ten-thousandth part. In this way it has been ascertained, not only that molecules taken from every specimen of hydrogen in our laboratories have the same set of periods of vibration, but that light, having the same set of periods of vibration, is emitted from the sun and from the fixed stars.

We are thus assured that molecules of the same nature as those of our hydrogen exist in those distant regions, or at least did exist when the light by which we see them was emitted.

From a comparison of the dimensions of the buildings of the Egyptians with those of the Greeks, it appears that they have a common measure. Hence, even if no ancient author had recorded the fact that the two nations employed the same cubit as a standard of length, we might prove it from the buildings themselves. We should also be justified in asserting that at some time or other a material standard of length must have been carried from one country to the other, or that both countries had obtained their standards from a common source.

But in the heavens we discover by their light, and by their light alone, stars so distant from each other that no material thing can ever have passed from one to another, and yet this light, which is to us the sole evidence of the existence of these distant worlds, tells us also that each of them is built up of molecules of the same kinds as those which we find on earth. A molecule of hydrogen, for example, whether in Sirius or in Arcturus, executes its vibrations in precisely the same time.

Each molecule, therefore, throughout the universe, bears impressed on it the stamp of a metric system as distinctly as does the metre of the Archives at Paris, or the double royal cubit of the Temple of Karnac.

同样地，我们可以从水、煤炭或者像格雷姆那样从陨铁中制备出氢气样品。任取两升氢气样品，它都可以和任取的一升氧气样品反应，正好生成两升水蒸气。

如果在任何一个样品的整个历史中，不论它是固锁在岩石中，还是漂流在海洋里，或者是跟着陨石穿越着未知的区域，只要分子发生了任何变化，上面的关系都不能再保持。

我们有另一种完全不同的办法来比较分子的性质。分子虽然不能被毁灭，但也不是刚性实体，它有内部运动，当内部运动被激发时，分子发出射线，这个射线的波长表征了分子振动的周期。

光谱仪可以比较不同光的波长差别，精度可达万分之一。用这个办法，我们可以确认，不但我们实验室中每一个氢气样品中的分子具有同样的振动周期组合，就连太阳和其他恒星发射的光也存在同样的振动周期组合。

这样我们就确信了在茫茫的宇宙中也存在和我们的氢分子一样的分子，或者说至少在那些我们用来观察分子的光线放射出时，它们是存在的。

比较古埃及和古希腊的建筑规模，可以感觉到它们似乎使用了同样的度量标准。因此，尽管历史上没有留下两个国家都使用同样的肘长作为长度标准的记载，我们也许可以通过建筑本身证明这一点。我们也有理由认为，肯定是某个实物长度标准在某个时候被人从一个国家带到了另一个国家，或者两个国家从同一个来源得到了这个标准。

对天空中的星体，我们只是通过它们发射的光线证实它们的存在，它们之间的距离如此遥远，物质从一个恒星到另一个恒星的传递是完全不可能的，但是它们发射的光线，除了作为证明这些遥远天体存在的唯一证据以外，还告诉我们，组成每一个天体的分子都和我们在地球上发现的分子相同。比如，不论是天狼星还是大角星，上面的氢分子都以同样的周期振动。

宇宙中的每一个分子，都铭刻着一个度量系统的痕迹，它的独特清晰，就如同巴黎档案局的米原尺，或者卡尔纳克神庙的皇家双肘尺一样。

No theory of evolution can be formed to account for the similarity of molecules, for evolution necessarily implies continuous change, and the molecule is incapable of growth or decay, of generation or destruction.

None of the processes of Nature, since the time when Nature began, have produced the slightest difference in the properties of any molecule. We are therefore unable to ascribe either the existence of the molecules or the identity of their properties to the operation of any of the causes which we call natural.

On the other hand, the exact quality of each molecule to all others of the same kind gives it, as Sir John Herschel has well said, the essential character of a manufactured article, and precludes the idea of its being eternal and self existent.

Thus we have been led, along a strictly scientific path, very near to the point at which Science must stop. Not that Science is debarred from studying the internal mechanism of a molecule which she cannot take to pieces, any more than from investigating an organism which she cannot put together. But in tracing back the history of matter Science is arrested when she assures herself, on the one hand, that the molecule has been made, and on the other that it has not been made by any of the processes we call natural.

Science is incompetent to reason upon the creation of matter itself out of nothing. We have reached the utmost limit of our thinking faculties when we have admitted that because matter cannot be eternal and self-existent it must have been created.

It is only when we contemplate, not matter in itself, but the form in which it actually exists, that our mind finds something on which it can lay hold.

That matter, as such, should have certain fundamental properties—that it should exist in space and be capable of motion, that its motion should be persistent, and so on, are truths which may, for anything we know, be of the kind which metaphysicians call necessary. We may use our knowledge of such truths for purposes of deduction but we have no data for speculating as to their origin.

But that there should be exactly so much matter and no more in every molecule of hydrogen is a fact of a very different order. We have here a particular distribution of matter—a *collocation*—to use the expression of Dr. Chalmers, of things which we have no difficulty in imagining to have been arranged otherwise.

The form and dimensions of the orbits of the planets, for instance, are not determined by any law of nature, but depend upon a particular collocation of matter. The same is the case with respect to the size of the earth, from which the standard of what is called the metrical system has been derived. But these astronomical and terrestrial magnitudes are far inferior in scientific importance to that most fundamental of all standards which forms

任何进化论都不能解释分子的相似性，因为进化隐含着不断的变化，而分子既不生长也不腐朽，不能被制造也不能被摧毁。

自从自然界形成以来，自然界的所有过程都没能使任何分子的性质有丝毫的改变。所以我们不能把分子的存在或其性质的同一性归结为任何自然原因的作用。

另一方面，正像约翰·赫歇尔爵士说的那样，同类分子精确相同的性质是其所构成的物质的根本特征，这就排除了分子的永恒自存在性。

这样一来，我们就被引领着，沿着一条严格的科学道路，走到了科学的尽头。这不是说科学手段不能分解分子就不能用来研究分子的内部机制，就像不能因为科学不能组合成生物就不用她来研究生物一样。但在追溯物质的历史时，科学被困住了，她一方面确认分子已被创造，另一方面又确认任何自然界的过程都不能创造分子。

科学不能理解物质如何被无中生有地创造出来。当我们承认由于物质不能永恒自存在，因此它一定已经被创造出来的时候，我们已经到了自己思维能力的极限。

只有当我们思考物质的存在形式，而不是物质本身的时候，我们的思维才算找到了一些可以解决问题的线索。

我们知道，形而上学者认为必要的真理是：物质必须具有一些基本的性质——它应该存在于空间中，能够运动，运动应该是持续的等等。我们可以利用已知的与这些真理有关的知识进行推导，但是并无数据可以用来猜测这些真理的本源。

但是每一个氢分子的质量都正好是这么多，这个事实就完全是另一个性质了。这种质量的分布，或者用查默斯博士的词语——这种**布置**，实在是太特别了，让我们都不习惯。

比如，行星轨道的形式和尺度，也不是由任何自然定律决定的，而是取决于特殊的质量的布置。作为米制计量体系标准的地球的尺寸也是这样。但是这些天文和地理的量值在科学上的重要性都远不能和构成分子系统基础的那些最基本的标准相比。我们知道自然界过程一直在起作用，就算最后不摧毁，它们至少也要改变地球

the base of the molecular system. Natural causes, as we know, are at work, which tend to modify, if they do not at length destroy, all the arrangements and dimensions of the earth and the whole solar system. But though in the course of ages catastrophes have occurred and may yet occur in the heavens, though ancient systems may be dissolved and new systems evolved out of their ruins, the molecules out of which these systems are built—the foundation stones of the material universe—remain unbroken and unworn.

They continue this day as they were created, perfect in number and measure and weight, and from the ineffaceable characters impressed on them we may learn that those aspirations after accuracy in measurement, truth in statement, and justice in action, which we reckon among our noblest attributes as men, are ours because they are essential constituents of the image of Him Who in the beginning created, not only the heaven and the earth, but the materials of which heaven and earth consist.

Table of Moncular Date

		Hydrogen	Oxygen	Carbonic oxide	Carbonic acid
Rank I	Mass of molecule (hydrogen = 1)	1	16	14	22
	Velocity (of mean square), metres per second at 0°C	1859	465	497	396
Rank II	Mean path, tenth-metres.	965	560	482	379
	Collisions in a second, (millions)	17750	7646	9489	9720
Rank III	Diameter, tenth-metre	5.8	7.6	8.3	9.3
	Mass, twenty-fifth-grammes.	46	736	644	1012

Table of Diffusion: $\frac{(\text{centimetre})^2}{\text{second}}$ measure

	Calculated	Observed	
H & O	0.7086	0.7214	Diffusion of matter observed by Loschmidt.
H & CO	0.6519	0.6422	
H & CO_2	0.5575	0.5558	
O & CO	0.1807	0.1802	
O & CO_2	0.1427	0.1409	
CO & CO_2	0.1386	0.1406	
H	1.2990	1.49	Diffusion of momentum Graham and Meyer.
O	0.1884	0.213	
CO	0.1748	0.212	
CO_2	0.1087	0.117	
Air		0.256	Diffusion of temperature observed by Stefan.
Copper		1.077	
Iron		0.183	
Cane sugar in water		0.00000365	Voit
Diffusion in a day		0.3144	
Salt in water		0.00000116	Fick

(**8**, 437-441; 1873)

和整个太阳系的秩序和尺寸。然而，尽管在历史上宇宙曾经发生过灾变，以后也可能再发生，虽然旧的体系可能被消灭，新的体系在旧体系的废墟上演化发展，但作为这些系统组成基础的分子——物质宇宙的基石，却不会被破坏和磨损。

它们被创造的时候是什么样，现在还是什么样，数量、大小和质量都没有丝毫改变，从铭记在它们身上的不可磨灭的特征，我们也许能够明白，我们对测量的精准、言语的真切、行为的正义这些人类最高尚品质的追求，是因为它们也是造物主形象的根本组成部分。他当初不仅创造了天和地，也创造了组成天和地的所有物质。

分子数据表

		氢	氧	一氧化碳	二氧化碳
第一等级	分子质量（氢=1）	1	16	14	22
	0℃下的速度（均方根），米/秒	1859	465	497	396
第二等级	平均自由程，10^{-10}米	965	560	482	379
	每秒碰撞次数（百万次）	17750	7646	9489	9720
第三等级	直径，10^{-10}米	5.8	7.6	8.3	9.3
	质量，10^{-25}克	46	736	644	1012

扩散数据表：单位是 $\frac{(\text{厘米})^2}{\text{秒}}$

	计算值	测量值	
氢 & 氧	0.7086	0.7214	洛施密特观察的物质扩散
氢 & 一氧化碳	0.6519	0.6422	
氢 & 二氧化碳	0.5575	0.5558	
氧 & 一氧化碳	0.1807	0.1802	
氧 & 二氧化碳	0.1427	0.1409	
一氧化碳 & 二氧化碳	0.1386	0.1406	
氢	1.2990	1.49	格雷姆和迈耶的动量扩散
氧	0.1884	0.213	
一氧化碳	0.1748	0.212	
二氧化碳	0.1087	0.117	
空气		0.256	斯特藩观察的温度扩散
铜		1.077	
铁		0.183	
蔗糖在水溶液中		0.00000365	沃伊特
一天后扩散的距离		0.3144	
食盐在水溶液中		0.00000116	菲克

（何钧 翻译；鲍重光 审稿）

On the Dynamical Evidence of the Molecular Constitution of Bodies*

J. C. Maxwell

Editor's Note

Here James Clerk Maxwell offers an update on efforts to understand the properties of gases and liquids from first principles—what today is called statistical mechanics. Maxwell had made a decisive contribution with his kinetic theory of gases. Here he reports on other work, such as that of Rudolph Clausius, who introduced a theoretical quantity known as the "virial", representing an average over all particle pairs of the product of the distance between the particles and the force of attraction or repulsion. Maxwell also mentions the work of two other figures, an unknown graduate student named Johannes Diderik van der Waals and an American physicist, Josiah Willard Gibbs. Both later became leaders in understanding the fundamental behavior of material systems.

I

WHEN any phenomenon can be described as an example of some general principle which is applicable to other phenomena, that phenomenon is said to be explained. Explanations, however, are of very various orders, according to the degree of generality of the principle which is made use of. Thus the person who first observed the effect of throwing water into a fire would feel a certain amount of mental satisfaction when he found that the results were always similar, and that they did not depend on any temporary and capricious antipathy between the water and the fire. This is an explanation of the lowest order, in which the class to which the phenomenon is referred consists of other phenomena which can only be distinguished from it by the place and time of their occurrence, and the principle involved is the very general one that place and time are not among the conditions which determine natural processes. On the other hand, when a physical phenomenon can be completely described as a change in the configuration and motion of a material system, the dynamical explanation of that phenomenon is said to be complete. We cannot conceive any further explanation to be either necessary, desirable, or possible, for as soon as we know what is meant by the words configuration, motion, mass, and force, we see that the ideas which they represent are so elementary that they cannot be explained by means of anything else.

The phenomena studied by chemists are, for the most part, such as have not received a complete dynamical explanation.

* A lecture delivered at the Chemical Society, Feb. 18, by Prof. Clerk Maxwell, F. R. S.

关于物体分子构成的动力学证据*

麦克斯韦

编者按

在这篇文章中，克拉克·麦克斯韦介绍了从基本原理（现在的名字是统计力学）出发研究气体和液体性质的最新进展。此前，麦克斯韦提出了他的气体动力学理论，这是他作出的决定性的贡献。在本文中他报道了鲁道夫·克劳修斯等其他研究人员的工作。克劳修斯引入了"位力"这个理论物理量，用来表示所有粒子对的粒子间距离与粒子间引力或斥力乘积的总平均。麦克斯韦在这篇文章中还提到了另外两个人的工作，其中一个是当时还没什么名气的研究生约翰内斯·迪德里克·范德瓦尔斯，另一个是美国物理学家约西亚·威拉德·吉布斯，这两人后来都成了研究物质体系基本行为特征的领袖人物。

I

任何一种现象，一旦可以被视为某种适用于其他现象的普遍原理的实例，我们就可以说这种现象得到了解释。然而，解释包括很多不同层面，要视所采用原理的普适程度而定。因此，第一个观察到将水泼入火中这种现象的人，在发现结果总是相似的而且这些结果并不取决于水和火之间那瞬息万变、反复无常的不相容性时，便会获得一定程度的精神满足。这是一种最低层次的解释，在这种解释中，以上现象所涉及的分类还包含其他现象，而这些现象只是发生的时间和地点不同，所涉及的原理却是非常普适的，因此时间和地点并不属于能够决定自然过程的条件。另一方面，当某一物理现象能够被完全描述为物质系统结构或运动的变化时，我们就可以说这种现象得到了完整的动力学解释。我们想不出还有什么更进一步的解释是必要的、有用的或者可能的，因为一旦我们了解了结构、运动、质量和力这些词的意义，便会发现它们所表示的意义如此基本，以至于无法再用其他概念来解释它们。

化学家们研究的现象，大部分还没有得到完整的动力学解释。

* 2月18日，克拉克·麦克斯韦教授（皇家学会会员）在化学学会所作的讲座。

Many diagrams and models of compound molecules have been constructed. These are the records of the efforts of chemists to imagine configurations of material systems by the geometrical relations of which chemical phenomena may be illustrated or explained. No chemist, however, professes to see in these diagrams anything more than symbolic representations of the various degrees of closeness with which the different components of the molecule are bound together.

In astronomy, on the other hand, the configurations and motions of the heavenly bodies are on such a scale that we can ascertain them by direct observation. Newton proved that the observed motions indicate a continual tendency of all bodies to approach each other, and the doctrine of universal gravitation which he established not only explains the observed motions of our system, but enables us to calculate the motions of a system in which the astronomical elements may have any values whatever.

When we pass from astronomical to electrical science, we can still observe the configuration and motion of electrified bodies, and thence, following the strict Newtonian path, deduce the forces with which they act on each other; but these forces are found to depend on the distribution of what we call electricity. To form what Gauss called a "construirbar Vorstellung" of the invisible process of electric action is the great desideratum in this part of science.

In attempting the extension of dynamical methods to the explanation of chemical phenomena, we have to form an idea of the configuration and motion of a number of material systems, each of which is so small that it cannot be directly observed. We have, in fact, to determine, from the observed external actions of an unseen piece of machinery, its internal construction.

The method which has been for the most part employed in conducting such inquiries is that of forming an hypothesis, and calculating what would happen if the hypothesis were true. If these results agree with the actual phenomena, the hypothesis is said to be verified, so long, at least, as some one else does not invent another hypothesis which agrees still better with the phenomena.

The reason why so many of our physical theories have been built up by the method of hypothesis is that the speculators have not been provided with methods and terms sufficiently general to express the results of their induction in its early stages. They were thus compelled either to leave their ideas vague and therefore useless, or to present them in a form the details of which could be supplied only by the illegitimate use of the imagination.

In the meantime the mathematicians, guided by that instinct which teaches them to store up for others the irrepressible secretions of their own minds, had developed with the utmost generality the dynamical theory of a material system.

很多化合物分子的图示或模型已经建立。它们都是化学家根据可以说明或解释的化学现象的几何关系努力猜想物质系统的结构而得到的结果。不过，这些图示只不过是对构成分子的不同成分束缚在一起的各种不同的紧密程度的符号表示而已，没有一位化学家敢说他能从中看出更多的东西。

另一方面，在天文学中，天体的结构和运动所具有的尺度是我们可以通过直接观测来确定的。牛顿证明了观测到的运动显示出所有物体之间都有不断彼此靠拢的趋势，他建立的万有引力定律不仅解释了观测到的我们自身所在系统的运动，而且使我们能够计算出具有任意量值的天文学对象所组成的系统的运动情况。

从天文学转向电学，我们仍然可以观测带电体的结构和运动，并遵循严格的牛顿式方法来推导出它们的相互作用力。但是我们发现这种力依赖于所谓的电的分布。因此，建立一种关于电作用不可见过程的、被高斯称为“统一表象”的理论，就成为这门科学的迫切需要。

在试图将动力学方法推广以用于对化学现象的解释时，我们必须形成关于大量物质系统的结构和运动的观念，而其中每一个物质系统都小到无法直接观测的程度。事实上，我们只能通过观测那些看不见的结构部分的外部作用来确定其内部结构。

主要用来进行此类研究的方法是先建立一个假说，然后计算如果该假说成立会产生什么结果。若这些结果与实际现象相吻合，我们就说这个假说得到了确证，至少在没有其他人提出与客观现象更一致的另一种假说之前，是可以这样说的。

之所以有如此多的物理理论依靠假说的方式建立起来，是因为理论家早期没有获得具有足够普适性的方法和术语来表达他们归纳出的结果。受此限制，他们要么将观点表达得含糊不清以至于毫无用处，要么用另一种形式表达观点，而这种形式的细节只能靠非常规的想象力来弥补。

与此同时，数学家在其本能（这种本能使他们将自身才智孕育出的丰硕果实供给他人）的驱使下已经发展出了具有充分普适性的关于物质系统的动力学理论。

Of all hypotheses as to the constitution of bodies, that is surely the most warrantable which assumes no more than that they are material systems, and proposes to deduce from the observed phenomena just as much information about the conditions and connections of the material system as these phenomena can legitimately furnish.

When examples of this method of physical speculation have been properly set forth and explained, we shall hear fewer complaints of the looseness of the reasoning of men of science, and the method of inductive philosophy will no longer be derided as mere guess-work.

It is only a small part of the theory of the constitution of bodies which has as yet been reduced to the form of accurate deductions from known facts. To conduct the operations of science in a perfectly legitimate manner, by means of methodised experiment and strict demonstration, requires a strategic skill which we must not look for, even among those to whom science is most indebted for original observations and fertile suggestions. It does not detract from the merit of the pioneers of science that their advances, being made on unknown ground, are often cut off, for a time, from that system of communications with an established base of operations, which is the only security for any permanent extension of science.

In studying the constitution of bodies we are forced from the very beginning to deal with particles which we cannot observe. For whatever may be our ultimate conclusions as to molecules and atoms, we have experimental proof that bodies may be divided into parts so small that we cannot perceive them.

Hence, if we are careful to remember that the word particle means a small part of a body, and that it does not involve any hypothesis as to the ultimate divisibility of matter, we may consider a body as made up of particles, and we may also assert that in bodies or parts of bodies of measurable dimensions, the number of particles is very great indeed.

The next thing required is a dynamical method of studying a material system consisting of an immense number of particles, by forming an idea of their configuration and motion, and of the forces acting on the particles, and deducing from the dynamical theory those phenomena which, though depending on the configuration and motion of the invisible particles, are capable of being observed in visible portions of the system.

The dynamical principles necessary for this study were developed by the fathers of dynamics, from Galileo and Newton to Lagrange and Laplace; but the special adaptation of these principles to molecular studies has been to a great extent the work of Prof. Clausius of Bonn, who has recently laid us under still deeper obligations by giving us, in addition to the results of his elaborate calculations, a new dynamical idea, by the aid of which I hope we shall be able to establish several important conclusions without much symbolical calculation.

在所有关于物体构成的假说中，最可信的无疑是这样一种假说：它除了假定物体构成物质系统之外绝无更多假定，并仅就观测到的现象在合理范围内所提供的有关物质系统的条件和联系等信息进行推导。

在合理地阐明和解释这种物理假说方法的实例之后，我们受到的对科学界人士推理不严谨的抱怨将会变得很少，并且相应的归纳哲学方法也不会再被人嘲笑为凭空臆测了。

迄今为止，物体构成理论中只有一小部分已经通过已知的事实被归纳为精确的结论。要通过合理的实验和严格的论证引导科学活动沿着十分合理的道路前进，就需要一种战略技巧，这种技巧可遇不可求，即使对于那些凭借原始观测与丰富联想进行科学研究的人来说也是一样。下面的事实并不会贬低科学先驱的成就，即他们在未知领域的开拓工作，这些开拓工作曾经一度被隔离于有着已确定的作用基础的交流体系之外，而这个交流体系正是任何科学持续发展的唯一保障。

要研究物体构成，我们从一开始就不得不处理看不见的微粒。因为不管我们对于分子和原子得到的最终结论将是什么，现在我们已经得到了实验上的证据，可以证明物体可以被分割成小到无法察觉的部分。

由此，如果我们细心地注意到微粒这个词只表示物体的一小部分，而且毫不涉及关于物质终极可分性的假设，我们就可以认为物体是由微粒组成的，我们还可以断言，在物体或者具有可观测尺度的物体的某一部分中，微粒的数量是非常大的。

下面我们还需要一种可以研究由大量微粒组成的物质系统的动力学方法，这就要形成一种关于微粒的结构、运动以及作用在粒子上的力的概念，并从动力学理论中推导出在系统的可见部分中可观测到的现象，尽管这些现象是由不可见微粒的运动和结构决定的。

进行这种研究必不可少的动力学原理是由多位动力学奠基人发展起来的，从伽利略、牛顿到拉格朗日和拉普拉斯；而将这些原理应用于分子研究，则很大程度上是波恩的克劳修斯教授的贡献，除了复杂的计算结果之外，他最近提出的一种新的动力学观点，给我们的研究工作带来了很大的帮助。借助这种观点，我希望我们可以不用大量的符号演算也能得到一些重要的结论。

The equation of Clausius, to which I must now call your attention, is of the following form:—

$$pV = \frac{2}{3}T - \frac{2}{3}\Sigma\Sigma\left(\frac{1}{2}Rr\right)$$

Here p denotes the pressure of a fluid, and V the volume of the vessel which contains it. The product pV, in the case of gases at constant temperature, remains, as Boyle's Law tells us, nearly constant for different volumes and pressures. This member of the equation, therefore, is the product of two quantities, each of which can be directly measured.

The other member of the equation consists of two terms, the first depending on the motion of the particles, and the second on the forces with which they act on each other.

The quantity T is the kinetic energy of the system, or, in other words, that part of the energy which is due to the motion of the parts of the system.

The kinetic energy of a particle is half the product of its mass into the square of its velocity, and the kinetic energy of the system is the sum of the kinetic energy of its parts.

In the second term, r is the distance between any two particles, and R is the attraction between them. (If the force is a repulsion or a pressure, R is to be reckoned negative.)

The quantity $\frac{1}{2}Rr$, or half the product of the attraction into the distance across which the attraction is exerted, is defined by Clausius as the virial of the attraction. (In the case of pressure or repulsion, the virial is negative.)

The importance of this quantity was first pointed out by Clausius, who, by giving it a name, has greatly facilitated the application of his method to physical exposition.

The virial of the system is the sum of the virials belonging to every pair of particles which exist in the system. This is expressed by the double sum $\Sigma\Sigma(\frac{1}{2}Rr)$, which indicates that the value of $\frac{1}{2}Rr$ is to be found for every pair of particles, and the results added together.

Clausius has established this equation by a very simple mathematical process, with which I need not trouble you, as we are not studying mathematics tonight. We may see, however, that it indicates two causes which may affect the pressure of the fluid on the vessel which contains it: the motion of its particles, which tends to increase the pressure, and the attraction of its particles, which tends to diminish the pressure.

We may therefore attribute the pressure of a fluid either to the motion of its particles or to a repulsion between them.

现在我必须提醒大家注意，克劳修斯方程具有如下形式：

$$pV = \frac{2}{3}T - \frac{2}{3}\Sigma\Sigma\left(\frac{1}{2}Rr\right)$$

其中，p 表示某种流体的压力，V 表示盛放流体的容器的体积。由玻意耳定律可知，对于恒温条件下的气体，乘积 pV 在不同的体积和压力下基本保持不变。可见，方程的左边，是两个直接观测量的乘积。

方程的另一边由两项组成，第一项由微粒的运动决定，第二项由微粒间的相互作用力决定。

T 这个量指系统的动能，或者换一种说法，是由于系统各部分的运动而具有的那部分能量。

微粒的动能是其质量与速率平方的乘积的一半，而系统的动能则是其各部分动能的总和。

在第二项中，r 表示任意两个微粒之间的距离，R 表示它们之间的吸引力。（如果作用力为排斥力或者压力，则 R 取负值。）

$\frac{1}{2}Rr$ 这个量，即引力与彼此吸引的微粒之间距离的乘积的一半，克劳修斯将其定义为引力的位力。（对于压力和排斥力，位力取负值。）

克劳修斯首次指出了这一物理量的重要性。通过给这一物理量指定名字，克劳修斯大大推动了他的方法在物理解释中的应用。

系统的位力是系统中每一对微粒之间位力的总和，可以表示为双重求和 $\Sigma\Sigma(\frac{1}{2}Rr)$，意思是要得到每一对微粒之间的 $\frac{1}{2}Rr$ 值，并将结果累加起来。

克劳修斯通过简单的数学过程建立了这一方程，这里我不需要以此来烦扰各位，因为今晚我们不是在研究数学问题。不过我们还是可以看到，方程指出有两个因素可能会影响流体对容纳它的容器的压力：流体中微粒的运动倾向于增大压力，而微粒间的吸引力倾向于减小压力。

由此，我们可以将流体的压力归结为流体中微粒的运动或者微粒间的排斥力。

Let us test by means of this result of Clausius the theory that the pressure of a gas arises entirely from the repulsion which one particle exerts on another, these particles, in the case of gas in a fixed vessel, being really at rest.

In this case the virial must be negative, and since by Boyle's Law the product of pressure and volume is constant, the virial also must be constant, whatever the volume, in the same quantity of gas at constant temperature. It follows from this that Rr, the product of the repulsion of two particles into the distance between them, must be constant, or in other words that the repulsion must be inversely as the distance, a law which Newton has shown to be inadmissible in the case of molecular forces, as it would make the action of the distant parts of bodies greater than that of contiguous parts. In fact, we have only to observe that if Rr is constant, the virial of every pair of particles must be the same, so that the virial of the system must be proportional to the number of pairs of particles in the system—that is, to the square of the number of particles, or in other words to the square of the quantity of gas in the vessel. The pressure, according to this law, would not be the same in different vessels of gas at the same density, but would be greater in a large vessel than in a small one, and greater in the open air than in any ordinary vessel.

The pressure of a gas cannot therefore be explained by assuming repulsive forces between the particles.

It must therefore depend, in whole or in part, on the motion of the particles.

If we suppose the particles not to act on each other at all, there will be no virial, and the equation will be reduced to the form

$$Vp = \frac{2}{3}T$$

If M is the mass of the whole quantity of gas, and c is the mean square of the velocity of a particle, we may write the equation—

$$Vp = \tfrac{1}{3}Mc^2,$$

or in words, the product of the volume and the pressure is one-third of the mass multiplied by the mean square of the velocity. If we now assume, what we shall afterwards prove by an independent process, that the mean square of the velocity depends only on the temperature, this equation exactly represents Boyle's Law.

But we know that most ordinary gases deviate from Boyle's Law, especially at low temperatures and great densities. Let us see whether the hypothesis of forces between the particles, which we rejected when brought forward as the sole cause of gaseous pressure, may not be consistent with experiment when considered as the cause of this deviation from Boyle's Law.

When a gas is in an extremely rarefied condition, the number of particles within a given distance of any one particle will be proportional to the density of the gas. Hence the virial

让我们用克劳修斯的结果来验证一下下面的理论：气体压力完全源于一个微粒施加在另一个微粒上的排斥力，而就恒容容器中的气体而言，这些微粒是真正静止的。

在上述情况中，位力必定为负值，又由玻意耳定律可知压力与体积的乘积为常数，则等量恒温的气体，不论其体积是多少，位力也一定是常数。由此可知，两个微粒之间的排斥力与距离的乘积 Rr 必为常数，换言之，排斥力必定与距离成反比。然而，牛顿已经指出这一定律不适用于分子间作用力的情况，因为它将使物体中相隔较远的部分间的相互作用比相邻的部分间的相互作用更强。实际上，我们只需注意到，如果 Rr 是常数，那么每一对微粒的位力都一定是相同的，于是系统的位力一定会正比于系统中微粒的对数——即正比于微粒数的平方，或者说正比于容器中气体数量的平方。根据这一定律，相同密度的气体在不同容器中的压力也是不同的，在较大容器中的压力会比在较小容器中的压力更大，而在开放的大气中的压力比在任何普通容器中的压力都大。

因此，气体的压力无法通过假定微粒间存在排斥力进行解释。

这样一来，气体的压力就一定是完全地或部分地依赖于微粒的运动。

如果我们假定微粒间没有任何相互作用，那么就没有位力存在，方程可简化为如下形式：

$$Vp = \frac{2}{3}T$$

如果用 M 表示气体的总质量，c^2 表示微粒速率的均方，我们就可以将方程写作——

$$Vp = \frac{1}{3}Mc^2,$$

或者用文字表述为，体积与压力的乘积是质量与速率均方乘积的 1/3。现在我们先假设，速率的均方仅取决于温度，那么这个方程恰好就是玻意耳定律，后面我们将通过独立的过程来证明。

但是我们知道大部分常见气体都会偏离玻意耳定律，尤其是在低温和高密度的状态下。现在我们再来看看刚才在作为产生气体压力的唯一原因被提出时，已经被我们否定的微粒间存在作用力的假说，在作为偏离玻意耳定律的原因时是否会与实验不一致。

当气体处于极稀薄的状态时，在任一微粒周围给定距离之内的微粒数量将正比于气体密度。因此，一个微粒作用于其他微粒而产生的位力将随密度变化，单位体

arising from the action of one particle on the rest will vary as the density, and the whole virial in unit of volume will vary as the square of the density.

Calling the density ρ, and dividing the equation by V, we get—

$$p = \frac{1}{3}\rho c^2 - \frac{2}{3}A\rho^2,$$

where A is a quantity which is nearly constant for small densities.

Now, the experiments of Regnault show that in most gases, as the density increases the pressure falls below the value calculated by Boyle's Law. Hence the viral must be positive; that is to say, the mutual action of the particles must be in the main attractive, and the effect of this action in diminishing the pressure must be at first very nearly as the square of the density.

On the other hand, when the pressure is made still greater the substance at length reaches a state in which an enormous increase of pressure produces but a very small increase of density. This indicates that the virial is now negative, or, in other words, the action between the particles is now, in the main, repulsive. We may therefore conclude that the action between two particles at any sensible distance is quite insensible. As the particles approach each other the action first shows itself as an attraction, which reaches a maximum, then diminishes, and at length becomes a repulsion so great that no attainable force can reduce the distance of the particles to zero.

The relation between pressure and density arising from such an action between the particles is of this kind.

As the density increases from zero, the pressure at first depends almost entirely on the motion of the particles, and therefore varies almost exactly as the pressure, according to Boyle's Law. As the density continues to increase, the effect of the mutual attraction of the particles becomes sensible, and this causes the rise of pressure to be less than that given by Boyle's Law. If the temperature is low, the effect of attraction may become so large in proportion to the effect of motion that the pressure, instead of always rising as the density increases, may reach a maximum, and then begin to diminish.

At length, however, as the average distance of the particles is still further diminished, the effect of repulsion will prevail over that of attraction, and the pressure will increase so as not only to be greater than that given by Boyle's Law, but so that an exceedingly small increase of density will produce an enormous increase of pressure.

Hence the relation between pressure and volume may be represented by the curve *A B C D E F G*, where the horizontal ordinate represents the volume, and the vertical ordinate represents the pressure.

积内的总位力将随密度的平方变化。

设密度为ρ，将方程两边同除以V，我们得到——

$$p = \frac{1}{3}\rho c^2 - \frac{2}{3}A\rho^2,$$

其中A这个量在密度很小时近似为常数。

勒尼奥的实验表明，大多数气体在密度增大时，其压力会下降到低于玻意耳定律的计算值。因此位力必定是正的；也就是说，微粒间的相互作用必定以引力为主，而且这种作用减小压力的效果在开始的时候一定是近似与密度的平方有关。

另一方面，当压力继续增大，物质最终将达到另一种状态，即压力急剧增大但密度只有很小的增加。这表明此时位力为负值，或者换句话说，此时微粒间的相互作用主要为斥力。我们可以由此确定，在任何可感知的尺度中，两个微粒之间的相互作用都是十分微弱的。随着微粒彼此靠近，它们之间的相互作用首先表现为引力，引力达到最大值后开始减小，最终转变为极大的斥力，以至于没有任何可获得的力能够将微粒间的距离减小到零。

源于微粒间这种相互作用的压力与密度之间的关系就是这样的。

随着密度从零开始逐渐增大，一开始压力几乎完全取决于微粒的运动，因此几乎与玻意耳定律计算得到的压力精确地吻合。随着密度继续增加，微粒间相互吸引力的影响逐渐体现出来，这使得压力的增大小于根据玻意耳定律预计的值。如果温度很低，吸引力的影响会大到可以与运动的影响相抗衡的程度，那么压力就不再总是随着密度的增加而增大，而是在达到一个最大值后开始减小。

最终，随着微粒的平均距离继续减小，斥力的影响将胜过引力的影响，而压力将会增大到超出玻意耳定律所预计的值，而且此时很小的密度增加也会导致压力的急剧增大。

由此，压力与体积之间的关系可以用曲线$A\ B\ C\ D\ E\ F\ G$来表示，其中横坐标代表体积，纵坐标代表压力。

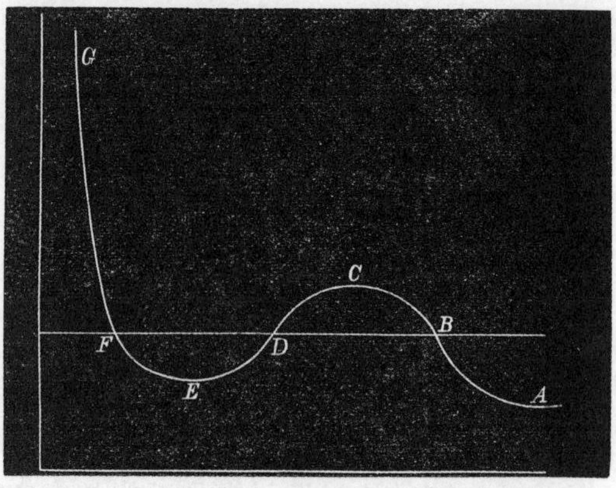

Fig. 1

As the volume diminishes, the pressure increases up to the point *C*, then diminishes to the point *E*, and finally increases without limit as the volume diminishes.

We have hitherto supposed the experiment to be conducted in such a way that the density is the same in every part of the medium. This, however, is impossible in practice, as the only condition we can impose on the medium from without is that the whole of the medium shall be contained within a certain vessel. Hence, if it is possible for the medium to arrange itself so that part has one density and part another, we cannot prevent it from doing so.

Now the points *B* and *F* represent two states of the medium in which the pressure is the same but the density very different. The whole of the medium may pass from the state *B* to the state *F*, not through the intermediate states *C D E*, but by small successive portions passing directly from the state *B* to the state *F*. In this way the successive states of the medium as a whole will be represented by points on the straight line *B F*, the point *B* representing it when entirely in the rarefied state, and *F* representing it when entirely condensed. This is what takes place when a gas or vapour is liquefied.

Under ordinary circumstances, therefore, the relation between pressure and volume at constant temperature is represented by the broken line *A B F G*. If, however, the medium when liquefied is carefully kept from contact with vapour, it may be preserved in the liquid condition and brought into states represented by the portion of the curve between *F* and *E*. It is also possible that methods may be devised whereby the vapour may be prevented from condensing, and brought into states represented by points in *B C*.

The portion of the hypothetical curve from *C* to *E* represents states which are essentially unstable, and which cannot therefore be realised.

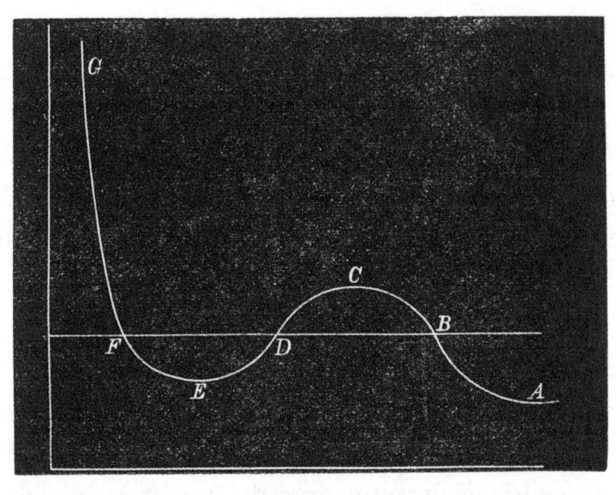

图1

随着体积缩小，压力先增大到 C 点，然后减小到 E 点，最终将随体积缩小而无限增大。

到目前为止，我们始终假设在实验过程中介质的每一部分都具有相同的密度。不过，这在实际中是不可能的，我们能够从外部施加于介质的唯一限制，仅仅是确保整个介质都置于确定的容器之中。因此，如果介质有可能自身调整为各处密度不同，我们也无法阻止。

B 点和 F 点代表介质的两种状态，其压力相同但密度相差甚远。全部介质可从状态 B 到达状态 F，不经过中间状态 CDE，而是直接由状态 B 一点一点地转化为状态 F。这样，整个介质的连续变化状态对应于图中直线 BF 上的点，B 点代表介质完全处于稀薄状态，F 则代表介质完全处于凝聚状态。这就是气体或者蒸气在液化过程中发生的现象。

因此，一般情况下，恒定温度时压力与体积的关系对应于图中的折线 $A\ B\ F\ G$。不过，如果在液化过程中小心地避免介质与蒸气接触，就可能可以使介质保持在液态，并达到图中 E 点和 F 点之间的曲线部分所代表的状态。也有可能可以设计出某种阻止蒸气凝聚的方法从而达到曲线 BC 上的点所代表的状态。

假设的曲线中从 C 点到 E 点的部分代表本质上不稳定的状态，因此是无法实现的。

Now let us suppose the medium to pass from *B* to *F* along the hypothetical curve *B C D E F* in a state always homogeneous, and to return along the straight line *F B* in the form of a mixture of liquid and vapour. Since the temperature has been constant throughout, no heat can have been transformed into work. Now the heat transformed into work is represented by the excess of the area *F D E* over *B C D*. Hence the condition which determines the maximum pressure of the vapour at given temperature is that the line *B F* cuts off equal areas from the curve above and below.

The higher the temperature, the greater the part of the pressure which depends on motion, as compared with that which depends on forces between the particles. Hence, as the temperature rises, the dip in the curve becomes less marked, and at a certain temperature the curve, instead of dipping, merely becomes horizontal at a certain point, and then slopes upward as before. This point is called the critical point. It has been determined for carbonic acid by the masterly researches of Andrews. It corresponds to a definite temperature, pressure and density.

At higher temperatures the curve slopes upwards throughout, and there is nothing corresponding to liquefaction in passing from the rarest to the densest state.

The molecular theory of the continuity of the liquid and gaseous states forms the subject of an exceedingly ingenious thesis by Mr. Johannes Diderik van der Waals*, a graduate of Leyden. There are certain points in which I think he has fallen into mathematical errors, and his final result is certainly not a complete expression for the interaction of real molecules, but his attack on this difficult question is so able and so brave, that it cannot fail to give a notable impulse to molecular science. It has certainly directed the attention of more than one inquirer to the study of the Low-Dutch language in which it is written.

The purely thermodynamical relations of the different states of matter do not belong to our subject, as they are independent of particular theories about molecules. I must not, however, omit to mention a most important American contribution to this part of thermodynamics by Prof. Willard Gibbs†, of Yale College, U. S., who has given us a remarkably simple and thoroughly satisfactory method of representing the relations of the different states of matter by means of a model. By means of this model, problems which had long resisted the efforts of myself and others may be solved at once.

(**11**, 357-359; 1875)

* Over de continuiteit van den gas en vloeistof toestand. Leiden: A. W. Sijthoff, 1873.

† "A method of geometrical representation of the thermodynamic properties of substances by means of surfaces." *Transactions of the Connecticut Academy of Arts and Sciences*, vol. II, Part 2.

现在，让我们设想介质始终以均匀的状态沿着假设的曲线$B\ C\ D\ E\ F$从B变化到F，并以蒸气和液体混合的形式沿着直线$F\,B$返回。因为温度一直是恒定的，所以其间不会有热能转化为功，而转化为功的热能对应于区域$F\ D\ E$与$B\ C\ D$的面积差。因此，在给定温度下使蒸气压力达到最大的条件，就是直线$B\,F$与上方和下方曲线围成的面积相等。

温度越高，由运动决定的那部分压力与由微粒间作用力决定的部分相比就越大。因此，随着温度升高，曲线的凹陷变得不明显，当温度达到某一个值时，曲线不再凹陷而在某一点达到水平线的位置之后曲线就像前面那样上升了。这个点称为临界点。安德鲁斯已经用巧妙的研究方法确定了碳酸的临界点。临界点对应于确定的温度、压力和密度。

在更高的温度下，曲线始终是向上倾斜的，在从最稀薄到最稠密的状态变化过程中不发生液化。

关于液态与气态连续性的分子理论构成了一篇极具才华的论文的主题，论文的作者是莱顿大学的毕业生约翰内斯·迪德里克·范德瓦尔斯先生*。尽管我觉得他在某些地方犯了一些数学错误，而且他的最终结果也确实不是对真实分子间相互作用的完整表达，但是他对这一难题所作的努力表现了他非凡的才华和勇气，以至于他的工作不可能不对分子科学产生重要的推动作用。确实有不止一位研究者受此驱使而去学习这篇文章撰写时所用的荷兰语。

物质不同状态间纯粹的热力学联系不属于我们的主题，因为它与这些关于分子的具体理论无关。不过，我决不能忘记提及一项极为重要的，由美国耶鲁大学的威拉德·吉布斯教授作出的热力学方面的贡献†，他利用一个模型给我们提供了一种极其简单又十分令人满意的表示不同物态间关联的方法。借助这个模型，那些长期以来阻碍我本人和其他人努力的难题可能立刻就会迎刃而解。

---

*《论液态和气态的连续性》，莱登赛特霍夫出版社，1873 年。

†《借助曲面表示物质热力学性质的几何描述法》，《康涅狄格艺术与科学学会学报》，第 2 卷，第 2 部分。

II

Let us now return to the case of a highly rarefied gas in which the pressure is due entirely to the motion of its particles. It is easy to calculate the mean square of the velocity of the particles from the equation of Clausius, since the volume, the pressure, and the mass are all measurable quantities. Supposing the velocity of every particle the same, the velocity of a molecule of oxygen would be 461 metres per second, of nitrogen 492, and of hydrogen 1,844, at the temperature 0 °C.

The explanation of the pressure of a gas on the vessel which contains it by the impact of its particles on the surface of the vessel has been suggested at various times by various writers. The fact, however, that gases are not observed to disseminate themselves through the atmosphere with velocities at all approaching those just mentioned, remained unexplained, till Clausius, by a thorough study of the motions of an immense number of particles, developed the methods and ideas of modern molecular science.

To him we are indebted for the conception of the mean length of the path of a molecule of a gas between its successive encounters with other molecules. As soon as it was seen how each molecule, after describing an exceedingly short path, encounters another, and then describes a new path in a quite different direction, it became evident that the rate of diffusion of gases depends not merely on the velocity of the molecules, but on the distance they travel between each encounter.

I shall have more to say about the special contributions of Clausius to molecular science. The main fact, however, is, that he opened up a new field of mathematical physics by showing bow to deal mathematically with moving systems of innumerable molecules.

Clausius, in his earlier investigations at least, did not attempt to determine whether the velocities of all the molecules of the same gas are equal, or whether, if unequal, there is any law according to which they are distributed. He therefore, as a first hypothesis, seems to have assumed that the velocities are equal. But it is easy to see that if encounters take place among a great number of molecules, their velocities, even if originally equal, will become unequal, for, except under conditions which can be only rarely satisfied, two molecules having equal velocities before their encounter will acquire unequal velocities after the encounter. By distributing the molecules into groups according to their velocities, we may substitute for the impossible task of following every individual molecule through all its encounters, that of registering the increase or decrease of the number of molecules in the different groups.

By following this method, which is the only one available either experimentally or mathematically, we pass from the methods of strict dynamics to those of statistics and probability.

II

现在让我们回过头来看看极稀薄状态的气体，在这种状态下压力完全来源于气体微粒的运动。由于体积、压力和质量都是可观测量，因此很容易利用克劳修斯方程计算出微粒速率的均方。假设每种微粒具有相同的速率，则在 0℃时，氧分子的速率为每秒 461 米，氮分子为每秒 492 米，而氢分子为每秒 1,844 米。

已经有多位学者多次提出过用气体微粒对容器表面的撞击来解释气体对容器的压力。不过，我们一直不能解释为什么从未观测到气体以接近于上面所提到的速率散布到大气中的现象，直到克劳修斯通过对大量微粒的运动的全面研究发展了现代分子科学的方法和观念。

我们得益于他提出的一个概念，即一个气体分子在与其他分子相继发生两次碰撞之间所经过路程的平均长度。一旦看到每个气体分子是如何在经历了一段极短的路程后与另一个分子相撞，随后气体分子又沿着完全不同的方向开始新的旅程，那么很明显的就是：气体扩散的速率不仅依赖于分子的速度，还与分子在相继发生的碰撞之间途经的距离有关。

关于克劳修斯对分子科学的特殊贡献，我还有很多要说。最主要的是，他通过展示如何用数学方式处理含无数分子的运动系统从而开创了数学物理的新领域。

至少在早期的研究中，克劳修斯并没有试图去确定同一气体中的所有分子是否都具有相同的速率，或者如果它们的速率不等，其分布是否应遵循某种规律。他的第一个假说似乎假定所有分子的速率都是相等的。不过显而易见的是，如果碰撞发生在大量分子之间，那么即使它们的速率开始时是相等的，之后也会变得不相等，因为除了在某些几乎无法满足的条件下，两个碰撞前具有相同速率的分子在碰撞后总会获得不同的速率。通过将分子按其速率分组，我们就只需记录不同组内分子数量的增加或减少，而不必完成跟踪每一个分子的所有碰撞这项不可能完成的任务。

利用这种无论从实验角度还是数学角度来说都是唯一可行的方法，我们从严格的动力学方法转到了统计和概率的方法上。

When an encounter takes place between two molecules, they are transferred from one pair of groups to another, but by the time that a great many encounters have taken place, the number which enter each group is, on an average, neither more nor less than the number which leave it during the same time. When the system has reached this state, the numbers in each group must be distributed according to some definite law.

As soon as I became acquainted with the investigations of Clausius, I endeavoured to ascertain this law.

The result which I published in 1860 has since been subjected to a more strict investigation by Dr. Ludwig Boltzmann, who has also applied his method to the study of the motion of compound molecules. The mathematical investigation, though, like all parts of the science of probabilities and statistics, it is somewhat difficult, does not appear faulty. On the physical side, however, it leads to consequences, some of which, being manifestly true, seem to indicate that the hypotheses are well chosen, while others seem to be so irreconcilable with known experimental results, that we are compelled to admit that something essential to the complete statement of the physical theory of molecular encounters must have hitherto escaped us.

I must now attempt to give you some account of the present state of these investigations, without, however, entering into their mathematical demonstration.

I must begin by stating the general law of the distribution of velocity among molecules of the same kind.

If we take a fixed point in this diagram and draw from this point a line representing in direction and magnitude the velocity of a molecule, and make a dot at the end of the line, the position of the dot will indicate the state of motion of the molecule.

If we do the same for all the other molecules, the diagram will be dotted all over, the dots being more numerous in certain places than in others.

The law of distribution of the dots may be shown to be the same as that which prevails among errors of observation or of adjustment.

The dots in the diagram before you may be taken to represent the velocities of molecules, the different observations of the position of the same star, or the bullet-holes round the bull's eye of a target, all of which are distributed in the same manner.

The velocities of the molecules have values ranging from zero to infinity, so that in speaking of the average velocity of the molecules we must define what we mean.

当两个分子发生碰撞时，它们就会从原本所在组的一对转变成另一组的一对，而且在同时发生大量碰撞时，平均来看，同一时间段内进入某一个组的分子数量，不会多于或少于离开该组的分子数量。当系统达到这种状态时，分子在各组的分布一定符合某种确定的规律。

我在了解了克劳修斯的研究工作之后，便立即努力探求这一规律。

我在 1860 年发表的结果后来由路德维希·玻尔兹曼博士进行了更为严格的研究，他还将他的方法应用于对化合物分子运动的研究。数学研究，例如所有关于概率统计的科学部分，虽然有些困难，却不无裨益。从物理角度来说，数学方法的某些推论是非常正确的，似乎可以说明所选假说是正确的，而另一些却与已知的实验结果相违背，这使我们不得不承认，到目前为止，我们遗漏了关于分子碰撞物理理论完整表述中的某些核心部分。

现在我必须努力为各位介绍这方面研究的现状，不过，我们不会涉及其中的数学证明。

我必须从同种分子中速度分布的一般规律开始说起。

如果从图中选择一个固定的点，从这点出发画一条线来表示速度的方向和大小，并在线段末端画出端点，那么端点的位置就表示分子的运动状态。

如果我们对所有分子作同样处理，那么整张图上将画满了点，某些位置的点会比其他位置的点多。

可以看出，这些点的分布规律与观测或调节中经常出现的误差的分布规律是相同的。

你面前这张图中的点可以用来表示分子的速度，或者对同一颗恒星位置的不同观测结果，或者位于目标靶心周围的弹洞，上述这些具有相同的分布方式。

分子速率的取值范围是从零到无限大，因此说到分子的平均速率我们必须先定义我们指的是什么。

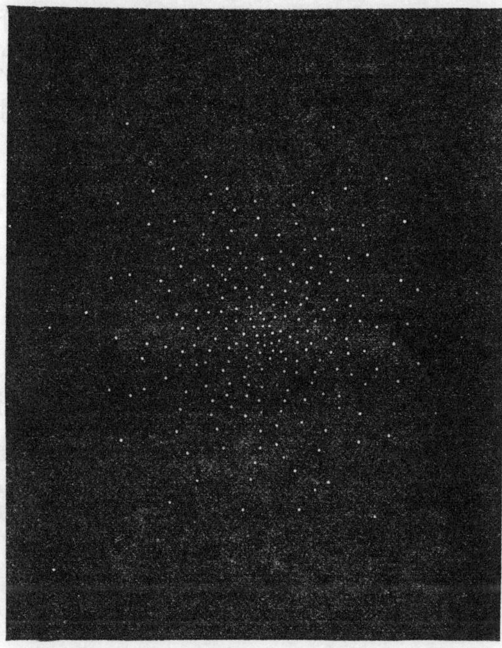

Fig. 2. Diagram of Velocities

The most useful quantity for purposes of comparison and calculation is called the "velocity of mean square". It is that velocity whose square is the average of the squares of the velocities of all the molecules.

This is the velocity given above as calculated from the properties of different gases. A molecule moving with the velocity of mean square has a kinetic energy equal to the average kinetic energy of all the molecules in the medium, and if a single mass equal to that of the whole quantity of gas were moving with this velocity, it would have the same kinetic energy as the gas actually has, only it would be in a visible form and directly available for doing work.

If in the same vessel there are different kinds of molecules, some of greater mass than others, it appears from this investigation that their velocities will be so distributed that the average kinetic energy of a molecule will be the same, whether its mass be great or small.

Here we have perhaps the most important application which has yet been made of dynamical methods to chemical science. For, suppose that we have two gases in the same vessel. The ultimate distribution of agitation among the molecules is such that the average kinetic energy of an individual molecule is the same in either gas. This ultimate state is also, as we know, a state of equal temperature. Hence the condition that two gases shall have the same temperature is that the average kinetic energy of a single molecule shall be the same in the two gases.

图 2. 速度图像

为了便于比较和计算，最有用的量叫作“均方速率”。这个量的平方是所有分子速率平方的平均值。

这就是上面给出的利用不同气体性质计算得到的速度。以均方速率运动的分子具有的动能等于介质内全体分子的平均动能，并且若以此速率运动的单个物体的质量等于气体的总质量，那么该物体与气体具有相同的动能，只是前者具有可见的形式并可直接用于做功。

如果同一容器中包含不同种类的分子，其中一些具有比较大的质量，那么根据这项研究，这些分子的速度分布方式应满足每种分子具有相同的平均动能，不管其质量是大还是小。

这里我们用到的可能是动力学方法最重要的应用，即将它应用于化学科学。比如，假定我们将两种气体置于同一容器中。分子运动的最终分布形式应满足两种气体中每一种的单个分子的平均动能相等。正如我们所知，这种最终态也是等温态。由此，两种气体具有相同温度的条件，就是两种气体中单个分子的平均动能相同。

Now, we have already shown that the pressure of a gas is two-thirds of the kinetic energy in unit of volume. Hence, if the pressure as well as the temperature be the same in the two gases, the kinetic energy per unit of volume is the same, as well as the kinetic energy molecule. There must, therefore, be the same number of molecules in unit of volume in the two gases.

This result coincides with the law of equivalent volumes established by Gay Lussac. This law, however, has hitherto tested on purely chemical evidence, the relative masses of the molecules of different substances having been deduced from the proportions in which the substances enter into chemical combination. It is now demonstrated on dynamical principles. The molecule is defined as that small portion of the substance which moves as one lump during the motion of agitation. This is a purely dynamical definition, independent of any experiments on combination.

The density of a gaseous medium, at standard temperature and pressure, is proportional to the mass of one of its molecules as thus defined.

We have thus a safe method of estimating the relative masses of molecules of different substances when in the gaseous state. This method is more to be depended on than those founded on electrolysis or on specific heat, because our knowledge of the conditions of the motion of agitation is more complete than our knowledge of electrolysis, or of the internal motions of the constituents of a molecule.

I must now say something about these internal motions, because the greatest difficulty which the kinetic theory of gases has yet encountered belongs to this part of the subject.

We have hitherto considered only the motion of the centre of mass of the molecule. We have now to consider the motion of the constituents of the molecule relative to the centre of mass.

If we suppose that the constituents of a molecule are atoms, and that each atom is what is called a material point, then each atom may move in three different and independent ways, corresponding to the three dimensions of space, so that the number of variables required to determine the position and configuration of all the atoms of the molecule is three times the number of atoms.

It is not essential, however, to the mathematical investigation to assume that the molecule is made up of atoms. All that is assumed is that the position and configuration of the molecule can be completely expressed by a certain number of variables.

Let us call this number n.

现在，我们已经指出，气体的压力是单位体积动能的 2/3。因此，如果两种气体具有相同的压力和温度，那么它们每单位体积的动能以及每个分子的动能也是相同的。由此，两种气体单位体积内必定含有相同数量的分子。

这一结果与盖·吕萨克提出的等容定律是一致的。不过，到目前为止该定律只基于化学证据，不同物质的相对分子质量已经从物质参加化合反应时所占据的比例中推导出来了。现在已经用动力学原理证明了该定律。分子被定义为扰动运动中作为一个整体运动的物体的一小部分。这是一个纯粹的动力学定义，与任何化合实验无关。

在标准的温度和压力下，气态介质的密度正比于该气体按以上定义的一个分子的质量。

这样，我们就有一种可以估计气态时不同物质的相对分子质量的可靠方法。这种方法比那些建立在电解或者比热基础上的方法更可信赖，因为我们已经掌握的关于扰动运动条件方面的知识比关于电解或者分子结构内部运动的知识更完备。

现在我必须谈谈这些分子的内部运动，因为目前气体动力学理论遇到的最大困难就在这部分。

迄今为止我们考虑的只是分子质心的运动。现在我们不得不考虑分子组分相对于质心的运动。

如果我们假定分子是由原子构成的，并且每个原子都是所谓的质点，那么每个原子都可以在 3 个不同且独立的方向上运动，这对应于空间的 3 个维度，因此要确定分子中所有原子结构和位置所需的变量数就是原子个数的 3 倍。

不过，对数学研究来说，假定分子是由原子构成的并不是一个基本假设。全部的前提假设只是，分子的位置和结构可以用一定数量的变量完整地表达。

让我们将这个数设为 n。

Of these variables, three are required to determine the position of the centre of mass of the molecule, and the remaining n–3 to determine its configuration relative to its centre of mass.

To each of the n variables corresponds a different kind of motion.

The motion of translation of the centre of mass has three components.

The motions of the parts relative to the centre of mass have n–3 components.

The kinetic energy of the molecule may be regarded as made up of two parts—that of the mass of the molecule supposed to be concentrated at its centre of mass, and that of the motions of the parts relative to the centre of mass. The first part is called the energy of translation, the second that of rotation and vibration. The sum of these is the whole energy of motion of the molecule.

The pressure of the gas depends, as we have seen, on the energy of translation alone. The specific heat depends on the rate at which the whole energy, kinetic and potential, increases as the temperature rises.

Clausius had long ago pointed out that the ratio of increment of the whole energy to that of the energy of translation may be determined if we know by experiment the ratio of the specific heat at constant pressure to that at constant volume.

He did not, however, attempt to determine *à priori* the ratio of the two parts of the energy, though he suggested, as an extremely probable hypothesis, that the average values of the two parts of the energy in a given substance always adjust themselves to the same ratio. He left the numerical value of this ratio to be determined by experiment.

In 1860 I investigated the ratio of the two parts of the energy on the hypothesis that the molecules are elastic bodies of invariable form. I found, to my great surprise, that whatever be the shape of the molecules, provided they are not perfectly smooth and spherical, the ratio of the two parts of the energy must be always the same, the two parts being in fact equal.

This result is confirmed by the researches of Boltzmann, who has worked out the general case of a molecule having n variables.

He finds that while the average energy of translation is the same for molecules of all kinds at the same temperature, the whole energy of motion is to the energy of translation as n to 3.

For a rigid body n=6, which makes the whole energy of motion twice the energy of energy of translation.

在这些变量中，有 3 个是用来确定分子质心位置的，而其余 n–3 个则用来确定相对于质心的分子结构。

n 个变量的每一个都对应于一种不同的运动。

质心的平移运动有 3 个分量。

各部分相对于质心的运动则包含 n–3 个分量。

分子的动能可以看作是由两部分组成——其中一部分是假定整个分子的质量集中于其质心所产生的，另一部分是各部分相对质心运动所产生的。第一部分称为平动动能，第二部分称为转动和振动动能。上述量的总和就是分子运动的总能量。

正如我们所知，气体的压力只由平动动能决定。比热则依赖于温度升高时动能与势能的总增加量与温度增量的比值。

克劳修斯早就指出，如果我们能够通过实验获得恒压比热与恒容比热的比值，就能确定总能量增量与平动动能增量的比值。

但是，他并没有试图**先验性地**确定两部分能量的比值，尽管他提出了一个极有可能成立的假说：对于一种给定的物质，两部分能量的平均值总是具有相同的比值。他将这一比值的具体数值留待实验确定。

1860 年，我基于分子是不变形的弹性体的假说研究了两部分能量的比值。我惊讶地发现，不论分子是何种形状，只要它们不是绝对光滑和完美球状，那么两部分能量的比值就一定总是相同的，实际上这两部分能量是相等的。

这一结果被玻尔兹曼的研究工作所证实，他给出了具有 n 个变量的分子在一般情况下的结果。

他发现，相同温度下任何种类的分子都具有相同的平均平动动能，运动的总能量与平动动能之比为 n：3。

对于刚体来说 n=6，这使得运动的总能量为平动动能的 2 倍。

But if the molecule is capable of changing its form under the action of impressed forces, it must be capable of storing up potential energy, and if the forces are such as to ensure the stability of the molecule, the average potential energy will increase when the average energy of internal motion increases.

Hence, as the temperature rises, the increments of the energy of translation, the energy of internal motion, and the potential energy are as 3, (n–3), and e respectively, where e is a positive quantity of unknown value depending on the law of the force which binds together the constituents of the molecule.

When the volume of the substance is maintained constant, the effect of the application of heat is to increase the whole energy. We thus find for the specific heat of a gas at constant volume—

$$\frac{1}{2J}\frac{p_0V_0}{273^{\circ}}(n+e)$$

where p_0 and V_0 are the pressure and volume of unit of mass at zero centigrade, or 273° absolute temperature, and J is the dynamical equivalent of heat. The specific heat at constant pressure is

$$\frac{1}{2J}\frac{p_0V_0}{273^{\circ}}(n+2+e)$$

In gases whose molecules have the same degree of complexity the value of n is the same, and that of e *may* be the same.

If this is the case, the specific heat is inversely as the specific gravity, according to the law of Dulong and Petit, which is, to a certain degree of approximation, verified by experiment.

But if we take the actual values of the specific heat as found by Regnault and compare them with this formula, we find that $n + e$ for air and several other gases cannot be more than 4.9. For carbonic acid and steam it is greater. We obtain the same result if we compare the ratio of the calculated specific heats

$$\frac{2+n+e}{n+e}$$

with the ratio as determined by experiment for various gases, namely, 1.408.

And here we are brought face to face with the greatest difficulty which the molecular theory has yet encountered, namely, the interpretation of the equation $n + e = 4.9$.

If we suppose that the molecules are atoms—mere material points, incapable of rotatory energy or internal motion—then n is 3 and e is zero, and the ratio of the specific heats is 1.66, which is too great for any real gas.

但是，如果分子能够在外力作用下改变形状，它就一定可以储存势能，而如果外力是能保证分子的稳定性的那种，当内部运动的平均能量增加时，平均势能也会增加。

因此，随着温度升高，平动动能、内部运动能量和势能的增量分别为 3，(n–3) 和 e，其中 e 是取值为正的未知量，其数值取决于将分子各组成部分束缚在一起的力的定律。

当物体的体积保持不变时，加热的效果是增加总能量。我们由此给出恒容气体的比热——

$$\frac{1}{2J}\frac{p_0V_0}{273^\circ}(n+e)$$

其中 p_0和 V_0是零摄氏度或绝对温度 273° 时单位质量的压力和体积，而 J 则是热的动力学当量。恒压条件下的比热为

$$\frac{1}{2J}\frac{p_0V_0}{273^\circ}(n+2+e)$$

对于那些有相同复杂程度的分子，n 值是相同的，e 值**可能**是相同的。

如果情况就是如此，那么根据杜隆–珀蒂定律，比热与比重成反比，这是一条在一定精度范围内已经被实验证明了的定律。

但是，如果我们采用勒尼奥所发现的实际比热值并将其与此公式进行比较就会发现，对空气和其他几种气体，$n+e$ 的值不会超过 4.9。而对碳酸和水蒸气来说这个值则大一些。如果将计算所得的比热之比值

$$\frac{2+n+e}{n+e}$$

与实验确定的比值相比，我们得到相同的结果，即 1.408。

此时，我们将面临分子理论中最大的困难，也就是，如何解释等式 $n+e=4.9$。

如果我们假设分子是不可分割的质点——没有转动能量或者内部运动——那么 n 为 3 而 e 为 0，比热的比值是 1.66，而这个值对于任何实际气体来说都太大了。

But we learn from the spectroscope that a molecule can execute vibrations of constant period. It cannot therefore be a mere material point, but a system capable of changing its form. Such a system cannot have less than six variables. This would make the greatest value of the ratio of the specific heats 1.33, which is too small for hydrogen, oxygen, nitrogen, carbonic oxide, nitrous oxide, and hydrochloric acid.

But the spectroscope tells us that some molecules can execute a great many different kinds of vibrations. They must therefore be systems of a very considerable degree of complexity, having far more than six variables. Now, every additional variable introduces an additional amount of capacity for internal motion without affecting the external pressure. Every additional variable, therefore, increases the specific heat, whether reckoned at constant pressure or at constant volume.

So does any capacity which the molecule may have for storing up energy in the potential form. But the calculated specific heat is already too great when we suppose the molecule to consist of two atoms only. Hence every additional degree of complexity which we attribute to the molecule can only increase the difficulty of reconciling the observed with the calculated value of the specific heat.

I have now put before you what I consider to be the greatest difficulty yet encountered by the molecular theory. Boltzmann has suggested that we are to look for the explanation in the mutual action between the molecules and the etherial medium which surrounds them. I am afraid, however, that if we call in the help of this medium, we shall only increase the calculated specific heat, which is already too great.

The theorem of Boltzmann may be applied not only to determine the distribution of velocity among the molecules, but to determine the distribution of the molecules themselves in a region in which they are acted on by external forces. It tells us that the density of distribution of the molecules at a point where the potential energy of a molecule is ψ, is proportional to $e^{-\frac{\psi}{\kappa\theta}}$ where θ is the absolute temperature, and κ is a constant for all gases. It follows from this, that if several gases in the same vessel are subject to an external force like that of gravity, the distribution of each gas is the same as if no other gas were present. This result agrees with the law assumed by Dalton, according to which the atmosphere may be regarded as consisting of two independent atmospheres, one of oxygen, and the other of nitrogen; the density of the oxygen diminishing faster than that of the nitrogen, as we ascend.

This would be the case if the atmosphere were never disturbed, but the effect of winds is to mix up the atmosphere and to render its composition more uniform than it would be if left at rest.

Another consequence of Boltzmann's theorem is, that the temperature tends to become equal throughout a vertical column of gas at rest.

但是，通过光谱仪我们了解到，分子可以进行周期固定的振动。因此分子不可能只是一个质点，而应是一个结构可以改变的体系。这一体系的变量不可能少于6个。这使得比热比值的最大值是1.33，而这个结果对于氢气、氧气、氮气、碳氧化物、氮氧化物和氢氯酸来说都太小了。

但是光谱仪告诉我们某些分子可以进行多种不同类型的振动。因此它们必然是具有相当复杂程度的体系，具有的变量数远大于6。那么，每增加一个变量，都会在不影响外部压力的前提下引入一些附加的内部运动的能力。因此，不论是在恒压还是恒容条件下计算，每增加一个变量都会使比热增加。

分子以势能形式储存能量的能力也是如此。但是，当我们假定分子仅由两个原子组成，计算得到的比热值就已经太大了。于是，分子的复杂程度每增加一点，都只会加大使比热的观测值与计算值相一致的难度。

现在我已将我认为的分子理论遇到的最大困难呈现在各位面前。玻尔兹曼曾经建议，应该从分子与围绕在其周围的以太介质的相互作用中寻求解释。不过，要是我们借助于这种介质的话，恐怕只会使已经过大的比热计算值变得更大。

玻尔兹曼定理不仅可以用于确定分子的速度分布，还可以用于确定在外力作用下分子自身在一个区域中的分布。它告诉我们，在分子势能为ψ的一点，分子的分布密度正比于$e^{-\frac{\psi}{\kappa\theta}}$，其中$\theta$为绝对温度，而$\kappa$对于所有气体都是常数。由此可知，如果同一容器中的几种气体受到外力（例如重力）作用，每一种气体的分布与没有其他气体存在时是一样的。这一结果与道尔顿提出的定律是一致的，根据这个定律，大气可以被视为是由两种独立的气体组成，一种为氧气，另一种为氮气；随着海拔升高，氧气的浓度比氮气的浓度减小得更快些。

以上是大气丝毫不受扰动的情况，而风的影响会将大气混匀，使其组成比保持静止时更均匀。

玻尔兹曼定理的另一推论是，静止状态的气体在垂直方向上温度趋于一致。

In the case of the atmosphere, the effect of wind is to cause the temperature to vary as that of a mass of air would do if it were carried vertically upwards, expanding and cooling as it ascends.

But besides these results, which I had already obtained by a less elegant method and published in 1866, Boltzmann's theorem seems to open up a path into a region more purely chemical. For if the gas consists of a number of similar systems, each of which may assume different states having different amounts of energy, the theorem tells us that the number in each state is proportional to $e^{-\frac{\psi}{\kappa\theta}}$ where ψ is the energy, θ the absolute temperature, and κ a constant.

It is easy to see that this result ought to be applied to the theory of the states of combination which occur in a mixture of different substances. But as it is only during the present week that I have made any attempt to do so, I shall not trouble you with my crude calculations.

I have confined my remarks to a very small part of the field of molecular investigation. I have said nothing about the molecular theory of the diffusion of matter, motion, and energy, for though the results, especially in the diffusion of matter and the transpiration of fluids are of great interest to many chemists, and though from them we deduce important molecular data, they belong to a part of our study the data of which, depending on the conditions of the encounter of two molecules, are necessarily very hypothetical. I have thought it better to exhibit the evidence that the parts of fluids are in motion, and to describe the manner in which that motion is distributed among molecules of different masses.

To show that all the molecules of the same substance are equal in mass, we may refer to the methods of dialysis introduced by Graham, by which two gases of different densities may be separated by percolation through a porous plug.

If in a single gas there were molecules of different masses, the same process of dialysis, repeated a sufficient number of times, would furnish us with two portions of the gas, in one of which the average mass of the molecules would be greater than in the other. The density and the combining weight of these two portions would be different. Now, it may be said that no one has carried out this experiment in a sufficiently elaborate manner for every chemical substance. But the processes of nature are continually carrying out experiments of the same kind; and if there were molecules of the same substance nearly alike, but differing slightly in mass, the greater molecules would be selected in preference to form one compound, and the smaller to form another. But hydrogen is of the same density, whether we obtain it from water or from a hydrocarbon, so that neither oxygen nor carbon can find in hydrogen molecules greater or smaller than the average.

对于大气来说，风的影响导致温度变化，例如，若一定质量的气体垂直向上运动，那么它在上升过程中会膨胀并会冷却。

然而，除了给出这些我已经于1866年就以不那么精巧的方法获得并发表的结果之外，玻尔兹曼的定理似乎还开辟了一条通向更纯粹的化学领域的道路。因为，如果气体是由大量相似的系统构成，每一个系统都可以假定为具有不同能量的不同状态，定理告诉我们每一状态中系统的数量正比于$e^{-\frac{\psi}{\kappa\theta}}$，其中$\psi$表示能量，$\theta$表示绝对温度，而$\kappa$是一个常数。

容易看出，这一结果应当被用于发生在不同物质混合体中的组合的状态理论。不过，由于这项工作是我这个星期才刚刚开始的，我还不想以我粗糙的计算结果来搅扰诸位。

我一直将论题限制在分子研究领域的一个很小的范围内。我并没有提及关于物质扩散、运动以及能量的分子理论，因为，尽管这些结果，尤其是关于物质扩散和液体蒸发的结果，对很多化学家来说极有吸引力，并且从这些结果中我们可以推导出重要的分子数据，但这些数据依赖于两个分子碰撞的条件，因而属于我们研究工作中必然具有很多假设性的部分。我认为，还是展示一下能表明流体各部分运动的证据，并且描述一下此运动在不同质量的分子中的分布情况更好些。

为表明同种物质的所有分子都具有相同的质量，我们就要谈到格雷姆引入的透析法，用这种方法可以通过一个多孔塞进行过滤，将两种不同密度的气体分离。

如果在一种气体中存在不同质量的分子，将同样的透析过程足够多次地重复之后，我们就可以得到两部分气体，其中一部分的平均分子质量比另一部分的大。这两部分气体有不同的密度和化合量。现在，也许可以说，没有谁曾以足够精细的方式对每一种化学物质进行过这一实验。然而自然过程在持续不断地进行着这种类型的实验。如果同种物质具有十分相似但质量略有差别的分子，那么其中质量较大的分子将会被优先选择来形成一种化合物，而质量较小的分子则形成另一种。但是，无论是从水中还是从碳氢化合物中得到的氢都具有相同的密度，因此，氧和碳都不能在氢分子中找到大于或小于平均质量的个体。

The estimates which have been made of the actual size of molecules are founded on a comparison of the volumes of bodies in the liquid or solid state, with their volumes in the gaseous state. In the study of molecular volumes we meet with many difficulties, but at the same time there are a sufficient number of consistent results to make the study a hopeful one.

The theory of the possible vibrations of a molecule has not yet been studied as it ought, with the help of a continual comparison between the dynamical theory and the evidence of the spectroscope. An intelligent student, armed with the calculus and the spectroscope, can hardly fail to discover some important fact about the internal constitution of a molecule.

The observed transparency of gases may seem hardly consistent with the results of molecular investigations.

A model of the molecules of a gas consisting of marbles scattered at distances bearing the proper proportion to their diameters, would allow very little light to penetrate through a hundred feet.

But if we remember the small size of the molecules compared with the length of a wave of light, we may apply certain theoretical investigations of Lord Rayleigh's about the mutual action between waves and small spheres, which show that the transparency of the atmosphere, if affected only by the presence of molecules, would be far greater than we have any reason to believe it to be.

A much more difficult investigation, which has hardly yet been attempted, relates to the electric properties of gases. No one has yet explained why dense gases are such good insulators, and why, when rarefied or heated, they permit the discharge of electricity, whereas a perfect vacuum is the best of all insulators.

It is true that the diffusion of molecules goes on faster in a rarefied gas, because the mean path of a molecule is inversely as the density. But the electrical difference between dense and rare gas appears to be too great to be accounted for in this way.

But while I think it right to point out the hitherto unconquered difficulties of this molecular theory, I must not forget to remind you of the numerous facts which it satisfactorily explains. We have already mentioned the gaseous laws, as they are called, which express the relations between volume, pressure, and temperature, and Gay Lussac's very important law of equivalent volumes. The explanation of these may be regarded as complete. The law of molecular specific heats is less accurately verified by experiment, and its full explanation depends on a more perfect knowledge of the internal structure of a molecule than we as yet possess.

目前已有的对分子实际大小的估计都建立在将物体固态或液态时的体积与气态时的体积相比较的基础之上。在对分子体积的研究中我们遇到了很多困难，但同时也得到了很多可靠的结果，这使研究充满了希望。

对于分子可能的振动，相关理论的研究尚未开始，它本应在不断将动力学理论与光谱证据相比较的帮助下展开。用光谱工具和计算法武装起来的才智之士，一定会发现一些与分子内部结构有关的重要事实。

观测到的气体的透明度看来似乎很难与分子研究的结果一致。

一种认为气体是由间隔距离与其自身直径符合适当比例的分散小球组成的气体分子模型，只能允许极少量的光穿透 100 英尺的距离。

不过，要是我们还记得与光的波长相比分子的尺寸很微小，我们就会采用瑞利勋爵关于波与微小球体间相互作用的某些理论研究。他的研究表明，气体的透明度如果只受分子存在的影响，就会比我们有任何理由能够去相信的还要大得多。

一项几乎还没有人尝试过的更加困难的研究，是与气体的电学性质有关。还没有人能够解释为什么稠密的气体是非常良好的绝缘体，为什么气体在稀释或加热时可以放电，而完全的真空却是最好的绝缘体。

的确，稀薄气体中的分子扩散会进行得更快一些，因为分子运动的平均路程反比于密度。但是稠密气体与稀薄气体的电学差异似乎非常大以至于无法从这个角度来解释。

虽然我认为指出这种分子理论目前尚无法克服的困难是应该的，但我绝对不会忘记提醒诸位它已经令人满意地解释了大量问题。我们已经提到过气体定律，正如我们所说，它表达了气体的体积、压力与温度之间的关系，以及盖·吕萨克的极其重要的等容定律。可以认为关于这些定律的解释是完整的。分子比热定律已被实验相对粗略地验证了，而对它的完整解释则有赖于拥有比今天更完备的关于分子内部结构的知识。

But the most important result of these inquiries is a more distinct conception of thermal phenomena. In the first place, the temperature of the medium is measured by the average kinetic energy of translation of a single molecule of the medium. In two media placed in thermal communication, the temperature as thus measured tends to become equal.

In the next place, we learn how to distinguish that kind of motion which we call heat from other kinds of motion. The peculiarity of the motion called heat is that it is perfectly irregular; that is to say, that the direction and magnitude of the velocity of a molecule at a given time cannot be expressed as depending on the present position of the molecule and the time.

In the visible motion of a body, on the other hand, the velocity of the centre of mass of all the molecules in any visible portion of the body is the observed velocity of that portion, though the molecules may have also an irregular agitation on account of the body being hot.

In the transmission of sound, too, the different portions of the body have a motion which is generally too minute and too rapidly alternating to be directly observed. But in the motion which constitutes the physical phenomenon of sound, the velocity of each portion of the medium at any time can be expressed as depending on the position and the time elapsed; so that the motion of a medium during the passage of a sound-wave is regular, and must be distinguished from that which we call heat.

If, however, the sound-wave, instead of traveling onwards in an orderly manner and leaving the medium behind it at rest, meets with resistances which fritter away its motion into irregular agitations, this irregular molecular motion becomes no longer capable of being propagated swiftly in one direction as sound, but lingers in the medium in the form of heat till it is communicated to colder parts of the medium by the slow process of conduction.

The motion which we call light, though still more minute and rapidly alternating than that of sound, is, like that of sound, perfectly regular, and therefore is not heat. What was formerly called Radiant Heat is a phenomenon physically identical with light.

When the radiation arrives at a certain portion of the medium, it enters it and passes through it, emerging at the other side. As long as the medium is engaged in transmitting the radiation it is in a certain state of motion, but as soon as the radiation has passed through it, the medium returns to its former state, the motion being entirely transferred to a new portion of the medium.

Now, the motion which we call heat can never of itself pass from one body to another unless the first body is, during the whole process, hotter than the second. The motion of radiation, therefore, which passes entirely out of one portion of the medium and enters another, cannot be properly called heat.

不过，由这些探索得到的最重要的结果是关于热现象更加明确的概念。首先，介质的温度由介质中单个分子的平均平动动能来度量。在两种介质有热交换时，以此确定的温度有变得一致的倾向。

其次，我们知道了如何将所谓的热运动与其他类型的运动区分开。我们所谓的热运动的特性是完全无规则；也就是说，在任一给定时刻，一个分子的速度的方向和大小不可能用时间和现在分子所处的位置表示。

另一方面，在物体的可视运动中，物体的任一可视部分所含全部分子的质心速度，就是所观测到的该部分的速度。不过，分子也可能由于被加热而具有不规则的扰动。

同样地，在声音的传播过程中，物体不同部分的运动通常由于过于细微和过于快速变化而难以被直接观测。不过，在产生声音这一物理现象的运动过程中，介质中每一部分在任意时刻的速度都可以用位置和所经过的时间表示出来；因此在声波传播过程中介质的运动是规则的，必然不同于我们所说的热运动。

然而，如果声波不是以整齐有序的方式向前传播并使其所经之处的介质归于静止，而是遇到阻力从而将运动耗损在无规则的激发中，这种无规则的分子运动就无法再以声音的形式在一个方向上快速地传播，而只能以热的形式留在介质中，直到通过缓慢的传导过程流向介质较冷的部分。

被我们称为光的那种运动，尽管比声音运动更加细微和快速变化，但也和声音一样是非常规则的运动，因此不是热运动。以前被称为热辐射的物理现象在物理本质上与光是一样的。

辐射在到达介质的某一部分后，就会进入并穿过该部分，再从另一端出现。该部分介质在传导辐射时处于一种特定的运动状态，而一旦辐射穿过该部分介质，该部分介质便会回到先前的状态，而运动完全转移到了介质中新的部分。

现在，我们称为热的这种运动，不会自发地从一个物体进入另一个物体，除非在整个传导过程中，第一个物体总是比第二个物体的温度高一些。因此，从介质的一部分完全流出再进入另一部分的辐射运动，就不适宜被称作热了。

We may apply the molecular theory of gases to test those hypotheses about the luminiferous ether which assume it to consist of atoms or molecules.

Those who have ventured to describe the constitution of the luminiferous ether have sometimes assumed it to consist of atoms or molecules.

The application of the molecular theory to such hypotheses leads to rather startling results.

In the first place, a molecular ether would be neither more nor less than a gas. We may, if we please, assume that its molecules are each of them equal to the thousandth or the millionth part of a molecule of hydrogen, and that they can traverse freely the interspaces of all ordinary molecules. But, as we have seen, an equilibrium will establish itself between the agitation of the ordinary molecules and those of the ether. In other words, the ether and the bodies in it will tend to equality of temperature, and the ether will be subject to the ordinary gaseous laws as to pressure and temperature.

Among other properties of a gas, it will have that established by Dulong and Petit, so that the capacity for heat of unit of volume of the ether must be equal to that of unit of volume of any ordinary gas at the same pressure. Its presence, therefore, could not fail to be detected in our experiments on specific heat, and we may therefore assert that the constitution of the ether is not molecular.

(**11**, 374-377; 1875)

我们可以用气体分子理论去检验那些假定光以太是由原子或分子组成的假说。

那些敢于描绘光以太组成的人们也曾假定它是由原子或分子组成的。

将分子理论应用于这类假说会导致相当令人惊讶的结果。

首先，以太分子只会是气体本身。如果我们愿意的话，我们可以假定每个以太分子都是一个氢分子的千分之一或百万分之一，并且它们可以自由穿越任何常见分子的间隙。但是，正如我们已经看到的，普通分子的扰动与以太分子的扰动之间会自发地建立平衡。换句话说，以太与其中的物体会趋于温度相同的状态，那么以太就会服从诸如压力和温度等常见的气体定律。

以太除了具有气体的性质以外，根据杜隆和珀蒂建立的定律，相同压力下，单位体积以太的热容与单位体积任何普通气体的热容相等。因此，在比热实验中不可能检测不到以太的存在，从而我们可以断言以太不是由分子构成的。

（王耀杨 翻译；李芝芬 审稿）

Maxwell's Plan for Measuring the Ether

Editor's Note

One of the most remarkable of Maxwell's scientific exploits was a scheme for telling the velocity of the Earth and the Solar System as a whole relative to the luminiferous ether, supposed at the time to be necessary for the propagation of electromagnetic waves. Maxwell's proposal to D. P. Todd, director of the Nautical Almanac office in Washington, D.C., was that accurate measurements of the rotation of Jupiter's satellites around their planet would allow this relative velocity through the ether to be derived. If Maxwell had been able to execute this plan (for which in Todd's opinion the data were not yet sufficiently accurate), he would have discovered that the ether is irrelevant to the propagation of electromagnetic waves and indeed does not exist—the foundation for Einstein's Theory of Special Relativity.

"On a Possible Mode of Detecting a Motion of the Solar System through the Luminiferous Ether". By the late Prof. J. Clerk Maxwell. In a letter to Mr. D. P. Todd, Director of the *Nautical Almanac* Office, Washington, U. S. Communicated by Prof. Stoke, Sec. R. S.

Mr. Todd has been so good as to communicate to me a copy of the subjoined letter, and has kindly permitted me to make any use of it.

As the notice referred to by Maxwell in the *Encyclopaedia Britannica* is very brief, being confined to a single sentence, and as the subject is one of great interest, I have thought it best to communicate the letter to the Royal Society.

From the researches of Mr. Huggins on the radial component of the relative velocity of our sun and certain stars, the coefficient of the inequality which we might expect as not unlikely, would be only something comparable with half a second of time. This, no doubt, would be a very delicate matter to determine. Still, for anything we know *à priori* to the contrary, the motion might be very much greater than what would correspond to this; and the idea has a value of its own, irrespective of the possibility of actually making the determination.

In his letter to me Mr. Todd remarks, "I regard the communication as one of extraordinary importance, although (as you will notice if you have access to the reply which I made) it is likely to be a long time before we shall have tables of the satellites of Jupiter sufficiently accurate to put the matter to a practical test."

麦克斯韦测量以太的计划

编者按

麦克斯韦最突出的科学成就之一是，他制定了地球和太阳系作为一个整体相对于以太的速度的测量计划。当时，以太被认为是电磁波传播的必要条件。麦克斯韦向华盛顿特区航海历书处的办公室主任托德建议，通过精确测量木星卫星围绕木星的公转可以得到太阳系整体相对于以太的速度。如果麦克斯韦当时真的实施了这个计划（托德认为该计划所需的数据还不够精确），那么他会发现，以太和电磁波的传播没有任何关系，而且实际上以太根本就不存在——这便是爱因斯坦的狭义相对论的基础。

《论一种探测太阳系在以太中运动的可能方式》，作者是已故的克拉克·麦克斯韦教授。这篇文章出现在麦克斯韦写给美国华盛顿**航海历书处**的办公室主任托德先生的一封信中，由皇家学会的秘书斯托克教授宣读。

托德先生十分慷慨地给了我一份附信的拷贝，并且大方地允许我使用它。

因为麦克斯韦提到的《大英百科全书》中的参考资料十分简略，仅仅只有一句话。而这个题目又十分吸引人，所以我觉得最好还是在皇家学会宣读这封信。

根据哈金斯先生对太阳和某些星体的相对速度径向分量的研究，系数上的差别（我们暂且认为结果可靠）大约只有半秒。毫无疑问，这对实验测量来说是一个很精细的问题。而且这和我们**先验的**观点相反，我们先验的观点中的运动比这里涉及到的大很多。如果不考虑实际测量的可操作性，这个想法还是有它自身的价值的。

在托德先生给我的信中，他评论到，“虽然（如您将在我的答复中看到的那样）将这个计划付诸实施需要足够精确的木星卫星的数据表，而且这恐怕需要很长时间才能获得，但是我仍然认为这封信极其重要。”

I have not thought it expedient to delay the publication of the letter on the chance that something bearing on the subject might be found among Maxwell's papers.

(Copy)

Cavendish Laboratory,

Cambridge,

19th March, 1879

Sir,

I have received with much pleasure the tables of the satellites of Jupiter which you have been so kind as to send me, and I am encouraged by your interest in the Jovial system to ask you if you have made any special study of the apparent retardation of the eclipses as affected by the geocentric position of Jupiter.

I am told that observations of this kind have been somewhat put out of fashion by other methods of determining quantities related to the velocity of light, but they afford the *only* method, so far as I know, of getting any estimate of the direction and magnitude of the velocity of the sun with respect to the luminiferous medium. Even if we were sure of the theory of aberration, we can only get differences of position of stars, and in the terrestrial methods of determining the velocity of light, the light comes back along the same path again, so that the velocity of the earth with respect to the ether would alter the time of the double passage by a quantity depending on the square of the ratio of the earth's velocity to that of light, and this is quite too small to be observed.

But if J E is the distance of Jupiter from the earth, and l the geocentric longitude, and if l' is the longitude and λ the latitude of the direction in which the sun is moving through ether with velocity v, and if V is the velocity of light and t the time of transit from J to E,

$$\mathrm{JE} = [\ \mathrm{V} - v\cos\lambda\cos(l - l')\ ]\ t$$

By a comparison of the values of t when Jupiter is in different signs of the zodiac, it would be possible to determine l' and $v\cos\lambda$.

I do not see how to determine λ, unless we had a planet with an orbit very much inclined to the ecliptic. It may be noticed that whereas the determination of V, the velocity of light, by this method depends on the differences of J E, that is, on the diameter of the earth's orbit, the determination of $v\cos\lambda$ depends on J E itself, a much larger quantity.

But no method can be made available without good tables of the motion of the satellites, and as I am not an astronomer, I do not know whether, in comparing the observations with the tables of Damoiseau, any attempt has been made to consider the term in $v\cos\lambda$.

将这封信推迟到可以在麦克斯韦的论文中找到相关内容的时候再发表，我认为并不适宜。

（原信的拷贝）

卡文迪什实验室，

剑桥，

1879 年 3 月 19 日

先生：

我非常高兴能收到您寄给我的木星卫星的数据表。受您对木星系统的兴趣的启发，我想问您有没有对由木星相对于地心的位置而造成的木星卫星蚀的明显延迟作过专门的研究。

我得知，在其他测量与光速有关的物理量的方法面前，这种观测已经有些过时了。但是，据我所知，这种观测给我们提供了估算太阳相对于以太介质的速度的方向和大小的**唯一**方法。即便我们坚信像差理论，我们得到的也只是恒星位置的差别。并且按照在地球上测量光速的方法，光按原路返回，地球相对于以太的速度会改变这一往返的时间。但是，这个时间上的改变量依赖于地球速度与光速之比的平方，因此，这个量太小，以至于无法观测。

但是，如果 JE 表示木星与地球之间的距离，l 表示地心经度，l' 和 λ 分别表示太阳在以太中运动方向的经度和纬度，v 表示太阳在以太中运动的速度，V 表示光速，t 表示光从木星到地球的传播时间，

$$\mathrm{JE} = [\mathrm{V} - v\cos\lambda\cos(l - l')]\, t$$

通过比较木星在黄道不同位置时的 t，就有可能确定 l' 和 $v\cos\lambda$。

除非有一个行星的轨道向黄道极大倾斜，不然我就无法知道怎样确定 λ。我们应该注意到，尽管用这种方法测量的光速 V 也与 JE 的变化有关，也就是说，依赖于地球轨道的直径，但是 $v\cos\lambda$ 却依赖于一个更大的量——JE 本身。

然而，没有精确的卫星运动数据就不会有任何行之有效的方法。我本人并不是天文学家，我不知道这些观测与达穆瓦索的数据表相比，是否曾经尝试过将 $v\cos\lambda$ 中的各项考虑在内。

I have, therefore, taken the liberty of writing to you, as the matter is beyond the reach of any one who has not made a special study of the satellites.

In the article E [ether] in the ninth edition of the "Encyclopaedia Britannica", I have collected all the facts I know about the relative motion of the ether and the bodies which move in it, and have shown that nothing can be inferred about this relative motion from any phenomena hitherto observed, except the eclipses, &c., of the satellites of a planet, the more distant the better.

If you know of any work done in this direction, either by yourself or others, I should esteem it a favour to be told of it.

Believe me,

Yours faithfully,

(Signed) J. Clerk Maxwell

(**21**, 314-315; 1880)

因此，我十分冒昧地给您写这封信。这个问题对没有专门研究过卫星的人来说，实在是勉为其难。

在《大英百科全书》第9版关于以太的文章中，我收集了所有我知道的关于以太以及在其中运动的物体的相对运动的资料，从这些资料中我发现，由目前观测到的实验现象，除了行星离我们越远，越有利于卫星蚀等现象以外，从其他任何已观测到的相关现象中都不能推断出有关这种相对运动的结论。

如果您知道有任何人，不论是您还是别人，做过这方面的工作，都请您告诉我，我将不胜感激。

相信我，

您忠实的，

（签名）克拉克·麦克斯韦

（王静 翻译；鲍重光 审稿）

Clerk Maxwell's Scientific Work

P. G. Tait

Editor's Note

James Clerk Maxwell died in November 1879. In this essay four months later, Scottish physicist Peter Guthrie Tait, who had known Maxwell from childhood, paid tribute to his accomplishments. Maxwell was producing influential, original work before the age of twenty. In 1864 he published a landmark paper giving the first complete statement of his theory of electricity and magnetism. It explained electromagnetic phenomena without recourse to action at a distance, and provided a unified view of what light is. Guthrie notes that the facility of Maxwell's thinking did not always translate into effective lectures. While the treatises he wrote were models of clarity, his extemporaneous lectures gave free rein to his imagination in a way that taxed his audiences.

AT the instance of Sir W. Thomson, Mr. Lockyer, and others I proceed to give an account of Clerk Maxwell's work, necessarily brief, but I hope sufficient to let even the non-mathematical reader see how very great were his contributions to modern science. I have the less hesitation in undertaking this work that I have been intimately acquainted with him since we were schoolboys together.

If the title of mathematician be restricted (as it too commonly is) to those who possess peculiarly ready mastery over symbols, whether they try to understand the significance of each step or no, Clerk Maxwell was not, and certainly never attempted to be, in the foremost rank of mathematicians. He was slow in "writing out", and avoided as far as he could the intricacies of analysis. He preferred always to have before him a geometrical or physical representation of the problem in which he was engaged, and to take all his steps with the aid of this: afterwards, when necessary, translating them into symbols. In the comparative paucity of symbols in many of his great papers, and in the way in which, when wanted, they seem to grow full-blown from pages of ordinary text, his writings resemble much those of Sir William Thomson, which in early life he had with great wisdom chosen as a model.

There can be no doubt that in this habit, of constructing a mental representation of every problem, lay one of the chief secrets of his wonderful success as an investigator. To this were added an extraordinary power of penetration, and an altogether unusual amount of patient determination. The clearness of his mental vision was quite on a par with that of Faraday; and in this (the true) sense of the word he was a mathematician of the highest order.

But the rapidity of his thinking, which he could not control, was such as to destroy, except for the very highest class of students, the value of his lectures. His books and his written

克拉克·麦克斯韦的科学工作

泰特

编者按

1879 年 11 月，詹姆斯·克拉克·麦克斯韦逝世。在 4 个月后的这篇短文中，苏格兰物理学家彼得·格思里·泰特（他从小就认识麦克斯韦）热情称颂了麦克斯韦的成就。麦克斯韦在 20 岁之前就开始发表有影响力的原创研究论文。1864 年，他发表了一篇首次完整阐述其电磁理论的论文。在这篇里程碑式的论文中，麦克斯韦抛开一定距离外的实际效应去解释电磁现象，并就光的本质提出了一个统一性的观点。格思里认为，麦克斯韦深刻的思考并没有通过他那些颇有影响的演讲全部体现出来。不过，与他那堪称逻辑清晰之典范的专著不同，他的即席演讲在某种程度上则更自由地展现了他那对听众来说过于跳跃的想象力。

应汤姆孙爵士、洛克耶先生以及其他一些人士的要求，我将对麦克斯韦的科学工作进行介绍。介绍必然是简略的，但我希望足以让即使没有数学背景的读者也能了解他对现代科学的伟大贡献。我与麦克斯韦在学生时代就已经熟识，因此，我毫不犹豫地接受了这个任务。

如果说数学家的头衔只属于那些不论是否试图理解每一步的意义，都对符号了如指掌的人（事实通常就是这样），那么克拉克·麦克斯韦就不是、也从来不试图成为这样的一流数学家。他总是不慌不忙地“完稿”，并且尽可能避免复杂的分析。他更喜欢以几何或物理的形式表示自己所研究的问题，并且总是借助于以下方式来完成下一步的研究：在必要时，将这些表示转化成符号的形式。在他众多的伟大著作中所出现的较少的符号，在需要的时候似乎又能拓展成为直接触及主题的成熟的篇章，从这些方面来看，他的著作与威廉·汤姆孙爵士的十分相似，他在早年就很有远见卓识地把爵士当作自己的榜样。

毫无疑问，为每个问题建立思维上的表示，这个习惯是麦克斯韦作为一名研究者能够获得巨大成功的主要秘诀之一。除此之外，他还拥有卓越的洞察力和非凡的意志力。他的思路清晰，堪比法拉第。从数学家一词的这种（真正的）意义来说，麦克斯韦就是一位最高层次的数学家。

然而，麦克斯韦的思维之敏捷，连他自己都无法控制，这使得他的讲座只能被极高层次的学生所接受，而其他人则很难从中受益。他的著作和演讲稿（通常是对

addresses (always gone over twice in MS.) are models of clear and precise exposition; but his *extempore* lectures exhibited in a manner most aggravating to the listener the extraordinary fertility of his imagination.

His original work was commenced at a very early age. His first printed paper, "*On the Description of Oval Curves, and those having a Plurality of Foci*", was communicated for him by Prof. Forbes to the Royal Society of Edinburgh, and inserted in the "*Proceedings*" for 1846, before he reached his fifteenth year. He had then been taught only a book or two of Euclid, and the merest elements of Algebra. Closely connected with this are three unprinted papers, of which I have copies (taken in the same year), on "*Descartes' Ovals*", "*The Meloid and Apioid*", and "*Trifocal Curves*". All of these, which are drawn up in strict geometrical form and divided into consecutive propositions, are devoted to the properties of plane curves whose equations are of the form

$$mr + nr' + pr'' + \cdots = \textit{constant}$$

r, r', r'', &c., being the distances of a point on the curve from given fixed points, and m, n, p, &c., mere numbers. Maxwell gives a perfectly general method of tracing all such curves by means of a flexible and inextensible cord. When there are but two terms, if m and n have the same sign we have the ordinary Descartes' Ovals, if their signs be different we have what Maxwell called the Meloid and the Apioid. In each case a simple geometrical method is given for drawing a tangent at any point, and some of the other properties of the curves are elegantly treated.

Clerk Maxwell spent the years 1847–1850 at the University of Edinburgh, without keeping the regular course for a degree. He was allowed to work during this period, without assistance or supervision, in the Laboratories of Natural Philosophy and of Chemistry: and he thus experimentally taught himself much which other men have to learn with great difficulty from lectures or books. His reading was very extensive. The records of the University Library show that he carried home for study, during these years, such books as Fourier's *Théorie de la Chaleur*, Monge's *Géometrie Descriptive*, Newton's *Optics*, Willis' *Principles of Mechanism*, Cauchy's *Calcul Différentiel*, Taylor's *Scientific Memoirs*, and others of a very high order. These were *read through*, not merely consulted. Unfortunately no list is kept of the books consulted in the Library. One result of this period of steady work consists in two elaborate papers, printed in the *Transactions of the Royal Society of Edinburgh*. The first (dated 1849) "*On the Theory of Rolling Curves*", is a purely mathematical treatise, supplied with an immense collection of very elegant particular examples. The second (1850) is "*On the Equilibrium of Elastic Solids*". Considering the age of the writer at the time, this is one of the most remarkable of his investigations. Maxwell reproduces in it, by means of a special set of assumptions, the equations already given by Stokes. He applies them to a number of very interesting cases, such as the torsion of a cylinder, the formation of the large mirror of a reflecting telescope by means of a partial vacuum at the back of a glass plate, and the theory of Örsted's apparatus for the compression of water. But he

手稿的的再一次重温）是表述清晰精确的典范；但是他的即席演讲却因为极富想象力的风格而使听众难以理解。

麦克斯韦最初的工作在他年轻时就已经着手展开。他发表的第一篇论文《论椭圆曲线及多焦点椭圆曲线》由福布斯教授代他在爱丁堡皇家学会的会议上宣读，并收录在《爱丁堡皇家学会会刊》中。当时是 1846 年，麦克斯韦尚不满 15 岁。那时的他只学过一两本欧几里德的书和最基本的代数基础。紧接着，他又写了另外 3 篇没有发表的文章，我有这几篇文章的拷贝，它们分别是：《笛卡尔椭圆》、《芫菁科昆虫形曲线和芹亚科植物形曲线》、《三焦点曲线》。这些建立在严格的几何形式上并且被分成论题连贯的文章，可以用来研究具有以下形式的方程所描述的平面曲线的性质：

$$mr + nr' + pr'' + \cdots = \text{常量}$$

r，r'，r'' 等是从一个给定的固定点到曲线上某一点的距离，m，n，p 等仅仅是一些数字。麦克斯韦用容易弯曲但不能伸展的绳子，给出了一种理想的绘制这样曲线的一般方法。当方程中只有 m 和 n 两项时，如果两者同号，就得到普通的笛卡尔椭圆，如果两者异号，就得到麦克斯韦所谓的芫菁科昆虫形曲线和芹亚科植物形曲线。在每种情况下，麦克斯韦都给出了在任意一点作切线的简单几何方法，并且很好地处理了曲线其他方面的性质。

1847~1850 年，克拉克·麦克斯韦就读于爱丁堡大学，在此他并不需要去学那些学位要求的常规课程，而是被允许在既无人帮助也无人指导的条件下在自然哲学与化学实验室工作。因此，他在实验中自学了很多东西，而这些是其他人很难从书本中或课堂上学到的。他的阅读非常广泛。大学图书馆的记录显示，他在这些年中借回家研读的书有傅立叶的《热力学理论》、蒙日的《几何学说明》、牛顿的《光学》、威利斯的《力学原理》、柯西的《微分计算》、泰勒的《科学回忆录》等高水平著作。这些书他全部通读，而不是仅仅翻阅一下。很可惜图书馆没有保存他在馆内阅览的书单。麦克斯韦这段时间持续学习的成果之一是发表在《爱丁堡皇家学会会报》上的两篇详细论文。第一篇《论曲线滚动理论》（1849 年），这是一篇纯数学的论文，文章给出了大量简洁而恰当的例子。第二篇是《论弹性固体的平衡》（1850 年）。考虑到作者当时的年龄，这可以被认为是他最不寻常的研究之一。麦克斯韦运用一系列特殊的假设，重新构造了已经由斯托克斯给出的方程。他将这些方程应用到许多非常有趣的情况中，比如圆柱体的扭曲、通过在玻璃板后形成局部真空的方法实现反射式望远镜的巨大镜面的构造、奥斯特的水压缩装置的理论。此外，他还将其方程应用于张力（向一个垂直穿过透明板的圆柱体施加力偶后在透明平板中产

also applies his equations to the calculation of the strains produced in a transparent plate by applying couples to cylinders which pass through it at right angles, and the study (by polarised light) of the doubly-refracting structure thus produced. He expresses himself as unable to explain the permanence of this structure when once produced in isinglass, gutta percha, and other bodies. He recurred to the subject twenty years later, and in 1873 communicated to the Royal Society his very beautiful discovery of the *temporary* double refraction produced by shearing in viscous liquids.

During his undergraduateship in Cambridge he developed the germs of his future great work on "Electricity and Magnetism" (1873) in the form of a paper "On Faraday's Lines of Force", which was ultimately printed in 1856 in the "Trans. of the Cam. Phil. Soc." He showed me the MS. of the greater part of it in 1853. It is a paper of great interest in itself, but extremely important as indicating the first steps to such a splendid result. His idea of a fluid, incompressible and without mass, but subject to a species of friction in space, was confessedly adopted from the analogy pointed out by Thomson in 1843 between the steady flow of heat and the phenomena of statical electricity.

Other five papers on the same subject were communicated by him to the *Philosophical Magazine* in 1861–1862, under the title *Physical Lines of Force*. Then in 1864 appeared his great paper "*On a Dynamical Theory of the Electromagnetic Field*". This was inserted in the *Philosophical Transactions*, and may be looked upon as the first complete statement of the theory developed in the treatise on *Electricity and Magnetism*.

In recent years he came to the conclusion that such analogies as the conduction of heat, or the motion of the mass-less but incompressible fluid, depending as they do on Laplace's equation, were best symbolised by the quaternion notation with Hamilton's ∇ operator; and in consequence, in his work on electricity, he gives the expressions for all the more important physical quantities in their quaternion form, though without employing the calculus itself in their establishment. I have discussed in another place (*Nature*, vol. VII, p. 478) the various important discoveries in this remarkable work, which of itself is sufficient to secure for its author a foremost place among natural philosophers. I may here state that the main object of the work is to do away with "action at a distance," so far at least as electrical and magnetic forces are concerned, and to explain these by means of stresses and motions of the medium which is required to account for the phenomena of light. Maxwell has shown that, on this hypothesis, the velocity of light is the ratio of the electro-magnetic and electro-static units. Since this ratio, and the actual velocity of light, can be determined by absolutely independent experiments, the theory can be put at once to an exceedingly severe preliminary test. Neither quantity is yet fairly known within about 2 or 3 percent, and the most probable values of each certainly agree more closely than do the separate determinations of either. There can now be little doubt that Maxwell's theory of electrical phenomena rests upon foundations as secure as those of the undulatory theory of light. But the life-long work of its creator has left it still in its infancy, and it will probably require for its proper development the services of whole generations of mathematicians.

生的）的计算，以及对同样产生的双折射结构的研究（用偏振光）。他无法解释为什么在云母、杜仲胶以及其他物体中一旦产生了这种结构就会持续存在。20 年以后，他又重新回到这个题目，并于 1873 年向皇家学会宣读了他非常美妙的发现——由粘性液体中的切变造成的**暂时**双折射。

在剑桥读本科时，麦克斯韦完成了最终于 1856 年发表在《剑桥哲学学会学报》上的论文《论法拉第力线》，这篇文章是他后来的伟大著作《电磁学》（1873 年）的雏形。1853 年，他曾给我看过这篇论文的大部分手稿。这篇论文本身就十分有趣，但更重要的是它显示了走向未来辉煌成就的第一步。麦克斯韦关于流体的观点是，流体不可压缩，没有质量，但是要克服空间中的某种摩擦力，这无疑采用了汤姆孙于 1843 年从稳定热流和静电现象之间类推出的结果。

关于这个主题还有另外 5 篇文章，以《论物理力线》为题发表在 1861~1862 年间的《哲学杂志》上。1864 年，他的伟大著作《电磁场的动力学理论》问世了。这篇文章被收录在《自然科学会报》上，可以将其看作对《电磁学》专著中所阐述理论的第一次完整论述。

近年，他得出了以下结论：对于这些遵循拉普拉斯方程的类似物理量，比如热传导或没有质量却不可压缩的流体运动来说，含有哈密顿算子 ∇ 的四维表示法是最佳的表示方式。接着，他在电学研究中给出了所有更重要的物理量的四维形式，然而他并没有在建立表示方法的同时进行计算。我在别处（《自然》，第 7 卷，第 478 页）讨论过这项伟大工作中的多个重要发现，这本身就足以使其作者跻身最伟大的科学家行列。我可以在这里说，这项工作的目标就是，至少在涉及电磁力时，弄清“超距作用”，并且要通过光现象所需的介质应力和运动来解释它们。麦克斯韦指出，在这种假设下，光速为电磁单元与静电单元之比。因为这个比值和光速都可以由完全独立的实验确定，所以上面的理论可以用极为严格的初级实验来检验。但是，目前这两个物理量都还没能在 2% ~ 3% 的误差范围内被清楚地认识，每个量的最可几数值当然比分散的数值更加集中。现在毫无疑问，麦克斯韦电现象理论建立的基础和光的波动学说的基础一样可信。但是，这个理论在其创建者的毕生努力下也还是处在初级阶段，它的合理发展可能还需要整整一代数学家的努力。

This was not the only work of importance to which he devoted the greater part of his time while an undergraduate at Cambridge. For he had barely obtained his degree before he read to the Cambridge Philosophical Society a remarkable paper *On the Transformation of Surfaces by Bending*, which appears in their *Transactions* with the date March 1854. The subject is one which had been elaborately treated by Gauss and other great mathematicians, but their methods left much to be desired from the point of view of simplicity. This Clerk Maxwell certainly supplied; and to such an extent that it is difficult to conceive that any subsequent investigator will be able to simplify the new mode of presentation as much as Maxwell simplified the old one. Many of his results, also, were real additions to the theory; especially his treatment of the *Lines of Bending*. But the whole matter is one which, except in its almost obvious elements, it is vain to attempt to popularise.

The next in point of date of Maxwell's greatest works is his "Essay on the Stability of the Motion of Saturn's Rings", which obtained the Adam's Prize in the University of Cambridge in 1857. This admirable investigation was published as a pamphlet in 1859. Laplace had shown in the *Mécanique Céleste* that a uniform solid ring cannot revolve permanently about a planet; for, even if its density were so adjusted as to prevent its splitting, a slight disturbance would inevitably cause it to fall in. Maxwell begins by finding what amount of *want* of uniformity would make a solid ring stable. He finds that this could be effected by a satellite rigidly attached to the ring, and of about $4\frac{1}{2}$ times its mass:—but that such an arrangement, while not agreeing with observation, would require extreme artificiality of adjustment of a kind not elsewhere observed. Not only so, but the materials, in order to prevent its behaving almost like a liquid under the great forces to which it is exposed, must have an amount of rigidity far exceeding that of any known substance.

He therefore dismisses the hypothesis of solid rings, and (commencing with that of a ring of equal and equidistant satellites) shows that a continuous liquid ring cannot be stable, but may become so when broken up into satellites. He traces in a masterly way the effects of the free and forced waves which must traverse the ring, under various assumptions as to its constitution; and he shows that the only system of rings which can dynamically exist must be composed of a very great number of separate masses, revolving round the planet with velocities depending on their distances from it. But even in this case the system of Saturn cannot be permanent, because of the mutual actions of the various rings. These mutual actions must lead to the gradual spreading out of the whole system, both inwards and outwards:—but if, as is probable, the outer ring is much denser than the inner ones, a very small increase of its external diameter would balance a large change in the inner rings. This is consistent with the progressive changes which have been observed since the discovery of the rings. An ingenious and simple mechanism is described, by which the motions of a ring composed of equal satellites can be easily demonstrated.

Another subject which he treated with great success, as well from the experimental as from the theoretical point of view, was the Perception of Colour, the Primary Colour Sensations, and the Nature of Colour Blindness. His earliest paper on these subjects bears

以上并不是麦克斯韦利用在剑桥读本科期间的大部分时间来完成的唯一重要的工作。因为直到他在剑桥哲学学会的会议上宣读了一篇著名的论文《论弯曲引起的表面变换》，并于 1854 年 3 月发表在《剑桥哲学学会学报》上，他才获得了学位。高斯和其他一些伟大的数学家都曾经详细研究过这个课题，但是从简洁的角度来看，他们的方法还有很多需要改进的地方。克拉克·麦克斯韦无疑完成了后续的简化工作，并且将其简化到了这样的程度：很难想象任何一个后继研究者在简化现有的新模型时，能达到像麦克斯韦简化旧模型时那样的程度。他的很多结果，是对原有理论的丰富，尤其是他对**弯曲线**所作的处理。但问题是，除了其中几乎显而易见的部分，这些结果都没有实现普及。

麦克斯韦的下一个伟大著作是《关于土星环运动稳定性的评论》，这篇文章在 1857 年获得了剑桥大学的亚当斯奖。这项令人称赞的研究在 1859 年被印成了小册子。拉普拉斯在他的《天体力学》中指出，均匀的固体环不可能持久地围绕行星转动，因为，即使环的密度调整到可以使其避免分裂的程度，一个微小的扰动也会不可避免地导致其塌陷。麦克斯韦从寻找使一个固体环稳定**所需的**均匀度开始。他发现，可以通过将一个卫星与这个环作刚性连接来实现环的稳定，其中卫星的质量等于环质量的 $4\frac{1}{2}$ 倍。但是由于这种方法与观测不符，于是就需要极精巧的调节方式，而这种调节方式也没有在其他地方看到过。不但如此，为了避免环出现类似暴露在强力下的液体那样的表现，环的材料必须具有足够的硬度，而这种硬度远远超过了任何已知材料的硬度。

因此，麦克斯韦放弃了固体环的假设，开始设想一个由等距的相同行星组成的环。他指出，连续的液体环无法保持稳定，但是当它破裂成多个卫星的时候就可能达到稳定状态。他极为巧妙地论述了在各种假设的环结构中必须穿过环的或自由或受迫的波的影响，并且指出，唯一一种能够动态稳定存在的环形结构必须由大量分离的物质组成，这些物质围绕行星转动，其速度取决于它们到行星的距离。然而，即使在这种情况下，由于不同环之间的相互作用，土星系统也不可能持久稳定。因为这种相互作用必然会导致整个系统向内外两个方向扩散。但是，如果外环密度比内环密度大，外环直径很小的增大就能平衡内环直径很大的改变。这与环被发现以来所观察到的不断变化是一致的。麦克斯韦描述了一个独特而又简洁的机制，用它可以很容易地说明由相同卫星组成的环的运动。

麦克斯韦的另一个研究课题是颜色的感知、基础色觉以及色盲的本质。这项工作无论是从实验角度还是从理论角度来看都极为成功。他最早关于这些问题的文章诞生于 1855 年，第 7 篇则发表于 1872 年。“由于他在颜色组成方面的研究和其他光

date 1855, and the seventh has the date 1872. He received the Rumford Medal from the Royal Society in 1860, "For his Researches on the Composition of Colours and other optical papers". Though a triplicity about colour had long been known or suspected, which Young had (most probably correctly) attributed to the existence of three sensations, and Brewster had erroneously* supposed to be objective, Maxwell was the first to make colour-sensation the subject of actual measurement. He proved experimentally that any colour C (given in intensity of illumination as well as in character) may be expressed in terms of three arbitrarily chosen standard colours, X, Y, Z, by the formula

$$\mathrm{C} = a\mathrm{X} + b\mathrm{Y} + c\mathrm{Z}$$

Here a, b, c are numerical coefficients, which may be positive or negative; the sign = means "matches", + means "superposed", and – directs the term to be taken to the other side of the equation.

These researches of Maxwell's are now so well known, in consequence especially of the amount of attention which has been called to the subject by Helmholtz' great work on Physiological Optics, that we need not farther discuss them here.

The last of his greatest investigations is the splendid Series on the Kinetic Theory of Gases, with the closely connected question of the sizes, and laws of mutual action, of the separate particles of bodies. The Kinetic Theory seems to have originated with D. Bernoulli; but his successors gradually reverted to statical theories of molecular attraction and repulsion, such as those of Boscovich. Herapath (in 1847) seems to have been the first to recall attention to the Kinetic Theory of gaseous pressure. Joule in 1848 calculated the average velocity of the particles of hydrogen and other gases. Krönig in 1856 (*Pogg. Ann.*) took up the question, but he does not seem to have advanced it farther than Joule had gone; except by the startling result that the weight of a mass of gas is only half that of its particles when at rest.

Shortly afterwards (in 1859) Clausius took a great step in advance, explaining, by means of the kinetic theory, the relations between the volume, temperature and pressure of a gas, its cooling by expansion, and the slowness of diffusion and conduction of heat in gases. He also investigated the relation between the length of the mean free path of a particle, the number of particles in a given space, and their least distance when in collision. The special merit of Clausius' work lies in his introduction of the processes of the theory of probabilities into the treatment of this question.

Then came Clerk Maxwell. His first papers are entitled "Illustrations of the Dynamical Theory of Gases", and appeared in the *Phil. Mag.* in 1860. By very simple processes he treats the collisions of a number of perfectly elastic spheres, first when all are of the same mass, secondly when there is a mixture of groups of different masses. He thus verifies

* All we can positively say to be erroneous is some of the principal arguments by which Brewster's view was maintained, for the subjective character of the triplicity has not been absolutely *demonstrated*.

学方面的论文”，麦克斯韦在1860年获得了皇家学会颁发的拉姆福德奖章。虽然人们知道或者猜测出颜色的三原色已经有很长一段时间了，例如杨曾经（很可能正确地）把三原色归因于存在三种主观色觉，而布鲁斯特错误地*猜测三原色是客观的，但是麦克斯韦却是第一个将实际测量引入色觉这个课题的人。他在实验中证明了，任何颜色C（以照度和特性的形式给出）都可以用三种任选的标准颜色X，Y，Z按照下面的公式表示出来：

$$\mathrm{C} = a\mathrm{X} + b\mathrm{Y} + c\mathrm{Z}$$

其中，a，b，c是数值系数，可以取正也可以取负；等号表示“匹配”，加号表示“叠加”，减号表示把这一项移到方程的另一边。

现在麦克斯韦的这些研究已经是众所周知的了，特别是后来亥姆霍兹关于生理光学的伟大工作使得这个研究课题吸引了很多注意力，因此，我们就不需要在这里进一步详细讨论它们了。

麦克斯韦最后一项伟大的研究是他那套卓越的关于气体动力学的丛书。这套丛书的内容与物体离散粒子的大小和相互作用定律等问题密切相关。动力学理论似乎最早起源于伯努利，但是他的后继者逐渐回归到关于粒子吸引和排斥的统计理论上，就如博斯科维克所做的那样。赫拉帕斯（1847年）似乎是第一个重新注意到气体压力动力学理论的人。焦耳在1848年计算了氢气和其他气体微粒的平均速度。科隆尼格在1856年（《波根多夫年鉴》）考虑到了这个问题，但是除了得到气体的重量只有静止气体微粒的一半这一惊人结果之外，他似乎并没有进一步发展焦耳的结果。

在不久之后（1859年），克劳修斯取得了巨大的进展。他用动力学理论成功地解释了气体体积、温度、压强之间的关系，膨胀造成的冷却，以及气体中缓慢的热扩散和热传导。他还研究了气体微粒的平均自由程长度、给定空间中的粒子数以及粒子碰撞过程中的最小距离这三者之间的关系。克劳修斯工作的最大价值在于，他在处理这个问题时，引入了概率论的方法。

然后就是克拉克·麦克斯韦。他第一篇论文的题目是《气体动力学理论图示》，于1860年发表在《哲学杂志》上。他通过一个非常简单的过程来处理许多完全弹性小球的碰撞。他首先研究了所有小球的质量都相等的情况，然后研究了含有不同质

---

* 我们只能说布鲁斯特的观点所使用的主要论据是错误的，而三原色的主观性还没有完全被证实。

Gay-Lussac's law, that the number of particles per unit volume is the same in all gases at the same pressure and temperature. He explains gaseous friction by the transference to and fro of particles between contiguous strata of gas sliding over one another, and shows that the coefficient of viscosity is independent of the density of the gas. From Stokes' calculation of that coefficient he gave the first deduced approximate value of the mean length of the free path; which could not, for want of data, be obtained from the relation given by Clausius. He obtained a closely accordant value of the same quantity by comparing his results for the kinetic theory of diffusion with those of one of Graham's experiments. He also gives an estimate of the conducting power of air for heat; and he shows that the assumption of non-spherical particles, which during collision change part of their energy of translation into energy of rotation, is inconsistent with the known ratio of the two specific heats of air.

A few years later he made a series of valuable experimental determinations of the viscosity of air and other gases at different temperatures. These are described in *Phil. Trans.* 1866; and they led to his publishing (in the next volume) a modified theory, in which the gaseous particles are no longer regarded as perfectly elastic, but as repelling one another according to the law of the inverse fifth power of the distance. This paper contains some very powerful analysis, which enabled him to simplify the mathematical theory for many of its most important applications. Three specially important results are given in conclusion, and they are shown to be independent of the particular mode in which gaseous particles are supposed to act on one another. These are:—

1. In a mixture of particles of two kinds differing in amounts of mass, the average energy of translation of a particle must be the same for either kind. This is Gay Lussac's Law already referred to.
2. In a vertical column of mixed-gases, the density of each gas at any point is ultimately the same as if no other gas were present. This law was laid down by Dalton.
3. Throughout a vertical column of gas gravity has no effect in making one part hotter or colder than another; whence (by the dynamical theory of heat) the same must by true for all substances.

Maxwell has published in later years several additional papers on the Kinetic Theory, generally of a more abstruse character than the majority of those just described. His two latest papers (in the *Phil. Trans.* and *Camb. Phil. Trans.* of last year) are on this subject:—one is an extension and simplification of some of Boltzmann's valuable additions to the Kinetic Theory. The other is devoted to the explanation of the motion of the radiometer by means of this theory. Several years ago (*Nature*, vol. XII, p. 217), Prof. Dewar and the writer pointed out, and demonstrated experimentally, that the action of Mr. Crookes' very beautiful instrument was to be explained by taking account of the increased length of the mean free path in rarefied gases, while the then received opinions ascribed it either to evaporation or to a quasi-corpuscular theory of radiation. Stokes extended the explanation to the behaviour of disks with concave and convex surfaces, but the subject was not at all fully investigated from the theoretical point of view till Maxwell took it up. During the last ten years of his life he had no rival to claim concurrence with him in the whole wide domain

量小球的情况。他由此证明了盖·吕萨克定律，即在同温同压下，所有气体单位体积内的粒子数相等。他把气体的摩擦力解释为相邻气层之间微粒的相对运动，并且指出粘滞系数和气体的密度无关。他从斯托克斯对粘滞系数的计算出发，第一次推出了气体平均自由程的近似值，由于缺少数据，这个数值是无法从克劳修斯给出的关系中得到的。他把自己用扩散动力学理论计算出的结果和格雷厄姆一个实验的结果进行比较，发现二者能够很好地吻合。麦克斯韦还估算出了空气的热导率，并且指出，非球形粒子在碰撞中能将部分平动动能转化为转动动能的假设与已知的空气的两种比热之比不符。

几年后，他做了一系列很有价值的实验，来确定不同温度下空气和其他气体的粘性。这些工作的结果发表在 1866 年的《自然科学会报》上。在该杂志接下来的一卷上，他又发表了修正后的理论，此理论中不再把气体粒子视为完全弹性小球，而是认为粒子之间存在着与相互距离的五次方成反比的排斥作用。这篇论文包含一些非常有力的分析，这使得他可以为了理论的重要应用而进行数学理论上的简化。结论中给出了 3 个特别重要的结果，并且这些结果和气体粒子之间的相互作用模型无关。它们是：

1. 在由两种质量不同的粒子组成的混合物中，两种粒子的平均平动动能一定相等。这是盖·吕萨克定律已经提到过的。
2. 在装有混合气体的立柱容器中，每一种气体在任一点的密度最终都将相同，就好像没有其他气体存在一样。这个定律是道尔顿建立的。
3. 在整个装有气体的立柱容器中，重力并没有使某一部分的温度高于或低于另外一部分。因此（根据热动力学理论），这个规律应该适用于所有物质。

在之后的几年中，麦克斯韦发表了另外几篇关于动力学理论的文章，这些文章比上面提到的工作中的大多数都更加深奥。他最近的两篇论文发表在去年的《自然科学会报》和《剑桥哲学学报》上：一篇是对玻耳兹曼在动力学理论上的重要补充的推广和简化；另一篇文章中，他用这个理论解释了辐射计的运转。几年之后，杜瓦教授等人（《自然》，第 12 卷，第 217 页）指出并且通过实验证实了：考虑到稀薄气体中平均自由程的增加，无论是采用蒸气辐射理论还是准颗粒辐射理论都可以解释克鲁克斯先生非常精巧的仪器的作用。斯托克斯将其推广，用于解释具有凹凸表面的圆盘的行为，但是从理论的角度来看，这个课题在麦克斯韦着手之前并没有得到充分的研究。在麦克斯韦生命的最后十年里，他没有遇到能够在分子力学的广阔领域内和他平起平坐的对手，然而在更深奥的电学领域，倒是有两三个人能与他

of molecular forces, and but two or three in the still more recondite subject of electricity.

"Every one must have observed that when a slip of paper falls through the air, its motion, though undecided and wavering at first, sometimes becomes regular. Its general path is not in the vertical direction, but inclined to it at an angle which remains nearly constant, and its fluttering appearance will be found to be due to a rapid rotation round a horizontal axis. The direction of deviation from the vertical depends on the direction of rotation… These effects are commonly attributed to some accidental peculiarity in the form of the paper…" So writes Maxwell in the *Cam. and Dub. Math. Jour*. (May, 1854), and proceeds to give an exceedingly simple and beautiful explanation of the phenomenon. The explanation is, of course, of a very general character, for the complete working out of such a problem appears to be, even yet, hopeless; but it is thoroughly characteristic of the man, that his mind could never bear to pass by any phenomenon without satisfying itself of at least its general nature and causes.

In the same volume of the *Math. Journal* there is an exceedingly elegant "problem" due to Maxwell, with his solution of it. In a note we are told that it was "suggested by the contemplation of the structure of the crystalline lens in fish". It is as follows:—

A transparent medium is such that the path of a ray of light within it is a given circle, the index of refraction being a function of the distance from a given point in the plane of the circle. Find the form of this function, and show that for light of the same refrangibility—

1. The path of *every ray within the medium* is a circle.
2. All the rays proceeding from any point in the medium will meet accurately in another point.
3. If rays diverge from a point without the medium and enter it through a spherical surface having that point for its centre, they will be made to converge accurately to a point within the medium.

Analytical treatment of this and connected questions, by a novel method, will be found in a paper by the present writer (*Trans. R. S. E.* 1865).

Optics was one of Clerk Maxwell's favourite subjects, but of his many papers on various branches of it, or subjects directly connected with it, we need mention only the following:—

"On the General Laws of Optical Instruments" (*Quart. Math. Jour.* 1858)
"On the Cyclide" (*Quart. Math. Journal*, 1868)
"On the best Arrangement for Producing a Pure Spectrum on a Screen" (*Proc. R. S. E.* 1868)
"On the Focal Lines of a Refracted Pencil" (*Math. Soc. Proc.* 1873)

A remarkable paper, for which he obtained the Keith Prize of the *Royal Society of Edinburgh*, is entitled "On Reciprocal Figures, Frames, and Diagrams of Forces." It is published in the *Transactions* of the Society for 1870. Portions of it had previously appeared in the *Phil. Mag.* (1864).

相提并论。

“每个人都会注意到，一张纸片在空气中飘落时，虽然一开始摇摆不定，但是它的运动会趋于规则。它通常的路径并不是沿垂直方向，而是与垂直方向成一个角度，这个角度基本是一个常数。我们会发现纸片一开始的飘动是围绕一条水平轴快速转动。偏离垂直轴的方向取决于转动的方向……。这些结果通常被归因于纸张形状的某些偶然特性……”麦克斯韦在 1854 年 5 月发表于《剑桥与都柏林数学杂志》上的文章中这样写到，他想对这个现象作出非常简洁而漂亮的解释。当然，这个解释十分笼统，因为即使现在看来，完全解决这个问题也是希望渺茫。但这正是麦克斯韦的性格，他的思想决不容忍自己与任何连一般性质及成因都得不到满意解释的现象擦肩而过。

在《剑桥与都柏林数学杂志》的同一卷中，麦克斯韦提出了一个极其精彩的“问题”，并且自己作出了解答。我们从一则记录中得知，这个工作是“在思考鱼的晶状体结构时受到的启发”。内容如下：

所谓介质是透明的就是指光线在此介质中的传播路径是一个特定的圆，介质某一点的折射率是这一点到圆平面中给定点距离的函数。我找到了这个函数的形式，并且发现对于具有相同折射性质的光线来说——

1. **每一条光线在介质中的**路径都是一个圆。
2. 从介质中任意一点发出的光线，都会在另外一点精确相遇。
3. 如果光线在介质外的某一点发散，并且经由一个以此发散点为球心的球面进入介质，那么这些光线将精确地会聚到介质中的某一点上。

我本人在一篇文章（《爱丁堡皇家学会会报》，1865 年）中，用一种新颖的方法对这个问题及相关问题进行了分析。

光学是克拉克·麦克斯韦最喜欢的课题之一，但是在他关于光学不同分支或者直接与光学相关的大量文章中，我们只需提及下面这些：

《论光学仪器的普遍规律》（《数学季刊》，1858 年）
《论四次圆纹曲面》（《数学季刊》，1868 年）
《论在屏幕上生成纯光谱的最佳方案》（《爱丁堡皇家学会会刊》，1868 年）
《论折射光束的焦线》（《数学学会会刊》，1873 年）

麦克斯韦还有一篇意义重大的文章，题目是《论力的对应线图、框架和图解》，为此他获得了爱丁堡皇家学会的基思奖。文章于 1870 年发表在《爱丁堡皇家学会会报》上，文章中的一部分之前已经在《哲学杂志》（1864 年）上出现过。

The triangle and the polygon of forces, as well as the funicular polygon, had long been known; and also some corresponding elementary theorems connected with hydrostatic pressure on the faces of a polyhedron; but it is to Rankine that we owe the full principle of diagrams, and reciprocal diagrams, of frames and of forces. Maxwell has greatly simplified and extended Rankine's ideas: on the one hand facilitating their application to practical problems of construction, and on the other hand extending the principle to the general subject of stress in bodies. The paper concludes with a valuable extension to three dimensions of Sir George Airy's "Function of Stress".

His contributions to the *Proceedings of the London Mathematical Society* were numerous and valuable. I select as a typical specimen his paper on the forms of the stream-lines when a circular cylinder is moved in a straight line, perpendicular to its axis, through an infinitely extended, frictionless, incompressible fluid (vol. III, p. 224). He gives the complete solution of the problem; and, with his usual graphical skill, so prominent in his great work on Electricity, gives diagrams of the stream-lines, and of the paths of individual particles of the fluid. The results are both interesting and instructive in the highest degree.

In addition to those we have mentioned we cannot recall many pieces of *experimental* work on Maxwell's part:—with two grand exceptions. The first was connected with the determination of the British Association Unit of Electric Resistance, and the closely associated measurement of the ratio of the electrokinetic to the electrostatic unit. In this he was associated with Professors Balfour Stewart and Jenkin. The Reports of that Committee are among the most valuable physical papers of the age; and are now obtainable in a book-form, separately published. The second was the experimental verification of Ohm's law to an exceedingly close approximation, which was made by him at the Cavendish Laboratory with the assistance of Prof. Chrystal.

In his undergraduate days he made an experiment which, though to a certain extent physiological, was closely connected with physics. Its object was to determine why a cat always lights on its feet, however it may be let fall. He satisfied himself, by pitching a cat gently on a mattress stretched on the floor, giving it different initial amounts of rotation, that it instinctively made use of the conservation of Moment of Momentum, by stretching out its body if it were rotating so fast as otherwise to fall head foremost, and by drawing itself together if it were rotating too slowly.

I have given in this journal (vol. XVI, p. 119) a detailed account of his remarkable elementary treatise on "Matter and Motion", a work full of most valuable materials, and worthy of most attentive perusal not merely by students but by the foremost of scientific men.

His "Theory of Heat", which has already gone through several editions, is professedly elementary, but in many places is probably, in spite of its admirable definiteness, more difficult to follow than any other of his writings. In intrinsic importance it is of the same high order as his "Electricity", but as a whole it is *not* an elementary book. One of the

力的三角法则、多边形法则、以及索状多边形法则早已为人们所熟知，一些和多面体表面静压有关的基础理论也是如此。但是兰金认为，我们缺少一套关于力的线图、框架、图解以及力本身的完整法则。麦克斯韦极大地简化并推广了兰金的观点：一方面，他使这个理论在建筑学实际问题上的应用更加方便；另一方面，他将这个理论推广到了物体中的压力这一一般主体。在文章结尾，麦克斯韦将乔治·艾里爵士的“压力函数”推广到了三维的情形，这个推广很有意义。

他为《伦敦数学学会会刊》贡献了大量有价值的稿件。我选取他的一篇文章作为其中的典范。这篇文章讨论了当一个圆柱体沿着垂直于轴的直线穿过一个不可压缩、无摩擦且可无限扩展的液体时液体中流线的形式（第 3 卷，第 224 页）。他给出了这个问题的完整解答，并且利用他在电学巨著中常用的图解技巧给出了流线和液体中单个粒子路径的图形。

除了上面提到的那些工作，麦克斯韦没有多少**实验性的**工作能被我们铭记，但是有两个工作是例外。第一个是关于确定电阻英制单位以及与其密切相关的电动力学单位与静电学单位之比的测量。这个工作是麦克斯韦和鲍尔弗·斯图尔特教授、詹金教授合作完成的。相关委员会的报告是这个时期最有价值的物理学论文之一，现在这些报告已经集结成册并且单独出版了。第二项工作是在实验上以极高的精度验证了欧姆定律。这项工作是他在卡文迪什实验室由克里斯托尔教授协助完成的。

麦克斯韦在读本科期间做了一个实验，虽然从某种程度上说这是一个生理学实验，但是它也和物理学密切相关。实验的目的是解释为什么猫在落地的时候总是能保持脚先着地。实验中他把猫轻轻地抛到毯子上，抛掷时，让猫具有不同的初始转动。麦克斯韦得到了满意的结果：猫在空中的时候，本能地利用了角动量守恒。如果给它的初始转动过快，它就会把身体伸展开，避免头先着地；相反地，如果给它的初始转动过慢，它就会缩成一团，最后总是能避免头先着地。

我曾经在贵刊上（第 16 卷，第 119 页）详细介绍了麦克斯韦著名的关于《物质和运动》的基础论文。他的这项工作有许多重要的结果，因此值得每一个人——不仅仅是学生，还包括一流的科学家——仔细研读。

他的《热学理论》已经出了好几版。尽管他自称这部书很基础，尽管书的思路的确非常清晰，但是其中许多地方可能比作者的其他任何著作都更难理解。这部书本身的重要性堪比他的《电学》，但是总体上却**不是**一本基础读物。克拉克·麦克斯

few knowable things which Clerk Maxwell did not know, was the distinction which most men readily perceive between what is easy and what is hard. What *he called* hard, others would be inclined to call altogether unintelligible. In the little book we are discussing there is matter enough to fill two or three large volumes without undue dilution (perhaps we should rather say, *with the necessary dilution*) of its varied contents. There is nothing flabby, so to speak, about anything Maxwell ever wrote: there is splendid muscle throughout, and an adequate bony structure to support it. "Strong meat for grown men" was one of his favourite expressions of commendation; and no man ever more happily exposed the true nature of the so-called "popular science" of modern times than he did when he wrote of "the forcible language and striking illustrations by which those who are past hope of being even beginners [in science] are prevented from becoming conscious of intellectual exhaustion before the hour has elapsed."

To the long list of works attached to Maxwell's name in the Royal Society's Catalogue of Scientific Papers may now be added his numerous contributions to the latest edition of the "Encyclopaedia Britannica"—Atom, Attraction, Capillarity, &c. Also the laborious task of preparing for the press, with copious and very valuable original notes, the "Electrical Researches of the Hon. Henry Cavendish." This work has appeared only within a month or two, and contains many singular and most unexpected revelations as to the early progress of the science of electricity. We hope shortly to give an account of it.

The works which we have mentioned would of themselves indicate extraordinary activity on the part of their author, but they form only a fragment of what he has published; and when we add to this the further statement, that Maxwell was always ready to assist those who sought advice or instruction from him, and that he has read over the proof-sheets of many works by his more intimate friends (enriching them by notes, always valuable and often of the quaintest character), we may well wonder how he found time to do so much.

Many of our readers must remember with pleasure the occasional appearance in our columns of remarkably pointed and epigrammatic verses, usually dealing with scientific subjects, and signed $\frac{dp}{dt}$*. The lines on Cayley's portrait, where determinants, roots of −1, space of *n* dimensions, the 27 lines on a cubic surface, &c., fall quite naturally into rhythmical English verse; the admirable synopsis of Dr. Ball's Treatise on Screws; the telegraphic love-letter with its strangely well-fitting *volts* and *ohms*; and specially the "Lecture to a Lady on Thomson's Reflecting Galvanometer", cannot fail to be remembered. No living man has shown a greater power of condensing the whole marrow of a question into a few clear and compact sentences than Maxwell shows in these verses. Always having a definite object, they often veiled the keenest satire under an air of charming innocence and *naïve* admiration. Here are a couple of stanzas from unpublished pieces of a similar kind:—first, some ghastly thoughts by an excited evolutionist—

* This *nom de plume* was suggested to him by me from the occurrence of his initials in the well-known expression of the second Law of Thermodynamics (for whose establishment on thoroughly valid grounds he did so much) $\frac{dp}{dt}$ = J. C. M.

韦不知道的少数几个显而易见的事情之一就是，难与易的区别，而这是多数人都能分清的。**他所谓的**难事，其他人会认为是根本无法理解的。我们正在谈论的这本薄薄的书的内容不用过度展开（可能我们应该说，**经过必要的展开**），其内容就足以写满两三卷书。可以说，麦克斯韦写的东西从不松散拖沓：他的文章只有健美的肌肉和适量的用以支撑的骨架。“成人的强健肌肉”是麦克斯韦最喜欢的表达称赞的说法。“过来人总是希望自己保持那种［科学上的］初学者的状态，对他们来说，有力的语言和精彩的图示就是能在时间消逝之前避免灵感和智慧枯竭的灵药。”当麦克斯韦写出上面这段话时，恐怕没有人比他更乐于揭示现代所谓“流行科学”的真实本质了。

在皇家学会科学文献目录中，有麦克斯韦署名的工作已经可以列出一长串了。现在，应该还要加上他对《大英百科全书》中原子、吸引作用、毛细作用等方面所作的很多贡献。另外，他还为编写出版《亨利·卡文迪什电学研究》付出了辛勤的劳动，整理了丰富而珍贵的手稿。这项工作是在最近的一两个月才问世的，书中记录着电学早期发展历程带给我们的很多意想不到的非凡启迪。我们希望不久之后能介绍一下这项工作。

上面提到的工作已经显示了作者超常的科研能力，然而这些只是他已发表著作的冰山一角。如果我还补充说，麦克斯韦总是乐于帮助那些寻求建议或指引的人，而且他通读过很多好友的著作的校样（在上面所作的注释使其更加丰富，往往起到画龙点睛的作用），大家可能会觉得十分惊讶，他哪来那么多时间完成这么多事情。

很多读者一定还记得在我们的专栏中偶尔出现的那些非常尖锐的讽刺小诗，通常都是关于科学主题的，并且署名$\frac{dp}{dt}$*；在凯莱肖像画上的诗句中由行列式、−1 的根、n 维空间以及三次曲面上的 27 条线等很自然地组成的一首充满韵律的英文小诗；为鲍尔博士关于旋量的论文写的令人赞叹的简介；用出奇得体的**伏特**和**欧姆**组成的电报情书；特别是那篇《为女士所作的关于汤姆孙反射检流计的演讲》，所有这些都让人无法忘怀。当今世上没有人可以超越麦克斯韦在小诗中表现出的用几个清晰而简洁的句子就把问题的精髓概括出来的能力。这些小诗总是具有明确的目标，但是又把最尖锐的讽刺隐藏在迷人的纯真和**纯朴的**赞美之中。这里有两段没有发表的类似这种风格的片段：首先，是一个狂热的革命者的可怕想法——

---

* 这个笔名是我建议他取的，因为我在著名的热力学第二定律$\frac{dp}{dt}$ = J. C. M.中发现了他名字的首字母缩写（而他本人也为这个定律能够建立在坚实的基础之上做了许多工作）。

To follow my thoughts as they go on,
Electrodes I'd place in my brain;
Nay, I'd swallow a live entozöon,
New feelings of life to obtain—

next on the non-objectivity of Force—

Both Action and Reaction now are gone;
Just ere they vanished
Stress joined their hands in peace, and made them one,
Then they were banished.

It is to be hoped that these scattered gems may be collected and published, for they are of the very highest interest, as the work during leisure hours of one of the most piercing intellects of modern times. Every one of them contains evidence of close and accurate thought, and many are in the happiest form of epigram.

I cannot adequately express in words the extent of the loss which his early death has inflicted not merely on his personal friends, on the University of Cambridge, on the whole scientific world, but also, and most especially, on the cause of common sense, of true science, and of religion itself, in these days of much vain-babbling, pseudo-science, and materialism. But men of his stamp never live in vain; and in one sense at least they cannot die. The spirit of Clerk Maxwell still lives with us in his imperishable writings, and will speak to the next generation by the lips of those who have caught inspiration form his teachings and example.

(**21**, 317-321; 1880)

我跟随着自己的感觉，
我要把电极放进我的脑子里；
要不我就吞下活生生的寄生虫，
我的生命会有崭新的感受——

另一个，是关于力的非客观性——

作用力和反作用力都消失了；
就在他们消失之前，
压力让他们静静地携起手来，合二为一，
然后，他们被放逐天涯。

人们希望能够把这些散落的宝石集结出版，因为它们是现代最敏锐的智者中的一员闲暇时完成的作品，而又是如此有趣。每一首诗都证明了作者缜密的思维，并且很多都是以讽刺诗那种诙谐的手法写成的。

我实在无法用语言表达麦克斯韦的早逝是多么巨大的损失，受到损失的不仅仅是他的朋友、剑桥大学和整个科学界，特别是，在充满空谈、伪科学和物质主义的今天，人们对常识、真科学以及宗教本身的探究也会因此受到巨大的损失。然而，脚踏实地的人决不会生活在空谈中，至少从某种意义上讲，这样的人不会从世界上消失。麦克斯韦的精神会在他不朽的著作中与我们同在，并且这种精神会由那些受过他的教诲并以他为榜样的人，传承给下一代。

（王静 翻译；鲍重光 审稿）

On a New Kind of Rays*

W. C. Röntgen

Editor's Note

This is an English translation of Wilhelm Conrad Röntgen's German report, in December 1895, of the discovery of X-rays. While experimenting with a cathode ray tube (also called a Crookes' or Lenard's tube, after earlier investigators), in which electrons or "kathode rays" are accelerated by electric fields, Röntgen found that the tube emits radiation that penetrates black paper and induces fluorescence in a screen on the other side. The rays also can penetrate matter and produce photographic images of "buried" objects such as bones. Röntgen deduces that these rays are not cathode rays, but seem instead to be akin to ultraviolet rays, yet with much greater penetrating power. He called them X-rays simply "for the sake of brevity".

(1) A discharge from a large induction coil is passed through a Hittorf's vacuum tube, or through a well-exhausted Crookes' or Lenard's tube. The tube is surrounded by a fairly close-fitting shield of black paper; it is then possible to see, in a completely darkened room, that paper covered on one side with barium platinocyanide lights up with brilliant fluorescence when brought into the neighbourhood of the tube, whether the painted side or the other be turned towards the tube. The fluorescence is still visible at two metres distance. It is easy to show that the origin of the fluorescence lies within the vacuum tube.

(2) It is seen, therefore, that some agent is capable of penetrating black cardboard which is quite opaque to ultra-violet light, sunlight, or arc-light. It is therefore of interest to investigate how far other bodies can be penetrated by the same agent. It is readily shown that all bodies possess this same transparency, but in very varying degrees. For example, paper is very transparent; the fluorescent screen will light up when placed behind a book of a thousand pages; printer's ink offers no marked resistance. Similarly the fluorescence shows behind two packs of cards; a single card does not visibly diminish the brilliancy of the light. So, again, a single thickness of tinfoil hardly casts a shadow on the screen; several have to be superposed to produce a marked effect. Thick blocks of wood are still transparent. Boards of pine two or three centimetres thick absorb only very little. A piece of sheet aluminium, 15 mm thick, still allowed the X-rays (as I will call the rays, for the sake of brevity) to pass, but greatly reduced the fluorescence. Glass plates of similar thickness behave similarly; lead glass is, however, much more opaque than glass free from lead. Ebonite several centimetres thick is transparent. If the hand be held before the fluorescent screen, the shadow shows the bones darkly, with only faint outlines of the surrounding tissues.

* By W. C. Röntgen. Translated by Arthur Stanton from the *Sitzungsberichte der Würzburger Physik-medic. Gesellschaft*, 1895.

论一种新型的射线*

伦琴

编者按

此文译自威廉·康拉德·伦琴 1895 年 12 月的一份关于发现了 X 射线的德文报告。当伦琴用阴极射线管（早期的科研人员也把它称作克鲁克斯管或莱纳德管，管中的电子或“阴极射线”被电场加速）进行实验时，他发现阴极射线管发射的射线能够穿透黑色的纸并在其另一侧的屏幕上显示出荧光。这种射线还可以穿透物质，人们可以利用它拍摄出像骨骼这样的“被遮挡的”物质的照片。伦琴推测这种射线不属于阴极射线，它似乎更接近紫外线，但具有更强的穿透力。“为了简便起见”，他把这类射线称为“X 射线”。

(1) 让大号感应线圈中产生的放电通过希托夫真空管，或者通过抽成真空的克鲁克斯管或莱纳德管。管子用黑纸包裹严实。在完全黑暗的房间里，将一面涂有铂氰酸钡的纸放在管子旁边，不论朝向管子的是涂有铂氰酸钡的一面还是没有涂的那面，纸都会被鲜艳的荧光照亮。在 2 米之外，这种荧光依然可见。很显然，荧光来源于真空管中。

(2) 由此可见，某些射线能够穿透这种紫外光、太阳光和电弧光都几乎不能透过的黑纸板。这就引起了人们研究这种射线到底能够多大程度地穿透其他物质的兴趣。很容易就能证明，所有物质对这种射线都是透明的，只是透明的程度大不相同。比如，纸是非常透明的，即使将荧光屏置于一本 1,000 页厚的书后面，我们仍然会在荧光屏上看到亮光，印刷油墨也不会造成明显的阻挡。类似地，单独一张卡片不会明显减弱光的强度，即使在两叠卡片后面我们也仍然能看到亮光。同样，单张锡箔纸的遮挡几乎不会使荧光屏上出现阴影，要产生明显的遮挡效果就必须重叠许多张锡箔纸。厚木块对于这种射线也是透明的。2~3 厘米厚的松木板的吸收效果非常微弱。15 毫米厚的铝板也能使 X 射线（为了简便起见，我将称这种射线为 X 射线）透过，但是能够大幅度地减弱荧光。玻璃板的作用与厚度相近的铅板类似，不过，含铅的玻璃对这种射线的阻挡效果比不含铅的玻璃更强。几厘米厚的硬质橡胶也是透明的。如果把手放在荧光屏前，屏幕上就会显示出骨骼的黑影，而周围组织则只有模糊的轮廓。

---

* 作者为伦琴。由阿瑟·斯坦顿译自 1895 年的《维尔茨堡物理学医学学会会刊》。

Water and several other fluids are very transparent. Hydrogen is not markedly more permeable than air. Plates of copper, silver, lead, gold, and platinum also allow the rays to pass, but only when the metal is thin. Platinum 0.2 mm thick allows some rays to pass; silver and copper are more transparent. Lead 1.5 mm thick is practically opaque. If a square rod of wood 20 mm in the side be painted on one face with white lead, it casts little shadow when it is so turned that the painted face is parallel to the X-rays, but a strong shadow if the rays have to pass through the painted side. The salts of the metals, either solid or in solution, behave generally as the metals themselves.

(3) The preceding experiments lead to the conclusion that the density of the bodies is the property whose variation mainly affects their permeability. At least no other property seems so marked in this connection. But that the density alone does not determine the transparency is shown by an experiment wherein plates of similar thickness of Iceland spar, glass, aluminium, and quartz were employed as screens. Then the Iceland spar showed itself much less transparent than the other bodies, though of approximately the same density. I have not remarked any strong fluorescence of Iceland spar compared with glass (see below, No. 4).

(4) Increasing thickness increases the hindrance offered to the rays by all bodies. A picture has been impressed on a photographic plate of a number of superposed layers of tinfoil, like steps, presenting thus a regularly increasing thickness. This is to be submitted to photometric processes when a suitable instrument is available.

(5) Pieces of platinum, lead, zinc, and aluminium foil were so arranged as to produce the same weakening of the effect. The annexed table shows the relative thickness and density of the equivalent sheets of metal.

	Thickness	Relative thickness	Density
Platinum	0.018 mm	1	21.5
Lead	0.050 mm	3	11.3
Zinc	0.100 mm	6	7.1
Aluminium	3.500 mm	200	2.6

From these values it is clear that in no case can we obtain the transparency of a body from the product of its density and thickness. The transparency increases much more rapidly than the product decreases.

(6) The fluorescence of barium platinocyanide is not the only noticeable action of the X-rays. It is to be observed that other bodies exhibit fluorescence, *e.g.* calcium sulphide, uranium glass, Iceland spar, rock-salt, &c.

水和其他几种液体对于这种射线都是非常透明的。氢气的透明度并没有明显强于空气。铜、银、铅、金和铂质的金属板只有在很薄的时候才能使这种射线透过。0.2 毫米的铂能使这种射线部分透过，银和铜则更透明一些。1.5 毫米厚的铅板基本上是不透明的。将一根边长为 20 毫米的方木棒的一个侧面涂上铅白，当木棒涂有铅白的面与射线平行时，几乎不会产生阴影，但是当射线必须穿过涂有铅白的一面时，就会产生明显的阴影。不论是固态的金属盐还是金属盐溶液，一般都能像金属本身一样阻挡该射线。

(3) 根据上述实验我们可以得出结论：物质的密度是这样一种性质，它的变化主要影响射线在该物质中的透过程度。至少其他性质的影响看起来都不如密度明显。不过，单是密度还不能完全决定物质对该射线的透明度。我用厚度相近的冰洲石板、玻璃板、铝板和石英板作为样品进行的实验表明，尽管这些物质具有近似相同的密度，但冰洲石对该射线的透明度却比其他物质小得多。在用冰洲石进行的实验中，我从来没有观察到像用玻璃进行的实验中出现的那样明显的荧光（见下文，第 4 部分）。

(4) 对于所有物体，增加厚度都会提高其对 X 射线的阻挡程度。我们已经在照相底片上对阶梯状叠放的多层锡箔进行了成像，得到的图像表现出了厚度的这种有规律的增加。如果根据此原理制成适当的仪器，则可以作为光度计使用。

(5) 为了得到对 X 射线相同的减弱效果，我将铂、铅、锌和铝分别制成如下规格的金属片。附表给出了具有相同减弱效果的各种金属片的密度和相对厚度。

	厚度	相对厚度	密度
铂	0.018 毫米	1	21.5
铅	0.050 毫米	3	11.3
锌	0.100 毫米	6	7.1
铝	3.500 毫米	200	2.6

从这些数据中可以清楚地看出，我们不可能根据金属密度与其厚度的乘积来确定其透明度。透明度增加的速度比该乘积减少的速度快很多。

（6）X 射线所产生的显著作用并不是只能使铂氰酸钡发出荧光。可以观测到，X 射线也能使其他一些物质发出荧光，例如硫化钙、铀玻璃、冰洲石和岩盐等。

Of special interest in this connection is the fact that photographic dry plates are sensitive to the X-rays. It is thus possible to exhibit the phenomena so as to exclude the danger of error. I have thus confirmed many observations originally made by eye observation with the fluorescent screen. Here the power of the X-rays to pass through wood or cardboard becomes useful. The photographic plate can be exposed to the action without removal of the shutter of the dark slide or other protecting case, so that the experiment need not be conducted in darkness. Manifestly, unexposed plates must not be left in their box near the vacuum tube.

It seems now questionable whether the impression on the plate is a direct effect of the X-rays, or a secondary result induced by the fluorescence of the material of the plate. Films can receive the impression as well as ordinary dry plates.

I have not been able to show experimentally that the X-rays give rise to any calorific effects. These, however, may be assumed, for the phenomena of fluorescence show that the X-rays are capable of transformation. It is also certain that all the X-rays falling on a body do not leave it as such.

The retina of the eye is quite insensitive to these rays: the eye placed close to the apparatus sees nothing. It is clear from the experiments that this is not due to want of permeability on the part of the structures of the eye.

(7) After my experiments on the transparency of increasing thicknesses of different media, I proceeded to investigate whether the X-rays could be deflected by a prism. Investigations with water and carbon bisulphide in mica prisms of 30° showed no deviation either on the photographic or the fluorescent plate. For comparison, light rays were allowed to fall on the prism as the apparatus was set up for the experiment. They were deviated 10 mm and 20 mm respectively in the case of the two prisms.

With prisms of ebonite and aluminium, I have obtained images on the photographic plate, which point to a possible deviation. It is, however, uncertain, and at most would point to a refractive index 1.05. No deviation can be observed by means of the fluorescent screen. Investigations with the heavier metals have not as yet led to any result, because of their small transparency and the consequent enfeebling of the transmitted rays.

On account of the importance of the question it is desirable to try in other ways whether the X-rays are susceptible of refraction. Finely powdered bodies allow in thick layers but little of the incident light to pass through, in consequence of refraction and reflection. In the case of the X-rays, however, such layers of powder are for equal masses of substance equally transparent with the coherent solid itself. Hence we cannot conclude any regular reflection or refraction of the X-rays. The research was conducted by the aid of finely-powdered rock-salt, fine electrolytic silver powder, and zinc dust already many times employed in chemical work. In all these cases the result, whether by the fluorescent screen

在这方面，特别让人感兴趣的是照相干版对 X 射线是敏感的。这就使我们可以将实验现象记录下来以避免出现错误。利用照相的方法，我已经确认了很多最初通过肉眼在荧光屏上观测得到的实验结果。X 射线穿透木块或纸板的能力很有用处。在对照相干版进行曝光时，可以不用除去遮光板或者其他保护盒，因此实验就不必在暗室中进行。当然，千万不要把装有未曝光照相干版的盒子放在真空管附近。

干版上留下的影像到底是 X 射线的直接效应，还是由干版材料发出的荧光引起的次级效应，现在看来还是一个令人疑惑的问题。和普通的干版一样，胶片也可以记录到影像。

我还没能通过实验证明 X 射线是否可以产生热效应，不过我们猜测它可以，因为荧光现象表明 X 射线可以引起能量转移。而且可以肯定的是，照射在物体上的 X 射线并没有全部以荧光形式离开物体。

人眼的视网膜对这种射线非常不敏感：即使眼睛离装置很近也看不见任何东西。实验结果清楚地表明，这并不是因为该射线在眼睛这部分结构中的透过程度不够。

(7) 在通过实验研究了不同介质随着厚度增加对该射线的透明度的变化之后，我又研究了 X 射线是否会被棱镜偏转。在顶角为 30° 的云母棱镜中分别装入水和二硫化碳进行实验，结果发现照相干版和荧光板上都没有显示出偏移。为了对照，我也用可见光在相同的实验装置上进行了实验，结果发现可见光在穿过上述两种棱镜时分别偏转了 10 毫米和 20 毫米。

在使用硬质橡胶和铝制成的棱镜进行实验时，我得到的照相干版上的影像显示射线可能发生了偏转，不过这一点还不能确定，而且偏转所对应的折射率最多也只有 1.05。使用荧光屏时则观测不到偏转。用较重金属进行的研究目前还没有任何结果，这是因为它们的透明度都很小，因而透射的射线非常微弱。

考虑到 X 射线能否被偏转这个问题的重要性，我们就有必要尝试用其他方法来研究 X 射线能否发生折射。由于反射和折射的原因，微细粉末形成的厚层几乎不能使入射光透过。不过，这种多层粉末对于 X 射线的透明度与同质量同组成的整块固体是一样的。因此我们不能得出 X 射线具有常规的反射或折射特性的结论。我又对细粉末状的岩盐、电解得到的细银粉和已经在化学实验中使用了多次的锌粉进行了研究。所有研究结果都表明，不论是用荧光屏还是用照相的方法，粉末与相应的固

or the photographic method, indicated no difference in transparency between the powder and the coherent solid.

It is, hence, obvious that lenses cannot be looked upon as capable of concentrating the X-rays; in effect, both an ebonite and a glass lens of large size prove to be without action. The shadow photograph of a round rod is darker in the middle than at the edge; the image of a cylinder filled with a body more transparent than its walls exhibits the middle brighter than the edge.

(8) The preceding experiments, and others which I pass over, point to the rays being incapable of regular reflection. It is, however, well to detail an observation which at first sight seemed to lead to an opposite conclusion.

I exposed a plate, protected by a black paper sheath, to the X-rays so that the glass side lay next to the vacuum tube. The sensitive film was partly covered with star-shaped pieces of platinum, lead, zinc, and aluminium. On the developed negative the star-shaped impression showed dark under platinum, lead, and, more markedly, under zinc; the aluminium gave no image. It seems, therefore, that these three metals can reflect the X-rays; as, however, another explanation is possible, I repeated the experiment with this only difference, that a film of thin aluminium foil was interposed between the sensitive film and the metal stars. Such an aluminium plate is opaque to ultra-violet rays, but transparent to X-rays. In the result the images appeared as before, this pointing still to the existence of reflection at metal surfaces.

If one considers this observation in connection with others, namely, on the transparency of powders, and on the state of the surface not being effective in altering the passage of the X-rays through a body, it leads to the probable conclusion that regular reflection does not exist, but that bodies behave to the X-rays as turbid media to light.

Since I have obtained no evidence of refraction at the surface of different media, it seems probable that the X-rays move with the same velocity in all bodies, and in a medium which penetrates everything, and in which the molecules of bodies are embedded. The molecules obstruct the X-rays, the more effectively as the density of the body concerned is greater.

(9) It seemed possible that the geometrical arrangement of the molecules might affect the action of a body upon the X-rays, so that, for example, Iceland spar might exhibit different phenomena according to the relation of the surface of the plate to the axis of the crystal. Experiments with quartz and Iceland spar on this point lead to a negative result.

(10) It is known that Lenard, in his investigations on kathode rays, has shown that they belong to the ether, and can pass through all bodies. Concerning the X-rays the same may be said.

体对 X 射线的透明度没有任何差别。

因此，很明显透镜是不能会聚 X 射线的。事实上，大尺寸的玻璃透镜和硬质橡胶透镜对 X 射线都没有会聚作用，这已经得到了证明。圆柱的透视影像显示，中间部分的阴影比边缘更深一些；如果在圆柱内部填入透明度比柱体材料更高的物质，那么在得到的影像中，中间部分会比边缘更亮一些。

(8) 上述实验和另外一些我未提及的实验，都表明这种射线不具备常规的反射能力。不过，我还是要详细介绍一个乍看上去似乎会使人们得出相反结论的实验。

实验中，我将一块用黑纸套保护起来的玻璃干版置于 X 射线中，使其玻璃面靠近真空管，并用铂、铅、锌和铝质的星形金属片部分地遮挡干版的感光膜。在显影后的负片上，铂片和铅片下方出现了黑色的星形影像，锌片下方的影像更加清晰，而铝片并没有产生阴影。由此看来，前三种金属可以反射 X 射线。不过也可能有另外的解释。我又重复了这一实验，这次唯一的不同之处是，我在感光膜和星形金属片之间插入了一块薄薄的铝箔。这块铝箔对紫外线是不透明的，但对 X 射线是透明的。结果出现了和以前一样的影像。这再次表明 X 射线在金属表面发生了反射。

如果综合考虑这一观测结果和其他一些结果，包括关于粉末透明度的结果以及关于表面状态不能有效改变 X 射线穿过物体的路径的结果，我们就会得出这样一个可能的结论：对于 X 射线来说，并不存在普通意义上的反射，物体对于 X 射线的作用，就像混浊介质对于可见光一样。

我还没有得到任何可以表明在不同介质表面 X 射线会发生折射的证据，这样看来，X 射线在所有物质中的传播速度可能都相同，而且在一种渗透一切物质、包容各种物质分子的介质中也是一样的。随着物质密度的增大，其分子对 X 射线的阻挡效果也变得更加明显。

(9) 分子的几何构型看起来可能会影响物质对 X 射线的阻挡作用，例如，对于冰洲石晶体而言，表面与晶轴之间相对取向的不同可能就会导致不同的现象。但是，为此而用石英和冰洲石进行的实验却得到了阴性的结果。

(10) 我们知道，莱纳德在对阴极射线的研究中已经指出，阴极射线属于以太，可以穿透任何物体。估计 X 射线可能也是这样的。

In his latest work, Lenard has investigated the absorption coefficients of various bodies for the kathode rays, including air at atmospheric pressure, which gives 4.10, 3.40, 3.10 for 1 cm, according to the degree of exhaustion of the gas in discharge tube. To judge from the nature of the discharge, I have worked at about the same pressure, but occasionally at greater or smaller pressures. I find, using a Weber's photometer, that the intensity of the fluorescent light varies nearly as the inverse square of the distance between screen and discharge tube. This result is obtained from three very consistent sets of observations at distances of 100 and 200 mm. Hence air absorbs the X-rays much less than the kathode rays. This result is in complete agreement with the previously described result, that the fluorescence of the screen can be still observed at 2 metres from the vacuum tube. In general, other bodies behave like air; they are more transparent for the X-rays than for the kathode rays.

(11) A further distinction, and a noteworthy one, results from the action of a magnet. I have not succeeded in observing any deviation of the X-rays even in very strong magnetic fields.

The deviation of kathode rays by the magnet is one of their peculiar characteristics; it has been observed by Hertz and Lenard, that several kinds of kathode rays exist, which differ by their power of exciting phosphorescence, their susceptibility of absorption, and their deviation by the magnet; but a notable deviation has been observed in all cases which have yet been investigated, and I think that such deviation affords a characteristic not to be set aside lightly.

(12) As the result of many researches, it appears that the place of most brilliant phosphorescence of the walls of the discharge-tube is the chief seat whence the X-rays originate and spread in all directions; that is, the X-rays proceed from the front where the kathode rays strike the glass. If one deviates the kathode rays within the tube by means of a magnet, it is seen that the X-rays proceed from a new point, *i.e.* again from the end of the kathode rays.

Also for this reason the X-rays, which are not deflected by a magnet, cannot be regarded as kathode rays which have passed through the glass, for that passage cannot, according to Lenard, be the cause of the different deflection of the rays. Hence I conclude that the X-rays are not identical with the kathode rays, but are produced from the kathode rays at the glass surface of the tube.

(13) The rays are generated not only in glass. I have obtained them in an apparatus closed by an aluminium plate 2 mm thick. I purpose later to investigate the behaviour of other substances.

(14) The justification of the term "rays", applied to the phenomena, lies partly in the regular shadow pictures produced by the interposition of a more or less permeable body between the source and a photographic plate or fluorescent screen.

莱纳德在最近的工作中研究了各种物体对阴极射线的吸收系数，比如一个大气压下的空气的吸收系数，根据放电管抽真空程度的不同，每一厘米对应的吸收系数分别是 4.10、3.40 和 3.10。为了根据放电的本质来作出判断，我在基本相同的压强下进行了研究，不过偶尔也会用更高一点或更低一点的压强。利用韦伯光度计，我发现荧光的强度近似与屏幕到放电管距离的平方成反比。这个结论是根据三组非常一致的观测结果得到的，其观测距离分别为 100 毫米和 200 毫米。因此，空气对 X 射线的吸收比对阴极射线的吸收低很多。这一结果与前述的在距真空管 2 米处的屏幕上仍会出现荧光的结果是完全一致的。大体上，其他物质的性质与空气类似，它们对 X 射线比对阴极射线更加透明。

(11) 另一个更明显也更值得关注的区别是磁场的作用。即使是在非常强的磁场中，我也没有观测到 X 射线的任何偏转。

阴极射线在磁场作用下会发生偏转，这是它的独特性质之一。赫兹和莱纳德曾经观测到存在好几种阴极射线，它们的区别在于激发磷光的能力不同、被吸收的容易程度不同以及在磁场作用下的偏转不同。但是对于所有已经被研究过的阴极射线，人们都观测到了显著的偏转，我认为这种偏转代表了阴极射线的一种绝不该被忽视的特性。

(12) 很多研究结果表明，放电管管壁上磷光最强的位置是在 X 射线产生并向四周各个方向发散的那个源头处，也就是说，X 射线产生于阴极射线轰击玻璃的前沿位置。如果利用磁场使管中的阴极射线偏转，就会看到 X 射线从另一个位置上产生，但仍然是在阴极射线的终端位置。

基于这一原因，我们不能把在磁场作用下并不偏转的 X 射线看作是已经穿透玻璃的阴极射线，因为按照莱纳德的说法，这条通道不可能是由阴极射线的不同偏转造成的。由此我断定，X 射线与阴极射线是不同的，它是阴极射线作用于真空管的玻璃表面而产生的。

(13) 并不是只有用玻璃才能产生 X 射线。我曾利用一种被 2 毫米厚的铝板包裹起来的装置得到了 X 射线。以后我将研究其他物质是否也能产生 X 射线。

(14) 在描述这种现象时我使用了“射线”这个词，这在一定程度上是因为，将不太透明的物体插入到源和照相干版或荧光屏之间时会产生规则的阴影。

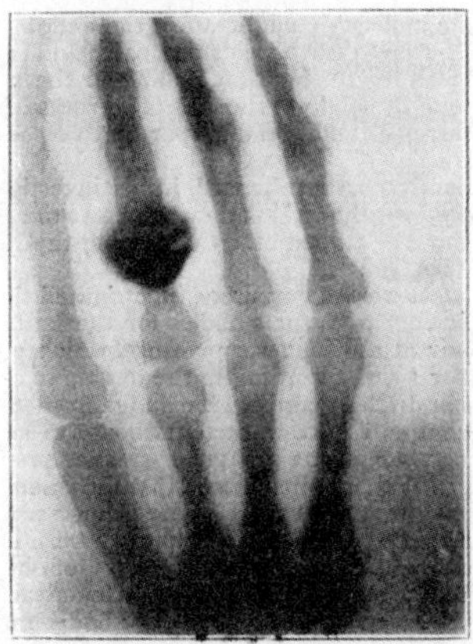

Fig. 1. Photograph of the bones in the fingers of a living human hand. The third finger has a ring upon it.

I have observed and photographed many such shadow pictures. Thus, I have an outline of part of a door covered with lead paint; the image was produced by placing the discharge-tube on one side of the door, and the sensitive plate on the other. I have also a shadow of the bones of the hand (Fig. 1), of a wire wound upon a bobbin, of a set of weights in a box, of a compass card and needle completely enclosed in a metal case (Fig. 2), of a piece of metal where the X-rays show the want of homogeneity, and of other things.

Fig. 2. Photograph of a compass card and needle completely enclosed in a metal case

For the rectilinear propagation of the rays, I have a pin-hole photograph of the discharge apparatus covered with black paper. It is faint but unmistakable.

(15) I have sought for interference effects of the X-rays, but possibly, in consequence of their small intensity, without result.

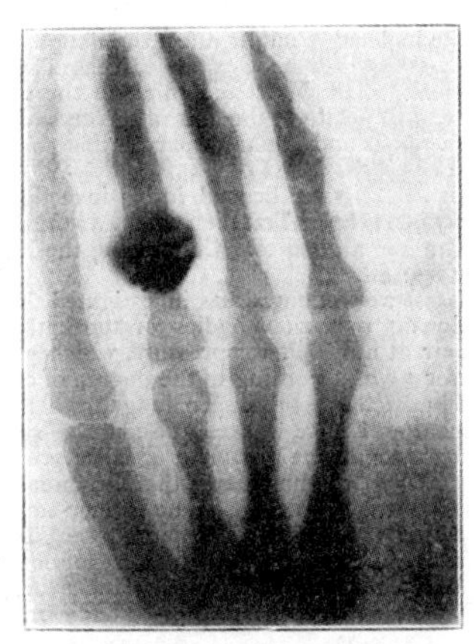

图 1. 活人手指（第三指上戴着一枚戒指）骨骼的影像

我已经观察并用照相记录了很多这样的阴影。由此，我记录下了门的局部轮廓。我是用含铅涂料刷了门的轮廓，然后把放电管放置在门的一侧，而把光敏照相干版放置在另一侧，这样就得到了门的局部轮廓的影像。我还记录了其他许多物体的阴影，这包括手掌骨骼（图 1）、缠在绕线筒上的导线、一套装在盒子里的砝码、完全密封于金属盒子中的罗经刻度盘和指针（图 2）、一块在 X 射线下显示出具有不均匀缺陷的金属片，以及其他一些物品。

图 2. 完全密封于金属盒子中的罗经刻度盘和指针的影像

为了说明射线的直线传播，我用针孔照相的方法拍摄了用黑纸覆盖的放电装置。照片虽然有些模糊，但却可以明白无误地分辨出装置。

(15) 我曾经试图寻找 X 射线的干涉效应，但并没有检测到，这可能是由于强度太低的缘故。

(16) Researches to investigate whether electrostatic forces act on the X-rays are begun but not yet concluded.

(17) If one asks, what then are these X-rays; since they are not kathode rays, one might suppose, from their power of exciting fluorescence and chemical action, them to be due to ultra-violet light. In opposition to this view a weighty set of considerations presents itself. If X-rays be indeed ultra-violet light, then that light must possess the following properties.

(*a*) It is not refracted in passing from air into water, carbon bisulphide, aluminium, rock-salt, glass or zinc.

(*b*) It is incapable of regular reflection at the surfaces of the above bodies.

(*c*) It cannot be polarised by any ordinary polarising media.

(*d*) The absorption by various bodies must depend chiefly on their density.

That is to say, these ultra-violet rays must behave quite differently from the visible, infra-red, and hitherto known ultra-violet rays.

These things appear so unlikely that I have sought for another hypothesis.

A kind of relationship between the new rays and light rays appears to exist; at least the formation of shadows, fluorescence, and the production of chemical action point in this direction. Now it has been known for a long time, that besides the transverse vibrations which account for the phenomena of light, it is possible that longitudinal vibrations should exist in the ether, and, according to the view of some physicists, must exist. It is granted that their existence has not yet been made clear, and their properties are not experimentally demonstrated. Should not the new rays be ascribed to longitudinal waves in the ether?

I must confess that I have in the course of this research made myself more and more familiar with this thought, and venture to put the opinion forward, while I am quite conscious that the hypothesis advanced still requires a more solid foundation.

(**53**, 274-276; 1896)

(16) 关于静电力对 X 射线是否有作用的研究工作正在进行，但目前尚无结论。

(17) 也许有人会问 X 射线到底是什么。既然这种射线不是阴极射线，有人可能就会根据其激发荧光和引发化学反应的能力猜想它是紫外光。然而，一系列认真的思考都是反对这种观点的。如果 X 射线真是一种紫外光，那么这种紫外光就必须具有如下性质：

(*a*) 它在由空气进入水、二硫化碳、铝、岩盐、玻璃或锌时，不会发生折射。

(*b*) 它在上述物质的表面不会发生常规的反射。

(*c*) 任何普通的偏振介质都不能使它偏振。

(*d*) 不同物质对它的吸收主要取决于该物质的密度。

也就是说，这种紫外线必须具有与可见光、红外线以及迄今为止已知的紫外线都十分不同的性质。

看起来这些是很难成立的，因此我想到了另一种假说。

这种新型的射线与普通光之间看起来应该存在着某种关联，至少在形成阴影、激发荧光以及引发化学反应这些方面都是相似的。长期以来我们都知道，除了能够解释光现象的横向振动外，在以太中可能存在纵向振动，某些物理学家甚至认为纵向振动是必定存在的。尽管目前人们还不完全清楚纵向振动是否存在，也没有通过实验论证这种纵向振动的性质，但是，难道我们就不能认为这种新型的射线属于以太中的纵波吗？

我必须要承认的是，在研究过程中我越来越倾向于这一观点。另外我也十分清楚这一新假说还需要更为可靠的证据，我承认在目前的情况下抛出这一观点是比较冒昧的。

（王耀杨 翻译；江丕栋 审稿）

Professor Röntgen's Discovery

A. A. C. Swinton

Editor's Note

One of the most sobering things about this verification of Röntgen's discovery of X-rays, less than a month after they were first reported, is that it shows how tepid the reception of great discoveries can be among scientific peers. Campbell-Swinton hints that the newspapers have been getting excited over a phenomenon that is not "entirely novel". But that is because he somewhat misinterprets Röntgen's results. Swinton insists on regarding the X-rays as "some portion of the kathode radiations", and points out that cathode rays are already known to produce photographic images—missing Röntgen's claim that his X-rays are not cathode rays at all.

THE newspaper reports of Prof. Röntgen's experiments have, during the past few days, excited considerable interest. The discovery does not appear, however, to be entirely novel, as it was noted by Hertz that metallic films are transparent to the kathode rays from a Crookes or Hittorf tube, and in Lenard's researches, published about two years ago, it is distinctly pointed out that such rays will produce photographic impressions. Indeed, Lenard, employing a tube with an aluminium window, through which the kathode rays passed out with comparative ease, obtained photographic shadow images almost identical with those of Röntgen, through pieces of cardboard and aluminium interposed between the window and the photographic plate.

Prof. Röntgen has, however, shown that this aluminium window is unnecessary, as some portion of the kathode radiations that are photographically active will pass through the glass walls of the tube. Further, he has extended the results obtained by Lenard in a manner that has impressed the popular imagination, while, perhaps most important of all, he has discovered the exceedingly curious fact that bone is so much less transparent to these radiations than flesh and muscle, that if a living human hand be interposed between a Crookes tube and a photographic plate, a shadow photograph can be obtained which shows all the outlines and joints of the bones most distinctly.

Working upon the lines indicated in the telegrams from Vienna, recently published in the daily papers, I have, with the assistance of Mr. J. C. M. Stanton, repeated many of Prof. Röntgen's experiments with entire success. According to one of our first experiments, an ordinary gelatinous bromide dry photographic plate was placed in an ordinary camera back. The wooden shutter of the back was kept closed, and upon it were placed miscellaneous articles such as coins, pieces of wood, carbon, ebonite, vulcanised fibre, aluminium, &c., all being quite opaque to ordinary light. Above was supported a Crookes tube, which was excited for some minutes. On development, shadows of all the articles

伦琴教授的发现

斯温顿

编者按

距离伦琴最初宣布发现 X 射线还不到一个月，就有了这篇对伦琴的发现的查证，这是一个应该引起人们警醒的事例，它表明科学界同行对重大发现的态度也会是冷淡的。坎贝尔–斯温顿含蓄地指出，报业为之感到兴奋的现象实际上并不是一个“全新的”现象。但他之所以这样说是因为他在某种程度上误会了伦琴得到的结论。斯温顿坚持把 X 射线看作是“阴极射线的一部分”，并指出人们早就知道阴极射线能够用于拍摄照片，可他没有注意到伦琴所称的 X 射线根本就不是阴极射线。

前一段时间，关于伦琴教授的实验的新闻报道引起了相当广泛的关注。不过，这一发现似乎并不是全新的，因为赫兹就曾注意到从克鲁克斯管或希托夫管中发射出来的阴极射线能够穿透金属薄片，而大约两年前莱纳德就在其发表的研究报告中明确地指出这种射线可以产生影像。莱纳德使用了一个带铝窗的管子，阴极射线可以比较容易地从此窗中穿出，伦琴的实验中则是射线穿过了插在管窗与照相干版之间的纸板和铝片。实际上，莱纳德得到了与伦琴的结果几乎完全一样的阴影图像。

不过，伦琴教授已经阐明这个铝窗并不是必需的，因为一部分能够引起成像的阴极辐射是从玻璃管壁中穿出的。此外，他还以一种能给人们留下深刻印象的方式推广了莱纳德得到的结果，而也许最为重要的是，他发现了一个极为新奇的现象，即这种辐射穿透骨骼的能力比穿透肌肉的能力差很多，如果将活人的手置于克鲁克斯管与照相干版之间，就能得到一张非常清晰地显示出骨骼关节轮廓的阴影图像。

最近，许多日报都刊载了来自维也纳的电报，根据其中提供的线索，在斯坦顿先生的协助下，我已经完全成功地重复了伦琴教授的很多实验。在最初的一次实验中，我们将一张普通的凝胶溴化物照相干版放置在普通相机后面。背面的木质快门始终保持关闭状态，并紧接着放置各种物品，诸如硬币、木块、炭、硬质橡胶、硬化纤维和铝等，所有这些物品对于普通的可见光都是完全不透明的。在这些物品上方固定一个已经激发了几分钟的克鲁克斯管。显影后，放置的所有物品的阴影都清

placed on the slide were clearly visible, some being more opaque than others. Further experiments were tried with thin plates of aluminium or of black vulcanised fibre interposed between the objects to be photographed and the sensitive surface, this thin plate being used in place of the wood of the camera back. In this manner sharper shadow pictures were obtained. While most thick metal sheets appear to be entirely opaque to the radiations, aluminium appears to be relatively transparent. Ebonite, vulcanised fibre, carbon, wood, cardboard, leather and slate are all very transparent, while, on the other hand, glass is exceedingly opaque. Thin metal foils are moderately opaque, but not altogether so.

As tending to the view that the radiations are more akin to ultraviolet than to infra-red light, it may be mentioned that a solution of alum in water is distinctly more transparent to them than a solution of iodine in bisulphide of carbon.

So far as our own experiments go, it appears that, at any rate without very long exposures, a sufficiently active excitation of the Crookes tube is not obtained by direct connection to an ordinary Rhumkorff induction coil, even of a large size. So-called high frequency currents, however, appear to give good results, and our own experiments have been made with the tube excited by current obtained from the secondary circuit of a Tesla oil coil, through the primary of which were continuously discharged twelve half-gallon Leyden jars, charged by an alternating current of about 20,000 volts pressure, produced by a transformer with a spark-gap across its high-pressure terminals.

For obtaining shadow photographs of inanimate objects, and for testing the relative transparency of different substances, the particular form of Crookes tube employed does not appear to greatly signify, though some forms are, we find, better than others. When, however, the human hand is to be photographed, and it is important to obtain sharp shadows of the bones, the particular form of tube used and its position relative to the hand and sensitive plate appear to be of great importance. So far, owing to the frequent destruction of the tubes, due to overheating of the terminals, we have not been able to ascertain exactly the best form and arrangement for this purpose, except that it appears desirable that the electrodes in the tube should consist of flat and not curved plates, and that these plates should be of small dimensions.

The accompanying photograph of a living human hand (Fig. 1) was exposed for twenty minutes through an aluminium sheet 0.0075 in thickness, the Crookes tube, which was one of the kind containing some white phosphorescent material (probably sulphide of barium), being held vertically upside down, with its lowest point about two inches above the centre of the hand.

晰可见，其中一些物品的阴影比其他的更加明显。在后来的实验中，我们尝试着将薄铝板或黑色硬化纤维薄板插入待成像的物体与感光表面之间，这一薄板用来代替相机后的木质快门。用这种方式我们得到了更加清晰的阴影图像。大部分厚金属板对于这种辐射似乎都是完全不透明的，而铝板则似乎比较透明。硬质橡胶、硬化纤维、炭、木块、纸板、皮革以及石板都是非常透明的，相反，玻璃则是非常不透明的。薄金属板是中等透明的，不过也不全是这样。

为了支持该辐射更类似于紫外线而不是红外线的观点，我们要说明一下，与碘的二硫化碳溶液相比，明矾的水溶液对该辐射的透明度明显好得多。

就我们的实验情况来说，如果不进行很长时间的曝光，单靠将一个普通的拉姆科夫感应线圈与克鲁克斯管直接相连，即使是用大号的线圈，看起来似乎也无法使克鲁克斯管产生足够引起成像活性的激发辐射。不过，我们常说的高频电流看来能够给出好的结果。我们在实验中使用的克鲁克斯管，是由特斯拉油线圈的次级电路产生的电流来激发的，通过其初级电路的是连续放电的 12 个半加仑莱顿瓶，这些莱顿瓶通过电压约为 20,000 伏特的交流电进行充电，此交流电由一高压端带有放电间隙的变压器产生。

要获得无生命物体的阴影或检验不同物质对此种辐射的相对透明度，使用哪种结构的克鲁克斯管看起来并不是非常要紧，尽管我们发现某些结构的管子比另外一些好一点。但是，在获取人手的影像时，重要的是得到骨骼的清晰阴影，那么所使用的管子的特殊结构以及管子相对于人手和感光干版的位置就显得至关重要了。到目前为止，由于管子经常因其末端过热而毁坏，我们还没能弄清楚对于上述目的来说什么样的管子结构和摆放位置是最好的，不过能够确定的是，管子中的电极应该采用平板电极而不是曲面电极，并且应该用尺寸较小的电极板。

本文所附的活人手掌的影像（图 1）是该辐射穿过厚度为 0.0075 的铝片持续曝光 20 分钟而得到的，实验中使用的克鲁克斯管中包含某种白色磷光物质（可能是硫化钡），管子颠倒后垂直放置，其最低点位于掌心上方大约 2 英寸处。

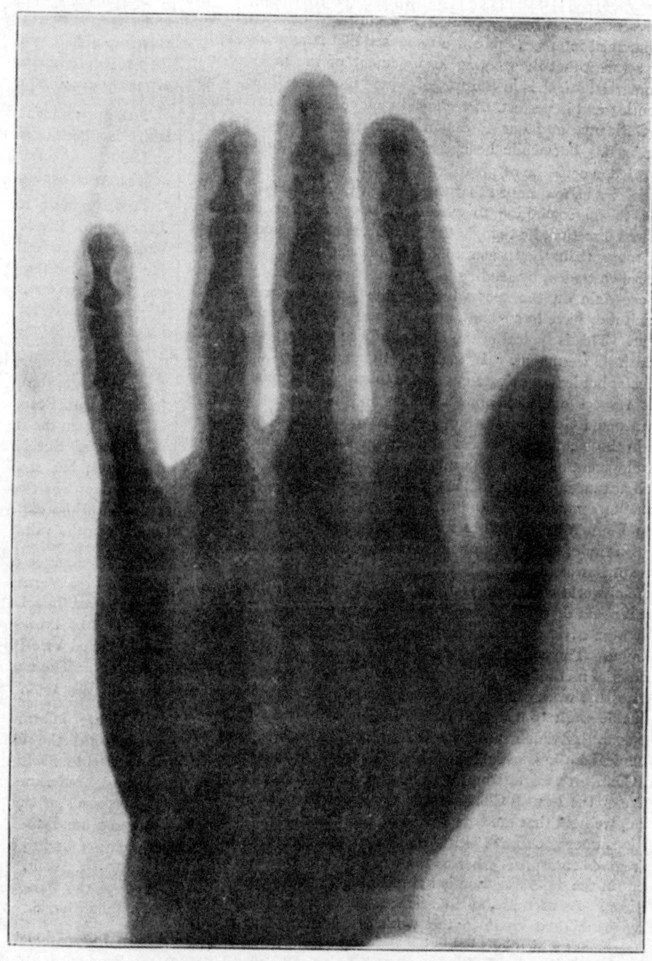

Fig. 1. Photograph of a living human hand

By substituting a thin sheet of black vulcanised fibre for the aluminium plate, we have since been able to reduce the exposure required to four minutes. Indeed with the aluminium plate, the twenty minutes' exposure appears to have been longer than was necessary. Further, having regard to the great opacity of glass, it seems probable that where ordinary Crookes tubes are employed, a large proportion of the active radiations must be absorbed by the glass of the tube itself. If this is so, by the employment of a tube partly constructed of aluminium, as used by Lenard, the necessary length of exposure could be much reduced.

(**53**, 276-277; 1896)

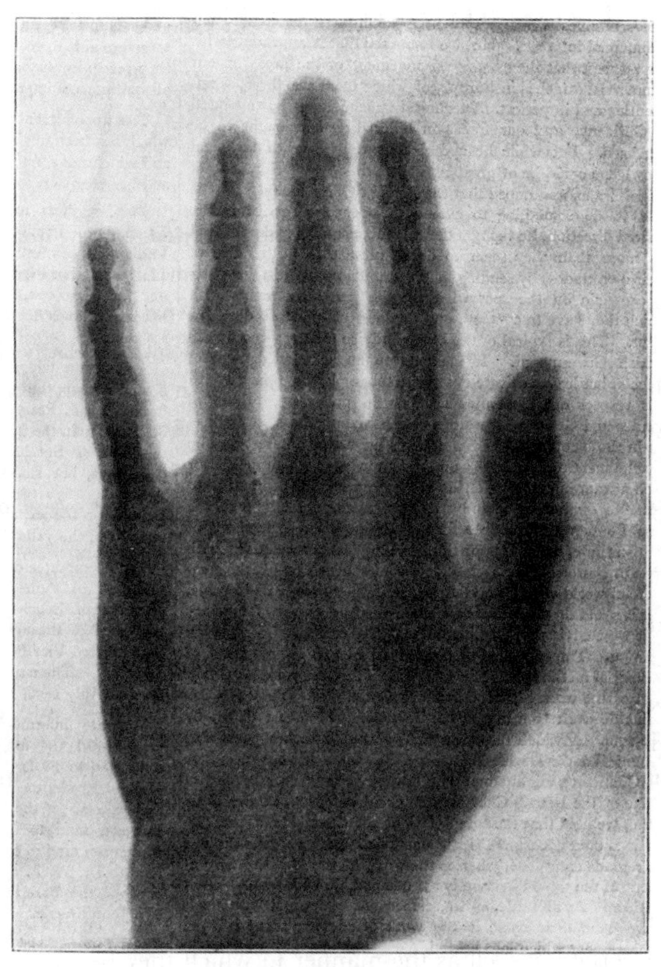

图 1. 活人手掌的影像

用一块黑色硬化纤维薄板代替铝板后，我们发现可以把曝光时间缩短到 4 分钟。实际上，在使用铝板的情况下，20 分钟的曝光时间似乎也比必需的曝光时间长一些。此外，考虑到玻璃很显著的不透明性，看来在使用普通克鲁克斯管时，大部分能够形成影像的辐射一定被管子自身的玻璃壁吸收了。如果确实如此，那么使用部分为铝质材料的管子（如同莱纳德所用的那样）的话，必需的曝光时间应该会大大缩短。

（王耀杨 翻译；江丕栋 审稿）

New Experiments on the Kathode Rays*

M. J. Perrin

Editor's Note

Physicists were puzzled by cathode rays, which carried energy from a negative electrode (cathode) toward a positive electrode inside a vacuum tube. Experimenters had determined that they originated at the cathode, but could not say what they were. In this classic paper the French physicist Jean Perrin reports experiments that helped to clarify the mystery. He placed into a cathode ray tube a metal cylinder linked to an electroscope, which would measure any charge deposited into it. When the cathode rays were directed into the tube, Perrin detected a significant negative charge. Perrin speculated that the charge carriers were negative ions created near the cathode. In fact they were electrons, as J. J. Thomson discovered one year later.

(1) Two hypotheses have been propounded to explain the properties of the kathode rays.

Some physicists think with Goldstein, Hertz, and Lenard, that this phenomenon is like light, due to vibrations of the ether †, or even that it is light of short wavelength. It is easily understood that such rays may have a rectilinear path, excite phosphorescence, and affect photographic plates.

Others think, with Crookes and J. J. Thomson, that these rays are formed by matter which is negatively charged and moving with great velocity, and on this hypothesis their mechanical properties, as well as the manner in which they become curved in a magnetic field, are readily explicable.

This latter hypothesis has suggested to me some experiments which I will now briefly describe, without for the moment pausing to inquire whether the hypothesis suffices to explain all the facts at present known, and whether it is the only hypothesis that can do so. Its adherents suppose that the kathode rays are negatively charged; so far as I know, this electrification has not been established, and I first attempted to determine whether it exists or not.

(2) For that purpose I had recourse to the laws of induction, by means of which it is possible to detect the introduction of electric charges into the interior of a closed electric conductor, and to measure them. I therefore caused the kathode rays to pass into a

---

* Translation of a paper by M. Jean Perrin, read before the Paris Academy of Sciences on December 30, 1895.

† These vibrations might be something different from light; recently M. Jaumann, whose hypotheses have since been criticised by M. H. Poincaré, supposed them to be longitudinal.

关于阴极射线的新实验*

佩兰

编者按

物理学家们对在真空管中把能量从负极（阴极）带到正极的阴极射线感到迷惑不解。实验可以证明它们来自阴极，但不能说明它们到底是什么。法国物理学家让·佩兰在这篇经典论文中用实验揭开了阴极射线的神秘面纱。他在阴极射线管内放置了一个与验电器相连的金属圆柱体，验电器可以测量进入其中的电荷。当阴极射线进入真空管时，佩兰检测到了大量的负电荷。佩兰推测这些载流子是在阴极附近产生的负离子。实际上它们就是一年之后汤姆逊发现的电子。

（1）现在可以解释阴极射线性质的假说有两种。

一部分物理学家和戈尔德施泰因、赫兹、莱纳德的意见一致，认为这种现象和光一样，是由以太的振动引起的†，或者它就是一种短波长的光。很容易就可以理解，这种射线可能是沿直线传播的，能激发磷光，而且可以使照相干版感光。

另一部分人则与克鲁克斯、汤姆逊持相同的观点，认为这种射线是由带负电的物质组成，并以极快的速度运动。用这种假说可以解释它们的力学性质，也可以很容易地说明为什么它们在磁场中的路径会弯曲。

后一种假说启发我进行了一些实验，我将在这里简单地描述这些实验，暂时先不管这个假说是否能解释目前所有已知的现象，或者是否只有这一种假说可以解释这些现象。它的支持者认为阴极射线是带负电的，而据我所知，这种带电性还没有被确认，我首先要确定它是否带电。

（2）为了达到这个目的，我将借助电磁感应定律，用这个定律，我们可以检测引入闭合导电体内部的电量，并且进行定量测量。因此，我让阴极射线通过法拉第

* 这篇文章翻译自让·佩兰在1895年12月30日向巴黎科学院宣读的论文。

† 这种振动可能与光有些不同。最近，尧曼（其猜想曾经被普安卡雷批判过）提出这种振动可能是纵向的。

Faraday's cylinder. For this purpose I employed the vacuum tube represented in Fig. 1. A B C D is a tube with an opening α in the centre of the face B C. It is this tube which plays the part of a Faraday's cylinder. A metal thread soldered at S to the wall of the tube connects this cylinder with an electroscope.

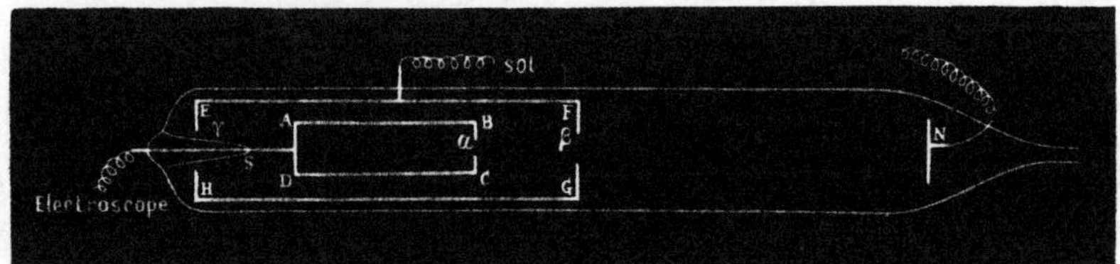

Fig. 1

E F G H is a second cylinder in permanent communication with the earth, and pierced by two small openings at β and γ; it protects the Faraday's cylinder from all external influence. Finally, at a distance of about 0.10 m in front of F G , was placed an electrode N. The electrode N served as kathode; the anode was formed by the protecting cylinder E F G H; thus a pencil of kathode rays passed into the Faraday's cylinder. This cylinder invariably became charged with negative electricity.

The vacuum tube could be placed between the poles of an electro-magnet. When this was excited, the kathode rays, becoming deflected, no longer passed into the Faraday's cylinder, and this cylinder was then not charged; it, however, became charged immediately the electromagnet ceased to be excited.

In short, the Faraday's cylinder became negatively charged when the kathode rays entered it, and only when they entered it; *the kathode rays are then charged with negative electricity*.

The quantity of electricity which these rays carry can be measured. I have not finished this investigation, but I shall give an idea of the order of magnitude of the charges obtained when I say that for one of my tubes, at a pressure of 20 microns of mercury, and for a single interruption of the primary of the coil, the Faraday's cylinder received a charge of electricity sufficient to raise a capacity of 600 C. G. S. units to 300 volts.

(3) The kathode rays being negatively charged, the principle of the conservation of electricity drives us to seek somewhere the corresponding positive charges. I believe that I have found them in the very region where the kathode rays are formed, and that I have established the fact that they travel in the opposite direction, and fall upon the kathode. In order to verify this hypothesis, it is sufficient to use a hollow kathode pierced with a small opening by which a portion of the attracted positive electricity might enter. This electricity could then act upon a Faraday's cylinder inside the kathode.

圆筒。为此我设计了如图 1 所示的真空管。A B C D 是一根管子，在 B C 面中心的 α 处有一个小孔。正是这根管子起到了法拉第圆筒的作用。一根焊接在管壁 S 处的金属线将圆筒和外部的验电器连接起来。

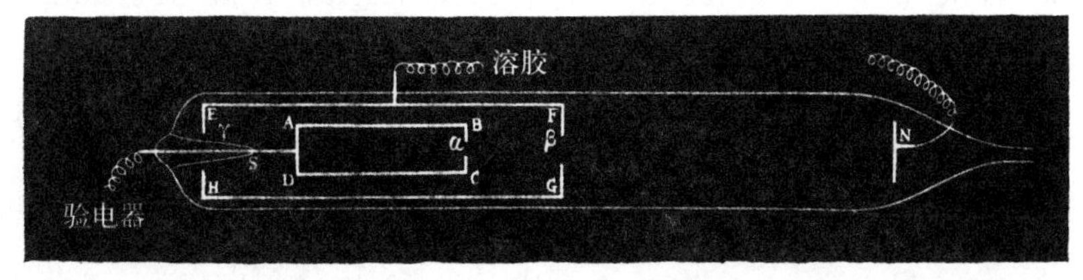

图 1

E F G H 是另一个圆筒，永久接地，并且在 β、γ 处穿有两个小孔；这个圆筒可以屏蔽外界对法拉第圆筒的干扰。最后，在 F G 前面大约 0.1 米的地方有一个电极 N。电极 N 作为阴极，屏蔽圆筒 E F G H 作为阳极。在这样的条件下，将一束阴极射线通入法拉第圆筒，这个圆筒将一直带负电。

真空管可以放置在电磁铁的两极之间。当电磁铁通电时，阴极射线将发生偏转，不能再通入法拉第圆筒，这个圆筒也将不再带电，而当电磁铁断电之后，法拉第圆筒马上又带电了。

简单地说就是，当且仅当有阴极射线进入时，法拉第圆筒带负电，**所以阴极射线一定带负电**。

射线所带的电量可以被测量出来。我还没有完成这项研究，但是我可以给出一个有关所获电量的数量级的概念，对于一个压力为 20 微米汞柱的真空管，将初级线圈截断，法拉第圆筒接收的电量足以使 600 单位（厘米克秒制）的电容器的电势差提高到 300 伏特。

（3）阴极射线带负电，根据电荷守恒定律，我们应该能在某处找到相应的正电荷。我确信我已经在阴极射线产生的地方找到了正电荷，我认为它们向相反的方向运动，而后撞在了阴极上。为了证明这种说法，只要用一个中空的阴极就行，在阴极上穿一个小孔，以使一部分被吸引过来的正电荷可以由此通过。进入阴极的正电荷会影响阴极内部的法拉第圆筒。

The protecting cylinder E F G H with its opening β fulfilled these conditions, and this time I therefore employed it as the kathode, the electrode N being the anode. The Faraday's cylinder is then invariably charged with *positive electricity*. The positive charges were of the order of magnitude of the negative charges previously obtained.

Thus, at the same time as negative electricity is *radiated* from the kathode, positive electricity travels towards that kathode.

I endeavoured to determine whether this positive flux formed a second system of rays absolutely symmetrical to the first.

(4) For that purpose I constructed a tube (Fig. 2) similar to the preceding, except that between the Faraday's cylinder and the opening β was placed a metal diaphragm pierced with an opening β', so that the positive electricity which entered by β could only affect the Faraday's cylinder if it also traversed the diaphragm β'. Then I repeated the preceding experiments.

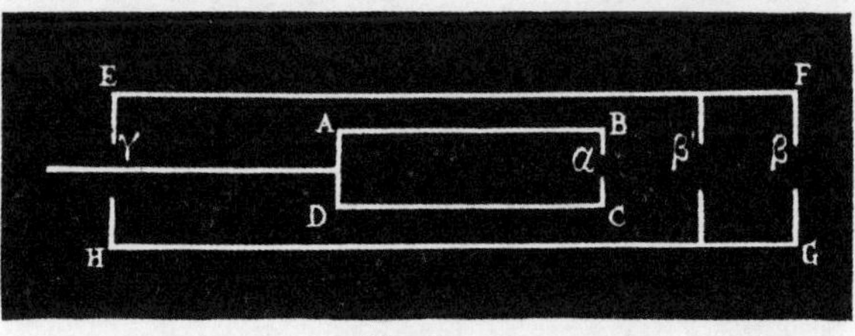

Fig. 2

When N was the kathode, the rays emitted from the kathode passed through the two openings β and β' without difficulty, and caused a strong divergence of the leaves of the electroscope. But when the protecting cylinder was the kathode, the positive flux, which, according to the preceding experiment, entered at β, did not succeed in separating the gold leaves except at very low pressures. When an electrometer was substituted for the electroscope, it was found that the action of the positive flux was real but very feeble, and increased as the pressure decreased. In a series of experiments at a pressure of 20 microns, it raised a capacity of 2,000 C. G .S. units to 10 volts; and at a pressure of 3 microns, during the same time, it raised the potential to 60 volts.*

By means of a magnet this action could be entirely suppressed.

(5) These results as a whole do not appear capable of being easily reconciled with the theory which regards the kathode rays as an ultra-violet light. On the other hand, they agree well with the theory which regards them as a material radiation, and which, as it appears to me, might be thus enunciated.

* The breaking of the tube has temporarily prevented me from studying the phenomenon at lower pressures.

带有小孔 β 的屏蔽圆筒 E F G H 满足以上条件，因此这一回我用 E F G H 作阴极，电极 N 作阳极。这样，法拉第圆筒就会一直带**正电**。其所带正电荷和前面所测的负电荷的数量级相同。

这说明，在阴极**发射**负电荷的同时，正电荷也在向阴极运动。

我下决心要确定这种正电流是否能形成另一个和阴极射线完全对称的射线系统。

(4) 为此我构造了一个和前面类似的管子（见图 2），与前面管子唯一的不同之处是，在法拉第圆筒和小孔 β 之间放置了一个金属膜片，膜片上有一个小孔 β'，这样，从 β 进入的正电荷只有在也通过 β' 的情况下才能作用于法拉第圆筒。我用这个装置重复了上面的实验。

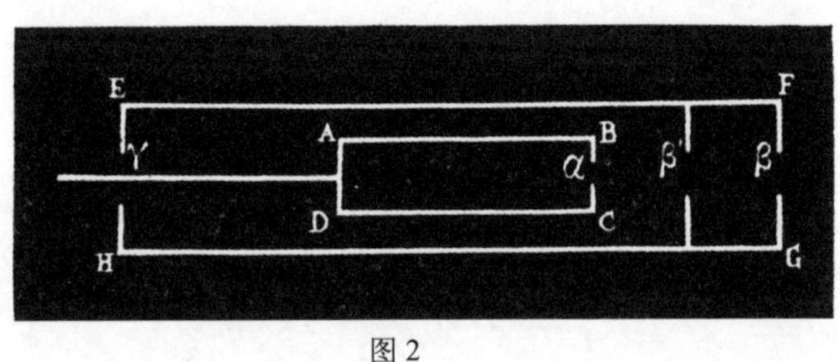

图 2

当 N 作为阴极时，由阴极发射的射线可以顺利地通过 β 和 β' 两个孔，使验电器的两个叶片张得很大。但是当屏蔽圆筒作为阴极时，根据前述的实验，通过 β 进入的正电流并没有使验电器的金箔张开，除非是在压力很低的情况下。当用静电计代替验电器时，可以看到正电流的确存在但非常微弱，并随着压力的减小而增大。在 20 微米汞柱条件下进行的一系列实验中，这一正电流可以把 2,000 单位（厘米克秒制）的电容器的电势差提高到 10 伏特；在压力为 3 微米汞柱时，同样时间内电势差被提高到了 60 伏特。*

利用磁铁可以完全地抑制这种作用。

(5) 总的来说，把阴极射线看作是紫外线的假说似乎不太容易解释这些实验结果。另一方面，这些实验结果与认为阴极射线是一种物质辐射的假说符合得很好，在我看来，这些实验结果恐怕只能这么解释。

* 真空管的爆裂使我暂时无法再在低压条件下研究这一现象。

In the neighbourhood of the kathode, the electric field is sufficiently intense to break into pieces (*into ions*) certain of the molecules of the residual gas. The negative ions move towards the region where the potential is increasing, acquire a considerable speed, and form the kathode rays; their electric charge, and consequently their mass (at the rate of one valence-gramme for 100,000 Coulombs) is easily measurable. The positive ions move in the opposite direction; they form a diffused brush, sensitive to the magnet, and not a radiation in the correct sense of the word.*

(**53**, 298-299; 1896)

* This work has been carried out in the laboratory of the Normal School, and in that of M. Pellat at the Sorbonne.

在阴极附近，电场强度强到足以把一定量的残余气体分子打成碎片（**变成离子**）。负离子向着电势增加的方向运动，速度很大，这形成了阴极射线，它们的电量很容易被测定，从而其质量（100,000 库仑对应 1 克当量）也很容易得到。正离子向相反方向运动，形成一个发散的尾巴，对磁场非常敏感，准确地说这就不是辐射了。*

（王锋 翻译；江丕栋 审稿）

* 这项研究是在师范学院的实验室和索邦大学佩拉的实验室中进行的。

The Effect of Magnetisation on the Nature of Light Emitted by a Substance*

P. Zeeman

Editor's Note

Does a magnetic field influence the light emitted by an atom? Here Pieter Zeeman reports the first evidence that it does. Zeeman heated sulphur in a ceramic chamber with transparent ends, and placed the chamber in a magnetic field. With the light of an arc lamp, he then measured the absorption spectrum and found a broadening of certain lines, attributing this to a change in the frequency of the absorbed light. Zeeman noted that the polarization of light emitted in the presence of a field behaves as predicted by Lorentz, owing to the circular motion of charged particles within the atom. He estimates the charge/mass ratio for these particles as being about 10^7.

IN consequence of my measurements of Kerr's magneto-optical phenomena, the thought occurred to me whether the period of the light emitted by a flame might be altered when the flame was acted upon by magnetic force. It has turned out that such an action really occurs. I introduced into an oxyhydrogen flame, placed between the poles of a Ruhmkorff's electromagnet, a filament of asbestos soaked in common salt. The light of the flame was examined with a Rowland's grating. Whenever the circuit was closed both D lines were seen to widen.

Since one might attribute the widening to the known effects of the magnetic field upon the flame, which would cause an alteration in the density and temperature of the sodium vapour, I had resort to a method of experimentation which is much more free from objection.

Sodium was strongly heated in a tube of biscuit porcelain, such as Pringsheim used in his interesting investigations upon the radiations of gases. The tube was closed at both ends by plane parallel glass plates, whose effective area was 1 cm. The tube was placed horizontally between the poles, at right angles to the lines of force. The light of an arc lamp was sent through. The absorption spectrum showed both D lines. The tube was continuously rotated round its axis to avoid temperature variations. Excitation of the magnet caused immediate widening of the lines. It thus appears very probable that the period of sodium light is altered in the magnetic field. It is remarkable that Faraday, as early as 1862, had made the first recorded experiment in this direction, with the incomplete resources of that period, but with a negative result (Maxwell, "Collected Works", vol. II, p. 790).

---

* Translated by Arthur Stanton from the *Proceedings of the Physical Society of Berlin.*

磁化对物质发射的光的性质的影响*

塞曼

编者按

磁场会影响原子发射的光吗？彼得·塞曼在这篇报告中首次证明这种效应是存在的。塞曼在两端透明的陶瓷真空室中加热硫磺，并把这个真空室放入磁场中。在弧光灯的照射下，他测量了吸收光谱并发现某些特定的谱线出现了加宽的现象，他把这归因于被吸收光线的频率的改变。塞曼特别提到，有场存在时发射出来的光的偏振与洛仑兹预言的一样，是由原子内带电粒子的圆周运动产生的。他估计这些粒子的荷质比约为 10^7。

我在对克尔磁光效应进行测量时突然产生了这样的想法：当磁力作用于火焰时，火焰发射出的光的周期是否会发生变化。结果证实这样的作用确实存在。我把浸泡在普通食盐中的石棉丝放在置于鲁姆科夫电磁体两极之间的氢氧焰中。火焰光用罗兰光栅检验。每当电路接通时都能看到两条 D 线的加宽。

鉴于也许有人会将谱线加宽归因于磁场对火焰的某种已知作用使钠蒸气的密度和温度发生了变化，我已采用了更加没有异议的实验方法进行了确证。

我们在素瓷管（与普林斯海姆在他著名的气体辐射实验中所用的一样）内对钠进行高温加热。管的两端用两块相互平行的平玻璃板密封，其有效区域为 1 厘米。该管被水平地置于两极之间，与磁力线垂直。弧光灯的光穿过其中，吸收光谱中显示出两条 D 线。管子不停地绕着它的轴自转以保持各处温度均衡，磁作用使谱线迅速加宽。很可能是因为钠光的周期在磁场中发生了变化。值得注意的是，这方面第一个有记录的实验是法拉第早在 1862 年进行的，那时的资源并不完备，得到的是阴性的结果（麦克斯韦，《文集》，第 2 卷，第 790 页）。

---

* 由阿瑟·斯坦顿翻译自《柏林物理学会会刊》。

It has been already stated what, in general, was the origin of my own research on the magnetisation of the lines in the spectrum. The possibility of an alteration of period was first suggested to me by the consideration of the accelerating and retarding forces between the atoms and Maxwell's molecular vortices; later came an example suggested by Lord Kelvin, of the combination of a quickly rotating system and a double pendulum. However, a true explanation appears to me to be afforded by the theory of electric phenomena propounded by Prof. Lorentz.

In this theory, it is considered that, in all bodies, there occur small molecular elements charged with electricity, and that all electrical processes are to be referred to the equilibrium or motion of these "ions". It seems to me that in the magnetic field the forces directly acting on the ions suffice for the explanation of the phenomena.

Prof. Lorentz, to whom I communicated my idea, was good enough to show me how the motion of the ions might be calculated, and further suggested that if my application of the theory be correct there would follow these further consequences: that the light from the edges of the widened lines should be circularly polarised when the direction of vision lay along the lines of force; further, that the magnitude of the effect would lead to the determination of the ratio of the electric charge the ion bears to its mass. We may designate the ratio e/m. I have since found by means of a quarter-wave length plate and an analyser, that the edges of the magnetically-widened lines are really circularly polarised when the line of sight coincides in direction with the lines of force. An altogether rough measurement gives 10^7 as the order of magnitude of the ratio e/m when e is expressed in electromagnetic units.

On the contrary, if one looks at the flame in a direction at right angles to the lines of force, then the edges of the broadened sodium lines appear plane polarised, in accordance with theory. Thus there is here direct evidence of the existence of ions.

This investigation was conducted in the Physical Institute of Leyden University, and will shortly appear in the "Communications of the Leyden University".

I return my best thanks to Prof. K. Onnes for the interest he has shown in my work.

(**55**, 347; 1897)

P. Zeeman: Amsterdam.

前面已经介绍了我对谱线磁化进行研究的起因。周期变化的可能性使我首先想到的是原子与麦克斯韦分子涡旋之间的加速和减速作用力；然后想到的是开尔文勋爵提出的一个快速旋转体系与双摆复合体的例子。然而，使我受到启发并最终得出正确结论的是洛仑兹教授提出的关于电现象的理论。

这个理论认为：在所有物体中，都存在小的、带电的分子单元，所有电的过程都与这些“离子”的平衡或运动有关。在我看来，只要认为在磁场中力直接作用于这些离子上，就足以解释这些现象。

我向洛仑兹教授阐述了我的观点，他友好地告诉我离子如何运动也许是可以计算的，并进一步建议说，如果我对该理论的应用是正确的，那么就会出现以下结果：当沿磁力线方向观察时，从加宽谱线边缘发出的光应该是圆偏振光；此外，这个效应的大小将能使人们测定离子所带电荷与其质量的比值。我们可以用 e/m 表示这个比值。后来我用四分之一波片和检偏器测量发现，当观测方向与磁力线一致时，磁场加宽谱线的边缘果然是圆偏振的。粗略的测定表明，如果用 e 来表示电磁单位，e/m 这一比值的数量级大约为 10^7。

反之，如果观察火焰的方向与磁力线垂直，加宽的钠线边缘出现的是平面偏振光，这与理论相符。这些都是离子存在的直接证据。

这项研究是在莱顿大学物理研究所进行的，不久之后研究报告将刊登在《莱顿大学学报》上。

非常感谢昂内斯教授对我的工作的重视。

(沈乃澂 翻译；赵见高 审稿)

Intra-atomic Charge

F. Soddy

Editor's Note

What was the internal structure of an atom? While Ernest Rutherford's experiments in 1911 had convinced him that the atom contained a dense, positively charged nucleus, others were not so sure. Here Rutherford's sometime collaborator Frederick Soddy suggested that the nucleus must also contain negative charges, expelled during so-called radioactive beta decay. Soddy introduces the term "isotope": atoms essentially identical in their chemical properties but with differing nuclei. For any given nuclear charge, he asserted, an atom may have any number of electrons in an "outer ring system". Changes in this number are a consequence of chemical action, with no effect on the nucleus. Clarification of this view awaited the discovery of the proton and neutron.

THAT the intra-atomic charge of an element is determined by its place in the periodic table rather than by its atomic weight, as concluded by A. van der Broek (*Nature*, November 27, p. 372), is strongly supported by the recent generalisation as to the radio-elements and the periodic law. The successive expulsion of one α and two β particles in three radio-active changes in any order brings the intra-atomic charge of the element back to its initial value, and the element back to its original place in the table, though its atomic mass is reduced by four units. We have recently obtained something like a direct proof of van der Broek's view that the intra-atomic charge of the nucleus of an atom is not a purely positive charge, as on Rutherford's tentative theory, but is the difference between a positive and a smaller negative charge.

Fajans, in his paper on the periodic law generalisation (*Physikal. Zeitsch.*, 1913, vol. XIV, p. 131), directed attention to the fact that the changes of chemical nature consequent upon the expulsion of α and β particles are precisely of the same kind as in ordinary electrochemical changes of valency. He drew from this the conclusion that radio-active changes must occur in the same region of atomic structure as ordinary chemical changes, rather than with a distinct inner region of structure, or "nucleus", as hitherto supposed. In my paper on the same generalisation, published immediately after that of Fajans (*Chem. News*, February 28), I laid stress on the absolute identity of chemical properties of different elements occupying the same place in the periodic table.

A simple deduction from this view supplied me with a means of testing the correctness of Fajans's conclusion that radio-changes and chemical changes are concerned with the same region of atomic structure. On my view his conclusion would involve nothing else than that, for example, uranium in its tetravalent uranous compounds must be chemically

原子内的电荷

索迪

编者按

原子的内部结构是怎样的？1911 年，当欧内斯特·卢瑟福用实验验证了原子包含一个致密的带正电的核时，其他人并没有表示十分肯定。在这篇文章中，曾经与卢瑟福合作过的弗雷德里克·索迪提出原子核中必须同时也包含负电荷，这些负电荷会在放射性 β 衰变时发射出去。索迪引入了"同位素"的概念，即原子核结构不同但化学性质基本一致的原子。他认为，对于任意给定的核电荷，原子"外层系统"可以排布任意的电子数量。化学反应能引起外层电子数的变化，但对原子核没有影响。后来质子和中子的发现证实了他的观点。

范德布鲁克断言（《自然》，11 月 27 日，第 372 页），一种元素原子内的电荷是由它在周期表中的位置而不是它的原子量确定的，这一论断受到最近一些关于放射性元素与周期律的结论的强力支持。如果在 3 次放射性变化中相继发射出 1 个 α 粒子和 2 个 β 粒子，那么不管这 3 次衰变的次序如何，都会使该元素的原子内电荷回到初始数值，元素也回到了它在周期表中的原始位置，但是它的原子质量却减少了 4 个单位。最近我们获得了一些证据，能够直接支持范德布鲁克的观点：就像卢瑟福的初步理论所指出的，原子核内的电荷并不是单纯的正电荷，而是正电荷与较小的负电荷的差值。

法扬斯在他那篇关于周期律的一般法则的论文（《物理学杂志》，1913 年，第 14 卷，第 131 页）中，特别指出了如下事实：由于 α 粒子和 β 粒子的发射而引起的元素化学性质的改变，与普通的会发生价态变动的由电化学变化引起的物质化学性质的改变完全属于同一类型。他由此得出的结论是，放射性变化必定与普通的化学变化一样发生在原子结构的同一区域，而不是像我们目前所假设的——放射性变化发生在被称为"核"的一个完全不同的内部区域。在法扬斯的论文发表后不久，我也针对同一主题发表了一篇论文（《化学新闻》，2 月 28 日），文中我将重点放在了周期表中处于同一位置的不同元素具有完全相同的化学性质这一点上。

此观点的一个简单推论为我提供了一种检验法扬斯认为的放射性变化与化学变化发生于原子结构中同一区域这一观点是否正确的方法。我认为，他的结论其实就是下面的意思：举例来说，铀的四价化合物中的铀元素必定与钍化合物中的钍元

identical with and non-separable from thorium compounds. For uranium X, formed from uranium I by expulsion of an α particle, is chemically identical with thorium, as also is ionium formed in the same way from uranium II. Uranium X loses two β particles and passes back into uranium II, chemically identical with uranium. Uranous salts also lose two electrons and pass into the more common hexavalent uranyl compounds. If these electrons come from the same region of the atom uranous salts should be chemically non-separable from thorium salts. But they are not.

There is a strong resemblance in chemical character between uranous and thorium salts, and I asked Mr. Fleck to examine whether they could be separated by chemical methods when mixed, the uranium being kept unchanged throughout in the uranous or tetravalent condition. Mr. Fleck will publish the experiments separately, and I am indebted to him for the result that the two classes of compounds can readily be separated by fractionation methods.

This, I think, amounts to a proof that the electrons expelled as β rays come from a nucleus not capable of supplying electrons to or withdrawing them from the ring, though this ring is capable of gaining or losing electrons from the exterior during ordinary electro-chemical changes of valency.

I regard van der Broek's view, that the number representing the net positive charge of the nucleus is the number of the place which the element occupies in the periodic table when all the possible places from hydrogen to uranium are arranged in sequence, as practically proved so far as the relative value of the charge for the members of the end of the sequence, from thallium to uranium, is concerned. We are left uncertain as to the absolute value of the charge, because of the doubt regarding the exact number of rare-earth elements that exist. If we assume that all of these are known, the value for the positive charge of the nucleus of the uranium atom is about 90. Whereas if we make the more doubtful assumption that the periodic table runs regularly, as regards numbers of places, through the rare-earth group, and that between barium and radium, for example, two complete long periods exist, the number is 96. In either case it is appreciably less than 120, the number were the charge equal to one-half the atomic weight, as it would be if the nucleus were made out of α particles only. Six nuclear electrons are known to exist in the uranium atom, which expels in its changes six β rays. Were the nucleus made up of α particles there must be thirty or twenty-four respectively nuclear electrons, compared with ninety-six or 102 respectively in the ring. If, as has been suggested, hydrogen is a second component of atomic structure, there must be more than this. But there can be no doubt that there must be some, and that the central charge of the atom on Rutherford's theory cannot be a pure positive charge, but must contain electrons, as van der Broek concludes.

素具有完全相同的化学性质且不可区分。因为，由铀 I 发射 1 个 α 粒子而形成的铀 X 与钍，以及由铀 II 以同样方式形成的“镤”，在化学性质上是一致的。铀 X 发射 2 个 β 粒子就又回到铀 II，铀 II 与铀的化学性质是一致的。亚铀盐也可以失去 2 个电子并形成更为常见的六价铀的化合物。如果这些电子来自原子中的同一区域，那么亚铀盐和钍盐应该具有相同的化学性质并且不可区分。然而事实并非如此。

亚铀盐与钍盐的化学性质非常相似，我已经请弗莱克先生研究是否可以在始终保持铀元素的四价或六价状态不发生改变的前提下用化学方法将铀盐与钍盐的混合物中的铀和钍分离开来。弗莱克先生将独立发表他的实验，我很感激能够使用他取得的结果，即通过分馏的方法可以很顺利地分离这两类化合物。

我认为这足以证明，以 β 射线形式发射出的电子来自核，尽管核无法为核外圈层提供电子或从中取走电子，但核外圈层可以在普通的会发生价态变动的电化学变化中从外部获取电子或失去电子。

在研究从铊到铀这些处在周期表序列尾部的成员的电荷相对值时，我把范德布鲁克的观点当作是已经被证实的，即从氢到铀按顺序排好周期表中所有可能的位置后，代表核所具有的净正电荷的数值就正好是元素在周期表中所处位置对应的数值。不过我们仍旧不能确定电荷的绝对值，因为对到底存在多少种稀土元素还存在疑问。如果我们假定这些都是已知的，那么铀原子核中正电荷的数值大约是 90。如果我们采取另一个更不确定的假设，即从周期表中位置对应的数值的角度来看，包括全部稀土元素以及介于钡与镭之间的元素在内的周期表是规则排布的，整个周期表有两个完整的长周期，这样得到的数值就会是 96。不管是哪种情况，该数值都明显小于 120，即便假设其电荷数值就是 120，那也只是其原子量的一半，如果真是这样，那核可能就只有 α 粒子了。已知铀原子中存在 6 个核电子，在放射性变化中这些核电子以 6 次 β 射线的形式发射出来。如果核是由 α 粒子构成的话，那么与核外圈层中具有 96 个或者 102 个电子相对应，就必须有 30 个或者 24 个核电子。另外已经有人提出氢是原子结构的另一种构件，如果考虑上这一点的话，那么其核电子数就不止于此了。但是有一点毋庸置疑，那就是必定存在一些核电子，而且卢瑟福理论中原子的中心电荷也不是单纯的正电荷，而是必定包含负电子，就像范德布鲁克断定的那样。

So far as I personally am concerned, this has resulted in a great clarification of my ideas, and it may be helpful to others, though no doubt there is little originality in it. The same algebraic sum of the positive and negative charges in the nucleus, when the arithmetical sum is different, gives what I call "isotopes" or "isotopic elements", because they occupy the same place in the periodic table. They are chemically identical, and save only as regards the relatively few physical properties which depend upon atomic mass directly, physically identical also. Unit changes of this nuclear charge, so reckoned algebraically, give the successive places in the periodic table. For any one "place," or any one nuclear charge, more than one number of electrons in the outer-ring system may exist, and in such a case the element exhibits variable valency. But such changes of number, or of valency, concern only the ring and its external environment. There is no in- and out-going of electrons between ring and nucleus.

(**92**, 399-400; 1913)

Frederick Soddy: Physical Chemistry Laboratory, University of Glasgow.

就目前我个人的思考结果来说，以上是对我的观点的一个很明晰的解释。虽然很明显其中并无多少创见，但也许会对其他人有所帮助吧。根据核内正负电荷的代数和相同而算数和不同的现象，我提出了“同位素”或“同位置元素”的概念，因为它们在周期表中处于相同的位置。它们在化学性质上是完全相同的，除了很有限的一些直接决定于原子量的物理性质外，其他大部分物理性质也是完全相同的。从数学角度来看，核电荷数会单位递增地发生变化，这使得周期表中的位置也连续地变化。对于周期表中任意一个位置，或者说任意一个确定的核电荷数，可以存在不止一种外层电子数，在这种情况下元素就表现出了不同的价态。不过，这种电子数或者价态的变化只是考虑了核外圈层及其外部环境，并没有把核外圈层与核之间的电子进出过程考虑进去。

（王耀杨 翻译；汪长征 审稿）

The Structure of the Atom

E. Rutherford

Editor's Note

Responding to a comment by Frederick Soddy, Ernest Rutherford here clarifies his view on the structure of the atomic nucleus. Soddy had suggested that Rutherford believed the nucleus to contain positive charges only. On the contrary, Rutherford insists, he believes only that the atomic nucleus is small, dense, and positively charged overall. Moreover, he thinks that two of the key products of radioactive decay—alpha and beta particles—might both originate from the nucleus. Rutherford supports a recent suggestion that the charge on the atomic nucleus is equal to the atomic number, and not to half the atomic weight. This observation prefigured the revelation that the atomic number is the equal to the number of protons in the nucleus.

IN a letter to this journal last week, Mr. Soddy has discussed the bearing of my theory of the nucleus atom on radio-active phenomena, and seems to be under the impression that I hold the view that the nucleus must consist entirely of positive electricity. As a matter of fact, I have not discussed in any detail the question of the constitution of the nucleus beyond the statement that it must have a resultant positive charge. There appears to me no doubt that the α particle does arise from the nucleus, and I have thought for some time that the evidence points to the conclusion that the β particle has a similar origin. This point has been discussed in some detail in a recent paper by Bohr (*Phil. Mag.*, September, 1913). The strongest evidence in support of this view is, to my mind, (1) that the β ray, like the α ray, transformations are independent of physical and chemical conditions, and (2) that the energy emitted in the form of β and γ rays by the transformation of an atom of radium C is much greater than could be expected to be stored up in the external electronic system. At the same time, I think it very likely that a considerable fraction of the β rays which are expelled from radio-active substances arise from the external electrons. This, however, is probably a secondary effect resulting from the primary expulsion of a β particle from the nucleus.

The original suggestion of van der Broek that the charge on the nucleus is equal to the atomic number and not to half the atomic weight seems to me very promising. This idea has already been used by Bohr in his theory of the constitution of atoms. The strongest and most convincing evidence in support of this hypothesis will be found in a paper by Moseley in *The Philosophical Magazine* of this month. He there shows that the frequency of the X radiations from a number of elements can be simply explained if the number of unit charges on the nucleus is equal to the atomic number. It would appear that the charge on the nucleus is the fundamental constant which determines the physical and chemical properties of the atom, while the atomic weight, although it approximately follows the order of the nucleus charge, is probably a complicated function of the latter depending on the detailed structure of the nucleus.

(**92**, 423; 1913)

E. Rutherford: Manchester, December 6, 1913.

原子结构

卢瑟福

编者按

为了回应弗雷德里克·索迪的意见，欧内斯特·卢瑟福在这篇文章中进一步解释了他对原子核结构的观点。索迪曾指出卢瑟福认为原子核只包含正电荷，而卢瑟福却强调说他只承认原子核很小、很致密，以及整体带正电。此外，卢瑟福认为，放射性衰变的两个主要产物，即 α 粒子和 β 粒子，可能都源自原子核。卢瑟福同意最近有人提出的关于核电荷等于原子序数而非原子量的一半的观点。这个结果预示了原子序数与原子核中的质子数相等这一关系。

在上周致贵刊的一封信中，索迪先生对我关于放射性现象中原子核的理论进行了相关讨论，他似乎以为，我认定原子核必须完全由带正电荷的粒子构成。事实上，我只是认为原子核必定具有总和为正的电荷，而对它的具体构成并未发表看法。在我看来，α 粒子无疑是产生于核的，而且经过一段时间的思考，我认为有证据表明 β 粒子也源自核。玻尔在最近的一篇论文（《哲学杂志》，1913 年 9 月）中对这一点进行了较为详细的讨论。支持这一观点的最强有力的证据是：(1) 与 α 射线一样，β 射线的衰变也是与物理和化学条件无关的；(2) 镭原子发生衰变时以 β 射线或 γ 射线的形式放出的能量 C，比预想的外部电子系统所存储的能量大得多。不过，放射性物质发射出的 β 射线可能有相当一部分来源于外部电子。这也许是原子核中的 β 粒子的初级辐射引起的一种次级效应。

范德布鲁克最先提出，核所带的基本电荷数应等于原子序数而不是原子量的一半，我认为这是非常有可能的。玻尔在他的原子结构理论中已经应用了这种观点。在莫塞莱本月发表于《哲学杂志》上的文章中，可以找到支持这个假说的最强有力也最令人信服的证据。文中表明，如果核所带的基本电荷数等于原子序数，就能方便地对多种元素所发出的 X 辐射的频率进行解释。看起来，核所带的基本电荷数是确定原子物理和化学性质的基本常数，尽管原子量的顺序与相应原子核电荷数的顺序基本一致，但原子量并非如人们以前所设想的那样是其核电荷数的两倍，可能是核电荷数的复杂函数，其函数关系与核的具体结构有关。

（王耀杨 翻译；鲍重光 审稿）

The Reflection of X-rays

Editor's Note

The German physicist Max von Laue demonstrated the phenomenon of X-ray diffraction in 1912. This high-energy form of light, with wavelengths comparable to the spacing between molecules in crystalline solids, could be used to reveal crystal structures. Other physicists had deduced the relationship between the lattice spacing and the angles at which bright diffraction spots should occur. Here Maurice de Broglie, the brother of physicist Louis de Broglie, introduces what came to be known as the rotating-crystal method for recording X-ray diffraction from a single crystal. The technique detects X-rays reflected along the surface of a series of so-called "Laue cones". It became the standard method of X-ray diffraction for many years.

IN view of the great interest of Prof. Bragg's and Messrs. Moseley and Darwin's researches on the distribution of the intensity of the primary radiation from X-ray tubes, it may be of interest to describe an alternate method which I have found very convenient (*Comptes rendus*, November 17, 1913).

As we know, the wavelength of the reflected ray is defined by the equation $n\lambda = 2d\sin\theta$, where n is a whole number, d the distance of two parallel planes, and θ the glancing angle. If one mounts a crystal with one face in the axis of an instrument that turns slowly and regularly, such as, for instance, a registering barometer, the angle changes gradually and continuously.

If, therefore, one lets a pencil of X-rays, emerging from a slit, be reflected from this face on to a photographic plate, one finds the true spectrum of the X-rays on the plate, supposing intensity of the primary beam to have remained constant. (This can be tested by moving another plate slowly before the primary beam during the exposure.)

The spectra thus obtained are exactly analogous to those obtained with a diffraction grating, and remind one strongly of the usual visual spectra containing continuous parts, bands, and lines.

So far I have only identified the doublet, $11°17'$ and $11°38'$, described by Messrs. Moseley and Darwin. The spectra contain also a number of bright lines about two octaves shorter than these, and the continuous spectrum is contained within about the same limits. These numbers may be used in the interpretations of diffraction Röntgen patterns, as they were obtained with tubes of the same hardness as those used for producing these latter.

X 射线的反射

编者按

1912 年，德国物理学家马克斯·冯·劳厄证明了 X 射线的衍射现象。这种形式的光能量很高，其波长与晶格中分子之间的距离相近，因而可以用于研究晶体的结构。其他物理学家推算出了晶格间距与预计会出现明亮衍射斑点的角度之间的关系。在这篇文章中，物理学家路易斯·德布罗意的哥哥莫里斯·德布罗意介绍了记录单晶 X 射线衍射的方法，后来被称作旋转晶体法。该技术探测沿着一组被称作“劳厄锥”的晶面反射的 X 射线。许多年来这种方法一直是人们研究 X 射线衍射的标准方法。

鉴于大家对布拉格教授、莫塞莱先生以及达尔文先生关于从 X 射线管发射的初级辐射强度分布的研究有极大的兴趣，我已发现的另一种很方便的方法可能也会引起大家的兴趣（《法国科学院院刊》，1913 年 11 月 17 日）。

如我们所知，反射射线的波长由方程 $n\lambda = 2d\sin\theta$ 确定，式中 n 是一个整数，d 是两个平行平面间的距离，θ 是掠射角。如果我们将一块晶体放在一台缓慢而有规律旋转的仪器（例如，记录式气压计）上，使晶体的一个表面沿仪器的轴向，那么反射的角度将连续不断地发生变化。

如果我们使一束 X 射线从狭缝中射出，并被此晶面反射到照相干版上，假定初级光束的强度保持不变，即可在干版上得到 X 射线的真实光谱。（可以通过曝光时在初级光束前方缓慢移动另一块照相干版来检验光束强度是否保持不变。）

这样得到的光谱与用衍射光栅得到的光谱非常类似，而且很容易使人想起那些含有连续区、谱带和谱线的普通可见光谱。

至今，我仅确认了莫塞莱先生和达尔文先生所描述的双线：11° 17′ 和 11° 38′。这个光谱还包含比双线短大约两个倍频程的许多亮线，还有在大致相同范围内的连续谱。这些光谱线也许可用于解释伦琴的衍射图样，因为产生伦琴的衍射图样时使用的管子与产生这些谱线时使用的管子具有相同的硬度。

The arrangement described above enables us to distinguish easily the spectra of different orders, as the interposition of an absorbing layer cuts out the soft rays, but does not weaken appreciably the hard rays of the second and higher orders.

It is convenient also for absorption experiments; thus a piece of platinum foil of 0.2 mm thickness showed transparent bands. The exact measurements will be published shortly, as well as the result of some experiments I am engaged upon at present upon the effect of changing the temperature of the crystal.

Maurice de Broglie

* * *

As W. L. Bragg first showed, when a beam of soft X-rays is incident on a cleavage plane of mica, a well-defined proportion of the beam suffers a reflection strictly in accordance with optical laws. In addition to this generally reflected beam, Bragg has shown that for certain angles of incidence, there occurs a kind of selective reflection due to reinforcement between beams incident at these angles on successive parallel layers of atoms.

Experiments I am completing seem to show that a generally reflected beam of rays on incidence at a second crystal surface again suffers optical reflection; but the degree of reflection is dependent on the orientation of this second reflector relative to the first.

The method is a photographic one. The second reflector is mounted on a suitably adapted goniometer, and the photographic plate is mounted immediately behind the crystal. The beam is a pencil 1.5 mm in diameter. When the two reflectors are parallel the impression on the plate, due to the two reflections, is clear. But as the second reflector is rotated about an axis given by the reflected beam from the first and fixed reflector, the optically reflected radiation from the second reflector—other conditions remaining constant—diminishes very appreciably. As the angle between the reflectors is increased from 0° to 90°, the impression recorded on the photographic plate diminishes in intensity. For an angle of 20° it is still clear; for angles in the neighbourhood of 50° it is not always detectable; and for an angle of 90° it is very rarely detectable in the first stages of developing, and is then so faint that it never appears on the finished print.

These results, then, would show that the generally reflected beam of X-rays is appreciably polarised in a way exactly analogous to that of ordinary light. Owing to the rapidity with which the intensity of the generally reflected beam falls off with the angle of incidence of the primary beam, it has not been possible to work with any definiteness with angles of incidence greater than about 78°, and this is unfortunately a considerably larger angle than the probable polarising angle. Experiments with incidence in the neighbourhood of 45° should prove peculiarly decisive, for whereas ordinary light cannot as a rule be completely polarised by reflection, the reflection of X-rays, which occurs at

上述装置能使我们很容易地区别不同级的谱线，因为插入吸收层可以截断软射线，但并不会太明显地减弱次级和更高级的硬射线。

进行吸收实验也是很方便的。用这样的方法，一片 0.2 毫米厚的铂箔会显示出透射带。至于确切的测量方法以及目前我在改变晶体温度方面所做的实验得到的结果，都将在我即将发表的文章中进行介绍。

莫里斯·德布罗意

* * *

正如布拉格首先指出的，当一束软 X 射线入射到云母的一个解理面上时，光束反射部分所占的比例严格遵照光学定律。除了这种普通的反射光束，布拉格指出，对于特定的入射角，由于以这些角度入射到原子中连续的平行层上的光束彼此相互增强，因而会产生一种选择性的反射。

我正在进行的实验似乎表明，当一个普通的反射束入射到下一个晶体表面上时，还会再次产生光学反射，但是，反射的程度取决于第二个反射面与第一个反射面之间的夹角。

我用的是照相记录法。第二个反射面被安装在调整好的测角仪上，照相干版紧贴在晶体后面。光束直径为 1.5 毫米。当两个反射面平行时，两次反射在照相干版上产生的影像是清晰的。但是，当以第一个固定反射面上反射的射线为轴旋转第二个反射面时，假如其他条件不变，则从第二个反射面上反射的光学辐射明显减弱。当反射面之间的夹角从 0° 增加到 90° 时，照相干版上记录的影像的清晰度不断降低。在角度为 20° 时，影像仍然很清晰；角度为 50° 左右时，经常检测不到影像；角度为 90° 时，由于影像非常微弱，在第一步显影阶段已经很难看到，而在最终冲洗出的照片上则从来没有出现过。

以上结果表明，X 射线普通反射束的偏振特性在某种程度上完全类似于普通光的偏振。由于普通反射束的强度随着初级光束入射角的增加而迅速下降，当入射角超过 78° 后就很难得到确定无疑的影像了。遗憾的是，这个角度比可能的偏振角大很多。入射角在 45° 附近的实验尤其具有决定意义，虽然通常不能通过反射使普通光完全偏振，但在原子平面上 X 射线的反射并不受被辐照的晶体表面上任何污染的影响，一旦出现偏振，那么偏振角处被反射的辐射将是完全偏振的。被选择性反射

planes of atoms, is independent of any contamination of the exposed crystal surface, and polarisation, once established, should prove complete for radiation reflected at the polarising angle. The selectively reflected X-rays seem to show the same effects as does the generally reflected beam. Selectively reflected radiation is always detectable after the second reflection, but this seems due to the selectively reflected radiation produced at the second reflector by the unpolarised portion of the beam generally reflected at the first reflector.

The application of a theory of polarisation to explain the above results is interestingly supported by the fact that in the case of two reflections by parallel reflectors, the proportion of X-rays reflected at the second reflector is invariably greater than the proportion of rays reflected at the first; that is, the ratio of reflected radiation to incident radiation at the second reflector is always greater than the same ratio at the first reflector. This might be expected if vibrations perpendicular to the plane of incidence are to be reflected to a greater extent than those in the plane of incidence. The proportion of such vibrations is larger in the beam incident on the second reflector than in the original beam, and a greater proportion of radiation would be reflected at the second reflector than could be at the first. For the case of parallel reflectors and incidence of a primary beam on the first at the polarising angle, the reflection at the second should be complete.

E. Jacot

(**92**, 423-424; 1913)

Maurice de Broglie: 29, Rue Chateaubriand, Paris, December 1.
E. Jacot: South African College, Cape Town, November 14.

的X射线与普通的反射有相同的效应。在第二次反射后总能检测到被选择性反射的辐射，但这似乎是由第一个反射面处被反射的那部分非偏振辐射在第二个反射面上发生选择性反射而造成的。

偏振理论可以用来解释上述结果，这得到了以下事实的强烈支持，在由两个平行反射面引起的两次反射中，第二个反射面上被反射的X射线总是多于第一个反射面上被反射的X射线；即在第二个反射面上被反射的辐射相对于入射辐射的比例，总是大于第一个反射面。这就可以预期，垂直于入射平面的振动被反射的量比在入射平面内的振动更大。与原光束相比，入射到第二个反射面上的光束中垂直于入射平面的振动所占比例更大一些，因此在第二个反射面上被反射的辐射的比例大于第一个反射面。在两个反射面相互平行的情况下，如果初始光束以偏振角入射到第一个反射面上，则在第二个反射面上它将被完全反射。

贾科

（沈乃澂 翻译；江丕栋 审稿）

Einstein's Relativity Theory of Gravitation

E. Cunningham

Editor's Note

Arthur Eddington had recently announced his measurements of the deflection of starlight during a solar eclipse, in apparent agreement with Einstein's general theory of relativity. Here Ebenezer Cunningham surveys Einstein's ideas. Einstein's 1905 work had established a link between inertia and energy, and his new work pursued the question of whether gravity too might be linked to energy. What emerges from the theory, Cunningham argues, is a view in which there is no ultimate criterion for the equality of space or time intervals, but only the equivalence of an infinite number of ways of mapping out physical events. All this has been made possible, he notes, by Einstein adopting mathematics already developed by Riemann, Levi-Civita and others.

I

THE results of the Solar Eclipse Expeditions announced at the joint meeting of the Royal Society and Royal Astronomical Society on November 6 brought for the first time to the notice of the general public the consummation of Einstein's new theory of gravitation. The theory was already in being before the war; it is one of the few pieces of pure scientific knowledge which have not been set aside in the emergency; preparations for this expedition were in progress before the war had ceased.

Before attempting to understand the theory which, if we are to believe the daily Press, has dimmed the fame of Newton, it may be worth while to recall what it was that he did. It was not so much that he, first among men, used the differential calculus. That claim was disputed by Leibniz. Nor did he first conceive the exact relations of inertia and force. Of these, Galileo certainly had an inkling. Kepler, long before, had a vague suspicion of a universal gravitation, and the law of the inverse square had, at any rate, been mooted by Hooke before the "Principia" saw the light. The outstanding feature of Newton's work was that it drew together so many loose threads. It unified phenomena so diverse as the planetary motions, exactly described by Kepler, the everyday facts of falling bodies, the rise and fall of the tides, the top-like motion of the earth's axis, besides many minor irregularities in lunar and planetary motions. With all these drawn into such a simple scheme as the three laws of motion combined with the compact law of the inverse square, it is no wonder that flights of speculation ceased for a time. The universe seemed simple and satisfying. For a century at least there was little to do but formal development of Newton's dynamics. In the mid-eighteenth century Maupertuis hinted at a new physical doctrine. He was not content to think of the universe as a great clock the wheels of which turned inevitably and irrevocably according to a fixed rule. Surely there must be some purpose, some divine economy in all its motions. So he propounded a principle of least action. But it soon appeared that this was only Newton's laws in a new guise; and so the eighteenth century closed.

爱因斯坦关于万有引力的相对论

坎宁安

编者按

阿瑟·爱丁顿最近宣布，他在日食期间对星光偏转的测量明确地证实了爱因斯坦的广义相对论。在这篇文章中，埃比尼泽·坎宁安简单描述了爱因斯坦的理论。爱因斯坦在1905年就已经建立了惯性和能量之间的联系，他现在的工作主要是考查重力是否也有可能与能量相关联。坎宁安指出，这个理论说明了这样一个观点：虽然不存在衡量空间间隔和时间间隔完全相等的绝对标准，但可以设计无限多种方式以保证多个物理事件的等同性。他说，爱因斯坦利用黎曼、列维齐维塔和其他人的数学理论已经验证了这些结论。

I

在11月6日皇家学会和皇家天文学会共同举办的会议上宣布的日食观测结果，使得爱因斯坦关于万有引力的新理论受到了公众的广泛关注。这个理论在战争之前就已经被提出来了；它是在战争中极少数没有被丢到一边的纯科学工作之一；日食观测的准备工作在战争结束之前就已经着手进行了。

如果在试图理解这个新理论之前，我们就已经像日报上说的那样，认为牛顿在其面前也会黯然失色，那么最好首先回顾一下牛顿所做的工作。并不能说牛顿是第一个使用微分计算的人。有些议论认为这是莱布尼茨的首创。牛顿也不是第一个认识到惯性和力之间关系的人。这方面，肯定是伽利略首先对此进行了初步的设想。开普勒在很久以前就对万有引力有过模糊的猜想，但是，引力与距离平方成反比的规律是在该“原理”建立之前，由胡克首先发现的。牛顿所做工作的杰出之处在于，他把这么多松散的线索整合到了一起。行星运动（已被开普勒精确描述过）、日常生活中的落体运动、潮汐的涨落、地轴的进动以及在月球和行星运动中出现的许多小的不规则性——牛顿将所有这些现象都统一了起来。牛顿把所有这些都归入了一个包括力学三大定律和简洁的平方反比关系的基本框架中，也难怪在后来很长一段时间内，科学上的思索都停滞了。整个宇宙都看似简单而圆满。在后来至少一个世纪的时间里，除了在形式上发展一下牛顿力学，没有其他工作可做。在18世纪中叶，莫佩尔蒂暗示了一种新的物理学说。他不满足于认为宇宙是一个在某种确定的法则下永不停止、永不倒退的钟表。宇宙的运动一定是有目的的，一定有一种神圣的力量在支配着它的运动。因此，他提出了最小作用原理。但是人们很快发现这只不过是牛顿定律的一种新的外在形式而已，然后18世纪就这样结束了。

The nineteenth saw great changes. When it closed, the age of electricity had come. Men were peering into the secrets of the atom. Space was no longer a mighty vacuum in the cold emptiness of which rolled the planets. It was filled in every part with restless energy. Ether, not matter, was the last reality. Mass and matter were electrical at bottom. A great problem was set for the present generation: to reconcile one with the other the new laws of electricity and the classical dynamics of Newton. At this point the principle of least action began to assume greater importance; for the old and the new schemes of the universe had this in common, that in each of them the time average of the difference between the kinetic and the potential energies appears to be a minimum.

One of the main difficulties encountered by the electrical theory of matter has been the obstinate refusal of gravitation to come within its scope. Quietly obeying the law of the inverse square, it heeded not the bustle and excitement of the new physics of the atom, but remained, independent and inevitable, a constant challenge to rash claimants to the key of the universe. The electrical theory seemed on the way to explain every property of matter yet known, except the one most universal of them all. It could trace to its origins the difference between copper and glass, but not the common fact of their weight; and now the ether began silently to steal away.

One matter that has seriously troubled men in Newton's picture of the universe is its failure to accord with the philosophic doctrine of the relativity of space and time. The vital quantity in dynamics is the acceleration, the change of motion of a body. This does not mean that Newton assumed the existence of some ultimate framework in space relative to which the actual velocity of a body can be uniquely specified, for no difference is made to his laws if any arbitrary constant velocity is added to the velocity of every particle of matter at all time. The serious matter is that the laws cannot possibly have the same simplicity of form relative to two frameworks of which one is in rotation or non-uniform motion relative to the other. It seems, for instance, that if Newton were right, the term "fixed direction" in space means something, but "fixed position" means nothing. It seems as if the two must stand or fall together. And yet the physical relations certainly make a distinction. Why this should be so has not yet been made known to us. Whatever new theory we adopt must take account of the fact.

It was with some feeling of relief that men hailed the advent of the ether as a substitute for empty space, though we may note in passing that some philosophers—Comte, for example—have held that the concept of an ether, infinite and intangible, is as illogical as that of an absolute space. But, jumping at the notion, physicists proposed to measure all velocities and rotations relative to it. Alas! the ether refused to disclose the measurements. Explanations were soon forthcoming to account for its reluctance; but these were so far-reaching that they explained away the ether itself in the sense in which it was commonly understood. At any rate, they proved that this creature of the scientific imagination was not one, but many. It quite failed to satisfy the cravings for a permanent standard against which motion might be measured. The problem was left exactly where it was before. This was prewar relativity, summarised by Einstein in 1905. The physicists complained loudly that he was taking away their ether.

19 世纪发生了重大的变化。在这个世纪末，电的时代到来了。人类要揭示原子中的秘密。太空也不再是有行星在其中运行的寒冷而空洞的广袤真空。宇宙中的每个部分都注满了运动着的能量。以太，而不是物质，才是最终的存在形式。物质实际上都是带电的。一个重大的问题摆在了当代人的面前：如何将新的电学理论和经典的牛顿力学原理统一起来。这样，最小作用原理就变得重要起来；因为它是新旧两种宇宙观相通的部分，在两者中，动能和势能之差的时间平均都应该取最小值。

物质的电学理论遇到的一个最主要的问题是，无法把万有引力引入到这个理论框架之中。在电学理论遵从平方反比定律的情况下，新兴的原子物理的蓬勃发展却被忽视掉了，电学独自向破解宇宙之谜的目标发起了挑战。电学理论力图说明物质的所有其他性质，而唯独将人们最为熟知的属性排除在外。它可以说清铜和玻璃之间存在差别的原因，但是不能解释它们共同的特性：重量。至此，以太学说也开始默默地销声匿迹了。

在牛顿的宇宙框架中最令人困扰的一点是：它无法与时空相对性的哲学学说达成一致。加速度在力学中是一个关键的量，它反映了物体运动状态的变化。但这并不意味着牛顿假设了空间中存在一个终极的参考系，一个物体的实际速度相对于这个参考系是唯一确定的，因为在他的理论中，任意一个质点的运动速度总可以加上一个常数速度。这种处理方法的严重缺陷是，当处理两个参考系的问题时，如果其中一个参考系相对于另一个做转动或者非匀速运动，那么前面所说的法则就不可能保持如此简单的形式。这样看来，如果牛顿是正确的，那么在空间中，“固定方向”是有意义的，但“固定位置”却没有任何意义。两者看似应该同时成立或者同时不成立，但是物理上的关系显然是有区别的。为什么会这样呢？我们还不知道。无论我们采用什么样的新理论，都要考虑到上面的问题。

当我们引入了以太的概念，用它来代替空无一物的空间时，问题看似得到了解决，尽管我们也许注意到，在传播这一概念的时候，一些哲学家，比如孔德，认为无限且无形的以太和绝对空间一样不合逻辑。但让我们先把这些看法抛在一边，物理学家们提出要测量所有相对于以太的平动和转动。唉！可惜以太却拒绝我们的测量。很快就出现了对这一难题的解释，但是这与能够用一种大家可以接受的方式来为以太辩解还有非常遥远的距离。无论如何，它们证明了这种科学想象的创造并不是唯一的，而是有很多种。我们渴望找到一种可以用来测量运动的永久标准，而以太的概念是非常失败的。问题和从前一样没有得到解决。还是战争之前由爱因斯坦于 1905 年总结出来的相对论解决了这个问题。物理学家们则强烈地抱怨爱因斯坦摒弃了他们的以太。

Let it not be thought, however, that the results of the hypothesis then advanced were purely negative. They showed quite clearly that many current ideas must be modified, and in what direction this must be done. Most notably it emphasised the fact that inertia is not a fundamental and invariable property of matter; rather it must be supposed that it is consequent upon the property of energy. And, again, energy is a relative term. One absolute quantity alone remained; one only stood independent of the taste or fancy of the observer, and that was "action". While the ether and the associated system of measurement could be selected as any one of a legion, the principle of least action was satisfied in each of them, and the magnitude of the action was the same in all.

But, still, gravitation had to be left out; and the question from which Einstein began the great advance now consummated in success was this. If energy and inertia are inseparable, may not gravitation, too, be rooted in energy? If the energy in a beam of light has momentum, may it not also have weight?

The mere thought was revolutionary, crude though it be. For if at all possible it means reconsidering the hypothesis of the constancy and universality of the velocity of light. This hypothesis was essential to the yet infant principle of relativity. But if called in question, if the velocity of light is only approximately constant because of our ordinary ways of measuring, the principle of relativity, general as it is, becomes itself an approximation. But to what? It can only be to something more general still. Is it possible to maintain anything at all of the principle with that essential limitation removed?

Here was exactly the point at which philosophers had criticised the original work of Einstein. For the physicist it did too much. For the philosopher it was not nearly drastic enough. He asked for an out-and-out relativity of space and time. He would have it that there is no ultimate criterion of the equality of space intervals or time intervals, save complete coincidence. All that is asked is that the order in which an observer perceives occurrences to happen and objects to be arranged shall not be disturbed. Subject to this, any way of measuring will do. The globe may be mapped on a Mercator projection, a gnomonic, a stereographic, or any other projection; but no one can say that one is a truer map than another. Each is a safe guide to the mariner or the aviator. So there are many ways of mapping out the sequences of events in space and time, all of which are equally true pictures and equally faithful servants.

This, then, was the mathematical problem presented to Einstein and solved. The pure mathematics required was already in existence. An absolute differential calculus, the theory of differential invariants, was already known. In pages of pure mathematics that the majority must always take as read, Riemann, Christoffel, Ricci, and Levi-Civita supplied him with the necessary machinery. It remained out of their equations and expressions to select some which had the nearest kinship to those of mathematical physics and to see what could be done with them.

(**104**, 354-356; 1919)

但是，我们不要以为后来提出的假说都是不正确的。这些假说明确地表明现有的许多观点需要修正，并且说明了修正的方向应该在哪里。尤其是它强调了一个事实：惯性不是物质的一个不会发生变化的基本属性，而应该被看作是随着能量的变化而变化的。我们要再次强调，能量是一个相对量。一个绝对量是独立不变的；它不依赖于观察者的体验和想象，这就是“作用”。当以太及与它相关的测量系统被选择作为大量作用中的任意一员时，它们都满足最小作用原理，并且作用量的总和不变。

但是，万有引力仍然没有被考虑进去，爱因斯坦正是从这个问题出发，现在已成功地获得了巨大的进展。如果能量和惯性是不可分割的，那么重力难道就不能建立在能量的基础之上吗？如果能量是一束具有动量的光束，那么它为什么不能有重量呢？

这样的想法是革命性的，尽管它还不够成熟。因为，如果这是可能的，那么它就意味着需要重新考虑光速不变性和普适性的假说。这个假说是尚不成熟的相对论的基本原则。但是如果它被质疑，如果光速只是因为我们通常的测量方法不够精确才大致不变，那么，被大家普遍接受的相对论法则就只是一种近似，就像它本身的系统一样。但这是对什么的近似呢？只能是对一种更加普遍的原理的近似。当消除了那种根本上的限制以后，原来的法则中还有没有什么东西可以保留下来呢？

就是在这个问题上，哲学家们批判了爱因斯坦早期的工作。对物理学家来说，这样的批评太偏激了。对哲学家来说，这种批评还远算不上严厉。爱因斯坦开始寻找一种彻底的时空相对论。他认为，除了完全重合之外，不存在衡量空间间隔和时间间隔完全相等的绝对标准。他只要求观察者观察到的事件的发生顺序和物体摆放的顺序不被打乱。在这个前提下，任何测量方法都将是可行的。这就好像我们可以用墨卡托投影法、心射切面投影法、立体投影法或者任何一种其他的方法去画地球，而没有人会说其中哪一种地图较之其他地图更准确。飞行员和海员使用任何一种地图都是安全可靠的。因此，也有很多方法可以标定时空中事件发生的顺序，每一种都描述了真实的情况，每一种都同样可信。

这样，爱因斯坦接下来就只需要解决那些数学上的问题了。他所需要的纯数学方法已经存在。绝对微分、微分不变量理论，这些都是已知的。大多数人经常研读的纯数学著作，如黎曼、克里斯托弗尔、里奇和列维齐维塔的著作，都为爱因斯坦提供了必要的数学工具。剩下的工作就是从那些方程和表达式中选出最接近数学物理的部分，并想办法把它们解出来。

II. The Nature of the Theory

In the first article an attempt was made to show the roads which led to Einstein's adventure of thought. On the physical side briefly it was this. Newton associated gravitation definitely with mass. Electromagnetic theory showed that the mass of a body is not a definite and invariable quantity inherent in matter alone. The energy of light and heat certainly has inertia. Is it, then, also susceptible to gravitation, and, if so, exactly in what manner? The very precise experiments of Eötvös rather indicated that the mass of a body, as indicated by its inertia, is the same as that which is affected by gravitation.

Also, how must the expression of Newton's law of gravitation be modified to meet the new view of mass? How, also, must the electromagnetic theory and the related pre-war relativity be adapted to allow of the effect of gravitation? With the relaxation of the stipulation that the velocity of light shall be constant, will the principle of relativity become more general and acceptable to the philosophic doctrine of relativity, or will it, on the other hand, become completely impossible?

One point arises immediately. The out-and-out relativist will not admit an absolute measure of acceleration any more than of velocity. The effect, however, of an accelerated motion is to produce an apparent change in gravitation; the measure of gravitation at any place must therefore be a relative quantity depending upon the choice which the observer makes as to the way in which he will measure velocities and accelerations. This is one of Einstein's fundamental points. It has been customary in expositions of mechanics to distinguish between so-called "centrifugal force" and "gravitational force". The former is said to be fictitious, being simply a manifestation of the desire of a body to travel uniformly in a straight line. On the other hand, gravitation has been called a real force because associated with a cause external to the body on which it acts.

Einstein asks us to consider the result of supposing that the distinction is not essential. This was his so-called "principle of equivalence". It led at once to the idea of a ray of light being deviated as it passes through a field of gravitational force. An observer near the surface of the earth notes objects falling away from him towards the earth. Ordinarily, he attributes this to the earth's attraction. If he falls with them, his sense of gravitation is lost. His watch ceases to press on the bottom of his pocket; his feet no longer press on his boots. To this falling observer there is no gravitation. If he had time to think or make observations of the propagation of light, according to the principle of equivalence he would now find nothing gravitational to disturb the rectilinear motion of light. In other words, a ray of light propagated horizontally would share in his vertical motion. To an observer not falling, and, therefore, cognisant of a gravitational field, the path of the ray would therefore be bending downward towards the earth.

The systematic working out of this idea requires, as has been remarked, considerable mathematics. All that can be attempted here is to give a faint indication of the line of attack, mainly by way of analogy.

II. 理论的本质

第一篇文章旨在说明爱因斯坦的思考方法，从物理学的角度来看，简要的说明就是这样。牛顿明确地将万有引力和质量联系起来。电磁学理论表明，一个物体的质量并不是物质确定不变的内在属性。光能和热能当然都具有惯性，那么，它也会受到万有引力的影响吗？如果确实如此，确切的作用方式又是怎样的呢？厄缶的精确实验更表明了由其惯性所表示的物体质量同样受到万有引力的影响。

牛顿万有引力定律的描述要怎样修正才能符合关于质量的新观点呢？电磁学理论和战争前提出的相对论要怎样调整才能允许万有引力效应的存在呢？在解除了光速不变的约束之后，相对论原理会不会成为一种更加普遍且被相对性的哲学理论所接受的法则呢？或者说，另一方面，它会不会被证明完全不可行呢？

这里马上就引出了一个观点。彻底的相对论者只能对速度进行绝对测量，却不能对加速度进行绝对测量。然而，加速运动在万有引力场中会发生明显的变化；因此在任意地点，万有引力的测量值肯定都是相对的，它取决于测量者所选择的测量速度和加速度的方式。这是爱因斯坦的主要观点之一。在力学上，对所谓“离心力”和“万有引力”的解释通常是有区别的。前者是一个虚拟的力，仅仅表现了物体要做匀速直线运动的趋势。另一方面，万有引力被认为是一个真实的力，因为它和作用于物体的外界因素有关。

爱因斯坦让我们考虑，如果这种区别不是本质的，结果会怎样。这就是他所说的“等效原理”。它立刻就引出了这样的设想：一束光穿过万有引力场时会发生弯曲。一个在地球表面附近的观察者会看到物体远离自己落向地球。一般来说，他会把这归结为地球的引力。如果他和物体一起下落，那么他对万有引力的感觉就会消失。他的怀表不再压在衣袋底部，他的脚也不再压在靴子上。对于这位正在下落的观察者来说，他是观察不到万有引力存在的。如果他有时间观察和思考光的传播，那么按照等效原理，他将看不到光线的直线运动被万有引力所干扰。换句话说，一束沿水平方向传播的光线，将与观察者一起同时做垂直运动。因此，一个没有下落的观察者可以感觉到万有引力的存在，所以光线在向地球运动的过程中会发生弯曲。

就像前面提到的那样，要把这样一个想法系统地求解出来，需要大量的数学运算。这里我们所能做的只是用模糊的示意来说明这个原理，主要通过类比法。

It is no new discovery to speak of time as a fourth dimension. Every human mind has the power in some degree of looking upon a period of the history of the world as a whole. In doing this, little difference is made between intervals of time and intervals of space. The whole is laid out before him to comprehend in one glance. He can at the same time contemplate a succession of events in time, and the spatial relations of those events. He can, for instance, think simultaneously of the growth of the British Empire chronologically and territorially. He can, so to speak, draw a map, a four-dimensional map, incapable of being drawn on paper, but none the less a picture of a domain of events.

Let us pursue the map analogy in the familiar two-dimensional sense. Imagine that a map of some region of the globe is drawn on some material capable of extension and distortion without physical restriction save that of the preservation of its continuity. No matter what distortion takes place, a continuous line marking a sequence of places remains continuous, and the places remain in the same order along that line. The map ceases to be any good as a record of distance travelled, but it invariably records certain facts, as, for example, that a place called London is in a region called England, and that another place called Paris cannot be reached from London without crossing a region of water. But the common characteristic of maps of correctly recording the shape of any small area is lost.

The shortest path from any place on the earth's surface to any other place is along a great circle; on all the common maps, one series of great circles, the meridians, is mapped as a series of straight lines. It might seem at first sight that our extensible map might be so strained that all great circles on the earth's surface might be represented by straight lines. But, as a matter of fact, this is not so. We might represent the meridians and the great circles through a second diameter of the earth as two sets of straight lines, but then every other great circle would be represented as a curve.

The extension of this to four dimensions gives a fair idea of Einstein's basic conception. In a world free from gravitation we ordinarily conceive of free particles as being permanently at rest or moving uniformly in straight lines. We may imagine a four-dimensional map in which the history of such a particle is recorded as a straight line. If the particle is at rest, the straight line is parallel to the time axis; otherwise it is inclined to it. Now if this map be strained in any manner, the paths of particles are no longer represented as straight lines. Any person who accepts the strained map as a picture of the facts may interpret the bent paths as evidence of a "gravitational field", but this field can be explained right away as due to his particular representation, for the paths can all be made straight.

But our two-dimensional analogy shows that we may conceive of cases where no amount of straining will make all the lines that record the history of free particles simultaneously straight; pure mathematics can show the precise geometrical significance of this, and can write down expressions which may serve as a measure of the deviations that cannot be removed. The necessary calculus we owe to the genius of Riemann and Christoffel.

把时间作为第四维并不是一个新的发现。每一个人在某种程度上都会把世界历史的一段时期当作一个整体来看待。在这样做的时候，时间段和空间段没有什么区别。在匆匆一瞥之中所有的东西都呈现到他面前要他去了解。他在仔细考虑一系列事件发生时间的同时，还要将它们和发生地点联系起来。比如，他能同时从时间顺序和疆域范围两方面来考虑大英帝国的扩张。所以可以这样说，他可以画一个地图，一个四维的地图，尽管不能画在纸上，但依然可以描述一系列事件。

让我们用我们所熟悉的二维地图进行类比。我们可以想象有一个描述世界上某个区域的地图，用来制作这个地图的材料可以不受物理限制，随意延展和扭曲以保持它的连续性。不管这种材料如何扭曲变形，它上面表示地点次序的连续直线依然保持连续，沿着这条直线各个地点的排列顺序不变。这种地图无法记录旅行的距离，但是它可以忠实地记录某些特定的事实，比如，伦敦位于英国境内，从伦敦到另一个地方——巴黎，不可能不跨越海洋。但是一般地图所具有的记录任意一小块地方的功能在这种地图中完全丧失了。

在地球表面从一地到另一地的最短路径是沿着大圆的路径；在普通的地图上，一系列的大圆，即经线，是用一系列的直线来表示的。初看起来，我们的可伸缩地图可以被拉伸开，这样，地球表面的所有大圆都可以呈直线。可事实并不是这样。我们可以把经线和另外一组由地球的另外一个直径确定的大圆表示成两组直线，但是这样做之后，所有其他的大圆就只能表示为曲线了。

将上述观点扩展到四维时空，就构成了爱因斯坦的基本概念。我们通常认为在没有万有引力的世界里，自由粒子将永远静止或者做匀速直线运动。我们可以想象一下，在四维的地图中，粒子的历史被记录为一条直线。如果这个粒子是静止的，那么这条直线就平行于时间轴，否则就是倾斜的。现在，如果地图以任意方式被拉伸，那么粒子的路径就不再被表示为直线。如果我们能接受可伸缩地图作为表征事实的方式，就可以把这些弯曲的路径看作是“万有引力场”的证据，但是，这种引力场也可以马上被解释成是由这种特殊的表示方法造成的，因为所有路径都可以变成直线。

但是类似的二维地图告诉我们，没有任何一种变形方式可以使所有记录自由粒子历史的线都同时呈直线；关于这一点，纯数学可以给出它在几何学上的精确证明，还可以给出表达式，用来度量那些不可消除的弯曲。天才的黎曼和克里斯托弗尔给出了我们所需的微积分算法。

Einstein now identifies the presence of curvatures that cannot be smoothed out with the presence of matter. This means that the vanishing of certain mathematical expressions indicates the absence of matter. Thus he writes down the laws of the gravitational field in free space. On the other hand, if the expressions do not vanish, they must be equal to quantities characteristic of matter and its motion. These equalities form the expression of his law of gravitation at points where matter exists.

The reader will ask: What are the quantities which enter into these equations? To this only a very insufficient answer can here be given. If, in the four-dimensional map, two neighbouring points be taken, representing what may be called two neighbouring occurrences, the actual distance between them measured in the ordinary geometrical sense has no physical meaning. If the map be strained, it will be altered, and therefore to the relativist it represents something which is not in the external world of events apart from the observer's caprice of measurement. But Einstein assumes that there is a quantity depending on the relation of the points one to the other which is invariant—that is, independent of the particular map of events. Comparing one map with another, thinking of one being strained into the other, the relative positions of the two events are altered as the strain is altered. It is assumed that the strain at any point may be specified by a number of quantities (commonly denoted g_{rs}), and the invariable quantity is a function of these and of the relative positions of the points.

It is these quantities g_{rs} which characterise the gravitational field and enter into the differential equations which constitute the new law of gravitation.

It is, of course, impossible to convey a precise impression of the mathematical basis of this theory in non-mathematical terms. But the main purpose of this article is to indicate its very general nature. It differs from many theories in that it is not devised to meet newly observed phenomena. It is put together to satisfy a mental craving and an obstinate philosophic questioning. It is essentially pure mathematics. The first impression on the problem being stated is that it is incapable of solution; the second of amazement that it has been carried through; and the third of surprise that it should suggest phenomena capable of experimental investigation. This last aspect and the confirmation of its anticipations will form the subject of the next article.

(**104**, 374-376; 1919)

III. The Crucial Phenomena

In the article last week an attempt was made to indicate the attitude of the complete relativist to the laws which must be obeyed by gravitational matter. The present article deals with particular conclusions.

现在，爱因斯坦认为，无法消除的弯曲表示物质的存在。这意味着，如果某个数学表达式为零则表示没有物质存在。于是他写出了自由空间中万有引力场的定律。另一方面，如果表达式不为零，那么它们一定等于描述物质及其运动的物理量。这些方程就是爱因斯坦在有物质存在的点上构筑的万有引力定律表达式。

读者可能会问：这些方程中都有哪些物理量？在这里我们只能给出一个非常不充分的回答。如果在四维地图中，两个相邻的点被认为代表两个相邻的事件，那么用普通几何方法测量的两点之间的实际距离将没有任何物理意义。如果这个地图发生变形，它就会被改变，所以对于相对论者来说，撇开观察者反复无常的测量结果，它代表了某种不存在于由事件组成的外部世界中的东西。但是爱因斯坦认为有一个物理量依赖于两个点之间的关系，具有不变性，也就是说，它不依赖于某种事件地图。比较一个地图和另一个地图，设想其中一个发生变形而成为另一个，代表两个事件的点的相对位置会随着变形方式的变化而变化。可以假设，任意点的变形可以用一些物理量来表示（一般记做 g_{rs}），而不变量是这些物理量和事件点相对位置的函数。

这个表征万有引力场的物理量 g_{rs} 被引入构成万有引力新定律的微分方程中。

当然，我们不可能用非数学语言将这个理论用数字精确地表达出来。但是这篇文章的主要目的是说明它的普遍特征。这个理论与许多其他理论的不同之处在于，它不是为了解释某个新发现而被构建的。构建它的目的是为了满足精神上的渴望和应对哲学上的质疑。它在本质上是纯数学问题。这个问题给我们的第一印象是，它是不可能被攻破的；第二点出人意料的是，它居然被攻破了；第三个令人惊奇的是，它竟然预言了可以用实验研究的现象。关于最后一方面以及对其预言的证实将是下一篇文章的主题。

III. 关键的现象

上周的文章旨在说明一个完全的相对论者对万有引力物质必须遵循的法则的态度。而本文是要介绍一些特定的结论。

As Minkowski remarked in reference to Einstein's early restricted principle of relativity: "From henceforth, space by itself and time by itself do not exist; there remains only a blend of the two" ("Raum und Zeit", 1908). In this four-dimensional world that portrays all history let (x_1, x_2, x_3, x_4) be a set of coordinates. Any particular set of values attached to these coordinates marks an event. If an observer notes two events at neighbouring places at slightly different times, the corresponding points of the four-dimensional map have coordinates slightly differing one from the other. Let the differences be called (dx_1, dx_2, dx_3, dx_4). Einstein's fundamental hypothesis is this: there exists a set of quantities g_{rs} such that

$$g_{11}dx_1^2 + 2g_{12}dx_1dx_2 + \cdots + g_{44}dx_4^2$$

has the same value, no matter how the four-dimensional map is strained. In any strain g_{rs} is, of course, changed, as are also the differences *dx*.*

If the above expression be denoted by $(ds)^2$, *ds* may conveniently be called the *interval* between two events (not, of course, in the sense of time interval). In the case of a field in which there is no gravitation at all, if dx_4 is taken to be *dt*, it is supposed that ds^2 reduces to the expression $dx_1^2 + dx_2^2 + dx_3^2 - c^2dt^2$, where *c* is the velocity of light. If this is put equal to zero, it simply expresses the condition that the neighbouring events correspond to two events in the history of a point travelling with the velocity of light.

Einstein is now able to write down differential equations connecting the quantities g_{rs} with the coordinates (x_1, x_2, x_3, x_4), which are in complete accord with the requirement of complete relativity. † These equations are assumed to hold at all points of space unoccupied by matter, and they constitute Einstein's law of gravitation.

Planetary Motion

The next step is to find a solution of the equations when there is just one point in space at which matter is supposed to exist, one point which is a singularity of the solution. This can be effected completely ‡ : that is, a unique expression is obtained for the interval between two neighbouring events in the gravitational field of a single mass. This mass is now taken to be the sun.

It is next assumed that in the four-dimensional map (which, by the way, has now a bad twist in it, that cannot be strained out, all along the line of points corresponding to the

* The gravitational field is specified by the set of quantities g_{rs}. When the gravitational field is small, these are all zero, except for g_{44}, which is approximately the ordinary Newtonian gravitational potential.

† These equations take the place of the old Laplace equation $\nabla^2V=0$. Just as that equation is the only differential equation of the second order which is entirely independent of any change of ordinary space coordinates, so Einstein equations are uniquely determined by the condition of relativity.

‡ The result is that the invariant interval *ds* is given by $ds^2 = (1 - 2m/r)(dt^2 - dr^2) - r^2(d\theta^2 + \sin^2\theta d\phi^2)$, the four coordinates being now interpreted as time and ordinary spherical polar coordinates.

闵可夫斯基这样评论爱因斯坦早期的狭义相对论："从今以后，单独的空间和单独的时间都将不存在；二者只能作为一个复合体而存在"（《空间和时间》，1908 年）。在这个描述了所有历史事件的四维空间中，(x_1, x_2, x_3, x_4) 被视为一组坐标。把任意一组特定数值代入坐标中，都能表示一个事件。如果一个观察者观察到发生地点和时间都很接近的两个事件，那么四维地图上相应两点的坐标也区别不大。我们把它们之间的差别表示为 (dx_1, dx_2, dx_3, dx_4)。爱因斯坦的基本假设是这样的：存在一组 g_{rs}，无论四维地图发生什么样的变形，

$$g_{11}dx_1^2 + 2g_{12}dx_1dx_2 + \cdots + g_{44}dx_4^2$$

都具有相同的值。在发生变形时，g_{rs} 的值当然会发生变化，差值 dx 也同样会改变。*

如果上面的表达式被记作 $(ds)^2$，为方便起见，我们可以把 ds 称作两个事件的**间隔**（当然不是一般观念中的时间间隔）。在万有引力场不存在的情况下，如果 dx_4 用 dt 来代替，那么 ds^2 就会退化成表达式 $dx_1^2 + dx_2^2 + dx_3^2 - c^2dt^2$，这里 c 是光速。如果这个表达式等于零，则表示一个以光速运动的点所经历的两个相邻的事件。

现在，爱因斯坦就可以写出将物理量 g_{rs} 与坐标 (x_1, x_2, x_3, x_4) 相联系的微分方程了，这与完善相对论的要求完全一致。† 这些方程包含了空间中所有未被物质占据的点，它们构成了爱因斯坦的万有引力定律。

行星的运动

下一步的任务是找到一个空间中只有一个点被物质占据的方程解，而这个点是解的奇点。这完全可以做到‡：也就是说，对于单一质点在引力场中的两个相邻事件的间隔，我们可以得到唯一的表达式。太阳可以当作这样的一个质点。

接下来假设在四维地图（现在这个地图严重扭曲，不能把对应于太阳每个时刻位置的点组成的线拉伸开）中，在太阳引力场中运动的质点的路径将是图上任意两

* 万有引力场由一组物理量 g_{rs} 来说明，当万有引力场很小的时候，这些值除了 g_{44} 以外都为零，这就近似成为牛顿的万有引力势场。

† 这些方程代替了旧的拉普拉斯方程 $\nabla^2 V=0$。正如拉普拉斯方程是唯一一个完全不受普通空间坐标体系变化影响的二阶微分方程一样，爱因斯坦方程是唯一一个由相对论条件确定的方程。

‡ 结果是：不变的间隔 ds 可以由下式确定，$ds^2 = (1 - 2m/r)(dt^2 - dr^2) - r^2(d\theta^2 + \sin^2\theta d\phi^2)$，这四个坐标可以解释为时间和普通的球面极坐标。

positions of the sun at every instant of time) the path of a particle moving under the gravitation of the sun will be the most direct line between any two points on it, in the sense that the sum of all the intervals corresponding to all the elements of its path is the least possible.* Thus the equations of motion are written down. The result is this:

The motion of a particle differs only from that given by the Newtonian theory by the presence of an additional acceleration towards the sun equal to three times the mass of the sun (in gravitational units) multiplied by the square of the angular velocity of the planet about the sun.

In the case of the planet Mercury, this new acceleration is of the order of 10^{-8} times the Newtonian acceleration. Thus up to this order of accuracy Einstein's theory actually arrives at Newton's laws: surely no dethronement of Newton.

The effect of the additional acceleration can easily be expressed as a perturbation of the Newtonian elliptic orbit of the planet. It leads to the result that the major axis of the orbit must rotate in the plane of the orbit at the rate of 42.9″ per century.

Now it has long been known that the perihelion of Mercury does actually rotate at the rate of about 40″ per century, and Newtonian theory has never succeeded in explaining this, except by *ad hoc* assumptions of disturbing matter not otherwise known.

Thus Einstein's theory almost exactly accounts for the one outstanding failure of Newton's scheme, and, we may note, does not introduce any discrepancy where hitherto there was agreement.

The Deflection of Light by Gravitation

The new theory having justified itself so far, it was thought worth while for British astronomers to devote their main energies at the recent solar eclipse to testing its prediction of an entirely new phenomenon.

As was remarked above, the propagation of light in the ordinary case of freedom from gravitational effect is represented by the equation $ds = 0$.

This Einstein boldly transfer to his generalised theory. After all, it is quite a natural assumption. The propagation of light is a purely objective phenomenon. The emission of a disturbance from one point at one moment, and its arrival at another point at another moment, are events distinct and independent of the existence of an observer. Any law that connects them must be one which is independent of the map the observer uses; ds being an invariant quantity, $ds = 0$ expresses such an invariant law.

---

* This corresponds to the fact that in a field where there is no acceleration at all the path of a particle is the shortest distance between two points.

点之间最直的线，即与路径上所有组成部分对应的所有间隔之和尽可能最小。* 这样，就可以写出运动方程。结果如下：

一个质点的运动与牛顿理论给出的运动的不同之处在于多出了一个朝向太阳的加速度，这个加速度的值等于三倍的太阳质量（万有引力单位）乘以行星绕太阳运动的角速度的平方。

对于水星，这个新加速度的量级是牛顿加速度的 10^{-8}。这样，低于这个精确度，爱因斯坦理论就还原成了牛顿理论：牛顿理论当然不会失效。

这个多出来的加速度可以被看作是牛顿椭圆行星轨道的一种扰动。这就造成了轨道主轴在轨道平面中以每世纪 42.9 角秒的速度进动。

很久以前我们就知道，水星的近日点的确在以每世纪约 40 角秒的速度进动，牛顿理论从来没有成功地解释过这个现象，除非特意假设有一个在其他情况下未曾出现的干扰物体。

这样，爱因斯坦的理论就彻底解决了牛顿理论框架中的一个重要不足，并且，我们注意到，爱因斯坦理论和牛顿理论没有矛盾，至今它们仍然是统一的。

光在万有引力作用下的偏转

到目前为止，这个新理论的正确性已经得到了证明，英国天文学家们正将他们的主要精力放在近期的日食上，为的是检验此理论所预言的一个全新的现象，大家都认为这是一件值得做的事情。

正如上面所说的那样，在没有万有引力效应的情况下，光的传播可以表示为方程 $ds = 0$。

爱因斯坦大胆地将它移植到了他的广义理论中。毕竟，这是一个很自然的假设。光的传播是一个纯客观的现象。在某一时刻某一点发生的干扰，以及这个干扰在另一时刻到达另一点，这两者是不同的事件，与观察者是否存在无关。任何将它们联系起来的法则都一定不依赖于观察者所使用的地图；ds 是一个不变的量，$ds = 0$ 表示出了这样一个不变的法则。

* 它对应于这样一个事实：在一个没有加速度的场中，粒子运动的路径是两点之间的最短距离。

This leads at once to a law of variation of the velocity of light in the gravitational field of the sun.

$$v = c(1 - 2m/r)$$

Here m, as before, is the mass of the sun in gravitational units, and is equal to 1.47 kilometres, while c is the velocity of light at a great distance from the sun. Thus the path of a ray is the same as that if, on the ordinary view, it were travelling in a medium the refractive index of which was $(1 - 2m/r)^{-1}$. In this medium the refractive index would increase in approaching the sun, so that the rays would be bent round towards the sun in passing through it. The total amount of the deflection for a ray which just grazes the sun's surface works out to be 1.75″, falling off as the inverse of the distance of nearest approach.

The apparent position of a star near to the sun is thus further from the sun's centre than the true position. On the photographic plate in the actual observations made by the Eclipse Expedition the displacement of the star image is of the order of a thousandth of an inch. The measurements show without doubt such a displacement. The stars observed were, of course, not exactly at the edge of the sun's disc; but on reduction, allowing for the variation inversely as the distance, they give for the bend of a ray just grazing the sun the value 1.98″, with a probable error of 6 percent, in the case of the Sobral expedition, and of 1.64″ in the Principe expedition.

The agreement with the theory is close enough, but, of course, alternative possible causes of the shift have to be considered. Naturally, the suggestion of an actual refracting atmosphere surrounding the sun has been made. The existence of this, however, seems to be negatived by the fact that an atmosphere sufficiently dense to produce the refraction in question would extinguish the light altogether, as the rays would have to travel a million miles or so through it. The second suggestion, made by Prof. Anderson in *Nature* of December 4, that the observed displacement might be due to a refraction of the ray in travelling through the earth's atmosphere in consequence of a temperature gradient within the shadow cone of the moon, seems also to be negatived. Prof. Eddington estimates that it would require a change of temperature of about 20°C per minute at the observing station to produce the observed effect. Certainly no such temperature change as this has ever been noted; and, in fact, in Principe, at which the Cambridge expedition made its observations, there was practically no fall of temperature.

Gravitation and the Solar Spectrum

It was suggested by Einstein that a further consequence of his theory would be an apparent discrepancy of period between the vibrations of an atom in the intense gravitational field of the sun and the vibrations of a similar atom in the much weaker field of the earth. This is arrived at thus. An observer would not be able to infer the intensity of the gravitational field in which he was placed from any observations of atomic vibrations in the same field: that is, an observer on the sun would estimate the period of vibration of

这立刻就引出了在太阳引力场中光速不变的定律。

$$v = c(1-2m/r)$$

这里的 m 和前面一样，是太阳在万有引力单位下的质量，它相当于 1.47 千米，c 是远离太阳处的光速。这样，从通常的观点上看，一束光的行进路线和它在折射率为 $(1-2m/r)^{-1}$ 的介质中传播一样。在这个介质中，越接近太阳折射率就越大，所以光线在经过太阳的时候就会发生弯曲。一束刚好掠过太阳表面的光的偏转角度是 1.75 角秒，偏转角度会随着光线与太阳之间最小距离的倒数的减小而减小。

靠近太阳的恒星的视位置比它的真实位置离日心更远。在观测日食时所拍的照相底片上，恒星图像位移的数量级为千分之一英尺。测量结果毫无疑问地显示了这样的一个位移。当然，观测的恒星并非恰好在日面的边缘；但是可以利用偏转量和距离成反比的关系进行化规，在索布拉尔的观测队测算出一束刚好掠过太阳表面的光线的偏转角度是 1.98 角秒，允许的误差范围是 6%；在普林西比岛的观测队得到的结果是 1.64 角秒。

实验和理论已经足够吻合，但是，也必须考虑到引起位移的其他可能原因。很自然的，有人怀疑在太阳周围有一个能发生真正的折射效应的大气层。但是，它的存在却可以被以下因素否定，即产生这样的折射作用要求大气层足够厚，而这么厚的大气层会使光线消失，因为光线需要经过 100 万公里左右才能穿过去。第二种可能性是安德森教授在 12 月 4 日的《自然》上提出的。他认为，观察到的位移可能是由光线穿过地球大气层时的折射造成的，因为在月影锥内存在温度梯度，温度梯度可以引发折射现象，这个猜测也不成立。爱丁顿教授估算过，要产生我们观测到的效应，观测站的温度变化应达到每分钟 20℃之多。这样的温度变化当然从未有过；事实上，在剑桥考察队进行观测的普林西比岛，温度并未下降过。

万有引力和太阳光谱

爱因斯坦指出，他的理论还会进一步导出这样的结果：原子在太阳的强引力场中的振动周期和其在弱得多的地球引力场中的振动周期有明显的差别。这个结果是这样得到的：当一个观察者与他所观察的振动原子处在同一个引力场中时，他不可能通过观察原子的振动来判断这个引力场的强度。也就是说，一个在太阳上的观察者测量到的原子振动周期与一个相似原子在地球上的振动周期相同，前提是他本人

an atom there to be the same that he would find for a similar atom in the earth's field if he transported himself thither. But on transferring himself he automatically changes his scale of time; in the new scale of time the solar atom vibrates differently, and, therefore, is not synchronous with the terrestrial atom.

Observations of the solar spectrum so far are adverse to the existence of such an effect. What, then, is to be said? Is the theory wrong at this point? If so, it must be given up, in spite of its extraordinary success in respect of the other two phenomena.

Sir Joseph Larmor, however, is of opinion that Einstein's theory itself does not in reality predict the displacement at all. The present writer shares his opinion. Imagine, in fact, two identical atoms originally at a great distance from both sun and earth. They have the same period. Let an observer A accompany one of these into the gravitational field of the sun, and an observer B accompany the other into the field of the earth. In consequence of A and B having moved into different gravitational fields, they make different changes in their scales of time, so that actually the solar observer A will find a different period for the solar atom from that which B, on the earth, attributes to his atom. It is only when the two observers choose so to measure space and time that they consider themselves to be in identical gravitational fields that they will estimate the periods of the atoms alike. This is exactly what would happen if B transferred himself to the same position as A. Thus, though an important point remains to be cleared up, it cannot be said that it is one which at present weighs against Einstein's theory.

(**104**, 394-395; 1919)

来到地球。但是在他转换观测地点的时候，他会自动调整时间尺度；在新的时间尺度中，太阳引力场中的原子振动周期将发生变化，所以就与地球上的原子不同步了。

迄今为止，太阳光谱的观测结果并不支持这种效应的存在。接下来我们该说些什么呢？这个理论在这一点上是否错了？如果是这样，它必须被放弃，尽管它成功地解释了另外两个现象。

然而，约瑟夫·拉莫尔爵士认为，事实上爱因斯坦的理论本身并没有预言过这种移位效应。本文作者也同意他的观点。事实上，我们可以想象，两个相同的原子最初处于既远离太阳又远离地球的某地。它们具有相同的周期。让观察者 A 伴随着其中一个原子来到太阳引力场中，让观察者 B 伴随着另一个原子来到地球引力场中。由于 A 和 B 进入的引力场不同，他们的时间尺度发生的变化也不同，因此，太阳上的观察者 A 将发现太阳上原子的振动周期与地球上的观察者 B 看到的地球上原子的振动周期不同。只有当两个观察者在同一个引力场中来测量时空时，他们才会判断出两原子的周期相同。如果 B 来到 A 的位置，就会发生以上所说的情况。所以，尽管还有一个要点有待澄清，但我们目前还不能说这是一个与爱因斯坦理论相悖的现象。

（王静 翻译；鲍重光 审稿）

A Brief Outline of the Development of the Theory of Relativity*

A. Einstein

Editor's Note

By 1921, Einstein's theory of relativity was widely accepted. Here he describes the theory's historical development, starting from the aim to rid physics of reliance on action at a distance. Maxwell had achieved this for electricity and magnetism, and his mathematical formulation led others to suppose that all space is filled with an ether that carried the electric and magnetic fields. This led to difficulties, which Einstein overcame by abandoning the belief that events may be simultaneous regardless of an observer's motion. But this new understanding didn't encompass gravity. The supposition that gravity and inertia are identical prompted the general theory of relativity. Einstein wonders if gravitational and electrical phenomena might be unified in a theory of all nature's forces—a theory physicists still seek today.

THERE is something attractive in presenting the evolution of a sequence of ideas in as brief a form as possible, and yet with a completeness sufficient to preserve throughout the continuity of development. We shall endeavour to do this for the Theory of Relativity, and to show that the whole ascent is composed of small, almost self-evident steps of thought.

The entire development starts off from, and is dominated by, the idea of Faraday and Maxwell, according to which all physical processes involve a continuity of action (as opposed to action at a distance), or, in the language of mathematics, they are expressed by partial differential equations. Maxwell succeeded in doing this for electro-magnetic processes in bodies at rest by means of the conception of the magnetic effect of the vacuum-displacement-current, together with the postulate of the identity of the nature of electro-dynamic fields produced by induction, and the electro-static field.

The extension of electro-dynamics to the case of moving bodies fell to the lot of Maxwell's successors. H. Hertz attempted to solve the problem by ascribing to empty space (the ether) quite similar physical properties to those possessed by ponderable matter; in particular, like ponderable matter, the ether ought to have at every point a definite velocity. As in bodies at rest, electro-magnetic or magneto-electric induction ought to be determined by the rate of change of the electric or magnetic flow respectively, provided that these velocities of alteration are referred to surface elements moving with the body. But the theory of Hertz was opposed to the fundamental experiment of Fizeau on the

* Translated by Dr. Robert W. Lawson.

相对论发展概述*

爱因斯坦

编者按

爱因斯坦的相对论在1921年得到了大家的广泛认可。在这篇演讲稿中，爱因斯坦从物理学力图摆脱超距作用的影响开始，对这个理论的历史沿革进行了回顾。麦克斯韦的电磁理论完成了这一使命，其他物理学家根据他的数学公式提出了以太假说，即认为所有的空间中都充满了作为电场和磁场媒介的以太。这使相对性原理遇到了困难，而爱因斯坦通过放弃事件的同时性可能与观测者的运动无关的观点摆脱了这个困境。但是这种新的理解没有考虑到重力。爱因斯坦猜测重力和惯性可能具有同一性，这一构想促使他提出了广义相对论。爱因斯坦设想重力和电现象或许也可以用一种适用于所有自然力的理论统一起来，这也是今天的物理学家们正在寻找的理论。

用尽可能简练的语言来阐述一系列观念的演变，但仍充分完整地把这种演变的连续性保留下来，这是一件很吸引人的事。在讲述相对论的发展时，我们将尽力做到这一点，并说明其整个发展过程是由一系列细微而又不言而喻的思维过程构成的。

整个发展历程始于并受制于法拉第和麦克斯韦的观念，按照他们的观念，所有的物理过程都包含连续作用（与超距作用相反），或者用数学语言来表示就是利用偏微分方程来描述物理过程。麦克斯韦利用真空位移电流的磁效应概念以及感生电动力场和静电场在本质上完全相同这一假定，成功地构筑了描述静止介质中电磁过程的偏微分方程。

把电动力学理论推广到运动物体的重任落在了麦克斯韦的后继者的身上。赫兹试图通过赋予虚空（以太）与一般有重物质颇类似的物理性质来解决这个问题。特别是，与有重物质一样，以太在空间的每一点上都应该有确定的速度。正如静止物体那样，如果电流或磁流的变化速度是以随物体一起运动的曲面元作为参考的话，那么电磁感应或者磁电感应应当分别由电流或磁流的变化率决定。但是赫兹的理论与斐索有关光在流动液体中传播的基本实验相矛盾。就是说，麦克斯韦理论对运动

* 由罗伯特·劳森博士翻译。

propagation of light in flowing liquids. The most obvious extension of Maxwell's theory to the case of moving bodies was incompatible with the results of experiment.

At this point, H. A. Lorentz came to the rescue. In view of his unqualified adherence to the atomic theory of matter, Lorentz felt unable to regard the latter as the seat of continuous electro-magnetic fields. He thus conceived of these fields as being conditions of the ether, which was regarded as continuous. Lorentz considered the ether to be intrinsically independent of matter, both from a mechanical and a physical point of view. The ether did not take part in the motions of matter, and a reciprocity between ether and matter could be assumed only in so far as the latter was considered to be the carrier of attached electrical charges. The great value of the theory of Lorentz lay in the fact that the entire electro-dynamics of bodies at rest and of bodies in motion was led back to Maxwell's equations of empty space. Not only did this theory surpass that of Hertz from the point of view of method, but with its aid H. A. Lorentz was also pre-eminently successful in explaining the experimental facts.

The theory appeared to be unsatisfactory only in *one* point of fundamental importance. It appeared to give preference to one system of coordinates of a particular state of motion (at rest relative to the ether) as against all other systems of coordinates in motion with respect to this one. In this point the theory seemed to stand in direct opposition to classical mechanics, in which all inertial systems which are in uniform motion with respect to each other are equally justifiable as systems of coordinates (Special Principle of Relativity). In this connection, all experience also in the realm of electro-dynamics (in particular Michelson's experiment) supported the idea of the equivalence of all inertial systems, *i.e.* was in favour of the special principle of relativity.

The Special Theory of Relativity owes its origin to this difficulty, which, because of its fundamental nature, was felt to be intolerable. This theory originated as the answer to the question: Is the special principle of relativity really contradictory to the field equations of Maxwell for empty space? The answer to this question appeared to be in the affirmative. For if those equations are valid with reference to a system of coordinates K, and we introduce a new system of coordinates K′ in conformity with the—to all appearances readily establishable—equations of transformation

$$\left.\begin{aligned} x' &= x - vt \\ y' &= y \\ z' &= z \\ t' &= t \end{aligned}\right\} \text{(Galileo transformation)},$$

then Maxwell's field equations are no longer valid in the new coordinates (x', y', z', t'). But appearances are deceptive. A more searching analysis of the physical significance of space and time rendered it evident that the Galileo transformation is founded on arbitrary assumptions, and in particular on the assumption that the statement of simultaneity has a meaning which is independent of the state of motion of the system of coordinates used. It was shown that the field equations for *vacuo* satisfy the special principle of relativity,

物体的这种最直接的推广与实验结果不符。

正在这时，洛仑兹进行了补救。由于洛仑兹是物质原子理论的忠实支持者，所以他觉得不能把物质看成是连续电磁场的所在地。因此，他设想这些场是连续的以太的某种状态。洛仑兹认为，从力学和物理学两方面的观点来看，以太在本质上与物质无关。以太不参与物质的运动，以太和物质之间的相互关系仅在于，物质被看成是所附电荷的载体。洛仑兹理论的重要价值在于，它使包括静止物体和运动物体在内的整个电动力学回归到了真空中的麦克斯韦方程。该理论不仅在方法论上超越了赫兹的理论，而且洛仑兹还利用它非常成功地解释了许多实验事实。

这个理论似乎仅仅在一个重要的基本点上不能令人满意。这就是，似乎某个具有特殊运动状态的坐标系（它相对于以太是静止的）要比相对于这个坐标系运动的所有其他坐标系更加优越。从这一点上来看，这个理论好像违背了经典力学，因为在经典力学中，所有相互间做匀速运动的惯性系都同样有理由被用来当作坐标系（狭义相对性原理）。在这一点上，包括电动力学领域在内的所有经验（尤其是迈克尔逊实验）都支持所有惯性系均等价这一观点，即都支持狭义相对性原理。

狭义相对论就是为解决这一困难而诞生的，这个困难由于它具有的根本性而无法让人容忍。狭义相对论最初被用于解答下述问题：狭义相对性原理真的与真空中的麦克斯韦场方程矛盾吗？答案似乎是肯定的。因为，如果某些方程对于坐标系 K 是成立的，而且我们引进一个新的坐标系 K′，使它符合于（显然容易做到）如下的变换方程：

$$\left.\begin{aligned} x' &= x - vt \\ y' &= y \\ z' &= z \\ t' &= t \end{aligned}\right\}\text{（伽利略变换），}$$

那么麦克斯韦场方程组在新的坐标系 (x', y', z', t') 中不再成立。但表面现象是靠不住的，在更透彻地分析时间和空间的物理意义后发现，伽利略变换是建立在几个相当任意的假设上面的，尤其是假设同时性的陈述与所使用的坐标系的运动状态无关。研究表明，如果我们利用下面的变换方程，则真空中的场方程可以满足狭义

provided we make use of the equations of transformation stated below:

$$\left.\begin{aligned} x' &= \frac{x - vt}{\sqrt{1 - v^2/c^2}} \\ y' &= y \\ z' &= z \\ t' &= \frac{t - vx/c^2}{\sqrt{1 - v^2/c^2}} \end{aligned}\right\} \text{(Lorentz transformation)}$$

In these equations x, y, z represent the coordinates measured with measuring-rods which are at rest with reference to the system of coordinates, and t represents the time measured with suitably adjusted clocks of identical construction, which are in a state of rest.

Now in order that the special principle of relativity may hold, it is necessary that all the equations of physics do not alter their form in the transition from one inertial system to another, when we make use of the Lorentz transformation for the calculation of this change. In the language of mathematics, all systems of equations that express physical laws must be co-variant with respect to the Lorentz transformation. Thus, from the point of view of method, the special principle of relativity is comparable to Carnot's principle of the impossibility of perpetual motion of the second kind, for, like the latter, it supplies us with a general condition which all natural laws must satisfy.

Later, H. Minkowski found a particularly elegant and suggestive expression for this condition of co-variance, one which reveals a formal relationship between Euclidean geometry of three dimensions and the space-time continuum of physics.

Euclidean Geometry of Three Dimensions.	*Special Theory of Relativity.*
Corresponding to two neighbouring points in space, there exists a numerical measure (distance ds) which conforms to the equation $ds^2 = dx_1^2 + dx_2^2 + dx_3^2$	Corresponding to two neighbouring points in space-time (point events), there exists a numerical measure (distance ds) which conforms to the equation $ds^2 = dx_1^2 + dx_2^2 + dx_3^2 + dx_4^2$
It is independent of the system of coordinates chosen, and can be measured with the unit measuring-rod.	It is independent of the inertial system chosen, and can be measured with the unit measuring-rod and a standard clock. x_1, x_2, x_3 are here rectangular coordinates, whilst $x_4 = \sqrt{-1}ct$ is the time multiplied by the imaginary unit and by the velocity of light.
The permissible transformations are of such a character that the expression for ds^2 is invariant, *i.e.* the linear orthogonal transformations are permissible.	The permissible transformations are of such a character that the expression for ds^2 is invariant, *i.e.* those linear orthogonal substitutions are permissible which maintain the semblance of reality of x_1, x_2, x_3, x_4. These substitutions are the Lorentz transformations.
With respect to these transformations, the laws of Euclidean geometry are invariant.	With respect to these transformations, the laws of physics are invariant.

From this it follows that, in respect of its *rôle* in the equations of physics, though not with regard to its physical significance, time is equivalent to the space coordinates (apart from the relations of reality). From this point of view, physics is, as it were, a Euclidean geometry of four dimensions, or, more correctly, a statics in a four-dimensional Euclidean continuum.

相对性原理：

$$\left.\begin{aligned} x' &= \frac{x - vt}{\sqrt{1 - v^2/c^2}} \\ y' &= y \\ z' &= z \\ t' &= \frac{t - vx/c^2}{\sqrt{1 - v^2/c^2}} \end{aligned}\right\} \text{（洛仑兹变换）}$$

在上述方程中 x、y、z 表示位置坐标，用相对于坐标系静止的量尺来测量；t 表示时间，用处于静止状态的经过适当校准并且具有相同构造的时钟来测量。

现在，为了使狭义相对性原理成立，要求所有的物理方程在使用洛仑兹变换来计算它们从一个惯性系到另一个惯性系的转换时其形式保持不变。用数学语言来描述就是，所有描述物理定律的方程相对于洛仑兹变换必须是协变的。因此，从方法论的角度来看，狭义相对性原理可以与第二种永恒运动不能实现的卡诺定理相比拟，因为正如卡诺定理一样，狭义相对性原理为我们提供了所有自然规律必须遵守的一般法则。

随后，闵可夫斯基找到了一个特别简洁而又极具启发性的方式来表述这个协变条件，揭示了三维欧几里德几何学与物理学中时空连续统之间的对应关系。

三维欧几里德几何学	狭义相对论
对应于空间中两个相邻的点，存在一种按如下方程计算的数值度量（距离 ds）： $ds^2 = dx_1^2 + dx_2^2 + dx_3^2$	对应于时空中两个相邻的点（点事件），存在一种按如下方程计算的数值度量（距离 ds）： $ds^2 = dx_1^2 + dx_2^2 + dx_3^2 + dx_4^2$
该距离与所选的坐标系无关，并可以用单位量尺测量。	该距离与所选择的惯性系无关，并可以用单位量尺和标准时钟测量。这里的 x_1，x_2，x_3 是直角坐标系中的坐标，而 $x_4=\sqrt{-1}ct$ 是时间和虚数单位以及光速的乘积。
保持 ds^2 表达式不变的变换是允许的，即线性正交变换是允许的。	保持 ds^2 表达式不变的变换是允许的，即那些线性正交变换是允许的，它们保持了 x_1，x_2，x_3，x_4 表面上的实数性，这些变换就是洛仑兹变换。
相对于这些变换，欧几里德几何学中的定律保持不变。	相对于这些变换，物理学中的定律保持不变。

由此可以看出，时间坐标在物理方程中的**作用**与空间坐标等价（除了实数性之外），虽然不是就其物理意义而言。按照这种观点，物理学过去和现在都是一种四维欧几里德几何学，或者，更确切地说，是四维欧几里德连续统中的一种静力学。

The development of the special theory of relativity consists of two main steps, namely, the adaptation of the space-time "metrics" to Maxwell's electro-dynamics, and an adaptation of the rest of physics to that altered space-time "metrics". The first of these processes yields the relativity of simultaneity, the influence of motion on measuring-rods and clocks, a modification of kinematics, and in particular a new theorem of addition of velocities. The second process supplies us with a modification of Newton's law of motion for large velocities, together with information of fundamental importance on the nature of inertial mass.

It was found that inertia is not a fundamental property of matter, nor, indeed, an irreducible magnitude, but a property of energy. If an amount of energy E be given to a body, the inertial mass of the body increases by an amount E/c^2, where c is the velocity of light *in vacuo*. On the other hand, a body of mass m is to be regarded as a store of energy of magnitude mc^2.

Furthermore, it was soon found impossible to link up the science of gravitation with the special theory of relativity in a natural manner. In this connection I was struck by the fact that the force of gravitation possesses a fundamental property, which distinguishes it from electro-magnetic forces. All bodies fall in a gravitational field with the same acceleration, or—what is only another formulation of the same fact—the gravitational and inertial masses of a body are numerically equal to each other. This numerical equality suggests identity in character. Can gravitation and inertia be identical? This question leads directly to the General Theory of Relativity. Is it not possible for me to regard the earth as free from rotation, if I conceive of the centrifugal force, which acts on all bodies at rest relatively to the earth, as being a "real" field of gravitation, or part of such a field? If this idea can be carried out, then we shall have proved in very truth the identity of gravitation and inertia. For the same property which is regarded as *inertia* from the point of view of a system not taking part in the rotation can be interpreted as *gravitation* when considered with respect to a system that shares the rotation. According to Newton, this interpretation is impossible, because by Newton's law the centrifugal field cannot be regarded as being produced by matter, and because in Newton's theory there is no place for a "real" field of the "Koriolis-field" type. But perhaps Newton's law of field could be replaced by another that fits in with the field which holds with respect to a "rotating" system of coordinates? My conviction of the identity of inertial and gravitational mass aroused within me the feeling of absolute confidence in the correctness of this interpretation. In this connection I gained encouragement from the following idea. We are familiar with the "apparent" fields which are valid relatively to systems of coordinates possessing arbitrary motion with respect to an inertial system. With the aid of these special fields we should be able to study the law which is satisfied in general by gravitational fields. In this connection we shall have to take account of the fact that the ponderable masses will be the determining factor in producing the field, or, according to the fundamental result of the special theory of relativity, the energy density—a magnitude having the transformational character of a tensor.

狭义相对论的发展经历了两个主要阶段，这就是使时空“度规”适合于麦克斯韦电动力学，以及使物理学中的其余部分适合于这个新的时空“度规”。第一个阶段的成果有同时性的相对性、运动对量尺及时钟的影响、运动学的修正，特别是还有新的速度相加定理。第二个阶段给我们提供了在高速情况下对牛顿运动定律的修正，以及关于惯性质量本质的具有基本重要性的知识。

研究表明惯性不是物质的基本属性，也不是一个不能分解的基本量，而只是能量的一个属性。如果我们赋予一个物体大小为 E 的能量，则此物体的惯性质量将增加 E/c^2，这里 c 是光在真空中的传播速度。同样，一个质量为 m 的物体将被认为具有 mc^2 的能量。

此外，我们很快发现很难把引力科学同狭义相对论以自然的方式联系起来。这种情况使我意识到引力具有一种不同于电磁力的基本性质。在引力场中，所有物体都以相同的加速度下落，或者说，一个物体的引力质量和惯性质量在数值上是相等的（这只不过是同一个事实的另一种表达方式）。这种数值上的相同暗示着两者本质上的等同。引力和惯性能够等同吗？这个问题直接导致了广义相对论的产生。如果我把作用于所有相对于地球静止的物体上的离心力想象成是一个“真实的”引力场，或者是这种引力场的一部分，那我难道不能认为地球是不转动的吗？如果这个想法能够实现，那么我们已经真正证明了引力和惯性的等同性。在不随地球转动的参考系里看来是**惯性**的这一特性，在随地球一起转动的参考系里可以被解释为**引力**。按照牛顿的观点，这样的解释是说不通的，因为牛顿定律告诉我们，离心力场不能被看作是由物质产生的，而且因为在牛顿的理论中，没有把“科里奥利场”这种类型的场当成是“真实的”场。但是，或许牛顿的有关场的定律可以用场的另外一种定律取代，这种定律既适合于这种场又在“转动”坐标系中成立？我坚信惯性质量与引力质量的等同性，这使我有绝对的信心认为上述解释是正确的。就这一点来说，我从以下观点中受到了鼓舞。我们熟悉“表观的”场，这些场在那些相对于一个惯性系做任意运动的坐标系中是有效的。借助于这些特殊的场，我们应当有可能研究引力场通常所满足的定律。关于这一点，我们不得不考虑这样的事实，即有重物质是产生场的决定性因素。或者可以这样表达，按照狭义相对论的基本结果，能量密度这个具有张量变换特性的物理量是产生场的决定因素。

On the other hand, considerations based on the metrical results of the special theory of relativity led to the result that Euclidean metrics can no longer be valid with respect to accelerated systems of coordinates. Although it retarded the progress of the theory several years, this enormous difficulty was mitigated by our knowledge that Euclidean metrics holds for small domains. As a consequence, the magnitude *ds*, which was physically defined in the special theory of relativity hitherto, retained its significance also in the general theory of relativity. But the coordinates themselves lost their direct significance, and degenerated simply into numbers with no physical meaning, the sole purpose of which was the numbering of the space-time points. Thus in the general theory of relativity the coordinates perform the same function as the Gaussian coordinates in the theory of surfaces. A necessary consequence of the preceding is that in such general coordinates the measurable magnitude *ds* must be capable of representation in the form

$$ds^2 = \sum_{uv} g_{uv}\, dx_u\, dx_v ,$$

where the symbols g_{uv} are functions of the space-time coordinates. From the above it also follows that the nature of the space-time variation of the factors g_{uv} determines, on one hand the space-time metrics, and on the other the gravitational field which governs the mechanical behaviour of material points.

The law of the gravitational field is determined mainly by the following conditions: First, it shall be valid for an arbitrary choice of the system of coordinates; secondly, it shall be determined by the energy tensor of matter; and thirdly, it shall contain no higher differential coefficients of the factors g_{uv} than the second, and must be linear in these. In this way a law was obtained which, although fundamentally different from Newton's law, corresponded so exactly to the latter in the deductions derivable from it that only very few criteria were to be found on which the theory could be decisively tested by experiment.

The following are some of the important questions which are awaiting solution at the present time. Are electrical and gravitational fields really so different in character that there is no formal unit to which they can be reduced? Do gravitational fields play a part in the constitution of matter, and is the continuum within the atomic nucleus to be regarded as appreciably non-Euclidean? A final question has reference to the cosmological problem. Is inertia to be traced to mutual action with distant masses? And connected with the latter: Is the spatial extent of the universe finite? It is here that my opinion differs from that of Eddington. With Mach, I feel that an affirmative answer is imperative, but for the time being nothing can be proved. Not until a dynamical investigation of the large systems of fixed stars has been performed from the point of view of the limits of validity of the Newtonian law of gravitation for immense regions of space will it perhaps be possible to obtain eventually an exact basis for the solution of this fascinating question.

(**106**, 782-784; 1921)

另一方面，基于对狭义相对论度规结果的考虑导致了这样的结论，即欧几里德度规不再适用于加速参考系。尽管它使理论的进程延迟了几年，但是这个巨大的困难在我们认识到欧几里德度规对于小的区域依然适用之后就变得比较容易解决了。结果，ds 这个迄今在狭义相对论中定义的物理量，在广义相对论中其物理含义仍保持不变。但是坐标本身失去了直接的意义，而完全退化成没有物理意义的数字，它们的唯一用途是标记时空点，因此广义相对论中的坐标与曲面论中高斯坐标的作用相同。以上的叙述必然给出这样一个结论：可测量的量 ds 必定可以用这种广义坐标表达为以下形式：

$$ds^2 = \sum_{uv} g_{uv}\, dx_u\, dx_v \text{ ,}$$

这里符号 g_{uv} 是时空坐标的函数。以上的分析还表明，因子 g_{uv} 随时空变化的特性，一方面决定了时空度规，另一方面还决定了支配质点力学行为的引力场。

引力场的定律主要取决于以下几个条件：第一，它应该在任意选择的坐标系中都是有效的；第二，它应当由物质的能量张量决定；第三，它所包含的因子 g_{uv} 的微分系数最高不超过二阶，并且它相对于这些微分都是线性的。这样我们就得到了一个定律，虽然它在本质上不同于牛顿定律，但是在由它给出的推论中它与牛顿定律吻合得很好，以至于发现只有很少几个判据能够用来对它进行决定性的实验检验。

下面是一些目前尚待解决的重要问题。电场和引力场在特性上真的有那么不同以至于不能用一个形式上的统一体把它们都包含进去吗？引力场对物质的结构起一定作用吗？原子核内部的连续统在相当程度上可以被看作是非欧氏的吗？最后一个问题涉及到宇宙论，即惯性是否源自于远距离物质之间的相互作用？与后者相关的是：宇宙的空间范围是有限的吗？在这一点上我的观点和爱丁顿的观点相悖。与马赫一样，我感到迫切需要一个肯定的答案，但暂时还找不到证据证明。只有从这样的观点（即认为在广袤宇宙空间使用牛顿引力定律具有局限性的观点）出发来对庞大恒星系的动力学进行研究之后，才有可能最终获得解决这一令人困惑的问题的确切依据。

（王锋 翻译；张元仲 审稿）

Atomic Structure

N. Bohr

Editor's Note

Here Danish physicist Niels Bohr ponders the possibility of explaining atomic properties on the basis of the new quantum theory. Physicists had conjectured about how the grouping of elements in the periodic table might reflect specific patterns of electrons arranged around the nucleus, yet without explaining how such patterns arise. As Bohr notes, his model of the hydrogen atom suggested that electrons may fall into distinct shells. Much of the periodic table, says Bohr, can be understood as the successive filling of these shells. The stability of filled shells would explain the chemical inactivity of inert gases such as helium and argon. Bohr's model was supplanted by a more mathematically rigorous quantum theory, yet much of his qualitative picture remains intact today.

IN a letter to *Nature* of November 25 last Dr. Norman Campbell discusses the problem of the possible consistency of the assumptions about the motion and arrangement of electrons in the atom underlying the interpretation of the series spectra of the elements based on the application of the quantum theory to the nuclear theory of atomic structure, and the apparently widely different assumptions which have been introduced in various recent attempts to develop a theory of atomic constitution capable of accounting for other physical and chemical properties of the elements. Dr. Campbell puts forward the interesting suggestion that the apparent inconsistency under consideration may not be real, but rather appear as a consequence of the formal character of the principles of the quantum theory, which might involve that the pictures of atomic constitution used in explanations of different phenomena may have a totally different aspect, and nevertheless refer to the same reality. In this connection he directs attention especially to the so-called "principle of correspondence", by the establishment of which it has been possible—notwithstanding the fundamental difference between the ordinary theory of electromagnetic radiation and the ideas of the quantum theory—to complete certain deductions based on the quantum theory by other deductions based on the classical theory of radiation.

In so far as it must be confessed that we do not possess a complete theory which enables us to describe in detail the mechanism of emission and absorption of radiation by atomic systems, I naturally agree that the principle of correspondence, like all other notions of the quantum theory, is of a somewhat formal character. But, on the other hand, the fact that it has been possible to establish an intimate connection between the spectrum emitted by an atomic system—deduced according to the quantum theory on the assumption of a certain type of motion of the particles of the atom—and the constitution of the radiation, which, according to the ordinary theory of electromagnetism, would result from the same type of motion, appears to me do afford an argument in favour of the reality of the assumptions of the spectral theory of a kind scarcely compatible with Dr. Campbell's suggestion. On

原子结构

玻尔

编者按

在这篇文章中丹麦物理学家尼尔斯·玻尔反复思索是否可以用新的量子理论来解释原子的性质。物理学家们猜测周期表中的元素分组可能在某种程度上反映了核外电子的排布情况，但没有说明这种排布模式是如何形成的。玻尔指出，他的氢原子模型可以说明电子为什么会填入不同的壳层。玻尔说，周期表中大部分元素的核外电子都可以被认为是连续填充这些壳层的。充满电子的壳层结构比较稳定，因而表现出化学反应上的惰性，如氦、氩等惰性气体。虽然玻尔的模型已经被数学上更为严密的量子理论取代，但他对核外电子排布的大部分定性描述至今仍然有效。

诺曼·坎贝尔博士在去年 11 月 25 日写给《自然》的信中，谈到有关原子中电子运动和排布的假说是否可能保持前后一致的问题，这关系到量子理论能否作为原子结构的核心理论对一系列元素的光谱进行解释的问题，为了建立一个能够解释元素其他物理化学性质的原子组成理论，人们在最近的研究中提出了许多差异很大的假设。坎贝尔博士的观点令人振奋，他提出，表面上的矛盾可能不是真的，而是因为量子理论的原理具有表观化的特征，其对原子构成的描述用于解释不同的现象时可能使用完全不同的形式。他把人们对这个问题的注意力转向了所谓的“对应原理”，尽管普通电磁辐射理论与量子理论存在本质上的不同，但在建立了“对应原理”之后，就可能可以用经典辐射理论的结论推导出量子力学理论中的某些推论。

必须承认，我们至今尚未建立起一个完整的理论来详细地描述原子系统发射辐射和吸收辐射的机制，我当然同意“对应原理”像所有其他的量子理论概念一样，在某种意义上也带有形式化的特征。但是，从另一个角度上说，现在已经有可能在一个原子系统——根据基于假设原子中的粒子做某种形式的运动的量子理论推导得到——的发射光谱和由同样类型的运动产生的放射物的构成之间建立一种紧密的关系，根据普通的电磁学理论，我认为它提供了一个论点，这个论点支持一种与坎贝尔博士的建议几乎完全不相符的光谱理论假说。相反，如果我们承认用量子理论解

the contrary, if we admit the soundness of the quantum theory of spectra, the principle of correspondence would seem to afford perhaps the strongest inducement to seek an interpretation of the other physical and chemical properties of the elements on the same lines as the interpretation of their series spectra; and in this letter I should like briefly to indicate how it seems possible by an extended use of this principle to overcome certain fundamental difficulties hitherto involved in the attempts to develop a general theory of atomic constitution based on the application of the quantum theory to the nucleus atom.

The common character of theories of atomic constitution has been the endeavour to find configurations and motions of the electrons which would seem to offer an interpretation of the variations of the chemical properties of the elements with the atomic number as they are so clearly exhibited in the well-known periodic law. A consideration of this law leads directly to the view that the electrons in the atom are arranged in distinctly separate groups, each containing a number of electrons equal to one of the periods in the sequence of the elements, arranged according to increasing atomic number. In the first attempts to obtain a definite picture of the configuration and motion of the electrons in these groups it was assumed that the electrons within each group at any moment were placed at equal angular intervals on a circular orbit with the nucleus at the centre, while in later theories this simple assumption has been replaced by the assumptions that the configurations of electrons within the various groups do not possess such simple axial symmetry, but exhibit a higher degree of symmetry in space, it being assumed, for instance, that the configuration of the electrons at any moment during their motions possesses polyhedral symmetry. All such theories involve, however, the fundamental difficulty that no interpretation is given why these configurations actually appear during the formation of the atom through a process of binding of the electrons by the nucleus, and why the constitution of the atom is essentially stable in the sense that the original configuration is reorganized if it be temporarily disturbed by external agencies. If we reckon with no other forces between the particles except the attraction and repulsion due to their electric charges, such an interpretation claims clearly that there must exist an intimate interaction or "coupling" between the various groups of electrons in the atom which is essentially different from that which might be expected if the electrons in different groups are assumed to move in orbits quite outside each other in such a way that each group may be said to form a "shell" of the atom, the effect of which on the constitution of the outer shells would arise mainly from the compensation of a part of the attraction from the nucleus due to the charge of the electrons.

These considerations are seen to refer to essential features of the nucleus atom, and so far to have no special relation to the character of the quantum theory, which was originally introduced in atomic problems in the hope of obtaining a rational interpretation of the stability of the atom. According to this theory an atomic system possesses a number of distinctive states, the co-called "stationary states", in which the motion can be described by ordinary mechanics, and in which the atom can exist, at any rate for a time, without emission of energy radiation. The characteristic radiation from the atom is emitted only during a transition between two such states, and this process of transition cannot be described by ordinary mechanics, any more than the character of the emitted radiation

释光谱是合理的，那么“对应原理”在解释元素的系列光谱以及元素的其他物理化学性质方面也许能够得出最可信的推论；在这封信中，我将简单地介绍当我们在试图把量子理论应用于原子核系统以构建一个普适的原子组成理论时，是如何广泛地运用这一原则来克服目前遇到的主要难题的。

一系列原子组成理论的共性是都在极力寻找电子的排列和运动规律以解释元素的化学性质为什么会随原子序数的增加而发生周期性变化，正如众所周知的周期率明确指出的那样。周期率使人们猜想，原子中的电子可以被分为不同的组，每一组包括的电子数等于元素序列的一个周期。在早先试图明确描述各组中电子排布和运动的理论中，人们假设每一组中的电子在任一时刻都以相等的角间距排布在以原子核为中心的圆形轨道上，而在后来的理论中，这一简单的假设被新的假设取代，即认为各组中的电子排布不具有这种单一的轴对称，但在空间上则表现出更高程度的对称。例如，假设在电子运动的任一时刻它们的排布呈多面体对称。但是，所有这些理论都难以解释为什么通过原子核束缚电子而生成原子的过程中会出现这样的排布，也不能解释当外部介质暂时侵入时为什么原子的结构在原始排列重组后依然保持基本稳定。如果我们设想，除了由于所带电荷产生的吸引力或排斥力之外，粒子之间再无其他作用力，那么这样的解释就明确要求在原子中不同组的电子之间必须存在一种紧密的相互作用或“耦合”，这完全不同于认为各组电子运动轨道之间的距离也许远到足以使每一组电子单独形成原子的一个“壳层”的假设，耦合效应对外层电子的影响主要是部分地抵消由于电子带负电荷而受到的来自原子核的吸引力。

上述论点被认为涉及到核原子的基本特征，与量子理论的特殊性没有关系，人们把量子理论引入原子体系的初衷是希望它能够对原子的稳定性给出合理的解释。根据量子理论，原子体系中存在若干种不同的状态，即所谓的“定态”，电子在定态中的运动不能用一般理论来解释，在某一段时间内，原子可以以任意速率运动，但不会以辐射方式释放能量。只有在两个定态之间发生跃迁时原子才会发射特征辐射，而这种跃迁过程不能用一般理论来描述，更不用说利用普通的电磁理论根据运动规律计算发射的特征辐射了。与普通电磁理论形成鲜明的对照，量子理论假设跃迁通

can be calculated from the motion by the ordinary theory of electro-magnetism, it being, in striking contrast to this theory, assumed that the transition is always followed by an emission of monochromatic radiation the frequency of which is determined simply from the difference of energy in the two states. The application of the quantum theory to atomic problems—which took its starting point from the interpretation of the simple spectrum of hydrogen, for which no *a priori* fixation of the stationary states of the atoms was needed—has in recent years been largely extended by the development of systematic methods for fixing the stationary states corresponding to certain general classes of mechanical motions. While in this way a detailed interpretation of spectroscopic results of a very different kind has been obtained, so far as phenomena which depend essentially on the motion of one electron in the atom were concerned, no definite elucidation has been obtained with regard to the constitution of atoms containing several electrons, due to the circumstance that the methods of fixing stationary states were not able to remove the arbitrariness in the choice of the number and configurations of the electrons in the various groups, or shells, of the atom. In fact, the only immediate consequence to which they lead is that the motion of every electron in the atom will on a first approximation correspond to one of the stationary states of a system consisting of a particle moving in a central field of force, which in their limit are represented by the various circular or elliptical stationary orbits which appear in Sommerfeld's theory of the fine structure of the hydrogen lines. A way to remove the arbitrariness in question is opened, however, by the introduction of the correspondence principle, which gives expression to the tendency in the quantum theory to see not merely a set of formal rules for fixing the stationary states of atomic systems and the frequency of the radiation emitted by the transitions between these states, but rather an attempt to obtain a rational generalization of the electromagnetic theory of radiation which exhibits the discontinuous character necessary to account for the essential stability of atoms.

Without entering here on a detailed formulation of the correspondence principle, it may be sufficient for the present purpose to say that it establishes an intimate connection between the character of the motion in the stationary states of an atomic system and the possibility of a transition between two of these states, and therefore offers a basis for a theoretical examination of the process which may be expected to take place during the formation and reorganisation of an atom. For instance, we are led by this principle directly to the conclusion that we cannot expect in actual atoms configurations of the type in which the electrons within each group are arranged in rings or configurations of polyhedral symmetry, because the formation of such configurations would claim that all the electrons within each group should be originally bound by the atom at the same time. On the contrary, it seems necessary to seek the configurations of the electrons in the atoms among such configurations as may be formed by the successive binding of the electrons one by one, a process the last stages of which we may assume to witness in the emission of the series spectra of the elements. Now on the correspondence principle we are actually led to a picture of such a process which not only affords a detailed insight into the structure of these spectra, but also suggests a definite arrangement of the electrons in the atom of a type which seems suitable to interpret the high-frequency spectra and the chemical properties of the elements. Thus from a consideration of the possible transitions

常伴随着发出单色辐射的过程，频率只取决于两个状态之间的能量差。最初应用量子理论解决原子中的问题是从解释简单的氢原子光谱开始的，因为氢原子不需要**先验地**确定定态，现在人们根据特定力学运动的类别已经得到了确定定态的系统化方法，因而量子力学理论已经被广泛地用于解决原子中的问题。虽然通过这种方式我们已经获得了许多差异很大的对光谱结果的详细解释，但只局限于研究原子中与单电子运动有关的现象。对于包括几个电子的原子，则由于确定定态的方法不能够排除在原子内不同组（或称壳层）中选择电子数量及其排布时的随意性，而无法得到明确的解释。实际上，这导致的唯一的直接后果是：原子中每个电子的运动情况都可以用一个在中心力场中运动的粒子的一个定态来近似，这种近似只限于用索末菲关于氢线精细结构理论中的圆形或椭圆形轨道来描述定态。但在引入了“对应原理”之后，人们就有了消除随意性的方法，这表明量子理论的发展趋势不仅在于建立一套正式的规则以确定原子体系中的定态和在这些定态之间发生跃迁时发射的辐射频率，而且还要努力总结出合理的辐射电磁理论法则，这一法则对不连续特性的解释必须能够说明原子是稳定的。

在这里不用详述“对应原理”的具体内容，为了解决现在的矛盾，也许可以认为它足以在原子体系不同定态的运动特征与两个定态发生跃迁的可能性之间建立很强的相关性，因而为用理论检验一个原子形成或重组时可能发生的过程提供了依据。例如，根据这个原理我们得出的直接结论是：在实际原子中每一组电子的排布都不可能是环形也都不是多边形对称，因为形成这样的排布要求各组中所有的电子都必须在开始的时候同时被原子束缚住。恰恰相反，我们有必要在原子中寻找使电子有可能是一个一个连续地被原子束缚住的排布方式，我们也许能在元素系列发射光谱中观察到这个过程的最后阶段。现在利用“对应原理”，我们对该过程的描述不仅能够精确地解析这些光谱的结构，还能明确地提出原子中电子的排布方式，既适合解释高频光谱，又能说明元素的化学性质。因此从考虑定态之间可能出现的跃迁入手，根据束缚每一个电子的不同步骤，我们首先假设只有最开始的 2 个电子在可以被称为 1–量子的轨道上运动，1–量子轨道近似于一个中心体系的定态，即一个电子绕一个原子核旋转的体系的基准状态。在最开始的 2 个电子之后被束缚的电子将不能通

between stationary states, corresponding to the various steps of the binding of each of the electrons, we are led in the first place to assume that only the two first electrons move in what may be called one-quantum orbits, which are analogous to that stationary state of a central system which corresponds to the normal state of a system consisting of one electron rotating round a nucleus. The electrons bound after the first two will not be able by a transition between two stationary states to procure a position in the atom equivalent to that of these two electrons, but will move in what may be called multiple-quanta orbits, which correspond to other stationary states of a central system.

The assumption of the presence in the normal state of the atom of such multiple-quanta orbits has already been introduced in various recent theories, as, for instance, in Sommerfeld's work on the high-frequency spectra and in that of Landé on atomic dimensions and crystal structure; but the application of the correspondence principle seems to offer for the first time a rational theoretical basis for these conclusions and for the discussion of the arrangement of the orbits of the electrons bound after the first two. Thus by means of a closer examination of the progress of the binding process this principle offers a simple argument for concluding that these electrons are arranged in groups in a way which reflects the periods exhibited by the chemical properties of the elements within the sequence of increasing atomic numbers. In fact, if we consider the binding of a large number of electrons by a nucleus of high positive charge, this argument suggests that after the first two electrons are bound in one-quantum orbits, the next eight electrons will be bound in two-quanta orbits, the next eighteen in three-quanta orbits, and the next thirty-two in four-quanta orbits.

Although the arrangements of the orbits of the electrons within these groups will exhibit a remarkable degree of spatial symmetry, the groups cannot be said to form simple shells in the sense in which this expression is generally used as regards atomic constitution. In the first place, the argument involves that the electrons within each group do not all play equivalent parts, but are divided into sub-groups corresponding to the different types of multiple-quanta orbits of the same total number of quanta, which represents the various stationary states of an electron moving in a central field. Thus, corresponding to the fact that in such a system there exist two types of two-quanta orbits, three types of three-quanta orbits, and so on, we are led to the view that the above-mentioned group of eight electrons consists of two sub-groups of four electrons each, the group of eighteen electrons of three sub-groups of six electrons each, and the group of thirty-two electrons of four sub-groups of eight electrons each.

Another essential feature of the constitution described lies in the configuration of the orbits of the electrons in the different groups relative to each other. Thus for each group the electrons within certain sub-groups will penetrate during their revolution into regions which are closer to the nucleus than the mean distances of the electrons belonging to groups of fewer-quanta orbits. This circumstance, which is intimately connected with the essential features of the processes of successive binding, gives just that expression for the "coupling" between the different groups which is a necessary condition for the

过两个定态之间的跃迁进入原子中与最开始的 2 个电子等同的位置，但将在可以被称作多量子的轨道上运动，相当于一个中心体系中的其他定态。

最近的一些理论已经假定过这样的多量子轨道存在于正常状态的原子中，如在索末菲关于高频光谱的论文中，以及在朗代关于原子大小和晶体结构的著作中；但正是对应原理第一次为这些推论提供了合理的理论依据，也为在最开始的 2 个电子之后被束缚的电子轨道如何分配这一问题提供了解答。因此通过对束缚过程的进一步研究，对应原理提出了一个简单的规则，认为这些电子分组排列的方式也是元素化学性质随原子序数递增而表现出周期性的反映。实际上，如果我们考虑的是大量电子被一个带较多正电荷的原子核束缚，该规则指出：最开始的 2 个电子位于 1–量子轨道，后 8 个电子被束缚于 2–量子轨道，然后 18 个电子在 3–量子轨道，再后面 32 个电子位于 4–量子轨道。

尽管各组中电子轨道的排列呈现出惊人的空间对称性，但不能因此而认为这些组在原子构造上显示出大家普遍接受的简单壳层结构。首先，该规则认为，同一壳层中的电子并不都扮演同样的角色，总量子数相等的多量子轨道有不同的类型，对应于这些不同的类型，电子又被归入不同的子壳层，子壳层也是描绘电子在中心力场中运动的定态。因此，根据一个体系中存在 2 类 2–量子轨道，3 类 3–量子轨道，以此类推，我们可以得到这样的结论——上面提到的 8 电子壳层由 2 个 4 电子子壳层构成，18 电子壳层由 3 个 6 电子子壳层构成，32 电子壳层由 4 个 8 电子子壳层构成。

原子组成的另一个基本特征是：不同壳层中电子轨道的构型相互关联。因而对每一个壳层来说，其中某些子壳层中的电子在绕核旋转过程中会进入离核距离比量子数较低轨道（即更内层的轨道）中电子离核的平均距离更近的区域。该现象与连续成键过程的基本特征联系紧密，说明不同壳层之间存在“耦合”现象，这种耦合正是使原子构型保持稳定的必要条件。事实上，这样的耦合是整个理论的主要特征，

stability of atomic configurations. In fact, this coupling is the predominant feature of the whole picture, and is to be taken as a guide for the interpretation of all details as regards the formation of the different groups and their various sub-groups. Further, the stability of the whole configuration is of such a character that if any one of the electrons is removed from the atom by external agencies not only may the previous configuration be reorganised by a successive displacement of the electrons within the sequence in which they were originally bound by the atom, but also the place of the removed electron may be taken by any one of the electrons belonging to more loosely bound groups or sub-groups through a process of direct transition between two stationary states, accompanied by an emission of a monochromatic radiation. This circumstance—which offers a basis for a detailed interpretation of the characteristic structure of the high-frequency spectra of the elements—is intimately connected with the fact that the electrons in the various sub-groups, although they may be said to play equivalent parts in the harmony of the inter-atomic motions, are not at every moment arranged in configurations of simple axial or polyhedral symmetry as in Sommerfeld's or Landé's work, but that their motions are, on the contrary, linked to each other in such a way that it is possible to remove any one of the electrons from the group by a process whereby the orbits of the remaining electrons are altered in a continuous manner.

These general remarks apply to the constitution and stability of all the groups of electrons in the atom. On the other hand, the simple variations indicated above of the number of electrons in the groups and sub-groups of successive shells hold only for that region in the atom where the attraction from the nucleus compared with the repulsion from the electrons possesses a preponderant influence on the motion of each electron. As regards the arrangements of the electrons bound by the atom at a moment when the charges of the previously bound electrons begin to compensate the greater part of the positive charge of the nucleus, we meet with new features, and a consideration of the conditions for the binding process forces us to assume that new, added electrons are bound in orbits of a number of quanta equal to, or fewer than, that of the electrons in groups previously bound, although during the greater part of their revolution they will move outside the electrons in these groups. Such a stop in the increase, or even decrease, in the number of quanta characterising the orbits corresponding to the motion of the electrons in successive shells takes place, in general, when somewhat more than half the total number of electrons is bound. During the progress of the binding process the electrons will at first still be arranged in groups of the indicated constitution, so that groups of three-quanta orbits will again contain eighteen electrons and those of two-quanta orbits eight electrons. In the neutral atom, however, the electrons bound last and most loosely will, in general, not be able to arrange themselves in such a regular way. In fact, on the surface of the atom we meet with groups of the described constitution only in the elements which belong to the family of inactive gases, the members of which from many points of view have also been acknowledged to be a sort of landmark within the natural system of the elements. For the atoms of these elements we must expect the constitutions indicated by the following symbols:

Helium	(2_1),	Krypton	$(2_1 8_2 18_3 8_2)$,
Neon	$(2_1 8_2)$	Xenon	$(2_1 8_2 18_3 18_3 8_2)$,
Argon	$(2_1 8_2 8_2)$,	Niton*	$(2_1 8_2 18_3 32_4 18_3 8_2)$,

* "Niton" was the provisional name in 1921 for the radioactive gas now called "radon".

也是理解不同壳层及其子壳层结构所有细节的基础。另外，整个原子结构具有保持稳定的特点——如果原子中的任何一个电子由于外界因素被移走，不仅被原子束缚的电子会在以前原子构型的基础上通过连续位移而进行重新组合，而且那些受束缚较弱的壳层或子壳层中的电子可以通过两个定态之间的直接跃迁而占据被移走的电子的位置，同时发出单色辐射。这个能为详细解释元素高频光谱特征结构提供依据的现象与以下的事实密切相关：尽管我们认为不同子壳层中的电子在原子内的简谐运动中起着同样的作用，但它们并非每时每刻都在外形上保持索末菲和朗代的著作中所说的简单轴对称或者多面体对称，相反，这些电子的行为是通过这样的方式相互联系的——当外力从壳层中移走任何一个电子后，其余电子的轨道将发生连续的变化。

这些观点可用于解释原子中所有壳层电子的组成方式和稳定性。另外，在连续的壳层和子壳层中，上述电子数目的简单变化只会发生在原子中的某个区域，在这个区域内，原子核的吸引力对每个电子运动的影响远远大于电子之间的排斥力。考虑到之前被原子核束缚的电子的电荷开始抵消掉原子核的大部分正电荷时的电子排布情况，我们得到了一些新的观点，成键过程要求的条件使我们不得不假设新增电子所在轨道的量子数等于或小于以前该壳层被束缚的电子，尽管它们在绕原子核旋转的大部分时间内都在该壳层中其他电子的外侧运动。一般说来，表征连续壳层中电子运动状态的轨道量子数或许在一半以上的电子被束缚后就不再增减了。在成键过程中，电子仍会首先排布在确定结构的壳层中，因此 3–量子轨道壳层还会有 18 个电子，而 2–量子轨道壳层还会有 8 个电子。然而，在中性原子中，最后成键且受束缚最弱的电子通常不会按照这样的规则排布。事实上，在原子表层，我们仅在惰性气体元素中观察到上面描述的壳层结构，因此，从各方面看，惰性气体家族都可以称得上是人们认识天然元素体系的里程碑。我们料想这些元素的原子组成可以用符号表示如下：

氦 (2_1),	氪 ($2_1 8_2 18_3 8_2$),
氖 ($2_1 8_2$),	氙 ($2_1 8_2 18_3 18_3 8_2$),
氩 ($2_1 8_2 8_2$),	氡 *($2_1 8_2 18_3 32_4 18_3 8_2$),

* “Niton” 是现在被称为 “radon” 的放射性气体在 1921 年的临时名称。

where the large figures denote the number of electrons in the groups starting from the innermost one, and the small figures the total number of quanta characterising the orbits of electrons within each group.

These configurations are distinguished by an inherent stability in the sense that it is especially difficult to remove any of the electrons from such atoms so as to form positive ions, and that there will be no tendency for an electron to attach itself to the atom and to form a negative ion. The first effect is due to the large number of electrons in the outermost group; hence the attraction from the nucleus is not compensated to the same extent as in configurations where the outer group consists only of a few electrons, as is the case in those families of elements which in the periodic table follow immediately after the elements of the family of the inactive gases, and, as is well known, possess a distinct electro-positive character. The second effect is due to the regular constitution of the outermost group, which prevents a new electron from entering as a further member of this group. In the elements belonging to the families which in the periodic table precede the family of the inactive gases we meet in the neutral atom with configurations of the outermost group of electrons which, on the other hand, exhibit a great tendency to complete themselves by the binding of further electrons, resulting in the formation of negative ions.

The general lines of the latter considerations are known from various recent theories of atomic constitution, such as those of A. Kossel and G. Lewis, based on a systematic discussion of chemical evidence. In these theories the electro-positive and electro-negative characters of these families in the periodic table are interpreted by the assumption that the outer electrons in the atoms of the inactive gases are arranged in especially regular and stable configurations, without, however, any attempt to give a detailed picture of the constitution and formation of these groups. In this connection it may be of interest to direct attention to the fundamental difference between the picture of atomic constitution indicated in this letter and that developed by Langmuir on the basis of the assumption of stationary or oscillating electrons in the atom, referred to in Dr. Campbell's letter. Quite apart from the fact that in Langmuir's theory the stability of the configuration of the electrons is considered rather as a postulated property of the atom, for which no detailed *a priori* interpretation is offered, this difference discloses itself clearly by the fact that in Langmuir's theory a constitution of the atoms of the inactive gases is assumed in which the number of electrons is always largest in the outermost shell. Thus the sequence of the number of electrons within the groups of a niton atom is, instead of that indicated above, assumed to be 2, 8, 18, 18, 32, such as the appearance of the periods in the sequence of the elements might seem to claim at first sight.

The assumption of the presence of the larger groups in the interior of the atom, which is an immediate consequence of the argument underlying the present theory, appears, however, to offer not merely a more suitable basis for the interpretation of the general properties of the elements, but especially an immediate interpretation of the appearance of such families of elements within the periodic table, where the chemical properties of

其中较大的数字代表从最内层开始每一壳层上的电子数，较小的数字代表每一壳层中电子轨道的总量子数。

惰性气体的原子结构具有很强的内在稳定性，从某种意义上说，很难从这样的原子中移走任何一个电子使其变成正离子；同样，一个电子与这类原子结合而使其变成负离子也是不可能的。前者是由于最外层有大量的电子，因此对原子核吸引力的抵消程度大于那些外层只有较少电子的原子，如元素周期表中紧跟在惰性气体之后的那些元素，正如大家所知道的，这种结构具有明显的正电特性。后者是由于最外层电子的规则排列阻止了一个外来电子加入该壳层成为新成员。另一方面，我们发现在元素周期表中排在惰性气体之前的那些族的元素，它们的中性原子的最外层电子结构非常倾向于吸引外来电子以填满该壳层从而形成负离子。

后面提到的种种看法都是根据最近关于原子组成的各种理论得到的，如科塞尔和刘易斯建立在化学实验系统分析基础上的理论。这些理论认为，如果假定惰性气体原子的外层电子结构非常规则和稳定，就可以理解为什么元素周期表中某些族的元素带有正电特性而另一些族的元素带有负电特性，但没有人试图对原子壳层的结构和成因给予详细解释。关于这一点也许更值得关注的是：本文中所阐述的原子结构理论与坎贝尔博士在来信中提到的朗缪尔的理论有本质的不同，朗缪尔假设电子在原子中处于定态或振荡态。本文提到的理论与朗缪尔的理论的不同之处在于：电子排布的稳定性被视为是原子的必要属性，但关于这一点没有给出详细的**先验的**解释，这种不同本身很明确地表明，朗缪尔的理论认为惰性气体原子的最外层电子数总是最大的。因此，氡原子每一壳层上的电子数不再是上文提到的，而是 2、8、18、18、32，乍一看似乎在表面上仍然保留了元素的周期性。

然而，作为当前理论的一个直接推论，原子内部存在较大壳层的假设不仅更适合于解释各种元素的一般性质，而且能马上解释元素周期表中为什么存在相邻元素的化学性质相差非常小的族。事实上，这些族的存在是因为随着原子序数的递增，增加的电子填充到了原子内部的可以容纳大量电子的壳层中。如此说来，我们也许

successive elements differ only very slightly from each other. The existence of such families appears, in fact, as a direct consequence of the formation of groups containing a larger number of electrons in the interior of the atom when proceeding through the sequence of the elements. Thus in the family of the rare earths we may be assumed to be witnessing the successive formation of an inner group of thirty-two electrons at that place in the atom where formerly the corresponding group possessed only eighteen electrons. In a similar way we may suppose the appearance of the iron, palladium, and platinum families to be witnessing stages of the formation of groups of eighteen electrons. Compared with the appearance of the family of the rare earths, however, the conditions are here somewhat more complicated, because we have to do with the formation of a group which lies closer to the surface of the atom, and where, therefore, the rapid increase in the compensation of the nuclear charge during the progress of the binding process plays a greater part. In fact, we have to do in the cases in question, not, as in the rare earths, with a transformation which in its effects keeps inside one and the same group, and where, therefore, the increase in the number in this group is simply reflected in the number of the elements within the family under consideration, but we are witnesses of a transformation which is accompanied by a confluence of several outer groups of electrons.

In a fuller account which will be published soon the questions here discussed will be treated in greater detail. In this letter it is my intention only to direct attention to the possibilities which the elaboration of the principles underlying the spectral applications of the quantum theory seems to open for the interpretation of other properties of the elements. In this connection I should also like to mention that it seems possible, from the examination of the change of the spectra of the elements in the presence of magnetic fields, to develop an argument which promises to throw light on the difficulties which have hitherto been involved in the explanation of the characteristic magnetic properties of the elements, and have been discussed in various recent letters in *Nature*.

(**107**, 104-107; 1921)

N. Bohr: Copenhagen, February 14.

可以认为在稀土族元素中，电子在最多可以容纳 32 个电子而原来仅含有 18 个电子的内壳层中连续地填充。我们同样可以猜测到铁、钯、铂族元素是在逐步填充最多可容纳 18 个电子的壳层。然而，与稀土族相比，铁、钯、铂族元素的情况更复杂一些，因为在靠近原子表面的壳层填充电子时我们不得不面临的难题是，在成键过程中对核电荷的补偿的增加速度非常快。事实上，我们还必须用一种在同一壳层内部发生的转变来解决这些在稀土元素中不存在的难题，因此该壳层中电子数目的增加只是反映了被考虑的族中原子序数的增加，但是我们发现了伴随着几个外层电子汇合的转化过程。

我在即将发表的报告中将更详细地说明这里提到的问题。我写这封信的目的只是让大家注意，应用量子理论解释光谱的那些基本原理的详细阐述，看起来也有可能同时解释了元素的其他性质。就此而言，我还想提醒大家，通过研究磁场中元素光谱的变化，也许可以提出一种有望解决迄今为止人们在解释元素磁特性时遇到的困难的观点，《自然》近期的几篇快报已经对此问题进行了讨论。

（王锋 翻译；李淼 审稿）

The Dimensions of Atoms and Molecules

W. L. Bragg and H. Bell

Editor's Note

In the early 1920s, physicists were struggling to understand the physical structure of atoms using a model introduced by Bohr in 1913, in which electrons filled up shells around the nucleus. Bohr's arguments predicted that elements within any particular period (row) of the periodic table should have roughly the same atomic size, but that atomic size should jump markedly when moving from one period to the next. Here Lawrence Bragg and H. Bell report data on the dimensions of atoms that support Bohr's ideas. Estimating these dimensions from crystal densities and liquid viscosity, they find that elements near the end of rows in the table have almost identical dimensions, while a definite increase happens from one period to the next.

CERTAIN relations which are to be traced between the distances separating atoms in a crystal make it possible to estimate the distance between their centres when linked together in chemical combination. On the Lewis-Langmuir theory of atomic constitution, two electro-negative elements when combined hold one or more pairs of electrons in common, so that the outer electron shell of one atom may be regarded as coincident with that of the other at the point where the atoms are linked together. From this point of view, estimates may be made (W. L. Bragg, *Phil. Mag.*, vol. XI, August, 1920) from crystal data of the diameters of these outer shells. The outer shell of neon, for example, was estimated from the apparent diameters of the carbon, nitrogen, oxygen, and fluorine atoms, which show a gradual approximation to a minimum value of 1.30×10^{-8} cm. The diameters of the inert gases as found in this way are given in the second column of the following table:

Gas	Diameter 2σ (Crystals)	Diameter $2\sigma'$ (Viscosity)	Difference $2\sigma'-2\sigma$
Helium	—	1.89	—
Neon	1.30	2.35	1.05
Argon	2.05	2.87	0.82
Krypton	2.35	3.19	0.84
Xenon	2.70	3.51	0.81

In the third column are given Rankine's values (A. O. Rankine, *Proc. Roy. Soc.*, A, vol. XCVIII, 693, pp. 360–374, February, 1921) for the diameters of the inert gases calculated from their viscosities by Chapman's formula (S. Chapman, *Phil. Trans. Roy. Soc.*, A, vol. CCXVI, pp. 279–348, December, 1915). These are considerably greater than the diameters calculated from crystals, but this is not surprising in view of our ignorance both of the field of force surrounding the outer electron shells and of the nature of the

原子和分子的尺度

布拉格，贝尔

编者按

20 世纪 20 年代初期的物理学家总想应用 1913 年玻尔提出的模型来努力理解原子的物理结构，在玻尔的模型中，电子填充原子核外的壳层。玻尔的理论预言，在周期表中任意一个周期（一行）的元素应该具有大致相同的原子尺寸，但是从一个周期变化到下一个周期时，原子的大小将发生显著的变化。在这篇文章中，劳伦斯·布拉格和贝尔列举了各种原子的尺寸数据，这些数据支持玻尔的理论。根据这些由晶体密度和液体粘度估算得到的数据，他们发现周期表中每行靠近行尾的元素具有几乎相同的尺寸，而当元素从一个周期过渡到下一个周期时，原子尺寸出现了一定的增加。

通过探索晶体中原子间距离之间的特定关系，可以估算出它们在化学结合中彼此连接时两者中心之间的距离。根据原子结构的刘易斯–朗缪尔理论，两种负电性的元素在结合时共用一对或若干对电子，这可以看作是一个原子的外层电子与另一个原子的外层电子在两个原子连接处重合。从这种观点出发，我们可以利用某些外壳直径的晶体数据作出估计（布拉格，《哲学杂志》，第 11 卷，1920 年 8 月）。例如，氖的外壳直径就是根据碳、氮、氧和氟原子的表观直径估计得到的——估算逐渐逼近于一个最小值 1.30×10^{-8} cm。下表中的第二列给出了以上述方式得到的惰性气体的直径：

气体	直径2σ（晶体）	直径$2\sigma'$（粘度）	差值 $2\sigma' - 2\sigma$
氦	——	1.89	——
氖	1.30	2.35	1.05
氩	2.05	2.87	0.82
氪	2.35	3.19	0.84
氙	2.70	3.51	0.81

第三列中给出了兰金的值（兰金，《皇家学会学报》，A 辑，第 98 卷，第 693 期，第 360 ~ 374 页，1921 年 2 月），这是根据查普曼公式由粘度计算出来的惰性气体的直径（查普曼，《皇家学会自然科学会报》，A 辑，第 216 卷，第 279~348 页，1915 年 12 月），它们明显大于利用晶体数据计算出的直径。但是，考虑到我们忽略了外部电子层周围的力场以及将原子连接在一起的共用电子的性质，这就不足为奇了，

electron-sharing which links the atoms together, for it is quite possible that their structures might coalesce to a considerable extent. The constancy of the differences between the two estimates given in the fourth column shows that the *increase* in the size of the atom as each successive electron shell is added is nearly the same (except in the case of neon), whether measured by viscosity or by the crystal data. Further, Rankine has shown that the molecule Cl_2 behaves as regards its viscosity like two argon atoms with a distance between their centres very closely equal to that calculated from crystals, and that the same is true for the pairs Br_2 and krypton, I_2 and xenon.

We see, therefore, that the evidence both of crystals and viscosity measurements indicates that (*a*) the elements at the end of any one period in the periodic table are very nearly identical as regards the diameters of their outer electron shells, and (*b*) in passing from one period to the next there is a definite increase in the dimensions of the outer electron shell, the absolute amount of this increase estimated by viscosity agreeing closely with that determined from crystal measurements.

A further check on these measurements is afforded by the infra-red absorption spectra of HF, HCl, and HBr. The wave-number difference δv between successive absorption lines determines the moment of inertia I of the molecule in each case, the formula being

$$\delta v = \frac{h}{4\pi^2 c\mathrm{I}},$$

where h is Planck's constant and c the velocity of light.

It is therefore possible to calculate the distances between the centres of the nuclei in each molecule, for

$$s^2 = \frac{m + m'}{mm'} \cdot \frac{h}{4\pi^2 c m_{\mathrm{H}} \delta v},$$

where m and m' are the atomic weights relative to hydrogen and m_{H} the mass of the hydrogen atom. The following table gives these distances (E. S. Imes, *Astroph. Journal*, vol. 1, p. 251, 1919). It will be seen that there are again increases in passing from F to Cl and Cl to Br, which agree closely with the increases in the radii σ of the electron shells given by the crystal and viscosity data.

	$s \times 10^8$			$\sigma \times 10^8$ (Crystals)		$\sigma' \times 10^8$ (Viscosity)	
HF	0.93		Neon (=F)	0.65		1.17	
		0.35			*0.37*		*0.26*
HCl	1.28		Argon (=Cl)	1.02		1.43	
		0.15			*0.15*		*0.15*
HBr	1.43		Krypton (= Br)	1.17		1.58	
					0.18		*0.17*
HI	—		Xenon (=I)	1.35		1.75	

The increase from fluorine to chlorine of 0.35×10^{-8} cm confirms the estimate given by crystals of 0.37×10^{-8} cm, as against the estimate 0.26×10^{-8} cm given by viscosity data.

因为很有可能它们的结构会有一定程度的重叠。第四列显示了两种估计值之间差值的恒定性，这表明，不论结果是通过粘度还是通过晶体数据测得的，在相继加入各个电子层时，原子尺寸的**增加**几乎是一样的（氖的情况例外）。兰金还进一步指出，Cl_2 在粘度性质方面的表现如同两个氩原子，其中心间距与根据晶体数据计算出来的几乎相等，对于 Br_2 与氪以及 I_2 与氙来说也是如此。

由此我们看到，晶体和粘度的测量结果都指出：(*a*) 就其外部电子层的直径来说，周期表中任一周期末尾的元素几乎是一样的；(*b*) 从一个周期过渡到下一个周期时，外部电子层尺寸会有一个确定的增加量，通过粘度估计出的这一增量的绝对数值与利用晶体数据确定的结果非常接近。

HF、HCl 和 HBr 的红外吸收光谱可以进一步证实上述测量结果。相继的吸收谱线的波数差 δv 决定了各种情况下分子的转动惯量 I，公式为

$$\delta v = \frac{h}{4\pi^2 c\mathrm{I}},$$

其中，h 为普朗克常数，c 为光速。

由此就有可能计算出每个分子中核心间的距离，因为

$$s^2 = \frac{m + m'}{mm'} \cdot \frac{h}{4\pi^2 cm_{\mathrm{H}}\delta v},$$

其中，m 和 m' 为相对于氢的原子量，m_{H} 为氢原子的质量。下面的表格给出了这些距离（艾姆斯，《天体物理学杂志》，第 1 卷，第 251 页，1919 年）。我们将看到，从 F 过渡到 Cl 以及从 Cl 过渡到 Br，该距离都有所增加，并与通过晶体数据和粘度给出的电子壳层半径 σ 的增量非常接近。

$s\times10^8$					$\sigma\times10^8$（晶体）		$\sigma'\times10^8$（粘度）	
HF	0.93		氖	(=F)	0.65		1.17	
		0.35				*0.37*		*0.26*
HCl	1.28		氩	(=Cl)	1.02		1.43	
		0.15				*0.15*		*0.15*
HBr	1.43		氪	(=Br)	1.17		1.58	
						0.18		*0.17*
HI	—		氙	(=I)	1.35		1.75	

从氟到氯的增量 0.35×10^{-8}cm 肯定了晶体测量给出的估计值 0.37×10^{-8} cm，但与粘度数据给出的估计值 0.26×10^{-8} cm 不一致。根据上述结果可知，要得到氢原子

It follows from the above that the distance between the hydrogen nucleus and the centre of an electro-negative atom to which it is attached is obtained by adding 0.26×10^{-8} cm to the radius of the electro-negative atom as given by crystal structures. The radius of the inner electron orbit, according to Bohr's theory, is 0.53×10^{-8} cm, double this value. The crystal data, therefore, predict the value $\delta v = 13.0\ \mathrm{cm}^{-1}$ for the HI molecule, corresponding to a distance 1.61×10^{-8} cm between their atomic centres.

This evidence is interesting as indicating that the forces binding the atoms together are localised at that part of the electron shell where linking takes place.

(**107**, 107; 1921)

W. L. Bragg, H. Bell: Manchester University, March 16.

核与附着其上的负电性原子中心之间的距离，只需将该原子根据晶体结构得到的半径加上 0.26×10^{-8} cm 即可。根据玻尔理论，内部电子轨道半径为 0.53×10^{-8} cm，是这一数值的两倍。因此，晶体数据预言 HI 分子的 $\delta v = 13.0 \text{ cm}^{-1}$，与其原子中心间的距离 1.61×10^{-8} cm 相符。

这个证据是引人关注的，因为它表明，将原子束缚在一起的力就位于电子层中发生连接的地方。

（王耀杨 翻译；李芝芬 审稿）

Waves and Quanta

L. de Broglie

Editor's Note

By 1923, physicists were facing up to the implications of Planck's and Einstein's discoveries about the quantized nature of light. Although a wave phenomenon, light also seemed particulate. Bohr's model of the atom had exploited the quantization principle for electrons. Here Louis de Broglie suggests that particles with mass, such as electrons, may also have associated waves, and that this idea could put Bohr's view on firmer ground. De Broglie says that Bohr's results can be obtained by demanding that an integral number of such electron waves must fit into its orbit around the nucleus. De Broglie's suggestion was confirmed in dramatic fashion by the discovery in 1927 of electron diffraction by crystals.

THE quantum relation, energy = $h \times$ frequency, leads one to associate a periodical phenomenon with any isolated portion of matter or energy. An observer bound to the portion of matter will associate with it a frequency determined by its internal energy, namely, by its "mass at rest." An observer for whom a portion of matter is in steady motion with velocity βc, will see this frequency lower in consequence of the Lorentz-Einstein time transformation. I have been able to show (*Comptes rendus*, September 10 and 24, of the Paris Academy of Sciences) that the fixed observer will constantly see the internal periodical phenomenon in phase with a wave the frequency of which $v = \frac{m_0 c^2}{h\sqrt{1-\beta^2}}$ is determined by the quantum relation using the whole energy of the moving body—provided it is assumed that the wave spreads with the velocity c/β. This wave, the velocity of which is greater than c, cannot carry energy.

A radiation of frequency v has to be considered as divided into atoms of light of very small internal mass ($<10^{-50}$ gm) which move with a velocity very nearly equal to c given by $\frac{m_0 c^2}{\sqrt{1-\beta^2}} = hv$. The atom of light slides slowly upon the non-material wave the frequency of which is v and velocity c/β, very little higher than c.

The "phase wave" has a very great importance in determining the motion of any moving body, and I have been able to show that the stability conditions of the trajectories in Bohr's atom express that the wave is tuned with the length of the closed path.

The path of a luminous atom is no longer straight when this atom crosses a narrow opening; that is, diffraction. It is then *necessary* to give up the inertia principle, and we must suppose that any moving body follows always the ray of its "phase wave"; its path will then bend by passing through a sufficiently small aperture. Dynamics must undergo

波与量子

德布罗意

编者按

物理学家们在1923年的主要工作是探讨普朗克和爱因斯坦关于光的量子性质的发现。虽然光是一种波，但它似乎也具有粒子的特性。玻尔的原子模型已经用到了电子的量子化原则。路易斯·德布罗意在本文中提出像电子这样有质量的粒子可能也有相对应的波，这一观点为玻尔的理论奠定了更坚实的基础。德布罗意认为，玻尔理论中的结果可以由电子波波长取整数以适合核外电子轨道得到。1927年，德布罗意的假设就被电子在晶体中的衍射实验成功证实了。

量子关系式，即能量 = 普朗克常数 h × 频率，使人们可以把一个周期现象与任一孤立的物质或能量联系起来。与物体一起运动的观察者观测到的周期现象的频率将由物体的内部能量，即“静止质量”决定。当物体相对于观察者以 βc 的速度匀速运动时，由洛仑兹–爱因斯坦时间变换公式，观察者观测到的频率变低。我已经指出（巴黎科学院的《法国科学院院刊》，9月10日和24日），这位固定不动的观察者总能通过波的相位看到频率为 $v = \frac{m_0 c^2}{h\sqrt{1-\beta^2}}$ 的内禀周期性现象，该公式是依据量子关系式推导出来的，利用了运动物体的总能量，并假设波以 c/β 的速度传播。这个速度大于 c 的波不能携带能量。

频率为 v 的辐射必须被看成由内部质量极小（小于 10^{-50} 克）的光原子组成，其运动速度（由公式 $\frac{m_0 c^2}{\sqrt{1-\beta^2}} = hv$ 决定）很接近于 c。这些光原子沿着频率为 v，速度为 c/β（仅比 c 略高一点）的非实物波缓慢行进。

“相位波”对于确定任何物体的运动都是至关重要的。我已经指出，玻尔原子轨道的稳定条件表明相位波的波长应与闭合轨道的长度相匹配。

当光原子穿过一个小孔时，它的路径就不再是一条直线，即产生了衍射现象。因此**必须**摒弃惯性原理，我们必须假定，任何运动物体总沿着它的“相位波”的放射路径行进。因此，当它经过一个足够小的孔时，其轨迹会发生弯曲。如同几何光

the same evolution that optics has undergone when undulations took the place of purely geometrical optics. Hypotheses based upon those of the wave theory allowed us to explain interferences and diffraction fringes. By means of these new ideas, it will probably be possible to reconcile also diffusion and dispersion with the discontinuity of light, and to solve almost all the problems brought up by quanta.

(**112**, 540; 1923)

Louis de Broglie: Paris, September 12.

学被波动光学取代一样，实物粒子的动力学也应经历相应的变革。我们可以利用基于波动理论的假设解释干涉和衍射产生的条纹。借助这些新思想，还有可能把漫射和散射现象与光的不连续性联系起来，解决由量子引出的几乎所有问题。

（王锋 翻译；刘纯 审稿）

Thermal Agitation of Electricity in Conductors

J. B. Johnson

Editor's Note

In the early days of electronics and telephone communications, engineers aimed to reduce the noise in their circuits to a minimum. But there are fundamental limits, as physicist and engineer John Johnson of Bell Telephone Laboratories here reports. Johnson shows that ordinary electrical conductors exhibit spontaneous, thermally induced fluctuations of voltage, even when not being driven by external currents. This voltage noise depends directly on the temperature of the sample, leading Johnson to surmise that it must arise from continual agitation and excitation by random thermal energy. Thus, he concludes, the minimum voltage that can be usefully amplified in electronic circuits is limited by the very matter from which they are built.

ORDINARY electric conductors are sources of spontaneous fluctuations of voltage which can be measured with sufficiently sensitive instruments. This property of conductors appears to be the result of thermal agitation of the electric charges in the material of the conductor.

The effect has been observed and measured for various conductors, in the form of resistance units, by means of a vacuum tube amplifier terminated in a thermocouple. It manifests itself as a part of the phenomenon which is commonly called "tube noise". The part of the effect originating in the resistance gives rise to a mean square voltage fluctuation V^2 which is proportional to the value R of that resistance. The ratio V^2/R is independent of the nature or shape of the conductor, being the same for resistances of metal wire, graphite, thin metallic films, films of drawing ink, and strong or weak electrolytes. It does, however, depend on temperature and is proportional to the absolute temperature of the resistance. This dependence on temperature demonstrates that the component of the noise which is proportional to R comes from the conductor and not from the vacuum tube.

A similar phenomenon appears to have been observed and correctly interpreted in connexion with *a current sensitive* instrument, the string galvanometer (W. Einthoven, W. F. Einthoven, W. van der Horst, and H. Hirschfeld, *Physica*, 5, 358–360, No. 11/12, 1925). What is being measured in these cases is the effect upon the measuring device of continual shock excitation resulting from the random interchange of thermal energy and energy of electric potential or current in the conductor. Since the effect is the same for different conductors, it is evidently not dependent on the specific mechanism of conduction.

The amount and character of the observed noise depend upon the frequency-characteristic of the amplifier, as would be expected from experience with the small-shot

导体中电的热扰动

约翰逊

编者按

在电子通讯和电话通信的早期，工程师们的目标是将电路中的噪声降至最小。但贝尔电话实验室的物理学家、工程师约翰·约翰逊认为，噪声的减少是有基本极限的。约翰逊指出：即使在没有加载外部电流的情况下，导电体也会表现出由自发热扰动导致的电压波动。这种电压噪声与样品的温度直接相关，这使约翰逊联想到它应该是由随机热能的连续激励和扰动造成的。因此他得出以下结论：在电路中能够被有效放大的最小电压受到组成电路的材料的限制。

普通的导电体是电压自发涨落的来源，而这种波动电压可以用足够灵敏的仪器进行测量。导体的这种性质是导体材料中电荷热扰动的结果。

在各种导体中都已经观测到这类效应，并以电阻的形式，通过终端连接温差电偶的真空管放大器对此进行了测量。它通常作为所谓的“电子管噪声”现象的一部分而表现出来。源于电阻波动的这部分效应造成了电压均方值 V^2 的波动，此波动与电阻值 R 成正比。比率 V^2/R 与导体的性质和形状无关，对金属丝、石墨、金属薄膜、绘图墨水薄膜及强电解质或弱电解质来说，该比率都是相同的。然而，该比率却与温度有关，并与电阻的绝对温度成正比。这种温度依赖性表明，与 R 成比例的噪声分量来自导体，而不是来自真空管。

目前已经有人用**对电流灵敏**的仪器——弦线电流计观测到类似现象并正确地进行了解释（老艾因特霍芬、小艾因特霍芬、范德霍斯特以及赫希菲尔德，《物理学》，第 5 卷，第 358~360 页，第 11/12 期，1925 年）。在这些情况下测量到的是，导体中的热能与电势或电流能量的随机交换引起的持续冲击激发对测量装置的影响。因为该效应对不同导体是相同的，所以显然它与传导的特定机制无关。

正如从散粒效应的经验可以预期的那样，观测到的噪声的量及其特性与放大器的频率特性有关。在室温下，来自电阻的表观输入功率在 10^{-18} 瓦的量级。至少在声

effect. The apparent input power originating in the resistance is of the order 10^{-18} watt at room temperature. The corresponding output power is proportional to the area under the graph of *power amplification–frequency*, at least in the range of audio frequencies. The magnitude of the "initial noise", when the quietest tubes are used without input resistance, is about the same as that produced by a resistance of 5,000 ohms at room temperature in the input circuit. For the technique of amplification, therefore, the effect means that the limit to the smallness of voltage which can be usefully amplified is often set, not by the vacuum tube, but by the very matter of which electrical circuits are built.

(**119**, 50-51; 1927)

J. B. Johnson: Bell Telephone Laboratories, Inc., New York, N. Y., Nov. 17.

频范围内，相应的输出功率与**功率放大–频率**关系图中曲线下的面积成正比。当采用无输入电阻的最平稳的真空管时，“初始噪声”的量级约与在室温下输入回路中5,000欧姆的电阻产生的噪声量级相同。因此，对于放大技术而言，这种效应意味着可以被有效放大的电压的最小值与构成电路的材料有关，而不由真空管决定。

（沈乃澂 翻译；赵见高 审稿）

The Scattering of Electrons by a Single Crystal of Nickel

C. Davisson and L. H. Germer

Editor's Note

In his PhD thesis of 1924, Louis de Broglie hypothesized that electrons and other massive particles might behave as waves, just as photons reveal the particle-like aspects of otherwise wave-like electromagnetic radiation. This implied that material particles should exhibit wave-like phenomena such as interference or diffraction. Here Clinton Davisson and Lester Germer of the Bell Telephone Laboratories verified this prediction by measuring the diffraction of an electron beam from a nickel crystal. Using the known spacing of atomic planes in nickel, they were able to calculate the expected angles where diffracted beams should occur if the electrons were indeed acting as waves in the manner de Broglie suggested. The experimental results agreed with these predictions, and they won Davisson the 1937 Nobel Prize in physics.

IN a series of experiments now in progress, we are directing a narrow beam of electrons normally against a target cut from a single crystal of nickel, and are measuring the intensity of scattering (number of electrons per unit solid angle with speeds near that of the bombarding electrons) in various directions in front of the target. The experimental arrangement is such that the intensity of scattering can be measured in any latitude from the equator (plane of the target) to within 20° of the pole (incident beam) and in any azimuth.

The face of the target is cut parallel to a set of {111}-planes of the crystal lattice, and etching by vaporisation has been employed to develop its surface into {111}-facets. The bombardment covers an area of about 2 mm^2 and is normal to these facets.

As viewed along the incident beam the arrangement of atoms in the crystal exhibits a threefold symmetry. Three {100}-normals equally spaced in azimuth emerge from the crystal in latitude 35°, and, midway in azimuth between these, three {111}-normals emerge in latitude 20°. It will be convenient to refer to the azimuth of any one of the {100}-normals as a {100}-azimuth, and to that of any one of the {111}-normals as a {111}-azimuth. A third set of azimuths must also be specified; this bisects the dihedral angle between adjacent {100}- and {111}-azimuths and includes a {110}-normal lying in the plane of the target. There are six such azimuths, and any one of these will be referred to as a {110}-azimuth. It follows from considerations of symmetry that if the intensity of scattering exhibits a dependence upon azimuth as we pass from a {100}-azimuth to the next adjacent {111}-azimuth (60°), the same dependence must be exhibited in the reverse

镍单晶对电子的散射

戴维森，革末

编者按

1924年，路易斯·德布罗意在他的博士论文中猜测，电子以及其他一些有质量的粒子的运动方式可能与波类似，就像光子除了具有电磁辐射的波动行为以外，还显示出与此不同的粒子行为。这一假说意味着，物质粒子会表现出干涉或者衍射这样的波动特征。在这篇文章中，贝尔电话实验室的克林顿·戴维森和莱斯特·革末通过对从镍晶中发出的电子束的衍射效应进行测量证实了这一假说。如果正如德布罗意猜想的那样电子具有像波一样的性质，那么根据镍晶中已知的原子平面之间的间距，他们就能计算出应该出现衍射束的预期角度。实验结果与这些预测是相吻合的。戴维森因此获得了1937年的诺贝尔物理学奖。

现在正在进行的一系列实验中，我们用一窄束电子垂直轰击一块从镍单晶上切下来的靶子，并测量了靶前方不同方向上的电子散射强度（与每单位立体角中速度和轰击电子的速度相似的电子数量）。我们的实验装置可以测量从赤道（靶平面）到与轴（入射束）成20°范围内在任意纬度处和在任意方位角上的散射强度。

被切割下来的靶面平行于镍单晶点阵的一组{111}面，切下来之后用蒸气进行蚀刻，以使其表面形成一系列{111}小晶面。轰击的电子束打在约2平方毫米的范围内，并与以上的小晶面垂直。

沿着入射电子束的方向观察，我们发现原子在晶体中的排列呈现三重对称。从晶体出发，在纬度为35°处有三条相隔同样方位角的{100}法线，并且在这些方位角中间，有三条{111}法线出现在纬度20°处。为了方便起见，任何一条{100}法线的方位角都可以被看作是一个{100}方位角，任何一条{111}法线的方位角也可以被看作是一个{111}方位角。我们还必须指定第三组方位角；它将平分相邻的{100}方位角和{111}方位角之间的二面角，并且包含一条位于靶平面上的{110}法线。有6个这样的方位角，它们当中的任意一个都可以被看作是一个{110}方位角。考虑到对称性的要求，当我们从{100}方位角转到下一个相邻的{111}方位角(60°)时，如果散射强度的变化依赖于方位角的话，那么当我们从60°转到下一个相邻的

order as we continue on through 60° to the next following {100}-azimuth. Dependence on azimuth must be an even function of period $2\pi/3$.

In general, if bombarding potential and azimuth are fixed and exploration is made in latitude, nothing very striking is observed. The intensity of scattering increases continuously and regularly from zero in the plane of the target to highest value in co-latitude 20°, the limit of observations. If bombarding potential and co-latitude are fixed and exploration is made in azimuth, a variation in the intensity of scattering of the type to be expected is always observed, but in general this variation is slight, amounting in some cases to not more than a few percent of the average intensity. This is the nature of the scattering for bombarding potentials in the range from 15 volts to near 40 volts.

At 40 volts a slight hump appears near 60° in the co-latitude curve for azimuth-{111}. This hump develops rapidly with increasing voltage into a strong spur, at the same time moving slowly upward toward the incident beam. It attains a maximum intensity in co-latitude 50° for a bombarding potential of 54 volts, then decreases in intensity, and disappears in co-latitude 45° at about 66 volts. The growth and decay of this spur are traced in Fig. 1.

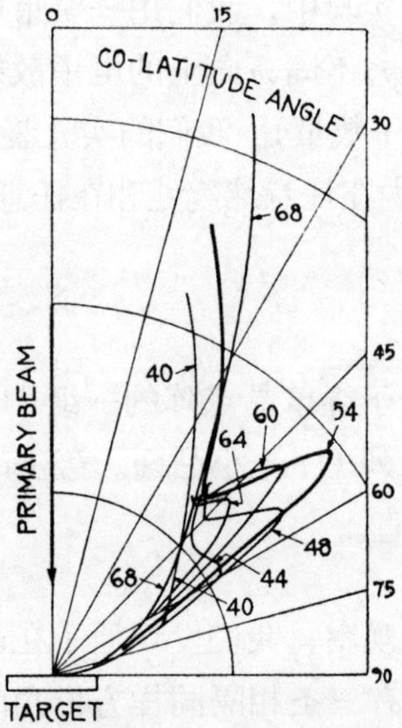

Fig. 1. Intensity of electron scattering vs. co-latitude angle for various bombarding voltages—azimuth-{111}-330°

A section in azimuth through this spur at its maximum (Fig. 2—Azimuth-330°) shows that it is sharp in azimuth as well as in latitude, and that it forms one of a set of three such spurs, as was to be expected. The width of these spurs both in latitude and in azimuth is almost completely accounted for by the low resolving power of the measuring device. *The spurs are due to beams of scattered electrons which are nearly if not quite as well defined as the primary*

{100} 方位角时，相同的依赖关系会按相反的次序再现。散射强度对方位角的依赖关系肯定是一个周期为 $2\pi/3$ 的偶函数。

一般来说，如果轰击电压和方位角固定，当沿着纬度测量的时候，不会观察到任何令人惊奇的现象。散射强度连续而有规律地从靶平面处的零增加到余纬 20° 时的最大值，余纬 20° 是我们的观测极限。如果轰击电压和余纬固定，而沿着方位角进行探测，则通常可以观察到预期之中的散射强度变化，但总的来说这种差别很小，在某些情况下总的变化量不会大于平均强度的百分之几。这就是散射在轰击电压从 15 伏特到接近 40 伏特之间变动时的特性。

当电压为 40 伏特时，在方位角为 {111} 的余纬曲线上接近 60° 处出现了一个小峰。随着电压的增加，这个峰迅速变大形成一个很强的尖峰，同时朝着入射电子束的方向缓慢向上提升。当轰击电压为 54 伏特，余纬为 50° 时，峰值达到最大，然后强度逐渐减小，在电压约为 66 伏特，余纬为 45° 时消失。图 1 描述了这个尖峰从增大到衰减的过程。

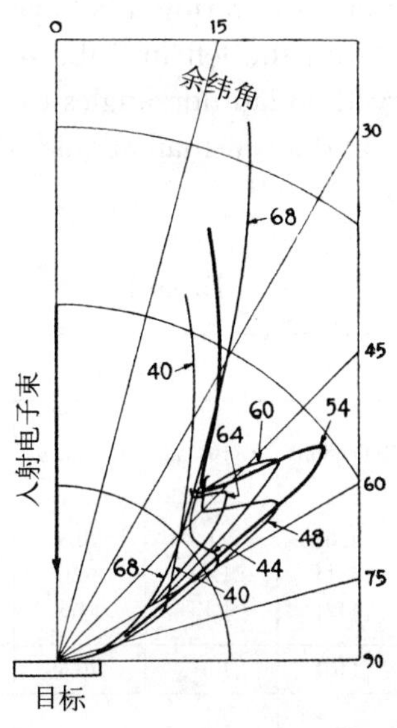

图 1. 不同轰击电压下电子散射强度随余纬角的变化——方位角{111}-330°

在最大值处穿过这个尖峰的方位角截面图（图 2 之方位角 330°）表明，它以方位角为变量的时候和以纬度为变量的时候一样尖锐，并且，如我们所料，它形成了 3 个尖峰组合中的一个。这些尖峰的宽度在随方位角变化和随纬度变化时都较大，几乎可以肯定地说这是由测量仪器的低分辨率造成的。**尖峰是由散射的电子束引起**

beam. The minor peaks occurring in the {100}-azimuth are sections of a similar set of spurs that attains its maximum development in co-latitude 44° for a bombarding potential of 65 volts.

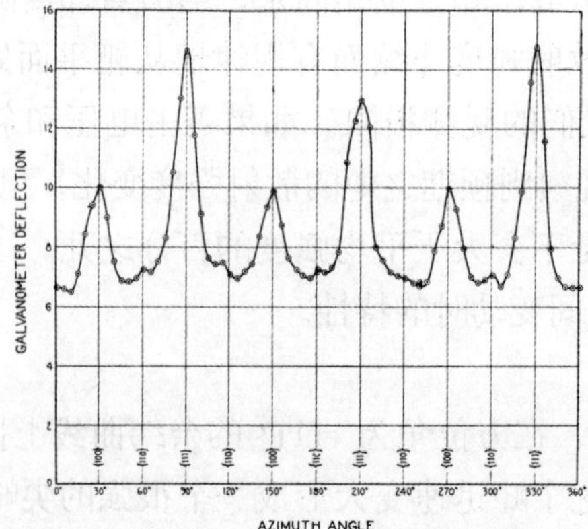

Fig. 2. Intensity of electron scattering vs. azimuth angle—54 volts, co-latitude 50°

Thirteen sets of beams similar to the one just described have been discovered in an exploration in the principal azimuths covering a voltage range from 15 volts to 200 volts. The data for these are set down on the left in Table I (columns 1–4). Small corrections have been applied to the observed co-latitude angles to allow for the variation with angle of the "background scattering", and for a small angular displacement of the normal to the facets from the incident beam.

Table I

Azimuth	Electron Beams			X-ray Beams						
	Bomb. Pot (volts)	Co-lat. θ	Intensity	Reflections	$\lambda\times10^8$ cm	Co-lat. θ	Co-lat. θ'	$v\times10^{-8}$ cm/sec	$n\lambda\times10^8$ cm	$n\left\{\frac{\lambda mv}{h}\right\}$
{111}	54	50°	0.5	{220}	2.03	70.5	52.7	4.36	1.65	0.99
	100	31	0.5	{331}	1.49	44.0	31.6	5.94	1.11	0.91
	174	21	0.9	{442}	1.13	31.6	22.4	7.84	0.77	0.83
	174	55	0.15	{440}	1.01	70.5	52.7	7.84	1.76	2(0.95)
{100}	65	44	0.5	{311}	1.84	59.0	43.2	4.79	1.49	0.98
	126	29	1.0	{422}	1.35	38.9	27.8	6.67	1.04	0.95
	190	20	1.0	{533}	1.04	28.8	20.4	8.19	0.74	0.83
	159	61	0.4	{511}	1.05	77.9	59.0	7.49	1.88	2(0.97)
{110}	138	59	0.07	{420}	1.22	78.5	59.5	6.98	1.06	1.02
	170	46	0.07	{531}	1.04	57.1	41.7	7.75	0.89	0.95
{111}	110	58	0.15	..	..	..	..	6.23	1.82	1.56
{100}	110	58	0.15	..	..	..	..	6.23	1.82	1.56
{110}	110	58	0.25	..	..	..	..	6.23	1.05	0.90

的，对这些散射电子束的界定即使不能像对原射线束的界定那样精确，至少也能相差不多。在 {100} 方位角上出现的次级峰是另一组类似的尖峰，它们在余纬 44°，轰击电压 65 伏特时达到最大值。

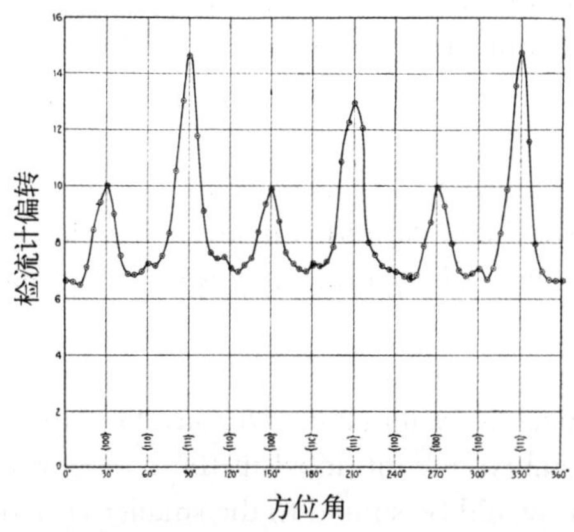

图 2. 电子散射强度随方位角的变化——54伏特，余纬50°

电压在从 15~200 伏特的范围内变动时，我们在主方位角方向发现了 13 组与上面所描述的电子束类似的射束。结果见表 I 左侧（第 1~4 列）。考虑到“背景散射”随角度的变化，以及晶面法线与入射电子束之间的角位移，我们对观察到的余纬角进行了微小的修正。

表 I

方位角	电子束			X 射线束						
	轰击电压（伏特）	余纬度 θ	强度	反射	$\lambda\times10^{8}$厘米	余纬度 θ	余纬度 θ'	$v\times10^{-8}$ 厘米/秒	$n\lambda\times10^{8}$ 厘米	$n\left\{\frac{\lambda mv}{h}\right\}$
{111}	54	50°	0.5	{220}	2.03	70.5	52.7	4.36	1.65	0.99
	100	31	0.5	{331}	1.49	44.0	31.6	5.94	1.11	0.91
	174	21	0.9	{442}	1.13	31.6	22.4	7.84	0.77	0.83
	174	55	0.15	{440}	1.01	70.5	52.7	7.84	1.76	2(0.95)
{100}	65	44	0.5	{311}	1.84	59.0	43.2	4.79	1.49	0.98
	126	29	1.0	{422}	1.35	38.9	27.8	6.67	1.04	0.95
	90	20	1.0	{533}	1.04	28.8	20.4	8.19	0.74	0.83
	159	61	0.4	{511}	1.05	77.9	59.0	7.49	1.88	2(0.97)
{110}	138	59	0.07	{420}	1.22	78.5	59.5	6.98	1.06	1.02
	170	46	0.07	{531}	1.04	57.1	41.7	7.75	0.89	0.95
{111}	110	58	0.15	..	..	..	..	6.23	1.82	1.56
{100}	110	58	0.15	..	..	..	..	6.23	1.82	1.56
{110}	110	58	0.25	..	..	..	..	6.23	1.05	0.90

If the incident electron beam were replaced by a beam of monochromatic X-rays of adjustable wave-length, very similar phenomena would, of course, be observed. At particular values of wave-length, sets of three or of six diffraction beams would emerge from the incident side of the target. On the right in Table I (columns 5, 6 and 7) are set down data for the ten sets of X-ray beams of longest wave-length which would occur within the angular range of our observations. Each of these first ten occurs in one of our three principal azimuths.

Several points of correlation will be noted between the two sets of data. Two points of difference will also be noted; the co-latitude angles of the electron beams are not those of the X-ray beams, and the three electron beams listed at the end of the Table appear to have no X-ray analogues.

The first of these differences is systematic and may be summarised quantitatively in a simple manner. If the crystal were contracted in the direction of the incident beam by a factor 0.7, the X-ray beams would be shifted to the smaller co-latitude angles θ' (column 8), and would then agree in position fairly well with the observed electron beams—the average difference being 1.7°. Associated in this way there is a set of electron beams for each of the first ten sets of X-ray beams occurring in the range of observations, the electron beams for 110 volts alone being unaccounted for.

These results are highly suggestive, of course, of the ideas underlying the theory of wave mechanics, and we naturally inquire if the wave-length of the X-ray beam which we thus associate with a beam of electrons is in fact the h/mv of L. de Broglie. The comparison may be made, as it happens, without assuming a particular correspondence between X-ray and electron beams, and without use of the contraction factor. Quite independently of this factor, the wave-lengths of all possible X-ray beams satisfy the optical grating formula $n\lambda = d\sin\theta$, where d is the distance between lines or rows of atoms in the surface of the crystal—these lines being normal to the azimuth plane of the beam considered. For azimuths-{111} and-{100}, $d = 2.15\times10^{-8}$ cm and for azimuth-{110}, $d = 1.24\times10^{-8}$ cm. We apply this formula to the electron beams without regard to the conditions which determine their distribution in co-latitude angle. The correlation obtained by this procedure between wave-length and electron speed v is set down in the last three columns of Table I.

In considering the computed values of $n(\lambda mv/h)$, listed in the last column, we should perhaps disregard those for the 110-volt beams at the bottom of the Table, as we have had reason already to regard these beams as in some way anomalous. The values for the other beams do, indeed, show a strong bias toward small integers, quite in agreement with the type of phenomenon suggested by the theory of wave mechanics. These integers, one and two, occur just as predicted upon the basis of the correlation between electron beams and X-ray beams obtained by use of the contraction factor. The systematic character of the departures from integers may be significant. We believe, however, that this results from

如果入射电子束被一束波长可调的单色 X 射线取代，我们当然可以观察到非常类似的现象。当波长为一些特定值时，在靶的入射面一侧就会出现 3 条一组或 6 条一组的衍射光束。表 I 右侧（第 5、6、7 列）列出了在我们能观测到的角度范围内由波长最长的 10 组 X 射线束得到的数据。前 10 组中每组数据都来自我们先前设定的 3 个主方位角中的一个。

在以上表中关于电子束和 X 射线束的两部分数据中有几点关系值得我们注意。同时还要注意两个区别：电子束的余纬角不是 X 射线束的余纬角；对于表格底部列出的 3 行电子束数据，没有 X 射线束数据与之对应。

第一个区别是系统上的，可以用一种简单的方式定量地总结出来。如果晶体沿着入射束方向被压缩至 70%，即压缩因子为 0.7，那么 X 射线束将移至更小的余纬角 θ'（第 8 列），这样它的位置将与观测到的电子束的情况精确地吻合——平均差异为 1.7°。同理，对于我们能观察到的前 10 组 X 射线束中的每一个，都存在一组电子束与之对应，只有 110 伏特的电子束不能应用这种方法。

当然，这些结果完全证实了波动力学理论的基本观点，我们自然很想知道，与电子束相关联的 X 射线束的波长是否就是德布罗意所说的 h/mv。也许不用假设 X 射线束和电子束之间的特定关系，也不用使用压缩因子，我们就可以作出比较。所有的 X 射线束都满足光栅公式 $n\lambda = d\sin\theta$，这与压缩因子完全不相干，在这个公式中，d 是位于晶体表面的原子行或列的间距——这些行列都垂直于我们粒子束所在的方位角平面。对于方位角 {111} 和方位角 {100}，$d = 2.15 \times 10^{-8}$ 厘米；对于方位角 {110}，$d = 1.24 \times 10^{-8}$ 厘米。我们可以把这个公式应用于电子束中而无需考虑支配电子束沿余纬度分布情况的因素。用这种方式得到的波长和电子速度 v 之间的关系列于表 I 的最后 3 列中。

至于在最后一列中列出的计算值 $n(\lambda mv/h)$，我们可能应该不考虑表格底部 110 伏特的电子束数据，因为我们已经有理由把这些电子束看作是异常情况。对于其他电子束，$n(\lambda mv/h)$ 值趋向于一个比较小的整数，与波动力学理论预言的现象十分吻合。使用压缩因子对电子束和 X 射线束之间的差别进行校正之后，也得到了预想的结果，即整数 1 和 2。对整数的偏离可能主要是由系统特性造成的。我们认为误差来

imperfect alignment of the incident beam, or from other structural deficiencies in the apparatus. The greatest departures are for beams lying near the limit of our co-latitude range. The data for these are the least trustworthy.

(**119**, 558-560; 1927)

C. Davisson, L. H. Germer: Bell Telephone Laboratories, Inc., New York, N.Y., Mar. 3.

自入射束不严格准直或者仪器的其他结构缺陷。最大的偏离出现在我们能观察到的余纬度极限附近，那里的数据是最不可靠的。

（王静 翻译；赵见高 审稿）

The Continuous Spectrum of β-rays

C. D. Ellis and W. A. Wooster

Editor's Note

Since the discovery of radioactivity in 1895, three different mechanisms had been identified. In some cases an atom would disintegrate by shedding an α-particle (the nucleus of a helium atom), in others by emitting a γ-ray (a high-frequency X-ray). In both these mechanisms, the amount of energy lost in the disintegration was found to be the same for each kind of disintegrating atom. In the third mechanism of radioactive decay, however, in which the particle shed during the disintegration is an electron, the energy carried away by the particle is indeterminate, ranging from zero to a certain maximum characteristic of the atom concerned. This raises problems referred to in the letter below.

THE continuous spectrum of the β-rays arising from radio-active bodies is a matter of great importance in the study of their disintegration. Two opposite views have been held about the origin of this continuous spectrum. It has been suggested that, as in the α-ray case, the nucleus, at each disintegration, emits an electron having a fixed characteristic energy, and that this process is identical for different atoms of the same body. The continuous spectrum given by these disintegration electrons is then explained as being due to secondary effects, into the nature of which we need not enter here. The alternative theory supposes that the process of emission of the electron is not the same for different atoms, and that the continuous spectrum is a fundamental characteristic of the type of atom disintegrating. Discussion of these views has hitherto been concerned with the problem of whether or not certain specified secondary effects could produce the observed heterogenity, and although no satisfactory explanation has yet been given by the assumption of secondary effects, it was most important to clear up the problem by a direct method.

There is a ready means of distinguishing between the two views, since in one case a given quantity of energy would be emitted at each disintegration equal to or greater than the maximum energy observed in the electrons escaping from the atom, whereas in the second case the average energy per disintegration would be expected to equal the average energy of the particles emitted. If we were to measure the total energy given out by a known amount of material, as, for example, by enclosing it in a thick-walled calorimeter, then in the first case the heating effect should lead to an average energy per disintegration equal to or greater than the fastest electron emitted, no matter in what way this energy was afterwards split up by secondary effects. Since on the second hypothesis no secondary effects are presumed to be present, the heating effect should correspond simply to the average kinetic energy of the particles forming the continuous spectrum.

To avoid complications due to α-rays or to γ-rays from parent or successive atoms, we

β 射线的连续谱

埃里斯，伍斯特

编者按

自从 1895 年发现放射性以来，人们已经确认放射性存在三种不同的机制。在某些情况下，原子会通过放射出一个 α 粒子（氦原子的原子核）而衰变，而在另一些情况下，原子会通过放射 γ 射线（一种高频 X 射线）而衰变。在这两种机制中，每一种衰变原子在衰变过程中的能量损失都是相同的。然而在放射性衰变的第三种机制中，原子衰变时放射出的粒子是电子，这个电子携带的能量是不确定的，其范围是从零到相关原子的最大特征能量。这就引发了下文中提到的问题。

在物质衰变的研究中，放射性物质产生的 β 射线连续谱是一个非常重要的问题。关于这种连续谱的起源有两种相反的观点。一种观点认为，和放射出 α 射线的衰变一样，在每一次衰变中原子核都发射出一个具有固定的特征能量的电子，并且对同一物体的不同原子这个过程都是相同的。于是，这些衰变电子的连续谱就被归因为二次效应，而有关二次效应的本质我们不必在这里进行讨论。另外一种观点认为，发射电子的过程对不同原子是不同的，并且连续谱是衰变原子类型的基本特征。关于这两种观点的讨论至今停留在是否有某种具体的二次效应能够产生可观测的连续谱这一问题上，尽管二次效应的假设还没能给出一个令人满意的解释，但最重要的是，我们应该用一种直接的方法来解释以上的问题。

有一个简易的方法可以区分以上两种观点，因为从第一个观点来看，每次衰变中放出的特定能量应该等于或者大于观察到的从原子中逃逸出的电子的最大能量，而从第二种观点来看，每次衰变的平均能量应该等于发射的电子的平均能量。如果我们测量一定量的物质发射出的电子的总能量，例如将这些物质放在一个厚壁量热计中，那么按照第一种观点，由热效应就能得出每次衰变的平均能量，这个值应该是等于或大于发射出的最快的电子的能量，不管这个能量随后以什么方式被二次效应分解。而在第二种观点中，由于不存在二次效应，所以热效应应该简单地对应于形成连续谱的电子的平均动能。

为了避免衰变中母原子或后续原子产生的 α 射线或者 γ 射线的干扰，我们在

measured the heating effect in a thick-walled calorimeter of a known quantity of radium E. This measurement proved difficult because of the small rate of evolution of heat, but by taking special precautions it has been possible to show that the average energy emitted at each disintegration of radium E is 340,000 ± 30,000 volts. This result is a striking confirmation of the hypothesis that the continuous spectrum is emitted as such from the nucleus, since the average energy of the particles as determined by ionisation measurements over the whole spectrum gives a value about 390,000 volts, whereas if the energy emitted per disintegration were equal to that of the fastest β-rays, the corresponding value of the heating would be three times as large—in fact, 1,050,000 volts.

Many interesting points are raised by the question of how a nucleus, otherwise quantised, can emit electrons with velocities varying over a wide range, but consideration of these will be deferred until the publication of the full results.

(**119**, 563-564; 1927)

C. D. Ellis, W. A. Wooster: Cavendish Laboratory, Cambridge, Mar.23.

厚壁量热计中测量了一定量的镭 E 的热效应。由于放热速度很慢，这个测量很难进行，但是通过采用特别的措施，我们得到镭 E 每次衰变辐射出的平均能量是 340,000 ± 30,000 电子伏。这个结果对于连续谱来源于原子核中的发射这一假设是个有力的确证。因为通过电离测量整个能谱而得到的电子平均能量大约是 390,000 电子伏，如果每次衰变放出的能量等于最快的 β 射线的能量，那么相应的热效应的值就应该是当前测量值的 3 倍，即 1,050,000 电子伏。

一个量子化的原子核怎么能发射速度变化范围如此广泛的电子？这个问题引发了很多有趣的观点，不过还是等到详细的结果发表之后再来考虑这些。

（王锋 翻译；江丕栋 审稿）

A New Type of Secondary Radiation

C. V. Raman and K. S. Krishnan

Editor's Note

Physics in the early twentieth century was dominated by scientists in Europe and the United States. Yet a landmark discovery in quantum physics is reported here by two Indian physicists in Calcutta. As hitherto understood, light scattering from a stationary material object should preserve its frequency. But Chandrasekhara Venkata Raman and Kariamanickam Srinivasa Krishnan demonstrate that a small part of the scattered light can significantly change frequency. This "Raman effect" involves an exchange of energy between the scattered photons and the internal degrees of freedom of atoms or molecules. The effect is used today to probe molecular structure and motion, and the chemical nature of materials. For his discovery, Raman was awarded the 1930 Nobel Prize in physics.

IF we assume that the X-ray scattering of the "unmodified" type observed by Prof. Compton corresponds to the normal or average state of the atoms and molecules, while the "modified" scattering of altered wave-length corresponds to their fluctuations from that state, it would follow that we should expect also in the case of ordinary light two types of scattering, one determined by the normal optical properties of the atoms or molecules, and another representing the effect of their fluctuations from their normal state. It accordingly becomes necessary to test whether this is actually the case. The experiments we have made have confirmed this anticipation, and shown that in every case in which light is scattered by the molecules in dust-free liquids or gases, the diffuse radiation of the ordinary kind, having the same wave-length as the incident beam, is accompanied by a modified scattered radiation of degraded frequency.

The new type of light scattering discovered by us naturally requires very powerful illumination for its observation. In our experiments, a beam of sunlight was converged successively by a telescope objective of 18 cm aperture and 230 cm focal length, and by a second lens of 5 cm focal length. At the focus of the second lens was placed the scattering material, which is either a liquid (carefully purified by repeated distillation *in vacuo*) or its dust-free vapour. To detect the presence of a modified scattered radiation, the method of complementary light-filters was used. A blue-violet filter, when coupled with a yellow-green filter and placed in the incident light, completely extinguished the track of the light through the liquid or vapour. The reappearance of the track when the yellow filter is transferred to a place between it and the observer's eye is proof of the existence of a modified scattered radiation. Spectroscopic confirmation is also available.

Some sixty different common liquids have been examined in this way, and every one of them showed the effect in greater or less degree. That the effect is a true scattering and not a fluorescence is indicated in the first place by its feebleness in comparison with the

一种新型的二次辐射

拉曼，克里希南

编者按

20 世纪早期的物理学主要是由欧美科学家主导的，然而，一项在量子物理方面具有里程碑意义的发现却是由两名印度物理学家在加尔各答作出的。就当时人们所知，从一个静止实物散射出来的光线应该保持频率不变，这一点大家都可以理解，但钱德拉塞卡拉·文卡塔·拉曼和卡瑞马尼卡姆·斯里尼瓦桑·克里希南却用实验证实，有一小部分散射光频率变化很大。“拉曼效应”包含散射光子与原子或分子的内自由度之间能量的交换过程。现在利用这个效应可以检测分子结构和分子运动以及材料的化学性质。拉曼因发现了这个效应而获得了 1930 年的诺贝尔物理学奖。

如果我们假定，康普顿教授观察到的“不变”的 X 射线散射对应于原子和分子的正常态或平均态，而波长发生改变的“变”散射对应于原子和分子相对于正常态或平均态的涨落，那么我们就可以预测，普通光的散射应该也存在两种类型，一种取决于原子或分子的正常光学性质，另一种则代表了它们相对于正常态的涨落效应。因此有必要检验真实的情况是否确实如此。我们的实验证实了上述预测。实验表明，由任何一种无尘的液体或气体分子造成的光散射，都不仅包含了与入射光波长相同的正常漫射辐射，同时也伴随频率发生变化的变散射辐射。

要观察到我们发现的这种新型光散射，自然就需要非常强的光照。在我们的实验中，一束太阳光依次通过口径为 18 厘米、焦距为 230 厘米的望远镜物镜和一个焦距为 5 厘米的透镜而被会聚。在第二个透镜的焦点处放置散射材料，这些材料或者是在真空中反复蒸馏得到的非常纯净的液体，或者是无尘的蒸气。为了探测变散射辐射的存在，我们使用了互补的滤光片。当把一个蓝紫色滤光片和一个黄绿色滤光片一起放置在入射光处时，透过液体或蒸气的光路会完全消失。而当把入射光处的黄色滤光片移置到散射材料和观测者的眼睛之间时，透过散射材料的光路就会重新出现。这就证实了变散射辐射的存在。光谱分析的结果也证实了这一点。

我们用这种方法检测了六十多种不同的常见液体，所有结果中都或多或少地出现了这种效应。这是一种真正的散射效应而不是一种荧光现象，因为与普通的散射

ordinary scattering, and secondly by its polarisation, which is in many cases quite strong and comparable with the polarisation of the ordinary scattering. The investigation is naturally much more difficult in the case of gases and vapours, owing to the excessive feebleness of the effect. Nevertheless, when the vapour is of sufficient density, for example with ether or amylene, the modified scattering is readily demonstrable.

(**121**, 501-502; 1928)

C. V. Raman, K. S. Krishnan: 210 Bowbazar Street, Calcutta, India, Feb. 16.

相比它的强度非常微弱，而且在很多情况下它具有与普通的散射相当的非常强的偏振性。这种效应的强度非常微弱，因此要在气体和蒸气中开展这项研究自然是非常困难的。不过，当蒸气浓度足够大时，例如乙醚或戊烯的蒸气，还是很容易观察到变散射的。

（王锋 翻译；李芝芬 审稿）

Wave Mechanics and Radioactive Disintegration

R. W. Gurney and E. U. Condon

Editor's Note

The "wave mechanics" description of quantum theory arose from the work of Erwin Schrödinger in the 1920s, which built on the suggestion of Louis de Broglie in 1924 that matter can possess wave-like properties. Schrödinger's description of the behaviour of quantum particles was formulated purely in terms of waves (or wavefunctions) whose amplitude in different parts of space specified the probability of the particle being there. Here Ronald Gurney and Edward Condon perceptively states what this implies for the decay of radioactive atomic nuclei by emission of alpha particles. They points out that the escape of the alpha particle can be regarded simply in terms of overlap between wavefunctions inside and outside the nucleus. This is a form of quantum "tunnelling" through an energy barrier.

AFTER the exponential law in radioactive decay had been discovered in 1902, it soon became clear that the time of disintegration of an atom was independent of the previous history of the atom and depended solely on chance. Since a nuclear particle must be held in the nucleus by an attractive field, we must, in order to explain its ejection, arrange for a spontaneous change from an attractive to a repulsive field. It has hitherto been necessary to postulate some special arbitrary "instability" of the nucleus; but in the following note it is pointed out that disintegration is a natural consequence of the laws of quantum mechanics without any special hypothesis.

It is well known that the failure of classical mechanics in molecular events is due to the fact that the wave-length associated with the particles is not small compared with molecular dimensions. The wave-length associated with α-particles is some 10^5 smaller, but since the nuclear dimensions are smaller than atomic in about the same ratio, the applicability of the wave mechanics would seem to be ensured.

In the classical mechanics, the orbit of a moving particle is entirely confined to those parts of space for which its potential energy is less than its total energy. If a ball be moving in a valley of potential energy and have not enough energy to get over a mountain on one side of the valley, it must certainly stay in the valley for all time, unless it acquire the deficiency in energy somehow. But this is not so on the quantum mechanics. It will always have a small but finite chance of slipping through the mountain and escaping from the valley.

In the diagram (Fig.1), let *O* represent the centre of a nucleus, and let *ABCDEFG* represent a simplified one-dimensional plot of the potential energy. The parts *ABC* and *GHK* represent the Coulomb field of repulsion outside the nucleus, and the internal part *CDEFG* represents the attractive field which holds α-particles in their orbits. Let *DF* be an allowed orbit the energy of which, say 4 million volts, is given by the height of *DF* above

波动力学和放射性衰变

格尼，康登

编者按

20 世纪 20 年代，埃尔温·薛定谔在路易斯·德布罗意于 1924 年提出的物质可能具有波动性这个观点的基础上进行了一系列研究工作，开启了用“波动力学”描述量子理论的先河。薛定谔完全从波（或者说是波函数）的角度系统地描述了量子化粒子的行为，空间中不同位置上波的振幅代表了粒子在该处出现的概率。在这篇文章中，罗纳德·格尼和爱德华·康登敏锐地指出了薛定谔的描述对于理解放射性原子核发射出 α 粒子的衰变的意义。他们认为，α 粒子的逃逸可以简单地理解成核内外波函数的重叠。这正是一种穿透能量壁垒的量子“隧道效应”。

在 1902 年放射性衰变的指数律被发现后，人们很快认识到一个原子衰变的时间与这个原子以前的历史无关，而完全是由概率决定的。因为核子一定是被引力场束缚在原子核中的，所以为了解释核子的发射，我们必须假设一个从引力场到斥力场的自发转变。目前，有必要假设原子核具有某种特别的任意“不稳定性”，但在下文中，我们会指出衰变是量子力学原理的固有结果，并不需要任何特别的假设。

众所周知，经典力学在分子事件中失效是由于与分子尺度相比粒子的波长较大。α 粒子的波长约为分子尺度的 $1/10^5$，但因为原子核的尺度比原子尺度小差不多同样的比率，波动力学的适用性似乎是有保证的。

在经典力学中，一个运动粒子的轨道被严格限制在势能小于总能量的区域内。如果一个球在势阱中运动，并且其能量不足以使它翻过势垒，那么它将一直呆在势阱中，除非它以某种方式获得足以翻越势垒的能量。但在量子力学中情况并非如此，球总是有一个很小但不为零的机会可以穿透势垒从势阱中逃逸出来。

如图 1 所示，设 O 表示一个原子核的中心，$ABCDEFG$ 表示一条简化了的一维势能曲线。其中，ABC 段和 GHK 段表示核外的库仑排斥场；中间的 $CDEFG$ 段表示将 α 粒子束缚在其轨道上的引力场。设 DF 是一个容许轨道，它的能量，比如说 400 万电子伏，由 OX 到 DF 的高度表示。我们可以近似地说，这个轨道对应的波函

OX. Approximately, we can say that with this orbit will be associated a wave-function which will die away exponentially from *D* to *B*. Again, corresponding to motion outside the nucleus along *BM*, there will be a wave-function which will die away exponentially from *B* to *D*. The fact that these two functions overlap in the region *BD* means that there is a small but finite probability that the particle in the orbit *DF* will escape from the nucleus along *BM*, acquiring kinetic energy equal to the height of *DFBM* above *OX*, say 4 million volts. This occurrence will be spontaneous and governed solely by chance.

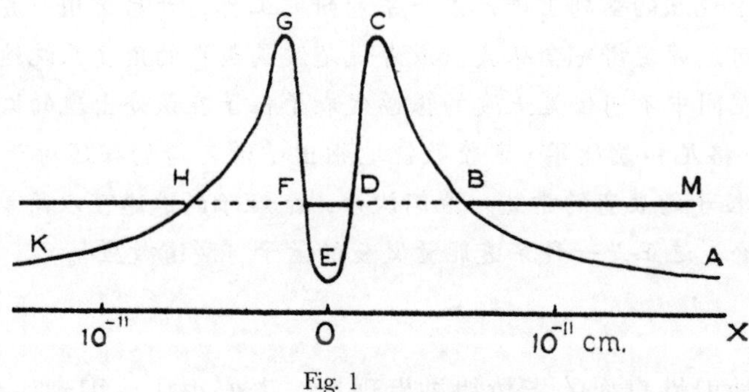

Fig. 1

The rate of disintegration, that is, the probability of escape, depends on the amount of overlapping of the wave-functions in the regions *DB* and *FH*, and this is extremely sensitive to the height to which the potential curve at *C* rises above *BDF*. By varying this height through a small range we can obtain all periods of radioactive decay from a fraction of a second, through the 10^9 years of uranium, to practical stability. (In considering the transmutation of a molecule into its isomer, Hund found a similar vast range of transformation periods, *Zeit. f. P.*, 43, 810; 1927) If the potential curves for the interaction of an α-particle with the various radioactive nuclei are similar, we can obtain a qualitative understanding of the Geiger-Nuttall relation between the rate of disintegration and the range of the emitted α-particles. For the α-particles of high energy the wave function for outside motion will overlap that for the inside motion more, and the rate of disintegration will be greater.

Besides obtaining a general idea of the mysterious instability of the nucleus, we can visualise in this way one of the most puzzling results of recent experimental work. An α-particle having the same range (2.7 cm) as those emitted by uranium should, if fired directly at the uranium nucleus, penetrate its structure; while faster α-particles should do so, even when not fired directly at the nucleus. It was therefore disconcerting when, on examining the scattering of fast α-particles fired at uranium, Rutherford and Chadwick (*Phil. Mag.*, 50, 904; 1925) could find no indication of any departure from the inverse square laws. But from the model outlined above, this is what would be expected. For if the height of *BM* above *OX* represents the energy of the uranium α-particles, then a faster particle fired at the nucleus will simply run part way up the hill *ABC* and return without having encountered any change in the repulsive field or any nuclear particles (which are describing orbits within the region *GEC*).

数从 D 到 B 呈指数衰减。此外，相应于核外运动的 BM 段中，粒子运动的波函数从 B 到 D 也呈指数衰减。这两个波函数在 BD 区域交叠的事实意味着，存在一个很小但不为零的概率，使得在 DF 轨道上的粒子能够沿着 BM 逃逸出原子核，同时获得 OX 到 $DFBM$ 的高度所表示的动能，比如说 400 万电子伏。这一事件是自发的并且只受概率控制。

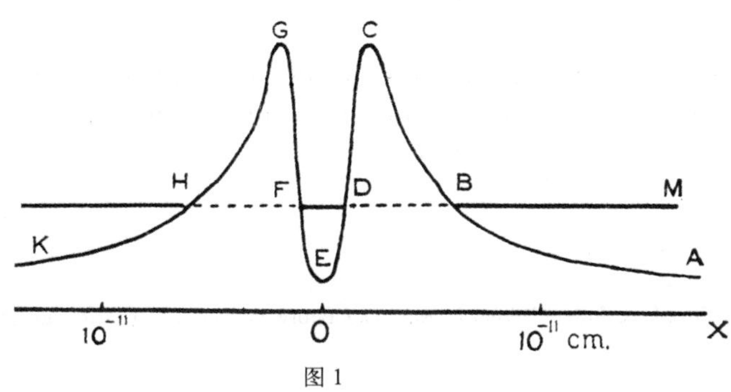

图 1

衰变速率，即逃逸概率，依赖于 DB 和 FH 区域中波函数的交叠量，并且对势能曲线中 C 点到 BDF 的高度十分敏感。通过在小范围内改变这个高度，我们可以获得所有放射性衰变元素的衰变周期，从几分之一秒、10^9 年（铀）到实质上是稳定的，各种情况都存在。（在考察分子的异构体之间的转化时，洪德发现了一个与此类似的转变周期的巨大范围，见《物理学杂志》，第 43 卷，第 810 页，1927 年）如果 α 粒子与各种放射性核相互作用的势能曲线是类似的，我们就能定性地理解衰变速率和所发射 α 粒子的射程之间的盖革–努塔耳关系。对于高能 α 粒子，核外运动波函数与核内运动波函数交叠得更多，从而衰变速率将会更大。

除了获得关于原子核神秘不稳定性的一般观点以外，以此方法我们还可以解释最近实验工作中最令人迷惑的一些结果中的一个。如果一个与铀发射出的 α 粒子射程相同（2.7 厘米）的 α 粒子直接射在铀核上，它将穿透铀核；而更快的 α 粒子即使不是直接射在铀核上，也会穿透铀核。因此，令人不解的是，在快 α 粒子射在铀上的散射实验中，卢瑟福和查德威克（《哲学杂志》，第 50 卷，第 904 页，1925 年）没能发现任何偏离库仑平方反比定律的迹象。但是从以上概述的模型来看，这应该是预期的结果。因为如果 OX 到 BM 的高度表示铀发射的 α 粒子的能量，那么射在核上的更快的粒子将仅仅沿着势垒 ABC 向上爬一段然后返回，其间没有经历排斥场的任何变化也没有遇到任何核子（核子在 GEC 区域内运动）。

The peculiar property of the wave mechanical equations which finds application here has also been applied to the theory of the emission of electrons from cold metals under the action of intense fields (Oppenheimer, *Proc. Nat. Acad. Sci.*, 14, 363; 1928; and Fowler and Nordheim, *Proc. Roy. Soc.*, A, 119, 173; 1928). Ordinarily, an atom does not lose its electrons because the attractive field of the atom remains attractive to all distances. But when an intense field is applied, then the attractive field is reversed in sign a short distance from the atom. This makes the resultant potential energy curve similar to that in the diagram, and so the atoms begin to shed their electrons.

Much has been written of the explosive violence with which the α-particle is hurled from its place in the nucleus. But from the process pictured above, one would rather say that the α-particle slips away almost unnoticed.

(**122**, 439; 1928)

Ronald W. Gurney, Edw. U. Condon: Palmer Physical Laboratory, Princeton University, July 30.

这里应用到的波动力学方程的特殊性质也曾被应用于强场作用下冷金属发射电子的理论中（奥本海默，《美国科学院院刊》，第14卷，第363页，1928年；福勒和诺德海姆，《皇家学会学报》，A辑，第119卷，第173页，1928年）。一般情况下，原子不会失去它的电子，因为不论相隔多远的距离，原子的引力场都吸引着电子。然而，当施加一个强场时，原子附近一个短距离范围之内，原子引力场将被颠倒。这使得合成的势能曲线类似于图1中的曲线，因此原子开始发射它们的电子。

很多文献都提到了将 α 粒子从原子核内抛出的一种爆炸性的力量。然而，从以上描绘的过程来看，人们宁愿说 α 粒子几乎是神不知鬼不觉地溜出原子核的。

（王锋 翻译；李军刚 审稿）

The "Wave Band" Theory of Wireless Transmission

A. Fleming

Editor's Note

Ambrose Fleming here takes issue with a way of understanding wave communications, for telephone or television. In both technologies, devices encode signals as amplitude modulations of a carrier wave. In terms of Fourier analysis, one can view the resulting wave as occupying a "band" of frequencies around the carrier frequency, and it had become common to consider how these bands should be apportioned, which bands were allowed and so forth. But Fleming argues that talk of bands obscures the role of the amplitude. Too large an amplitude could cause interference between different transmissions, much as speaking too loudly at the theatre can be disruptive. Focusing on amplitude rather than bands, Fleming suggests, will help avoid unnecessary restrictions on the new technologies.

IN scientific history we meet with many examples of scientific theories or explanations which have been widely adopted and employed, not because they can be proved to be true but because they provide a simple, easily grasped, plausible explanation of certain scientific phenomena. The majority of persons are not able to see their way through complicated phenomena and so thankfully adopt any short-cut to a supposed comprehension of them without objection.

Ease of comprehension is not, however, a primary quality of Nature, and it does not follow that because we can imagine a mechanism capable of explaining some natural phenomenon it is therefore accomplished in that way. There is a widely diffused belief in a certain theory of wireless telephonic transmission, and also of television, that for securing good effects it is necessary to restrict or include operations within a certain width of "wave band". But although this view has been very much adopted there is good reason to think that it is merely a kind of mathematical fiction and does not correspond to any reality in Nature.

Let us consider how it has arisen. We send out from all wireless telephone transmitters an electromagnetic radiation of a certain definite and constant frequency expressed in kilocycles. Thus 2LO London broadcasts on 842 kilocycles. This means that it sends out 842,000 electric vibrations or waves per second. Every broadcasting station has allotted to it a certain frequency of oscillation and it is not allowed to depart from it.

It is like a lighthouse which sends out rays of light of one pure colour or an organ which emits a single pure musical note. For most broadcasting stations this peculiar and individual frequency lies somewhere between a million and half a million per second, though for the long wave stations like Daventry it is so low as 193,000 or 193 kilocycles.

无线传输的“波带”理论

弗莱明

编者按

在本文中安布罗斯·弗莱明对应用于无线电话和电视的波通信有不同的理解。在这两项技术中，装置把信号编译成振幅调制的载波。在傅立叶分析中，我们可以认为得到的波占据了载波频率附近的一个频“带”。大家通常要考虑的是这些频带如何分配，哪些频带能够被允许等等。但弗莱明认为人们对频带的过分关注掩盖了振幅的作用。太大的振幅可能会导致不同传输过程之间的干扰，就像在戏院中大声讲话带来的麻烦一样。弗莱明指出，对振幅而不是频带的关注将会帮助我们避免在应用这项新技术时遇到不必要的麻烦或限制。

在科学史上，我们遇到的许多科学理论或解释被广泛接受并应用的情况，并不是因为它们能够被证明是真理，而是因为它们为某些科学现象提供了简单、易于被理解并且似乎合理的解释。大多数人并不能透过复杂现象发现真理，因此也就乐于不加质疑地接受某种能够便捷地解释复杂现象的假说。

然而，简单的理解并不是自然界的基本特征，也不能随即得到，即不会因为我们想象一种能够解释某些自然现象的机制，它就以那种方式来实现。在关于无线电话传输以及电视的某种理论中，存在一种广为流传的认识，即为了达到可靠的良好效果，必须限制在具有确定宽度的“波带”内操作。这种观点虽然已被广泛接受，但我们有理由认为，这只是一类数学虚构，并不能与自然界的任何现实相对应。

让我们考虑这是如何产生的。我们从所有的无线电话发射台发出一个具有确定的恒定频率（以千周表示）的电磁辐射。伦敦 2LO 电台按照这种方式以 842 千周的频率进行广播。这意味着它每秒内发出 842,000 次电子振荡或 842,000 个波。每个广播站已分配到一个确定的振荡频率，而且不允许偏离此频率。

这就像一座发出单色光的灯塔或一个发出单一音符的风琴。对于大多数广播站而言，这类特有的专用频率位于每秒 50 万~100 万之间，然而对于类似达文特里这样的长波站，它的频率低至 193,000，或 193 千周。

When we speak or sing or cause music to affect the microphone at a broadcasting studio the result is to cause the emitted vibrations, which are called the *carrier waves*, to fluctuate in height or wave amplitude, but does not alter the number of waves sent out per second. It is like altering the height or size of the waves on the surface of the sea without altering the distance from crest to crest which is called the wave-length.

Suppose the broadcasting station emits a carrier wave of frequency n and let $p = 2\pi n$. Then we may express the amplitude a of this wave at any time t by the function $a = A \sin pt$ where A is the maximum amplitude. If on this we impose a low frequency oscillation due to a musical note of frequency m and let $2\pi m = q$, then we can express the modulated vibration by the function

$$a = A \cos qt \sin pt$$

But by a well-known trigonometrical theorem this is equal to

$$\frac{A}{2}\{\sin(p+q)t + \sin(p-q)t\} ,$$

and thence may be supposed to be equivalent to the simultaneous emission of two carrier waves of frequency $n + m$ and $n - m$.

If the imposed note or acoustic vibration is very complex in form, then in virtue of Fourier's theorem it may be resolved into the sum of a number of simple harmonic terms of form cos qt, and each of these may be considered to be equivalent to a pair of co-existent carrier waves. Hence the complex modulation of a single frequency carrier wave might be imitated by the emission of a whole spectrum or multitude of simultaneous carrier waves of frequencies ranging between the limits $n + N$ and $n - N$, where n is the fundamental carrier frequency and N is the maximum acoustic frequency occurring and $2N$ is the width of the wave band. This, however, is a purely mathematical analysis, and this band of multiple frequencies does not exist, but only a carrier wave of one single frequency which is modulated in amplitude regularly or irregularly.

If the sounds made to the microphone at the broadcasting station are very complex, such as those due to instrumental music or speech, then in virtue of this mathematical theorem the very irregular fluctuations in amplitude of the single carrier wave can be imitated if we suppose the station to send out simultaneously a vast number of carrier waves of various frequencies lying between certain limits called the "width of the wave band".

This, however, is merely a mathematical artifice similar to that employed when we resolve a single force or velocity in imagination into two or more component forces. Thus, if we consider a ball rolling down an inclined plane and desire to know how far it will roll in one second, we can resolve the single vertical gravitational force on the ball into two components, one along the plane and one perpendicular to it. But this is merely an ideal division for convenience of solution of the problem; the actual force is one single force acting vertically downwards. Similar reasoning is true with regard to wireless telephony. What happens, as a matter of fact, is that the carrier wave of one single constant

当我们在播音室里对着麦克风说话、唱歌或放出音乐，都会导致发射振荡的产生，这被称作**载波**，它会使波的高度或波幅出现起伏，但并不改变每秒钟发出的波数。这类似于只改变海面上波浪的高度或大小，而不改变被称作波长的从波峰到波峰的距离。

假定广播站发射一个频率为 n 的载波，并令 $p = 2\pi n$。我们通过函数 $a = A \sin pt$ 表示在任意时刻 t 这个波的振幅 a，其中 A 是最大振幅。如按这种方式，我们施加一个由频率为 m 的音符引起的低频振荡，并令 $2\pi m = q$，则我们可以通过下面的函数方程表示调制振荡

$$a = A \cos qt \sin pt$$

通过人们熟知的三角定理，上式等于

$$\frac{A}{2}\left\{\sin(p+q)t + \sin(p-q)t\right\} ,$$

因此可以认为其等效于同时发射频率为 $n + m$ 和 $n - m$ 的两个载波。

如果施加的音律或声音振荡在形式上很复杂，那么利用傅立叶定理，我们可将其分解为如 $\cos qt$ 形式的许多简谐项之和，可以等效地认为其中每一项都是一对共存的载波。因此单频载波的复杂调制可以用整个谱的发射或大量频率在 $n + N$ 到 $n - N$ 之间的同步载波来模拟，其中 n 是基本载波频率，而 N 是出现的最大声波频率，$2N$ 是波带的宽度。然而，这是纯数学分析，这个多频波带并不存在，而只存在单频的载波，可以对其振幅进行规则或不规则的调制。

如果向广播站里的麦克风发出的声音非常复杂，例如乐器发出的声音或讲话的声音，那么根据这个数学原理，我们可以假定广播站里同时发出大量频率被限定在“波带宽度”之内的载波来模拟单个振幅不规则波动的载波。

然而，这仅是一个数学技巧，类似于我们在假想中将单个力或速度分解为两个或更多个分量。因此，如果我们考虑一个沿倾斜平面滚下的球，并要求知道球在一秒钟内滚动多远，我们就可以将球受到的单一的垂直引力分解为两个分量，分别与平面平行和垂直。但这只是便于解决问题的一个理想的分解；实际的力仍只是作用方向垂直向下的单个力。对于无线电话也存在着相似的思考过程。事实上，发生的情况是，一个单一恒定频率的载波按照某种规则或不规则的规律在振幅上发生变

frequency suffers a variation in amplitude according to a certain regular or irregular law. There are no multiple wave-lengths or wave bands at all.

The receiver absorbs this radiation of fluctuating amplitude and causes the direct current through the loud speaker to vary in accordance with the fluctuations of amplitude of the carrier wave; the carrier wave vibrations being rectified by the detector valve.

The same thing takes place in the case of wireless transmission in television. The scanning spot passes over the object and the reflected light falls on the photoelectric cells and creates in them a direct current which varies exactly in proportion to the intensity of the reflected light. This photoelectric current is employed to modulate the amplitude of a carrier wave, and the neon lamp at the receiving end translates back these variations of carrier wave amplitude into variations in the cathode light of the neon tube.

There is neither in wireless telephony nor in television any question of various bands of wavelength. There is nothing but a carrier wave of one single frequency which experiences change of amplitude. The whole question at issue then is, what range in amplitude is admissible?

In the case of television it is usual for critics of present achievements to say that good or satisfactory television cannot be achieved within the limits of the nine kilocycle band allowed. But there is in reality no wave band involved at all. It is merely a question of what change in amplitude in a given carrier wave can be permitted without creating a nuisance.

It is something like the question: How loud can you whisper to your next neighbour at a concert or theatre without being considered to be a nuisance? People do whisper in this way, and provided not too loudly, it is passed over. But if anyone is so ill-mannered as to speak too loudly he is quickly called to order, or turned out.

It is, however, not an easy thing to define a limit to wave amplitudes. They are measured in microvolts per metre and are difficult to measure. But a wave-length is easy to define in kilocycles or in metres, and hence the method has been adopted of limiting emission to an imaginary band of wavelengths which, however, do not exist.

The definition is imperfect or elusive. It is something like the old-fashioned definition of metaphysics as "a blind man in a dark room groping for a black cat which isn't there". Similarly, the supposed wave band is not there. All that is there is a change, gradual or sudden, in the amplitude of the carrier wave. It is clear, then, that sooner or later we shall have to modify our code of wireless laws.

We have no reason for limiting the output of our broadcasting stations to some imaginary wave band of a certain width, say nine kilocycles or whatever may be the limiting width, but we have reason for limiting the range of amplitude of the carrier waves sent out.

化。根本不存在多重的波长或波带。

接收器吸收了这类振幅振荡变化的辐射，并使通过扬声器的直流按照载波振幅的振荡变化而变化；载波的振动通过检测器的电子管进行整流。

对于电视的无线传输来说也存在相同的情况。扫描点在物体上扫描，而反射光落在光电元件上，并在其中产生了与反射光强度精确成正比变化的直流。这个光电流用于载波振幅的调制，在接收端的氖灯会将这些载波振幅的变化转换回氖管的阴极光的变化。

在无线电话和电视中均不存在各种不同波带的问题。只是存在振幅被调制的单频载波。那么争论的全部问题就是，可以允许振幅在什么范围内变化呢？

对于电视而言，批评现有成果的评论者们通常会提出，在允许的 9 千周的限制范围内，性能良好的或令人满意的电视不可能实现。但实际上根本不存在所谓的波带。这仅仅是一个在不产生干扰的情况下可以允许给定的载波中振幅如何变化的问题。

这就类似于如下的问题：在音乐厅或剧场中你可以用多大的声音悄悄对邻座说话而不至于影响其他人呢？人们以这种方式耳语，并保持声音不是很大，这样可以不被注意。但是如果任何人非常不礼貌地用很大的声音说话，他很快就会被要求保持安静或被逐出会场。

然而，确定波幅的调制范围并不是件容易的事情。它们以每米微伏的数量级来进行计量，并且是很难被测定的。但波长很容易用千周或米来定义，因此现已采用的方法是将发射限定到虚构的波带范围内，然而，这个波带实际上并不存在。

这样的定义是不完善的，或者说是难以理解的。这有点像形而上学的老式定义，如“一个在暗室里摸索一只并不存在的黑猫的盲人”。类似地，我们假设的波带并不存在。存在的只是载波中振幅逐步或突然的变化。显然，总有一天我们将不得不修改我们的无线通信的编码规则。

我们没有理由将我们的广播电台的输出限制在某种确定宽度的假想波带内，比如说 9 千周或其他任何限制范围，但我们有理由限定输出载波振幅的范围。

Some easily applied method will have to be found of defining and measuring the maximum permissible amplitude of the carrier waves as affected by the microphone or other variational appliance. It may perhaps be thought that an unnecessary fuss is here being made on what may be regarded as simply a way of explaining things, but experience in other arts shows how invention may be greatly retarded by unessential official restrictions. Consider, for example, the manner in which mechanical traction was retarded in Great Britain for years by ridiculous regulations limiting the speed of such vehicles on highway roads. The only restrictions that should be imposed are those absolutely necessary in the interests of public safety or convenience, and all else tend to throttle and retard invention and progress.

(**125**, 92-93; 1930)

我们必将找到某些易于应用的方法，用来定义及测量被麦克风或其他有变化的设备影响的载波的最大允许振幅。也许有人会认为，这只是事物的一种解释方式，没有必要像我们这样小题大做，但其他技术中的经验表明，不必要的官方限定是如何使可能的重大发明被推迟的。考虑一下这样的事例，机械牵引在英国的发展多年受阻，正是由于制定了荒谬的法规限制公路上的这类车辆的速度而导致的。唯一应该加以限制的，是那些出于公共安全性或便利性的考虑绝对必要的方面，而其他所有方面的限制往往会扼杀并妨碍发明和进步。

（沈乃澂 翻译；李军刚 审稿）

Electrons and Protons

Editor's Note

Paul Dirac was one of the most creative physicists of the early twentieth century. In 1932 he was appointed Lucasian professor at Cambridge University, and his theory unifying quantum mechanics with Einstein's theory of special relativity won him the 1933 Nobel Prize in physics. Here *Nature* reports on one of the implications of this theory, as Dirac outlined in a paper in the *Proceedings of the Royal Society*. His equations predicted "electrons of negative energy", which meant, of positive charge. The report echoes Dirac's initial suspicion that these predicted positively charged particles would behave like protons, but they soon proved to be "positive electrons" or positrons, made of antimatter.

A theory of positive electricity has been put forward by Dr. P. A. M. Dirac in the January number of the *Proceedings of the Royal Society*. The relativity quantum theory of an electron leads to a wave equation which possesses solutions corresponding to negative energies—the energy of the electron of ordinary experiment being reckoned as positive—and although there are serious difficulties encountered in any immediate attempt to associate these negative states with protons, the existence of positive electricity can be predicted by a fairly direct line of argument. Since the stable states of an electron are those of lowest energy, all the electrons would tend to fall into the negative energy states—with emission of radiation—were it not for the Pauli exclusion principle, which prevents more than one electron from going to any one state. If, however, it is assumed that "there are so many electrons in the world that... all the states of negative energy are occupied except perhaps a few...", it may be supposed that the infinite number of electrons present in any volume will remain undetectable if uniformly distributed, and only the few "holes", or missing states of negative energy will be amenable to observation. The step is then made of regarding these "holes" as "*things of positive energy*" which are identified with the protons. A difficulty now arises in ordinary electromagnetic theory which apparently has to cope with the presence of negative electricity of infinite density; this is met by supposing that for ordinary purposes volume-charges must be measured by departures from a "normal state of electrification", which is "the one where every electronic state of negative energy and none of positive energy is occupied." The problem of the large mass of the proton, as compared with that of the electron, is not discussed in detail, but a possible line of attack is indicated. Dr. Dirac has included the minimum of mathematical analysis in this paper, which can be followed in all essential points by anyone acquainted with the principles of the quantum theory.

(**125**, 182; 1930)

电子和质子

编者按

保罗·狄拉克是20世纪初最有建树的物理学家之一。1932年，他被聘任为剑桥大学的卢卡斯教授。他因提出能将量子力学与爱因斯坦的狭义相对论统一起来的理论而获得了1933年的诺贝尔物理学奖。在发表于《皇家学会学报》的一篇论文中，狄拉克对他的理论所蕴含的一个推断进行了概述，《自然》的这篇文章报道的正是这些。狄拉克的方程预测出"具有负能量的电子"，即带正电荷的电子。狄拉克最初认为，这些被预测到的带正电荷的粒子的行为方式与质子类似，这篇报道再次复述了这一观点，但不久之后，这些粒子就被证明是组成反物质的"正电子"。

在1月的《皇家学会学报》上，狄拉克博士提出了正电子理论。他通过电子的相对论量子理论导出了一个波动方程，而这个波动方程包含相应于负能量的解（普通实验中电子的能量被认为是正的）。尽管将这种负能态与质子联系起来的尝试遇到了很多困难，但是，通过非常直接的论证可以预言正电子的存在。因为电子的稳定态是能量最低的态，所以伴随着发射辐射，所有电子都将落入负能态，但是这与泡利不相容原理不同，这一原理规定不可能有多于一个的电子处于任何一个相同的状态。然而，如果假设"世界上有非常多的电子以至于除了极少的负能态以外，几乎所有的负能态都被占据了"，那么可以推测，这些存在于负能态体系中的无数个电子将不会被探测到（若电子是均匀分布的），而仅有极少的"空穴"或遗漏的负能态可以很容易地被观察到。接下来他把这些"空穴"视为与质子相同的"**具有正能量的物质**"。密度无限大的负电的出现使普通电磁学理论遇到了困难。不过，当假定一般情况下必须通过"带电的正常状态"的偏离来测量体电荷时，便可以解决这个难题，所谓带电的正常状态是指"每一个具有负能量的电子态均被占据而没有一个正能态被占据的状态"。在这篇文章中，他并没有详细讨论质子质量比电子质量大这一问题，但是却提出了一种可能的解决思路。狄拉克博士的论文仅涉及少量数学分析，这就使任何一个学习过量子理论基本原理的人都能看懂这些数学分析的要点。

（王锋 翻译；李淼 审稿）

Artificial Disintegration by α-particles

J. Chadwick and G. Gamow

Editor's Note

In 1919 Ernest Rutherford had shown that alpha particles fired into nitrogen gas could create hydrogen, apparently because the particles kick a proton out of the nucleus while being themselves captured. Rutherford called the process "artificial disintegration" (popularly, splitting) of the atom. Here James Chadwick and George Gamow suggest that protons might also be produced without the alpha particle entering the nucleus. Gamow went on to propose that protons, with only half the charge of alpha particles, might approach the positively charged nucleus more easily. That idea stimulated work at Cambridge on a high-voltage device to accelerate protons to high energies, leading to the first induced splitting of a lithium nucleus by proton bombardment by John Cockroft and Ernest Walton.

IT is commonly assumed that the process of artificial disintegration of an atomic nucleus by collision of an α-particle is due to the penetration of the α-particle into the nuclear system; the α-particle is captured and a proton is emitted.

On general grounds it seems possible that another process may also occur, the ejection of a proton without the capture of the α-particle.

Consider a nucleus with a potential field of the type shown in Fig. 1, where the potential barrier for the α-particle is given by the full line and that for the proton by the dotted line. Let the stable level on which the proton exists in the nucleus be $-E_p^\circ$ and the level on which the α-particle remains after capture be $-E_\alpha^\circ$.

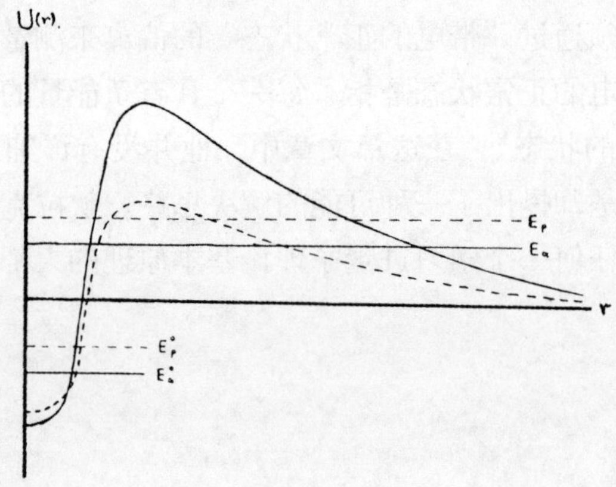

Fig. 1

α 粒子引发的人工衰变

查德威克，伽莫夫

编者按

1919 年，欧内斯特·卢瑟福发现 α 粒子射入氮气后可以产生氢，这显然是因为 α 粒子被俘获的同时把一个质子赶出了原子核。卢瑟福把这个过程称作原子的"人工衰变"（更通俗地说是裂变）。詹姆斯·查德威克和乔治·伽莫夫在本文中提出，在没有 α 粒子进入原子核的情况下我们也可以得到质子。伽莫夫还指出，质子所带电荷数只有 α 粒子的一半，它有可能更容易接近带正电的原子核。这种观点推动了剑桥大学的约翰·考克饶夫和欧内斯特·瓦耳顿的工作，他们用高压装置把质子加速到很高的能量，然后通过质子轰击首次实现了锂核的诱导裂变。

通常的假设是，一个 α 粒子与原子核发生碰撞的人工衰变过程，是由 α 粒子穿透到核系统内部导致的；在 α 粒子被俘获的同时，也会发射出一个质子。

按照通常的观点，另外一种过程也可能发生，即在没有俘获 α 粒子的情况下，释放出一个质子。

研究一个处于图 1 所示的势场条件下的核，其中实线代表 α 粒子的势垒，虚线代表质子的势垒。令原子核中质子的稳态能级为 $-E_p^\circ$，被俘获后的 α 粒子的能级为 $-E_\alpha^\circ$。

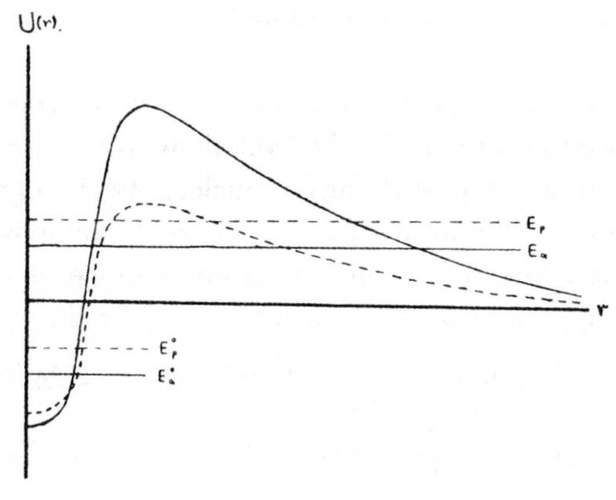

图 1

If an α-particle of kinetic energy E_α penetrates into this nucleus and is captured, the energy of the proton emitted in the disintegration will be $E_p = E_\alpha + E_\alpha^\circ - E_p^\circ$, neglecting the small kinetic energy of the recoiling nucleus. If the nucleus disintegrates without capture of the α-particle, the initial kinetic energy of the α-particle will be distributed between the emitted proton and the escaping α-particle (again neglecting the recoiling nucleus). The disintegration protons may have in this case any energy between $E_p = 0$ and $E_p = E_\alpha - E_p^\circ$.

Thus, if both these processes occur, the disintegration protons will consist of two groups: a continuous spectrum with a maximum energy less than that of the incident α-particles and a line spectrum with an energy greater or less than that of the original α-particles according as $E_\alpha^\circ > E_p^\circ$ or $E_\alpha^\circ < E_p^\circ$, but in either case considerably greater than the upper limit of the continuous spectrum (see Fig. 2).

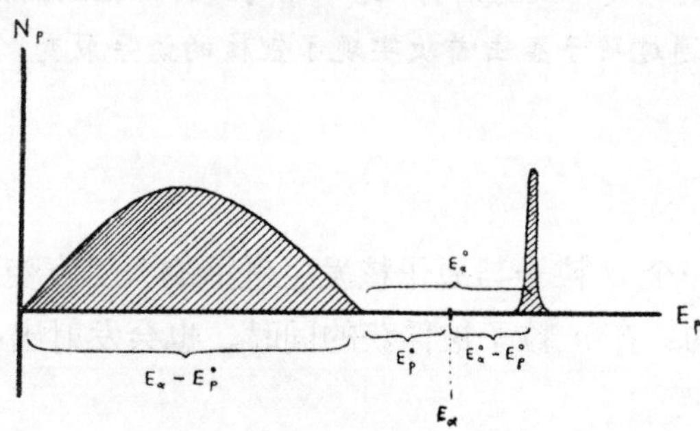

Fig. 2

In some experiments of one of us in collaboration with J. Constable and E. C. Pollard, the presence of these two groups of protons appears quite definitely in certain cases, for example, boron and aluminium. A full discussion of these and other cases of disintegration will be given elsewhere, but it may be noted that the existence of groups of protons has already been reported by Bothe and by Pose. In general the experimental results suggest that with incident α-particles of energy about 5×10^6 volts (α-particles of polonium) the process of non-capture is several times more frequent than the process of capture.

It is clear that, if our hypothesis is correct, accurate measurement of the upper limit of the continuous spectrum and of the line will allow us to estimate the values of the energy levels of the proton and α-particle in the nucleus. In the case of aluminium bombarded by the α-particles of polonium the protons in the continuous spectrum have a maximum range of 32 cm and those of the line spectrum a range of 64 cm. These measurements give the following approximate values for the energy levels:

$$E_p^\circ = 0.6 \times 10^6 \ e \text{ volts, and } E_\alpha^\circ = 2 \times 10^6 \ e \text{ volts.}$$

On the wave mechanics the probability of disintegration of both types is given by the square of the integral

$$W = \int f(r_{\alpha,p}) \cdot \psi_\alpha \cdot \psi_p \cdot \phi_\alpha \cdot \phi_p \cdot dV \cdot dV' \quad (1)$$

如果动能为 E_α 的 α 粒子穿透到这个核的内部并被俘获，那么衰变中发射出的质子的能量是 $E_p = E_\alpha + E_\alpha^o - E_p^o$，这里忽略了此过程中反冲核的微小动能。如果在核衰变的过程中没有俘获 α 粒子，α 粒子的初始动能将在发射出的质子与逃逸的 α 粒子之间分配（再次忽略反冲核）。在这种情况下，衰变质子的能量可能是 $E_p = 0$ 与 $E_p = E_\alpha - E_p^o$ 之间的任意值。

因此，如果这两种过程都存在，那么衰变的质子将由两部分组成：最大能量小于入射 α 粒子能量的连续谱和能量大于或小于初始 α 粒子能量的线状谱，相当于 $E_\alpha^o > E_p^o$ 或 $E_\alpha^o < E_p^o$，但对于这两种情况中的任何一种，能量都比连续谱的上限大很多（见图 2）。

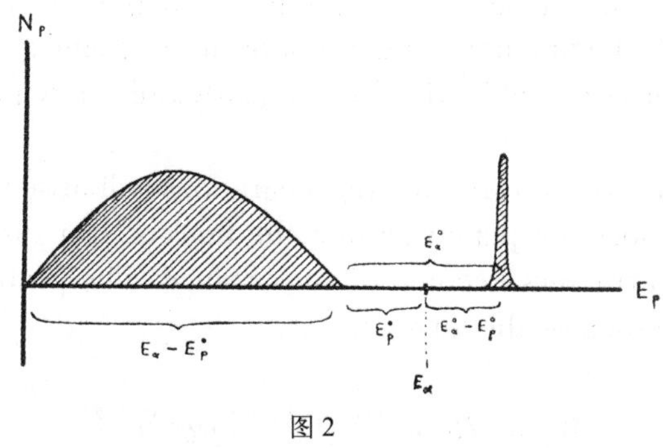

图 2

在我们中的一位与康斯特布尔和波拉德合作进行的一些实验中，在某些情况下这两组质子的存在会非常明确地表现出来，比如，硼和铝的情况。关于这些以及其他一些衰变的情况将会在别的地方给出全面的讨论，但可以注意到的是，博特和波泽已经报道了多组质子的存在。一般而言，实验结果显示，当入射 α 粒子（钋的 α 粒子）的能量约为 5×10^6 电子伏时，非俘获过程发生的频率比俘获过程高出几倍。

显然，如果我们的假设是正确的，那么对连续谱和线状谱上限的精确测量将使我们能对原子核中的质子和 α 粒子的能级数值进行估计。在用钋的 α 粒子轰击铝的情况下，在连续谱中质子的最大范围为 32 厘米，而在那些线状谱中的范围为 64 厘米。由这些测量得到的能级近似值如下：

$$E_p^o = 0.6 \times 10^6 \text{ 电子伏}，E_\alpha^o = 2 \times 10^6 \text{ 电子伏}。$$

根据波动力学，两种类型的衰变概率由积分的平方给出

$$W = \int f(r_{\alpha,p}) \cdot \psi_\alpha \cdot \psi_p \cdot \phi_\alpha \cdot \phi_p \cdot dV \cdot dV' \quad (1)$$

where $f(r_{\alpha,p})$ is the potential energy of an α-particle and a proton at the distance $r_{\alpha,p}$ apart, and the wave functions ψ_α, ψ_p represent the solutions for the α-particle and proton before and ϕ_α, ϕ_p after the disintegration. In calculating the integral (1) we must develop the incident plane wave of the α-particle into spherical harmonics corresponding to different azimuthal quantum numbers of the α-particle, and deal with each term separately.

In the case of capture of the α-particle the estimation of (1) can be carried out quite simply. It can be shown that the effect of the higher harmonics is very small, and that the disintegration is due almost entirely to the direct collisions. Thus we obtain for the probability of disintegration

$$W_1^2 = \frac{A}{v_\alpha^2} \cdot e^{-\frac{8\pi^2 e^2}{h} \cdot \frac{Z}{v_\alpha}} \cdot e^{-\frac{4\pi^2 e^2}{h} \cdot \frac{Z}{v_p}} \tag{2}$$

where v_α and v_p are the velocities of the initial α-particle and the ejected proton respectively. Since only the first harmonic is important in disintegration of this type, it is to be expected that the protons will be distributed nearly uniformly in all directions.

When the α-particle is not captured the disintegrations will arise mainly from collisions in which the α-particle does not penetrate into the nucleus. For disintegration produced in this way the higher harmonics become of importance. The probability of disintegration can be roughly represented by the formula

$$W_2^2 = B \cdot e^{-\frac{8\pi^2 e^2}{h} \cdot Z\left(\frac{1}{v'_\alpha} - \frac{1}{v_\alpha}\right)} \cdot e^{-\frac{4\pi^2 e^2}{h} \cdot \frac{Z}{v_p}} \tag{3}$$

where v'_α is the velocity of the α-particle after the collision, and B is a function of the angle of ejection of the proton. The protons of the continuous spectrum will not be emitted uniformly in all directions. According to the expression (3) the distribution with energy of the protons in the continuous spectrum will have a maximum value for an energy of ejection of about 0.3 of the upper limit, and will vanish for zero energy and at the upper limit.

More detailed accounts of the experimental results and of the theoretical calculations will be given shortly.

(**126**, 54-55; 1930)

J. Chadwick, G. Gamow: Cavendish Laboratory, Cambridge, June 18.

式中 $f(r_{\alpha,p})$ 是相距为 $r_{\alpha,p}$ 的一个 α 粒子和一个质子之间的势能，波函数 ψ_α 和 ψ_p 表示 α 粒子和质子在衰变前的解，ϕ_α 和 ϕ_p 表示它们在衰变后的解。在计算积分 (1) 时，我们必须将 α 粒子的入射平面波展开为对应于 α 粒子不同角量子数的球谐函数的形式，并对每一项分别处理。

在俘获 α 粒子的情况下，对 (1) 式进行估算非常简单。可以看到的是，高次谐波的效应很小，衰变几乎全部由直接碰撞产生。因此我们可以得到衰变概率

$$W_1^2 = \frac{A}{v_\alpha^2} \cdot e^{-\frac{8\pi^2 e^2}{h} \cdot \frac{Z}{v_\alpha}} \cdot e^{-\frac{4\pi^2 e^2}{h} \cdot \frac{Z}{v_p}} \tag{2}$$

式中 v_α 和 v_p 分别是初始 α 粒子和发射出的质子的速度。因为在这类衰变中，只有一次谐波起主要作用，所以可以认为质子在所有方向上都是均匀分布的。

当 α 粒子未被俘获时，衰变将主要通过 α 粒子并不穿透到核内的碰撞产生。在由这种方式产生的衰变过程中，高次谐波开始起主要作用。衰变概率可以通过下述公式近似地表示

$$W_2^2 = B \cdot e^{-\frac{8\pi^2 e^2}{h} \cdot Z\left(\frac{1}{v'_\alpha} - \frac{1}{v_\alpha}\right)} \cdot e^{-\frac{4\pi^2 e^2}{h} \cdot \frac{Z}{v_p}} \tag{3}$$

式中 v'_α 是碰撞后 α 粒子的速度，B 是质子发射角度的函数。连续谱中的质子在各个方向上的发射并不均匀。根据表达式（3)，连续谱中质子的能量分布，在约为上限的 0.3 处其发射能量达到最大值，而在上限处突然变为零。

不久之后，我们将给出关于实验结果和理论计算的更详细的解释。

（沈乃澂 翻译；江丕栋 审稿）

Fine Structure of α-rays

G. Gamow

Editor's Note

George Gamow was a Russian scientist who left the Soviet Union in 1932 and worked at several Western European universities until moving to the United States before the Second World War. He made a powerful impression by his versatility as a scientist, his capacity to write clearly for the general public and his engagement in public causes such as advocacy of building nuclear weapons in the United States. This brief letter offers an explanation of why γ-rays emitted by radioactive atoms may have a variety of energies.

IT is usually assumed that the long range α-particles observed in C'-products of radioactive series correspond to different quantum levels of the α-particle in the nucleus. If after the preceding β-disintegration the nucleus is left in an excited state with the α-particle on one of the levels of higher energy, one of the two following processes can take place: either the α-particle will cross the potential barrier surrounding the nucleus and will fly away with the total energy of the excited level (long range α-particle), or it will fall down to the lowest level, emitting the rest of its energy in the form of electromagnetic radiation (γ-rays), and will later fly away as an ordinary α-particle of the element in question. Thus there must exist a correspondence between the different long range α-particles and the γ-rays of the preceding radioactive body. If p is the relative number of nuclei in the excited state, λ the corresponding decay constant, and θ the probability of transition of the nucleus from the excited state to one of the states of lower energy with emission of energy (in form of γ-quanta or an electron from the electronic shells of the atom), the relative number of long range α-particles must be $N = p\frac{\lambda}{\theta}$. Knowing the number of α-particles in each long range group and calculating, from the wave mechanical theory of radioactive disintegration, the corresponding values of λ, we can estimate for each group the value θ/p, giving a lower limit for the probability of γ-emission. For example, for thorium-C' possessing besides the ordinary α-particles also two groups of long range α-particles, we have for transition probabilities from two excited states to the normal state $\theta_1 < 0.4 \times 10^{12}$ sec^{-1} and $\theta_2 < 2 \times 10^{12}$ sec^{-1}, which is the right order of magnitude for the emission of light quanta of these energies. With decreasing energy λ decreases much more rapidly (exponentially) than θ, so that the number of long range α-particles from the lower excited levels will be very small. (From this point of view we can also easily understand why the long range α-particles were observed only for C'-products for which the energy of normal α-particles is already much greater than for any other known radioactive element.)

α 射线的精细结构

伽莫夫

编者按

乔治·伽莫夫是一位俄国科学家，他于 1932 年离开苏联，在几所西欧的大学里工作。第二次世界大战爆发前夕，他去了美国。他是一位多才多艺的科学家，为公众撰写的普及读物非常清晰易懂，他在美国投身于公众事业，支持原子武器，这些都给人们留下了深刻的印象。在这篇简短的快报文章中，他解释了为什么放射性原子发射出的 γ 射线可能具有不同的能量。

我们通常假设，C' 放射系列产物中观测到的长程 α 粒子对应于原子核中 α 粒子的不同量子能级。如果在 β 衰变之后，原子核处于 α 粒子占据某个更高能级的激发态，那么就可能发生下面两个过程中的一个：或者是 α 粒子穿过原子核周围的势垒，携带激发态的所有能量而逃逸（长程 α 粒子）；或者是 α 粒子降至最低能级，将剩余能量以电磁辐射（γ 射线）的形式发射出去，然后再以普通 α 粒子的形式逃逸出该原子核。这样，在长程 α 粒子和之前的放射体放出的 γ 射线之间就应该存在某种关联。如果 p 是处于激发态的原子核的相对数量，λ 是相应的衰减常数，θ 是原子核从激发态跃迁到某个低能态并辐射出能量（以 γ 量子或者是从该原子电子壳层中发射出的电子的形式）的概率，那么长程 α 粒子的相对数量就是 $N = p\frac{\lambda}{\theta}$。我们已经知道了每一个长程组内的 α 粒子数量，并且可以通过辐射衰变的波动力学理论计算出相应的 λ 值，那么我们就可以估算出每一组的 θ/p 值，得到一个 γ 辐射发生概率的下限。比如，在钍 C' 的衰变中，除释放普通的 α 粒子之外，还有两组长程 α 粒子，对于从这两个激发态跃迁到正常态的概率，我们知道：$\theta_1 < 0.4 \times 10^{12}$ 秒 $^{-1}$，$\theta_2 < 2 \times 10^{12}$ 秒 $^{-1}$，这个数量级对于辐射这些能量的光量子来说是合适的。当能量减小时，λ（呈指数减小）比 θ 减小的速度快很多，因此来自较低激发态的长程 α 粒子数量将会非常少。（这样看来，就不难理解为什么长程 α 粒子只有在 C' 的衰变产物中才会出现了，C' 过程产物中的正常 α 粒子的能量远远高于其他已知的任何放射性元素产生的能量。）

A difficulty arises with the recent experiments of S. Rosenblum (*C. R.*, p. 1,549; 1929; p. 1,124; 1930), who found that the α-rays of thorium-*C* consist of five different groups lying very close together. The energy differences and intensities of the different groups relative to the strongest one (α_0) are, according to Rosenblum:

$$E\alpha_1 - E\alpha_0 = +40.6 \text{ kv} \quad I\alpha_1 = 0.3$$
$$E\alpha_2 - E\alpha_0 = -287 \ \ldots \quad I\alpha_2 = 0.03$$
$$E\alpha_3 - E\alpha_0 = -442 \ \ldots \quad I\alpha_3 = 0.02$$
$$E\alpha_4 - E\alpha_0 = -421 \ \ldots \quad I\alpha_4 = 0.005$$

If we suppose that these groups are due to α-particles escaping from different excited quantum levels in the nucleus, we meet with very serious difficulties. The decay constant λ for the energy of thorium-*C* fine structure particles is very small ($\lambda \sim 10^{-2}$ sec^{-1}), and in order to explain the relatively great number of particles in different groups we must assume also very small transition probabilities. We must assume that thorium-*C* nucleus can stay in an excited state without emission of energy for a period of half an hour!

We can, however, obtain the explanation of these groups by assuming that we have here a process quite different from the emission of long range α-particles. Suppose that two (or more) α-particles stay on the normal level of the thorium-*C* nucleus. It can happen that after one of the α-particles has escaped the nucleus will remain in an excited state with the other particle on a certain level of higher energy. (In this case the energy of the escaping α-particle will be smaller than the normal level and obviously will not correspond to any quantum level inside the nucleus.) From the excited state the nucleus (thorium-*C*″ now) can afterwards jump down to the normal level, emitting the energy difference in form of a γ-quantum.

Thus the relative number of different groups will not depend on the probability of γ-emission but only on the transition integral:

$$W = \int f(r_{1,2})\, \psi E_0(\alpha_1)\, \psi E_0(\alpha_2)\, \psi E_n(\alpha_1)\, \psi E\alpha_n(\alpha_2)\, dv_1\, dv_2$$

where $f(r)$ is the interaction energy of two α-particles at a distance r apart, ψE_0 and ψE_n the eigenfunctions of an α-particle in the normal and n^{th} excited states, and ψE_a the eigenfunction of an escaping α-particle with the energy: $E_{\alpha_n} = E_0 - (E_n - E_0)$.

According to this scheme, the γ-rays corresponding to different fine structure groups of thorium-*C* must be observed as γ-rays of thorium-*C* (ejecting electrons from *K*, *L*, *M*, ... shells of the thorium-C″-atom) and not as the rays of thorium-*B*, as we would expect in the case of long range particle explanation. The level scheme of the thorium-*C*″-nucleus as given by fine structure energies is represented in Fig. 1.

罗森布拉姆在最近的实验中遇到了一些困难（《法国科学院院刊》，1929 年第 1,549 页，1930 年第 1,124 页），他发现钍 C 放射的 α 射线是由非常接近的 5 个不同的组组成的。罗森布拉姆给出了其余各组相对于最强的那一组（α_0）的能量差和强度：

$$E\alpha_1 - E\alpha_0 = +40.6 \text{ 千电子伏} \quad I\alpha_1 = 0.3$$
$$E\alpha_2 - E\alpha_0 = -287 \text{ 千电子伏} \quad I\alpha_2 = 0.03$$
$$E\alpha_3 - E\alpha_0 = -442 \text{ 千电子伏} \quad I\alpha_3 = 0.02$$
$$E\alpha_4 - E\alpha_0 = -421 \text{ 千电子伏} \quad I\alpha_4 = 0.005$$

如果我们假设这些组分是由于逃逸的 α 粒子曾处于原子核内的不同激发量子能级而形成的，那么我们将遇到很大的麻烦。钍 C 精细结构粒子的能量衰减常数 λ 非常小（λ 约为 10^{-2} 秒 $^{-1}$），而且为了解释各组中何以有相对那么大数量的粒子，我们必须同时假设跃迁概率非常小。我们必须假设钍 C 原子核可以停留在激发态而不辐射能量长达半个小时！

然而，如果我们设想一个完全不同于发射长程 α 粒子的过程，就可以解释这 5 组 α 粒子了。假设有两个（或者更多的）α 粒子处于钍 C 原子核的正常能级上。其中一个 α 粒子逃逸到原子核外后，原子核有可能还保持在激发态，因为剩下的 α 粒子可能处于某个能量较高的能级上。（在这种情况下，逃逸 α 粒子的能量低于正常能级，而且它显然与原子核内的任何量子能级都不相等。）随后原子核（这里是钍 C''）可以从激发态跃迁到正常态，放出一个 γ 量子以释放两态之间的能量差。

这样，不同组的相对数量将与 γ 辐射的概率无关，而只与跃迁积分相关：

$$W = \int f(r_{1,2})\,\psi E_0(\alpha_1)\,\psi E_0(\alpha_2)\,\psi E_n(\alpha_1)\,\psi E\alpha_n(\alpha_2)\,dv_1\,dv_2$$

式中，$f(r)$ 是两个 α 粒子在相距为 r 时的相互作用能，ψE_0 和 ψE_n 分别是 α 粒子在正常态和第 n 个激发态的本征函数，ψE_a 是能量为 $E_{\alpha n} = E_0 - (E_n - E_0)$ 的逃逸 α 粒子的本征函数。

按照这种解释，对应于钍 C 不同精细结构组分的 γ 射线应该被看作是钍 C 的 γ 射线（发射出的电子来自于钍 C'' 原子的 K，L，M 等壳层），而不能被看作是钍 B 的 γ 射线，正如我们在解释长程粒子时预期的那样。图 1 中所示的钍 C'' 原子核能级图画出了能量的精细结构。

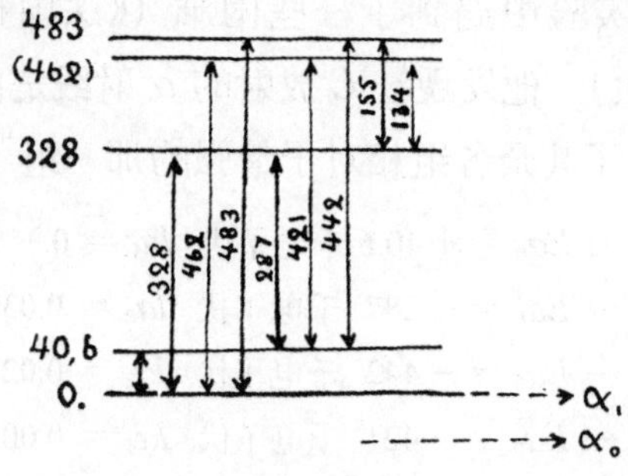

Fig. 1

In the observed γ-ray spectra of thorium-C + C'' (Black, *Proc. Roy. Soc.*, pp. 109–166; 1925) we can find lines with the energies: 40.8; 163.3; 279.4; 345.8; 439.0; 478.8; 144.6 kv fitting nicely with the energy differences in Fig. 1.

Thus we see that the fine structure group of highest energy corresponds to the normal level of the nucleus, while the other groups are due to the ordinary α-particles which have lost part of their energy, leaving the nucleus in an excited state.

I am glad to express my thanks to Dr. R. Peierls and Dr. L. Rosenfeld for the opportunity to work here.

(**126**, 397; 1930)

G. Gamow: Piz da Daint, Switzerland, July 25.

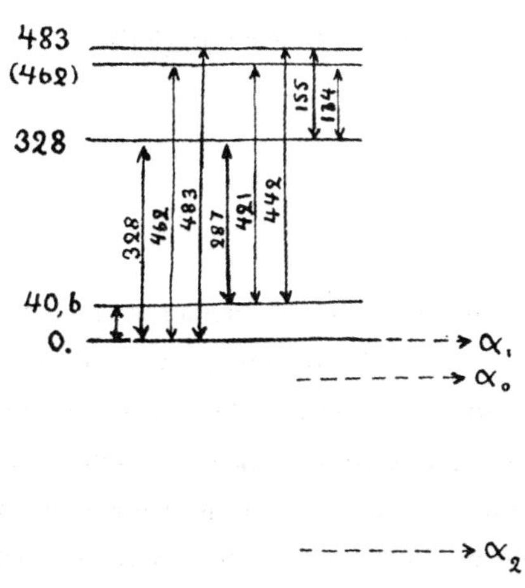

图 1

在实测的钍 C 和钍 C'' 的 γ 射线能谱中（布莱克，《皇家学会学报》，1925 年，第 109~166 页），我们可以找到对应于以下能量的谱线：40.8，163.3，279.4，345.8，439.0，478.8，144.6 千电子伏，这与图 1 中的能级差吻合得很好。

我们可以看到，能量最高的精细结构组分对应于原子核的正常态，而其他组分则是由损失掉部分能量的普通 α 粒子离开处于激发态的原子核造成的。

我非常感谢佩尔斯博士和罗森菲尔德博士给我提供在这里工作的机会。

（王静 翻译；江丕栋 审稿）

The Proton*

P. A. M. Dirac

Editor's Note

Here English physicist Paul Dirac describes his prediction, reported earlier in another journal, that electrons may have "negative energy" and consequently positive charge. This prediction followed from Dirac's "relativistic" quantum theory, which made quantum mechanics compatible with special relativity. The "positive electrons" appear as "holes" in a pervasive electron sea. Dirac suspects they behave as protons, but there are problems with that: protons have very different masses from electrons, and the electrons and holes were predicted to annihilate. Dirac alludes to an idea of J. Robert Oppenheimer that a positive electron may in fact be a different particle with the mass of an electron. And so it later proved: they were positrons, anti-matter versions of electrons.

MATTER is made up of atoms, each consisting of a number of electrons moving round a central nucleus. It is likely that the nuclei are not simple particles, but are themselves made up of electrons, together with hydrogen nuclei, or protons as they are called, bound very strongly together. There would thus be only two kinds of simple particles out of which all matter is built, the electrons, each carrying a charge $-e$, and the protons, each carrying a charge $+e$.

It should be mentioned here that there is a difficulty in this point of view provided by the nitrogen atom. One can infer from the charge and mass of the nitrogen nucleus that it should consist of 14 protons and 7 electrons, but it appears to have properties inconsistent with its being composed of an odd number of simple particles. However, very little is really known about nuclei, and the opinion is generally held by physicists that some way of evading this difficulty will be found and that all nuclei will ultimately be shown to be made up of electrons and protons.

It has always been the dream of philosophers to have all matter built up from one fundamental kind of particle, so that it is not altogether satisfactory to have two in our theory, the electron and the proton. There are, however, reasons for believing that the electron and proton are really not independent, but are just two manifestations of one elementary kind of particle. This connexion between the electron and proton is, in fact, rather forced upon us by general considerations about the symmetry between positive and negative electric charge, which symmetry prevents us from building up a theory of the negatively charged electrons without bringing in also the positively charged protons. Let us examine how this comes about.

---

* Based on a paper read before Section A (Mathematical and Physical Science) of the British Association at Bristol on Sept. 8.

质　子*

狄拉克

编者按

这篇文章报道了英国物理学家保罗·狄拉克就电子可能具有“负能量”从而带有正电荷这一预测所作的论述，该预测在更早些时候已发表在其他期刊上。狄拉克的“相对论性的”量子理论使量子力学与狭义相对论得以相容，前述的预测正是这一理论的结果。“带正电荷的电子”就像是无处不在的电子海中的“空穴”。狄拉克猜测，这些“带正电荷的电子”的行为方式与质子类似，不过这一猜测存在一些问题：质子与电子在质量上的差别非常大，而且据预测电子和这些空穴相遇会湮灭。在这里，狄拉克也提到了罗伯特·奥本海默的观点，即带正电荷的电子可能就是与电子的质量相同的另一种粒子。后来人们证明确实如此：这些粒子就是正电子，电子的反物质形式。

物质是由原子构成的，每一个原子是由若干个围绕中心原子核转动的电子组成的。原子核很可能不是基本粒子，而是由电子和氢原子核（或者所谓质子）紧密束缚在一起构成的。这样所有的物质都只由这两种基本粒子构成，其中每一个电子带电荷 $-e$，每一个质子带电荷 $+e$。

这里需要指出的是，氮原子的存在给这个观点提出了一个难题。由氮原子核的电荷和质量，我们可以推断出氮原子核是由 14 个质子和 7 个电子组成的，但是氮原子核表现出来的性质似乎与它是由奇数个基本粒子构成这一点不符。然而，关于原子核，我们知之甚少，而且物理学家们普遍认为将来总会有办法克服这个困难，并且最终将会证明所有的原子核都是由电子和质子构成的。

哲学家总是梦想所有的物质都是由一种基本粒子构成的，所以我们的理论——包含两种基本粒子（电子和质子）——并不能使所有人都满意。然而人们有理由相信电子和质子并不是毫无关系的，它们只是一种基本粒子的两种表现形式。而事实上，电子和质子之间的联系在某种程度上是关于正负电荷之间对称性的一般认识强加给我们的，这种对称性使我们不能构建一套只包含带负电的电子，而不包含带正电的质子的理论。下面让我们看看为什么会是这样。

* 基于 9 月 8 日在布里斯托尔向英国科学促进会的 A 分部（数学和物理科学）宣读的一篇论文。

The energy W of a particle in free space is determined in terms of its momentum p according to relativity theory by the equation

$$W^2/c^2 - p^2 - m^2c^2 = 0,$$

where m is the rest-mass of the particle and c is the velocity of light. This equation can easily be generalised to apply to a charged particle moving in an electromagnetic field and can be used as a Hamiltonian to give the equations of motion of the particle, and thus its possible tracks in space-time.

Now the above equation is quadratic in W, allowing of both positive and negative values for W. Thus for some of the tracks in space-time the energy W will have positive values and for the others negative values. Of course a particle with negative energy (kinetic energy is referred to throughout) has no physical meaning. Such a particle would have less energy the faster it is moving and one would have to put energy into it to bring it to rest, quite contrary to anything that has ever been observed.

The usual way of getting over this difficulty is to say that the tracks for which W is negative do not correspond to anything real in Nature and are to be simply ignored. This is permissible only provided that for every track W is either always positive or always negative, so that one can tell definitely which tracks are to be ignored. This condition is fulfilled in the classical theory, where W must vary continuously, since W can never be numerically less than mc^2 and is thus precluded from changing from a positive to a negative value. In the quantum theory, however, discontinuous variations in a dynamical variable such as W are permissible, and detailed calculation shows that W certainly will make transitions from positive to negative values. We can now no longer ignore the states corresponding to a negative energy and it becomes imperative to find some physical meaning for them.

We can deal with these states mathematically, in spite of their being physically nonsense. We find that an electron with negative energy moves in an electromagnetic field in the same way as an ordinary electron with positive energy would move if its charge were reversed in sign, so as to be $+e$ instead of $-e$. This immediately suggests a connexion between negative-energy electrons and protons. One might be tempted at first sight to say that a negative-energy electron *is* a proton, but this, of course, will not do, since protons certainly do not have negative kinetic energy. We must therefore establish the connexion on a different basis.

For this purpose we must take into consideration another property of electrons, namely, the fact that they satisfy the exclusion principle of Pauli. According to this principle, it is impossible for two electrons ever to be in the same quantum state. Now the quantum theory allows only a finite number of states for an electron in a given volume (if we put a restriction on the energy), so that if only one electron can go in each state, there is room for only a finite number of electrons in the given volume. We thus get the idea of a *saturated* distribution of electrons.

根据相对论，自由空间中粒子的能量 W 由它的动量 p 决定，即

$$W^2/c^2 - p^2 - m^2c^2 = 0 ,$$

其中 m 是粒子的静止质量，c 是光速。这个方程可以很容易地推广到带电粒子在电磁场中运动的情况，并且可以被用作哈密顿量，给出带电粒子的运动方程，从而得到带电粒子在时空中可能的径迹。

上面的方程中 W 项是二次的，所以 W 既可能是正的，也可能是负的。因此对时空中的一些径迹而言能量 W 是正的，而对其他一些则是负的。当然粒子具有负能量（动能总是会涉及到）是没有物理意义的。这样的粒子运动得越快，它的能量就越小，我们不得不给它能量使它静止，然而这与我们观察到的所有现象都是截然不同的。

通常克服这个困难的办法是认为具有负 W 的径迹不对应于任何真实的自然现象，而只需要简单地把它忽略掉。不过这只有在每一条径迹的 W 值恒正或者恒负的前提下才成立，因为只有这样我们才可以明确地判断哪一条径迹应该被忽略。这个条件在 W 连续变化的经典理论中是满足的，因为 W 在数值上不能小于 mc^2，所以排除了 W 从正值变化到负值的可能性。然而，在量子理论中，像 W 这样的动力学量可以不是连续变化的，并且详细的计算表明 W 确实可以从正值变化到负值。因此我们不能再忽略负能量对应的状态，而必须为它们寻找某种物理意义。

我们可以从数学上处理这些状态，而先不去管它们是否具有物理意义。我们发现，如果普通电子的电荷符号发生翻转，即从 $-e$ 变为 $+e$，那么一个具有负能量的电子在电磁场中的运动方式和一个普通的具有正能量的正电子一样。这就意味着负能电子和质子之间存在某种联系。乍一看这种情况，人们可能会说负能电子**就是**质子，但是这无疑是不成立的，因为质子的动能不可能是负的。因此，我们必须在另外的基础上构建它们的联系。

为此我们必须考虑电子的另外一个特性，即它们满足泡利不相容原理。根据这一原理，两个电子永远不可能处于同一个量子态。因为量子理论在给定的空间内只允许有限数目的电子态（如果我们给能量一个限制），所以如果每一个态只允许一个电子占据，那么在给定的空间内只能容纳有限数目的电子。这样我们就会得到电子**饱和**分布的概念。

Let us now make the assumption that almost all the states of negative energy for an electron are occupied, and thus the whole negative-energy domain is almost saturated with electrons. There will be a few unoccupied negative-energy states, which will be like holes in the otherwise saturated distribution. How would one of these holes appear to our observations? In the first place, to make the hole disappear, which we can do by filling it up with a negative-energy electron, we must put into it a negative amount of energy. Thus to the hole itself must be ascribed a positive energy. Again, the motion of the hole in an electromagnetic field will be the same as the motion of the electron that would fill up the hole, and this, as we have seen, is just the motion of an ordinary particle with a charge $+e$. These two facts make it reasonable to assert that *the hole is a proton*.

In this way we see the proper role to be played by the negative-energy states. There is an almost saturated distribution of negative-energy electrons extending over the whole of space, but owing to its uniformity and regularity it is not directly perceptible to us. Only the small departures from perfect uniformity, brought about through some of the negative-energy states being unoccupied, are perceptible, and these appear to us like particles of positive energy and positive charge and are what we call protons.

This theory of the proton involves certain difficulties, which will now be discussed. The theory postulates the existence everywhere of an infinite number of negative-energy electrons per unit volume, and thus an infinite density of electric charge. According to Maxwell's equations, this would give rise to an infinite electric field. We can easily avoid this difficulty by a re-interpretation of Maxwell's equations. A perfect vacuum is now to be considered as a region in which all the states of negative energy and none of those of positive energy are occupied. The electron distribution in such a region must be assumed to produce no field, and only the departures from this vacuum distribution can produce a field according to Maxwell's equations. Thus, in the equation for the electric field E

$$\operatorname{div} E = -4\pi\rho,$$

the electric density ρ must consist of a charge $-e$ for each state of positive energy that is occupied, together with a charge $+e$ for each state of negative energy that is unoccupied. This gives complete agreement with the usual ideas of the production of electric fields by electrons and protons.

A second difficulty is concerned with the possible transitions of an electron from a state of positive energy to one of negative energy, which transitions were the original cause of our having to give a physical meaning to the negative-energy states. These transitions are very much restricted when nearly all the negative-energy states are occupied, since an electron in a positive-energy state can then drop only into one of the unoccupied negative-energy states. Such a transition process would result in the simultaneous disappearance of an ordinary positive-energy electron and a hole, and would thus be interpreted as an electron and proton annihilating one another, their energy being emitted in the form of electromagnetic radiation.

现在我们假设电子的负能态几乎都被占据了，因此整个负能区域电子几乎是饱和的。有一些没被占据的负能态，它们就像饱和分布的负能态电子海中的一些空穴。在我们看来这些空穴是什么样的呢？首先，为了使这些空穴消失，我们需要填充一个负能量的电子，即放入一个负能量。这样空穴本身必须具有正能量。其次，空穴在电磁场中的运动方式和填充空穴的电子的运动方式一样，就像我们之前看到的那样，就是一个带 $+e$ 电荷的普通粒子的运动。这两个事实使我们有理由断言——**空穴就是质子**。

这样我们就看到了负能态起到的作用。整个空间中负能电子几乎处于饱和分布，但由于它们表现出来的均匀性和规律性，因而不能直接被我们觉察到。只有在完美的均匀性上出现一些小的偏离，即有一些负能态没被占据，才能被我们觉察到。这些偏离在我们看来就是些具有正能量和带正电荷的粒子，也就是我们所谓的质子。

这个关于质子的理论存在一些问题，下面我们就来讨论这些问题。这个理论假设空间每处单位体积内都存在无限数目的负能电子，这样电荷密度就是无穷大了。根据麦克斯韦方程，这将会产生一个无限大的电场。不过我们可以很容易地通过重新解释麦克斯韦方程来克服这个困难。理想的真空被认为是所有的负能态都被占据，而所有的正能态都没被占据的空间。我们认为在这样的空间中，电子分布不会产生任何场，只有当电子分布偏离真空分布时才会产生根据麦克斯韦方程得到的场。因此，在电场 E 的方程中

$$\text{div}\ E = -4\pi\rho\ ,$$

电荷密度 ρ 是由每一个被占据的正能态上的电荷 $-e$ 和每一个没被占据的负能态上的电荷 $+e$ 组成的。这和通常电子和质子产生电场的观点是完全一致的。

第二个困难涉及到电子可能存在从正能态向负能态的跃迁，这种跃迁是我们必须赋予负能态以物理意义的最初原因。当几乎所有的负能态都被占据时，这种跃迁是非常受限制的，因为处于正能态的电子只能落入没被占据的负能态。这样的跃迁过程导致一个普通的正能态电子和一个空穴同时消失，所以可以解释成一个电子和一个质子的互相湮灭，它们的能量以电磁辐射的形式发射出来。

There appears to be no reason why such processes should not actually occur somewhere in the world. They would be consistent with all the general laws of Nature, in particular with the law of conservation of electric charge. But they would have to occur only very seldom under ordinary conditions, as they have never been observed in the laboratory. The frequency of occurrence of these processes according to theory has been calculated independently by several investigators, with neglect of the interaction between the electron and proton (that is, the Coulomb force between them). The calculations give a result much too large to be true. In fact, the order of magnitude is altogether wrong. The explanation of this discrepancy is not yet known. Possibly the neglect of the interaction is not justifiable, but it is difficult to see how it could cause such a very big error.

Another unsolved difficulty, perhaps connected with the previous one, is that of the masses. The theory, when one neglects interaction, requires the electron and proton to have the same mass, while experiment shows the mass ratio to be about 1,840. Perhaps when one takes interaction into account the theoretical masses will differ, but it is again difficult to see how one could get the large difference required by experiment.

An idea has recently been put forward by Oppenheimer (*Phys. Rev.*, vol. 35, p. 562) which does get over these difficulties, but only at the expense of the unitary theory of the nature of electrons and protons. Oppenheimer supposes that all, and not merely nearly all, of the states of negative energy are occupied, so that a positive-energy electron can never make a transition to a negative-energy state. There being now no holes which we can call protons, we must assume that protons are really independent particles. The proton will now itself have negative-energy states, which we must again assume to be all occupied. The independence of the electron and proton according to this view allows us to give them any masses we please, and further, there will be no mutual annihilation of electrons and protons.

At present it is too early to decide what the ultimate theory of the proton will be. One would like, if possible, to preserve the connexion between the proton and electron, in spite of the difficulties it leads to, as it accounts in a very satisfactory way for the fact that the electron and proton have charges equal in magnitude and opposite in sign. Further advances in the theory of quantum electrodynamics will have to be made before one can deal accurately with the interaction and see whether it will settle the difficulties, or whether, perhaps, a new idea can be introduced which will answer this purpose.

(**126**, 605-606; 1930)

看起来没有什么理由可以说明为什么这样的过程不能在现实世界的某处发生。它们会遵守自然界所有的一般规律，特别是电荷守恒定律。但是它们在普通条件下必然很少发生，因为即使在实验室中它们也还没有被观察到。一些研究者已经独立地计算出这些过程发生的理论频率，计算中忽略了电子和质子之间的相互作用（即它们之间的库仑力）。计算给出的结果太大了，肯定是不正确的。事实上，结果的数量级都是完全错误的。现在还不知道为什么会出现这样的差异。可能忽略相互作用是不合理的，但是仍然很难理解为什么会导致这么大的错误。

另外一个没有解决的困难就是质量问题，这可能和前一个困难有关。如果忽略相互作用，这个理论就要求电子和质子具有相同的质量，然而实验表明它们的质量比约为 1,840。也许考虑相互作用后理论上的质量会有所不同，但还是很难理解怎样才能得到实验要求的那么大的质量差。

奥本海默最近提出的一个观点（《物理学评论》，第 35 卷，第 562 页）的确可以解决这个困难，但是它却牺牲了关于电子和质子本质的统一理论。奥本海默假定所有的 (不仅仅是几乎所有的) 负能态都被占据了 , 因此正能电子不能跃迁到负能态。这里没有我们可以称之为质子的空穴，所以我们必须假定质子是真正独立的粒子。这样，质子本身也有自己的负能态，而且我们必须假设它们也被完全占据了。根据这个观点，电子和质子的独立性允许我们随心所欲地给它们的质量赋值，而且它们之间也不会相互湮灭。

目前断定质子的最终理论还为时尚早。如果可能的话，人们愿意保留电子和质子之间的这种关系，而不管它带来的困难，因为它非常圆满地解释了这个事实——电子和质子携带的电荷大小相等，而符号相反。量子电动力学需要进一步的发展，人们才可以准确地计算相互作用，才可以知道我们的理论是否可以解决这些困难，或者是否会出现新的可以回答这个问题的观点。

（王锋 翻译；李淼 审稿）

The Ether and Relativity

J. H. Jeans

Editor's Note

Here the English physicist James Hopwood Jeans responds to a letter from Oliver Lodge criticizing Jeans' recent claim that the laws of the universe would only be penetrated by the use of mathematics. Jeans affirms his belief that "No one except a mathematician need ever hope fully to understand those branches of science which try to unravel the fundamental nature of the universe—the theory of relativity, the theory of quanta and the wave mechanics." Lodge suggested that the universe might ultimately turn out to have been created or designed on aesthetic, rather than mathematical lines. If so, one might expect artists, not mathematicians, to be best suited to fundamental science. But Jeans notices no such aptitude in his artist friends.

I obviously must not ask for space to discuss all the points raised in Sir Oliver Lodge's interesting letter in *Nature* of Nov. 22, and so will attempt no reply to those parts of it which run counter to the ordinarily accepted theory of relativity. For I am sure nothing I could say would change his views here. But I am naturally distressed at his thinking I have quoted him with a "kind of unfairness", and should be much more so, had I not an absolutely clear conscience and, as I think, the facts on my side.

In the part of my book to which Sir Oliver objects most, I explained how the hard facts of experiment left no room for the old material ether of the nineteenth century. (Sir Oliver explains in *Nature* that he, too, has abandoned this old material ether.) I then quoted Sir Oliver's own words to the effect that many people prefer to call the ether "space", and his sentence, "The term used does not matter much."

I took these last words to mean, not merely that the ether by any other name would smell as sweet to Sir Oliver, but also that he thought that "space" was really a very suitable name for the new ether. He now explains he was willing to call the ether "space", "for the sake of peace and agreement". If I had thought it was only *qua* pacifist and not *qua* scientist that he was willing to call the ether "space", I naturally would not have quoted him as I did, and will, of course, if he wishes, delete the quotation from future editions of my book. But I did not know his reasons at the time, and so cannot feel that I acted unfairly in quoting his own words verbatim from an Encyclopaedia article.

Against this, I seem to find Sir Oliver attributing things to me that, to the best of my belief, I did not say at all, as, for example, that a mathematician alone can hope to understand the universe. My own words were (p. 128):

以太与相对论

金斯

编者按

英国物理学家詹姆斯·霍普伍德·金斯写这篇文章是为了回应奥利弗·洛奇的一篇快报，奥利弗·洛奇在该快报文章中批评了金斯最近提出的观点，即宇宙中的定律只有用数学方法才能解释清楚。金斯坚持认为“除了数学家以外，没有人能完全理解那些试图揭示宇宙基本性质的科学分支——相对论、量子理论和波动力学。”洛奇认为宇宙的创造和设计最终有可能是按照美学原则而非数学原理进行的。如果事实真的如此，那么艺术家应该比数学家更适合研究基础科学。但金斯在他的艺术界朋友中没有发现这种特别的倾向。

显然，我没有必要要求一个很大的版面来讨论奥利弗·洛奇爵士发表在11月22日《自然》上的那篇引人注意的快报中提到的全部要点，因而也不会试图回复该快报文章中那些与人们普遍接受的相对论不相符合的内容。因为我知道无论我在这里说什么也不能改变他的观点。但是，如果他认为我是以“一种不公正的方式”引述他的话，我自然会感到不安，而且，如果我不曾拥有一个绝对清晰的意识并认为事实一定站在我这一边的信念，我将会感到更加不安。

在我的书中，奥利弗爵士最不赞同的部分是我关于铁一般的实验事实如何使得19世纪的旧的物质性以太再无容身之地的解释。（奥利弗爵士在《自然》上曾解释说他也已经抛弃了这种旧的物质性以太。）接着我引述了奥利弗爵士本人针对很多人更愿意称以太为“空间”这一现象所说的话，他的原话是，“用什么样的术语关系不大。”

我引用最后这句话是想说明，不仅以任何其他方式命名的以太一样合乎奥利弗爵士的胃口，而且他认为“空间”对于新的以太来说确实是一个很合适的名称。他现在解释说自己乐于称以太为“空间”，“为的是息事宁人和意见统一”。如果我以前认为他只是作为和平主义者而不是作为科学家才乐于称以太为“空间”，自然就不会像我之前那样引述他的话，当然，如果他希望的话，我将在我那本书的新版中删去那段引文。但当时我并不知道他是这样想的，所以我不知道我从一篇百科全书的文章中一字不差地引述他的原话是不公正的做法。

与此相反，我可以非常负责任地说，我发现奥利弗爵士把我根本没有说过的话强加在我身上，比如，单凭数学家就能理解宇宙。我的原话是这样的（第128页）：

"No one except a mathematician need ever hope fully to understand those branches of science which try to unravel the fundamental nature of the universe—the theory of relativity, the theory of quanta and the wave-mechanics."

This I stick to, having had much experience of trying to explain these branches of science to non-mathematicians. In the same way, if the material universe had been created or designed on aesthetic lines—a possibility which others have contemplated besides Sir Oliver Lodge—then artists ought to be specially apt at these fundamental branches of science. I have noticed no such special aptitude on the part of my artist friends. Incidentally, I think this answers the question propounded in the News and Views columns of *Nature* of Nov. 8, which was, in brief:—If the universe were fundamentally aesthetic, how could an aesthetic description of it possibly be given by the methods of physics? Surely the answer is that if the objective universe were fundamentally aesthetic in its design, physics (defined as the science which explores the fundamental nature of the objective universe) would be very different from what it actually is; it would be a *milieu* for artistic emotion and not for mathematical symbols. Of course, we may come to this yet, but if so, modern physics would seem rather to have lost the scent.

However, I am glad to be able to agree with much that Sir Oliver writes, including the quotations from Einstein which he seems to bring up as heavy artillery to give me the final *coup de grace*:—"In this sense, therefore, there exists an ether", and so on. On this I would comment that nothing in science seems to exist any more in the good old-fashioned sense—that is, without qualifications; and modern physics always answers the question, "To be or not to be?" by some hesitation compromise, ambiguity, or evasion. All this, to my mind, gives strong support to my main thesis.

(**126**, 877; 1930)

J. H. Jeans: Cleveland Lodge, Dorking, Nov. 23.

“除了数学家以外，没有人能完全理解那些试图揭示宇宙基本性质的科学分支——相对论、量子理论和波动力学。”

我仍然坚持这一点，因为在试图向非数学家解释上述科学分支方面我已经积累了许多经验。同样地，如果物质性宇宙的创造和设计是按美学原则进行的——除了奥利弗·洛奇爵士之外还有其他人也曾考虑过这种可能性，那么艺术家应该特别容易理解这些基础的科学分支。迄今为止，在我的艺术家朋友中我并没有发现这种特别的倾向。顺便提一句，我认为这解答了11月8日《自然》的“新闻与视点”栏目中提出的问题，简单地说就是：如果宇宙基本上是美学的，那么物理学方法又怎么能给出一个关于宇宙的美学描述呢？答案只能是这样的，如果客观的宇宙基本上是依美学观点设计的，那么物理学（定义为研究客观宇宙基本性质的科学）将会与它实际的样子大为不同；这种依美学观点设计的**宇宙环境**将适合于艺术情感而非数学符号。当然，我们可以这样做，但是果真如此的话，现代物理学似乎会迷失方向。

不过，我很高兴在很多方面与奥利弗爵士意见一致，包括他引用的爱因斯坦的话——“所以就这个意义而言，是有某种以太存在”等，看起来他要把这些当作重炮给我最终的**致命一击**。对此我的评论是，在科学中没有什么东西能以旧有的形式安然存在，这是绝对的；现代物理学经常会以某种迟疑不决的折衷、模棱两可或者遁辞来回答“存在还是不存在？”的问题。在我看来，所有这些都强有力地支持了我的主要观点。

（王耀杨 翻译；张元仲 审稿）

Unit of Atomic Weight

F. W. Aston

Editor's Note

Francis Aston, working at the Cavendish Laboratory of the University of Cambridge, had devised an instrument for measuring the atomic masses of individual atoms, now called the mass spectrometer. This led him earlier to postulate the notion of isotopes, which have identical chemical properties but different masses. Having used the device to identify the isotopes of more than 80 different chemical elements, Aston here advocates the need for a new standard of atomic mass, to replace the practice then current of referring all masses to that of oxygen—for this element, having several isotopes, is not an appropriate reference point.

THE discovery of the complexity of oxygen clearly necessitates a reconsideration of the scale on which we express the weights of atoms. Owing to the occurrence of O^{17} and O^{18}, now generally accepted, it follows that the mean atomic weight of this element, the present chemical standard, is slightly greater than the weight of its main constituent O^{16}. The most recent estimate of the divergence is 1.25 parts per 10,000.

This quantity, even apart from its smallness, is not of much significance to chemists, for the experience of the last twelve years has shown that complex elements do not vary appreciably in their isotopic constitution in natural processes or in ordinary chemical operations. Physics, on the other hand, is concerned with the weights of the individual atoms, and by the methods of the mass-spectrograph and the analysis of band spectra it is already possible to compare some of these with an accuracy of 1 in 10,000. Furthermore, the theoretical considerations of the structure of nuclei demand an accuracy of 1 in 100,000, which there is reasonable hope of attaining in the near future. The chemical unit is clearly unsuitable, and it seems highly desirable that a proper unit for expressing these quantities should be decided upon.

The proton, the neutral hydrogen atom, one-quarter of the neutral helium atom, one-sixteenth of the neutral oxygen atom 16, and several other possible units have been suggested. None of these is quite free from objection. It is desirable that this matter should be given attention, so that when a suitable opportunity occurs for a general discussion of the subject, each point of view may be afforded its proper weight in arriving at a conclusion.

(**126**, 953; 1930)

F. W. Aston: Trinity College, Cambridge, Dec. 4.

原子量的单位

阿斯顿

编者按

在剑桥大学卡文迪什实验室工作的弗朗西斯·阿斯顿设计了一种用来测量单个原子的原子质量的仪器，现在我们称之为质谱仪。这使得他更早地提出了同位素（化学性质完全相同但质量不同）的概念。利用该仪器鉴定了超过 80 种化学元素的同位素以后，在这篇文章中阿斯顿主张，有必要采用一种新的原子质量的标准，以取代当时测量所有其他原子质量时利用氧作为标准的做法——因为氧有多种同位素，不适合作为原子质量的参考标准。

氧元素具有多种同位素，这一发现无疑使我们必须重新考虑用以表述原子量的标度。目前在化学上是以氧元素的平均原子质量作为原子量的标准，但由于现在同位素 O^{17} 和 O^{18} 的发现已经得到了普遍的认可，所以氧元素的平均原子质量略大于氧元素中的主要组成部分 O^{16} 的原子质量。最新的估计表明这一差别是 0.125‰。

这个差值对于化学家来说没有太大的意义，更别说它还非常微小，因为最近 12 年的研究经验已经表明，对于同位素形式复杂多样的元素来说，在自然过程或者化学操作中其各种同位素的相对丰度并不发生明显改变。但是，物理学研究要考虑单个原子的质量，通过质谱仪和谱带分析的方法目前已经可以以万分之一的精度来分辨某些原子质量的差别。此外，对原子核结构的理论研究需要十万分之一的精度，在不久的将来对原子质量的测定有望能够达到这一精度。因此，目前化学上采用的原子量的单位显然是不合适的，对于物理学研究来说，似乎迫切需要确定一个能够描述这些量的合适的原子量单位。

质子的质量，中性氢原子的质量，氦原子质量的 1/4，中性氧同位素 O^{16} 原子质量的 1/16，以及其他几种可能的原子量单位都被提出来了。但所有这些无一例外都遭到了一些反对。这个问题应该受到关注，一旦出现一个对这一主题进行广泛讨论的合适时机，那么为了得出结论每一种观点都可以适当地发挥作用。

（王锋 翻译；李芝芬 审稿）

Evidence for a Stellar Origin of the Cosmic Ultra-penetrating Radiation

V. F. Hess

Editor's Note

Physicists were still pondering the nature of cosmic rays. Earlier studies failed to find any evidence that the Sun emitted such rays, but here Victor Hess reports new experiments showing that it does. As he notes, recent experiments at high altitude in the Swiss Alps found the average intensity of radiation to be higher during the day and lower at night. Further experiments with lead shielding showed that the Sun's light included a component of highly penetrating rays, with intensity equal to about 0.5 percent of the total observed cosmic ray intensity. Hess argues that cosmic rays most probably have a stellar origin, as all other stars probably emit them much as the Sun. The precise nature of these particles remained unknown.

WHILE in former years all observers were agreed that the sun does not contribute any noticeable amount to the total intensity of the cosmic ultra-radiation, the increase in the sensitivity of the apparatus used within recent years, and the increase in the number of observations made at different stations and under different experimental conditions, makes it possible to investigate once more whether the influence of the sun is altogether negligible.

Very accurate and trustworthy registrations of the cosmic radiation have been carried out with Prof. G. Hoffmann's high-pressure ionisation chamber at Muottas Muraigl (2,456 m. above sea-level) in the Engadine. These measurements show, beyond any doubt, that the average intensity of the radiation is somewhat greater in daytime than during the night. G. Hoffmann and F. Lindholm[1] give the average difference between day and night intensities as 0.12 mA., ~0.0125 ions per c.c. per sec. while the apparatus was unscreened from above, and 0.04 mA., ~0.0042 *I* with a lead-screening of 6 cm. and 9 cm. thickness. (The letter "*I*" always denotes "ions per c.c. and sec.".) F. Lindholm,[2] with the same apparatus, found from longer series of observations (8 months) the values in the accompanying table (see Table 6 of his paper).

In Hoffmann and Lindholm's apparatus a compensation current of one milliampere corresponds to an ionisation of 0.104 *I*. Therefore the total intensity of the ultra-radiation with the apparatus unscreened from above was about 2.50 *I* at Muottas Muraigl.

The difference between day and night intensity can be taken, provisionally at least, as the actual intensity of the solar penetrating radiation. One can see at once that at Muottas

宇宙超穿透性辐射起源于恒星的证据

维克托·赫斯

编者按

物理学家们仍在思考宇宙射线的性质。以前的研究未能找到任何证据证明太阳发射了这类射线，如今维克托·赫斯报告了他用新的实验结果说明确实如此。正如他所指出的，在瑞士阿尔卑斯山上的高海拔区进行的一项最新实验发现，辐射的平均强度白天比晚上高。采用铅屏蔽板以后再做的实验表明，太阳光中包含一个穿透力很强的射线成分，其强度约为宇宙射线总观测强度的0.5%。赫斯认为宇宙射线很可能起源于恒星，因为除太阳以外所有其他恒星发射的宇宙射线很可能与太阳发射的一样多。这些粒子的确切性质现在还不清楚。

在过去，所有的观测者一致认为，在宇宙超级辐射的总强度中，太阳没有任何值得注意的贡献。近年来，随着观测仪器灵敏度的不断增强，以及在不同国家、不同实验环境下进行的观测次数不断增多，于是有可能再一次研究由太阳造成的影响是不是可以完全忽略不计。

有人把霍夫曼教授的高压电离室放在瑞士恩加丁地区的穆拉古尔山(海拔2,456 m)上，由此得到了一些非常准确而且可靠的有关宇宙辐射的数据。这些测量结果毫无疑问地说明白天的平均辐射强度要略高于夜晚。霍夫曼和林霍尔姆[1]给出了昼夜间强度差异的平均值：当仪器上方没有屏蔽时，平均值为0.12 mA，或~0.0125个离子每立方厘米每秒；当使用6 cm和9 cm厚的铅板屏蔽时，平均值是0.04 mA，~0.0042 *I*（符号“*I*”通常表示“每立方厘米每秒的离子数”）。林霍尔姆[2]使用同样的仪器进行了更长期的观测（8个月），所得数据列于附表中（参见他文章中的表6）。

在霍夫曼和林霍尔姆使用的仪器中，一个1 mA的补偿电流相当于0.104 *I*的电离值。由此得出，在穆拉古尔山上由顶部没有铅板屏蔽的仪器测得的超级辐射的总强度大约为2.50 *I*。

我们至少可以暂时把昼夜间的强度差视为太阳贯穿辐射的实际强度。于是马上就可以看到在海拔2,456 m的穆拉古尔山上，大约有一半这类太阳辐射成分能够穿

Muraigl, 2,456 m. above sea-level, about one-half of this solar radiation component is able to penetrate through 10 cm. of lead. This component is therefore far more penetrating than the gamma rays from radioactive substances. If we assume that all of the above-mentioned 0.011 I is of solar origin, we can compute the absorption coefficient in lead μ_{Pb} (it will suffice to take the case of perpendicular incidence) from the equation $I = I_0 e^{-\mu_{Pb} d}$ taking $I_0 = 0.011$, $I = 0.0058$, and $d = 10$ cm.; thus we obtain $\mu_{Pb} = 0.064$ cm.$^{-1}$ and the mass absorption coefficient $\left(\frac{\mu}{\rho}\right)_{Pb} = 5.7\times10^{-3}$ cm.^{2}/gm.

This value is almost exactly equal to the mass absorption coefficient value of the total cosmic radiation at the same altitude ($(\mu/\rho)_{Pb} = 6.3\times10^{-3}$ cm.^{2}/sec. as found by Büttner on the Eiger glacier 2.3 km. above sea-level).[3] If we assume that part of the (0.011 I) difference between day and night values with unscreened apparatus is due to an increase in the average content of radium emanation and its products in the air during daytime, then we should get an even more pronounced hardness of the solar penetrating rays, that is, a smaller value for their mass absorption coefficient. Therefore we are justified in concluding that *the sun emits penetrating rays of at least the same penetrating power as the well-known cosmic ultra-radiation. The total amount of the solar penetrating rays (at 2,456 m. above sea-level) is about one-half percent of the total intensity of the cosmic radiation, as it is seen from the accompanying table*. Of course, one might think it possible to explain the increase in the total radiation during daytime as due to an indirect influence of the sun (that is, an increase in the scattering of the ultra-rays by the heating of the atmosphere during the day). In this case, however, one would expect that this scattered radiation, represented by the difference between the day and night values, would be much softer than the general cosmic radiation; but this is in contradiction to the experimental results analysed above.

Recent observations of R. Steinmaurer[4] on the summit of the Sonnblick (3,100 m. above sea-level) in the summer of 1929, made with three different instruments (two of the Kolhörster double loop-electrometer type and one of the Wulf–Kolhörster type), also show clearly that the total ultra-radiation in daytime is slightly higher than at night; the difference amounts to about 0.7 percent (0.06 I, average difference for the three forms of apparatus mentioned above, the total intensity on the Sonnblick being about 8.7 I with the screening open on the top). The increase of radiation was also observed with apparatus screened with 7 cm. iron all around, but the number of these observations on the Sonnblick is not sufficient for quantitative calculations. It may be mentioned that even in the old observations on the summit of the Obir (2,000 m. above sea-level), made by V. F. Hess and M. Kofler,[5] the solar influence is noticeable (the total intensity of the ultra-radiation plus earth-radiation during the day being 11.11, during the night 11.09 I, in the average for 13 months), although at that time the apparatus were not screened from the earth radiation. The difference of 0.02 I was—at that time—considered as practically amounting to zero.

Observations with apparatus of the Wulf– or Kolhörster type for shorter periods (like those of Kolhörster–v. Salis on the Jungfraujoch, on the Mönch, and of Büttner at other places

过 10 cm 厚的铅板。因此这部分辐射的穿透性大大高于放射性物质发出的 γ 射线。如果我们假设所有上述的 0.011 I 全部起源于太阳，我们就可以根据方程 $I = I_0 e^{-\mu_{Pb} d}$ 计算出铅的吸收系数 μ_{Pb}（只考虑垂直入射的情况已经足够），代入 $I_0 = 0.011$，$I = 0.0058$ 和 $d = 10$ cm，我们得到 $\mu_{Pb} = 0.064$ cm^{-1} 和质量吸收系数 $\left(\frac{\mu}{\rho}\right)_{Pb} = 5.7 \times 10^{-3}$ cm^2/g。

这个数值几乎精确地等于所有宇宙辐射在这个高度上的质量吸收系数（比特纳在海拔 2.3 km 的艾格尔冰川上的测量值为$(\mu/\rho)_{Pb} = 6.3 \times 10^{-3}$ cm^2/sec）。[3] 如果我们假设在没有屏蔽层的仪器上测量到的昼夜间差异（0.011 I）中，有一部分是由于白天空气中镭射气和其产物的平均含量上升引起的，那么我们观察到的太阳贯穿射线会更硬，即它们的质量吸收系数会更小。因此我们认为以下结论是合理的：**太阳发出的贯穿射线的穿透能力至少与著名的宇宙超穿透辐射相当。太阳贯穿射线的总量（在海拔 2,456 m 处）大约占宇宙辐射总强度的 0.5%，如附表所示。**当然，有人可能将白天辐射总量的升高解释为受太阳间接影响所致（即白天被加热的大气增加了对超穿透射线的散射）。然而在这种情况下，因散射造成的辐射——由昼夜间辐射量的差异表示，会比一般的宇宙辐射更软；而这与上面分析的实验结果是矛盾的。

1929 年夏天，斯坦莫勒 [4] 在松布利克山峰顶（海拔约 3,100 m）用三种不同仪器（两台柯尔霍斯特型双环静电计和一台伍尔夫–柯尔霍斯特型）的最新观测也明确显示出测量的超级辐射总量白天的数值略高于夜晚；差值大约为 0.7%（用上面提到的三种仪器测量的差值取平均后得到 0.06 I，顶部没有屏蔽的仪器在松布利克山测量的总强度约为 8.7 I。）当仪器四周用 7 cm 厚的铁板屏蔽时仍然可以观测到辐射量的增加，但是在松布利克山上的观测次数太少不足以作出定量计算。人们也许会提到即使从之前赫斯和科夫勒 [5] 在奥柏（海拔 2,000 m）山顶的观测数据中也可以看出太阳带来的影响（超级辐射加上地球辐射的总强度在 13 个月中的平均值：白天为 11.11 I，晚上为 11.09 I），尽管那时的仪器没有屏蔽掉地球辐射的影响。0.02 I 的差别在当时几乎可以被看作是零。

至于使用伍尔夫型或柯尔霍斯特型仪器进行的短周期观测（比如冯萨利斯在少女峰和修士峰以及比特纳在阿尔卑斯山其他地方用柯尔霍斯特型仪器所做的观测），

in the Alps) naturally do not show the influence of the solar component of the ultra-rays, on account of the lesser degree of accuracy of the means; therefore Corlin,[6] using the observations on the Mönch and the Zugspitze, came to negative conclusions as to the solar influence. From the data given below it is quite safe to conclude, according to the most accurate and most numerous observations at present available, that *the sun contributes an amount of about 0.5 percent to the total intensity of the cosmic ultra-radiation at 2.5 km. above sea-level. The penetrating power of the solar ultra-rays is at least as great as that of the total cosmic radiation.* There is no doubt that this solar component of the ultra-radiation is also present at lower levels; on account of its very small absolute intensity it will, of course, be far more difficult to prove its existence in these levels. An analysis of the very accurate registrations of the total radiation by Hoffmann and Steinke in Königsberg and in Halle in this direction might be successful.

Period	Number of Days	Armour open above		
		Mean Values		Difference (Day–Night)
		Day	Night	
1928 January–March	(32)	24.46 mA.	24.34 mA.	0.12 mA. = 0.0125 I
1928 June, July, October	(39)	23.98 mA.	23.88 mA.	0.10 mA. = 0.0104 I
1929 January–February	(11)	24.68 mA.	24.59 mA.	0.09 mA. = 0.0094 I
Weighted average difference			0.011 I	

Period	Number of Days	Armour closed (10 cm. lead screening all around)		
		Mean Values		Difference (Day–Night)
		Day	Night	
1928 March	(2)	19.54 mA.	19.50 mA.	0.04 mA. = 0.0042 I
1928 July	(8)	19.21 mA.	19.17 mA.	0.04 mA. = 0.0042 I
1929 February	(6)	19.46 mA.	19.38 mA.	0.08 mA. = 0.0084 I
Weighted average difference behind 10 cm. lead			0.0058 I (ions/c.c./sec.)	

If the sun, as the fixed star nearest to our planet, emits rays of about the same qualities as the total cosmic penetrating radiation, one cannot but assume that all fixed stars are sources of a radiation of similar qualities. The sun being a relatively old star of the yellow dwarf type may, of course, be expected to yield far less total quantity of the ultra-penetrating radiation than, for example, the younger giant stars. Naturally, the ultra-penetrating rays which we observe can only come from the outermost layers of the stars, since they are not able to penetrate material layers of more than a few hundred metres water equivalent.

It is not possible, at present, to say more about the nature of these stellar ultra-rays: whether they are electrons or protons accelerated in cosmic electric fields, or indeed photons (quanta) created by atomic mass shrinking or annihilation processes. This hypothesis of a partly stellar origin of the ultra-penetrating cosmic radiation does not necessarily exclude the possibility that another part of this radiation is created in interstellar space by the formation of certain elements out of hydrogen, according to

由于仪器精度不够高，自然也测不出太阳的超级射线成分的影响；因此科兰[6]使用在修士峰和楚格峰上的观测结果得出太阳对辐射量无影响的结论。从下表的数据中完全可以得到以下结论，根据目前能得到的所有最精确的观测结果，我们认为**在海拔 2.5 km 处，太阳辐射占宇宙超级辐射总强度的 0.5%。太阳超能射线的穿透能力至少与宇宙辐射的总体穿透能力相当**。毫无疑问，目前太阳的这一超穿透辐射成分的贡献仍然处于较低的水平。由于这一成分的绝对强度值很小，要想证实其存在非常困难，霍夫曼和施坦因克在柯尼斯堡和哈雷对总辐射量的非常精确的数据进行了分析，他们在这方面的努力也许会取得成功。

周期	天数	上方防护外壳打开		
		平均值		差值（日—夜）
		日	夜	
1928 年 1~3 月	(32)	24.46 mA	24.34 mA	0.12 mA = 0.0125 *I*
1928 年 6、7、10 月	(39)	23.98 mA	23.88 mA	0.10 mA = 0.0104 *I*
1929 年 1~2 月	(11)	24.68 mA	24.59 mA	0.09 mA = 0.0094 *I*
加权平均差值：0.011 *I*				

周期	天数	防护外壳关闭（四周用 10 cm 厚铅板屏蔽）		
		平均值		差值（日—夜）
		日	夜	
1928 年 3 月	(2)	19.54 mA	19.50 mA	0.04 mA = 0.0042 *I*
1928 年 7 月	(8)	19.21 mA	19.17 mA	0.04 mA = 0.0042 *I*
1929 年 2 月	(6)	19.46 mA	19.38 mA	0.08 mA = 0.0084 *I*
用 10 cm 铅板屏蔽后的加权平均差值：0.0058 *I*（离子数 /c.c./ 秒）				

作为离地球最近的恒星，如果太阳发出的射线具有与总的宇宙穿透性辐射大致相同的性质，则我们不得不假定所有恒星都是发出类似射线的辐射源。太阳是一颗年代比较久远的黄矮星，它释放出的超级穿透性射线在总量上自然会远远少于那些年轻一些的巨星。当然，我们观察到的超级穿透性射线只能来自恒星的最外层，因为它们不可能穿过厚度超过几百米水当量的物质层。

我们现在还不能对恒星超级射线的性质作更多的说明：不能判断它们到底是被宇宙电场加速的电子或质子，还是在原子质量减小或湮灭过程中放出的光子（量子）。按照爱丁顿和密立根的想法，虽然最小值原理假说更倾向于引导我们尝试用以太阳超穿透性辐射为实验证据的恒星起源假说来解释全部观察到的现象，但是一部分超穿透性宇宙辐射源自恒星的假说未必非要排除另一部分宇宙辐射来自星际空间

Eddington's and Millikan's ideas, although the principle of minimum hypothesis would rather induce us to try whether the stellar origin hypothesis, based on the experimental evidence of the solar ultra-penetrating rays, would suffice to explain the observed facts.

The conclusions put forward in this note certainly support the original ideas of Prof. Nernst first mentioned in 1921.[7] A few years ago, when the first results of observations on the daily period according to sidereal time were published, he wished that it were possible to increase the sensitivity of our apparatus until we could detect the ultra-rays from a single stellar nebula or a single star. I think the results put forward here indicate that a modest beginning has been made in this direction. At least it has been possible now to detect the influence and the penetrating power of the ultra-rays from the sun. It may be added that the evidence here brought forward for a stellar origin of the cosmic ultra-rays is completely independent of the existence of a daily period according to sidereal time, a subject which is still under discussion.

(**127**, 10-11; 1931)

Victor F. Hess: Institute of Experimental Physics, University of Graz, Austria, Nov. 4.

References:

1. *Gerlands Beitr. z. Geophysik*, **20**, 52 (1928).
2. *Gerlands Beitr. z. Geoph.*, **26**, 416-439 (1930).
3. *Zeitschr. f. Geophys.*, **3**, 179 (1927).
4. *Sitz. Ber. Akad. d. Wiss. Wien*, II. a. **139**, 281-318 (1930).
5. *Phys. Zeitschr.*, **18**, 585 (1917).
6. *Zeitschr. f. Physik*, **50**, 808-848 (1928).
7. *Das Weltgebäude im Lichte der neueren Forschung* (Verlag Springer, Berlin).

的可能性，星际空间中的宇宙辐射是在氢元素合成某些较重元素的过程中产生的。

这篇短文中的结论理所当然地支持了能斯特教授在 1921 年时就已经率先提出的想法[7]。几年前，当基于恒星时的日周期的首次观测结果发表出来时，他就希望能够提高仪器的灵敏度，直到我们能够分辨出来自单个恒星或单一恒星星云的超级射线。我认为本文提供的结果可以表明朝这方面的努力已经开始。至少现在检测出太阳超级射线的影响力和穿透力已经成为可能。需要补充说明的是，这里提出的有关宇宙超级射线起源于恒星的证据完全独立于基于恒星时的日周期的存在，后者是一个仍在讨论之中的课题。

（史春晖 翻译；马宇蒨 审稿）

Present Status of Theory and Experiment as to Atomic Disintegration and Atomic Synthesis*

R. A. Millikan

Editor's Note

In the mid-1920s, Robert Millikan coined the term "cosmic rays" for the high-energy radiation recently detected in the upper atmosphere. While some physicists suggested that the rays might come from atmospheric electricity, Millikan favoured an extraterrestrial origin. Here he describes recent evidence supporting this view. The discovery of unstable radioactive elements showed that some process must have created these atoms in the relatively recent past by building heavier elements up out of lighter ones. Perhaps cosmic-ray collisions are the cause, Millikan suggests. He proposes a broader process of heavy-element formation in space, but today's well-founded theory locates it in thermonuclear reactions within stars and supernovae, or during processes in the very early universe.

MY task is to attempt to trace the history of the development of scientific evidence bearing on the question of the origin and destiny of the physical elements. I shall list ten discoveries or developments, all made within the past hundred years, which touch in one way or another upon this problem and constitute indications or sign-posts on the road toward an answer.

Prior to the middle of the nineteenth century, little experimental evidence of any sort had appeared, so that the problem was wholly in the hands of the philosopher and the theologian. Then came, first, the discovery of the equivalence of heat and work, and the consequent formulation of the principle of the conservation of energy, probably the most far-reaching physical principle ever developed.

Following this, and directly dependent upon it, came, second, the discovery, or formulation, of the second law of thermodynamics, which was first interpreted, and is still interpreted by some, as necessitating the ultimate "heat-death" of the universe and the final extinction of activity of all sorts; for all hot bodies are observed to be radiating away their heat, and this heat after having been so radiated away into space apparently cannot be reclaimed by man. This is classically and simply stated in the humpty-dumpty rhyme. As a natural if not necessary corollary to this was put forward by some, in entire accord with the demands of medieval theology, a *Deus ex machina* initially to wind up or start off this running-down universe.

---

* Retiring presidential address to the American Association for the Advancement of Science, delivered at Cleveland on Dec. 29.

原子衰变与原子合成的理论和实验现状*

罗伯特·密立根

编者按

20 世纪 20 年代中期，罗伯特·密立根将最新在高层大气中发现的高能辐射命名为“宇宙射线”。虽然有些物理学家认为这种射线可能来自大气电，而密立根则赞成其起源于地球之外。他在这篇文章中介绍了近期得到的一些证据以支持这一观点。不稳定放射性元素的发现说明：在不久以前，一定有某种轻元素合成重元素的过程产生了这些原子。密立根认为宇宙射线的碰撞可能就是起因。他还提出了较为广泛地在宇宙空间中形成重元素的过程，但当今确立的理论则认为重元素的生成过程起源于恒星和超新星内部的热核反应，或者形成于早期宇宙的某些反应之中。

我的任务是试图追溯关于物质元素起源和命运问题的科学证据的发展历史。我将列出 10 个最近 100 年内的发现或进展，它们会以这种或那种方式与此问题相关联，并且充当着通向答案之路上的指示灯或者路标。

在 19 世纪中叶以前，人们几乎得不到任何实验上的证据，所以这个问题的解答完全由哲学家和神学家说了算。然后出现了第一个发现，即热和功的等价性，由此导致了能量守恒定律，后者或许能称得上是迄今为止影响最为深远的物理定律。

在这之后，并与此直接相关，热力学第二定律作为第二个发现或构想出现了，该定律最初被解释为（某些人至今仍作此解释）宇宙必然要走向最终的“热寂”，各种活动最终都将停止；因为人们发现所有热的物体都在向外辐射自己的热量，以这种方式散发到宇宙中去的热量显然无法被人类收回。这和经典而简单的《汉普蒂·邓普蒂》歌谣中描述的情况相符（译者注：humpty-dumpty 在歌谣中被比作“一经损坏就无法修复的东西”）。有些人提出这即使不是必然结果也是一种很自然的结果，完全适合了中世纪神学的要求，即最初是由**救世主**灭世或创世这个不断被耗尽的宇宙的。

* 卸任美国科学促进会会长的演说辞，12月29日发表于克利夫兰市。

Then came, third, the discovery, through studies both in geology and biology, of the facts of evolution—facts which showed that, so far as the biological field is concerned, the process of creation, or upbringing from lower to higher forms, has been continuously going on for millions upon millions of years and is presumably going on now. This tended to direct attention away from the *Deus ex machina*, to identify the Creator with his universe, to strengthen the theological doctrine of immanence, which represents substantially the philosophic position of Leonardo da Vinci, Galileo, Newton, Francis Bacon, and most of the great minds of history down to Einstein.

Neither evolution nor evolutionists have in general been atheistic—Darwin least of all—but their influence has undoubtedly been to raise doubts about the legitimacy of the dogma of the *Deus ex machina* and of the correlative one of the heat-death. This last dogma rests squarely on the assumption that we, infinitesimal mites on a speck of a world, know all about how the universe behaves in all its parts, or more specifically, that the radiation laws which seem to us to hold here cannot possibly have any exceptions anywhere, even though that is precisely the sort of sweeping generalisation that has led us physicists into error half a dozen times during the past thirty years, and also though we know quite well that conditions prevail outside our planet which we cannot here duplicate or even approach. Therefore the heat-death dogma has always been treated with reserve by the most thoughtful of scientific workers. No more crisp or more cogent statement of what seems to me to be the correct position of science in this regard has come to my attention than is found in the following recent utterance of Gilbert N. Lewis, namely, "Thermodynamics gives no support to the assumption that the universe is running down". "*Gain of entropy always means loss of information and nothing more.*"

The fourth discovery bearing on our theme was the discovery that the dogma of the immutable elements was definitely wrong. By the year 1900 the element radium had been isolated and the mean lifetime of its atoms found to be about two thousand years. This meant definitely that the radium atoms that are here now have been formed within about that time; and a year or two later the element helium was definitely observed to be growing out of radium here and now. This raised insistently the question as to whether the creation, or at least the formation, of all the elements out of something else may not be a continuous process—stupendous change in viewpoint the discovery of radioactivity brought about, and a wholesome lesson of modesty it taught to the physicist. But a couple of years later, uranium and thorium, the heaviest known elements, were definitely caught in the act of begetting radium, and all the allied chain of disintegration products. Since, however, the lifetime of the parent atom, uranium, has now been found to be a billion years or so, we have apparently ceased to inquire whence it comes. We are disposed to assume, however, that it is not now being formed on earth. Indeed, we have good reason to believe that the whole radioactive process is confined to a very few, very heavy elements which are now giving up the energy which was once stored up in them—we know not how—so that radioactivity, though it seemed at first to be pointing away from the heat-death, has not at all, in the end, done so. Indeed, it seems to be merely one mechanism by which stored-up energy is being frittered away into apparently unreclaimable radiant heat—another case of humpty-dumpty.

然后出现了第三个发现，地质学和生物学对进化现象的研究结果表明：就生物学领域而言，创造生命的过程或从低级形式向高级形式发展的过程，已经持续了亿万年而且据推测现在还在继续。这似乎使人们的视线从**救世主**中转移开来，认为造物主与他的天地万物是等同的，从而强化了上帝无所不在的神学理论，这一理论充分反映了历史上大多数伟大的思想家，从列奥纳多·达·芬奇、伽利略、牛顿、弗朗西斯·培根，一直到爱因斯坦的哲学立场。

一般而言，进化论和进化论者都不倾向于无神论——达尔文尤其如此——但是由于他们的影响，人们的确已开始怀疑**救世主**理论以及与之相联系的热寂说的合理性。这个最后的学说直接需要假设，我们人类，这些宇宙中某一点上的无限小微粒，已经了解了宇宙中所有组成部分的行为，或者更明确地说，我们现在认定的辐射定律不可能在任何地方出现例外情况，尽管它就是那个曾让我们这些物理学家在过去的 30 年里犯过 6 次错误的并非普适的定律，也尽管我们很清楚地知道我们所在行星之外的环境是我们无法复制甚至是无法介入的。因此，大多数有思想的科学工作者都对热寂说采取保留的态度。在我所注意到的关于该问题的正确科学观点方面，没有谁的观点比吉尔伯特·刘易斯在最近讲话中的下述论断更简明、更有说服力，即“热力学并不支持宇宙在退化的假设”。“**熵的增加通常只意味着信息的减少，仅此而已**”。

第四个与我们这个主题相关的发现是，元素不可改变的观点是完全错误的。到 1900 年，元素镭已经被分离出来，并得出镭原子的平均寿命大约为 2,000 年。这显然说明现在的镭原子大约是在那个时间范围内形成的；一两年以后，人们可以观测到从镭中生成的氦元素。这使人们急切地想知道，所有这些由其他元素生成新元素的创造过程，或至少可以说是形成过程，是否不会连续进行——放射性的发现带来了观念上的巨大变化，也有益地告诫了物理学家们要谦虚。然而几年之后，目前已知的两种最重的元素——铀和钍，已被确认都能在生成镭以及一系列相关的衰变产物的过程中得到。既然现在发现作为母原子铀的寿命在 10 亿年左右，显然我们不用再去追问它是从哪儿来的了。然而，我们倾向于假定它不是现在在地球上形成的。实际上，我们有足够的理由相信所有的放射性过程只局限于极少数很重的元素，它们目前正在释放曾经储存在其内部的能量——我们不知道这种储存是怎样实现的——所以虽然放射现象起初似乎远离了热寂说，可到了最后却不是完全如此。放射现象看起来其实仅仅是储存能慢慢转化为明显不可逆转的辐射热的一种机制——“汉普蒂·邓普蒂”的另一个实例。

The fifth significant discovery was the enormous lifetime of the earth—partly through radioactivity itself, which assigns at least a billion and a half years—and the still greater lifetime of the sun and stars—thousands of times longer than the periods through which they could possibly exist as suns if they were simply hot bodies cooling off. This meant that new and heretofore unknown sources of heat energy had to be found to keep the stars pouring out such enormous quantities of radiation for such ages upon ages.

The sixth discovery, and in many ways the most important of all, was the development of evidence for the interconvertibility of mass and energy. This came about in three ways. In 1901 Kaufman showed experimentally that the mass of an electron could be increased by increasing sufficiently its velocity: that is, energy could be definitely converted into mass. About the same time the pressure of radiation was experimentally established by Nichols and Hull at Dartmouth College, New Hampshire, and Lebedew at Moscow. This meant that radiation possesses the only distinguishing property of mass, the property by which we define it, namely, inertia. The fundamental distinction between radiation and matter thus disappeared. These were direct, experimental discoveries. Next, in 1905, Einstein developed the interconvertibility of mass and energy as a necessary consequence of the special theory of relativity. If, then, the mass of the sun could in any way be converted into radiant heat, there would be an abundant source of energy to keep the sun going so long as necessary, and all our difficulties about the lifetimes of the sun and stars would have disappeared. But what could be the mechanism of this transformation?

Then came the seventh discovery, which constituted a very clear finger-post, pointing to the possibility of the existence of an integrating or building-up process among the physical elements, as well as in biological forms, in the discovery that the elements are all definitely built up out of hydrogen; for they—the ninety-two different atoms—were all found, beginning about 1913 by the new method of so-called positive ray analysis, to be exact multiples of the weight of hydrogen within very small limits of uncertainty. This fact alone raises very insistently the query as to whether they are not being built up somewhere out of hydrogen now. They certainly were once so put together, and some of them, the radioactive ones, are now actually caught in the act of splitting up. Is it not highly probable, so would say any observer, that the inverse process is going on somewhere, especially since the process would involve no violation either of the energy principle or of the second law of thermodynamics; for hydrogen, the element out of which they all must be built, has not a weight exactly one in terms of the other ninety-two, but about 1 percent more than one, so that since mass or weight had been found in the sixth discovery to be expressible in terms of energy, the union of any number of hydrogen atoms into any heavier element, meant that 1 percent of the total available potential energy had disappeared and was therefore available for appearance as heat.

When, about 1914–15, this fact was fitted by MacMillan, Harkins, and others into the demand made above in the fifth discovery for a new source of energy to keep the sun pouring out heat so copiously for such great lengths of time, is seemed to the whole

第五个重大发现是发现地球已经存在了很长时间——部分是根据元素放射性确定的，地球年龄至少为 15 亿年，而太阳和其他恒星的年龄则更长——这个年龄要比假设恒星作为一个炽热的天体仅是单纯地不断冷却所耗的时间长几千倍。这意味着我们不得不去寻找能使恒星在漫长岁月里不断释放出大量辐射的热能的新来源，这种能源迄今为止尚不可知。

第六个发现是找到了质量和能量相互转化的证据，从许多方面来看，这个发现是所有发现中最重要的。这可以从三方面说明。1901 年考夫曼用实验证明，通过充分提高电子的速度可以增加电子的质量：也就是说，能量确实可以转化为质量。几乎是同时，新罕布什尔州达特茅斯学院的尼科尔斯、赫尔和莫斯科的列别捷夫用实验验证了辐射压的存在。这说明辐射具有质量的唯一与众不同的特征，即惯性。这样辐射和物质之间的根本区别就消失了。这些都是直接由实验发现的。紧接着在 1905 年，爱因斯坦利用狭义相对论推导出质量和能量之间的相互转化这一必然结果。假如太阳的质量可以通过某种方式转化为辐射热，那么就有了充足的能量来源以保持太阳在如此长的时间里放出热，而我们对太阳及其他恒星年龄的疑问也就不复存在了。但是这种转变机制又是什么呢？

于是就有了第七个发现，这个发现非常明确地指出：组成物质的元素之间有可能出现积聚或合成的过程，在生物形态中也是如此，人们发现元素无疑都是由氢元素构成的；因为自 1913 年以来用所谓的阳射线分析法发现：92 种不同原子的重量均为氢元素重量的精确倍数且误差极小。仅仅是这一事实就使人们迫切想知道它们是不是现在正在某个地方由氢合成。它们肯定曾被这样合成过，而且现在它们之中的那些放射性元素会在裂变的过程中被发现。任何观测者都会说，很有可能在某处正进行着裂变的逆过程，特别是当这个过程并不违反能量原理和热力学第二定律的时候。氢作为构建其他元素的必要成分，相对于其他 92 个元素来说，其重量并不正好是 1，而是比 1 大 1%。所以既然在第六个发现中质量或重量可以表现为能量的形式，则任意数目的氢原子合并成任何较重元素时意味着全部可利用势能的 1% 消失了，并转化成了可以释放的热能。

大约在 1914 至 1915 年，这个事实被麦克米伦、哈金斯等人用作解释第五个发现中所需的一种新能源，以保持太阳长期散发出巨大的热量。对整个物理学界来说，由氢构成更重元素的过程实际上已被证明在太阳和恒星内部环境下确实在发生。这

world of physics that the building up of the heavier elements out of hydrogen under the conditions existing within the sun and stars had been practically definitely proved to be taking place. This would not provide an escape from the heat-death, but it would enormously postpone it, that is, until all the hydrogen in the universe had been converted into the heavier elements.

By this process, however, the suns could stoke at most but 1 percent of their total mass, assuming they were wholly hydrogen to begin with, into their furnaces, and 99 percent of the mass of the universe would remain as cold, dead ash when the fires were all gone out and the heat-death had come. But about 1917 the astronomer began to chafe under the time-limitation thus imposed upon him, and this introduced the eighth consideration bearing upon our theme. He could get a hundred times more time—from now on, much more than that, because only a small fraction of the matter in the universe is presumably now hydrogen—by assuming that, in the interior of heavy atoms, occasionally a negative electron gets tired of life at the pace it has to be lived in the electron world, and decides to end it all and commit suicide; but, being paired by Nature in electron-fate with a positive, he has to arrange a suicide pact with his mate, and so the two jump into each other's arms in the nucleus, and the two complementary electron lives are snuffed out at once; but not without the letting loose of a terrific death-yell, for the total mass of the two must be transformed into a powerful ether pulse which, by being absorbed in the surrounding matter, is supposed to keep up the mad, hot pace in the interiors of the suns. This discovery, or suggestion, to account for the huge estimated stellar lifetimes, of the complete annihilation of positive and negative electrons within the nucleus, makes it unnecessary to assume, at least for stellar lifetime purposes, the building up of the heavier elements out of hydrogen. Indeed, it seems rather unlikely that both kinds of processes, atom-building and atom-annihilating, are going on together in the same spot under the same conditions, so we must turn to further experimental facts to get more light.

The ninth sign-post came into sight in 1927, when Aston made a most precise series of measurements on the relative masses of the atoms, which made it possible to subject to a new test the Einstein formula for the relation between mass and energy, namely, $E = Mc^2$. This Aston curve is one of the most illuminating finger-pointings we now have. It shows that:

1. Einstein's equation actually stands the quantitative test for radioactive or disintegrating processes right well, and therefore receives new experimental credentials.

2. The radioactive or disintegrating process with the emission of an alpha ray must be confined to a very few heavy elements, since these are the only ones so situated on the curve that mass can disappear, and hence heat energy appear, through such disintegration.

3. All the most common elements, except hydrogen, are already in their most stable condition, that is, their condition of minimum mass, so that if we disintegrate them we shall have to do work upon them, rather than get energy out of them.

并不能使宇宙摆脱热寂的结局，但可以极大地推迟它的到来，也就是说，可以推迟到宇宙中所有的氢都被转化成了较重的元素时才发生。

然而，假设这些像太阳一样的恒星全部由氢元素构成，并将它们投入其自身的熔炉之中，这一过程最多只能消耗掉其总质量的 1%，宇宙中 99% 的质量在燃料烧尽热寂来临时将变为冰冷的死灰。但在 1917 年前后，天文学家开始对这个强加给他的时间限制感到烦扰，并由此引出了与我们这个主题相关的第八个设想。他从现在起可以获得比那个时间限制长几百倍甚至更长的时间，因为据推测现在宇宙中的物质只有一小部分是氢——假设在重原子内部，偶尔一个负电子对自己不得不生活在电子世界中感到厌倦而决定通过自杀结束这一切；但自然界注定电子要和正电子配成对，它必须筹划与自己的搭档一起自杀，这样在原子核中它们两个分别跳到了对方的怀抱中，这两个正负互补的电子的生命同时结束；但在死亡时并非没有发出临死前那可怕的叫喊，因为它们两个的总质量必然转化成一个强大的以太脉冲，并被周围物质吸收，该过程被认为是保持恒星内部剧烈热反应的原因。这个发现或提议，即利用原子核内部正负电子的彻底湮灭来解决预测的恒星年龄大得不可思议的难题，使人们至少在恒星年龄问题上不需要假设由氢合成重元素的过程。的确，原子合成和原子湮灭这两类过程在同一地点同一条件下同时进行看起来不太可能，所以我们必须通过更多的实验得到更清楚的解释。

1927 年，当阿斯顿对原子的相对质量进行了一系列非常精确的测定，使爱因斯坦在质量和能量之间建立的公式，即 $E = Mc^2$ 得到了新的验证之后，第九个里程碑也就出现在了人们的视野中。阿斯顿曲线是我们目前拥有的最具启发性的指证之一。它说明了以下几点：

1. 爱因斯坦方程实际上可用作放射性或衰变过程的有效的定量检验，因而增加了实验的可信度。

2. 放出一个 α 射线的放射性或衰变过程肯定只局限于很少几个重元素之中，因为它们在曲线中处于质量可能减少的位置上，因而可以通过这样的衰变过程释放出热能。

3. 所有除氢以外的最常见的元素都处于最稳定的状态之中，也就是说，它们处于质量最小的状态，所以如果我们想让它们发生衰变必须对它们做功，而不是从中获取能量。

4. Therefore, man's only possible source of energy other than the sun is the upbuilding of the common elements out of hydrogen or helium, or else the entire annihilation of positive and negative electrons; and there is no likelihood that either of these processes is a possibility on earth.

5. If the foregoing upbuilding process is going on anywhere, the least penetrating and the most abundant radiation produced by it, that corresponding to the formation of helium out of hydrogen, ought to be about ten times as energetic as the hardest gamma rays, that is, it ought to correspond to about twenty-six million electron-volts in place of two and a half million.

6. Other radiations corresponding to the only other abundant elements, namely, oxygen (oxygen, nitrogen, carbon), silicon (magnesium, aluminium, silicon), and iron (iron group), should be found about four times, seven times, and fourteen times as energetic as the "helium rays".

7. The radiation corresponding to the smallest annihilation process that can take place—the suicide of a positive and negative electron—is three hundred and fifty times as energetic as the hardest gamma ray, or thirty-five times as energetic as the "helium ray".

This brings us to the tenth discovery, that of the cosmic rays. These reveal:*

1. A radiation, the chief component of which, according to our direct comparison, is five times as penetrating as the hardest gamma ray, which, with the best theoretical formula we have relating energy and penetrating power (Klein–Nishina), means a ray ten times as energetic as the hardest gamma ray, *precisely according to prediction.*

2. Special bands of cosmic radiation that are roughly where they should be to be due to the formation of the foregoing abundant elements out of hydrogen, though (for reasons to be given presently) no precise quantitative check is to expected except in the case of helium.

3. No radiation of significant amount anywhere near where it is to be expected from the annihilation hypothesis, thus indicating that at least 95 percent of the observed cosmic rays are due to some other less energetic processes.

4. A radiation that is completely independent of the sun, the great hot mass just off our bows, and not appreciably dependent on the Milky Way or the nearest spiral nebula, Andromeda, one that comes in to us practically uniformly from all portions of the celestial dome, and is so invariable with both time and latitude at a given elevation that the observed small fluctuations at a given station reflect with much fidelity merely the changes

* See articles by Millikan and by Millikan and Cameron, *Phys. Rev.*, Dec. 1, 1930, and in press.

4. 因此，除了太阳以外，人类可能获得能量的方式只有通过用氢或氦合成普通元素，或正负电子全部湮灭的过程；这两个过程都不可能在地球上发生。

5. 假如上述合成过程可以在任意地点进行，由此产生的穿透力最弱、强度最高的辐射，按照由氢合成氦的过程推算，应该能够达到最硬的 γ 射线能量的 10 倍左右，也就是说，大致相当于 26 百万电子伏特而不是 2.5 百万电子伏特。

6. 与其他含量丰富的元素，即氧（氧、氮、碳），硅（镁、铝、硅）和铁（铁族元素）相关联的辐射过程所对应的能量应当分别约为“氦射线”的 4 倍、7 倍和 14 倍。

7. 由可能发生的最小的湮灭过程——正负电子的自杀过程发出的辐射是最硬的 γ 射线能量的 350 倍，是“氦射线”能量的 35 倍。

这把我们引入了与宇宙射线相关的第十个发现。观测表明 *：

1. 我们根据直接对比发现：一个主体部分是穿透力为最硬的 γ 射线 5 倍的辐射，用能量与穿透力之间最好的理论公式——克莱因 – 仁科公式计算，该射线的能量为最硬的 γ 射线的 10 倍，**这是根据预测推算的精确结果**。

2. 宇宙辐射的特殊频带，大致位于由氢合成上述含量丰富元素时产生的辐射应在的位置，尽管（由于马上要提到的原因）除了氦元素以外，我们都不能进行精确的定量计算。

3. 在正负电子湮灭假设对应的位置附近没有观测到数量显著的辐射，这说明观测到的宇宙射线至少有 95% 是由其他一些能量较低的过程造成的。

4. 这种辐射与就在我们凸窗外面的那个巨大的发热体，即太阳完全不相干，而且与银河系或离我们最近的旋涡星云——仙女座星云也没有关系，它几乎均匀地从天穹的各个位置向我们而来，在某一特定的高度，其观测值不随时间和纬度而变，个别观测站观测值的微小涨落仅仅是由于射线在到达观测器之前必须要穿过的大气

* 见密立根以及密立根和卡梅伦发表在《物理学评论》上的文章，1930年12月1日，即将出版。

in the thickness of the absorbing air blanket through which the rays have had to pass to get to the observer.

This last property is the most amazing and the most significant property exhibited by the cosmic rays, and before drawing the final conclusions its significance will be discussed. For it means that at the time these rays enter the earth's atmosphere, they are practically pure ether waves or photons. If they were high-speed electrons or even had been appreciably transformed by Compton encounters in passing through matter into such high-speed electrons or beta rays, these electrons would of necessity spiral about the lines of force of the earth's magnetic field and thus enter the earth more abundantly near the earth's magnetic poles than in lower latitudes. This is precisely what the experiments made during the last summer at Churchill, Manitoba (lat. 59° N.), within 730 miles of the north magnetic pole, showed to be *not true*, the mean intensity of the rays there being not measurably different from that at Pasadena in lat. 34° N.

Nor is the conclusion that the cosmic rays enter the earth's atmosphere as a practically pure photon beam dependent upon these measurements of last summer alone. It follows also from the high altitude sounding-balloon experiments of Millikan and Bowen in April 1922, taken in connexion with the lower balloon flights of Hess and Kolhörster in 1911–14. For in going to an altitude of 15.5 km. we got but one-fourth the total discharge of our electroscope which we computed we should have obtained from the extrapolation of our predecessors' curves. This shows that somewhere in the atmosphere below a height of 15.5 km. the intensity of the ionisation within a closed vessel exposed to the rays goes through a maximum, and then decreases, quite rapidly, too, in going to greater heights. We have just taken very accurate observations up to the elevation of the top of Pike's Peak (4.3 km.), and found that within this range the rate of increase with altitude is quite as large as that found in the Hess and Kolhörster balloon flights, so that there can be no uncertainty at all about the existence of this maximum. Such a maximum, however, means that the rays, before entering the atmosphere, have not passed through enough matter to begin to get into equilibrium with their secondaries—beta rays and photons of reduced frequency—in other words, *that they have not come through an appreciable amount of matter in getting from their place of origin to the earth.*

This checks with the lack of effect of the earth's magnetic field on the intensity of the rays; and the two phenomena, of quite unrelated kinds and brought to light years apart, when taken together, prove most conclusively, I think, that the cosmic rays cannot originate even in the outer atmospheres of the stars, though these are full of hydrogen and helium in a high temperature state, but that they must originate rather in those portions of the universe from which they can come to the earth without traversing matter in quantity that is appreciable even as compared with the thickness of the earth's atmosphere—in other words, that *they must originate in the intensely cold regions in the depths of interstellar space.*

Further, the more penetrating the beta rays produced by Compton encounters, the greater the thickness of matter that must be traversed before the beam of pure photons

吸收层的厚度变化造成的，这种涨落是吸收层厚度变化的真实反映。

最后一个特性是宇宙射线最令人惊异也是最重要的特性，在下最终定论之前将对这个问题的重要性进行讨论。因为这意味着宇宙射线在进入地球大气层时基本上是由纯以太波或光子组成的。如果它们是高速运动的电子，或曾经在穿透物质的过程中因发生康普顿碰撞而转化成了高速电子或 β 射线，那么这些电子必然是以围绕地球磁力线的螺旋形轨迹前进的，因此进入地球两磁极的射线应该比低纬度地区的多。但去年夏天在距离北磁极 730 英里的马尼托巴省丘吉尔市（北纬 59°）进行的实验说明**事实并非如此**，在那里测量的射线平均强度值与在北纬 34° 的帕萨迪纳得到的测量值没有什么区别。

进入地球大气层的宇宙射线几乎全部由光子组成这一结论并不仅仅是根据去年夏季的测量结果，它也能从 1922 年 4 月密立根和鲍恩利用探空气球在高空处进行的实验和与之相关联的赫斯和柯尔霍斯特 1911~1914 年低空区的气球实验中找到根据。因为在到达 15.5 千米的高度时，验电器的放电量并没有达到我们根据前人实验曲线外推得到的计算值，而仅为它的四分之一。这说明在大气中某个低于 15.5 千米的地方，密闭容器内宇宙射线导致的电离程度先是达到了一个最大值，但当升至更高的高度后又开始急剧下降。我们刚刚对派克峰（4.3 千米）峰顶以下各个高度处的射线强度进行了精确的测量，发现在这个范围内，射线强度随高度上升的增加率与赫斯和柯尔霍斯特在气球飞行实验中得到的结果一致，所以存在一个最大值是毋庸置疑的。但这个最大值说明宇宙射线在到达地球大气层之前并没有穿过足够厚的物体以能与它们的次级粒子——β 射线和频率降低的光子达到平衡。也就是说，**这些射线在从它们的发源地到地球之间未曾穿过数量可观的物质。**

这证实了地球磁场对宇宙射线强度没有影响；这两个现象完全不相干，并且相距若干光年，当把它们联系在一起的时候，我认为可以下结论说：宇宙射线不可能发源于恒星，哪怕是在它的大气外层，尽管那里有大量处于高温中的氢和氦。不过它们肯定是来自宇宙中那些能让它们甚至不必穿透和地球大气层一般厚的物质就可以到达地球的地方——换句话说，**宇宙射线必定起源于星际空间深处的超低温区域。**

此外，由康普顿碰撞产生的 β 射线的穿透力越强，进入大气层的纯光子束在与它的次级粒子达到平衡之前需要穿透的物质厚度就越大；且当达到此种平衡时，测

which enters the atmosphere gets into equilibrium with its secondaries; and until such equilibrium is reached, the apparent absorption coefficient must be less than the coefficient computed with the aid of the Klein–Nishina formula from the energy released in the process from which the radiation arises. Now the Bothe–Kolhörster experiments of about a year ago show that when the energies of the incident photons are sufficiently high, the beta rays released by Compton encounters do indeed become abnormally penetrating: so that it is to be expected that, for the cosmic rays produced by the formation of the heavier of the common elements like silicon and iron out of hydrogen, the observed absorption coefficients will be somewhat smaller than those computed from the energy available for their formation. This is precisely the behaviour which our cosmic ray depth-ionisation curve actually reveals. At the highest altitudes at which we have recently observed (14,000 ft.), the helium rays have reached equilibrium with their secondaries, and the observed and computed coefficients agree as they should. For the oxygen rays the observed coefficent is a little lower than the computed value—about 17 percent lower; for the silicon rays still lower—about 30 percent; and for the iron rays considerably lower still—about 60 percent: all in beautiful qualitative agreement with the theoretical demands as outlined.

The foregoing results seem to point with much definiteness to the following conclusions:

1. The cosmic rays have their origin not in the stars but rather in interstellar space.

2. They are due to the building up in the depths of space of the commoner heavy elements out of hydrogen, which the spectroscopy of the heavens shows to be widely distributed through space. That helium and the common elements oxygen, nitrogen, carbon, and even sulphur, are also found between the stars is proved by Bowen's beautiful recent discovery that the "nebulium lines" arise from these very elements.

3. These atom-building processes cannot take place under the conditions of temperature and pressure existing in the sun and stars, the heats of these bodies having to be maintained presumably by the atom-annihilating process postulated by Jeans and Eddington as taking place there.

4. All this says nothing at all about the second law of thermodynamics or the *Wärme-Tod*, but it does contain a bare suggestion that if atom formation out of hydrogen is taking place all through space, as it seems to be doing, it may be that the hydrogen is somehow being replenished there, too, from the only form of energy that we know to be all the time leaking out from the stars to interstellar space, namely, radiant energy. This has been speculatively suggested many times before, in order to allow the Creator to be continually on his job. Here is, perhaps, a little bit of *experimental* finger-pointing in that direction. But it is not at all proved or even perhaps necessarily suggested. If Sir James Jeans prefers to hold one view and I another on this question, no one can say us nay. The one thing of which we may all be quite sure is that neither of us *knows* about it. But for the continuous building up of the common elements out of hydrogen in the depths of interstellar

得的表观吸收系数肯定小于根据辐射产生过程中释放的能量通过克莱因–仁科公式计算得到的系数。因为一年前的博思–柯尔霍斯特实验证明，当入射光子的能量足够高的时候，由康普顿碰撞产生的 β 射线确实表现出格外强的穿透力；所以可以预测，如果宇宙射线是在由氢合成像硅和铁这样较重的常见元素时产生的，则所观察到的吸收系数将比根据合成过程可获得的能量计算得到的系数低一些。这正是我们的宇宙射线深度–电离曲线揭示出来的特性。在我们最近测量的最大高度处（14,000 英尺），氦射线已经和它们的次级粒子达到了平衡，因而系数测量值和计算值达到了应有的一致。对于氧射线，系数测量值略低于计算值——约低 17%；硅射线更低一些——约 30%；而铁射线则低得更多——约 60%；但它们都与所述的理论要求定性地一致。

根据上述结果显然可以得到以下结论：

1. 宇宙射线的发源地不是恒星，而是在星际空间。

2. 宇宙射线起源于太空深处的氢合成较为常见的重元素的过程，太空的光谱表明这些重元素广泛地分布于宇宙中。鲍恩的最新发现证明氦和常见元素氧、氮、碳甚至硫都存在于恒星间的太空中，这个出色的发现即“氰线”就是由这些元素引起的。

3. 这样的原子合成过程不可能在太阳或其他恒星内的高温高压环境下进行，根据金斯和爱丁顿的设想，这些天体的热量也许只能靠发生在那里的原子湮灭过程来维持。

4. 所有这些都没有提及热力学第二定律或**热寂**，但它确实包含了这样一个假设：假如由氢合成其他原子的过程在宇宙空间中普遍存在，就像我们现在看到的这样，那么氢元素有可能也在同一地点以某种方式得到补充，即通过我们已知的从恒星向星际空间不断释放能量的唯一方式——辐射能。为了使造物主的工作得以延续下去，这一点以前已经被设想过多次了。在这方面也许有一点点**实验上的**迹象。但它根本就得不到验证，甚至可能连提出的必要都没有。如果在这个问题上詹姆斯·金斯爵士支持一个论点而我倾向于另一个，没有人能说我们不对。也许只有一件事情我们能完全确定，那就是我们俩对此都**一无所知**。不过宇宙射线就可以很好地为在广漠星际空间中存在由氢不断合成常见元素的过程提供实验上的证据。我不是不知道

space the cosmic rays furnish excellent experimental evidence. I am not unaware of the difficulties of finding an altogether satisfactory kinetic picture of how these events take place, but acceptable and demonstrable facts do not, in this twentieth century, seem to be disposed to wait on suitable mechanical pictures. Indeed, has not modern physics thrown the purely mechanistic view of the universe root and branch out of its house?

(**127**, 167-170; 1931)

Robert A. Millikan: California Institute of Technology, Pasadena, California.

要找到一个完全令人满意的动力学解释来说明这些现象的来龙去脉是困难的，但是在20世纪，可接受的明显事实似乎无意于长期等候合适的机械性理论解释的出现。事实上，当代物理学不是已经彻底把纯粹机械的宇宙观赶出门外了吗？

（王锋 翻译；朱永生 审稿）

New Aspects of Radioactivity*

C. D. Ellis

Editor's Note

With the quantum theory of atomic structure in place, physicists began to ponder the structure of the nucleus. Observations of high-energy electromagnetic radiation called gamma rays emitted by radioactive nuclei seemed to offer information on internal nuclear states. Here Charles D. Ellis discusses several of the most promising techniques for investigating these gamma rays. The most useful involved measuring the energies of secondary electrons ejected from an atom by the photoelectric effect, involving fine wires coated with an atomically thin layer of radioactive material. Although the gamma ray energies could only be measured to an accuracy of one part in 500, Ellis is hopeful that such data will clarify nuclear structure in the near future.

γ-Ray and Nuclear Structure

UNTIL a few years ago, the fundamental problems of physics were those concerned with the structure of the atom. The nucleus was necessarily often referred to, but only in relation to its effect on the behaviour of the electrons in the atom. It was found that for most purposes the net charge, *Ze*, was a sufficient description of the nucleus. Within, however, the last three years, the whole attitude of physicists to this problem has changed; on one hand, our knowledge of those phenomena which depend on the intimate structure of the nucleus has been greatly increased; on the other hand, wave mechanics has proved to be eminently suitable for a theoretical attack on this problem, and has already provided a solution of some of the outstanding problems.

Of the many lines of investigation which have been developed, not the least interesting is that of the characteristic electromagnetic radiation that can be emitted by radioactive nuclei. These radiations are termed the γ-rays and are in general of considerably shorter wave-length than the X-rays. They bear the same relation to the structure of the nucleus as do the ordinary optical and X-ray spectra to the structure of the electronic system of the atom, but there is this one point of difference. The optical and X-ray spectra can conveniently be studied for a series of elements because the process of excitation is under control, but it is only in a few isolated cases that it has yet been possible to excite a nucleus by external agencies to emit characteristic radiation. Some of the radioactive bodies, however, emit these radiations spontaneously, since the process of disintegration leaves the newly formed nucleus in an excited state and able to emit its characteristic radiation. The nuclear spectra have therefore only been examined in detail for those radioactive bodies which happen to emit them, and it has been impossible as yet to find any general laws

* Substance of two lectures delivered at the Royal Institution on Nov. 4 and 11.

放射性研究的新面貌*

查尔斯·埃利斯

编者按

考虑到原子结构在量子理论中的重要地位，物理学家们开始研究原子核的结构。通过观察从放射性原子核发射出的高能电磁辐射——γ 射线，人们也许可以得到关于原子核内部状态的信息。查尔斯·埃利斯在文中讨论了几种用以研究这种 γ 射线的最有前景的技术。其中最有价值的是测量次级电子能量的技术，在细金属丝表面涂上一层原子那么薄的放射性材料，发生光电效应的原子就会释放出这些次级电子。虽然这种测量 γ 射线能量的方法只能达到 1/500 的精确度，但埃利斯对用这种方法得到的数据能在不远的未来破解原子核的结构充满信心。

γ 射线和核结构

直至几年前，物理学的基本问题还是关于原子的结构。通常核是必然要涉及的，但仅涉及其对原子中电子行为的影响。人们发现，在大多数场合净电荷 Ze 足以描述原子核。然而在近三年内，物理学家们对此问题的总体看法已发生了变化，原因如下：一方面，我们对那些依赖于核的本质结构的有关现象的认识已大大增加；另一方面，波动力学已被证明非常适合从理论上攻破这个问题，并且已经为某些重要问题提供了一个解决方式。

在已经开展的许多研究中，比较令人感兴趣的是对放射性核发射的特征电磁辐射的研究。这些辐射被命名为 γ 射线，它们通常具有比 X 射线短得多的波长。γ 射线和核结构的关系就如同普通光及 X 射线光谱与原子中电子系统结构的关系一样，但是有一点不同：因为激发过程处于控制之下，一系列元素的普通光和 X 射线光谱很方便进行研究；然而，到目前为止，仅在个别情况下有可能用外部因素激发一个原子核来发射特征辐射。不过，某些放射体可以自发发射这些辐射，因为衰变过程使新形成的核处于激发态，并能发射其特征辐射。因此，只有那些碰巧发射辐射的放射体的核谱得到了详细的研究，迄今为止尚不太可能通过观察接连的一系列不同原子核核谱的相似性来找出支配这些核谱分布的任何普遍规律。

* 本文中的内容来自作者11月4日和11日发表于英国皇家研究院的两个报告。

governing the arrangement of these spectra by noting the similarities in the spectra from a succession of different nuclei.

The result of this was that, until a few years ago, while there was a great deal of information about the nuclear spectra of several radioactive bodies, it was still impossible to associate this with any definite feature of the structure. Recently the position has changed greatly, and it now seems possible to view in the nuclear level systems which can be deduced from the γ-ray measurements the characteristic stationary states of α-particles or protons in the nucleus, and to associate such level systems directly with the ground states deducible from other evidence.

Methods of Investigating the γ-Rays

A simple method that was of great importance in the early days of radioactivity was to investigate the absorption of the radiation emitted by a particular body by placing a radioactive source at some distance from an electroscope and observing how the ionisation decreased when successive sheets of some material such as aluminium or lead were interposed. It was frequently possible to analyse the resulting absorption curve into a series of simple exponential curves, and thus to obtain a general idea of the different components of the complex radiation. Methods such as this could never yield very precise information, and they have now been superseded by more accurate methods.

The crystal method, in the forms used for X-rays, has been applied with considerable success to γ-rays[1]. In one respect the technique is simpler, since in place of the X-ray tube with all the apparatus necessary to run it, it is only necessary to use a fine tube containing the radioactive material, but in other respects the experiments are far more difficult. Owing to the very short wave-length, of the order of 40 X.U. to 4 X.U., the glancing angles are extremely small, and not only is the adjustment of the apparatus considerably more difficult but it is also impossible to measure the wave-length with much accuracy. Further, in comparison with an X-ray tube, the normal amount of radioactive material constitutes an extremely weak source of radiation. As a result it has not yet been possible to push this method when using photographic registration beyond 16 X.U. Recently Steadman[2] has devised an arrangement, using an electrical counter in place of a photographic plate, which may overcome some of these difficulties.

The method which has given us most of our information is based on the photoelectric effect. The general principle is very simple and is as follows[3]. A tube containing the radioactive body, the γ-rays of which are under investigation, is placed inside a small tube of some material of high atomic weight, such as platinum. In their passage through the platinum, the γ-rays eject groups of photoelectrons the energies of which are connected with the frequency of the γ-rays by the Einstein law. Thus the γ-ray of frequency v will lead to the ejection of a series of groups of electrons of energies $hv\text{-}K_{Pt}$, $hv\text{-}L_{Pt}$, etc., according to whether the conversion occurs in the K, L, etc., state of the platinum atoms. This electronic emission can be separated out into a corpuscular spectrum by

直至几年前，尽管有了关于某些放射体核谱的大量信息，但是仍然不可能将其与原子结构的任何确定特性联系在一起。最近，情况发生了很大的变化，现在似乎可以把核中 α 粒子或质子的特征定态看作是由 γ 射线测量推导出的核能级系统中的一部分，并将此能级系统直接与由其他证据推断出的基态相联系。

研究 γ 射线的方法

在放射性研究早期，一个极为重要的简单方法是研究特殊物体所发射的辐射的吸收，具体做法是将放射源置于与验电器有一定距离的地方，然后观测在接连插入诸如铝或铅这些材料的薄片时，电离如何减小。通常可以将生成的吸收曲线分解成一系列简单的指数曲线，进而大致了解复合辐射中不同组分的情况。利用这类方法不可能得到很精确的信息，因而它们现在已被更精确的方法所代替。

在 X 射线研究中所用的晶体方法已经非常成功地应用到了 γ 射线的研究中[1]。从一方面来看技术上会变得更为简单，因为不再需要 X 射线管以及运行它所需的所有装置，只需要使用含有放射性材料的细管；但从其他方面来看，实验却要困难得多：由于 γ 射线的波长很短，数量级从 40 X.U. 到 4 X.U.，因此掠射角极小，这不仅使装置的调整更加困难，而且不可能很准确地测量波长。此外，与 X 射线管比较起来，正常量的放射性材料只是极弱的辐射源。因此，当使用的波长范围超出 16 X.U. 时，还不能使用这种方法。最近，斯特德曼[2]已设计了一台装置，用电子计数器来代替照相底片，这也许能克服某些困难。

基于光电效应的方法已经给我们带来了大部分信息。一般原理非常简单，叙述如下[3]：将一个装有能放出所要研究的 γ 射线的放射体的管，放入由某些高原子量材料，例如铂，制成的小管内。在它们通过铂时，γ 射线发射成组的光电子，它们的能量与 γ 射线频率的关系符合爱因斯坦定律。因此，频率为 v 的 γ 射线将按照转换是否发生在铂原子的 K、L 等能量态，发射一系列能量为 $hv\text{-}K_{Pt}$、$hv\text{-}L_{Pt}$ 等的光电子群。用一般的半圆磁聚焦方法，可以将这类电子发射分离为微粒谱，通常用照相法记录微粒谱。因为在大多数情况下，只有从 K 能级来的电子群才会有足够的强度产

the usual method of semicircular magnetic focusing. It is usual to register these spectra photographically, and there is not a great deal of difficulty in analysing them and deducing the corresponding γ-rays, since in most cases it is only the electronic group from the *K* level which is sufficiently intense to give a detectable effect. The general application of the method is greatly limited by the fact that the photographic impression of the groups of electrons always shows as a broad, rather diffuse band. The reason is that, although the photoelectrons are ejected from the platinum atoms with sharply defined energies, only those from the surface of the tube actually emerge with their full velocity. Those from the lower layers are retarded in their passage out, and cause the diffuse character of the band.

Fortunately, the radioactive atoms themselves provide us with much more favourable opportunities for observing this photoelectric conversion, by what is termed internal conversion. This is by itself an extremely interesting phenomenon, and will be referred to in detail later. For the present purpose it is convenient to describe it as follows. When a radioactive nucleus emits a quantum hv of radiation, this does not always escape as such from the atom but may be absorbed by the electronic structure of the atom in its passage out. This internal conversion follows the usual photoelectric laws, and thus a radioactive body which emits γ-rays will also emit a corpuscular spectrum similar in every respect to that coming from the platinum tube already mentioned, except that the energies are now $hv\text{-}K_{\text{rad}}$, $hv\text{-}L_{\text{rad}}$, ... The result is in principle in no way different from the previous case where the γ-rays were converted in the platinum, but the importance of this phenomenon for determining the wave-length of the γ-rays depends on the following facts. If a normal amount of radioactive material is deposited on the surface of a fine wire, the actual number of atoms is so small that the layer is in general less than one atom deep. The electrons liberated by this internal photoelectric effect therefore all escape with their full energy and give extremely sharp lines on a photographic plate, in striking contrast to the broad bands obtained by the normal external photoelectric effect. There is the further advantage that the probability of this internal conversion is so great that measurable lines can be obtained with far shorter exposures than by the other method, and the effects of γ-rays are detectable which are so weak as to be quite unattackable by the other method.

The γ-rays of many radioactive bodies have been analysed by this method, and the main features of the characteristic nuclear spectra are known. The accuracy with which the frequencies can be determined is, however, considerably lower than that realised with X-ray spectra. Even in the case of the bodies radium B and radium C, which have been extensively investigated, the relative frequencies are probably not known to much better than one part in five hundred, and the absolute error may be greater. The chief cause for this lies in the difficulty of obtaining a homogeneous magnetic field over a large area.

Intensities of the γ-Rays

An important method[4] of investigating the intensities has been developed by Skobeltzyn, based on the Compton effect of the γ-rays. A narrow pencil of γ-rays is allowed to pass

生可检测到的光电效应，所以分析这些谱线以及导出相应的 γ 射线并没有多大的困难。这个方法的普遍推广大大受限于以下事实：电子群的照相印记经常呈现为一个宽而弥散的条带。因为，尽管光电子是以确定的能量从铂原子上发射出来的，但实际上只有管表面的光电子才能全速发射。较底层的电子在发射穿透出金属表面的路径上受到阻滞减速，从而引起条带的弥散性。

所幸的是，放射性原子自身为我们观测这种被称作内转换的光电转换提供了更多有利的机会。这本身就是一个极有趣的现象，后面将详细提及。就目前的目的而言，可方便地将其描述如下：当放射性核发射一个辐射量子 $h\nu$ 时，这个量子并不总会从原子中逃逸，而是可能在逃逸路上被原子的电子结构所吸收。这种内转换遵从通常的光电定律，因此发射 γ 射线的放射体将也会发射微粒谱，各方面都类似于前面提到的来自铂管的微粒谱，只不过现在的能量是 $h\nu\text{-}K_{\text{rad}}$、$h\nu\text{-}L_{\text{rad}}$……这个结果基本上与以前 γ 射线在铂中转换的情况并无不同，但这类现象对于测定 γ 射线波长的重要性取决于以下事实：将放射性物质的正常量放在细金属丝的表面，那么原子的实际数目会很少，以至于该层通常不到一个原子的深度。因此，这种内部光电效应释放的电子都能以全部能量逃逸，并在照相底片上产生极其清晰的谱线，它与从正常的外部光电效应中得到的宽而弥散的谱线形成鲜明的对比。这种方法还有另一个好处，即这种内转换的概率如此之大，以至于在比其他方法短得多的曝光时间下就可以获得可测量的谱线。在这种情况下 γ 射线的效应是可以检测得到的，而用其他方法很难检测到这种非常微弱的效应。

人们已用这种方法对许多放射体的 γ 射线作了分析，从而了解了特征核谱的主要特性。然而，用此方法测定 γ 射线频率的准确度比用 X 射线谱法测得的要低很多。即使在已经进行过大量研究的镭 B 和镭 C 中，也没有发现相对频率会大大高于 1/500，绝对误差可能更大。其主要原因是，在一个较大的范围内，要获得均匀磁场是有困难的。

γ 射线的强度

斯科别利兹根据 γ 射线的康普顿效应已建立了一种研究 γ 射线强度的重要方法[4]。让一束 γ 射线的窄光锥通过膨胀室，并以常规方式观测由 γ 射线的康普顿效

through an expansion chamber and the recoil electrons liberated by the Compton effect of the γ-rays are observed in the usual manner. In addition, a magnetic field parallel to the axis of the chamber is applied at the moment of expansion, so that the tracks of the recoil electrons are curved by an amount depending on their velocity. By observing both the curvature and the direction of emission of the recoil electrons, it is possible to associate each electron with a γ-rays of definite frequency. A statistical study is made of the relative number of the recoil electron tracks, and from a knowledge of the general laws of scattering it is possible to deduce the relative intensities of the γ-rays.

Owing to a variety of experimental causes, the resolution of the method is not very high, and the effect of two neighbouring γ-rays cannot always be clearly separated. This disadvantage, however, is far outweighed by the definiteness of the results about the intensity distribution throughout the spectrum, and by the fact that the method detects weak γ-rays equally efficiently as strong γ-rays. The interpretation involves a knowledge of the laws of scattering, but there is both a reasonable theoretical foundation and internal evidence from these experiments which combine to render the uncertainties due to this cause of little importance at present.

The photoelectric method has been applied to determine the intensities of the γ-rays by Ellis and Aston[5]. The corpuscular spectra liberated from the radioactive atoms themselves by the internal conversion are clearly of no use in this connexion, since the relative intensities of the groups depend upon the unknown laws of internal conversion. If, however, the corpuscular spectrum ejected from platinum is observed, we are concerned only with the normal photoelectric effect. Supposing that the X-ray absorption results could be extrapolated to the γ-ray region, it would then be possible to deduce the intensities of the γ-rays from the intensities of the corresponding electronic groups. It is, however, precisely this point which is doubtful, and the accuracy of this method is at present limited by the accuracy of the empirical formula which it was necessary to assume for the photoelectric method. The method, however, has one extremely important advantage, which is, that if a γ-ray is sufficiently intense to give a measurable corpuscular group, then the intensity of this group can be determined independently of neighbouring weak γ-rays. It will be seen that these two methods are really complementary, one supplying the deficiencies of the other. The γ-rays of radium B and radium C are the only ones that have yet been intensively investigated, but the results seem consistent, and we know not only the general distribution throughout the spectrum but also the individual intensities of all the strong γ-rays.

The results that have just been mentioned referred to the relative intensities of the γ-rays, and in the analogous case of X-rays or optical spectra this would be all that could be stated. However, in the case of the radioactive bodies it is possible to define and to deduce the absolute intensities. This depends upon the fact that the process of excitation is due to the disintegration of the atom. When a nucleus disintegrates, the departure of the disintegration particle, α or β, may leave the nucleus in an excited state, and its subsequent return to its normal state is the cause of the emission of the γ-rays. The γ-rays

应释放的反冲电子。此外，在膨胀瞬间，施加一个方向平行于膨胀室轴线的磁场，于是反冲电子轨迹的弯曲度取决于其速度。通过观测反冲电子发射轨迹的曲率和方向，可以将每个电子与确定频率的 γ 射线联系起来。通过对反冲电子轨迹的相对数进行统计研究，并根据散射的基本定律，可以推导出 γ 射线的相对强度。

由于各种实验原因，这种方法的分辨率不是很高，并且也不是总能将两个相邻 γ 射线的效应清楚地分开。然而，以下两个优点的价值在很大程度上掩盖了这个缺点：其一是能得到整个谱线结果中 γ 射线强度的确定分布，其二是这种方法检测弱 γ 射线与检测强 γ 射线一样有效。数据的解释需要用到散射定律的知识，但是有合理的理论基础和来自这些实验本身的证据，这两者的结合使得目前散射知识带来的不确定性变得无关紧要了。

埃利斯和阿斯顿[5]已将光电方法应用于测定 γ 射线的强度。放射性原子本身通过内转换释放的微粒谱显然不能用于与此相关联的问题，因为成组光电子的相对强度与尚不为人所知的内转换定律有关。然而，如果观测从铂中发射的微粒谱，我们可以只关心正常的光电效应。假定 X 射线的吸收结果可以外推到 γ 射线区，那么从相应的电子群强度推导 γ 射线的强度就将成为可能。然而，不可靠的恰恰是这一点，目前这种方法的准确度受限于在光电方法中必须假设的经验公式的准确度。然而，这种方法具有一个极其重要的优点，即如果 γ 射线的强度强到足以获得可测的微粒电子群，那么这组电子强度的测定可与邻近的弱 γ 射线无关。由此可见，这两种方法实际上是互补的，一种方法弥补了另一种方法的缺陷。虽然镭 B 和镭 C 的 γ 射线是迄今为止唯一得到深入研究的 γ 射线，但结果似乎是一致的，我们不仅知道了 γ 射线整个光谱的大致分布，而且知道了所有强 γ 射线各自的强度。

上面所提到的结果指的是 γ 射线的相对强度。如果是在与 X 射线或光学光谱类似的情况下，相对强度将是所能说明的全部情况。不过，在放射体的情况下，要确定和推导出绝对强度是可能的。这是因为激发过程是由于原子的衰变造成的。当原子核衰变时，会释放出衰变粒子（α 粒子或者 β 粒子），从而有可能使核处于激发态，随后它又回到正常态，并发射 γ 射线。因此，γ 射线仅在这类衰变之后发射，

are, therefore, emitted only after this disintegration, and it is possible to define the absolute intensity of a γ-ray as the average number of quanta emitted per disintegration. It follows that the absolute intensity of any γ-ray cannot be greater than unity. The simplest way of deducing these absolute intensities is to make use of the measurements of the total amount of energy emitted in the form of γ-rays. Knowing both the frequencies and the relative intensities of the γ-rays, it is easy to calculate the average number of quanta of each frequency emitted per disintegration. This further step has already been carried out for the γ-rays of radium B and radium C.

If we now review the information that we possess about the γ-rays of radium B and C and anticipate that which we shall no doubt in time possess about the rays of other bodies, it will be seen that on the whole it compares very favourably with that available about X-ray spectra. The accuracy of the wave-length determinations is certainly much lower, but we have this important information about the absolute intensities. For example, a prominent γ-ray of radium C has a wave-length of 20.2 X.U., which may be in error by one part in five hundred to even one part in three hundred, but on the other hand, we can say that a quantum of this radiation is emitted by the nucleus on the average twice in every three disintegrations.

Applications to the Structure of the Nucleus

The preceding account will have shown the extent to which the spectroscopy of the γ-rays has advanced. Its application to the problem of nuclear structure is only at the beginning, but it is already possible to indicate the possible lines of advance.

It has been realised for some time that there were many examples of combination differences between the frequencies of the γ-rays from any one body, and that this indicated, what was otherwise probable, that the γ-rays could be associated with a nuclear level system. Little progress, however, was made with this idea for several years, due to the realisation of the difficulty of associating such a level system with any specific part of the nucleus. In the nucleus there are α-particles, protons, and electrons, and in general any of these particles might be the emitters of the γ-rays. This question is still open, but there is now sufficient evidence to make it reasonable to try the hypothesis that the γ-rays are emitted by transitions of α-particles between stationary states in the nucleus.

The theories of Gamow and of Gurney and Condon[6] have shown that we may regard the process of emissions of an α-particle as due to the gradual leak of the wave function through a potential barrier. An extremely important result of this view is that the energy of the α-particle outside the atom, which can of course be measured, is the same as the energy of the α-particle in the stationary state in the nucleus which it occupied before the disintegration. For example, the α-particle from radium C is found to be emitted with an energy of 7.68 million volts. We therefore deduce that in the radium C nucleus there is an α-particle level with a positive energy of this amount. Such a level gives a natural basis on which to build the level system deducible from the γ-rays. We imagine that as a result of

可以将 γ 射线的绝对强度定义为每次衰变发射的平均量子数。由此可以断定，任何 γ 射线的绝对强度都不能大于 1 个单位。导出这些绝对强度的最简单的方法是利用对以 γ 射线形式发射的总能量的测量结果。知道了 γ 射线的频率和相对强度，就很容易计算出每次衰变时发射的对应于每个频率的平均量子数。人们已经对镭 B 和镭 C 的 γ 射线应用了这种深层次的研究方法。

如果现在回顾一下我们所拥有的关于镭 B 和镭 C 的 γ 射线的信息，并预期我们无疑将及时获得其他放射体 γ 射线的信息，就会发现，与从 X 射线光谱得到的可用信息相比，γ 射线的信息在总体上是非常令人满意的。测定波长的准确度确实会降低很多，但我们有了绝对强度的重要信息。例如，镭 C 的一种主要 γ 射线的波长为 20.2 X.U.，其误差可能为 1/500，甚至可以达到 1/300，但另一方面，我们也可以说，原子核在每三次衰变中平均辐射两次这样的量子。

在核结构研究中的应用

前面的叙述显示出 γ 射线光谱学已经发展到了怎样的程度。虽然将其应用于解决核结构问题仅仅是刚刚起步，但我们已经可以看到可能的发展方向。

之前人们就已经认识到，有许多例子显示来自任何一个放射体的 γ 射线在频率上存在多组差异，这表明 γ 射线可能与核能级系统相关，尽管别的解释也有可能。然而，多年以来，由于很难将这样一个核能级系统与核的任何特定部分联系起来，因而人们在这一构想上没有取得什么进展。在核中，存在 α 粒子、质子和电子，一般来说这些粒子中的任何一个都有可能发射 γ 射线，这个问题仍然是一个尚未解决的问题。但是现在有足够的证据说明这样去假设是合理的，即认为 γ 射线是因核内 α 粒子在定态之间的跃迁而发射的。

伽莫夫以及格尼和康登的理论[6]指出，我们可以认为 α 粒子的发射过程是由于波函数通过势垒的逐渐泄漏。由这个观点可以得到一个极为重要的结果，即原子外面 α 粒子的能量（当然可以测量）与核衰变前在核中处于定态的 α 粒子的能量相等。例如，我们发现镭 C 发射的 α 粒子能量是 7.68 百万电子伏特。于是可以推出，在镭 C 的核内，存在一个具有 7.68 百万电子伏特正能量的 α 粒子能级。这一能级给出了建立根据 γ 射线推导的能级系统的自然基础。我们可以设想，某些内核排列激发 α 粒子，使其到达某一个更高的激发态，然后通过发射频率相当于能量差的 γ

some internal nuclear arrangement an α-particle is excited to one of certain higher states, and that from these states it arrives at the ground state by emitting γ-rays of frequencies corresponding to the energy differences. It now follows, however, that if an α-particle can leak out through the potential barrier from the ground level, it can do so still more easily from the excited levels. We should therefore expect to find a certain number of high-speed α-particles corresponding to these modes of disintegration.

The existence of such long-range α-particles has of course been known for a long time, and in fact many tentative suggestions have been put forward associating the energy differences of the groups of α-particles with the frequencies of the γ-rays. The present-day point of view, however, goes much further than this, since it predicts definite relations between the intensities of the γ-rays and the number of long-range particles. That such a relation must exist can be easily seen in the following way. Suppose that on the average out of every thousand disintegrations there are n cases where an α-particle is excited to a certain state, the rate of leak through the potential barrier is given to a fair approximation by theory, and the probability of the nuclear transition can at least be estimated. We are therefore able in terms involving only the unknown quantity n to write down the number of long-range α-particles we should expect and the number of quanta of radiation. Both these quantities can also be measured, perhaps not with a very high accuracy, but yet sufficient to see whether there is an agreement with theory or not.

This is really a stringent test for the theory, because although the theories of the probabilities of nuclear transitions are necessarily tentative, any adjustment which proved necessary for one γ-ray must also apply to all the others. By arguments of this type Fowler[7] has been led to associate one excited α-particle level of the radium C nucleus with the corresponding nuclear transition formed from the β-ray spectrum. It seems likely that this line of investigation will lead to definite and valuable results. It is of course quite probable that several nuclear transitions will not be able to be associated with long-range α-particles, but it would then be possible to draw the important conclusion that these transitions were due to protons or α-particles of small positive or of negative energy.

Internal Conversion

Reference was made above to internal conversion and it was pointed out that groups of electrons are ejected from the K, L, M states of radioactive atoms with just those energies that they would have if radiation were emitted from the nucleus but was absorbed photoelectrically before it escaped. It has been frequently pointed out that there was no need and, in fact, no justification to assume that in this case the radiation was ever actually emitted at all[8]. All that could be truly inferred from the experimental results was that an excited nucleus could either emit its excess energy as radiation or had some means of transferring this energy to the electronic structure of the atom.

On the old quantum mechanics, it was difficult to imagine any method other than that of radiation transfer, but the wave mechanics suggests that there is a far more intimate

射线又从激发态回到基态。然而，如果一个 α 粒子在从基态能级穿过势垒时会泄漏，那么它就更容易从激发态能级泄漏。因此我可以预期将发现有一定数量的高速 α 粒子与这些衰变模式相对应。

诚然，很久以前我们就知道这类长程 α 粒子是存在的，事实上我们已提出过很多将 α 粒子群的能量差与 γ 射线的频率联系起来的初步建议。然而，目前的观点比这已更进了一步，因为它预测了 γ 射线强度与长程 α 粒子数量之间的确定关系。通过以下方式很容易看出这种确定关系是必然存在的：假定平均每千次衰变有 n 个 α 粒子被激发到某个激发态上，根据理论可以非常好地近似得到穿出势垒的泄漏率，因而至少可以估计出核跃迁的概率。因此我们能够用只包含未知数 n 的项写出我们所预期的长程 α 粒子的数量及 γ 辐射的量子数。这两个量都可以通过测量得到，也许准确度不是很高，但已足以了解它与理论是否一致。

这实际上是对理论的严格检验，因为虽然核跃迁概率理论目前还不够完备，但是，被一种 γ 射线证明是必要的调整也应适用于所有其他的 γ 射线。这种观点引导福勒[7]将镭 C 核 α 粒子的一个激发态能级与 β 射线谱形成的相应核跃迁联系起来。看起来这类研究路线有可能会导致确定和有价值的结果。当然完全有可能，一些核跃迁与长程 α 粒子不相联系，但由此可能得出如下重要结论：这些核跃迁是由具有微正能量或负能量的质子或 α 粒子产生的。

内转换

上面已经提到了内转换，并已指出从放射性原子的 K、L、M 态发射出来的电子群所具有的能量刚好等于如果这些从核中发出的辐射量子在逃逸前被光电吸收的能量。已屡次指出，假定在内转换情况下 γ 光子会完全发射出来是没有必要的，也是不正确的[8]。根据实验结果能够真正推出来的是，一个激发核或者会以辐射形式发射其过剩的能量，或者会以一些方式将这类能量传递给原子的电子结构。

按照旧的量子论，很难想象除了辐射转换之外还有其他方法，但波动力学提出，在核粒子与电子结构之间存在着更为密切的联系。核内粒子的波函数将在一定

connexion between the nuclear particles and the electronic structure. The wave functions of the particles in the nucleus will extend out to a certain extent into the electronic region of the atom, and conversely the electronic wave functions will exist throughout the nucleus. As a model, we may think that every electron in the atom occasionally passes right through the nucleus, and that a nuclear particle might sometimes for a very short time be found to be actually outside the nucleus.

We have thus no difficulty in seeing, in a general way, how the nuclear energy might be transferred to the electronic system by a direct collision process. Which process, radiation or collision, is predominant can only be settled by experiment, and the answer given by experiment in this case is fortunately unambiguous. The measurements of Ellis and Aston[5] of the extent of this internal conversion and of the way in which it depends on the frequency of the associated radiation show clearly that the behaviour is incompatible with the radiation hypothesis, and we are thus led to conclude that the collision process is the most important. It will be seen that this process is really a collision of the second kind, between an electron and an excited nucleus.

The peculiar interest of this phenomenon lies in the fact that it represents an easily measurable example of direct interaction between the nucleus and the electronic system. There are several other cases where the interaction between the nucleus and the electronic system must be taken into account, but only in order to give the finer details. The importance of the phenomenon of internal conversion is that the entire phenomenon, even to its first approximation, depends upon interaction, and that no approach can be made to it with a simple point nucleus.

However, quite apart from the intrinsic interest of this interaction, the phenomenon of internal conversion seems likely to provide valuable information about the stationary states in the nucleus. The quantity that can actually be measured, the internal conversion coefficient, is the ratio of the probabilities of occurrence of this collision of the second kind and of the nuclear radiation transition. The latter is determined mainly by the energy difference of the initial and final states, whilst the absolute energies are involved in the former. In a general way it can be seen that the internal conversion should lead to a classification of the levels responsible for the γ-rays, or, in other words, should enable the γ-rays to be associated with a definite part of the nucleus.

While but little has yet been accomplished along these various lines of investigation of the nuclear levels, it is certainly true that the most difficult step has already been made. The problem can now be clearly envisaged, and definite lines of work proposed which seem likely to lead to results. The way appears open to an experimental investigation of certain radioactive nuclei, and to an interpretation of the experimental results in terms of nuclear phenomena.

(**127**, 275-278; 1931)

程度上延伸到原子的电子区域，反过来，电子的波函数也将存在于整个核内。作为模型，我们可以认为，原子中的每一个电子都偶尔会穿过核，核粒子实际上有时也可能在核外出现很短的时间。

因此我们不难看到，在一般情况下，核能量可通过直接碰撞过程转移到电子系统。辐射还是碰撞，这两种过程哪一个更占优势只能通过实验来确定。幸运的是，在这种情况下由实验给出的回答是明确的。埃利斯和阿斯顿[5]对内转换的程度以及它对相关辐射频率的依赖方式进行了测量，结果明确显示出与辐射假说的不相容。因此，我们得出的结论是碰撞过程才是最重要的。我们将会看到这个过程实际上是电子与激发核的第二类碰撞。

这种现象之所以具有特殊意义是因为它代表了核与电子系统之间发生直接相互作用的一个易于测量的例子。还有其他几种情况必须考虑核与电子系统之间的相互作用，但只是为了给出更精细的细节。内转换现象的重要性是整个现象（即使在一级近似下）依赖于核与电子的相互作用，哪种方法都不可能仅仅把原子核看成点粒子。

然而，除了对这类相互作用本身的意义之外，内转换现象似乎还能提供关于核内定态的重要信息。实际上能被测量的量是内转换系数，它是第二类碰撞与核辐射跃迁出现概率的比值。后者主要由初态和终态的能量差决定，而前者涉及绝对能量。就通常意义而论，内转换将导致引起 γ 射线发射的能级分类，换言之，它能将 γ 射线与核的某个确定部分联系起来。

尽管人们以各种方式对核能级的研究尚未取得明显的进展，不过，可以肯定的是我们已迈出了最困难的一步。现在可以清晰地设想问题，并且可以提出有可能得到结果的明确工作路线。这种方式为用实验研究某些放射性核以及根据核现象解释实验结果提供了可能。

（沈乃澂 翻译；尚仁成 审稿）

References:

1. Rutherford and Andrade, *Phil. Mag.*, **27**, 854; **28**, 262 (1924). Thibaud, Thèse, Paris (1925). Frilly, Thèse, Paris (1928). Meitner, *Zeit. f. Physik*, **52**, 645 (1928).
2. Steadman, *Phys. Rev.*, **36**, 460 (1930).
3. Ellis, *Proc. Roy. Soc.*, A, **101**, 1 (1922). Thibaud, Thèse, Paris (1925).
4. Skobeltzyn, *Zeit. f. Physik.*, **43**, 354 (1927): **58**, 595 (1929).
5. Ellis and Aston, *Proc. Roy. Soc.*, A, **129**, 180 (1930).
6. Gamow, *Zeit. f. Physik*, **51**, 204 (1928). Gurney and Condon, *Nature*, **122**, 439 (1928).
7. Fowler, *Proc. Roy. Soc.*, A, **129**, 1 (1930).
8. Smekal, *Zeit. f. Physik.*, **10**, 275 (1922). *Ann. d. Phys.*, **81**, 399 (1926). Rosseland, *Zeit. f. Physik*, **14**, 173 (1923).

The End of the World: from the Standpoint of Mathematical Physics*

A. S. Eddington

Editor's Note

This supplement contains a somewhat light-hearted address on "the end of the world" that Arthur Eddington delivered at the Mathematical Association. One must first ask "which end?", he says. Space itself may have no end: current cosmology suggested a universe shaped like the surface of a sphere, finite but without edges. As for time, the second law of thermodynamics suggests that the entire universe will eventually reach a state of thermodynamic equilibrium marked by complete disorganisation. However, if time is infinite then every conceivable fluctuation in the universe's particles will happen, temporarily disturbing this equilibrium. Eddington ends by predicting how the world will really end: as a ball of radiation growing ever larger, roughly doubling its size every 1,500 million years.

THE world—or space-time—is a four-dimensional continuum, and consequently offers a choice of a great many directions in which we might start off to look for an end; and it is by no means easy to describe "from the standpoint of mathematical physics" the direction in which I intend to go. I have therefore to examine at some length the preliminary question, Which end?

Spherical Space

We no longer look for an end to the world in its space dimensions. We have reason to believe that so far as its space dimensions are concerned the world is of spherical type. If we proceed in any direction in space we do not come to an end of space, nor do we continue on to infinity; but, after travelling a certain distance (not inconceivably great), we find ourselves back at our starting-point, having "gone round the world". A continuum with this property is said to be finite but unbounded. The surface of a sphere is an example of a finite but unbounded two-dimensional continuum; our actual three-dimensional space is believed to have the same kind of connectivity, but naturally the extra dimension makes it more difficult to picture. If we attempt to picture spherical space, we have to keep in mind that it is the *surface* of the sphere that is the analogue of our three-dimensional space; the inside and the outside of the sphere are fictitious elements in the picture which have no analogue in the actual world.

We have recently learnt, mainly through the work of Prof. Lemaître, that this spherical space is expanding rather rapidly. In fact, if we wish to travel round the world and get

* Presidential address to the Mathematical Association, delivered on Jan. 5.

以数学物理的视角看宇宙的终点*

阿瑟·爱丁顿

编者按

本文的内容是阿瑟·爱丁顿在数学协会上发表的一篇关于“宇宙的终点”的非正式讲话。他说，人们肯定首先要问“宇宙的终点是时间方向上的终点还是空间方向上的终点？”空间本身也许不存在终点：现代宇宙学认为宇宙类似于一个球体的表面，有限但无界。至于时间，热力学第二定律预言整个宇宙最终将达到一个以完全无序为标志的热平衡状态。然而，如果时间是无限的，那么宇宙粒子的每一次可能的涨落都会短暂地打破这种平衡。爱丁顿在结束语中预言了宇宙终结的实际方式：宇宙作为一个充满辐射的球不断地变大，每15亿年其大小约膨胀一倍。

宇宙或者说时空是一个四维连续区，因此我们可以从很多不同的角度来讨论它的终点。毫无疑问“以数学物理的视角”来描述以下我试图展开讨论的内容绝非易事。所以我必须相当仔细地考虑这个最初的问题：宇宙的终点是时间方向上的终点还是空间方向上的终点？

球面空间

我们不再寻找宇宙在空间方向上的终点。因为我们有理由相信，宇宙的空间部分具有球面结构。如果朝着空间中的任何一个方向一直走下去，我们不会到达空间的终点，也不会走到无穷远处；但是，当我们走了一段距离（并不是无法想象的远）以后，我们发现自己又回到了原来的出发点，相当于“绕了宇宙一圈”。我们把具有这种特性的连续区说成是有限但无界。球的表面即为有限无界二维连续区的一个例子；我们生活的三维空间被认为具有同样的连通性，但是由于比二维空间多出了一个空间维度，这使得我们很难用图形把这样的三维连续体表示出来。如果我们试图画出一个三维球面空间的话，我们必须要记住，只是这个三维球**面**对应于我们的三维空间；而这个球面的里面和外面都是我们虚构出来的，因而并不与现实的宇宙相对应。

最近，主要通过勒迈特教授的工作，我们已经获知，我们生活的这个三维球面空间正在快速地膨胀着。事实上，如果我们想环绕宇宙空间一周而回到出发点的

* 这篇文章来自爱丁顿于1月5日在数学协会上发表的讲话。

back to our starting-point, we shall have to move faster than light; because, whilst we are loitering on the way, the track ahead of us is lengthening. It is like trying to run a race in which the finishing-tape is moving ahead faster than the runners. We can picture the stars and galaxies as embedded in the surface of a rubber balloon which is being steadily inflated; so that, apart from their individual motions and the effects of their ordinary gravitational attraction on one another, celestial objects are becoming farther and farther apart simply by the inflation. It is probable that the spiral nebulae are so distant that they are very little affected by mutual gravitation and exhibit the inflation effect in its pure form. It has been known for some years that they are scattering apart rather rapidly, and we accept their measured rate of recession as a determination of the rate of expansion of the world.

From the astronomical data it appears that the original radius of space was 1,200 million light years. Remembering that distances of celestial objects up to several million light years have actually been measured, that does not seem overwhelmingly great. At that radius the mutual attraction of the matter in the world was just sufficient to hold it together and check the tendency to expand. But this equilibrium was unstable. An expansion began, slow at first; but the more widely the matter was scattered the less able was the mutual gravitation to check the expansion. We do not know the radius of space today, but I should estimate that it is not less than ten times the original radius.

At present our numerical results depend on astronomical observations of the speed of scattering apart of the spiral nebulae. But I believe that theory is well on the way to obtaining the same results independently of astronomical observation. Out of the recession of the spiral nebulae we can determine not only the original radius of the universe but also the total mass of the universe, and hence the total number of protons in the world. I find this number to be either 7×10^{78} or 14×10^{78}.* I believe that this number is very closely connected with the ratio of the electrostatic and the gravitational units of force, and, apart from a numerical coefficient, is equal to the square of the ratio. If F is the ratio of the electrical attraction between a proton and electron to their gravitational attraction, we find $F^2 = 5.3\times10^{78}$. There are theoretical reasons for believing that the total number of particles in the world is αF^2, where α is a simple geometrical factor (perhaps involving π). It ought to be possible before long to find a theoretical value of α, and so make a complete connexion between the observed rate of expansion of the universe and the ratio of electrical and gravitational forces.

Signposts for Time

I must not dally over space any longer but must turn to time. The world is closed in its space dimensions but is open in both directions in its time dimension. Proceeding from "here" in any direction in space we ultimately come back to "here"; but proceeding from

* This ambiguity is inseparable from the operation of counting the number of particles in finite but unbounded space. It is impossible to tell whether the protons have been counted once or twice over.

话，我们的行进速度必须要比光速还快；这是因为我们在路上行进的同时，我们前面的路途也在变长。就好像参加了一场终点线向前移动得比运动员跑步速度还快的赛跑一样。想象一下恒星和星系都镶嵌在一个橡胶气球的表面，而这个气球正在持续不断地膨胀着；这样，如果不考虑恒星和星系各自的运动以及它们之间的万有引力，这些天体之间的距离将由于宇宙空间的膨胀而变得越来越大。由于旋涡星云彼此相距很远，以至于它们几乎不会受到相互作用的引力影响，因而它们的形状很可能就体现了宇宙膨胀的效果。几年前我们就已经知道它们正在迅速地相互分散远离，我们可以通过测量它们之间的退行速度来推算宇宙的膨胀速度。

天文学数据表明宇宙空间最初的半径大约为 12 亿光年。要知道，远至几百万光年的天体实际上已经被我们观测到了，所以 12 亿光年也并非是不可想象的距离。在那种尺度下，宇宙中物质间相互作用的引力刚好可以把物质聚合在一起与宇宙空间的膨胀趋势相平衡。但是这种平衡并不稳定。于是宇宙开始膨胀，最初膨胀得比较慢；但随着物质之间的距离越来越远，它们之间的引力也越来越小，以至于越来越没有办法抑制宇宙的膨胀。我们不知道宇宙现在的半径，但我估计当今宇宙的半径应该至少是最初半径的 10 倍。

现在，我们的计算结果依赖于旋涡星云间分离速度的天文观测数据。但是我相信，我们也能从理论上顺利地得出不依赖天文观测数据的相同计算结果。从旋涡星云的退行速度我们不但可以推算出宇宙最初的尺度，还可以估算出宇宙总的质量，进而得到宇宙中总的质子数。我得出的宇宙中的总质子数为 7×10^{78} 或 14×10^{78}。* 我相信这个数值与静电力和引力的比值有密切的关系，并且只与这个比值的平方相差一个常数因子。如果 F 是质子和电子之间的静电力与万有引力的比值，我们得到 $F^2 = 5.3 \times 10^{78}$。理论上我们有理由相信宇宙中的粒子总数是 αF^2，其中 α 是一个简单的几何因子（也许和 π 有关）。可能不久我们就能得到 α 的理论值，从而给出宇宙膨胀速度的观测值同静电力与引力的比值之间的完整关系。

时间的指示牌

我不能继续讨论跟空间有关的问题，而必须开始讨论与时间有关的内容了。宇宙在空间上是闭合的，但是在时间尺度的两个方向上却都是开放的。从“这里”朝

* 这个结果的不确定性与在有限无界的空间中计算粒子数有不可分割的联系。因为我们无法分辨这些质子到底被计算过一次还是两次。

"now" towards the future or the past we shall never come across "now" again. There is no bending round of time to bring us back to the moment we started from. In mathematics this difference is provided for by the symbol $\sqrt{-1}$, just as the same symbol crops up in distinguishing a closed ellipse and an open hyperbola.

If, then, we are looking for an end of the world—or, instead of an end, an indefinite continuation for ever and ever—we must start off in one of the two time directions. How shall we decide which of these two directions to take? It is an important question. Imagine yourself in some unfamiliar part of space-time so as not to be biased by conventional landmarks or traditional standards of reference. There ought to be a signpost with one arm marked "To the future" and the other arm marked "To the past". My first business is to find this signpost, for if I make a mistake and go the wrong way I shall lead you to what is no doubt an "end of the world", but it will be that end which is more usually described as the *beginning*.

In ordinary life the signpost is provided by consciousness. Or perhaps it would be truer to say that consciousness does not bother about signposts; but wherever it finds itself it goes off on urgent business in a particular direction, and the physicist meekly accepts its lead and labels the course it takes "To the future". It is an important question whether consciousness in selecting its direction is guided by anything in the physical world. If it is guided, we ought to be able to find directly what it is in the physical world which makes it a one-way street for conscious beings. The view is sometimes held that the "going on of time" does not exist in the physical world at all and is a purely subjective impression. According to that view, the difference between past and future in the material universe has no more significance than the difference between right and left. The fact that experience presents space-time as a cinematograph film which is always unrolled in a particular direction is not a property or peculiarity of the film (that is, the physical world) but of the way it is inserted into the cinematograph (that is, consciousness). In fact, the one-way traffic in time arises from the way our material bodies are geared on to our consciousness:

> "Nature has made our gears in such a way
> That we can never get into reverse".

If this view is right, "the going on of time" should be dropped out of our picture of the physical universe. Just as we have dropped the old geocentric outlook and other idiosyncrasies of our circumstances as observers, so we must drop the dynamic presentation of events which is no part of the universe itself but is introduced in our peculiar mode of apprehending it. In particular, we must be careful not to treat a past-to-future presentation of events as truer or more significant than a future-to-past presentation. We must, of course, drop the theory of evolution, or at least set alongside it a theory of anti-evolution as equally significant.

着空间中的任意方向出发，最后我们都会回到“这里”；但是从“现在”出发，无论是向着未来还是向着过去前进，我们将永远不可能再回到“现在”。在这里并没有弯曲的时间回路能把我们带回到原来出发的时刻。从数学上说这种差别是由记号 $\sqrt{-1}$ 造成的，正如闭合椭圆和开放双曲线之间的区别也是由这个记号造成的一样。

于是，如果我们要找寻宇宙的终点——其实并不是终点，而是一种永无休止的持续，那么我们就必须从时间的两个方向中选一个作为开始的方向。但是我们如何决定应该选取哪一个时间方向呢？这是个很重要的问题。想象一下你处在时空中的某个陌生的部分，在那里你不会受到常规标志或传统参考标准的影响。但是那里应该有一个时间的指示牌，一边写着“通向未来”，而另一边写着“通向过去”。我要做的第一件事情就是要先找出这个指示牌，否则一旦我弄错了方向，那么无疑，我还是会把你们带到“宇宙的终点”，不过这个“终点”更通常地是被描述为宇宙的**起点**。

在日常生活中，我们的意识充当了时间的指示牌。或者更准确地说，我们的意识根本不关心所谓的时间指示牌；但是无论什么地方，在紧急情况下意识总是朝着某一个特定的方向行进思考，而物理学家也温和地接受了意识的引导并把它的行进方向标记为“通向未来”的方向。这里有一个重要的问题，意识在选择它的行进方向时是否也受到物理世界中的某种事物的影响？如果确实如此，那么我们应该能够直接找出究竟是物理世界中的什么因素使得有意识的人类认为时间是沿单向车道行进的。有观点认为“时间的流逝”只是一种主观的想法而不是物理世界中真实存在的现象。按照这个观点，物质世界中过去和未来的区别只不过相当于左和右之间的区别而已。事实上我们的经验显示时空就像一组电影胶片，它总是按照某种特定的方向放映，但这并不是由胶片（物理时空）本身的性质或特点所决定的，而是与胶片插入放映机的方式（意识）有关。这种观点认为，把事物储存到我们意识中的方式使我们认为时间具有单向性：

> “大自然给我们配备了这样的意识思维方式
> 使得我们永远不能倒退”。

如果以上的观点是对的，那么在我们关于物理世界的图像中就应该抛弃“时间的流逝”这种观点。就像我们放弃旧的时空观以及其他对周围事物观察得到的个人看法一样，我们必须抛弃事物是动态呈现的这种观念，因为这并不是世界本身的性质而只是我们人类在认识物理世界时引入的一种人类特有的理解方式。我们要特别注意的是，如果真是这样，那么我们就不能先入为主地认为“过去到未来”的呈现比“未来到过去”的呈现更正确或更重要。当然，我们也必须放弃进化论，或者至少要同时建立一种具有同样重要性的反进化论。

If anyone holds this view, I have no argument to bring against him. I can only say to him, "You are a teacher whose duty it is to inculcate in youthful minds a true and balanced outlook. But you teach (or without protest allow your colleagues to teach) the utterly one-sided doctrine of evolution. You teach it not as a colourless schedule of facts but as though there were something significant, perhaps even morally inspiring, in the progress from formless chaos to perfected adaptation. This is dishonest; you should also treat it from the equally significant point of view of anti-evolution and discourse on the progress from future to past. Show how from the diverse forms of life existing today Nature anti-evolved forms which were more and more unfitted to survive, until she reached the sublime crudity of the palaeozoic forms. Show how from the solar system Nature anti-evolved a chaotic nebula. Show how, in the course of progress from future to past, Nature took a universe which, with all its faults, is not such a bad effort of architecture and—in short, made a hash of it."

Entropy and Disorganisation

Leaving aside the guidance of consciousness, we have found it possible to discover a kind of signpost for time in the physical world. The signpost is of rather a curious character, and I would scarcely venture to say that the discovery of the signpost amounts to the same thing as the discovery of an objective "going on of time" in the universe. But at any rate it serves to discriminate past and future, whereas there is no corresponding objective distinction of left and right. The distinction is provided by a certain measurable quantity called entropy. Take an isolated system and measure its entropy S at two instants t_1 and t_2. We want to know whether t_1 is earlier or later than t_2 without employing the intuition of consciousness, which is too disreputable a witness to trust in mathematical physics. The rule is that the instant which corresponds to the greater entropy is the later. In mathematical form

$$dS/dt \text{ is always positive.}$$

This is the famous second law of thermodynamics.

Entropy is a very peculiar conception, quite unlike the conceptions ordinarily employed in the classical scheme of physics. We may most conveniently describe it as the measure of disorganisation of a system. Accordingly, our signpost for time resolves itself into the law that disorganisation increases from past to future. It is one of the most curious features of the development of physics that the entropy outlook grew up quietly alongside the ordinary analytical outlook for a great many years. Until recently it always "played second fiddle"; it was convenient for getting practical results, but it did not pretend to convey the most penetrating insight. But now it is making a bid for supremacy, and I think there is little doubt that it will ultimately drive out its rival.

There are some important points to emphasise. First, there is no other independent signpost for time; so that if we discredit or "explain away" this property of entropy, the distinction of past and future in the physical world will disappear altogether. Secondly,

如果有谁坚持时间的行进具有单向性这种观点，我没有任何办法说他是错的。我只能对他说：“你是一个致力于把一个真实均衡的世界观灌输给年青学生的老师。但是你所教授的（或没有阻止你的同事去教授的）是完全片面的进化论学说。你并没有告诉学生进化论只是单纯事实的罗列，而是让他们认为似乎从混乱无形到完美有序的过程才是重要的，才真正会是鼓舞人心的。你这样做其实是不诚实的；你应该把反进化论也放在同样重要的位置并讲述从未来到过去的演化过程。你应该向他们展示：现今自然界存在的丰富多彩的生命形态是如何反进化到越来越不适合生存，直到古生代最原始的形态的；自然界是如何从太阳系反进化成混沌的星云；以及，随着从未来到过去的时间进程，自然界是如何选择了一个充满问题的宇宙，而这些问题并不是宇宙本身的构造不好，总而言之，最初的宇宙是一团糨糊。”

熵和无序度

撇开意识的引导性不谈，我们已经发现在物理世界里有可能找到一种时间的指示牌。这个指示牌具有非常奇特的性质，当然，我还不敢说这个指示牌的发现与当年在宇宙中发现客观的“时间流逝”有一样的重要性。但是不管怎样，这个指示牌可以用来区别过去和未来，而空间上的左和右却没有相应的客观区分。我们可以用某种称为熵的可测量量来标志这种时间方向上的差别。假设有一个孤立的系统，我们在两个不同的时刻 t_1 和 t_2 分别测量它的熵 S。因为意识这个概念在数学物理领域里并不是可靠的证据，所以我们想在不靠意识直觉的情况下知道 t_1 究竟比 t_2 早还是晚。这里用到的规则是：熵越大，时间越晚。在数学形式中为：

dS/dt 总是大于零。

这就是著名的热力学第二定律。

熵是个很特殊的概念，与经典物理学常用的概念很不一样。我们可以把熵简单地描述为：熵是衡量系统无序性的物理量。于是，根据熵这个时间的指示牌，我们得到了这样的规律：从过去到未来，无序性是增加的。物理学发展的最奇妙的特点之一就是：多年来，与其他常规的分析方法一样，我们对于熵的认知一直都在悄悄地发展着。直到最近，熵在科学研究中仍只是“居于次要位置”。由熵可以很方便地得到实用的结果，但并不能由此表明人们对熵有了深刻的洞察力。不过现在熵正在谋求更大的发展，并且我认为，毫无疑问，它最终会超过它的对手。

这里还需要强调几个重要的地方。首先，世界上没有其他独立的时间指示牌；所以如果我们不信任或者“通过辩解来消除”熵作为时间指示牌的这个性质，那么物理学关于过去和未来的差别也会随之消失。第二，检验熵的实验结果应该都是一

the test works consistently; isolated systems in different parts of the universe agree in giving the same direction of time. Thirdly, in applying the test we must make certain that our system is strictly isolated. Evolution teaches us that more and more highly organised systems develop as time goes on; but this does not contradict the conclusion that on the whole there is a loss of organisation. It is partly a question of definition of organisation; from the evolutionary point of view it is quality rather than quantity of organisation that is noticed. But, in any case, the high organisation of these systems is obtained by draining organisation from other systems with which they come in contact. A human being as he grows from past to future becomes more and more highly organised—at least, he fondly imagines so. But if we make an isolated system of him, that is to say, if we cut off his supply of food and drink and air, he speedily attains a state which everyone would recognise as "a state of disorganisation".

It is possible for the disorganisation of a system to become complete. The state then reached is called thermodynamic equilibrium. The entropy can increase no further, and, since the second law of thermodynamics forbids a decrease, it remains constant. Our signpost for time disappears; and so far as that system is concerned, time ceases to go on. That does not mean that time ceases to exist; it exists and extends just as space exists and extends, but there is no longer any one-way property. It is like a one-way street on which there is never any traffic.

Let us return to our signpost. Ahead there is ever-increasing disorganisation. Although the sum total of organisation is diminishing, certain parts of the universe are exhibiting a more and more highly specialised organisation; that is the phenomenon of evolution. But ultimately this must be swallowed up in the advancing tide of chance and chaos, and the whole universe will reach a state of complete disorganisation—a uniform featureless mass in thermodynamic equilibrium. This is the end of the world. Time will *extend* on and on, presumably to infinity. But there will be no definable sense in which it can be said to *go* on. Consciousness will obviously have disappeared from the physical world before thermodynamical equilibrium is reached, and dS/dt having vanished, there will remain nothing to point out a direction in time.

The Beginning of Time

It is more interesting to look in the opposite direction—towards the past. Following time backwards, we find more and more organisation in the world. If we are not stopped earlier, we must come to a time when the matter and energy of the world had the maximum possible organisation. To go back further is impossible. We have come to an abrupt end of space-time—only we generally call it the "beginning".

I have no "philosophical axe to grind" in this discussion. Philosophically, the notion of a beginning of the present order of Nature is repugnant to me. I am simply stating the

致的；宇宙中任何地方的孤立系统都应该给出相同的时间方向。第三，在进行这类实验时，我们必须保证我们的系统是严格孤立的系统。进化论告诉我们，随着时间的推移，越来越高的有序系统产生了；但是这与整体上有序性的减少并不矛盾。这在一定程度上是有序性的定义问题；从进化的角度看，需要注意的是有序性的质而非量。但无论如何，有序性很高的系统是通过吸取其他与之接触的系统的有序性来实现的。一个人从过去到未来总是变得越来越有序，至少他自己愿意这么认为。但是如果我们使他成为一个孤立的系统，即切断他的饮食和空气供应，那么他很快就会达到一种大家都认同的“无序状态”。

系统可能达到完全的无序状态，我们称之为热力学平衡态。此时，熵不能再继续增加了，而热力学第二定律又不允许它减少，于是熵就只能保持为一个常数。这时，我们的时间指示牌也消失了，因而对于这样的一个系统，时间变得固定不动了。但这并不表示时间不存在；时间仍然像空间一样存在并延续着，但是它已经不再具有任何单向性了。或者说，单行道仍然存在，但不再有汽车在上面行驶了。

让我们回到时间指示牌的讨论上来。通过前面的讨论我们已经知道，无序性是不断增加的。虽然总的有序性正在减少，但是宇宙中的某些部分却展示出越来越高的特殊有序性；这就是进化现象。但是这些有序的系统最终会被不断增加的机遇与混沌所吞没，然后整个宇宙会达到一种完全无序的状态，即变成热力学平衡态下的一堆毫无特征的均匀物质。这就是宇宙的终点。时间仍然会永远地**延续**下去，可能没有尽头。但是时间的这种**延续**没有了明确的意义。很明显，在热力学平衡态即将到来之前，物理世界中已经不存在意识了，dS/dt 也成为零，再也没有一种指示牌能告诉我们，哪个时间方向通向未来，而哪个又是通向过去的。

时间的起点

如果我们朝着“时间流逝”相反的方向，即朝着过去的方向看，我们将会发现更加有趣的现象。当我们沿着“时间流逝”的方向往回走时，会发现宇宙的有序性越来越大。如果我们一直不停地走下去，就会到达一个物质和能量的有序度都为允许的最高限的时刻。这时，再想进一步走下去已经不可能了。我们已经来到了时空戛然而止的终点——只不过我们通常称之为“起点”。

在这里的讨论中我没有“涉足哲学”的意思。我并不认同哲学中那些与现在自然界秩序的起点有关的概念。我只是想说明物理规律中现有的基本概念使我们陷入

dilemma to which our present fundamental conception of physical law leads us. I see no way round it; but whether future developments of science will find an escape I cannot predict. The dilemma is this:—Surveying our surroundings, we find them to be far from a "fortuitous concourse of atoms". The picture of the world, as drawn in existing physical theories, shows arrangement of the individual elements for which the odds are multillions to 1 against an origin by chance. Some people would like to call this non-random feature of the world purpose or design; but I will call it non-committally anti-chance. We are unwilling to admit in physics that anti-chance plays any part in the reactions between the systems of billions of atoms and quanta that we study; and indeed all our experimental evidence goes to show that these are governed by the laws of chance. Accordingly, we sweep anti-chance out of the laws of physics—out of the differential equations. Naturally, therefore, it reappears in the boundary conditions, for it must be got into the scheme somewhere. By sweeping it far enough away from the sphere of our current physical problems, we fancy we have got rid of it. It is only when some of us are so misguided as to try to get back billions of years into the past that we find the sweepings all piled up like a high wall and forming a boundary—a beginning of time—which we cannot climb over.

A way out of the dilemma has been proposed which seems to have found favour with a number of scientific workers. I oppose it because I think it is untenable, not because of any desire to retain the present dilemma. I should like to find a genuine loophole. But that does not alter my conviction that the loophole that is at present being advocated is a blind alley. I must first deal with a minor criticism.

I have sometimes been taken to task for not sufficiently emphasising in my discussion of these problems that the results about entropy are a matter of probability, not of certainty. I said above that if we observe a system at two instants, the instant corresponding to the greater entropy will be the later. Strictly speaking, I ought to have said that for a smallish system the chances are, say, 10^{20} to 1, that it is the later. Some critics seem to have been shocked at my lax morality in making such a statement, when I was well aware of the 1 in 10^{20} chance of its being wrong. Let me make a confession. I have in the past twenty-five years written a good many papers and books, broadcasting a large number of statements about the physical world. I fear that for not many of these statements is the risk of error so small as 1 in 10^{20}. Except in the domain of pure mathematics, the trustworthiness of my conclusions is usually to be rated at nearer 10 to 1 than 10^{20} to 1; even that may be unduly boastful. I do not think it would be for the benefit of the world that no statement should be allowed to be made if there were a 1 in 10^{20} chance of its being untrue; conversation would languish somewhat. The only persons entitled to open their mouths would presumably be the pure mathematicians.

Fluctuations

The loophole to which I referred depends on the occurrence of chance fluctuations. If we have a number of particles moving about at random, they will in the course of time go through every possible configuration, so that even the most orderly, the most non-

了困境。我没有办法解决这个矛盾；也不能预料未来科学的发展能否避开这个矛盾。这个矛盾是这样的：考虑我们周围的物质，我们发现它们远不是“原子的偶然集合”。现有的物理理论所描述的物理世界图像表明，这些单个元素的分布情况是偶然出现的概率仅仅是 10 的 10 次方的 10 次方分之一。有人喜欢把这种现象称之为宇宙意图或设计的非随机特性；但是，我要把它称为不受约束的反概率性。在物理学领域内，我们不愿意承认反概率性在我们所研究的包含数以十亿的原子和量子系统的相互作用中发挥了作用；而且实际上我们所有的实验结果也证实了所有这些现象都是由随机定律决定的。于是，我们就可以把反概率性从物理定律（微分方程）中剔除了。因为它总要出现在物理框架中的某个地方，所以它会很自然地在方程的边界条件里再次现身。通过把它置于我们现有的物理问题的范围之外，我们幻想着我们已经摆脱了这个困难。只有当我们中的某些人被误导着试图穿越几十亿年的时间回到过去时，我们才会发现被我们剔除的所有反概率性现象都堆在那里像一堵高墙一样形成了一个不可逾越的边界，这即是时间的起点。

目前已经有人提出了一种可以解决以上矛盾的方法，这个方法似乎还得到了一些科学工作者的支持。然而，我不同意那种解释，因为我觉得它是站不住脚的，并不是因为我想让这个矛盾继续存在下去。我更愿意去找一个确切的着眼点，但是我确信现在鼓吹的着眼点其实是个死胡同。我必须首先回应对我的一些小的批评意见。

有时，我会因为没有在我所讨论的内容中充分强调熵是概率性而非确定性的物理量而受到批评。如前所述，如果我们在不同的时刻观察同一个系统，那么熵较大的时刻将是较晚的时刻。严格来讲，我应该如此表述：对于一个较小的系统，上述结论正确的可能性为 10^{20} ： 1。虽然我清楚地知道这个结论出错的概率是 $1/10^{20}$，但是，由于我在陈述上述结论时不够严谨，有些批评者仍会为此感到震惊。坦白地说：在过去的 25 年里，我撰写了很多文章和书籍，对物理现象做了大量的解释，恐怕我所做的各种解释中没有多少结论的出错概率会小于 $1/10^{20}$。在纯数学领域之外，我的结论的正误比估计更接近 10 ： 1，而不是 10^{20} ： 1；尽管如此，还是自负不已。我认为出错率为 $1/10^{20}$ 的结论对我们认识世界并不会有什么坏处；而我们应该稍微搁置关于出错概率的讨论。如果按照反对者的说法，这个世界上唯一能发表言论的就只可能是纯粹的数学家了。

涨 落

我前面所说的着眼点依赖于概率的涨落。我们考虑一群随机运动的粒子，随着时间的推移它们会经历任何可能的状态，所以只要我们等待足够长的时间，即使

chance configuration, will occur by chance if only we wait long enough. When the world has reached complete disorganisation (thermodynamic equilibrium) there is still infinite time ahead of it, and its elements will thus have opportunity to take up every possible configuration again and again. If we wait long enough, a number of atoms will, just by chance, arrange themselves in systems as they are at present arranged in this room; and, just by chance, the same sound-waves will come from one of these systems of atoms as are at present emerging from my lips; they will strike the ears of other systems of atoms, arranged just by chance to resemble you, and in the same stages of attention or somnolence. This mock Mathematical Association meeting must be repeated many times over—an infinite number of times, in fact—before t reaches $+\infty$. Do not ask me whether I expect you to believe that this will really happen.*

"Logic is logic. That's all I say."

So, after the world has reached thermodynamical equilibrium the entropy remains steady at its maximum value, except that "once in a blue moon" the absurdly small chance comes off and the entropy drops appreciably below its maximum value. When this fluctuation has died out, there will again be a very long wait for another coincidence giving another fluctuation. It will take multillions of years, but we have all infinity of time before us. There is no limit to the amount of the fluctuation, and if we wait long enough we shall come across a big fluctuation which will take the world as far from thermodynamical equilibrium as it is at the present moment. If we wait for an enormously longer time, during which this huge fluctuation is repeated untold numbers of times, there will occur a still larger fluctuation which will take the world as far from thermodynamical equilibrium as it was one second ago.

The suggestion is that we are now on the downward slope of one of these fluctuations. It has quite a pleasant subtlety. Is it chance that we happen to be running down the slope and not toiling up the slope? Not at all. So far as the physical universe is concerned, we have *defined* the direction of time as the direction from greater to less organisation, so that, on whichever side of the mountain we stand, our signpost will point downhill. In fact, on this theory, the going on of time is not a property of time in general, but is a property of the slope of the fluctuation on which we are standing. Again, although the theory postulates a universe involving an extremely improbable coincidence, it provides an infinite time during which the most improbable coincidence might occur. Nevertheless, I feel sure that the argument is fallacious.

If we put a kettle of water on the fire there is a chance that the water will freeze. If mankind goes on putting kettles on the fire until $t = \infty$, the chance will one day come off and the individual concerned will be somewhat surprised to find a lump of ice in his kettle. But it will not happen to *me*. Even if tomorrow the phenomenon occurs before my eyes, I shall not explain it this way. I would much sooner believe in interference by a demon than

* I am hopeful that the doctrine of the "expanding universe" will intervene to prevent its happening.

是最有秩序的、概率上最不可能的状态也会偶然出现。当宇宙达到完全的无序状态（热力学平衡态）后，时间仍然会无限地存在，宇宙中的元素将有机会反复经历各种可能的状态。如果我们观察足够长的时间，我们会发现，系统中一些原子所处的状态结构可能会偶然地与我们现在这个空间中的原子一样；同样偶然地，原子所组成的系统中的某一个系统出现的声波将可能与我口中现在发出的声波一样；而不管你是处于清醒还是昏昏欲睡的状态，这些声波都将去冲击由其他原子所组成的偶然与你的耳朵类似的系统。这个模拟的数学协会报告会将会在时间 t 达到无穷大之前重复很多次，实际上是无穷多次。别问我是否期望你们相信这些真的会发生。*

“逻辑就是逻辑，这就是我所说的全部。”

因此，当宇宙到达热力学平衡时，熵就会稳定地处在它的最大值上，除非出现一个千载难逢的极小的概率，使得熵从最大值回落到明显比最大值小的值。这个涨落消失之后，要等很长时间才会碰巧发生下一次涨落。尽管也许要等 10 的 10 次方的 10 次方年的时间，但是不用担心，我们拥有无限延续的时间。这种涨落的大小并没有什么限制，如果我们等待足够久，也许我们可以碰上一次大的涨落，使宇宙远离热力学平衡态，变成和我们现在这个宇宙一样的状态。如果我们等候更长的时间，其间会有数不清的类似的大涨落发生，也许还会有一次较大的涨落使宇宙远离热力学平衡态，变回到一秒钟以前的状态。

有人提出，我们现在正处在某个涨落的下坡过程中。这种提法存在令人兴奋的微妙之处。我们的宇宙刚好处在涨落的下坡过程而非往上的爬坡过程，这是一种巧合吗？完全不是。就物理世界而言，我们已经**定义**了时间的方向为有序度减少的方向，因此，无论我们站在山坡的哪一边，我们的指示牌都是指向下坡的。事实上，在这个理论里，总的来说时间的流逝并不是时间本身的性质，而是我们所处的那个涨落的山坡的性质。尽管这个理论假设宇宙包含了一个极不可能发生的概率事件，但同时该理论却提供了无限长的时间使得最不可能发生的概率事件最终总能发生。无论怎样，我个人觉得以上的说法是不合理的。

如果我们把一壶水放到火上，这壶水有可能结冰吗？如果一个人把一壶水放到火上无限长的时间，某一天这个人可能会惊讶地发现，他壶里的水居然结冰了。但这类事情不可能发生在**我**身上。即使将来有一天这种事情真的在我眼前发生了，我也不会用上面那样的方式去解释它。我宁愿相信这是一个魔鬼干的，而不是小概率

---

* 我希望“宇宙膨胀”说会阻止它的发生。

in a coincidence of that kind coming off; and in doing so I shall be acting as a rational scientist. The reason why I do not at present believe that devils interfere with my cooking arrangements and other business, is because I have become convinced by experience that Nature obeys certain uniformities which we call laws. I am convinced because these laws have been tested over and over again. But it is possible that every single observation from the beginning of science which has been used as a test, has just happened to fit in with the law by a chance coincidence. It would be an improbable coincidence, but I think not quite so improbable as the coincidence involved in my kettle of water freezing. So if the event happens and I can think of no other explanation, I shall have to choose between two highly improbable coincidences: (*a*) that there are no laws of Nature and that the apparent uniformities so far observed are merely coincidences; (*b*) that the event is entirely in accordance with the accepted laws of Nature, but that an improbable coincidence has happened. I choose the former because mathematical calculation indicates that it is the less improbable. I reckon a sufficiently improbable coincidence as something much more disastrous than a violation of the laws of Nature; because my whole reason for accepting the laws of Nature rests on the assumption that improbable coincidences do not happen—at least, that they do not happen in my experience.*

Similarly, if logic predicts that a mock meeting of the Mathematical Association will occur just by a fortuitous arrangement of atoms before $t = \infty$, I reply that I cannot possibly accept that as being the explanation of a meeting of the Mathematical Association in $t = 1931$. We must be a little careful over this, because there is a trap for the unwary. The year 1931 is not an absolutely random date between $t = -\infty$ and $t = +\infty$. We must not argue that because for only $1/x$th of time between $t = -\infty$ and $t = \infty$ a fluctuation as great as the present one is in operation, therefore the chances are x to 1 against such a fluctuation occurring in the year 1931. For the purposes of the present discussion, the important characteristic of the year 1931 is that it belongs to a period during which there exist in the universe beings capable of speculating about the universe and its fluctuations. Now I think it is clear that such creatures could not exist in a universe in thermodynamical equilibrium. A considerable degree of deviation is required to permit of living beings. Therefore it is perfectly fair for supporters of this suggestion to wipe out of account all those multillions of years during which the fluctuations are less than the minimum required to permit of the development and existence of mathematical physicists. That greatly diminishes x, but the odds are still overpowering. The *crude* assertion would be that (unless we admit something which is not chance in the architecture of the universe) it is practically certain that at any assigned date the universe will be almost in the state of maximum disorganisation. The *amended* assertion is that (unless we admit something which is not chance in the architecture of the universe) it is practically certain that a universe containing mathematical physicists will at any assigned date be in the state of maximum disorganisation which is not inconsistent with the existence of such creatures. I think it is quite clear that neither the original nor the amended version applies. We are thus driven

* No doubt "extremely improbable" coincidences occur to all of us, but the improbability is of an utterly different order of magnitude from that concerned in the present discussion.

事件发生了；而这样做的时候，我总该更像是一个理性的科学家吧。我现在之所以不相信魔鬼会干预我的烹调事务或者其他事情，是因为日常经验使我相信自然界会遵循一定的统一性，即物理定律。我相信这些定律是因为它们已经被反复验证过。当然，也有可能所有这些用来验证物理定律的实验一开始就恰好满足这些定律只是一种巧合。这或许是个难以置信的巧合，但是我相信这种巧合不会像我壶里的水发生结冰的巧合那样难以置信。一旦这种事件发生了，而我又找不到其他合适的解释，那么我只能在以下两种极为难以置信的巧合之间做出选择：(*a*) 这个世界上并不存在物理定律，迄今为止观察到的一致性仅仅是巧合。(*b*) 这个事件是符合物理定律的，只不过难以置信的巧合事件发生了。我更愿意接受前一种解释，因为数学计算的结果表明它的可能性更高一点。相比于违反物理定律而言，我把难以置信的巧合事件的发生看成是更加糟糕的事情；因为我之所以接受存在于自然界中的物理定律只是因为我相信难以置信的巧合事件不会发生——至少在我有生之年不会发生。*

同样，如果逻辑上预言在 $t=\infty$ 之前仅仅是由于原子的随机排列而出现了一个模拟的数学协会的报告会，那我要说的是，我不能接受这就是今天 $t=1931$ 年的这个数学协会报告会发生的理由。我们必须小心这一点，因为这里有个不小心就容易陷入的圈套。在 $t=-\infty$ 到 $t=+\infty$ 之间，1931 年并不是一个完全随机的时间。我们不能因为在 $t=-\infty$ 到 $t=+\infty$ 之间只是在第 $1/x$ 个时间里有一个与今天的数学报告会一样的涨落发生了，就说在 1931 年里发生这种涨落的概率是 $1/x$。从我们现在的讨论情况来看，1931 年的重要性就在于在这段时间里宇宙中出现了能够思考宇宙以及它的涨落的人类。现在我认为有一点是很清楚的，就是人类不可能在处于热力学平衡态下的宇宙中生存。要让生物能够生存下去，宇宙就必须在很大程度上偏离平衡态。因此，对于涨落论的支持者来说，不对在 10 的 10 次方的 10 次方年这段时间中发生涨落的概率比数学物理学家成长和存在的最小概率还小的情况进行说明是完全合理的。这就大大缩小了 x 的值，但是问题仍然存在。**最初**的假设应该是这样的（除非我们承认某些状态在宇宙的结构中不可能发生）：几乎可以肯定地说宇宙在给定的任一时间都将近似处于最无序的状态。**改进**后的说法是（除非我们承认某些状态在宇宙的结构中不可能发生）：几乎可以肯定地说有数学物理学家存在的宇宙在给定的任一时间都将处于最无序的状态，而这个状态也适合人类的生存。我想，我们已经很清楚地看出以上两种说法都是站不住脚的。于是我们被迫接受反概率论；而对待它的最好办法显然是，正如之前所说过的，把所有反概率性现象整理在一起并堆

* 毫无疑问，“极不可能”发生的巧合会在我们面前发生，只是这种巧合事件的发生概率与我们在此讨论的巧合事件有完全不同的数量级。

to admit anti-chance; and apparently the best thing we can do with it is to sweep it up into a heap at the beginning of time, as I have already described.

The connexion between our entropy signpost and that dynamic quality of time which we describe as "going on" or "becoming" leads to very difficult questions which I cannot discuss here. The puzzle is that the signpost seems so utterly different from the thing of which it is supposed to be the sign. The one thing on which I have to insist is that, apart from consciousness, the increase of entropy is the only trace that we can find of a one-way direction of time. I was once asked a ribald question: How does an electron (which has not the resource of consciousness) remember which way time is going? Why should it not inadvertently turn round and, so to speak, face time the other way? Does it have to calculate which way entropy is increasing in order to keep itself straight? I am inclined to think that an electron does do something of that sort. For an electric charge to face the opposite way in time is the same thing as to change the sign of the charge. So if an electron mistook the way time was going it would turn into a positive charge. Now, it has been one of the troubles of Dr. P. A. M. Dirac that in the mathematical calculations based on his wave equation the electrons do sometimes forget themselves in this way. As he puts it, there is a finite chance of the charge changing sign after an encounter. You must understand that they only do this in the mathematical problems, not in real life. It seems to me there is good reason for this. A mathematical problem deals with, say, four electric charges at the most; that is about as many as a calculator would care to take on. Accordingly, the unfortunate electron in the problem has to make out the direction of past to future by watching the organisation of three other charges. Naturally, it is deceived sometimes by chance coincidences which may easily happen when there are only three particles concerned; and so it has a good chance of facing the wrong way and becoming a positive charge. But in any real experiment we work with apparatus containing billions of particles—ample to give the electron its bearings with certainty. Dirac's theory predicts things which never happen, simply because it is applied to problems which never occur in Nature. When it is applied to four particles alone in the universe, the analysis very properly brings out the fact that in such a system there could be no steady one-way direction of time, and vagaries would occur which are guarded against in our actual universe consisting of about 10^{79} particles.

Heisenberg's Principle

A discussion of the properties of time would be incomplete without a reference to the principle of indeterminacy, which was formulated by Heisenberg in 1927 and has been generally accepted. It had already been realised that theoretical physics was drifting away from a deterministic basis; Heisenberg's principle delivered the knock-out blow, for it actually postulated a certain measure of indeterminacy or unpredictability of the future as a fundamental law of the universe. This change of view seems to make the progress of time a much more genuine thing than it used to be in classical physics. Each passing moment brings into the world something new—something which is not merely a mathematical extrapolation of what was already there.

积在时间的起点上。

熵这个指示牌和被称作“流逝”或“发展”的时间动态特征之间的关系导致了一些非常困难的问题，在这里我不可能做详细的讨论。其中的困难之处在于熵这个指示牌似乎和我们预期的时间标记非常不同。我需要强调的是，除了意识以外，熵的增加是我们发现时间具有单向性的唯一线索。曾经有人向我提出一个粗鄙的问题：一个电子（它没有意识）是如何记住时间的流向的？为什么它不会无意中改变方向，即，朝向时间的另一个方向？它需要事先计算好朝哪个方向熵会增加，然后再决定往哪个方向前进吗？我倾向于认为电子确实会作这样的判断。对一个电荷来说，朝相反的时间方向意味着它的电荷符号要发生变化。所以，如果一个电子错误地选择了时间方向，那么它就会变成一个正电荷。这跟狄拉克博士在以电子波动方程为基础进行数学计算时碰到的一个麻烦是一样的，在他的计算里电子真的会走错方向。就像狄拉克所发现的，在一次碰撞之后，电子有一定的概率可以改变电荷的符号。你必须要明白的是，这仅仅是数学的计算结果，并未在现实生活中发现。之所以是这样，我认为似乎有很合理的理由可以解释。假设我们考虑一个最多涉及四个电荷的数学问题，这个问题是任何一个计算器都能够处理的。于是，其中一个倒霉的电子不得不通过观察其他三个电荷的有序性来判断时间从过去到将来的流向。很自然，当我们只考虑三个电子的情况时，很容易发生偶然的巧合使得电子弄错方向；所以电子有很好的机会可以因为弄错了时间的方向而变成一个正电荷。但是在现实实验中，我们所用的仪器会包含几十亿个粒子——多到足够让电子准确地确定出时间的方向。狄拉克理论预言的情况从来就没有发生过，因为与之相适的问题从来没有出现在现实世界中。如果将狄拉克理论应用于只有四个电子的宇宙中，那么经过分析很可能会给出这样的结论：在这个系统中不会存在一个稳定的单向的时间方向，奇特的事情会不断地发生，而这些在我们这个包含了约 10^{79} 个粒子的真实宇宙中是完全不可能发生的。

海森堡原理

如果我们不考虑不确定性原理的话，那么我们对时间的讨论将是不完整的。这个原理是海森堡在 1927 年用公式表达出来并已得到一致认可的。人们已经意识到理论物理学正慢慢地偏离确定性这个基础；海森堡不确定性原理对人们的思维产生了巨大的冲击，因为这个原理作为物理世界的基本定律实际上认为未来理应具有一定程度上的不确定性或不可预测性。相比于经典物理学，这种观念上的转变使得时间的进展具有了更加真实的意义。每一个瞬间过后，世界都可能增添一些新事物——这些事物不可能单纯使用数学方法从过去已经发生的事件中推测出来。

The deterministic view which held sway for at least two centuries was that if we had complete data as to the state of the whole universe during, say, the first minute of the year 1600, it would be merely a mathematical exercise to deduce everything that has happened or will happen at any date in the future or past. The future would be determined by the present as the solution of a differential equation is determined by the boundary conditions. To understand the new view, it is necessary to realise that there is a risk of begging the question when we use the phrase "complete data". All our knowledge of the physical world is inferential. I have no direct acquaintance with my pen as an object in the physical world; I infer its existence and properties from the light waves which fall on my eyes, the pressure waves which travel up my muscles, and so on.

Precisely the same scheme of inference leads us to infer the existence of things in the past. Just as I infer a physical object, namely, my pen, as the cause of certain visual sensations now, so I may infer an infection some days ago as the cause of an attack of measles. If we follow out this principle completely we shall infer causes in the year 1600 for all the events which we know to have happened in 1930. At first sight it would seem that these inferred causes have just as much status in the physical world as my fountain pen, which is likewise an inferred cause. So the determinist thinks he has me in a cleft stick. If the scientific worker poking about in the universe in 1600 comes across these causes, then he has all the data for making a correct prediction for 1930; if he does not, then he clearly has not complete knowledge of the universe in 1600, for these causes have as much right to the status of physical entities as any of our other inferences.

I need scarcely stop to show how this begs the question by arbitrarily prescribing what we should deem to be complete knowledge of the universe in 1600, irrespective of whether there is any conceivable way in which this knowledge could be obtained at the time. What Heisenberg discovered was that (at least in a wide range of phenomena embracing the whole of atomic physics and electron theory) there is a provision of Nature that just half of the data demanded by our determinist friend might with sufficient diligence be collected by the investigators in 1600, and that complete knowledge of this half would automatically exclude all knowledge of the other half. It is an odd arrangement, because you can take your choice which half you will find out; you can know either half but not both halves. Or you can make a compromise and know both halves imperfectly, that is, with some margin of uncertainty. But the rule is definite. The data are linked in pairs and the more accurately you measure one member of the pair the less accurately you can measure the other member.

Both halves are necessary for a complete prediction of the future, although, of course, by judiciously choosing the type of event we predict we can often make safe prophecies. For example, the principle of indeterminacy will obviously not interfere with my prediction that during the coming year zero will turn up approximately $\frac{1}{37}$ of the total number of times the roulette ball is spun at Monte Carlo. All our successful predictions in physics and astronomy are on examination found to depend on this device of eliminating the inherent uncertainty of the future by averaging.

确定性的观点占主导地位至少有两个世纪了，确定论认为如果我们知道了某个时刻宇宙状态的完整数据，比如说 1600 年第一分钟的完整数据，那么单纯的数学推算就可以告诉我们这个世界以前是什么样子以及未来将会是什么样子。未来是由现在所决定的，因为一个微分方程的解由边界条件决定。要理解这个全新的观点，我们必须意识到我们一开始提出的“完整数据”这个概念是存在回避问题实质的风险的。而我们对物理世界的所有认识都是推论性的。我的钢笔并不是作为一个在物理世界中存在的物体使我直接认识到的；我是通过从笔上反射进我眼睛里的光以及握笔时在我肌肉中传播的压力波等感知它的存在和性质的。

我们对过去事物的感知也遵循着完全一样的模式。就像现在我可以通过我的视觉感受去推断一个客观的物体，即我的钢笔一样，我也可以从患上麻疹这个情况推断出几天前应该受到了感染。如果我们完全依照这个规律，就可以就 1930 年已经发生的全部事件去推断在 1600 年使它们发生的原因。乍一看似乎这些推断出来的原因与钢笔（也是一种推断出来的原因）的存在状态一样对我们认识物理世界具有同样重要的作用。于是确定论者们认为这令我陷入了进退两难的境地。如果某个科学工作者在 1600 年的世界中四处探寻并且恰好找到了这些事情的起因，那么他就有完整的数据来准确地预测 1930 年发生的事情；如果他做不到，那么他肯定没有完全认识 1600 年的宇宙，因为这些起因跟我们在其他推论过程中得出的起因一样都是些物理实体。

我还需要继续说明他们是如何为了回避问题的实质而武断地规定哪些内容应该被视为是 1600 年的世界的完整数据，而不考虑那时是否存在可以想象的方式去获取这些数据。海森堡发现（至少在包括整个原子物理学和电子理论的广泛物理现象中），按照自然界的规律，在我们支持确定论的朋友们所需的数据中，只有一半有可能被 1600 年的研究者通过不懈的努力收集到，而对这一半数据的完整认识将使我们自动失去了得到另一半数据的机会。这是个很奇怪的约定，因为你可以选择要了解哪一半数据；你可以知道任何一半的数据，但就是不能同时知道全部的数据。或者你也可以做出妥协而选择不完全地知道这两个一半的全部数据，即有不确定的部分。这个物理规律是确定的：数据是成对出现的，你对一对数据中的一个测量得越精确，则对另一个的测量结果就会越不精确。

当然，想要预测未来，全部数据的两个部分都是必要的，但是如果我们谨慎地选择预测的对象，我们还是可以做出一些可靠的预言的。例如，不确定性原理不会对我预测未来一年蒙特卡洛轮盘赌上的数字零出现的次数造成影响，这个次数大概是小球总的旋转次数的$\frac{1}{37}$。实践证明，所有物理学和天文学上的成功预测都是通过这种取平均的办法来消除未来的内在不确定性的。

As an illustration, let us consider the simplest type of prediction. Suppose we have a particle, say an electron, moving undisturbed with uniform velocity. If we know its position now and its velocity, it is a simple matter to predict its position at some particular future instant. Heisenberg's principle asserts that the position and velocity are paired data; that is to say, although there is no limit to the accuracy with which we might get to know the position and no limit to the accuracy with which we might get to know the velocity, we cannot get to know both. So our attempt at an accurate prediction of the future position of the particle is frustrated. We can, if we like, observe the position now and the position at the future instant with the utmost accuracy (since these are not paired data) and then calculate what has been the velocity in the meantime. Suppose that we use this velocity together with the original position to compute the second position. Our result will be quite correct, and we shall be true prophets—after the event.

This principle is so fully incorporated into modern physics that in wave mechanics the electron is actually pictured in a way which exhibits this "interference" of position and velocity. To attribute to it exact position and velocity simultaneously would be inconsistent with the picture. Thus, according to our present outlook, the absence of one half of the data of prediction is not to be counted as ignorance; the data are lacking because they do not come into the world until it is too late to make the prediction. They come into existence when the event is accomplished.

I suppose that to justify my title I ought to conclude with a prophecy as to what the end of the world will be like. I confess I am not very keen on the task. I half thought of taking refuge in the excuse that, having just explained that the future is unpredictable, I ought not to be expected to predict it. But I am afraid that someone would point out that the excuse is a thin one, because all that is required is a computation of averages and that type of prediction is not forbidden by the principle of indeterminacy. It used to be thought that in the end all the matter of the universe would collect into one rather dense ball at uniform temperature; but the doctrine of spherical space, and more especially the recent results as to the expansion of the universe, have changed that. There are one or two unsettled points which prevent a definite conclusion, so I will content myself with stating one of several possibilities. It is widely thought that matter slowly changes into radiation. If so, it would seem that the universe will ultimately become a ball of radiation growing ever larger, the radiation becoming thinner and passing into longer and longer wavelengths. About every 1,500 million years it will double its radius, and its size will go on expanding in this way in geometrical progression for ever.

(**127**, 447-453; 1931)

作为一个例证，让我们来考虑一个最简单的预测。假设我们有一个粒子，例如电子，没有受到任何扰动而做匀速直线运动。如果我们知道它现在的位置和速度，那么预测它未来某个时刻的位置是一件很简单的事情。海森堡不确定性原理告诉我们，位置和速度是结成对的数据；也就是说，我们可以无限精确地知道它的位置，也可以无限精确地知道它的速度，但是我们就是不能同时精确地知道这两个量。所以我们在试图准确预测这个粒子未来某个时刻对应的位置时遇到了困难。如果我们愿意的话，我们可以观察这个粒子现在时刻和未来时刻的精确位置（因为它们不是结成对的数据），然后计算出这段时间的平均速度。假如当我们由这个平均速度和最初时刻的位置来计算未来时刻的位置时，我们发现结果完全正确，于是我们成了真正的预言家——其实只是事后诸葛亮。

这个不确定性原理已经完全融入了现代物理学，在波动力学中的确可以认为在电子的位置和速度之间就表现出了那种“相干关系”。由于这种相干关系，同时精确地知道电子的位置和速度是与以上的物理图像相矛盾的。因此，按照我们现在的观点，缺少预测未来所需数据中的一半算不上是信息不灵通；数据不完整是因为在我们做出预测之前它们并不存在。它们是在我们的测量行为发生之时才出现的。

我认为为了充分说明我的主题，我应当用一个预言来作为对宇宙终点可能会是什么样子这个问题的总结。我承认我并不十分情愿回答这个问题。我甚至想，既然我刚才已经解释了未来是不可预测的，我就可以以此为借口拒绝回答这个问题。但恐怕有人会说，你这个借口太勉强，你所需要做的只是计算一下平均值，而那种类型的预言并不违背不确定性原理。过去我们认为宇宙中的所有物质最终会聚集到一起变成一个温度均匀、密度极大的圆球；但是球面空间的学说，尤其是由此得到的宇宙膨胀的最新结论改变了以前的看法。现在还存在一两个未得到解决的问题，这使得我不能得出明确的结论，所以我在这里只是给出其中的一种可能性。大家普遍认为物质会慢慢地转变成辐射。如果是那样的话，整个宇宙就会变成一个不断膨胀的充满辐射的球，辐射会变得越来越弱，相应的波长也变得越来越长。大概每过 15 亿年，宇宙的半径就会增加一倍，宇宙的大小将会以这种几何级数的增长方式永远地膨胀下去。

（沈乃澂 翻译；张元仲 审稿）

Obituary

Editor's Note

In the nineteenth century, the propagation of light raised considerable difficulty in the minds of scientists. As Maxwell had demonstrated, light consists of electromagnetic waves, but people found it difficult to visualise the occurrence of wave motion in an entirely empty space. So there became established the idea that light travels through an insubstantial (that is, massless) medium called the lumeniferous ether. Albert Michelson and his chemist colleague Edward Morley carried out an experiment using a specially constructed interferometer designed by Michelson to compare the velocity of light in two directions perpendicular to each other. The outcome of the experiment suggested that the behaviour of light in the two perpendicular arms of the equipment was identical, thus throwing doubt on the whole concept of the lumeniferous ether—a point from which Einstein's Special Theory of Relativity was founded in 1905.

Prof. A. A. Michelson, For Mem. R. S.

WE much regret to announce the death, which occurred on May 9, of Prof. A. A. Michelson, the distinguished physicist of the University of Chicago. Prof. Michelson was probably best known for his wonderful experimental work to detect any effect of the earth's rotation on the velocity of light. At the end of 1929, he resigned his position at the University of Chicago and went to Pasadena, where he proposed to carry out further work on this subject, and it is reported that preliminary measurements have already been made. Prof. Michelson had worked previously at Mount Wilson Observatory, Pasadena, and a brief account of repetitions of the famous Michelson–Morley experiment, as it is generally called, with a diagram of the apparatus, was contributed to *Nature* of Jan. 19, 1929, by him and his collaborators. The results then obtained showed no displacement of the interferometer fringes so great as one-fifteenth of that to be expected on the supposition of an effect due to a motion of the solar system of three hundred kilometres per second through the ether. Since then, Prof. Michelson has been awarded the Duddell Medal for 1929 of the Physical Society of London for his work on interferometry.

In *Nature* of Jan. 2, 1926, we were fortunate in being able to publish, as one of our series of "Scientific Worthies", an appreciation of Prof. Michelson and his work by Sir Oliver Lodge. We print below extracts from that article.

"Albert Abraham Michelson was born in Strelno, Poland, on Dec. 19, 1852. In 1854 his parents migrated to the United States. After emerging from High School in San Francisco, young Michelson was appointed to the Naval Academy, from which he graduated in 1873, and two years later became instructor in physics and chemistry under Admiral

讣 告

编者按

在 19 世纪，科学家们对光的传播的理解仍面临着很大的困难。麦克斯韦指出，光是由电磁波构成的，但人们觉得很难想象波会在真空中传播。所以有人提出光在质量为零的以太中传播。阿尔伯特·迈克尔逊与他的化学家同事爱德华·莫雷用迈克尔逊本人设计的干涉仪测量并比较了互相垂直的两束光在传播速度上的差异。实验结果说明，光在干涉仪两个相互垂直的臂上的行为是相同的，这使人们对传光以太的概念产生了怀疑——而基于这一点爱因斯坦在 1905 年提出了狭义相对论。

英国皇家学会为纪念迈克尔逊教授所写的悼词

我们非常遗憾地告诉大家：芝加哥大学杰出的物理学家迈克尔逊教授在 5 月 9 日与世长辞了。迈克尔逊教授最著名的工作大概要算他检测地球旋转对光速影响的完美实验。1929 年末，他放弃了在芝加哥大学的职位，赴帕萨迪纳继续开展这方面的研究，据报道他已经完成了前期的测量工作。迈克尔逊教授先前曾在帕萨迪纳的威尔逊山天文台工作过，他和他的合作者们在 1929 年 1 月 19 日的《自然》杂志上发表了有关迈克尔逊－莫雷实验的文章，他们用装置图简单再现了这个为大家所公认的著名实验。后来得到的结果表明，假设太阳系以每秒 300 公里的速度在以太中运动，干涉条纹的位移不会超过预期值的 1/15。1929 年，迈克尔逊教授因在干涉测量法方面的贡献而被伦敦物理学会授予达德尔奖章。

我们很荣幸能在 1926 年 1 月 2 日《自然》杂志的“杰出科学家”系列介绍中发表奥利弗·洛奇爵士对迈克尔逊教授及其研究工作的评论。以下内容是我们从那篇文章中摘录出来的。

“阿尔伯特·亚伯拉罕·迈克尔逊于 1852 年 12 月 19 日出生于波兰的斯特列罗。1854 年，他随父母移居美国。年轻的迈克尔逊在旧金山读完高中后被指派到海军学院学习一直到 1873 年毕业，两年后他成为桑普森上将手下的一名物理和化学教员，

Sampson, continuing this work until 1879. After a year in the Nautical Almanac Office at Washington, Michelson, now an ensign, went abroad for further study at the Universities of Berlin and Heidelberg, and at the Collège de France and the École Polytechnique in Paris. Upon his return to the United States in 1883 he became professor of physics in the Case School of Applied Science, Cleveland, Ohio; whence, after six years, he was called to Clark University, where he remained as professor until 1892, when the University of Chicago opened its doors. Prof. Michelson went to this new institution as professor of physics and head of the department. In June 1925 he was honoured by being appointed to the first of the Distinguished Service Professorships made possible by the new development programme of the University.

"It was while he was at Cleveland that Prof. Michelson collaborated with Prof. Morley in their joint experiment; and it may have been for the purpose of that experiment that he invented his particular form of interferometer, with the to-and-fro beams at right angles. Later, he applied it in Paris to the determination of the metre, with an estimated accuracy of about one part in two million.

"During the War, Prof. Michelson re-entered the Naval Service with the rank of Lieutenant-Commander, giving his entire time to seeking new devices for naval use, especially a range-finder, which became part of the U.S. Navy Equipment.

"A Nobel Prize was awarded to Prof. Michelson in 1907, the first American to get one for science; and the Copley Medal, the most distinguished honour of the Royal Society of London, was awarded him in the same year.

"The gold medal of the Royal Astronomical Society was presented to Prof. Michelson on Feb. 9, 1923; and the compact exposition of the reasons for that award, by the president, Prof. Eddington, on that occasion will be found in *Nature*, vol.111, p. 240.

"Michelson touched on many departments of physics, but in optics, the highest optics, he excelled. In this subject he can be regarded as the most fertile and brilliant disciple of the late Lord Rayleigh, for his inventions are based on a thorough assimilation of the principles of diffraction, interference, and resolving power; and his great practical achievements are the outcome of this knowledge. Michelson seemed to have a special instinct for all phenomena connected with the interference of light, with a taste for exact measurement surpassed by none in this particular region. The interferometer with which he began became in his hands much more than an interferometer. He applied it to the determination of the standard metre in terms of the wave-length of light, with exact results which will enable remote posterity millions of years hence to reconstruct, if they want to, the standard measures in vogue at this day. He applied it also to analyse the complex structure of spectrum lines, and with remarkable completeness to determine the shape and size of invisible objects, such as to ordinary vision, however much aided by telescopic power, will probably remain mere points of light.

并在此一直工作到 1879 年。后来他又为华盛顿的航海天文历编制局工作了一年，此时已是海军少尉的迈克尔逊走出了国门，先后到柏林大学、海德堡大学、法兰西学院和巴黎理工学校深造。他在 1883 年回到美国，随即成为位于俄亥俄州克利夫兰的凯斯应用科学学院的物理教授；6 年后，克拉克大学邀请他担任该校的教授，他在那里一直工作到 1892 年芝加哥大学成立。迈克尔逊教授在这个新建的大学中担任物理学教授和物理系系主任。1925 年 6 月，由于该大学新发展规划的需要，他荣幸地被提名为首位对社会有杰出贡献的教授。

“正是在克利夫兰工作的时候，迈克尔逊教授与莫雷教授合作进行了他们的实验；为完成这个实验，迈克尔逊教授发明了可以在相互垂直方向上使光线来回反射的干涉仪。后来他在巴黎用这个仪器测量了标准米，估计精度在 1/2,000,000 上下。

“在第一次世界大战期间，迈克尔逊教授重返海军服役，军衔为少校，他把自己全部的时间都用于研制适用于海军需要的新仪器，尤其是已经成为美国海军装备组成部分的测距仪。

“迈克尔逊教授获得了 1907 年的诺贝尔奖，他是第一位获此殊荣的美国科学家；在同一年，他还荣获了代表伦敦皇家学会最高荣誉的科普利奖章。

“1923 年 2 月 9 日，迈克尔逊教授被授予皇家天文学会金奖；《自然》杂志在第 111 卷的第 240 页上刊登了皇家天文学会主席爱丁顿教授在颁奖仪式上对其获奖原因的简短说明。

“迈克尔逊曾经涉足过物理学的许多领域，但他在光学领域，最前沿的光学领域所取得的成就最为突出。在这个领域中，他可以被认为是已故瑞利勋爵的所有弟子中成果最丰硕和最优秀的一位，因为他的发明是建立在透彻理解衍射、干涉原理以及分辨率的基础上的；而他在实践上的伟大成就就是能够充分运用这些知识的结果。迈克尔逊似乎对所有与光干涉相关的现象都有一种特别的直觉，在这个特定领域中他对精确测量的判断力是无人能及的。在他眼中，自己早先发明的干涉仪远远不只是一部干涉仪。他利用干涉仪根据光的波长测定了标准米，得到的结果非常精确，可以让几百万年以后的子孙重新构建当今流行的计量标准，只要他们愿意。他还利用干涉仪分析了谱线的复杂结构，并且完美地测定了不可见物体的形状和大小，这些不可见物体对于视力正常的人来说，即使借助大倍率望远镜也只能看到几个光点。

"In a magnificent paper in the *Phil. Mag.* of July 1890, Michelson suggested the application of interference methods to astronomy. He knew well that the resolving power of a telescope depended on the diameter of its aperture, and that the formation of an image was essentially an interference phenomenon; the minuteness of a point image, and therefore the clearness of definition, depending on the size of the object-glass. But he pointed out that if the aperture was limited to slits at opposite edges—so that no actual image anything like the object would be formed, but only the interference bands which the beams from the two slits could produce—a study of those bands would enable us to infer about the source of light very much more than we could get by looking at its image. For example, suppose it was a close double star, and suppose the slits over the object-glass were movable, so that they could be approached nearer together, or separated the whole distance of the aperture apart. A gradual separation of the slits would now cause the fringes to go through periods of visibility and invisibility; and the first disappearance of the fringes would tell us that the distance apart of the two components of the star (multiplied by the distance between the slits and divided by the distance of the star) would equal half a wave-length of light. The two components might be far too near together ever to be seen separately, and yet we could infer that the star was a double one; and by further attention to the visibility curve we could infer the relative brightness of the two components and their position relative to our line of sight.

"Furthermore, if, instead of looking at a star, we turned the slit-provided telescope on a planet with a disc too small for ordinary measurement, the size of that disc could be estimated from the behaviour of the interference fringes produced by its light in a suitable interferometer, or by the telescope converted into one.

"In view of the great interest aroused by the application of this method by Michelson himself, with the aid of collaborators at Mount Wilson Observatory, Pasadena, California, and with the hundred-inch telescope established there, it may be interesting to quote here part of the conclusion of his paper of date 1890:

" '(1) Interference phenomena produced under appropriate conditions from light emanating from a source of finite magnitude become indistinct as the size increases, finally vanishing when the angle subtended by the source is equal to the smallest angle which an equivalent telescope can resolve, multiplied by a constant factor depending on the shape and distribution of light in the source and on the order of the disappearance.

" '(2) The vanishing of the fringes can ordinarily be determined with such accuracy that single readings give results from fifty to one hundred times as accurate as can be obtained with a telescope of equal aperture.'

" 'If among the nearer fixed stars there is any as large as our sun, it would subtend an angle of about one hundredth of a second of arc; and the corresponding distance required to observe this small angle is ten metres, a distance which, while utterly out of question as

“1890 年 7 月，迈克尔逊在一篇发表于《哲学杂志》上的优秀论文中提到了干涉法在天文学中的应用。他非常清楚望远镜的分辨力取决于其孔径的大小，而从本质上看，像的形成是一种干涉现象；点像的精确性以及与此相关的轮廓清晰度取决于物镜的尺寸。但他指出，如果孔径受到对面边缘狭缝的限制，则不能形成与物体类似的实像，而来自两个狭缝的光束只能形成干涉带，通过研究这些干涉带推出的有关光源的信息要比我们从观察图像本身得到的信息多很多。例如，假设有一对相离很近的双星，并假设物镜上的狭缝是可移动的，它们可以靠得更近，也可以相距整个孔径那么大的距离。使两个狭缝逐渐远离就能实现条纹从可见到不可见的周期变化；条纹的首次消失将告诉我们，恒星两组分之间的距离（乘以狭缝间的距离，除以双星到地球的距离）等于光的半波长。也许双星之间的距离过于接近，以至于我们从未观测到它们是分开的两颗星，但我们已经可以推测出该恒星是一对双星；通过对可见度曲线的进一步研究，我们还可以推断出双星的相对亮度以及它们相对于我们视线的位置。

“此外，除了观察恒星以外，我们还可以将装有狭缝的望远镜对准一颗星盘很小、用普通测量方法难以观测到的行星，来自星盘的光在适当的干涉仪中或由望远镜改装成的干涉仪中形成干涉条纹，根据这些条纹的变化情况可以估算出星盘的尺寸。

“迈克尔逊本人把自己的方法应用于天文学领域激起了人们极大的兴趣，他得到了加州帕萨迪纳威尔逊山天文台的工作人员的帮助，并使用了安装在那里的孔径可达百英寸的望远镜，考虑到这一点，在此引述他在 1890 年论文中的部分结论也许是一件有意思的事情：

“‘(1) 在适当条件下，如果干涉现象的产生源于有限强度光源发出的光，那么随着光源尺寸的增加，干涉现象就会逐渐变模糊，当光源所张的角度等于等效孔径望远镜所能分辨的最小角度乘上一个常数因子时，干涉现象将最终消失，这个常数因子取决于光源中光的状态和频率分布以及干涉现象消失的量级。

“‘(2) 通常用于测量条纹相消的方法有很高的精确度，其单次测量所得结果的精度是等孔径望远镜的 50 至 100 倍。’

“‘如果在比较靠近我们的恒星中存在着一个与太阳大小差不多的恒星，那么它所张的角度将大约是 1/100 角秒；而观测这么小的角度需要 10 米的距离，望远镜物

regards the diameter of a telescope-objective, is still perfectly feasible with a refractometer. There is, however, no inherent improbability of stars presenting a much larger angle than this; and the possibility of gaining some positive knowledge of the real size of these distant luminaries would more than repay the time, care, and patience which it would be necessary to bestow on such a work.'

"There seemed little hope at that time, and certainly no reasoned expectation, that any stars, except perhaps some of the very nearest, could have discs big enough for perception and measurement even by this virtual telescope of thirty feet aperture. The possibility of giant stars came, however, above our mental horizon; and Eddington made the notable prediction that a star like Betelgeuse must be in a highly rarefied state at a tremendously high temperature, and that it would be swollen out by the pressure of light to a size almost comparable with the dimensions of a solar system, although it could not contain very much more matter than, say, two or five times our sun. His argument, in brief, is that the spectrum of a young red star like Betelgeuse shows that it cannot be radiating furiously. Why then is it so conspicuous an object to our vision? It can only be because it is of enormous size, its density perhaps a thousand times less than atmospheric air. By utilisation of the data available in the light of his theory of stellar constitution, Eddington made an estimate of the diameter of the star.

"So with great skill Michelson and his collaborators got the interferometer to work. After many preliminary adjustments, on Dec. 13, 1920, Dr. F. G. Pease at Mount Wilson, with Michelson's apparatus, measured the diameter of a star for the first time, using Betelgeuse for the purpose. The interference-fringes formed by the star were observed, the object mirrors were gradually separated, and it must have been a joyful moment when, as they grew farther and farther and farther apart, the fringes at the eye end became less distinct and ultimately disappeared. The distance apart of the mirrors now, multiplied by the proper fraction, gave the angular dimensions of the star—a thing which had never before been observed in the history of the world. An estimate of the star's distance gave its actual diameter, and confirmed Eddington's prediction!

"Other stars have since been measured, and the giant stars well deserve their name. Moreover, an instrument has been put in the hands of posterity to the power of which we can scarcely set a limit in investigating utterly invisible details, both about the heavenly bodies and about atoms, by the new and powerful method of analysing the radiation which they emit.

"The form of instrument adapted to the heavens is, however, not applicable to the atoms. The spectrum of atomic radiation is formed by a grating; and Rayleigh showed that the power of a prism spectroscope is expressed approximately by the number of centimetres of available thickness of glass, which is one form of saying that, to get high definition or separating power, we must use interference depending on a great number of wavelengths retardation. Michelson perceived that the retardation principle might be employed so as to make a grating which combined with its own effect the resolving power of a prism.

镜的直径完全可以达到这么长的距离，这个距离对于折光仪来说也毫无问题。然而，恒星所张角度大大高于 1/100 角秒的情况未必不可能出现；获得这些遥远发光体实际尺寸的有用信息是有可能的，而且值得人们投入时间、精力和耐心，而完成该项工作也需要这样的投入。'

“在当时几乎无法希望，当然也没有理由期望，恒星的星盘能够大到足以感知和测量的地步，即便是采用孔径为 30 英尺的虚拟望远镜，不过一些最接近我们的恒星也许可以除外。然而，可能存在巨星的想法进入了我们的视野；爱丁顿提出了那个著名的猜想，即像参宿四这样的恒星必然会处在极高温度下的高度稀薄状态，它在光压的作用下将膨胀到几乎与整个太阳系相近的尺寸，尽管它含有的物质不可能比太阳上的物质多多少，也就是太阳的 2 倍或 5 倍。简言之，他的结论是，从一颗类似于参宿四的年轻红巨星的光谱判断，它不可能在强烈地辐射能量。然而为什么我们很容易看到这样的天体呢？原因只能是它具有特别大的尺寸，而它的密度可能仅为空气密度的 1/1000。爱丁顿利用根据自己提出的恒星构造理论得到的数据估算出了这个恒星的直径。

“于是迈克尔逊和他的合作者们开始熟练地操作干涉仪进行研究。1920 年 12 月 13 日，在经过了多次调试之后，皮斯博士在威尔逊山上用迈克尔逊的装置首次测出了一颗恒星的直径，测量的恒星正是参宿四。可以观测到由恒星形成的干涉条纹，令物镜逐渐分离，当物镜之间的距离越来越远时，肯定会出现那个令人欣喜的时刻，目端的干涉条纹逐渐变模糊直至最终消失。这时用物镜之间的距离乘以一定的比例就得到了这颗恒星的角度大小——这在世界历史上还从未被人观测到过。根据对恒星距离的估计可以算出它的实际直径，从而验证了爱丁顿的预言！

“后来又测量了其他的恒星，发现那些巨星果然具有很大的体积。再者，这种已经传到后人手中的仪器功能非常了得，在用这种功能十分强大的新方法分析天体和原子发出的辐射时，我们简直不知道它是否有做不到的事情。

“然而，适用于观察天体的仪器规格并不适用于原子。原子辐射光谱是由光栅得到的；瑞利发现棱镜分光镜的分辨率可近似表示为玻璃可用厚度的厘米数，也就是说，为了达到更高的清晰度或分辨率，我们必须用到与大量波长延迟有关的干涉现象。迈克尔逊意识到也许可以利用延迟原理制成一个光栅，这个光栅既具有自己的特殊功用又具有棱镜的分辨能力。也许可以用一片厚度为 1 厘米或更厚的玻璃实现

A slab of glass, a centimetre or more thick, might be used to give the necessary lag in phase of many thousand wavelengths, and thereby secure a definition and resolving power unthought of before. So Michelson designed the Echelon spectroscope, consisting of thick slabs of glass, each protruding a millimetre or so beyond the other—a staircase spectroscope—which is now a regular instrument in the examination of the minute structure of spectrum lines.

"What, however, is popularly the best-known work of Michelson is the application of his interferometer to determine if possible the motion of the earth through the ether. The speed expected was of the order one-ten-thousandth of the velocity of light; but since the journey of the light in the instrument is a to-and-fro journey—one half-beam going as nearly as possible with and against the hypothetical stream of ether, while the other half-beam goes at right angles to that direction—the amount to be measured was not one-ten-thousandth but the square of that quantity; that is to say, the observer had to measure one part in a hundred million—no easy matter. The interferometer was mounted on a stone slab floating in mercury, and the whole observation conducted with great care. The result was zero; and that zero was used afterwards as the corner-stone of the great and beautiful edifice of relativity."

(**127**, 751-753; 1931)

所需的几千个波长的相位滞后，由此获得以前无法想象的清晰度和分辨率。于是迈克尔逊设计出了由厚玻璃片组成的阶梯光栅分光仪，每一个玻璃片都比另一个约长出 1 毫米，这种阶梯状的分光镜现在已经成为检测谱线精细结构的常规仪器。

“然而，大家普遍认为迈克尔逊最著名的成就是利用干涉仪来测定地球在以太中的运动，如果这种运动是存在的。人们预期其速度的数量级大概为光速的 1/10,000；但由于光在仪器中所走的路线是连续往返的，有一半的光束可以尽可能地与假想的以太流方向一致或相反，而另一半光束的行进方向与此垂直，测量出来的结果不该是 1/10,000，而应是 1/10,000 的平方；换言之，观测者要测量的是这个量的 1/100,000,000——这绝非易事。干涉仪被放在浮于水银表面的石板上，在整个观测过程中的操作都非常小心。最后得到的结果是零；后来这个零成为壮观美丽的相对论大厦的基石。”

（沈乃澂 翻译；张泽渤 审稿）

The Annihilation of Matter*

J. Jeans

Editor's Note

James Jeans notes here that one of the sacred principles of science was giving way. It had been a bedrock belief of physicists for centuries that matter can be neither created nor destroyed. Yet now, he points out, it looked increasingly certain that the process by which stars generate energy could only be explained through the annihilation of matter. Evidence showed that the average star has already emitted many times its own mass in radiation. The life history of a star seemed to be a continual annihilation of its substance, as massive particles give up their energy to produce radiation. The explanation for this is now seen to come from nuclear fusion, in which mass and energy are interconverted via Einstein's $E = mc^2$.

THROUGHOUT the greater part of the history of science, matter was believed to be permanent, incapable either of annihilation or of creation. Yet a large amount of astronomical evidence now seems to point to the annihilation of matter as the only possible source of the energy radiated by the stars. A position has thus been reached in which the majority of astronomers think it probable that annihilation of matter constitutes one of the fundamental processes of the universe, while many, and perhaps most, physicists look on the possibility with caution and even distrust. I have thought it might be of interest to attempt a survey of the present situation in respect to this question.

The Astronomical Evidence

The astronomical argument for the annihilation of matter is based, not on the intensity of stellar radiation, but on its duration. No transformation of a less drastic nature than complete annihilation is found capable of providing continuous radiation for the immense periods of time throughout which the stars have, to all appearances, lived. For, with one conspicuous exception, to be discussed later, all available methods of estimating stellar ages are found to indicate that the stars, as a whole, have already lived through periods of millions of millions of years.

Some of these methods depend on the rate of gravitational interaction between adjacent stars; for example, the velocities with which the stars move through space show an approximation to equipartition of energy, such as must have required millions of millions of years for its establishment. The individual members of the groups of stars known as moving star clusters appear to have had their courses changed by the gravitational pull of passing stars to an extent which again indicates action extending over millions of millions

* The substance of lectures delivered before the Universities of Princeton, Yale, and Harvard on May 23, 26, and 27 respectively, under the auspices of the Franklin Institute of Pennsylvania.

物质的湮灭*

詹姆斯·金斯

编者按

詹姆斯·金斯在文中指出：科学发展最神圣的原则之一是新旧理论的更迭。几百年来，物理学家们一直对物质既不能产生也不能灭亡这一原则深信不疑。可是现在，他指出，人们越来越相信恒星只有通过物质的湮灭过程才能获取能量。有证据表明：一般的恒星早已在过去的岁月里辐射掉了现有质量的好多倍。恒星存在的过程似乎就是其物质不断湮灭的过程，因为有质量的粒子会将其能量用于产生辐射。现在人们利用核聚变原理来解释这一现象，质量和能量的相互转化关系可以由爱因斯坦方程 $E = mc^2$ 表示。

在贯穿整个科学史的大部分时间里，物质都被看作是守恒的，它们既不会凭空产生，也不会凭空消失。但是现在大量天文学的证据似乎表明，物质的湮灭是恒星辐射能量的唯一来源。从而出现了这样一种局面：大多数天文学家认为物质湮灭可能是宇宙中的一个基本过程，而同时，许多也可能是大部分的物理学家对这种可能性持谨慎态度，甚至干脆不相信。我一直认为试着评述与这个问题相关的现状会是一件有意义的事情。

天文学的证据

天文学上关于物质湮灭的论证，不是以恒星辐射的强度为依据，而是基于恒星存在的时间。尚未发现一种不如完全湮灭那么剧烈的能量转化，能够在恒星显然存在的漫长岁月里，始终为其提供持续辐射所需要的能量。所有可以用来估算恒星寿命的方法都表明，整体看来，恒星已经存在了万亿年，只有一个方法明显例外，我将在后面讨论它。

有些方法取决于临近恒星的引力作用速度；例如，恒星穿越空间的速度表明它的能量是近似均分的，而要实现这样的均分必须经过数以万亿年的时间。被称为运动星团的星群中的个体成员，由于在其旁边经过的恒星的一定程度上的引力作用，它们的路径已经出现了改变，这再次表明活动过程已延续了数万亿年。目视双星的轨道也是如此。在这三个例子中，我们所使用的时钟的单位时间类似于在气体理论

* 这个由美国宾夕法尼亚州富兰克林科学馆赞助的演讲分别于5月23日、26日和27日发表于普林斯顿大学、耶鲁大学和哈佛大学。

of years. The same is true of the orbits of visual binary stars. In each of these three cases, the clock we use has for its unit a time analogous to what is called the "time of relaxation" in the theory of gases; in this comparison the single stars correspond to monatomic molecules, and binary stars to diatomic molecules, while the disintegration of a moving star cluster provides the counterpart of the process of gaseous diffusion.

These estimates of stellar ages are, of course, valid only if we assume that the changes in stellar motions and arrangements are produced solely by the gravitational pulls of other stars. Other causes are conceivable, and must indeed contribute something—pressure of radiation, bombardment by stray matter in space, or by the atoms of cosmic clouds diffused through space. But calculation shows that the contributions from these sources are quite negligible. Indeed, when we take them into account, the discussion of stellar movements is no longer a problem of astronomy, but of physics; we have to treat the stars as Brownian "particles" in a physical medium. When they are so treated, we find that the starry medium has a temperature—in the sense in which we speak of the temperature of moving Brownian particles—of the order of 10^{62} degrees. Both individual and binary stars exhibit the equipartition of energy which corresponds to a temperature of this order, whence it is obvious that physical agencies such as pressure of radiation and atomic pressures, which are in equilibrium with far lower temperatures of the order only of 10^4 degrees, cannot have made any appreciable contributions to the establishment of this equipartition; they act as mere drags on the stellar motions, tending on the average to check their speed.

In a second class of binary stars, the spectroscopic binaries, the two components are so close together that the gravitational pull from passing stars is approximately the same on each, and so cannot exert the differential action which would change the relative orbits of the constituent masses. Clearly there can be no question of any approximation to equipartition of energy in the internal motions of these systems. Nevertheless, it is possible to trace a steady sequence of configurations, beginning with almost circular orbits in which the two constituents are practically in contact—this being probably the condition of a system which has just formed by fission—and proceeding to orbits which are far from circular in shape, in which the components are at a substantial distance apart. It seems likely, although not certain, that this sequence is one of advancing age; when the parent star first breaks up to form a binary, the newly formed system starts at the first-mentioned end and moves gradually along the sequence. Now observation shows, beyond all doubt, that the stars at the far end of this sequence are substantially less massive than those at the beginning. We know the rate at which the various types of stars are radiating their mass away in the form of radiation, and from this we can calculate the time needed to produce the difference of mass which is observed to exist between the two ends of the sequence; again it proves to be a matter of millions of millions of years. Here the clock we use is the rate of outflow of radiation from a star, or its equivalent, the rate of loss of mass.

Against these various estimates must be set one piece of evidence which, if interpreted in the most obvious way, seems to point in exactly the opposite direction. This is, that the

中所称的“弛豫时间”；在这个类比中，单个恒星被当作单原子分子处理，双星被当作双原子分子，而运动星团的瓦解对应于气体的扩散过程。

当然，只有当我们假定其他恒星的引力作用是引起恒星运动和排列改变的唯一原因时，这些对恒星年龄的估计才成立。可以想象其他原因也是存在的，并且必然在某些方面有一定的贡献——辐射压力，由宇宙空间中的游离物质或弥散在空间中的宇宙云原子造成的轰击。但是计算表明源于这些过程的贡献是相当微不足道的。实际上，当我们把这些因素考虑进去时，对恒星运动的讨论就不再是一个天文学问题，而变为一个物理问题；我们必须把恒星当作是在物理介质中作布朗运动的“粒子”。做这样的处理之后，我们发现这个布满恒星的介质，以我们谈及运动着的布朗粒子的温度的方式而论，它的温度可达 10^{62} 度数量级。单个恒星和双星都表现出与此数量级温度相对应的能量均分，显而易见，物理中介作用，比如辐射压力和原子压力等建立平衡的温度要低得多，只有 10^{4} 度的量级，不可能对建立均分的过程有任何可观的贡献；它们不过是作为恒星运动的阻力罢了，平均而言往往是在减慢它们的速度。

在第二类双星系统，即光谱双星中，两颗子星是如此之近，以至于从它们附近经过的恒星对它们两个的吸引效果几乎相同，因此就不能以不同的作用效果改变双星系统组成物的相对轨道。显然对于这种系统的内部运动，毫无疑问可以采取任意近似方式以达到其能量均分的状态。尽管如此，仍然可以描绘出双星系统结构所固有的演变过程，从仍相互接触的两组分所处的近似圆轨道开始，这很可能就是一个系统刚刚分裂形成的状态；然后继续发展到形状远不是圆的轨道，此时这两个组分产生了实质上的分离。尽管还不能确定，但似乎可以认为这种演变过程就是一种老化的过程。当母星刚刚分裂形成双星系统时，新形成的系统从刚提到的结构变化的末端，逐渐沿着一系列变化过程演化。现在观测表明，毫无疑问，处于该演化过程末端的恒星所具有的质量大大低于初始阶段的质量。我们知道不同类型恒星以辐射形式损失质量的速率，由此可以计算出为产生存在于演化过程两端的质量差而需要的时间。结果再一次证明这个过程需要数万亿年。在这里我们所用的度量标准是恒星辐射的流失率，或与之相对应的，物质损失率。

反对这几种估算方法的人一定会提到这样一个证据，如果用最一目了然的方式解释的话，它似乎会得出完全相反的结论。即遥远的河外星云的谱线都发生了红移，

remote extra-galactic nebulae all show a shift of their spectral lines to the red, the amount of shift being approximately, although not exactly, proportional to the distance of the nebula. If this is interpreted in the most direct way, as a Doppler effect, the nebulae must all be scattering away from us and from one another in space, at so great a speed that the whole universe doubles its size about once in every 1,400 million years. Such a rate of increase seems quite inconsistent with the estimate which assigns ages of millions of millions of years to the stars. Calculation suggests that the original radius of the universe must have been of the order of 1,200 million light-years (Eddington), while the present radius of the universe appears to be only of the order of 2,000 million light-years (de Sitter). If these estimates could be treated as exact, we could fix the age of the universe definitely at just more than 1,000 million years, which is substantially less even than the age of the earth as indicated by its radioactive rocks. No one would claim any great degree of exactness for either of these estimates, especially the second, yet the general situation seems to forbid that the universe can have been doubling in size every 1,400 million years throughout a period of millions of millions of years.

Although alternative interpretations are tenable, none of them seems entirely convincing, and the present situation is extremely puzzling. While there is obviously room for much difference of opinion, many astronomers consider it likely that some other explanation of the apparent recessions of the nebulae will be found in time, in which event the road will be clear for the acceptance of ages of millions of millions of years for the stars, as suggested by the main bulk of astronomical evidence.

If such ages are provisionally accepted, calculation shows that the average star has already emitted many times its total mass in radiation; in other words, the average star must have started life with many times its present mass. Indeed, the sequence of spectroscopic binaries gives us a sort of picture of the life-history of a typical star. It starts with anything from ten to a hundred times the mass of the sun, and ends with a mass comparable to, or even less than, that of the sun. It is difficult to see where the enormous weight of the newly-born star can have been stored if not in the form of material atoms, or at any rate of material electrons and protons. Thus we are led to suppose that the life-history of the star is one of continual annihilation of its substance, the electrons and protons annihilating one another, and providing the energy for the star's radiation in so doing. Such, at least, is the conjecture suggested to us by astronomy; the testing of the conjecture rests with physics.

Highly Penetrating Radiation

If any direct evidence of this process of annihilation is to be obtained, it seems most likely that it will be found in the highly penetrating radiation which McLennan, Rutherford, and others discovered in the earth's atmosphere at the beginning of the present century. The reason, as we shall see later, is that here, and here alone in the whole of physics, we are dealing with photons of radiation whose mass is comparable with that to be expected in

红移量与星云的距离近似成正比，但不精确。如果这个现象被解释成最直接的原因，即多普勒效应，则这些星云必须全部从我们身边分散开去，并且在空间中互相远离。远离的速度是如此之大以至于大约每过 14 亿年宇宙的尺度就要增加一倍。这样的膨胀率似乎完全无法与之前对恒星存在了万亿年的估计相符。计算表明宇宙的初始半径必须在 12 亿光年量级（爱丁顿），而现在的宇宙半径似乎只有 20 亿光年量级（德西特）。如果可以把以上的这些估计看作是准确的，那么我们就可以确定宇宙的年龄只比 10 亿年多一些，甚至比由地球上岩石的放射性推得的地球年龄小很多。没有人对这两种估计的精确性提出太高的要求，尤其是对第二种，但总的情况似乎不允许宇宙在这万亿年间每 14 亿年就膨胀一倍。

尽管还有其他站得住脚的解释可供选择，但它们中没有一个可以让人完全信服，因而目前的局面令人十分困惑。尽管现在还有很多意见分歧，许多天文学家认为，其他一些关于星云存在明显退行的解释将很可能被适时发现。到那时，由大量天文学证据所支持的恒星已存在万亿年的观点就可以毫无阻碍地得到人们的认可了。

如果暂时接受这样的恒星年龄，那么通过计算可以得到，普通恒星已经辐射出相当于自身质量很多倍的能量了。换言之，普通恒星在诞生之初的质量必定是现在的很多倍。确实，光谱双星的演化过程为我们描画了一颗典型恒星的生命历程。它最初的质量是太阳质量的 10 到 100 倍，而终结时，它的质量与太阳的质量不相上下，甚至比太阳的质量还要小。很难想象，若不是以实物原子或者至少以实物电子和质子的形式，这些新生恒星的巨大能量能存储在哪里。因此我们就得到了这样的结论，恒星的生命历程就是其自身物质不断湮灭的过程，电子和质子相互湮灭，并以这种方式为恒星辐射提供能量。至少，从天文学上我们可以得到这样的推测，而检验这些推测则要有赖于物理学了。

高穿透性辐射

如果要找出湮灭过程的直接证据，那么该证据最有可能从本世纪初麦克伦南、卢瑟福和其他科学家在地球大气中发现的穿透力很强的辐射中获得。这是因为在整个物理学中只有这里涉及的辐射光子的质量与期望中由电子和质子湮灭产生的光子质量相当，这些我们在后面还会谈到。最近几年，赫斯、密立根和雷格纳等人十分

photons resulting from the annihilation of electrons and protons. In the last few years, this radiation has been studied in great detail by Hess, Millikan, Regener, and many others. Their investigations scarcely leave room for doubt that the radiation enters the earth's atmosphere from outer space; for which reason it is often described as "cosmic radiation".

It was at first taken for granted that this radiation must be of the nature of γ-radiation, since its penetrating power was greater than seemed possible for any kind of corpuscular radiation. This reason is now known to be inadequate, theoretical investigations having shown that corpuscular radiation, consisting of either α- or β-particles, might conceivably possess as high a penetrating power as the observed radiation.

Other arguments have, however, stepped into the breach, and show very convincingly that the radiation cannot be of the nature of either α or β radiation. The central fact is, in brief, that radiation which consisted of charged particles would be influenced by a magnetic field, whereas cosmic radiation is not. An electron or other charged particle in motion acquires magnetic properties in virtue of its motion; the faster it moves, the greater the force which a magnetic field exerts upon it. Now the penetrating power of the radiation under consideration is so great that it could only be attained by charged particles, if these were moving with very high speeds indeed. If a swarm of such particles became entangled in the earth's magnetic field, their high speed of motion would cause them to describe spiral paths coiled quite closely around the earth's lines of magnetic force, with the result that they would fall far more abundantly near the earth's magnetic poles than elsewhere. Epstein[1] estimates that for a shower of electrons to have the penetrating power of cosmic radiation, they would have to move with the energy produced by a fall through about 1,000 million volts, and has calculated that the incidence of electrons moving with this energy would be limited entirely to comparatively small circles surrounding the two magnetic poles. Actually the observed radiation falls so evenly on the different parts of the earth's surface that no variations have ever been detected. Members of the B.A.N.Z. Antarctic Expedition[2] found the same intensity of radiation within 250 miles of the south magnetic pole as they had previously measured in South Australia, and as others had found in the United States, Canada, and the North Atlantic. This seems to leave little room for doubt that the radiation is of the nature of very hard γ radiation.[3]

At first, some experiments by Bothe and Kohlhörster seemed to throw doubt on this conclusion. They had placed two Geiger counters, one vertically above the other, and found that the number of coincident discharges in the two counters was just about that which would be expected from purely geometrical considerations, if the radiation was corpuscular. Of course, the radiation which produced these ionisations was not necessarily the primary radiation which fell on the earth from outer space. Any primary radiation, as it traverses the atmosphere, is bound to produce secondary radiation of a variety of kinds, and any one of these might have been the immediate cause of the ionisation observed by Bothe and Kohlhörster. The primary radiation which first enters the earth's atmosphere might quite conceivably be electromagnetic, while the ionisation might be produced by a secondary corpuscular radiation.

细致地研究了这种辐射。他们的研究几乎不容置疑地表明这些辐射是从外太空进入到地球大气的；这就是这些辐射通常被称为“宇宙射线”的原因。

一开始人们理所当然地认为这种辐射本质上肯定是γ辐射，因为它的穿透力比被认为有可能的任何种类的粒子辐射都强。现在看来这个理由并不充分，理论研究已经证明粒子辐射，包括 α 粒子和 β 粒子的辐射，可能拥有和观测到的辐射一样强的穿透力。

然而其他的论证已经十分令人信服地表明这种辐射不可能是 α 或 β 辐射。简言之，其中心论点是，由带电粒子构成的辐射会受到磁场的影响，而宇宙辐射则不会。运动着的电子或其他带电粒子会受到磁场的作用，这是由于它在运动；它运动得越快，磁场对它施加的力也就越大。这种辐射的穿透力非常强，只有以很高速度运动的带电粒子才能达到这么强的穿透力。如果一大群这样的粒子进入地球磁场，高速运动会使它们绕着非常接近地球磁力线的轨道做螺旋运动，结果在地球磁极附近落下的粒子要远远多于其他地方的粒子。爱泼斯坦[1]估计对于一大批电子，若要具有宇宙辐射那样的穿透力，它们必须以大约穿过 10 亿伏特才能获得的能量运动，他还计算出以这个能量运动的入射电子将被完全限制在两磁极附近的相对较小的圆里。而实际上，观测到的辐射非常平均地落在地球表面的各个区域，并没有发现明显的差异。英国、澳大利亚、新西兰共同组成的南极探险队[2]的成员发现，在南磁极周围 250 英里区域内的辐射强度，与他们之前在南澳大利亚测得的，以及其他人在美国、加拿大和北大西洋得到的结果相同。此辐射具有极硬 γ 辐射的性质这一点似乎是无可置疑的[3]。

最初，博特和科尔霍斯特的一些实验结果似乎不支持这个推论。他们设置了两个盖革计数器，将其中一个垂直置于另一个的上面，并且发现如果这个辐射是粒子辐射，两个计数器记录的同时放电的粒子数几乎与用纯几何方式预计的结果相同。当然，产生这些电离的辐射不一定是从外太空进入地球的初级辐射。当任何初级辐射穿过大气层时，都必然会产生各种类型的次级辐射，其中任一种次级辐射都可能会直接导致博特和科尔霍斯特在实验中观察到的电离作用。首先进入地球大气的初级辐射非常可能是电磁辐射，而造成电离的辐射则是次级粒子辐射。

To examine this possibility, Bothe and Kohlhörster placed a block of gold between their two counters. This naturally caused a reduction in the number of coincidences, and from the amount of the reduction it was possible to calculate the penetrating power of the radiation which actually effected the ionisations. It was found to be approximately the same as that of the primary radiation. So far, then, everything could be explained by supposing that it was the primary radiation itself which produced the ionisations in the counters, and that this was corpuscular in its nature.

Recently this explanation has been tested by Moss–Smith[4] and found wanting. He extended the apparatus used by Bothe and Kohlhörster, by mounting yet a third counter vertically below the original two, and first verified that the number of coincident ionisations in all three counters was that which their geometrical arrangement would lead us to expect. Now if the radiation which produced these ionisations were corpuscular, it ought to be deflected by a magnetic field. For example, if a sufficiently strong magnetic field were inserted between the second and third counters, the third counter ought to be entirely shielded from the radiation which had passed through the first two counters, so that the number of coincident ionisations in the first two counters would remain as before, while the number in the third counter would fall to zero. Moss–Smith found that this did not happen. Although his magnetic field had many times the strength needed to shield the third counter completely, its insertion had no effect on the number of coincident ionisations. This showed that the ionising radiation was not corpuscular, and as Bothe and Kohlhörster had already shown that the ionising radiation was probably identical with the primary radiation, it confirmed the theoretical arguments of Millikan and Epstein, which proved the primary radiation to be of the nature of γ radiation.

The Mode of Production of the Radiation

If the primary radiation is of the nature of γ radiation, as these arguments and experiments seem to show, its origin ought to be disclosed by its penetrating power. Such radiation consists of photons, which may be compared to bullets, all moving with the same speed—the velocity of light. Their penetrating power accordingly depends solely on their mass, and a theoretical investigation enables us to deduce the one from the other. Every photon is, however, produced originally by an atomic upheaval, and its mass is exactly equal to the decrease of mass which the parent atom experienced as the result of this upheaval. For example, if the atom was one of hydrogen and the upheaval consisted of annihilation, the photon resulting from this annihilation must have a mass exactly equal to the original mass of the hydrogen atom, namely, 1.66×10^{-24} gm. Or again, if a proton and an electron mutually annihilate one another in any atom whatever, thus reducing its atomic weight by unity, the mass of the resulting photon must be equal to the combined masses of the proton and electron in situ in the atom, which again, except for a small "packing-fraction" mass, is equal to the mass of a hydrogen atom.

The most effective means of investigating the penetrating power of cosmic radiation is to sink suitable apparatus to varying depths below the surface of a lake, and observe the

为了检验这种可能性，博特和科尔霍斯特在两个计数器之间放置了一块金板。这自然就会造成同时发生放电的粒子数目减少，并且通过减少的数量就可以计算出真正引起电离的辐射的穿透力。结果发现它的穿透力和初级辐射的穿透力近似相同。假若是初级辐射本身造成了计数器中的电离作用，并且它本质上是粒子辐射，那么到目前为止的所有现象都可以得到解释。

最近莫斯–史密斯[4]对这个解释进行了验证并发现了不足之处。他改进了博特和科尔霍斯特使用的设备，在原有的两个计数器下面加入了垂直于它们的第三个计数器。他首先证实在三个计数器中同时发生的电离个数正是我们通过它们的几何排列得到的预期值。现在，如果引发这些电离的是粒子辐射，那么它应该在磁场中发生偏转。例如，如果一个足够强的磁场加到第二和第三个计数器之间，第三个计数器应该被完全屏蔽而免受从前两个计数器穿过的辐射，因此前两个计数器的计数将保持不变而第三个将减少为零。但是莫斯–史密斯发现这种情况并没有发生。尽管他的磁场强度已经是完全屏蔽第三个计数器所需强度的很多倍，但这个磁场的加入并没有对同时电离的计数情况产生任何影响。这表明电离辐射不是粒子性的，而博特和科尔霍斯特也已经指出电离辐射很可能就是初级辐射，这样就证实了密立根和爱泼斯坦的理论依据，从而得到初级辐射具有γ辐射的性质。

产生辐射的模式

如果就像前面的讨论和实验所表明的那样，初级辐射具有γ辐射的性质，那么它的起源就应该能够通过其穿透力揭示出来。这样的辐射由光子构成，可以把它们比作许多发以相同速度——光速运动着的子弹。因而它们的穿透能力也仅由其质量决定，并且我们可以通过理论研究从一个光子推导出另一个光子。然而每个光子都起源于原子激变，并且光子的质量精确地等于母原子在这个激变过程中减少的质量。例如，如果这里的原子是氢原子且激变由湮灭过程组成，那么湮灭产生的光子的质量必须精确等于氢原子的原始质量，即1.66×10^{-24} g。再比如，无论什么原子中的一个质子和一个电子相互之间发生了湮灭，这在整体上减少了原子量，那么产生的光子的质量必定等于原子中原来位置上的电子和质子质量之和。除去少量的“结合部分”的质量外，它也与氢原子的质量相等。

研究宇宙辐射穿透力最有效的方式，是把合适的仪器沉入到湖面以下不同深度，并观测入射的宇宙射线在不同深度被水吸收后引起的电离作用。密立根、雷格纳和

ionisation produced by the incidence of the cosmic rays after absorption by varying depths of water. Observations of this type have been performed with great care and skill by Millikan, Regener, and others.

Their results are none too easy of interpretation. L. H. Gray has shown[5] that there is a sort of softening effect continually in progress by which the absorption of a quantum of energy produces a recoil electron, which in turn produces radiation of energy comparable to, although somewhat lower than, the energy of the original quantum. After the radiation has travelled through a certain thickness of absorbing material, the observed ionisation no longer gives a true measure of the intensity of the primary radiation which has escaped absorption, but of this primary radiation in equilibrium with all its softer secondary components.

When this complication has been allowed for, the ionisation curve gives the intensity of the true primary radiation which remains after passing through varying thicknesses of absorbing matter. If this primary radiation consists of a mixture of constituents of different and clearly defined wavelengths, so that it has a line spectrum in the language of ordinary optics, these different constituents will have different coefficients of absorption. In such a case, it ought to be possible to analyse the observed curve into the superposition of a number of simple exponential curves, one for each constituent of the radiation.

Actually, it is found that this can be done. Different experimenters do not obtain results which are altogether accordant, but all agree in finding that there is a long stretch, near the end of the range of the radiation, over which its intensity decreases according to a simple exponential law. This can only mean that one particular constituent of the radiation is so much harder than the others that it persists in appreciable amount after traversing a thickness of matter which has completely absorbed all the softer constituents. Regener, who has studied the problem in great detail, finds that the hardest radiation of all has an absorption coefficient of 0.020 per metre of water. Other experimenters have found values which agree with this to within about 10 percent.

The mass of the photon can be deduced from the observed absorption coefficient μ of the radiation, by the use of a theoretical formula given by Klein and Nishina.[6] This can be written in the form

$$\mu = \frac{2\pi N e^4}{m^2 c^4} f\left(\frac{M}{m}\right),$$

where M is the mass of the photon, m of an electron, e, c have their usual meanings, and f represents a fairly complicated function of M/m. In all the applications of the formula to cosmic radiation, M/m is quite large, and for such values of M/m, f assumes the form

$$f\left(\frac{M}{m}\right) = \frac{1}{4}\left(\frac{M}{m} + 2\log\frac{2M}{m}\right).$$

其他一些人非常细心巧妙地完成了这样的实验观测。

他们的结果很难解释。格雷指出[5]有一种软化效应持续在起作用。由于这种效应持续在起作用，当一个能量量子被吸收时，就会产生一个反冲电子，这个电子反过来又引起了能量辐射。辐射的能量与原始量子的能量相比，尽管略低一些，但也大致相当。当辐射在吸收介质中传播一段距离之后，观察到的电离就不再能反映未被吸收的初级辐射的真实强度，而是对应于与所有较软的次级辐射部分保持平衡的初级辐射的值。

考虑到这个复杂的情况以后，就可以从电离曲线中得到穿过不同厚度吸收介质之后依然存在的真实的初级辐射强度。如果初级辐射包含波长不同的成分，那么它就会有像普通光学那样的线状谱，这些不同成分的吸收系数是不同的。在这种情况下，应该可以把观测到的曲线看作是多条简单指数曲线的叠加，每一个辐射成分都对应着一条指数曲线。

实际上人们发现这是可以做到的。不同的实验者并没有得到完全一致的结果，但是他们都一致认为在辐射区域的末端附近有一段很长的延伸范围，在这个范围内辐射强度呈简单的指数率衰减。这只能说明辐射中存在着一个特殊的成分，它比其他成分硬很多，以至于在经过一定厚度的介质后，所有较软的成分都已经被完全吸收，而这个成分的辐射仍然保持着相当的强度。雷格纳在对上述现象进行了仔细的研究之后发现，全部辐射中最硬的那个成分在被水吸收时的吸收系数为每米 0.020。其他实验者得到的数值与此一致，偏差不超过 10%。

利用克莱因–仁科给出的理论公式，我们可以从观测到的辐射吸收系数 μ 中推导出光子的质量[6]。这个公式可以由下面的形式给出：

$$\mu = \frac{2\pi N e^4}{m^2 c^4} f\left(\frac{M}{m}\right),$$

其中：M 为光子质量，m 为电子质量，e、c 为它们的通常定义，f 表示以 M/m 为变量的一个相当复杂的函数。对于所有的宇宙辐射，该公式中的 M/m 都非常大，而当 M/m 取这些值时，可以假设 f 有如下形式：

$$f\left(\frac{M}{m}\right) = \frac{1}{4}\left(\frac{M}{m} + 2\log\frac{2M}{m}\right)。$$

These formulae are calculated on the supposition that the absorption is caused by N electrons per unit volume, and that these are entirely free. This last condition can never be fully realised in Nature, since every electron is bound, more or less closely, to other electric charges. If an electron is bound to a system of mass m', we can allow for this binding by increasing m in the formula by a fraction of m', the fraction being large or small according as the coupling is tight or loose. Thus a loosely coupled electron behaves almost like a free electron, but an electron coupled tightly to a massive system, such, for example, as a proton or an atomic nucleus, behaves like an electron of very great mass, and the formula shows that this has no appreciable absorbing power.

The Klein–Nishina formula has been tested by comparing it with observation for γ-rays. In the case of the lighter elements, it gives values which agree well with the observed absorption, provided all the extra-nuclear electrons are treated as free, while the nuclear electrons are disregarded entirely. It is natural to disregard these, because the coupling of nuclear electrons in the lighter elements is known to be so close that even the hardest γ-rays make but little impression on them. This is true for the lighter elements only; in the case of lead, Chao[7] has found an additional scattering of the hardest γ-rays, which he believes to be of nuclear origin. In other words, he finds that some at least of the nuclear electrons in lead are not so closely coupled as to resist the onslaught of the hardest γ-radiation. Still less, then, can they be so closely coupled as to resist the incidence of the far more massive photons of cosmic radiation. From theoretical considerations of a very general nature[8] it appears probable that in dealing with cosmic radiation, the N in the Klein–Nishina formula should refer to all electrons, nuclear as well as extra-nuclear, and not merely to the latter. A further term ought also to be added to represent scattering by nuclear protons, but calculation shows that this is entirely insignificant in amount. The result of taking the nuclear electrons into account is to replace atomic number by atomic weight, so that the absorption by a given thickness of matter becomes strictly proportional to the mass of the matter, and absolutely independent of its nature, except possibly in so far as a further small absorption, caused by photoelectric action, may depend on the latter. The effect of this is to double, or more than double, the capacity of all atoms except hydrogen for absorbing cosmic radiation; it increases the absorbing power of water to 80 percent above the value usually calculated.

The following table shows the absorption coefficients (per metre of water) which I have calculated for the radiation produced by the synthesis of iron and by the annihilation of 1 and 4 protons respectively, with their accompanying electrons. The calculation is based on the Klein–Nishina formula, all electrons, including the nuclear electrons, being treated as absolutely free:

Process	$\frac{M}{m}$	Calculated μ (per metre, water)	Observed μ (Regener)
56H → Fe	876	0.136	..
+, − → 0	1845	0.071	0.073
4+, 4− → 0	7380	0.020	0.020

这些公式的计算基于这样的假设：吸收是由单位体积内 N 个完全自由的电子所引起的。电子是完全自由的这个条件在自然界中永远不可能完全满足，因为每个电子都会被其他电荷束缚着，不论束缚是强是弱。如果一个电子被一个质量为 m' 的系统所束缚，那么对于这种情况我们可以考虑把 m' 的一部分增加到公式中的质量 m 上，增加部分的大小取决于耦合的强弱。因此弱耦合电子的表现几乎与自由电子一样，但是对于一个与质子或原子核等大质量系统紧紧耦合的电子，它会表现得如同质量非常大的电子，公式表明在这种情况下它不具有明显的吸收能力。

通过与观测到的γ射线进行比较，人们已经检验了克莱因－仁科公式。对于较轻的元素，它给出的值与观测到的吸收曲线符合得相当好，条件是这里所有的核外电子都被当作自由电子处理，同时完全忽略核内电子。这样的忽略是自然的，因为我们知道较轻元素的核内电子耦合很强，即使是最硬的γ射线也只会对它们造成很小的影响。但是，这仅对较轻元素成立；就铅而言，赵[7]发现在最硬的γ射线下存在一个附加散射，他认为这来源于原子核。换句话说，他发现至少铅的某些核内电子的耦合强度还抵御不了最硬 γ 辐射的冲击。更不用说以它们的耦合强度去抵御数量巨大的宇宙辐射光子的冲击。从理论角度作非常一般性的考虑[8]，在处理宇宙辐射时，很可能克莱因－仁科公式中的 N 指的是包括核内和核外的所有电子，而不应该仅仅是后者。此外还应该增加一项，用来表示原子核内质子的散射，然而计算表明质子散射的作用量很小，完全可以忽略不计。把核内电子计算在内的结果就是要用原子量取代原子序数，因此对于给定厚度的介质，它的吸收就严格正比于射线所经过区域的介质质量，而与它的性质毫无关系，除非是由光电效应造成的更少量的吸收有可能与后者相关。这样就使除氢以外的所有原子在吸收宇宙辐射的能力上加倍，甚至大于原来的两倍。经过这样的处理，水的吸收能力比通常的计算值高出了80%。

对于由铁的合成以及由 1 个质子和 4 个质子分别与它们的伴随电子发生湮灭而产生的辐射，我分别进行了计算，并在下表中给出了吸收系数（水中，每米）。结果是根据克莱因－仁科公式计算出来的，计算过程中把所有的电子，包括核内电子，都视为自由电子：

过程	$\frac{M}{m}$	计算值 μ（水中，每米）	观测值 μ（雷格纳）
56H → Fe	876	0.136	..
+,– → 0	1845	0.071	0.073
4+,4– → 0	7380	0.020	0.020

The last column gives the absorption coefficients of the two most penetrating constituents of cosmic radiation, as analysed by Regener. Their agreement with the figures in the preceding column is probably well within errors of observation and analysis, and is rather too good to be attributed with much plausibility to mere accident; the odds against a double agreement, within 5 percent in one case and 2.7 percent in another, being about 3000 to 1. This seems to me to suggest quite strongly that the most penetrating constituent so far observed in cosmic radiation may originate in the annihilation of an α-particle and its two neutralising electrons (the components of a helium atom), while the next softer constituent may originate in the annihilation of a proton and its one neutralising electron (the components of a hydrogen atom).

An alternative possibility, which was first suggested by Millikan and has been championed mainly by him, is that the cosmic radiation may result from the building of electrons and protons into atoms. Yet the hardest constituents of the cosmic radiation appear to be far too hard to be produced by the synthesis of iron, while Millikan himself considers that the synthesis of heavier elements is probably ruled out by their rarity in the universe. If, as I have suggested, the annihilation of matter is the true origin of the two hardest constituents of the cosmic radiation, then it becomes possible to suppose, with Millikan, that the softer constituents are produced by the synthesis of simple atoms into complex. Many will, however, hesitate to accept such a mixed origin for the radiation. It certainly seems simpler to suppose that the two hardest constituents, and these alone, form the fundamental radiation, while all other constituents represent mere softened or degraded forms of these. Yet this supposition brings its own difficulties, since if we measure the intensity of the radiation by its ionising power, the supposed secondary radiation is found to have many times the ionising intensity of the primary. But whatever the origin of the softer constituents may be, the two hardest constituents, with their photons equal in mass to the atoms of hydrogen and helium respectively, appear to provide weighty evidence that matter can be, and is, annihilated somewhere out in the depths of space. If we can assume that this process occurs on a sufficiently large scale, this supposition brings order and intelligibility into a vast series of problems of astronomy and cosmogony in a way in which no other suppositions can.

The Place of Production of the Cosmic Radiation

Various suggestions have been made as to the place of origin of this highly penetrating radiation. Many of them are put out of court by the fact, which must now, I think, be regarded as well established, that the radiation is nearly constant in intensity at all times of day and night,[9] any variation being, at most, of the order of one part in 200. There seems to be a real variation of this amount, but in the main it appears to follow the variation of the barometer. Millikan considers that it is adequately explained by fluctuations in the absorbing power of the air blanket formed by the earth's atmosphere. It was at one time suggested that the radiation might consist of electrons ejected from thunder-clouds high up in the earth's atmosphere, or of electrons moving with enormous speeds acquired by drifting through electrostatic fields in space, the potential gradients in these fields being

最后一列给出的是宇宙辐射中两种穿透力最强的成分的吸收系数，这是雷格纳通过分析得到的。观测值与前一列中的计算值也许相符得还不错，其偏差在观测和分析所引起的误差范围内。很难把这么好的符合度仅仅归因于巧合。两次吻合，一次误差在 5% 以内，另一次在 2.7% 以内，出现错误的概率为 1/3,000。在我看来这个结果强烈地显示出：在迄今为止观测到的宇宙辐射中，穿透力最强的成分有可能起源于一个 α 粒子和它的两个使其变为电中性的电子（组成氦原子的成分）的湮灭，而另一个较软的成分可能来源于一个质子和一个使其变为电中性的电子（组成氢原子的成分）的湮灭。

由密立根首先提出并且主要由他倡导的另一种可能性是，宇宙辐射起源于电子和质子形成原子的过程。然而宇宙辐射中最硬的成分似乎比铁合成过程中产生的辐射硬很多，而密立根自己也认为更重元素的合成可以被排除掉，因为它们在宇宙中是很稀少的。如果按照我提出的，物质的湮灭是宇宙辐射中最硬的两个成分的真正来源，那么我和密立根设想，很可能较软的成分是在简单原子组成复合原子的过程中产生的。但是很多人会对宇宙辐射有多种起源表示怀疑。显然比较简单的假设是认为只有两种最硬的成分构成了基本辐射，而所有其他成分仅仅是这部分辐射的软化或降阶形式。但是这种假设给自己带来了一些难题，因为如果我们用电离能力去衡量宇宙辐射的强度，那么这个假设的次级辐射的电离强度会是初级辐射的很多倍。但是无论较软成分的来源是什么，这两个最硬成分中的光子所具有的质量分别与氢原子和氦原子的质量相同，这似乎为物质可以在遥远的空间中湮灭提供了有力的证据。如果我们假定这个过程发生在一个足够大的尺度内，那么这个假设就会使大量的天文学和宇宙学问题变得有条理并且易于理解，这是其他假设无法做到的。

宇宙辐射产生的位置

对于这种高穿透性辐射起源于何处人们有不同的看法。其中的很多观点都被以下这个在我看来现在已被牢固确立的事实排除掉了，即无论是白天还是黑夜，宇宙辐射的强度在任何时候几乎都是一个常数[9]，它的变化幅度至多为 1/200 量级。这个量似乎确实会发生一定的变化，但主要是随着气压而变化。密立根认为这些变化用地球大气层吸收能力的涨落就足以解释。曾经有人认为这些辐射可能包含那些由地球大气中的高空雷雨云发射出来的电子，或者包含因在空间静电场中漂移而获得极高速度的运动电子。这个静电场的电势梯度虽然微不足道，但是由于电场范围极为宽广，因而电势差非常大。即便仍然认为宇宙辐射由粒子辐射产生，也很难将这

slight, but the potential differences immense simply on account of the vast extent of the fields. Even if the radiation could still be treated as corpuscular, it would be very difficult to reconcile either of these suggested origins with the steadiness and uniformity with which the radiation falls on the earth's surface.

The fact that the intensity of the radiation is very approximately independent of both solar and sidereal time seems to show that no appreciable part of the radiation comes from the sun or stars. Counting the sun as a star, we receive more than 100,000,000 times as much starlight at midday as at midnight, yet apart from the purely local "barometer" effect just mentioned, we receive the same intensity of the radiation at both times. The fact that the intensity is approximately independent of the position of the Milky Way seems to show that the bulk at least of the radiation must come even from beyond the confines of the galactic system, thus justifying the name "cosmic radiation".

Where, then, does the radiation originate? For reasons which will be clear at the end of our quest, it is simpler to conduct our search in time rather than in space. The average density of matter in space is probably of the order of 10^{-30} gm. per c.c., and in each second of its existence, a beam of cosmic radiation passes through a layer of space 3×10^{10} cm. thick. Thus every second it passes on the average through 3×10^{-20} gm. per sq. cm. of its cross-section. We have, however, seen that the hardest constituent must pass through 50 gm. per sq. cm. before it is reduced in intensity by one percent, and this requires an average time of 16×10^{20} seconds, or about 5×10^{13} years—a period which, on any reckoning, is greater than the age of the stars; its intensity is reduced to $1/e$ times its original value after 5×10^{15} years, which is greater, so far as we know, than the age of the universe.

Thus, to an approximation, we may think of the hardest constituent of the cosmic radiation as indestructible, since the universe has not yet existed long enough for any appreciable amount of it to be absorbed. To a slightly less good approximation, the same is true of the softer constituents. This leads us to regard space as being permeated with all, or nearly all, of the cosmic radiation which has ever been generated since the world began. The rays come to us as messengers, not only from the farthest depths of space, but also from the remotest eras of time. And, since we cannot produce cosmic rays on earth, their message appears to be that the physics which prevails out in these far depths of space and time is something different from our terrestrial physics: different processes result in different products. So far as we can read the riddle of the rays, one at least of these processes appears to be the annihilation of matter, although whether this annihilation is taking place now, or occurred only in the remote past, or even only at the beginning of the world's history, we have no means of knowing; all that the rays show is that somewhere and sometime in the history of the universe, matter has been annihilated.

Similar remarks may be made with respect to the softer constituents. Millikan believes that these originate in the synthesis of complex atoms out of lighter ones, and so argues

些假定起源中的任何一个与落向地球表面的辐射所具有的稳定性和均匀性统一起来。

辐射强度与太阳时和恒星时近似无关这一事实似乎表明，可测到的辐射不会来源于太阳或者恒星。如果将太阳当作普通恒星处理，那么我们在正午接收到的星光是在午夜接收到的星光强度的 100,000,000 倍。然而排除刚才提到的纯粹由局部“气压”不同引起的作用之后，我们在这两个时刻得到的辐射强度仍然相同。而辐射强度与银河系的位置几乎无关这一事实也说明，宇宙辐射中至少有一大部分会来自于银河系范围以外，这也印证了“宇宙辐射”这个名字是恰如其分的。

那么这些辐射是从哪里来的？为了能够最终找到原因，从时间入手比从空间入手更简单。宇宙中的平均物质密度约为 10^{-30} g/cm^3，并且在宇宙存在的每一秒钟，一束宇宙射线会穿越 3×10^{10} cm 厚的宇宙空间层。因此每秒钟它穿越的辐射截面是 3×10^{-20} g/cm^2。然而我们已经发现，这个最硬成分的强度要减少 1%，就必须穿过 50 g/cm^2，这个过程所需的平均时间为 16×10^{20} 秒，或者约 5×10^{13} 年——以任何方式计算，这个时间都要比恒星的寿命长；辐射强度减小到初始值的 $1/e$ 需要 5×10^{15} 年，比我们现在认为的宇宙的年龄都要长。

因此，我们可以近似认为宇宙辐射中最硬的成分是不会衰减的，因为宇宙存在的时间长度还不足以使它被吸收的部分达到任何可观的量。也可作一个稍差一点的近似，认为较软的成分也是不会被吸收的。这使我们可以把宇宙空间看成是充满了自宇宙诞生以来到现在曾经产生的所有或几乎所有的宇宙辐射。我们接收到的这些射线不仅带来了太空最深处的信息，还带来了时间最久远的年代的信息。并且，因为我们不能在地球上制造出宇宙射线，所以它们所携带的信息显示出这些普遍适用于遥远空间和时间的物理定律有别于我们地球上的物理定律：不同的过程导致不同的结果。就我们现在对宇宙射线之谜的解读情况来看，其中至少有一个过程应该是物质的湮灭，尽管我们没有办法知道这种湮灭是发生在现在，还是发生在遥远的过去，或者甚至仅是在宇宙诞生之初；这些射线的存在已经能够充分说明，在宇宙演化史中的某个地点和某个时刻，物质发生了湮灭。

类似的说明也适用于较软的成分。密立根确信它们产生于较轻原子合成复杂原子的过程之中，并因此认为这种合成还在进行。但是由于这些较软的成分同样具有

that the act of creation is still in progress. But these softer constituents also have such high penetrating powers as to be virtually indestructible. Even if Millikan's interpretation of the origin of these rays were established, it would only prove that synthesis of matter had occurred somewhere and sometime during the long past history of the universe; it would not prove that any such synthesis was still in progress.

Indeed, the fact that the radiation does not vary in intensity with the position of the Milky Way may be thought to suggest that it is merely a relic of past eras in the history of the universe. It may be argued that if the radiation were still being generated, the huge mass of the Milky Way, comparatively close to our doors, would surely make its influence felt. It is, however, possible (and, I think, likely) that the radiation is still being generated in extra-galactic nebulae of earlier type than the galactic system; it may be that they only emit this radiation before they condense into stars; and that the atoms which can produce such radiation in the galactic system are all shut up inside the stars, so that the radiation is transformed into starlight before it reaches us.

Millikan has estimated that the total amount of cosmic radiation received on earth has about a tenth of the energy of starlight, sunlight not being counted in. Near the earth, the energy of radiation from the stars is intense enough to raise space to a temperature of about 3.5 degrees absolute, whereas the energy of cosmic radiation will raise this space only to about 2 degrees absolute. Out in the inter-galactic darkness the position is reversed. Here the feeble starlight and star-heat from distant galaxies can at most raise space to a fraction of a degree above absolute zero, but the intensity of cosmic radiation is probably the same as nearer home, corresponding to about 2 degrees absolute. Space as a whole appears likely to contain far more of cosmic radiation than of light and heat, although in assessing this fact, we must remember that cosmic radiation is virtually endowed with immortality, whereas ordinary radiation, in the form of light and heat, is not. The total annihilation of all the matter in the universe would raise space to about 10 degrees absolute, so that the cosmic radiation we observe could be produced by the annihilation of quite a small fraction of the universe.

This is not surprising, since the cosmic radiation which pervades space is necessarily quite distinct from the similar radiation which astronomers regard as the source of stellar light and heat. The annihilation of matter in stellar interiors would produce radiation of exactly the same high frequency as the observed cosmic radiation, but as this radiation fought its way outwards to lower temperatures, and finally to outer space, it would be continually softened, by a long succession of Compton encounters, until it finally emerged in the familiar form of starlight—ordinary temperature radiation at anything from 1,650° abs. to about 60,000° abs.; none of it could reach the earth in its original form.

The mere fact of its not having been completely absorbed shows that the cosmic radiation we receive on earth cannot have passed through more than a few kilometres of stellar matter at most; its penetrating power, high though it is, will not carry it through a greater thickness of matter than this. Consequently, it can scarcely have been generated at a

很强的穿透力，因此实际上也是不会衰减的。即便密立根对这些射线起源的解释可以成立，也只能证明在宇宙漫长的发展史中，物质的合成过程曾经在某个地点和某个时刻发生过，而不能证明这样的合成仍在进行。

实际上，辐射强度与银河系的位置无关这一事实也许可以这样来理解：宇宙辐射仅仅是宇宙发展史中的一个遗迹。如果这个辐射过程还在继续，而银河系的主体又离我们那么近，显然我们应该察觉到它所造成的影响。然而还有可能（我认为非常可能）这个辐射仍然在不断地从比银河系存在时间更长的河外星云中产生出来，也许它们只在凝聚成恒星之前发出这样的辐射，而银河系中能产生这种辐射的原子都被密闭在了恒星的内部，所以辐射在到达我们这里之前已经被转变为星光了。

密立根曾经估算过，在不把太阳光计算在内的情况下，地球上接收到的宇宙辐射的总能量大约为星光能量的 1/10。在地球附近，恒星发出的辐射能量足以使空间的绝对温度上升大约 3.5 度，而宇宙辐射的能量则只能使此空间的绝对温度升高 2 度。在星系间的黑暗区域，情况则截然相反。在那里，来自于遥远星系的微弱星光和恒星热量最多只能使空间温度稍稍升高到绝对零度以上不到 1 度，而宇宙辐射的强度则可能和地球附近的强度相同，对应的绝对温度大约为 2 度。总体来说，宇宙空间包含的宇宙辐射很可能会远远多于光和热，虽然在确定这一点的时候我们也必须记得，事实上宇宙辐射是永远不会消失的，而以光和热为呈现方式的普通辐射则会。宇宙中所有物质的完全湮灭将使空间的绝对温度上升 10 度左右，因此，产生我们观测到的宇宙辐射只需要湮灭掉极小部分的宇宙物质。

遍及空间的宇宙辐射和那些被天文学家们认作是恒星光和热来源的相似的辐射必然是明显不同的，这并不奇怪。恒星内部物质的湮灭会产生频率和观测到的宇宙辐射完全相同的高频辐射，但是当这种辐射向恒星外部温度较低的地方传播并最终到达外层空间时，它会因长期发生连续的康普顿散射而被不断软化，直到最终显现为我们所熟悉的星光形式——从绝对温度 1,650 度到绝对温度约 60,000 度之间的普通温度辐射，这种辐射不可能以原始形式到达地球。

宇宙辐射没有被完全吸收的事实足以说明：我们在地球上接收到的宇宙辐射，不可能曾穿透过超过几公里的恒星物质；尽管它的穿透能力很强，但也不能穿过比这更厚的物质。因此，在温度高于 100,000 度的地方几乎不可能产生这种辐射。我

place where the temperature was more than about 100,000 degrees. We must suppose that it originated fairly near to the surfaces of astronomical bodies, or, more probably still, in unattached atoms or molecules in free space. In contrast to this, the radiation which provides the energy poured out by the stars was probably generated in their central regions. Thus it must have been generated in matter at very high temperatures, while the similar radiation we receive on earth must have been generated at comparatively low temperatures.

Physical Principles

According to classical theories of electro-magnetism, any acceleration of a moving electron is accompanied by an emission of radiation, of amount given by the well-known formula of Larmor. Thus an electron, describing an orbit in an atom of, say, hydrogen, must continually radiate energy away, so that the orbit will continually shrink.

The quantum theory replaces this continuous emission of energy by a succession of discontinuous emissions; at each moment there is a definite calculable chance that the orbit will shrink in size by a finite amount, and emit a photon in the process. The orbit of lowest energy is anomalous; when an electron is describing this orbit, no further shrinkage in orbit or emission of radiation is possible.

The concept of annihilation of matter removes this anomaly by providing a state of still lower energy, in which proton and electron have both disappeared in radiation. The energy emitted in the process of annihilation corresponds, of course, to that which would be emitted continuously on the classical electro-dynamics while the orbit was shrinking to zero radius.

Although neither the new quantum theory nor the theory of wave mechanics in any way predicts that this process must actually happen, they are in no way definitely antagonistic to its occurrence. Certain forms of both, on the whole, seem rather to favour the possibility, but theoretical calculation based on these does not at present agree with numerical estimates derived from astronomical evidence. Dirac[10] has recently calculated the probability of annihilation given by the new quantum theory, and obtained a value which is substantially too large; according to his calculations, the universe ought to have dissolved into radiation long ago. Or, to put the same thing in another way, the stars ought to radiate energy far more furiously than they do.

The general principles of the quantum theory show that annihilation of matter might either occur spontaneously, after the manner of radioactive disintegration, or might be incited by a sufficiently high temperature, like the atomic changes which produce ordinary temperature radiation. The second process will only occur when the matter is traversed by photons with energy equal to that set free by annihilation of matter; the requisite temperature is found to be of the order of a million million degrees. Now it is quite impossible that the cosmic radiation we receive on earth can have originated in

们必须假设它产生于非常接近天体的表面的地方，或者，更有可能的是，产生于自由空间中的独立原子或分子。与此相反，能使恒星发出能量的辐射则可能来源于恒星的中心区域。因此它必定是在温度非常高的物质中产生的，而我们在地球上接收到的相似的辐射应该产生于相对较低的温度。

物理原理

根据经典电磁理论，任何作加速运动的电子都会产生辐射，其辐射量可由著名的拉莫尔公式给定。因此，某一个原子，比如说氢原子，其中沿一定轨道运动的电子必须持续地辐射出能量，因而其运动轨道将不断收缩。

量子理论用一系列不连续的辐射取代了这种连续的能量辐射，在每一个时刻都可以确切地计算出轨道收缩一个固定值并放出光子的概率。能量最低的轨道是反常的，当一个电子处于这个轨道时，它既不可能再降低轨道，也不可能再发出辐射。

物质湮灭的概念提出了一个能量更低的状态，消除了这个反常情况。在这种状态下，质子和电子同时消失并转化为辐射。当然，湮灭过程中放射的能量，应该与经典电动力学中轨道半径收缩为零时持续放射的能量相当。

尽管新的量子理论和波动力学理论都没有以任何方式预言这个过程真的一定会发生，但是它们也没有以任何确定的方式来否定它的出现。两种理论的特定形式总体上看来都颇为支持这种可能性，但是基于这两种理论的计算结果目前尚不能与通过天文学观测证据估算的数值相符。最近，狄拉克[10]根据新的量子理论计算了湮灭概率，但是得到的数值明显太大了。根据他的计算结果，宇宙早就应该变为辐射消失了。或者换个说法，恒星放出能量的方式应该比现在猛烈得多。

量子理论的普遍原理指出，物质的湮灭既可能是在放射性衰变之后自发发生的，也可能是被足够高的温度所激发，就像发出普通辐射的原子变化一样。第二种情况只能在当穿过物质的光子所具有的能量与物质湮灭释放的能量相等时才会发生，其需要的温度要达到万亿度的量级。而目前我们在地球上观测到的宇宙辐射根本不可能是从温度能达到这么高的区域中产生的。实际上我们发现它的温度很难超过

regions where the temperature approaches this; indeed, we have seen that it can scarcely have been more than 100,000° or so. Thus this radiation can only have originated from spontaneous annihilation. Cosmic radiation can, and very possibly does, provide evidence of the spontaneous annihilation of matter at low temperatures, but it cannot, from the nature of the case, give any evidence of annihilation being produced by high temperatures, since any radiation so produced could never get out to empty space.

There seem to me to be two strong reasons for supposing that this latter process is not operative in the stars, and that any radiation which is produced by annihilation inside the stars must be produced spontaneously, like the cosmic radiation which is produced outside.

In the first place, if the generation is not spontaneous, the temperature at the star's centre must be of the order of a million million degrees. An immensely steep temperature gradient would be needed to connect this temperature with that of a few thousand degrees at the surface of the star, and so steep a gradient can only be reconciled with the observed flow of heat out of the star by postulating a very high opacity for the stellar material. It has so far proved impossible to reconcile such a high value for the opacity with the theoretical value given by Kramers.

The second reason is as follows. If the generation of energy results from high temperature, the rate of generation will involve a factor of the usual type $e^{-Mc^2/RT}$, where M is the mass annihilated. As the temperature increases from zero up, this factor first becomes appreciable when RT begins to be appreciable in comparison with Mc^2. This happens at the temperature of about a million million degrees already mentioned. When this temperature is first approached, the exponential term is increasing very rapidly in comparison with the temperature T. But a dynamical investigation shows that when this happens, the star must be very unstable. In brief, the emission of appreciable radiation would be accompanied by instability in the star, so that the very stable structures we describe as stars cannot radiate by means of this mechanism. The dynamical result has, it is true, been rigorously proved only for a simple, and very idealised, model of stellar structure; but general thermodynamical principles show that any structure in which a small change of physical conditions results in a very great liberation of heat, is likely to be unstable—in brief, it is in an explosive state.

On the other side, there is one strong argument against supposing that stellar radiation is produced by spontaneous annihilation of matter; it is that if the sun's heat were produced by the spontaneous annihilation of its atoms, we might, expect that the earth's atoms would be subject to spontaneous annihilation at an equal or similar rate. Yet calculation shows that annihilation at even a ten-thousandth part of this rate would make the earth too hot for human habitation. Clearly, then, no appreciable annihilation of matter can occur inside the earth. This must be formed of atoms of a kind which do not undergo spontaneous annihilation, and if the sun derives its heat from the spontaneous annihilation of atoms, these must be of a different kind from the atoms of which our earth is formed.

100,000度左右。因此这个辐射只能来源于自发的湮灭。宇宙辐射能够，并且很可能已经，为物质在低温下的自发湮灭提供了证据，但是从这种情况的本质上说，它不能提供任何高温导致湮灭的证据，因为没有任何这样产生的辐射能逃逸到太空中。

在我看来，似乎有两个强有力的理由可以让我们认为后一种过程不会在恒星中发生，并且任何由恒星内部的湮灭过程所产生的辐射必定是自发发生的，就像恒星外产生的宇宙辐射一样。

首先，如果恒星中的湮灭不是自发的，那么恒星中心的温度就必须达到万亿度的量级。这就需要一个变化极大的温度梯度，以便将这个温度与恒星表面几千度的温度联系起来，只有在恒星物质不透明度非常高的条件下，这么大的梯度才可能与观测到的恒星外热流相一致。现在已经证明如此高的不透明度不可能与克拉默斯给出的理论计算值相符。

下面给出第二条理由。如果能量的产生是由高温引起的，那么产能率就会引入一个通常形式为 $e^{-Mc^2/RT}$ 的因子，其中 M 是湮灭的质量。随着温度由零度开始不断升高，当 RT 升至和 Mc^2 接近的量值时，这个因子的影响开始变得明显。当温度大约达到前面已经提到的万亿度时，这种情况就会发生。第一次达到这个温度时，这个指数项与温度 T 相比，增长得非常迅速。但动力学研究表明，当这种情况发生时恒星肯定是极不稳定的。简言之，明显的能量释放总会与恒星的不稳定性相伴，因此被我们认定为结构非常稳定的恒星是不会以这种机制发生辐射的。确实，只有在一个理想化程度非常高的简单恒星结构模型下，动力学结果才能得到严格的证明；但是普遍的热力学原理表明，若在一种结构中，物理状态很小的一个改变就会释放大量的热，那么任何这样的结构都很可能是不稳定的——简言之，它处在一个一触即发的状态。

另一方面，对于恒星辐射是由物质自发湮灭产生的猜想，存在一个有力论据可以反驳它，即如果是在太阳自身原子的自发湮灭中产生了太阳的热量，那么我们是否可以认为地球上的原子也会以相同或者相近的速率自发湮灭。然而计算表明，哪怕湮灭速率仅是这个速率的万分之一，也会使地球非常热，以至于人类无法居住。那么显然，地球内部不可能有明显的物质湮灭。地球必须由不会发生自发湮灭的原子所构成，而如果太阳的热量来自于原子的自发湮灭，那么这些原子与构成我们地球的原子一定不属于同样的类型。这并不是没有理由的，从地球的形成模式可知，

This is not in itself unreasonable; from the mode of the earth's formation, its atoms can be a sample only of those in the sun's outer layers. If we conjecture that those kinds of atoms which undergo spontaneous annihilation are of very great atomic weight, and so sink to the interiors of the stars, this difficulty disappears, and with it the problem of why no cosmic radiation is received directly from the Milky Way.

(**128**, 103-110; 1931)

References:

1. *Proceedings: National Academy of Sciences* (Oct. 1930).
2. *Nature*, 127, 924 (June 20, 1931).
3. Millikan, Dec. 29, 1930, Lecture at Pasadena, reprinted in *Nature*, 127, 167 (Jan. 31, 1931).
4. *Physical Review* (April 15, 1931).
5. *Proc. Roy. Soc.*, 122, 647.
6. *Nature*, 122, 398 (Sept. 15, 1928).
7. *Physical Review* (Nov. 15, 1930).
8. *Nature*, 127, 594 (April 18, 1931).
9. Hess, V. F., *Nature*, 127, 10 (Jan. 3, 1931).
10. *Proceedings of Cambridge Philosophical Society* (July 1930).

地球上的原子可能只与太阳外层的物质相同。如果我们推测，这种发生自发湮灭的原子具有很大的原子量，因而会陷入到恒星的内部，那么不但以上的问题不存在了，而且还能解释为什么没有接收到直接从银河系发出的宇宙辐射。

（周杰 翻译；尚仁成 审稿）

Quantum-mechanical Models of a Nucleus

R. H. Fowler

Editor's Note

By the 1930s, physicists believed they had a fair understanding of the structure of the atomic nucleus. Its mass was considered to be due to the number of protons it contained, while some of their charge was thought to be offset by the presence of negatively charged electrons involved in β-decay. But even so, it was not clear how so many protons, repelling one another, could be stably packed into so small a space. Here physicist Ralph Fowler suggests an explanation for the stability of nuclei which relies on the observation that radioactive atoms may emit α-particles. The real explanation invokes another nuclear particle, the neutron, not discovered until 1932.

IN their recent paper,[1] Lord Rutherford and Dr. Ellis have shown how the numerous γ-rays of radium C' can be arranged in a simple and orderly manner, which suggests, as they point out, that the multiplicity of the γ-rays is largely due to the excitation of several α-particles into the same excited level rather than to the excitation of one α-particle into several excited levels. Their arrangement of the lines of radium C' is probably not a unique scheme of this sort, but any reasonable scheme appears likely (they show) to present the same general features.

It seems desirable therefore to investigate theoretically in detail any simple model or models of a nucleus consisting of some fifty α-particles, which might show such general features. The main feature brought out by Rutherford and Ellis is that the γ-rays can be expressed in the form

$$hv = pE_0 - qE_1,$$

where p is an integer running from 1 to 4 and q an integer running from 0 up to perhaps 10; the value of E_1 is about $\frac{1}{16}E_0$ for radium C' and has much the same value for radium B. For radium C' more than one value of E_0 may be required.

There are two models which might be investigated with some chance of success; the first is a model in which each α-particle is considered to move independently in a central field (which is ultimately to be referred to the combined interactions with the other α-particles), but the whole family is affected by perturbing interactions of the form $V(r_{ij})$, between each pair i, j of all the α-particles, where r_{ij} is the distance between the α-particles i and j. Such a model is very like an atom of electrons, except that wave functions have to be symmetrical in the α-particles instead of antisymmetrical in the electrons, and this is the

原子核的量子力学模型

拉尔夫·福勒

编者按

到 20 世纪 30 年代时，物理学家们认为他们已经对原子核的结构非常了解了：原子核的质量由所包含的质子数决定，而质子的一部分电荷被 β 衰变中带负电的电子所中和。但是，即使这样，仍然不知道这么多相互排斥的质子怎样才能稳定地排布在一个非常狭小的空间中。在本文中，物理学家拉尔夫·福勒提出了一个能够解释这种稳定性的理论，该理论基于放射性原子会放出 α 粒子的实验证据。当然，正确的解释需要涉及另一种核粒子，即直到 1932 年才发现的中子。

卢瑟福勋爵和埃利斯博士在他们最近发表的文章 [1] 中指出，大量镭 C' 同位素放射出的 γ 射线可以用一种简单而有序的方法进行归纳，他们指出，γ 射线的多样性主要是由多个 α 粒子被激发到同一个激发态上，而不是由同一个 α 粒子被激发到多个激发态上造成的。他们对由镭 C' 放射出的 γ 射线进行归纳的方法也许并不是唯一可行的，但是任何适用的方法可能都会呈现出相同的特征。

这样，我们就迫切地需要建立具有这种特征的模型并对模型进行细致的理论研究，这个模型可以是一个只有简单相互作用的模型，也可以是包含了大约 50 个 α 粒子的原子核模型。卢瑟福和埃利斯得出的关于 γ 射线的最主要的特征可以用以下的形式来表示：

$$hv = pE_0 - qE_1,$$

p 是一个取值从 1 到 4 的整数，而 q 是一个从 0 开始的整数，最大可能只能取到 10；在镭 C' 同位素的衰变实验中，E_1 的取值大约是 E_0 取值的 1/16，与镭 B 同位素衰变实验中相应的 E_1 值大致相当。对于镭 C' 同位素来说，我们需要选取不同的 E_0 值。

我们可以研究以下两个可能成功解释原子核稳定性的模型。第一个模型认为每个 α 粒子都可以在中心势场中独立运动（这个中心势场指的是与其他 α 粒子之间的复杂相互作用），而整个原子核又受到形式为 $V(r_{ij})$ 的微扰势场的影响，其中 $V(r_{ij})$ 是每对粒子中第 i 个 α 粒子和第 j 个 α 粒子之间的相互作用，其中 r_{ij} 是第 i 个 α 粒子和第 j 个 α 粒子之间的距离。这样的模型与包含电子的原子模型非常类似，不同的是 α 粒子的波函数是对称的，而不像电子的波函数那样是反对称的。这种本质上的

essential difference which allows of the states of reduplicated excitation, which do not occur at all in atoms. This model can be still further simplified from a three-dimensional to a one-dimensional form for a first discussion.

The second model is one in which each pair i, j of α-particles act on one another with a potential energy $\frac{1}{2}\lambda r_{ij}^2$. This model is obviously a rather poor physical approximation to the type of force, but it has the advantage that it can be studied exactly and not merely by the approximations of a perturbation method. A discussion of both these models has been begun, but has as yet only been carried through for the first model simplified to one dimension.

Confining attention only to the most general features, likely to be true of any suitable similar model, the following results have been obtained, which are to a large extent in excellent accord with the scheme of Rutherford and Ellis, but also seem to indicate clearly that a rather more elaborate scheme should be adopted. The energy levels of the model which arise from excitation of more than one α-particle into a single excited state are of such a configuration that the corresponding γ-rays (if they could all be emitted) would be approximately of the frequencies

$$hv = p(E_0 - qE_1)$$

These frequencies agree with those of the proposed scheme of Rutherford and Ellis if the scheme is only very slightly modified, so that in place of the proposed single set of γ-rays of frequencies $2E_0 - qE_1$, we have the double set of frequencies $2(E_0 - qE_1)$ and $2(E_0' - qE_1)$ with E_0 and E_0' nearly equal, and in place of the single set $3E_0 - qE_1$, the triple set $3(E_0 - qE_1)$, $3(E_0' - qE_1)$, $3(E_0'' - qE_1)$ and so on. It is, moreover, clear that the reduplication of the upper levels is to be expected when we consider the three-dimensional version of the model. Further, the theory suggests that the ratio E_0/E_1 should be numerically somewhat less than $\frac{1}{2}n\,(= 26)$ in not too bad conformity with the observed value 16 for radium C'. The theory even suggests further that both E_0 and E_1, or perhaps rather E_1, will not vary very much between one radioactive nucleus and another. It is true that the observed values of E_1 (but not those of E_0) are much the same for radium C' and radium B. The γ-rays of other atoms have not yet been analysed in this way.

All these features are general and the conformity very reassuring. One can, however, further estimate the relative frequency of the emission of the various γ-rays corresponding to the transitions from a state of q-excited α-particles to states of $q-1$, $q-2$, $q-3$... excited α-particles. With an interaction energy of the proposed form, the transitions $q \rightarrow q-3$ should be absent, or at most very rare, and the transitions $q \rightarrow q-4$, $q \rightarrow q-5$, etc., entirely absent. The theory gives as a first approximation to R, the ratio of the frequency of occurrence of the transitions $q \rightarrow q-2$ and $q \rightarrow q-1$, the value

$$R\left(\frac{q \rightarrow q-2}{q \rightarrow q-1}\right) = \frac{q-1}{6.5}f$$

区别允许原子核中的粒子发生重复激发，而原子中的电子是不可能出现重复激发的情况的。在初步的讨论中，我们可以把三维情况进一步简化到一维情况。

第二个模型假设每对标记为 i，j 的 α 粒子对对彼此有势能为 $\frac{1}{2}\lambda r_{ij}^2$ 的相互作用。很明显，这个模型对作用力所做的近似是一个非常不好的物理近似，但是这个近似的优点在于我们可以对其进行严格的计算而不只是用微扰理论来进行近似研究。关于以上这两个模型的研究已经开始，但到目前为止只完成了被简化到一维情况下的第一个模型。

就像任何类似的合理模型那样，我们只把注意力集中到最普遍的特征上面来，并得到了以下的结果，这些结果在很大程度上和卢瑟福、埃利斯的构想十分吻合，但是似乎也明确说明我们需要建立一个更加精细的模型。模型中由多个 α 粒子跃迁到同一个激发态上所产生的能级具有这样的形式，即相应的 γ 射线（如果所有 γ 射线都被能放射出来）的频率大致为：

$$hv = p(E_0 - qE_1)$$

如果将卢瑟福与埃利斯的构想稍加修正，由此得出的射线频率将和上式中的射线频率吻合得很好。因此我们用频率关系对应于 $2(E_0 - qE_1)$ 和 $2(E_0' - qE_1)$ 的两组 γ 射线代替以前提出的频率关系对应于 $2E_0 - qE_1$ 的单组 γ 射线，其中，E_0 和 E_0' 几乎相等。我们还用频率关系对应于 $3(E_0 - qE_1)$、$3(E_0' - qE_1)$ 和 $3(E_0'' - qE_1)$ 的三组 γ 射线来代替频率关系对应于 $3E_0 - qE_1$ 的单组 γ 射线，由此类推。此外，很明显地，当我们考虑三维情况下的该模型时，将会出现高能级的简并情况。并且，理论表明，比值 E_0/E_1 在数值上应该略小于 $\frac{1}{2}n(=26)$，与我们在镭 C' 同位素中观察到的 E_0/E_1 值 16 相差得还不算太大。理论上还表明，不同的放射性原子核的 E_0 和 E_1 值（尤其是 E_1 的值）不会相差太多。在镭的 C' 和 B 同位素的观测结果中，E_1 的观测值的确相近，但是 E_0 的值却并非如此。而其他原子核放射出的 γ 射线还没有用这种方法分析过。

所有这些特性都是很普遍的，并且我们得到的一致性结果也很有说服力。然而我们还可以进一步估算各种不同的 γ 射线的相对放射频率，这些不同的放射频率对应 α 粒子从 q 激发态到 $q-1$, $q-2$, $q-3$ 等激发态的跃迁。按照前面所提到的相互作用能量的表达形式，$q \to q-3$ 的跃迁过程应该不存在，即使存在也是非常罕见的，而 $q \to q-4$，$q \to q-5$ 等情况的跃迁过程也都不存在。理论上给出了 $q \to q-2$ 和 $q \to q-1$ 这两个跃迁发生频率的之比 R 的第一级近似形式：

$$R\left(\frac{q \to q-2}{q \to q-1}\right) = \frac{q-1}{6.5}f$$

where f is a factor certainly less than unity and probably not so small as 1/10. The absolute value of the ratio R may be heavily affected by higher order terms, and we need not be concerned if the proposed scheme does not conform closely. The feature of R that is almost certainly of general importance is that R increases with q. This feature ought to be carefully borne in mind in the construction of any amended scheme. It is not yet possible to say whether these features can be incorporated in an otherwise satisfactory scheme, and a detailed re-examination must be undertaken.

The proposed scheme for radium C' is arranged to include values of p up to 4 and therefore transitions of the type $q \rightarrow q-4$. These certainly do not, and the transitions $q \rightarrow q-3$ probably do not, fit into the allowed transitions of the proposed model with the simple interactions proposed. But such transitions can be present if there are terms in the interactions depending essentially on the co-ordinates of three or more particles, not reducible to sums of terms depending on the co-ordinates of only two. Such terms are to be expected in such a close configuration, though one would scarcely expect their effect to be so large. If the proposed scheme proves ultimately to be correct, one may hope to work back from the γ-ray intensities to some knowledge of the magnitude of these triple and higher interactions.

To sum up, one may say that the scheme proposed by Rutherford and Ellis, so far as it has yet been closely analysed, *that is for frequencies only*, seems likely with trivial modifications to conform completely to the requirements of a simple quantum mechanical model so far as these requirements can yet be foreseen. Such a model, however, will make fairly stringent demands on intensity ratios, and as yet no scheme has been proposed and tested with these in mind. One may hope that further work on these lines will prove fruitful.

While these models may well be able to explain the complicated spectrum of radium C', it is well to remember that the corresponding spectrum of thorium C' is very much simpler and contains no families of γ-rays—except perhaps very faint ones—corresponding to those of radium C', which have been interpreted in the scheme as transitions $q \rightarrow q-2$, $q \rightarrow q-3$, and $q \rightarrow q-4$. It has of course, in addition, a very strong isolated γ-ray of very high frequency. If therefore in attempting to proceed with this analysis, which in any event I believe to be important, one is forced finally to conclude that such models will *not* explain the facts for radium C', there is no call for surprise or disappointment. It may still be that the proposed scheme of $q \rightarrow q-1$ transitions will account properly for the important *common* features of the γ-ray spectra of radium B, radium C', thorium C', and probably other nuclei. It is more than likely that the striking *differences* between the spectra of radium C' and thorium C' should be associated with the two extra free protons in radium C', the atomic weight of which is of the form $4n+2$, while that of thorium C', is $4n$.

In the models suggested above, the effect of the protons has been ignored primarily because there seems at present no simple way of incorporating them. But it is clear that the general effect of free protons present in normal and excited states will be to cause the

式中的f因子一定小于单位 1，但可能不太会小于 1/10。高阶项对比值 R 的绝对值有很大的影响，所以如果上面提到的构想中相应的 R 值与实验结果不能严格一致时，我们也不必太在意。R 的一个具有普遍重要性的特征是，它随着 q 的增大而增大。在对任何有效模型进行修正的时候，我们都必须记住 R 的这个特征。以上提到的所有特征到现在为止还不能确定是否可以整合到另一个令人满意的模型当中去，必须经过仔细地反复验证。

以上提出的关于镭 C' 同位素的构想中，p 的取值可以从 1 到 4，因此，构想中包含了 $q\rightarrow q-4$ 的跃迁。在前面提到的简单相互作用模型中 $q\rightarrow q-4$ 的跃迁肯定是被禁止的，而 $q\rightarrow q-3$ 的跃迁有可能被禁止。但是，如果相互作用项是由三个或者三个以上粒子的坐标所决定的，而且不能化简成只由两个粒子的坐标所决定的两体相互作用项的总和时，上面所说的 $q\rightarrow q-3$、$q\rightarrow q-4$ 的跃迁就可能发生了。以上所讨论的相互作用项在形式上都很相近，我们想象不到这些相互作用项的影响会如此之大。如果上面提出的考虑三个粒子相互作用的构想最终被证明是正确的，那么我们也许可以回过头来，从 γ 射线的强度出发来获知这些三重或者多重相互作用的大小。

总之，我们可以认为卢瑟福和埃利斯所提出的构想到现在为止已经被仔细地分析过了，**这是只从频率的角度进行研究的**，从这方面看，这一构想似乎经过细微的修正就可以完全符合可以预见的简单量子力学模型的要求。然而，这样的一个模型对强度比的要求非常严格，就我所知目前还没有人提出这方面的有关方案和检测方法。我们希望今后在这方面的研究工作能富有成效。

虽然这些模型也许可以很好地解释镭 C' 的复杂谱线，但不要忘记，钍 C' 的对应谱线要简单得多，而且不包括镭 C' 谱线中那些由 $q\rightarrow q-2$、$q\rightarrow q-3$、$q\rightarrow q-4$ 跃迁所产生的 γ 射线族——除了几条非常微弱谱线以外。当然，钍 C' 的相应谱线中还包含了另外一条非常强的独立的高频 γ 射线。如果我们试图对以上现象继续进行分析（我认为这个分析无疑重要的），我们最终不得不得出这样的结论，适用于钍 C' 的模型将**无法**用来解释镭 C' 的实验事实，对于这个结果，我们大可不必感到惊讶或失望。上面提到的 $q\rightarrow q-1$ 的跃迁方案，仍然可以很好地解释镭 B、镭 C'、钍 C'（可能还有其他的放射性原子核）衰变放射出的 γ 射线谱中**共有**的重要特征。更有可能发生的情况是，镭 C' 和钍 C' 谱线中惊人的**差异**应该与镭 C' 比钍 C' 多出两个自由质子有关。镭 C' 的原子量是 $4n+2$，而钍 C' 的原子量是 $4n$。

在上面所提出的模型中，质子的效应从一开始就被忽略了，因为目前似乎还没有将质子效应整合到模型中的简单方法。但是可以明确的是，在正常态和激发态，

set of low frequency transitions $q \rightarrow q-1$ to be repeated again at higher frequencies but with the same dependence on q, the constant shift between the two sets representing an excitation energy for a proton.

(**128**, 453-454; 1931)

R. H. Fowler: Cromwell House, Trumpington, Cambridge, Aug. 14.

Reference:

1. *Proc. Roy. Soc.*, A, 132, 667 (1931).

由自由质子所引发的普遍效应会使低频跃迁 $q \to q-1$ 在高频情况下重复发生，但是高频情况下的跃迁同样依赖于 q，两种情况下的常数变化代表了一个质子的激发能。

（王静 翻译；李军刚 审稿）

The Angular Momentum of Light

C. V. Raman

Editor's Note

Arthur Compton's pioneering work on the scattering of light by electrons showed how the process could be understood in quantum terms as a collision conserving energy and momentum. As Raman argues here, the scattering of light from molecules is considerably more complicated, since the photon may excite molecular internal degrees of freedom such as rotations or vibrations. If the intrinsic angular momentum of a photon may only take certain discrete values, this would account for a known "selection rule" governing the permitted changes in molecular angular momentum during light scattering. It also implies that any change in molecular angular momentum must be accompanied by a reversal of the circular polarisation of the photon, as confirmed by recent experiments in Calcutta.

THE work of Compton on X-ray scattering led to the general acceptance of the idea that the scattering of radiation by a material particle is a unitary process in which energy and linear momentum are conserved. A molecule is, however, a much more complicated structure than an electron, and the conservation principles by themselves would give us an erroneous idea of what we should expect in light-scattering. This follows from the fact that a molecule has in general three degrees of freedom of rotation, several degrees of freedom of vibration according to its complexity, and various possible modes of electronic excitation, and that each of these may correspond to one or other of an extended series of quantum numbers. Restricting ourselves to the cases in which the molecule takes up a part of the energy of the quantum, the conservation principles would indicate that the spectrum of the scattered light should contain an immense number of new lines.

Actually, a remarkable simplicity characterises the observed spectra of the light scattered by polyatomic molecules, a simplicity which is in striking contrast with the complexity of their absorption and emission spectra. It is clear that the Compton principles cannot be regarded as capable of *predicting* the observed phenomena of light-scattering, and that their utility lies solely in the *interpretation* of results discovered by experiment. These remarks seem necessary to correct an impression to the contrary which finds expression in some recent publications.

We may extend Compton's principle and add angular momentum to the quantities which we should expect to find conserved in the collision between a light-quantum and a molecule. The fact that, in liquids and solids, the mutual influence of the material particles is very considerable, attaches some uncertainty to the interpretation of the results obtained with them. The recent success of Bhagavantam at Calcutta in measuring the

光的角动量

钱德拉塞卡拉·文卡塔·拉曼

编者按

阿瑟·康普顿在研究电子对光的散射时发现这个过程可以用量子力学中保持能量和动量守恒的碰撞理论来解释。就像拉曼在这里指出的那样，分子对光的散射更加复杂，因为光子可以激发分子内部的自由度，如旋转或者振动。如果一个光子的内禀角动量只能取某些不连续的值，那么这就解释了为什么普遍认可的“选择定则”能够支配分子角动量在光散射过程中可能出现的变化。最近在加尔各答的实验证明：分子角动量的任何变化都将伴随着光子圆偏振的反转。

康普顿关于X射线散射的研究使人们普遍地接受了以下观点，即物质粒子对辐射的散射是一种维持能量和线性动量守恒的幺正过程。然而，一个分子的结构比一个电子的结构要复杂得多，而守恒定理本身使我们在预测光散射的结果时会出现错误。这是因为一个分子通常具有三个转动自由度、几个振动自由度（数量与它的复杂程度有关）以及多种电子激发的可能模式，每一种模式都可能对应于一个或另一个来自扩展系列的量子数。如果我们局限于认为分子占据量子能量的一部分，那么根据守恒原理，散射光的谱就应该包含数量巨大的新线。

实际上我们所观察到的多原子分子散射光的谱非常简单，要比它们的吸收谱和发射谱简单很多。显然，康普顿原理不能**预言**观察到的光散射现象，它们只能用于**解释**由实验得到的结果。上述意见对于修正最近一些刊物中出现的相反看法是很有必要的。

我们可以拓展康普顿原理，并把角动量加入到我们认为在光量子与分子发生碰撞时应该保持守恒的量当中。在液体和固体中，物质粒子之间的相互影响相当可观，因而对散射结果的解释会带有某种不确定性。巴加万塔姆最近在加尔各答成功地测量了光被气体散射时的偏振性和强度，该成果为这项课题的研究开辟了新的可能性。

polarisation and intensity of light scattered by gases, however, opens up new possibilities for the development of the subject.

As a working hypothesis, we may follow Dirac and assume that the angular momentum of a photon is *plus* or *minus* $h/2\pi$, intermediate values being inadmissible. This supposition enables us to interpret very simply the known selection rule $\Delta m = 0$ or ± 2 for the change of rotational quantum number of a diatomic molecule in light-scattering, which follows as a natural consequence of it. Further, it follows[1] that a change in rotational quantum number of the molecule should be accompanied by a reversal in the sign of circular polarisation of the photon, when the latter is scattered in the forward direction. This reversal has been actually observed by Bär and by Bhagavantam with the rotational wings accompanying the original mercury lines scattered in liquids, and the data obtained by Bhagavantam with hydrogen gas may also be interpreted as a confirmation of the same result.

It is remarkable that the latter result is also predicted by the classical electromagnetic theory of light for the case of a rotating anisotropic particle scattering circularly polarised radiation. Nevertheless, it is clear that the observed phenomena may be regarded as an experimental proof that radiation has angular momentum associated with it, and that it has the values $\pm h/2\pi$ for each quantum.

(**128**, 545; 1931)

C. V. Raman: 210 Bowbazar Street, Calcutta, Aug. 15.

Reference:

1. *Nature*, 128, 114 (July 18, 1931).

在对这一假设进行推导时，我们可以根据狄拉克的理论假定一个光子的角动量为 $\pm h/2\pi$，中间值的存在是不允许的。这个假定能使我们很简单地解释为什么双原子分子在光散射中转动量子数的改变要遵循选择定则 $\Delta m = 0$ 或 ± 2，这样选择定则的出现就成为了一个自然而然的结果。此外我们还发现[1]：当光子向前散射时，分子转动量子数的变化应伴随着光子圆偏振符号的反向。巴尔和巴加万塔姆用伴随在液体中散射的原来汞线的旋转两翼观测到了这一反向现象，巴加万塔姆用氢气得到的数据也可以用来证明同样的结论。

值得注意的是，后一个结果也可以在旋转的各向异性并具有圆偏振散射线的粒子的情况下由光的经典电磁理论推导出来。显然观测到的现象也可以作为辐射具有相关角动量的实验证据，而每个量子所具有的角动量大小为 $\pm h/2\pi$。

（沈乃澂 翻译；张泽渤 审稿）

Isolated Quantised Magnetic Poles

O. W. Richardson

Editor's Note

There were, in classical physics, no "magnetic charges"—no isolated magnetic poles without their complementary pole. But Paul Dirac suggested that quantum theory might allow the existence of such "monopoles", and here distinguished British physicist Owen Willans Richardson speculates on what their existence might entail for physics. While science focused on atoms made of electrically charged particles, a similar hierarchy of magnetic atoms might exist. Richardson argues that their properties would be unusual; in particular, the frequencies of their spectral lines would be some 10^{10} higher than for "ordinary" matter. Dirac had speculated that the forces between magnetic monopoles would prevent their separation in ordinary circumstances, but Richardson suggests this might happen for cosmic rays of extremely high energy.

IN the last number of the *Proceedings of the Royal Society*, Dirac has come to the conclusion that the quantum theory requires the existence of discrete magnetic poles of a strength equal to 137÷2 times the electronic charge. If such objects were common one might expect the universe to be a good deal different from what experimenters have found it to be, so far.

There seems no a priori reason why the whole theory of atom building which has been built up for electrons and nuclei—an electrostatic problem apart from details—should not be carried over bodily into the corresponding magneto-static problem of the attractions of the oppositely charged poles. In this way we might, at first, expect to get a set of "magnetic" atoms, similar to the electric atoms of which matter is generally supposed to be built up. These atoms would be a good deal different from those we think we are familiar with. How much different depends to some extent on what the mass of a magnetic pole is. The quantum theory does not tell this, but I think its value, if it exists, can be fixed by an argument based on classical ideas at about 500 times that of the corresponding electronic object. Following this general line of argument, the dimensions of these magnetic atoms come out at 10^{-14} cm. to 10^{-15} cm. compared with 10^{-7} cm. to 10^{-8} cm. for the atoms of the periodic table. The frequencies of the "spectral" lines emitted by these magnetically constructed atoms would run about 10^{10} times those of the corresponding lines of the electronic spectra; for example, the first line of the Lyman series would be raised from $v=2.5\times10^{15}$ to $v=3.1\times10^{25}$ sec.$^{-1}$ if the corresponding states are capable of existence. Even if quite large changes are made in the mass of the magnetic poles, which is the doubtful element, the corresponding numbers will still remain quite wide apart.

Dirac has suggested that the reason these magnetic poles have not been observed may be that the forces between them are so much larger than those between electrons and protons

孤立的量子化磁极

欧文·威兰斯·里查孙

编者按

在经典物理学中不存在“磁荷”这个概念——没有相反磁极的孤立磁极是不存在的。但保罗·狄拉克提出量子理论也许能允许这样的“磁单极”存在。在本文中，英国著名物理学家欧文·威兰斯·里查孙认为对于物理学来说磁单极的存在也许是必然的。既然科学主要研究由带电的粒子组成的原子，就应该有相应的带磁性的一类原子存在。里查孙指出它们的性质很不一般；尤其是它们的谱线频率是“普通”物质的 10^{10} 倍。狄拉克推测磁单极之间的强大作用力使它们在通常环境下无法分离，但里查孙认为在高能宇宙射线中这种现象就有可能会发生。

在最近一期的《皇家学会会刊》中，狄拉克已得出结论，量子理论要求存在强度相当于 137÷2 倍的电子电荷的单个磁极。假如这样的物质是普遍存在的，那么我们可以预期现实的宇宙与目前实验者们眼中的宇宙会有很大的不同。

似乎没有理由不能把建立在电子和原子核之上的一整套原子理论——除细节外基本属于静电问题——应用到以相反磁荷相互吸引为基础的相应的静磁问题上。按照这种方式，我们首先可以预期得到一组与带电原子类似的“磁”的原子，通常我们认为物质是由带电原子构成的。磁的原子与我们原有概念中所熟知的原子有很大的差别。差别的大小在某种程度上取决于磁极的质量。量子理论并未说明相应的具体数值，但我认为如果这个值存在，用经典方法论证则应该约为相应带电物的 500 倍。按照这个思路推算，这些磁的原子的尺度达到 10^{-14} cm 至 10^{-15} cm，而元素周期表中的原子尺度为 10^{-7} cm 至 10^{-8} cm。由这些磁性结构原子发射的“谱”线频率约为带电原子谱线相应频率的 10^{10} 倍；例如，假如以上磁性结构的原子是存在的，赖曼线系中的第一条线的频率将从 $v = 2.5 \times 10^{15}\ s^{-1}$ 增加至 $v = 3.1 \times 10^{25}\ s^{-1}$。即使对数值不确定的磁极质量作很大的调整，所得的结果仍与带电原子的情况相差很大。

狄拉克指出，这些磁极尚未被观测到的原因可能是因为它们之间的力远大于电子和质子之间的力，从而使它们不能分离。我们有理由相信它们不可能接近到上述

that they cannot be separated. There is reason for believing they could not get together to the extent indicated by the preceding numbers. The number of kinds of atoms with azimuthal quantum number 1 which can be formed from these magnetic units is much less than unity. This follows from Dirac's formula for the spectral terms for hydrogen, or alternatively, from the principle of minimum time. This may be forcing the required atoms too much into the pattern of those with which we are familiar. In any event, no atom with azimuthal quantum number less than $34\frac{1}{4}$ can be made out of these elements. Otherwise the time factors in the wave functions involve real exponentials and become infinite with lapse of time. However, even with such high quantum numbers the forces would still be enormous compared with those in corresponding electronic structures and the frequencies would still be quite high.

There may be an application of these products of the quantum theory in the field of "ultra-penetrating" radiations. I have no first-hand knowledge of the process of creation, but I should suspect it would be relatively difficult to create objects with the intrinsic energy of these magnetic poles. It seems likely, therefore, that their abundance would be very small compared with that of electrons and protons, but there might be enough in the universe to account for such ultra-penetrating radiations as are not capable of being accounted for otherwise. The possible existence of such isolated magnetic poles, with properties so very different from those of electrons and protons, obviously changes the basis for discussion of a good many cosmological questions.

(**128**, 582; 1931)

O. W. Richardson: King's College, London, W.C.2, Sept. 18.

数字表示的程度。这些磁单元形成的角量子数为1的原子种类数远小于1。这是根据与氢光谱项有关的狄拉克公式或者根据最小时间原理得出的。这也许是在迫使磁的原子过于符合我们所熟悉的模式。在任何情况下，磁的原子都不能构成角量子数小于 $34\frac{1}{4}$ 的原子。否则，波函数中的时间因子会含有实指数，并随时间的消逝而趋于无限大。然而，尽管有如此高的量子数，与相应的电子结构相比，磁单元之间的作用力仍然很大，频率也仍然很高。

这些量子理论的成果可能会在“超穿透”辐射领域中得到应用。我对孤立磁极的形成过程没有直接的了解，但我推测，产生这种具有磁极内能的物质是相当困难的。因此，地球上磁原子的数量似乎应该远小于电子和质子的数量，但在宇宙中磁原子的数量也许已经足以用来解释超穿透辐射这种不能用其他方法解释的现象。如果这类与电子和质子的性质差异很大的孤立磁极是存在的，那么将会使许多宇宙问题的研究思路发生明显的变化。

（沈乃澂 翻译；赵见高 审稿）

Maxwell and Modern Theoretical Physics*

N. Bohr

Editor's Note

On the occasion of the Maxwell Centenary Celebrations in Cambridge, October 1931, Niels Bohr presented an address on Maxwell and Modern Theoretical Physics. Bohr notes that Maxwell's theory of electricity and magnetism had made possible developments in the understanding of atomic matter, and also provided the foundations for the discovery of quantum theory. Although this theory moved beyond Maxwell's ideas, analysis in quantum theory still required Maxwell's theory, especially through the relations for the energy and momentum of light quanta. Bohr asserts that Maxwell's theory will always play a central role in physics, as the formulation of the laws of quantum physics depends on the existence of a classical world describable in Maxwell's terms.

I feel greatly honoured in being given this opportunity of paying a tribute of reverence to the memory of James Clerk Maxwell, the creator of the electromagnetic theory, which is of such fundamental importance to the work of every physicist. In this celebration we have heard the Master of Trinity and Sir Joseph Larmor speak, with the greatest authority and charm, of Maxwell's wonderful discoveries and personality, and of the unbroken tradition upheld here in Cambridge connecting his life and his work with our time. Although I have had the great privilege, in the years of my early studies, of coming under the spell of Cambridge and the inspiration of the great English physicists, I fear that it may not be possible for me to add anything of sufficient interest in this respect, but it gives me very great pleasure indeed to be invited to say a few words about the relation between Maxwell's work and the subsequent development of atomic physics.

I shall not speak of Maxwell's fundamental contributions to the development of statistical mechanics and of the kinetic theory of gases, which Prof. Planck has already discussed, especially as regards Maxwell's fruitful co-operation with Boltzmann. It is only my intention to make a few remarks about the application of the electromagnetic theory to the problem of atomic constitution, where Maxwell's theory, besides being extremely fruitful in the interpretation of the phenomena, has yielded the utmost any theory can do, namely, to be instrumental in suggesting and guiding new developments beyond its original scope.

I must, of course, be very brief in commenting upon the application of Maxwell's ideas to atomic theory, which in itself constitutes a whole chapter of physics. I shall just recall how successfully the idea of the atomic nature of electricity was incorporated into Maxwell's

---

* Address delivered on the occasion of the Maxwell Centenary Celebrations at Cambridge on Oct. 1.

麦克斯韦与现代理论物理*

尼尔斯·玻尔

编者按

1931 年 10 月，剑桥大学举行了麦克斯韦诞辰一百年纪念会，尼尔斯·玻尔在会上发表了一篇有关麦克斯韦和现代理论物理的演讲。玻尔指出麦克斯韦的电磁学理论为人类了解原子世界作出了贡献，也为创立量子理论提供了基础。虽然量子理论超越了麦克斯韦的学说，但量子力学中的分析手段离不开麦克斯韦的理论，尤其是在建立光量子能量和动量之间的关系时。玻尔认为麦克斯韦的理论在物理学中的核心地位不会丧失，因为量子物理定律中的公式都是按照麦克斯韦所描述的经典世界创建的。

我非常荣幸能有机会在此表达我对电磁学理论的创立者——詹姆斯·克拉克·麦克斯韦的敬意，因为电磁学理论对每一位物理学家的研究工作都是非常重要的。在这个纪念庆典上，三一学院院长和约瑟夫·拉莫尔爵士已经给我们作了最具权威性的精彩演讲，介绍了麦克斯韦的重要发现和伟大人格，还提到剑桥大学一直坚持把麦克斯韦的生活与工作和我们所处的时代紧密相连的传统。尽管我早年曾有幸在剑桥大学工作过一段时间，也接受过知名英国物理学家们的指点，但在这方面我恐怕讲不出更多吸引人注意的内容，不过我仍然非常高兴能应大会之邀来这里简单介绍一下麦克斯韦的研究与随后发展起来的原子物理之间的关系。

我不打算讲麦克斯韦对统计力学和气体动力学的重大贡献，因为普朗克教授已经在前面介绍过了，他还特别提到麦克斯韦和玻尔兹曼曾经一起进行过有效合作。在这里我只想谈一下如何应用电磁学理论解决原子的构成问题，在这个问题上，麦克斯韦的理论不仅很好地解释了相关的实验现象，而且已经达到了理论所能达到的极致，因为它可以启发和引导自身领域之外的研究工作的发展。

当然，我只能非常简短地介绍麦克斯韦理论在原子物理中的应用，因为仅这个问题本身就足以构成物理学中整整一章的内容。我要说洛伦兹和拉莫尔将电的原子本质与麦克斯韦理论结合在一起的思想非常成功，特别是由此可以解释包括塞曼效

---

* 这是 1931 年 10 月 1 日在剑桥举行的麦克斯韦诞辰一百周年庆祝大会上玻尔所作的演讲。

theory by Lorentz and Larmor, and especially how it furnished an explanation of the dispersion phenomena, including the remarkable features of the Zeeman effect. I would also like to allude to the important contribution to the electron theory of magnetism made by Prof. Langevin, whom we much regret not to be able to hear today. But above all, I think in this connexion of the inspiration given by Maxwell's ideas to Sir Joseph Thomson in his pioneer work on the electronic constitution of matter, from his early introduction of the fundamental idea of the electromagnetic mass of the electron, to his famous method, valid to this day, of counting the electrons in the atom by means of the scattering of Röntgen rays.

The developments of the atomic theory brought us soon, as everybody knows, beyond the limit of direct and consistent application of Maxwell's theory. I wish to emphasise, however, that it was just the possibility of analysing the radiation phenomena provided by the electromagnetic theory of light which led to the recognition of an essentially new feature of the laws of Nature. Planck's fundamental discovery of the quantum of action has necessitated, indeed, a radical revision of all our concepts in natural philosophy. Still, in this situation, Maxwell's theory continued to provide indispensable guidance. Thus the relation between energy and momentum of radiation, which follows from the electromagnetic theory, has found application even in the explanation of the Compton effect, for which Einstein's idea of the photon has been so appropriate a means of accounting for the marked departure from the classical ideas. The use of Maxwell's theory as a guide did not fail either in the later stage of atomic theory. Although Lord Rutherford's fundamental discovery of the atomic nucleus, which brought our picture of the atom to such wonderful completion, showed most strikingly the limitation of ordinary mechanics and electrodynamics, the only way to progress in this field has been to maintain as close contact as possible with the classical ideas of Newton and Maxwell.

At first sight it might perhaps look as if some essential modification of Maxwell's theory was needed here, and it has even been suggested that new terms should be added to his famous equations for electromagnetic fields in free space. But Maxwell's theory has proved far too consistent, far too beautiful, to admit of a modification of this kind. There could only be a question, indeed, of a generalisation of the whole theory, or rather of a translation of it into a new physical language, suited to take into account the essential indivisibility of the elementary processes in such a way that every feature of Maxwell's theory finds a corresponding feature in the new formalism. In the last few years, this aim has actually been attained to a large extent by the wonderful development of the new quantum mechanics or quantum electrodynamics, connected with the names of de Broglie, Heisenberg, Schrödinger, and Dirac.

When one hears physicists talk nowadays about "electron waves" and "photons", it might perhaps appear that we have completely left the ground on which Newton and Maxwell built; but we all agree, I think, that such concepts, however fruitful, can never be more than a convenient means of stating characteristic consequences of the quantum theory which cannot be visualised in the ordinary sense. It must not be forgotten that only the

应的显著特征在内的色散现象。我还想提及朗之万教授在磁性的电子理论方面所作的贡献。遗憾的是，他今天不能到场为我们作报告。但我认为最值得一提的是约瑟夫·汤姆逊爵士在物质电子结构方面的开拓性工作一直受到麦克斯韦理论的启发，从他早期引入电子电磁质量这一基本概念，到后来他发明用伦琴射线的散射计算原子中电子数的方法，这种方法一直沿用至今。

我们都知道，原子理论的迅速发展使我们现在已经不能再像原来那样直接利用麦克斯韦的理论来解决原子内部的问题。但我要强调的是，正是因为有了光的电磁学理论，我们才可能去分析辐射现象，才可能认识到自然法则的一个全新特征。的确，普朗克关于量子行为的重要发现使我们必须从根本上修正自然科学中的所有观念。但即便如此，麦克斯韦的理论仍然能为我们的研究工作提供必不可少的指导。由电磁学理论可以推出辐射能量与动量之间的关系，而我们可以利用这种关系来解释康普顿效应，虽然爱因斯坦的光子理论明确指出康普顿效应与经典理论之间存在着明显的区别。麦克斯韦的理论在原子理论后来的发展中仍然具有一定的指导意义。尽管卢瑟福勋爵对原子结构的完整描述说明普通力学和电动力学具有极大的局限性，但想在这一领域进一步拓展就必须尽可能紧密地联系牛顿和麦克斯韦的经典理论。

乍看起来，我们似乎需要从本质上对麦克斯韦的理论进行修正，甚至有人指出，应该在麦克斯韦著名的自由空间电磁场方程中加入新项。但麦克斯韦的理论高度自洽非常完美，人们无法对它进行这样的修正。实际上，我们只能对整个理论进行一致化处理，更确切地说就是把麦克斯韦的方程翻译成一种新的物理语言，这种新语言通过为麦克斯韦理论中的每一个特征找到其在新形式下的对应特征来反映基本物理过程中的不可分性。前一段时间，德布罗意、海森堡、薛定谔和狄拉克在量子力学或电动力学方面所取得的突破性进展已经在很大程度上实现了上述目标。

今天，当我们听到物理学家们谈论“电子波”和“光子”的时候，这些概念似乎已经完全脱离了牛顿和麦克斯韦建立起来的理论基础；但是我认为，大家都会同意这样一个观点：这些新观念虽然发挥了很大的作用，但它们不过是一种用来说明量子理论特殊性的便捷方法，缺乏一般意义上的形象化特征。不要忘记，只有关于

classical ideas of material particles and electromagnetic waves have a field of unambiguous application, whereas the concepts of photons and electron waves have not. Their applicability is essentially limited to cases in which, on account of the existence of the quantum of action, it is not possible to consider the phenomena observed as independent of the apparatus utilised for their observation. I would like to mention, as an example, the most conspicuous application of Maxwell's ideas, namely, the electromagnetic waves in wireless transmission. It is a purely formal matter to say that these waves consist of photons, since the conditions under which we control the emission and the reception of the radio waves preclude the possibility of determining the number of photons they should contain. In such a case we may say that all trace of the photon idea, which is essentially one of enumeration of elementary processes, has completely disappeared.

For the sake of illustration, let us imagine for a moment that the recent experimental discoveries of electron diffraction and photonic effects, which fall in so well with the quantum mechanical symbolism, were made before the work of Faraday and Maxwell. Of course, such a situation is unthinkable, since the interpretation of the experiments in question is essentially based on the concepts created by this work. But let us, nevertheless, take such a fanciful view and ask ourselves what the state of science would then be. I think it is not too much to say that we should be farther away from a consistent view of the properties of matter and light than Newton and Huygens were. We must, in fact, realise that the unambiguous interpretation of any measurement must be essentially framed in terms of the classical physical theories, and we may say that in this sense the language of Newton and Maxwell will remain the language of physicists for all time.

I do not think that this is a proper occasion to enter into further details regarding these problems, and to bring new views under discussion. In conclusion, however, I am glad to give expression to the great expectation with which the whole scientific world follows the exploration of an entirely new field of experimental physics, namely, the internal constitution of the nucleus, which is now carried on in Maxwell's laboratory, under the great leadership of the present Cavendish professor. In the fact that nobody here in Cambridge is likely to forget Newton's and Maxwell's work, we see perhaps the very best auguries for the continued success of these endeavours. Even if we must be prepared for a still further renunciation of ordinary visualisation, the basic concepts of physics which we owe to the great masters will certainly prove indispensable in this new field as well.

(**128**, 691-692; 1931)

物质粒子和电磁波的经典观念得到了实际的应用，而光子和电子波的概念还没有应用到实践中。应用受到限制的主要原因是由于存在量子效应，不可能认为观察到的现象与所用的仪器无关。作为一个例子，我想谈一下麦克斯韦理论最著名的应用，也就是在无线电传输领域中的应用。认为电磁波由光子组成只不过是形式上的说法而已，因为我们在发射和接收电磁波的时候是无法确定电磁波中包含的光子数目的。因此我们可以说，以计数为本质的光子概念已经彻底不复存在了。

为了进一步说明问题，我们可以想象一下如果严格符合量子力学规律的电子衍射和光子效应是在麦克斯韦和法拉第的工作问世之前就已经发现了，那将出现什么样的情况呢？当然，这种情况是不可能出现的，因为用来解释以上实验发现的理论是以麦克斯韦和法拉第这两人的理论成果为依据的。不过，我们还是可以发挥一下我们的想象力，问问自己，如果没有麦克斯韦和法拉第的理论，科学会处于一个什么样的水平。我想如果说我们应该在牛顿和惠更斯认为光是一种物质的理论上加以拓展，这并不为过。事实上，我们应该意识到的是，对任何测量结果的准确解释都必须建立在经典物理理论的框架之内，从这个意义上我们可以说，牛顿和麦克斯韦的语言仍将是物理学家们的通用语言。

我认为现在不是继续深入讨论这些问题并在讨论中引入新观点的最好时机。但在最后我想说的是，目前麦克斯韦实验室正在现任的卡文迪什教授的领导下开展原子核内部结构方面的研究，我希望整个科学界都去探索这一全新的实验物理领域。事实上，我相信在座的剑桥大学同仁们都不会忘记牛顿和麦克斯韦的工作，也许我们已经看到利用这些成果最有可能取得后续的成功。就算我们必须做好进一步放弃常规理论的准备，大师们为我们总结的这些基本物理观点在这个新领域中也将被证明是必不可少的。

（王静 翻译；鲍重光 审稿）

Experimental Proof of the Spin of the Photon

C. V. Raman and S. Bhagavantam

Editor's Note

Chandrasekhara Venkata Raman and S. Bhagavantam had recently presented experimental evidence suggesting that the photon carries one unit of intrinsic angular momentum. In subsequent weeks, great improvements in their technique led to much more accurate results, which they now report. Their experiments involved measuring the depolarisation of the light scattered in a fluid. By eliminated a source of error in the previous experiments, due to light scattered from the walls of the containing vessels, they were able to show that a theory based on photons carrying quantum-mechanical spin 1 accounted for the results far more accurately than an alternative theory for spinless photons.

IN a paper under this title which has recently appeared,[1] we have described and discussed observations which have led us to the conclusion that the light quantum possesses an intrinsic spin equal to one Bohr unit of angular momentum. In the four weeks which have elapsed since that paper was put into print, the experimental technique has been much improved in the direction of attaining greater precision. It appears desirable forthwith to report our newer results, which confirm the conclusion stated above.

As mentioned in earlier communications,[2] the experiment we set before ourselves was to determine the extent to which the depolarisation of Rayleigh scattering of monochromatic light is diminished when it is spectroscopically separated from the scattering of altered frequency arising from the molecular rotation in a fluid. An important improvement on our previous arrangements is the use of a pointolite mercury arc, which enables an intense beam of monochromatic light to be obtained which is rigorously transverse to the direction of observation. In the case of the feeble scattering by gases, a serious source of error is the parasitic illumination from the walls of the containing vessel. We have succeeded in eliminating this by using the gas under pressure in a steel cross with suitable arrangements for securing a dark background. The depolarisation of the scattered light is determined photographically with a spectrograph and a large nicol placed in front of the slit. The use of Schwarzchild's formula for photographic blackening enables the ratio of the horizontal and vertical components of scattered light to be calculated from the times of exposure in the two positions of the nicol which give equal densities in the spectra.

Using alternately a fine slit and a very broad slit on the spectrograph, the depolarisations of the Rayleigh scattering and of the total scattering respectively are determined. The following table gives the values for the case of oxygen, carbon dioxide, and nitrous oxide gases under pressure.

光子自旋的实验证据

拉曼，巴加万塔姆

编者按

钱德拉塞卡拉·文卡塔·拉曼和巴加万塔姆最近公布的实验结果证明光子带有一个单位的内禀角动量。在随后的几周内，他们通过改进实验技术得到了更为精确的结果，本文中报告了这些结果。他们用实验测量了光散射在一种液体中的解偏振现象。他们在去除了以前实验中由于容器壁的光散射所造成的误差之后指出：用假设光子在量子力学中的自旋为 1 的理论来解释实验结果要比假设光子没有自旋的理论更精确。

我们在最近发表的一篇同名文章中[1]，描述并讨论了我们的实验结果，即光量子具有一个玻尔角动量单位的内禀自旋。从那篇文章发表到现在已经过去了 4 周，我们的实验技术在得到更高的精确度方面又有了很大改进。我们似乎有必要立即把最新的实验结果发表出来，而这新的实验结果也证实了上述结论。

如从前的文章所提到的那样[2]，用光谱法可以将单色光的瑞利散射从液体中由分子转动产生的变频散射中分离出来，我们的实验就是要确定瑞利散射解偏振降低的程度。我们对以前实验装置的一个重要改进是使用了汞弧光点光源，这样我们就可以得到一束与观察者方向严格垂直的高强度单色光。在气体散射很弱的情况下，由容器壁产生的寄生亮度将导致严重的实验误差。我们通过在钢的十字形腔中对气体加压，而这种十字形腔可用适当方式确保处于黑暗的背景中，从而成功地降低了这种误差。利用放置于狭缝前的一个大号尼科尔棱镜和光谱仪，我们可以用照相法将散射光解偏振的数值确定下来。根据底片黑度的史瓦西公式，我们可用尼科尔棱镜上产生相同光谱强度的两个不同位置的曝光次数计算出散射光横向偏振分量和纵向偏振分量的比值。

在光谱仪上交替使用窄缝和宽缝，可以分别得到瑞利散射和总散射的解偏振百分比率。下表给出了 O_2、CO_2 以及 N_2O 气在加压条件下的相关数据：

Table: Depolarisation percent

Gas	Observed		Calculated	
	Total Scattering	Rayleigh Scattering	Kramers–Heisenberg Theory	Spin Theory
O_2	6.5	4.1	1.7	4.2
CO_2	10.3	6.3	2.8	6.7
N_2O	12.0	7.7	3.4	7.9

The depolarisations of the total scattering given in column 1 thus found spectroscopically are in good agreement with the best accepted values determined by other methods. Column 2 gives the observed depolarisations of the Rayleigh scattering, column 3 the values calculated from the Kramers–Heisenberg dispersion theory, and column 4 the values calculated from the theory of the spinning photons discussed in our paper. It will be seen that the values given by the latter are strikingly supported by the experimental results.

(**129**, 22-23; 1932)

C. V. Raman and S. Bhagavantam: 210 Bowbazar Street, Calcutta, India, Nov. 29.

References:

1. *Ind. Jour. Phy.*, **6**, 353 (Oct. 1931).

2. *Nature*, **128**, 576 and 727 (1931).

表：解偏振百分比率

气体	观测值		计算值	
	总散射	瑞利散射	克拉莫 – 海森堡理论	自旋理论
O_2	6.5	4.1	1.7	4.2
CO_2	10.3	6.3	2.8	6.7
N_2O	12.0	7.7	3.4	7.9

我们在第一列中给出了用光谱学方法观测得到的总散射解偏振比率，其数值与用其他方法得到的最佳结果十分吻合。第二列中给出的是观测到的瑞利散射解偏振比率，第三列是根据克拉莫 – 海森堡色散理论计算得到的数值，而第四列中的数据是由我们文章中讨论的光子自旋理论计算出来的。可以看到，第二种理论给出的数值与实验结果非常一致。

（王静 翻译；赵见高 审稿）

The Decline of Determinism*

A. Eddington

Editor's Note

Ten years ago, Arthur Eddington claims here, practically every notable physicist was a determinist, at least regarding physical processes. Physicists believed they possessed a powerful scheme of causal scientific law, embodied especially in classical mechanics, and saw the project of science as explaining ever more phenomena within this scheme. The quantum theory has now changed all this, stimulating reactions ranging from incredulity and cynicism to yawning indifference. Eddington argues, however, that the new indeterministic theory cannot be considered to be a rejection of the scientific method. He compares the older deterministic or "primary" laws to a scientific gold standard, now replaced by a "secondary" currency. Most other physicists now wonder why those primary laws were once accorded such reverence.

DETERMINISM has faded out of theoretical physics. Its exit has been commented on in various ways. Some writers are incredulous, and cannot be persuaded that determinism has really been eliminated. Some think that it is only a domestic change in physics, having no reactions on general philosophic thought. Some imagine that it is a justification for miracles. Some decide cynically to wait and see if determinism fades in again.

The rejection of determinism is in no sense an abdication of scientific method; indeed it has increased the power and precision of the mathematical analysis of observed phenomena. On the other hand, I cannot agree with those who belittle the general philosophical significance of the change. The withdrawal of physical science from an attitude it has adopted consistently for more than two hundred years is not to be treated lightly; and it involves a reconsideration of our views with regard to one of the perplexing problems of our existence. In this address, I shall deal mainly with the physical universe, and say very little about mental determinism or freewill. That might well be left to those who are more accustomed to arguing about such questions, if only they could be awakened to the new situation which has arisen on the physical side. At present I can see little sign of such an awakening.

Definitions of Determinism

Let us first be sure that we agree as to what is meant by determinism. I quote three definitions or descriptions for your consideration. The first is by a mathematician (Laplace):

* Presidential address to the Mathematical Association delivered on Jan. 4.

决定论的衰落*

阿瑟·爱丁顿

编者按

10 年前，阿瑟·爱丁顿在《自然》杂志上发表文章声称实际上所有的知名物理学家都是决定论者，至少他们面对物理过程时都是如此。物理学家们认为他们拥有一个因果相连的庞大科学体系——最明显的例子是经典力学领域，并且人们还在利用这种模式解释更多的现象。但现在量子理论改变了所有的一切，人们对此的反应各不相同：有人将信将疑，有人冷嘲热讽，还有些人漠不关心。然而，爱丁顿指出：不能认为新出现的非决定论是一种反科学的方法。他把旧有的决定论或称“基本定律”比作科学上的金本位，现在被“第二代”货币所取代。现在，大多数物理学家都不能理解为什么这些基本定律曾经受到过如此的推崇。

在理论物理领域里，决定论的身影渐渐淡去。人们对它的退出持有各种不同的观点。有些作者对此表示怀疑，不相信决定论真的被排除了；有些则认为它仅仅是物理学领域内部的变化，对一般的哲学思想没有任何影响；另一些人认为这是对奇怪现象的合理性解释；还有一部分人冷眼旁观，等着看决定论是否会卷土重来。

摒弃决定论绝不是要放弃科学方法；在对观测到的现象进行数学分析时，它也的确提高了分析的能力和精度。从另一方面来说，我无法认同那些低估了这个变化所具有的一般哲学意义的人。我们不能轻描淡写地摆脱两百多年来被自然科学所一致采取的看法；并且这使得我们必须重新审视一个有关存在物的复杂问题。在这里，我将主要就物质世界进行探讨，而几乎不涉及思想决定论和自由意志。最好把它们留给那些更加善于探讨此类问题的人，只要他们能意识到目前在物理学方面已经出现的新情况。而我目前几乎看不出任何意识到这种情况的迹象。

决定论的定义

让我们首先确保我们对决定论的含义有一致的认识。我要引用三种定义或描述给你们作参考。第一种是一位数学家（拉普拉斯）的叙述：

---

* 本文是 1 月 4 日爱丁顿对数学分会所作的演讲。

We ought then to regard the present state of the universe as the effect of its antecedent state and the cause of the state that is to follow. An intelligence, who for a given instant should be acquainted with all the forces by which Nature is animated and with the several positions of the entities composing it, if, further, his intellect were vast enough to submit those data to analysis, would include in one and the same formula the movements of the largest bodies in the universe and those of the lightest atom. Nothing would be uncertain for him; the future as well as the past would be present to his eyes. The human mind in the perfection it has been able to give to astronomy affords a feeble outline of such an intelligence. … All its efforts in the search for truth tend to approximate without limit to the intelligence we have just imagined.

The second is by a philosopher (C. D. Broad):

"Determinism" is the name given to the following doctrine. Let S be any substance, ψ any characteristic, and t any moment. Suppose that S is in fact in the state δ with respect to ψ at t. Then the compound supposition that everything else in the world should have been exactly as it in fact was, and that S should have been in one of the other two alternative states with respect to ψ is an impossible one. [The three alternative states (of which δ is one) are to have the characteristic ψ, not to have it, and to be changing.]

The third is by a poet (Omar Khayyám):

With Earth's first Clay They did the Last Man knead,
And there of the Last Harvest sow'd the Seed:
And the first Morning of Creation wrote What the Last Dawn of Reckoning shall read.

I propose to take the poet's description as my standard. Perhaps this may seem an odd choice; but there is no doubt that his words express what is in our minds when we refer to determinism. The other two definitions need to be scrutinised suspiciously; we are afraid there may be a catch in them. In saying that the physical universe as now pictured is not a universe in which "the first morning of creation wrote what the last dawn of reckoning shall read", we make it clear that the abandonment of determinism is no technical quibble, but is to be understood in the most ordinary sense of the words.

It is important to notice that all three definitions introduce the time-element. Determinism postulates not merely causes but pre-existing causes. Determinism means predetermination. Hence in any argument about determinism the dating of the alleged causes is an important matter; we must challenge them to produce their birth certificates.

Ten years ago, practically every physicist of repute was, or believed himself to be, a determinist, at any rate so far as inorganic phenomena are concerned. He believed that he had come across a scheme of strictly causal law, and that it was the primary aim of science to fit as much of our experience as possible into such a scheme. The methods, definitions, and conceptions of physical science were so much bound up with this assumption of

我们应把目前宇宙的状态看作是它先前状态的结果和未来状态的原因。假定有这么一位天才，他在某个确定的瞬间知道了所有驱动自然界运转的力以及组成宇宙的所有实体的几个位置，如果他还具有足够的才智来收集分析这些数据的话，他就能够把从最大物体到最小原子的运动都包括在同一个公式中。对他来说，没有什么是不确定的，过去和未来全在他的掌握之中。人类的完美主义思想为天文学描绘了一个如此聪慧的人的大致轮廓。……对我们刚才幻想中的这位天才而言，他为探寻到真理所做的努力几乎是不受限制的。

第二种定义来自于一位哲学家（布罗德）的观点：

“决定论”指的是下述学说：设 S 为任意一个物质，ψ 为任意一种特性，t 为任意一个时刻。假设 t 时刻特性为 ψ 的物质 S 实际上处于状态 δ，那么认为：世界上的任意其他物质都处于实际状态，并且物质 S 本来应该处于关于特性 ψ 的其他两种状态中的一种的复合推断是不可能成立的。[这三种状态（δ是其中一种）分别为：具备特性 ψ、不具备特性 ψ 和正在变化的状态。]

第三种说法则是一位诗人（奥马尔·海亚姆）提出来的：

最初的泥丸捏成了最终的人形，
最后的收成便是那最初的种子：
天地开辟时的老文章写就了天地掩闭时的字句。

我选择诗人的描述作为我对决定论的理解标准。看起来这可能是个奇怪的选择，但是毫无疑问他的描述正是我们提到决定论时脑海里所浮现的想法。其他的两个定义则需以怀疑的态度仔细研究，恐怕它们之中会存在某些陷阱。当我们提到现在所描绘的物质世界时，指的并不是那种“天地开辟时的老文章写就了天地掩闭时的字句”的宇宙。我们要让人明白放弃决定论并不是一种巧妙的推托，而是用一种最平实的语言就可以理解的。

值得特别注意的是三种定义都引入了时间元素。决定论不仅仅是要假定存在的起因，而是要假定起因在早先就已经存在了。决定论意味着预先决定。因此在任何有关决定论的争论中，确定所谓起因的时间是一个非常重要的问题，我们必须让它们为自身的起源提供证明。

十年前，至少是在考虑无机系统的物理现象时，几乎所有知名的物理学家都是，或者认为他们自己是，一个决定论者。他们确信自己已经发现了一个严格的因果律体系，并且相信科学的首要目的就是尽可能地把我们的日常体验纳入到这个体系内。物理科学中的方法、定义以及概念受到这种决定论假设过多的限制，以至于人们把

determinism that the limits (if any) of the scheme of causal law were looked upon as the ultimate limits of physical science.

To see the change that has occurred, we need only refer to a recent book which goes as deeply as anyone has yet penetrated into the fundamental structure of the physical universe, Dirac's "Quantum Mechanics". I do not know whether Dirac is a determinist or not; quite possibly he believes as firmly as ever in the existence of a scheme of strict causal law. But the significant thing is that in this book he has no occasion to refer to it. In the fullest account of what has yet been ascertained as to the way things work, causal law is not mentioned.

This is a deliberate change in the aim of theoretical physics. If the older physicist had been asked why he thought that progress consisted in fitting more and more phenomena into a deterministic scheme, his most effective reply would have been "What else is there to do?" A book such as Dirac's supplies the answer. For the new aim has been extraordinarily fruitful, and phenomena which had hitherto baffled exact mathematical treatment are now calculated and the predictions are verified by experiment. We shall see presently that indeterministic law is as useful a basis for practical predictions as deterministic law was. By all practical tests, progress along this new branch track must be recognised as a great advance in knowledge. No doubt some will say "Yes, but it is often necessary to make a detour in order to get round an obstacle. Presently we shall have passed the obstacle and be able to join the old road again." I should say rather that we are like explorers on whom at last it has dawned that there are other enterprises worth pursuing besides finding the North-West Passage; and we need not take too seriously the prophecy of the old mariners who regard these enterprises as a temporary diversion to be followed by a return to the "true aim of geographical exploration". But at the moment I am not concerned with prophecy and counter-prophecy; the important thing is to grasp the facts of the present situation.

Secondary Law

Let us first try to see how the new aim of physical science originated. We observe certain regularities in the course of Nature and formulate these as "laws of Nature". Laws may be stated positively or negatively, "Thou shalt" or "Thou shalt not". For the present purpose it is most convenient to formulate them negatively. Consider the following two regularities which occur in our experience:

(*a*) We never come across equilateral triangles whose angles are unequal.

(*b*) We never come across 13 trumps in our hand at bridge.

In our ordinary outlook we explain these regularities in fundamentally different ways. We

因果律体系的局限性（如果有的话）看作是物理科学的极限。

要了解已经出现的改变，我们仅仅需要参考最近出版的一本书——狄拉克所著的《量子力学》，它比任何人都更深层次地触及到物质世界的基本结构。我不知道狄拉克是否是一个决定论者，他很有可能一直都坚信存在着严格的因果律体系。但是值得注意的是，在这本书里他没有提及这个问题。在对已经可以确定的事物运行规律的最全面的叙述中，并没有提到因果律。

理论物理研究目标中的这样一个改变是经过深思熟虑的。如果年长些的物理学家被问及为什么科学的发展就在于把越来越多的现象纳入决定论的体系，那么他最有力的回应便是："除此之外还有什么别的办法吗？"狄拉克所写的那样一本书给出了答案。因为新的研究目标已经产生了非凡的效果，迄今仍难以给出精确数学处理的物理现象现在也可以计算出来了，并且很多预言得到了实验的验证。现在我们应该明白，非决定论的规律和决定论的规律一样有效，都可以作为实际预测的原则。所有实践都证明，我们必须把在这个新分支方向上取得的进展看作是知识的巨大进步。毫无疑问，有些人会说："对，但为了避开障碍物，通常会绕道而行。目前我们已经越过了障碍，可以重新回到以前的老路上去了。"不过我却这样认为：我们就像那些探险家，他们最后终于意识到，除了去探寻西北通道以外，还有其他值得追求的事业。我们不需要过分看重老水手的预言，他们把这些事业当作是一种暂时性的消遣，随后还是要回到"地理大发现这个终极目标上"。但此刻我不关心那些预言和反预言的论调，重要的是把握目前的实际情况。

次级定律

首先，让我们了解一下物理科学中的新目标是怎样出现的。我们观察自然过程中的某些规律，然后把它们明确表达为"自然定律"。这些定律既可以用肯定句式陈述也可以用否定句式陈述，比如"你应该"或者"你不应该"。对于当前的需要，用否定句式来阐明最为方便。在我们的生活经历中会出现以上两种规律：

(*a*) 我们绝不会见到一个等边三角形具有不相等的角。

(*b*) 在打桥牌时我们绝不会碰到手中有 13 张王牌的情况。

按照我们通常的观点，我们可以用根本不同的方法来解释这些规则。我们会说

say that the first occurs because the contrary experience is *impossible*; the second occurs because the contrary experience is *too improbable*.

This distinction is entirely theoretical; there is nothing in the observations themselves to suggest to which type a particular regularity belongs. We recognise that "impossible" and "too improbable" can both give adequate explanation of any observed uniformity of experience, and the older theory rather haphazardly explained some uniformities one way and other uniformities the other way. In the new physics we make no such discrimination; the union obviously must be on the basis of (*b*), not (*a*). It can scarcely be supposed that there is a law of Nature which makes the holding of 13 trumps in a properly dealt hand impossible; but it *can* be supposed that our failure to find equilateral triangles with unequal angles is only because such triangles are too improbable.

We must, however, first consider the older view which distinguished type (*a*) as a special class of regularity. Accordingly, there were two types of natural law. The earth keeps revolving round the sun, because it is *impossible* it should run away. Heat flows from a hot body to a cold, because it is *too improbable* that it should flow the other way. I call the first type *primary* law, and the second type *secondary* law. The recognition of secondary law was the thin end of the wedge that ultimately cleft the deterministic scheme.

For practical purposes primary and secondary law exert equally strict control. The improbability referred to in secondary law is so enormous that failure even in an isolated case is not to be seriously contemplated. You would be utterly astounded if heat flowed from you to the fire so that you got chilled by standing in front of it, although such an occurrence is judged by physical theory to be not impossible but improbable. Now it is axiomatic that in a deterministic scheme nothing is left to chance; a law which has the ghost of a chance of failure cannot form part of the scheme. So long as the aim of physics is to bring to light a deterministic scheme, the pursuit of secondary law is a blind alley since it leads only to probabilities. The determinist is not content with a law which prescribes that, given reasonable luck, the fire will warm me; he admits that that is the probable effect, but adds that somewhere at the base of physics there are other laws which prescribe just what the fire will do to me, luck or no luck.

To borrow an analogy from genetics, determinism is a *dominant character*. We can (and indeed must) have secondary indeterministic laws within any scheme of primary deterministic law—laws which tell us what is likely to happen but are overridden by the dominant laws which tell us what must happen. So determinism watched with equanimity the development of indeterministic law within itself. What matter? Deterministic law remains dominant. It was not foreseen that indeterministic law when fully grown might be able to stand by itself and supplant its dominant parent. There is a game called "Think of a number". After doubling, adding, and other calculations, there comes the direction "Take away the number you first thought of". We have reached that position in physics, and the time has come to take away the determinism we first thought of.

出现第一种情况是因为**不可能**发生相反的情况；出现第二种情况是因为相反情况发生的**概率极小**。

这完全是理论上的区分。观测本身不能说明某个特定的规律属于哪种类型。我们认为“不可能”和“概率极小”都可以对在实践中观测到的一致性做出合理的解释，早期的理论非常随意地对某些一致性以一种方式做出解释，而对其他的一致性又以另一种方式做出解释。在新物理学中，我们不作这样的区分；显然地，统一应基于标准 (*b*)，而不是 (*a*)。我们无法假设存在一种使手中有 13 张王牌的情况不可能发生的自然规律，但**可以**认为找不到不等角的等边三角形是因为这样的三角形出现的概率极小。

然而，我们必须首先顾及过去的观点，即是把 (*a*) 区分出来作为一种特殊规律。因此，我们有了两类自然定律。地球持续围绕太阳运行是因为它**不可能**逃离开。热量从高温物体流向低温物体，是因为它流向相反方向的**概率极小**。我把第一种类型叫做**基本**定律，而把第二种叫做**次级**定律。对次级定律的认识有如楔子的尖端终于劈开了决定论的框架。

对于实际的目的来说，基本定律和次级定律发挥了同样严格的控制作用。在次级定律中提到的不可能性发生的概率非常大，即使出现了个别反例，也不会得到认真对待。如果热量从你身体上流向了火，你会十分惊讶，因为这样一来你站在火前就会感到寒冷刺骨，尽管在物理理论中这种情况并非不可能发生，而不过是发生的概率极小罢了。在决定论体系中，显然任何事情都不是偶然发生的；该体系中的所有定律都不可能出现任何反例。只要物理学的目标是要建立一个决定论的体系，那么我们对次级定律的研究只能走进死胡同，因为这样的研究只会得到有可能发生的结果。决定论者不会认可这样的规则：只要运气还可以我就能感到火的温暖。他承认这是一种可能出现的结果，可是他还会补充道，其物理基础应该以另一条定律来规定，即无论有没有运气，火都会使我温暖。

借用一个遗传学的相似概念，决定论可以被看作是一种**显性性状**。我们能够（事实上是必须）把次级的非决定论置于任何一个基本的决定论体系之中，非决定论会告诉我们某件事情很有可能发生，然而那些告诉我们某件事情肯定会发生的显性定律优先于非决定论的定律。因此，决定论对其内部的非决定论定律的发展泰然视之。有什么关系呢？决定论的定律仍然是占优势地位的。没有人能预料见到当非决定论定律发展成熟后会独树一帜，取代占主导地位的母体。有一个游戏，名字叫做“猜数字”。在经过各种加倍、相加和其他运算后，最终达到“去掉最初所想的数字”的目的。在物理学中，我们已经实现了这样的一个目标，现在到了把我们最初想到的决定论去掉的时候了。

The growth of secondary law within the deterministic scheme was remarkable, and gradually sections of the subject formerly dealt with by primary law were transferred to it. There came a time when in some of the most progressive branches of physics secondary law was used exclusively. The physicist might continue to profess allegiance to primary law but he ceased to utilise it. Primary law was the gold to be kept stored in vaults; secondary law was the paper to be used for actual transactions. No one minded; it was taken for granted that the paper was backed by gold. At last came the crisis, and *physics went off the gold standard*. This happened very recently, and opinions are divided as to what the result will be. Prof. Einstein, I believe, fears disastrous inflation, and urges a return to sound currency—if we can discover it. But most theoretical physicists have begun to wonder why the now idle gold should have been credited with such magic properties. At any rate the thing has happened, and the immediate result has been a big advance in atomic physics.

We have seen that indeterministic or secondary law accounts for regularities of experience, so that it can be used for predicting the future as satisfactorily as primary law. The predictions and regularities refer to average behaviour of the vast number of particles concerned in most of our observations. When we deal with fewer particles the indeterminacy begins to be appreciable, and prediction becomes more of a gamble; until finally the behaviour of a single atom or electron has a very large measure of indeterminacy. Although some courses may be more probable than others, backing an electron to do anything is in general as uncertain as backing a horse.

It is commonly objected that our uncertainty as to what the electron will do in the future is due not to indeterminism but to ignorance. It is asserted that some character exists in the electron or its surroundings which decides its future, only physicists have not yet learned how to detect it. You will see later how I deal with this suggestion. But I would here point out that if the physicist is to take any part in the wider discussion on determinism as affecting the significance of our lives and the responsibility of our decisions, he must do so on the basis of what he has discovered, not on the basis of what it is conjectured he might discover. His first step should be to make clear that he no longer holds the position, occupied for so long, of chief advocate for determinism, and that he is *unaware* of any deterministic law in the physical universe. He steps aside and leaves it to others—philosophers, psychologists, theologians—to come forward and show, if they can, that they have found indications of determinism in some other way.* If no one comes forward, the hypothesis of determinism presumably drops; and the question whether physics is actually antagonistic to it scarcely arises. It is no use looking for an opposer until there is a proposer in the field.

---

* With the view of learning what might be said from the philosophical side against the abandonment of determinism, I took part in a symposium of the Aristotelian Society and Mind Association in July 1931. Indeterminists were strongly represented, but unfortunately there were no determinists in the symposium, and apparently none in the audience which discussed it. I can scarcely suppose that determinist philosophers are extinct, but it may be left to their colleagues to deal with them.

在决定论体系中，次级定律的发展是引人注目的，一部分原本利用基本定律来处理的问题逐渐开始偏向使用次级定律。有一段时间，物理学中的一些最前沿的分支只采用次级定律进行研究。可能会有这样的情况，物理学家在继续宣称自己忠实于基本定律的同时已经不再使用它了。基本定律有如存放在金库中的黄金，而次级定律则是真正用于交易的纸币。没有人会介意，大家都认为纸币以黄金为基础是理所当然的事情。最终危机出现了，**物理学放弃了金本位制**。这是最近才发生的，并且对结果将怎样人们说法不一。我确信，爱因斯坦教授对这种损失惨重的通货膨胀感到担忧，他强烈要求回归到健全的货币体系——当然，要以我们能找到这种货币体系为前提。不过大部分理论物理学家都已经开始质疑为什么应该认为现在闲置着的金子具有如此神奇的特性。无论如何，事情已经发生了，其直接的结果是使原子物理学取得了一个极大的进步。

我们已经知道，非决定论定律，或者说次级定律，解释了由经验总结的规律，因此它同样可以像基本定律一样很好地预测未来。这些预测和规律与我们在大多数情况下观测到的大量粒子的平均状态相关。当我们处理数目较少的粒子时，不确定性开始变得明显起来，而预测变得更像是在冒险。直到最后，单个原子或电子的行为在很大程度上是不确定的。虽然有些过程比另一些过程更有可能发生，但是在通常情况下，预测一个电子将会怎样就如同预测一匹马一样难以确定。

人们通常会对上述观点表示反对：我们不能确定电子将会如何运动的原因不是因为非决定论而是因为无知。有人声称，在电子内部或者它的周围环境中存在着确定其未来的某些特征因素，只是物理学家还不懂得如何才能检测到它。稍后你们就会知道我是如何看待这个观点的。但是这里我要指出，如果一个物理学家要在更广泛的意义上探讨决定论对我们的生活水平以及决策可靠性的影响，他就必须以他已经发现的物理现象为论据，而不能基于他推测出的可能会发现的现象。首先他应该搞清楚的是，他已经不再拥有占据了很久的决定论主要支持者的地位，并且他对物质世界中的任何决定论定律都**毫无所知**。他将退到一边，让其他人——哲学家、心理学家、神学家站出来说明他们以其他方式发现了决定论的迹象*，如果他们能找到的话。如果没有人挺身而出，那么决定论这个假设也许会被丢弃，也就不会引出物理学是否真的与决定论相抵触的问题。寻找反对者的努力都是徒劳的，除非有人自告奋勇。

* 因为想知道反对放弃决定论的哲学家有什么想法，我参加了亚里士多德学会和哲学学会在1931年7月举办的研讨会。但不幸的是与会者都是非决定论者，研讨会上一个决定论者也没有，在听众中也没有人讨论决定论。我不敢说持决定论观点的哲学家已经不复存在了，这也许应该由哲学界人士来了断。

Inferential Knowledge

It is now necessary to examine rather closely the nature of our knowledge of the physical universe.

All our knowledge of physical objects is by inference. We have no means of getting into direct contact with them; but they emit and scatter light waves, and they are the source of pressures transmitted through adjacent material. They are like broadcasting stations that send out signals which we can receive. At one stage of the transmission the signals pass along nerves within our bodies. Ultimately visual, tactual, and other sensations are provoked in the mind. It is from these remote effects that we have to argue back to the properties of the physical object at the far end of the chain of transmission. The image which arises in the mind is not the physical object, though it is a source of information about the physical object; to confuse the mental object with the physical object is to confuse the clue with the criminal. Life would be impossible if there were no kind of correspondence between the external world and the picture of it in our minds; and natural selection (reinforced where necessary by the selective activity of the Lunacy Commissioners) has seen to it that the correspondence is sufficient for practical needs. But we cannot rely on the correspondence, and in physics we do not accept any detail of the picture unless it is confirmed by more exact methods of inference.

The external world of physics is thus a universe populated with *inferences*. The inferences differ in degree and not in kind. Familiar objects which I handle are just as much inferential as a remote star which I infer from a faint image on a photographic plate, or an "undiscovered" planet inferred from irregularities in the motion of Uranus. It is sometimes asserted that electrons are essentially more hypothetical than stars. There is no ground for such a distinction. By an instrument called a Geiger counter, electrons may be counted one by one as an observer counts one by one the stars in the sky. In each case the actual counting depends on a remote indication of the physical object. Erroneous properties may be attributed to the electron by fallacious or insufficiently grounded inference, so that we may have a totally wrong impression of what it is we are counting; but the same is equally true of the stars.

In the universe of inferences, past, present, and future appear simultaneously, and it requires scientific analysis to sort them out. By a certain rule of inference, namely, the law of gravitation, we infer the present or past existence of a dark companion to a star; by an application of the same rule of inference we infer the existence on Aug. 11, 1999, of a configuration of the sun, earth, and moon, which corresponds to a total eclipse of the sun. The shadow of the moon on Cornwall in 1999 is already in the universe of inference. It will not change its status when the year 1999 arrives and the eclipse is observed; we shall merely substitute one method of inferring the shadow for another. The shadow will always be an inference. I am speaking of the object or condition in the external world which is called a shadow; our perception of darkness is not the physical shadow, but is one of the possible clues from which its existence can be inferred.

推理性的认识

现在有必要进一步审视我们对于物质世界的认识的本质。

我们对物理对象的所有认识都是通过推理获得的。我们没有办法与它们直接接触，但是它们发射或散射光波，并把压力传递给相邻的物质。它们就像广播电台一样发射我们可以接收到的信号。在传输的一个阶段信号沿着我们身体里的神经传递，最终在我们的头脑里形成视觉、触觉以及其他感觉。通过这些远程效应，我们必须反过来辨明在传输链的另一端的物质的性质。我们头脑中产生的图像并不是物质本身，尽管它是与物质相关的信息的来源；把头脑中的物质图像混同为物质的实体就等于把线索混同于罪犯。如果不存在外部世界与它在我们头脑中形成的图像之间的对应，就不可能有我们的生活。在日常生活中，自然选择（必要时需要补充精神病委员的鉴定结果）所得到的对应似乎就已经足够满足实际需要了。但是我们不能依赖这些对应，在物理领域内，我们不能接受根据图像得到的任何信息，除非经过更为精确的推理方法的确证。

物理学的外部世界就是这样一个充满**推理**的世界。这些推理只有程度上的差别，而没有本质上的区分。我手中的常见物体，与从照相底片上的模糊图像推断出的一颗遥远恒星，或从天王星运动的不规则性推出的“未知”行星一样，都可以由推论得到。有时候我们认为电子的不确定性高于恒星。但其实它们根本没有什么区别。利用一台名为盖革计数器的仪器，我们可以逐个将电子计数，犹如我们对天上的星星逐个计数一样。在上述两种情况下，计数实际上都取决于物质的远程影像。当然，对于电子错误的或根据不充分的推断会得到不正确的性质，所以对于我们所计量的对象，我们有可能得到一个完全错误的概念；然而我们在计数星星时，也同样会存在这样的情况。

在这个充满推理的宇宙里，过去、现在和未来似乎是同时存在的，需要我们用科学分析法去区分。利用特定的推理规律，即引力定律，我们可以推断出某个恒星周边在现在或过去存在一个不发光的伴星；利用这个定律我们同样可以推断，在 1999 年 8 月 11 日，由于太阳、地球和月亮的相对位置会出现一次日全食。1999 年将要在康沃尔出现的月亮阴影已经被推算了出来。这个推算在 1999 年到来和日食被观测到之前是不会改变的，我们只会不断替换推断出阴影的方法。阴影将永远是一个推断。我所说的阴影是外部世界中的物体或条件；其实我们对黑暗的感知并不是物理上的阴影，而是可以从黑暗的存在推断出来的其中一种可能的线索。

Of particular importance to the problem of determinism are our inferences about the past. Strictly speaking, our direct inferences from sight, sound, touch, all relate to a time slightly antecedent; but often the lag is more considerable. Suppose that we wish to discover the constitution of a certain salt. We put it in a test tube, apply certain reagents, and ultimately reach the conclusion that it *was* silver nitrate. It is no longer silver nitrate after our treatment of it. This is an example of retrospective inference: the property which we infer is not that of "being X" but of "having been X".

We noted at the outset that in considering determinism the alleged causes must be challenged to produce their birth certificates so that we may know whether they really were pre-existing. Retrospective inference is particularly dangerous in this connexion because it involves antedating a certificate. The experiment above mentioned certifies the chemical constitution of a substance, but the date we write on the certificate is earlier than the date of the experiment. The antedating is often quite legitimate; but that makes the practice all the more dangerous, it lulls us into a feeling of security.

Retrospective Characters

To show how retrospective inference might be abused, suppose that there were no way of learning the chemical constitution of a substance without destroying it. By hypothesis a chemist would never know until after his experiment what substance he had been handling, so that the result of every experiment he performed would be entirely unforeseen. Must he then admit that the laws of chemistry are chaotic? A man of resource would override such a trifling obstacle. If he were discreet enough never to say beforehand what his experiment was going to demonstrate, he might give edifying lectures on the uniformity of Nature. He puts a lighted match in a cylinder of gas, and the gas burns. "There you see that hydrogen is inflammable." Or the match goes out. "That proves that nitrogen does not support combustion." Or it burns more brightly. "Evidently oxygen feeds combustion." "How do you know it was oxygen?" "By retrospective inference from the fact that the match burned more brightly." And so the experimenter passes from cylinder to cylinder; the match sometimes behaves one way and sometimes another, thereby beautifully demonstrating the uniformity of Nature and the determinism of chemical law! It would be unkind to ask how the match must behave in order to indicate indeterminism.

If by retrospective inference we infer characters at an earlier date, and then say that those characters invariably produce at a future date the manifestation from which we inferred them, we are working in a circle. The connexion is not causation but definition, and we are not prophets but tautologists. We must not mix up the genuine achievements of scientific prediction with this kind of charlatanry, or the observed uniformities of Nature with those so easily invented by our imaginary lecturer. It is easily seen that to avoid vicious circles we must abolish purely retrospective characteristics—those which are never found as existing but always as having existed. If they do not manifest themselves until the moment that they cease to exist, they can never be used for prediction except by those who prophesy after the event.

在决定论问题中最重要的一点是我们对于过去的推断。严格说来，我们通过视觉、听觉、触觉直接得来的推断在时间上都有很短的滞后；但这种滞后往往是比较重要的。假如我们想知道某种盐类的构成，我们会把它放入试管内，然后加入特定的试剂，最后得到结论，原来它是硝酸银。在我们对它做了处理之后它已经不再是硝酸银了。这就是一个回顾性推理的例子：我们推出的结论不是“是 X”，而是“曾经是 X”。

我们从一开始注意到，在讨论决定论的时候，那些所谓的原因必须提供自己的出生证明，这样我们才能够知道它们是否的确是早已存在的。就此而论，回顾性推理是非常不可靠的，因为它需要提前提供证明。上面提到的实验证明了物质的化学成分，然而鉴定书上所证实的物质的存在日期要先于实验的日期。这种提前通常是非常合理的，但在实际操作中更加危险，因为它让我们产生了一种虚假的安全感。

回顾性的特征

为了表明回顾性的推理是如何被滥用的，假定如果不破坏物质就没有任何办法获知它的化学成分。根据这个假设，一个化学家在他进行实验前不会知道他所要测试的物质的化学成分，于是他所做的所有实验的结果都是不可预见的。那么他是否必须承认，所有的化学规律都是杂乱无章的？聪明人知道要先忽略掉这些不重要的障碍。如果他足够谨慎，从不在实验前说出自己的实验要去证明什么，那么他也许会对自然齐一性问题发表富有启发性的演讲。他把一根燃烧的火柴扔进一瓶气体中，气体燃烧起来。然后他说，“你们看，氢气是易燃的”。或者是火柴熄灭了。“这证明氮气是不能助燃的”。或者是火柴燃烧得更剧烈了。“很明显氧气是助燃的”。“你怎么知道它就是氧气呢？”“这是从火柴燃烧得更加剧烈的事实中推断出来的。”然后实验者就这样一瓶又一瓶地进行试验；火柴有时呈现出这样一种状态，有时是另一种状态，从而完美地证实了自然齐一性以及化学规律的决定论！要是提问说火柴应该怎样表现才能够表明非决定论，那将是一个很不客气的问题。

如果按照回顾性的推理，我们推断出的是较早以前的性质，然后说这些特性在以后会恒定不变地产生一种现象，而这种现象就是最初我们用来推断其性质的现象，我们就是在循环论证。这种关系并不是因果关系，而完全是在下定义。而我们也不是预言家，只是重复在使用同义词的人。我们决不能把真正的科学预言与此类骗术混同，也决不能把观察到的自然规律的一致性同那个虚构的演说家轻易编造出来的东西混同。很容易发现要避开这种恶性循环，我们就必须废除纯粹的回顾性特性——那些特性永远不是现有状态的特性，而总是曾经存在的某种状态的特性。如果这些性质在消逝的那一刻前不能表现出来，那么它们永远都不能用于预测，除非是在事后去做预测。

Chemical constitution is not a retrospective character, though it is often inferred retrospectively. The fact that silver nitrate can be bought and sold shows that there is a property of *being* silver nitrate as well as of *having been* silver nitrate. Apart from special methods of determining the constitution or properties of a substance without destroying it, there is one general method widely applicable. We divide the specimen into two parts, analyse one part (destroying it if necessary), and show that its constitution *has been X*; then it is usually a fair inference that the constitution of the other part is *X*. It is sometimes argued that in this way a character inferable retrospectively must always be also inferable contemporaneously; if that were true, it would remove all danger of using retrospective inference to invent fictitious characters as causes of the events observed. Actually the danger arises just at the point where the method of sampling breaks down, namely, when we are concerned with characteristics supposed to distinguish one individual atom from another atom of the same substance; for the individual atom cannot be divided into two samples, one to analyse and one to preserve. Let us take an example:

It is known that potassium consists of two kinds of atoms, one kind being radioactive and the other inert. Let us call the two kinds $K\alpha$ and $K\beta$. If we observe that a particular atom bursts in the radioactive manner, we shall infer that it was a $K\alpha$ atom. Can we say that the explosion was predetermined by the fact that it was a $K\alpha$ and not a $K\beta$ atom? On the information stated there is no justification at all; $K\alpha$ is merely an antedated label which we attach to the atom when we see that it has burst. We can always do that, however undetermined the event may be which occasions the label. Actually, however, there is more information which shows that the burst is not undetermined. Potassium is found to consist of two isotopes of atomic weights 39 and 41; and it is believed that 41 is the radioactive kind, 39 being inert. It is possible to separate the two isotopes and to pick out atoms known to be K^{41}. Thus, K^{41} is a contemporaneous character, and can legitimately predetermine the subsequent radioactive outburst; it replaces $K\alpha$ which was a retrospective character.

So much for the fact of outburst; now consider the time of outburst. Nothing is known as to the time when a particular K^{41} atom will burst except that it will probably be within the next thousand million years. If, however, we observe that it bursts at a time *t*, we can ascribe to the atom the retrospective character K^t, meaning that it had (all along) the property that it was going to burst at time *t*. Now, according to modern physics, the character K^t is not manifested in any way—is not even represented in our mathematical description of the atom—until the time *t* when the burst occurs and the character K^t, having finished its job, disappears. In these circumstances K^t is not a predetermining cause. Our retrospective labels and characters add nothing to the plain observational fact that the burst occurred without warning at the moment *t*; they are merely devices for ringing a change on the tenses.

The time of break-up of a radioactive atom is an example of extreme indeterminism; but it must be understood that, according to current theory, all future events are indeterminate in greater or lesser degree, and differ only in the margin of uncertainty. When the

化学组分并不具有回顾性的特性，尽管它总是在事后才被推断出来。硝酸银是可以被买卖的，这说明有这样一种表明它**现在是**硝酸银，而且它**曾经是**硝酸银的性质。除了一些特殊的方法可以测定物质的组成或性质而不用破坏它以外，还有一种得到广泛应用的一般方法。我们把这个样品分成两份，分析其中的一份（必要时可以去破坏它），并且表明它的成分**曾经是** $\boldsymbol{X}$；然后通常可以合理地推出，另一份的成分就是 X。于是时常有人认为，按照这种方式，一个可以在事后做回顾性推断得到的性质也可以实时推断出来；如果这是正确的话，那么就可以避开将回顾性推断得到的假想特性作为观测事件的原因所产生的全部危险。事实上，危险恰恰是在取样方法不再适用的情况下出现的，也就是当我们讨论能将同一个物质中单独的一个原子和另一个原子区分开的性质的时候。因为一个原子不能被分为两份，一份用于分析，另一份保留。让我们来看下面的例子：

我们知道，钾由两种类型的原子构成，一种是有放射性的，而另一种则是惰性的。我们把它们叫做 $\mathrm{K}\alpha$ 和 $\mathrm{K}\beta$。如果我们观察到一个原子以放射性的方式爆发，那么我们就推断它是 $\mathrm{K}\alpha$ 原子。我们是否可以根据爆发原子是 $\mathrm{K}\alpha$ 而不是 $\mathrm{K}\beta$ 预先断定爆炸的发生？这里提供的信息表明，根本就没有这样的道理；$\mathrm{K}\alpha$ 只是事先准备的一个标签，如果我们看见原子爆炸了就把这个标签贴在它上面。我们可以一直这样做下去，不管那个需要用到这个标签的事件是不是待定的。然而实际上，更多的信息表明爆发并不是未确定的。我们知道钾是由原子量分别为 39 和 41 的两种同位素组成的；而且我们确信原子量是 41 的那种具有放射性，而 39 的那种则是惰性的。于是可以将这两种同位素区分开，挑选出被称为 K^{41} 的原子。这样 K^{41} 就是一个实时性质，能用来合理地预测它后续的放射性爆发；因而可以用它来代替 $\mathrm{K}\alpha$ 这个回顾性的特性。

对于爆发的现象就讨论到这里，现在我们要考虑爆发的时间。我们根本不知道一个 K^{41} 原子会在何时爆发，除非说它或许在今后的十亿年内会爆发。然而如果我们在时刻 t 观察到了它的爆炸，那么我们就可以认为这是由于此原子具有回顾性特性 K^t 造成的，这就表明它（自始至终）具有在时刻 t 爆发的性质。现在，根据现代物理学，这个性质 K^t 没有以任何一种方式得到证明——甚至不能用我们对原子的数学描述来表示——直到 t 时刻那个原子爆发，这个性质 K^t 在完成了自己的使命后销声匿迹了。在这种情况下 K^t 并不是一个可预测的起因。我们的回顾性的标签及特性没有给这个简单的实测现象增加任何内容，爆发毫无征兆地在时刻 t 发生了；这些标签和特性只不过是变化发生时的报警装置。

放射性原子的衰变时刻就是一个极端的非决定论的例子；但是我们必须要明白，根据当前流行的理论，未来的事件，都或多或少是不可确定的，只是不确定度有所

uncertainty is below our limits of measurement, the event is looked upon as practically determinate; determinacy in this sense is relative to the refinement of our measurements. A being accustomed to time on the cosmic scale, who was not particular to a few hundred million years or so, might regard the time of break-up of the radioactive atom as practically determinate. There is one unified system of secondary law throughout physics and a continuous gradation from phenomena predictable with overwhelming probability to phenomena which are altogether indeterminate.

Criticism of Indeterminism

In saying that there is no contemporaneous characteristic of the radioactive atom determining the date at which it is going to break up, we mean that in the picture of the atom as drawn in present-day physics no such characteristic appears; the atom which will break up in 1960 and the atom which will break up in the year 150,000 are drawn precisely alike. But, it will be said, surely that only means that the characteristic is one which physics has not yet discovered; in due time it will be found and inserted in the picture either of the atom or its environment. If such indeterminacy were exceptional, that would be the natural conclusion, and we should have no objection to accepting such an explanation as a likely way out of a difficulty. But the radioactive atom was not brought forward as a difficulty; it was brought forward as a favourable illustration of that which applies in greater or lesser degree to all kinds of phenomena. There is a difference between explaining away an exception and explaining away a rule.

The persistent critic continues: "You are evading the point. I contend that there are characteristics unknown to you which completely predetermine not only the time of break-up of the radioactive atom but also all physical phenomena. How do you know there are not? You are not omniscient."

The curious thing is that the determinist who takes this line is under the illusion that he is adopting a more modest attitude in regard to our scientific knowledge than the indeterminist. The indeterminist is accused of claiming omniscience. I will not make quite the same countercharge against the determinist; but surely it is only a man who thinks himself *nearly* omniscient who would have the audacity to start enumerating all the things which (it occurs to him) might exist without his knowing it. I am so far from omniscient that my list would contain innumerable entries. If it is any satisfaction to the critic, my list does include deterministic characters—along with Martian irrigation works, ectoplasm, etc.—as things which might exist unknown to me.

It must be realised that determinism is a positive assertion about the behaviour of the universe. It is not sufficient for the determinist to claim that there is no fatal objection to his assertion; he must produce some reason for making it. I do not say he must prove it, for in science we are ready to believe things on evidence falling short of strict proof. If no reason for asserting it can be given, it collapses as an idle speculation. It is astonishing that even scientific writers on determinism advocate it without thinking it necessary to say

差别。如果这种不确定度在测量允许的范围以内，那么这件事情就可以被看作是可以确定的；在这种意义上确定性与我们的测量精度有关。一个习惯了以宇宙尺度度量时间的人，不会觉得几亿年上下的时间有什么特别了不起，那么他也许会认为放射性原子的衰变时间其实还是非常确定的。贯穿物理学的次级定律与从有绝对把握可以预测的现象到完全不能预测的现象之间的连续渐进是一个统一的体系。

对非决定论的批评

当谈及没有一个实时性质可以确定放射性原子将要爆发的时间时，我们的意思是在当前物理学所描述的原子图像中不存在这种性质；对一个将要在 1960 年爆发的原子和一个将要在 150,000 年爆发的原子的描述是完全相同的。但是有人会说，那只能说明当前物理学还没有发现这种性质，在适当的时候我们总会发现这个性质，并把它添加到原子或其周围结构的理论中去。如果这种不确定性是例外情况，那么就会很自然地得到上述结论，并且这种解释完全可以被我们拿来作为摆脱困难的可靠方法。但是放射性原子并不是作为一个难点提出的，提出它是为了在一定程度上能恰当地解释所有现象。对一个例外情况做出清楚的解释和对一个规则做出清楚的解释是不一样的。

但是固执的批评家还在继续："你在回避要点。我认为有一些你所不知道的性质不但能够完全预先确定放射性原子爆发的时间，还能够完全预先确定一切物理现象。你怎么知道没有那些性质呢？你又不是无所不知的。"

奇怪的是，持这种说法的决定论者错误地认为他们对待科学知识的态度比非决定论者更为谦虚。非决定论者因声称无所不知而受到指责。我并不打算采用完全一样的方式反击决定论者；但是可以肯定，只有自认为**接近**无所不知的人，才有胆量列举出所有（他想到的）可能存在但他自己不了解的东西。我远没有无所不知到可以列出无穷多条目的地步。如果能让那些批评家稍微满意一点的话，我承认我的列表中确实包括了一些确定性的特征，和火星灌溉工程、外质等一样，这些事物可能存在但非我所知。

必须认识到，决定论是关于宇宙性质的一个积极的论断。但是这并不足以让决定论者断言，他的论断没有致命的缺陷；他必须提出一些理由令它成立。我并不是说他必须去证明它，因为在科学上，我们宁愿去相信一些证据明显却缺乏严格证明的事情。如果无法给出其成立的理由，那么它就会成为一个空洞的想法。让人惊讶的是，即使是那些就决定论问题运用科学方法讨论的作者也认为要支持它并不必去

anything in its favour, merely pointing out that the new physical theories do not actually disprove determinism. If that really represents the status of determinism, no reputable scientific journal would waste space over it. Conjectures put forward on slender evidence are the curse of science; a conjecture for which there is no evidence at all is an outrage. So far as the physical universe is concerned, determinism appears to explain nothing; for in the modern books which go farthest into the theory of the phenomena no use is made of it.

Indeterminism is not a positive assertion. I am an indeterminist in the same way that I am an anti-moon-is-made-of-green-cheese-ist. That does not mean that I especially identify myself with the doctrine that the moon is not made of green cheese. Whether or not the green cheese lunar theory can be reconciled with modern astronomy is scarcely worth inquiring; the main point is that green-cheesism, like determinism, is a conjecture that we have no reason for entertaining. Undisprovable hypotheses of that kind can be invented *ad lib*.

Principle of Uncertainty

The mathematical treatment of an indeterminate universe does not differ much in form from the older treatment designed for a determinate universe. The equations of wave mechanics used in the new theory are not different in principle from those of hydrodynamics. The fact is that, since an algebraic symbol can be used to represent either a known or an unknown quantity, we can symbolise a definitely predetermined future or an unknown future in the same way. The difference is that whereas in the older formulae every symbol was theoretically determinable by observation, in the present theory there occur symbols the values of which are not assignable by observation.

Hence, if we use the equations to predict, say, the future velocity of an electron, the result will be an expression containing, besides known symbols, a number of undeterminable symbols. The latter make the prediction indeterminate. (I am not here trying to prove or explain the indeterminacy of the future; I am only stating how we adapt our mathematical technique to deal with an indeterminate future.) The indeterminate symbols can often (or perhaps always) be expressed as unknown phase-angles. When a large number of phase-angles are involved, we may assume in averaging that they are uniformly distributed from 0° to 360°, and so obtain predictions which could only fail if there has been an unlikely coincidence of phase-angles. That is the secret of all our successful prophecies; the unknowns are not eliminated by determinate equations but by averaging.

There is a very remarkable relation between the determined and the undetermined symbols, which is known as Heisenberg's Principle of Uncertainty. The symbols are paired together, every determined symbol having an undetermined symbol as partner. I think that this regularity makes it clear that the occurrence of undetermined symbols in the mathematical theory is not a blemish; it gives a special kind of symmetry to the whole picture. The theoretical limitation on our power of predicting the future is seen to be systematic, and it cannot be confused with other casual limitations due to our lack of skill.

提它的优越性，只需说明新的物理理论与决定论并不抵触就可以了。如果该理论代表了决定论的观点，那么任何著名的科学杂志都不会为此浪费空间了。在缺乏足够证据的情况下妄加推测是科学的祸根，而在没有任何证据的时候就下结论则是一种暴行。就整个物理世界而言，决定论似乎没有解释任何事情；因为在深入探讨现象背后的理论的现代书籍中，没有用到决定论的观点。

非决定论并不是一个积极的论断。我是一个非决定论者，同样也是一个反对月亮是由绿奶酪做成的人。这并不是说我想要特别强调自己赞同月亮不是由绿奶酪做成的学说。至于月亮是由绿奶酪做成的理论是否能被当代天文学所接受，完全不值得去调查；关键点在于绿奶酪论就像决定论，是一个我们没有理由考虑的假设。我们可以随意创造这种无需证伪的猜想。

不确定性原理

对一个不确定性宇宙的数学处理和以往对一个确定性宇宙所做的处理在形式上并没有太大区别。在这个新理论中，所用的波动力学方程跟流体力学方程没有原则上的差别。事实上，既然代数符号既可以用来代表一个已知的量也可以用来代表未知量，我们就可以用同样的方式来标识一个预先完全确定的未来或者一个不可确定的未来。不同之处是：按照以往的方式，在理论上每一个符号都是由观测确定的；而在当前的理论中，会有一些符号所代表的值不能通过观测给定。

于是，如果说我们要用方程预测一个电子未来的速度，那么我们最后的表达式除了包括已知的符号外，还包括一些不可确定的符号。后者使得预测变得不可确定。(这里我并不是想证明或者解释未来的不可确定性；我只是在表述如何使用我们的数学技巧去处理不可确定的未来。）那些不可确定的符号常常（或可能总是）被表示成未知的相角。当涉及很多相角的时候，我们可以假设从平均意义上说它们从 0°到 360°是均匀分布的，所得预测只有在相角不太可能同时出现的情况下才会出现错误。这就是我们能够成功预言的秘密所在；不是通过确定的方程，而是通过取平均的方法来消除那些未确定的量。

在确定的符号和不确定的符号之间有一个非常值得注意的关系，那就是海森堡测不准原理。这些符号都是成对的，每个确定的符号都对应着一个不确定的符号。我想这种规律性也清楚地告诉我们，在数学理论中出现不确定的符号并不是缺陷；它为整个物理图像带来了一种特殊的对称性。这个原理认为我们在预测未来能力上所能达到的极限取决于系统上的原因，而这种限制不会与因缺乏技巧而导致的偶然性限制混为一谈。

Let us consider an isolated system. It is part of a universe of inference, and all that can be embodied in it must be capable of being inferred from the influence which it broadcasts over its surroundings. Whenever we state the properties of a body in terms of physical quantities, we are imparting knowledge as to the response of various external indicators to its presence and nothing more. A knowledge of the response of all kinds of objects would determine completely its relation to its environment, leaving only its unget-at-able inner nature, which is outside the scope of physics. Thus, if the system is really isolated so that it has no interaction with its surroundings, it has no properties belonging to physics, but only an inner nature which is beyond physics. So we must modify the conditions a little. Let it for a moment have some interaction with the world exterior to it; the interaction starts a train of influences which may reach an observer; he can from this one signal draw an inference about the system, that is, fix the value of one of the symbols describing the system or fix one equation for determining their values. To determine more symbols there must be further interactions, one for each new value fixed. It might seem that in time we could fix all the symbols in this way, so that there would be no undetermined symbols in the description of the system. But it must be remembered that the interaction which disturbs the external world by a signal also reacts on the system.

There is thus a double consequence; the interaction starts a signal through the external world informing us that the value of a certain symbol p in the system is p_1, and at the same time it alters to an indeterminable extent the value of another symbol q in the system. If we had learned from former signals that the value of q was q_1, our knowledge will cease to apply, and we must start again to find the new value of q. Presently there may be another interaction which tells us that q is now q_2; but the same interaction knocks out the value p_1 and we no longer know p. It is of the utmost importance for prediction that a paired symbol and not the inferred symbol is upset by the interaction. If the signal taught us that at the moment of interaction p was p_1, but that p had been upset by the interaction and the value no longer held good, we should never have anything but retrospective knowledge—like the chemistry lecturer to whom I referred above. Actually we can have contemporaneous knowledge of the values of half the symbols, but never more than half. We are like the comedian picking up parcels who, each time he picks up one, drops another.

There are various possible transformations of the symbols and the condition can be expressed in another way. Instead of two paired symbols, one wholly known and the other wholly unknown, we can take two symbols each of which is known with some uncertainty; then the rule is that the product of the two uncertainties is fixed. Any interaction which reduces the uncertainty of determination of one increases the uncertainty of the other. For example, the position and velocity of an electron are paired in this way. We can fix the position with a probable error of 0.001 mm. and the velocity with a probable error of about 1 km. per sec.; or we can fix the position to 0.0001 mm. and the velocity to 10 km. per sec.; and so on. We divide the uncertainty how we like, but we cannot get rid of it. If current theory is right, this is not a question of lack of skill or a perverse delight of Nature in tantalising us; for the uncertainty is actually embodied in the theoretical picture of the

让我们来考虑一个孤立的系统。那是可推理的宇宙的一部分，宇宙中出现的一切事物都必须能够从其对周围环境所施加的影响中推断出来。每当我们用物理量描述一个物体的性质时，我们正在传递的信息只不过是用不同外部指示物的响应来判断它的存在。如果知道了各类物体的响应，我们就可以完全确定它与周围环境的联系，而仅剩下它的“不能得到的”内部性质，那就是物理范畴以外的东西了。因此，如果一个系统是完全孤立的，那么它就不会与其周围的环境产生相互作用，因而它不具备任何物理性质，而只具有超出物理范畴的内部性质。因此，我们必须把条件稍微修改一下。让它在一个短暂的时间内与外部世界产生一定的相互作用，这种作用引发一连串影响并被一个观测者所感知。观测者可以通过这个信号对上述系统进行推测，换言之，确定用来表述这个系统的某个符号的值，或者确定一个方程以便判定这些符号的值。要确定更多的符号，就必须有更进一步的相互作用，每一次相互作用都可以确定一个新符号的值。看起来通过这种方法我们总有一天会确定所有的符号，于是在描述这个系统的符号中就没有哪个是未确定的了。但是，一定要记住的是，这种通过信号对外部世界产生干扰的相互作用，同样也会对这个系统产生反作用。

于是，这里就会存在两个结论：这个在外部世界产生了一个信号的相互作用告诉我们，某个符号 p 在系统中的值是 p_1，同时它还把系统中另一个符号 q 值的大小改变到一个不确定的值。如果我们已经从以前的信号得知 q 的值是 q_1，那么我们的这个认识将不再适用，我们必须重新开始寻找符号 q 的新值。不久可能有另一个相互作用告诉我们，现在的 q 值是 q_2，但同时这个相互作用又改变了 p_1 的值，于是我们不再确知 p 了。预测中至关重要的问题就是：相互作用会扰动成对出现的符号，而不是那个要推断的符号。如果某个信号告诉我们：在当前的相互作用中 p 为 p_1，但那个 p 已经被相互作用所扰动，它的值不再准确，除了对先前的回顾性认识外我们绝不会得到任何东西，就和我前面提到的那个化学老师一样。事实上我们可以同时知道一半符号的值，但不会超过一半。我们就像拣包裹的喜剧演员，每拣起一个就会又丢掉一个。

这些符号可以有各种不同的变换，而条件也可以用另一种方式来表达。不采用这种其中一个完全确定而另一个完全不确定的成对符号，我们可以选取两个符号，其中每一个都带有一定的不确定性。于是就有了这样一个定律：两个不确定度的乘积是确定的。任何降低其中一个不确定度的相互作用都会增加另一个的不确定度。例如，一个电子的位置和速度以这种方式成对的。我们可以把位置的概率误差定为 0.001 mm，因而速度的概率误差大约是 1 km/s；或者我们可以把位置的概率误差定为 0.0001 mm，因而速度的概率误差是 10 km/s，依此类推。我们可以按照我们的喜好来分配这个不确定度，但却不能摆脱它。如果目前的理论是正确的，那么就不存在缺乏技巧或大自然偏偏要与我们作对的问题；这是因为这种不确定性实实在在地

electron; so that if we describe something as having exact position and velocity we cannot be describing an electron.

If we divide the uncertainty in position and velocity at time t_1 in the most favourable way, we find that the predicted position of the electron one second later (at time t_2) is uncertain to about five centimetres. That represents the extent to which the future position is not predetermined by anything existing one second earlier. If the position at time t_2 always remained uncertain to this extent, there would be no failure of determinism, for the thing we had failed to predict (exact position at time t_2) would be meaningless. But *when the second has elapsed* we can measure the position of the electron to 0.001 mm. or even more closely, as already stated. This accurate position is not predetermined; we have to wait until the time arrives and then measure it. It may be recalled that the new knowledge is acquired at a price. Along with our rough knowledge of position (to 5 cm.) we had a fair knowledge of the velocity; but when we acquire more accurate knowledge of the position, the velocity goes back into extreme uncertainty.

We might spend a long while admiring the detailed working of this cunning arrangement by which we are prevented from finding out more than we ought to know. But I do not think we should look on these as Nature's devices to prevent us from seeing too far into the future. They are the devices of the mathematician who has to protect himself from making impossible predictions. It commonly happens that when we ask silly questions, mathematical theory does not directly refuse to answer but gives a non-committal answer like $\frac{0}{0}$, out of which we cannot wring any definite meaning. Similarly, when we ask where the electron will be tomorrow, mathematical theory does not give the straightforward answer, "It is impossible to say, because it is not yet decided" , because that is beyond the resources of an algebraic vocabulary. It gives us an ordinary formula of *x*'s and *y*'s, but makes sure that we cannot possibly find out what the formula means—until tomorrow.

Mental Indeterminism

I have, perhaps fortunately, left myself no time to discuss the effect of indeterminacy in the physical universe on our general outlook. I will content myself with stating in summary form the points which seem to arise.

(1) If the whole physical universe is deterministic, mental decisions (or at least *effective* mental decisions) must also be predetermined. For if it is predetermined in the physical world (to which your body belongs) that there will be a pipe between your lips on Jan. 1, the result of your mental struggle on Dec. 31 as to whether you will give up smoking in the New Year is evidently predetermined. The new physics thus opens the door to indeterminacy of mental phenomena, whereas the old deterministic physics bolted and barred it completely.

(2) The door is opened slightly, but apparently the opening is not wide enough; for according to analogy with inorganic physical systems, we should expect the indeterminacy

蕴涵在电子的理论图像里，所以如果我们要用精确的位置和速度来描述某个物体的话，那么这在描述电子时是行不通的。

假如在 t_1 时刻，我们按照最适宜的方式把不确定度分配给位置和速度，那么我们会发现在一秒钟以后（t_2 时刻）电子位置的预测值具有约 5 cm 的不确定度。这就代表着通过任何存在于一秒钟之前的事物，也不可能把对未来位置的不确定度降到这个值以下。如果 t_2 时刻的位置不确定度一直保持这样的程度，那么我们可以说决定论并没有失效，因为那些我们所不能预测的事情（t_2 时刻电子的精确位置）是没有意义的。**但是当这一秒钟消逝之后**我们对电子位置的测量能够精确到 0.001 mm 甚至更高。这个精确的位置并不是预先知道的，我们必须要等到时机成熟时才能对它进行测量。还要回想起获得这个新的认识是需要付出代价的。除了我们对位置（达到 5 cm）的大概了解以外，我们还对速度有着相当不错的了解。但是当我们要对位置获得更加精确的认识时，速度就变得非常不确定了。

也许我们花了很长一段时间来欣赏这种巧妙处理的精细工作原理，正是它在防止我们去寻找比我们本该知道的还要多的事情。但是我并不认为我们应该把这当作是阻止我们去探索更加遥远的未来的自然机制。它们只是数学家们用来保护自己不至做出不可能的预测的手段。当我们提出一个很愚蠢的问题时，数学理论通常不会直接拒绝回答，但会给出一个类似 $\frac{0}{0}$ 的含糊答案，根据那个答案我们得不到任何明确的意思。同样，当我们提出电子明天会在什么地方出现的问题时，数学理论也不会给出直接的回答，“这不好说，因为现在还不能确定”，因为那超出了代数语言的范畴。它所给出的是一个关于 x 和 y 的一般表达式，但可以肯定的是，在明天到来之前，我们不可能知道这个表达式的含义。

精神非决定论

似乎很幸运，我没有给自己留出时间来讨论不确定性如何影响了我们对物理世界的总体认识。接下来我很乐意把似乎可以得出的要点做一个简单的概括。

（1）如果整个物理世界是可以确定的，那么心理决策（或者至少是**有效**的心理决策）也一定可以预先确定。因为假如在物理世界里（你的身体也是其中一部分）可以预先确定 1 月 1 日你嘴上会叼着一个烟斗，那么你在 12 月 31 日为新年是否要戒烟所做的心理斗争显然已经提前有了结果。因而这个新的物理学为我们打开了一扇通向心理现象的不确定性的大门，而旧的决定论的物理学则完全把它关在门里。

（2）这个门被轻轻地打开了，但显然打开得还不够宽。因为与一个无机的物理系统类比后我们会发现人类活动的不确定性在数量上是可以被忽略的。我们必须通

of human movements to be quantitatively insignificant. In some way we must transfer to human movements the wide indeterminacy characteristic of atoms, instead of the almost negligible indeterminacy manifested by inorganic systems of comparable scale. I think this difficulty is not insuperable, but it must not be underrated.

(3) Although we may be uncertain as to the intermediate steps, we can scarcely doubt what is the final answer. If the atom has indeterminacy, surely the human mind will have an equal indeterminacy; for we can scarcely accept a theory which makes out the mind to be more mechanistic than the atom.

(4) Is the human will really more free if its decisions are swayed by new factors born from moment to moment than if they are the outcome solely of heredity, training, and other predetermining causes? On such questions as these we have nothing new to say. Argument will no doubt continue "about it and about". But it seems to me that there is a far more important aspect of indeterminacy. It makes it possible that the mind is not utterly deceived as to the mode in which its decisions are reached. On the deterministic theory of the physical world, my hand in writing this address is guided in a predetermined course according to the equations of mathematical physics; my mind is unessential—a busybody who invents an irrelevant story about a scientific argument as an explanation of what my hand is doing—an explanation which can only be described as a downright lie. If it is true that the mind is so utterly deceived in the story it weaves round our human actions, I do not see where we are to obtain our confidence in the story it tells of the physical universe.

Physics is becoming difficult to understand. First relativity theory, then quantum theory, then wave mechanics have transformed the universe, making it seem ever more fantastic to our minds. Perhaps the end is not yet. But there is another side to this transformation. Naïve realism, materialism, the mechanistic hypothesis were simple; but I think that it was only by closing our eyes to the essential nature of experience, relating as it does to the reactions of a conscious being, that they could be made to seem credible. These revolutions of scientific thought are clearing up the deeper contradictions between life and theoretical knowledge, and the latest phase with its release from determinism marks a great step onwards. I will even venture to say that in the present theory of the physical universe we have at last reached something which a reasonable man might almost believe.

(**129**, 233-240; 1932)

过某种方法，把原子较大的不确定性特征转移到人的活动上来，以代替与无机系统相比几乎可以忽略不计的不确定性。我认为这个难点不是不能克服的，但也决不能低估它。

（3）虽然我们也许不能确定中间的步骤，但我们从不怀疑最后的解决方法。如果原子具有不确定性，那么人的思维肯定也会具有同样的不确定性；因为把思维说成比原子更机械的理论是绝大多数人都不愿意接受的。

（4）如果人的决定会受到不断出现的新因素的影响，而不仅仅是遗传、培养以及其他先决条件产生的结果，那么人的意识会更加自由吗？对于这样的问题，我们没有给出任何新的答案。这个争论肯定还会无休止地进行下去。就我看来，不确定性还有一个更加重要的方面。它使我们的思想在通过某种模式作决定的时候不会完全被蒙蔽。根据物理世界中的决定论，我的手在写这个讲稿，是由数学物理方程预先确定的进程决定的，我的思想在这里并不重要——一个爱管闲事的人想出了一个与科学问题有关的不相干的故事来解释我的手在做什么——这种解释只会被认为是一种彻头彻尾的谎言。如果思想真的是完全被蒙蔽的，在这个故事中它是围绕着我们人类的行为编排出来的，那么在这个故事中，我不认为我们会有信心描述物理世界。

物理学正在变得难以理解。首先是相对论，然后是量子理论，接着是波动力学，它们已经使宇宙发生了转变，使我们眼中的宇宙变得更加奇妙。也许这还不是终点。但是这种变化还有另外一面。朴素的现实主义、唯物主义以及机械论的假设都是较简单的；但是我认为，只有当我们不去理会经验的基本性质而把它跟有意识的生命的反应联系起来的时候，它们看起来才是可信的。这种科学思想上的变革扫清了存在于生命和理论认识之间的深刻矛盾，最近它从决定论中解放出来标志着往前迈进了伟大的一步。我甚至敢说，在物理世界目前的理论中，我们终于达到了能让一个明理的人几乎可以信任的程度。

（沈乃澂 翻译；葛墨林 审稿）

News and Views

Editor's Note

In the early 1930s, physicists were exploring the profound philosophical repercussions of quantum theory. Heisenberg's principle of indeterminacy apparently established clear limitations to determinism in physics. Did an absence of deterministic laws at the microscale undermine cause and effect? In a supplement to this issue, Arthur Eddington considered the "decline of determinism", insisting that indeterministic "secondary" law can still be useful for predicting the future. The author of this essay takes issue with that point, saying that Eddington proclaims the demise of determinism while at the same time supposing its implications for orderly causal links, and even the possibility of prediction, remain. But the essay applauds Eddington's statement that quantum indeterminacy does not imply a universe characterised by unrestrained caprice.

Determinism Defined

SIR Arthur Eddington's characteristically fascinating address on "The Decline of Determinism", which we publish as our Supplement this week, will be welcomed as a clear, unequivocal statement, by a leading authority, on a question which, even among the many revolutionary aspects of the new physics, holds a pre-eminent place for importance and interest. Such a statement is the more necessary because of the almost universal tendency for discussions of determinism to be concerned at bottom with words rather than ideas, and Sir Arthur has quite properly begun by stating definitely what he means by the determinism which he holds has declined. His thorough analysis leaves little room for disagreement, but many will wonder whether he has not achieved a Pyrrhic victory by conceding to the determinist the substance of his doctrine and destroying only the shadow. "The rejection of determinism is in no sense an abdication of scientific method", and "indeterministic or secondary law … can be used for predicting the future as satisfactorily as primary law". In other words, Sir Arthur does not allow that the first Morning of Creation wrote what the last Dawn of Reckoning shall read, but he allows that it might have read what the last Dawn shall write. Even the most perfervid determinist will scarcely ask more. Furthermore, he acknowledges that he does not know whether Dirac, whose book "goes as deeply as anyone has yet penetrated into the fundamental structure of the physical universe", is a determinist or not. It would seem, therefore, that the determinism in question cannot be of much importance even in physics.

新闻与观点

编者按

在20世纪30年代早期，物理学家们发现量子理论中蕴藏着深刻的哲学内涵。海森堡测不准原理已经证明物理学中的决定论思想存在着明显的局限性。在微观尺度上不存在决定论难道就会有损于因果推理？在为讨论这个问题特设的增刊上，阿瑟·爱丁顿考虑到了“决定论的衰落”，但他坚信非决定论的“次级”原理仍然可以用于预测未来。这篇短文的作者不支持爱丁顿的论点，说他一边宣布决定论的死亡一边又在鼓吹仍然存在有规律的因果联系甚至预测未来的可能性。但作者在文中对爱丁顿的另一个观点表示赞同，即认为量子理论的不确定性并不意味着宇宙处于完全无序的状态之中。

决定论的界定

在本周《自然》杂志的增刊上刊载了阿瑟·爱丁顿爵士的一篇极具吸引力的题为“决定论的衰落”的演说词。他用清晰准确的语言阐述了一个在新兴物理学各创新领域中显得十分突出的重要问题，他的观点为主要的物理学家所接受。因为大家对决定论的讨论越来越趋向于词藻之争而不是观点的表达，所以阿瑟爵士的这篇文章就显得更加难得了，一开始他就给出了他认为已经衰落了的决定论的明确含义。他的精辟论述没有给异见者留下多少反驳的余地，但很多人怀疑他还未取得皮洛士式的胜利（译者注：指付出重大代价才取得的胜利，古希腊国王皮洛士在公元前280~前279年间曾为打败罗马军队付出了惨重的代价），因为他的学说的基本内容对决定论者是认可的，仅仅否定了一些无关紧要的地方。“摒弃决定论绝不是要放弃科学方法”，“非决定论或次级定律……同样可以像基本定律一样很好地预测未来”。换言之，阿瑟爵士虽不允许在创世纪的第一个早晨就记录下在最后一次大灾难拂晓将看到的一切，但他却相信在创世纪的第一个早晨就能读到最后一次大灾难拂晓所记录下的一切。对于这一点，即使是最坚定的决定论者也不会再过分地要求什么了。阿瑟爵士还承认自己并不清楚狄拉克是不是决定论者，虽然狄拉克写了一本“比任何人都更深层次地触及到物质世界的基本结构”的书。看来决定论的观点即便在物理学中也算不上重要了。

Physical Inference and Prediction

Apparently, however, in spite of the unqualified statement concerning prediction quoted above, Sir Arthur denies that we can predict the behaviour of electrons more certainly than that of horses, and the importance, to all but the physicist, of the "decline of determinism" therefore depends on the recognition of electrons as bodies co-equal with ordinary physical objects. To establish this he claims that since physical objects, as well as electrons and such particles, are all "inferences", they differ only in degree and not in kind. We must not, however, be deceived by words. Objects which we see and handle may be, as he says, as inferential as an undiscovered planet inferred from irregularities in the motion of Uranus, but the inferences are of different kinds; otherwise, why, when a planet was seen in a different position from that inferred from the irregularities, was it *without question* preferred to the "undiscovered" inferential planet? There was not even an instinctive estimate of the "degree" of validity to be attributed to the two "inferences". Unless Sir Arthur assigns to "direct observation" a status essentially different from that of rational deduction, it is difficult to see how his position can be "in no sense an abdication of scientific method". All this, however, does not affect determinism in relation to physical objects, and it is to be hoped that Sir Arthur's plain statement will do much to remove the widespread delusion that modern physics has revealed a universe of unrestrained caprice.

(**129**, 228-229; 1932)

物理推论与预测

然而，尽管上面引用了有关预测问题的无保留的阐述，但阿瑟爵士认为我们对电子行为的预测并不会比对马的行为的预测更准确。因此，对于除物理学家以外的所有人而言，“决定论的衰落”的重要性在于它使人们认识到电子可以被看作是与普通物理对象等同的实体。为了证明这一点，阿瑟爵士声称由于物理对象，包括电子和类似电子的粒子，都是由“推论”得到的，它们只会有程度上的不同，而不会有本质上的差别。然而，我们千万不要被这些文字迷惑。正如他所述，我们可以看到和触摸到的物体也许和从天王星运动的不规则性能推出一颗未知行星的存在一样可以通过推断得到，不过这些推论所属的类别并不相同；否则，为什么当观测到的一颗行星并没有处于由无规则运动推断出的位置上时，人们还是**毫无异议**地倾向于把它看作就是那颗由推论得到的“未知”行星呢？我们甚至不能直观地估计出这两种“推论”的有效“程度”。除非阿瑟爵士给“直接观测”下一个本质上不同于理性推理的定义，否则很难想象他的立场怎么会是“决不能摈弃科学的方法”。然而所有这些都不会影响到与物理对象相关的决定论，我们希望阿瑟爵士的平实论述能在很大程度上消除人们普遍存在的错觉，即认为现代物理学揭示的是一个过于变幻无常的宇宙。

（金世超 翻译；赵见高 审稿）

Artificial Production of Fast Protons

J. D. Cockcroft and E. T. S. Walton

Editor's Note

To study the structure of the atomic nucleus, physicists in the early 1930s needed to accelerate probe particles to high energies. Among the first to develop such techniques were John Cockcroft and Ernest Walton in Cambridge. Here they report initial studies using a device for accelerating protons. An electrical discharge in hydrogen gas produced protons, which were then accelerated by high-voltage electrodes in a vacuum tube nearly one metre long, and detected at a fluorescent screen. Cockcroft and Walton measured proton velocities of up to 1.16×10^9 cm/s. The two physicists later used the device to achieve the first artificial splitting of the atom, or "transmutation" of the nucleus, for which they won the 1951 Nobel Prize for physics.

A high potential laboratory has been developed at the Cavendish Laboratory for the study of the properties of high speed positive ions. The potential from a high voltage transformer is rectified and multiplied four times by a special arrangement of rectifiers and condensers, giving a working steady potential of 800 kilovolts. Currents of the order of a milliampere may be obtained at a potential constant to 1–2 percent.

Protons from a discharge in hydrogen are directed down the axis of two glass cylinders 14 in. in diameter and 36 in. long, and accelerated by the steady potentials of the rectifier. They are then passed into an experimental chamber at atmospheric pressure through a mica window having a stopping power of about 1mm. air equivalent. Luminescence of the air can easily be observed.

The ranges of the protons in air and hydrogen have been measured using a fluorescent screen as a detector. The range in air at S.T.P. of a proton having a velocity of 10^9 cm./sec. is found to be 8.2 mm., whilst the corresponding range for hydrogen is 3.2 cm. The observed ranges support the general conclusions of Blackett on the relative ranges of protons and α-particles, although the absolute values of the ranges are lower for both gases. The ranges and stopping power will be measured more accurately by an ionisation method.

The maximum energy of the protons produced up to the present has been 710 kilovolts with a velocity of 1.16×10^9 cm./sec. and a corresponding range in air of 13.5 mm. at S.T.P. We do not anticipate any difficulty in working up to 800 kilovolts with our present apparatus.

(**129**, 242; 1932)

J. D. Cockcroft and E. T. S. Walton: Cavendish Laboratory, Cambridge, Feb. 2.

快质子的人工产生

约翰·考克饶夫，欧内斯特·瓦耳顿

编者按

为了研究原子核的结构，20 世纪 30 年代的物理学家需要把探测粒子加速到很高的能量。剑桥大学的约翰·考克饶夫和欧内斯特·瓦耳顿是首先利用这类技术的人之一。本文中介绍了他们用质子加速仪所作的研究。首先让氢气放电产生质子，随后质子在近一米长的真空管中被高压电极加速，最后在荧光屏上被检测出来。考克饶夫和瓦耳顿测得的质子速度高达 1.16×10^9 cm/s。后来这两位物理学家利用这个设备首次实现了原子的人工裂变，或称原子核的“蜕变”，他们因此获得了 1951 年的诺贝尔物理学奖。

为了研究高速正离子的性质，我们在卡文迪什实验室设立了一个高能实验室。由高压变压器输出的电压，经过特殊设置的整流器和电容器整流后，其值提高为 4 倍，最后可以得到 800 千伏的稳定工作电压。当电压的稳定度达到 1%~2% 时，我们就可以获得毫安级的电流。

我们使氢电离制得的质子沿着两个直径为 14 英寸，长为 36 英寸的玻璃圆管的轴线向下运动，质子被由整流器得到的稳定电压加速。然后它们会通过一个云母窗进入处于大气压下的实验靶室，云母片对质子的阻止作用相当于约 1 mm 厚的空气。在靶室中很容易观察到空气在质子的轰击下发出冷光的现象。

利用荧光屏作为探测器可以测量质子在空气和氢气中的射程。一个速度为 10^9 cm/s 的质子在处于标准温度和压强下的空气中的射程是 8.2 mm，而在相同条件下在氢气中的射程是 3.2 cm。我们测量的射程与布莱克特关于质子和 α 粒子相对射程的基本结论一致，但在两种气体中射程的绝对值都偏低。用电离方法测量射程和阻止本领，我们得到的结果会更精确。

目前能获得的质子的最高能量为 710 千伏，相应的速度为 1.16×10^9 cm/s，在标准条件空气中的射程为 13.5 mm。用我们现在的设备提高到 800 千伏是完全可以做到的。

（沈乃澂 翻译；尚仁成 审稿）

Possible Existence of a Neutron

J. Chadwick

Editor's Note

Physicists had recently noted that alpha particles hitting a beryllium target produced unknown secondary radiation with great penetrating power. When directed into any material containing hydrogen, this secondary radiation had been found to produce protons. Some physicists suspected that the unknown particles might be high-energy light quanta. But here James Chadwick reports that his studies on the ionisation of nitrogen atoms by collisions with the mysterious radiation point instead to the action of a hitherto unknown particle. Chadwick suggests that the particles have atomic mass one and charge zero: they are what we now call neutrons. It took further experiments to verify the hypothesis, but for this discovery Chadwick won the Nobel Prize in physics in 1935.

IT has been shown by Bothe and others that beryllium when bombarded by α-particles of polonium emits a radiation of great penetrating power, which has an absorption coefficient in lead of about 0.3 (cm.)$^{-1}$. Recently Mme. Curie-Joliot and M. Joliot found, when measuring the ionisation produced by this beryllium radiation in a vessel with a thin window, that the ionisation increased when matter containing hydrogen was placed in front of the window. The effect appeared to be due to the ejection of protons with velocities up to a maximum of nearly 3×10^9 cm. per sec. They suggested that the transference of energy to the proton was by a process similar to the Compton effect, and estimated that the beryllium radiation had a quantum energy of 50×10^6 electron volts.

I have made some experiments using the valve counter to examine the properties of this radiation excited in beryllium. The valve counter consists of a small ionisation chamber connected to an amplifier, and the sudden production of ions by the entry of a particle, such as a proton or α-particle, is recorded by the deflexion of an oscillograph. These experiments have shown that the radiation ejects particles from hydrogen, helium, lithium, beryllium, carbon, air, and argon. The particles ejected from hydrogen behave, as regards range and ionising power, like protons with speeds up to about 3.2×10^9 cm. per sec. The particles from the other elements have a large ionising power, and appear to be in each case recoil atoms of the elements.

If we ascribe the ejection of the proton to a Compton recoil from a quantum of 52×10^6 electron volts, then the nitrogen recoil atom arising by a similar process should have an energy not greater than about 400,000 volts, should produce not more than about 10,000 ions, and have a range in air at N.T.P. of about 1.3 mm. Actually, some of the recoil atoms in nitrogen produce at least 30,000 ions. In collaboration with Dr. Feather, I have observed the recoil atoms in an expansion chamber, and their range, estimated visually, was sometimes as much as 3 mm. at N.T.P.

可能存在中子

詹姆斯·查德威克

编者按

物理学家们最近指出：α 粒子在撞击铍靶之后能够产生一种穿透力很强的未知次级辐射。他们发现当这种次级辐射射向含氢物质时会有质子生成。一些物理学家怀疑未知粒子可能就是高能光量子。但在这篇文章中，詹姆斯·查德威克报告了氮原子与这种神秘射线碰撞后所产生的离子化现象，结果表明它更像是一种迄今未知的粒子。查德威克认为这种粒子的原子量为 1，电荷数为 0：我们现在把它们称为中子。虽然要验证这一假说还需要进一步的实验，不过查德威克因此而获得了 1935 年的诺贝尔物理奖。

博特等人曾指出，当用钋元素放射产生的 α 粒子轰击铍元素时，铍元素会释放出一种穿透力极强的射线，铅对这种射线的吸收系数约为 0.3 cm^{-1}。最近，约里奥–居里夫妇发现，在一个具有薄窗的容器中测量这种铍辐射导致的电离时，如果在薄窗前放置含氢的物质，电离就会增强。这种现象似乎是由实验中有速度最大值接近于 3×10^9 cm/s 的质子发射出来所致。他们认为能量是通过一个类似于康普顿效应的过程转移给质子的，并估算出这种铍辐射的量子的能量为 50×10^6 eV。

为了研究从铍元素中激发出的这种射线的性质，我利用真空管计数器做了一些实验。这种真空管计数器由一个连接到放大器的小电离腔组成。质子或 α 粒子进入电离腔后迅速产生离子的情况可以通过示波器显示的偏转来记录。这些实验表明，该射线轰击氢、氦、锂、铍、碳、空气、氩时都能发射出粒子。从射程和致电离能力来看，氢中发射出来的粒子类似于速度高达约 3.2×10^9 cm/s 的质子。其他元素中发射出来的粒子具有很高的致电离能力，看上去像是每种元素的反冲原子。

如果我们将这一质子发射过程视作能量为 52×10^6 eV 的量子的康普顿反冲过程，那么在类似的过程中产生的氮的反冲原子的能量应该不大于 400,000 eV，其产生的离子数应该不多于 10,000，常温常压下这种原子在空气中的射程应该约为 1.3 mm。而事实上，一些氮的反冲原子至少产生了 30,000 个离子。在与费瑟博士合作进行的实验中，我在膨胀室中观察到了反冲原子，其在常温常压下的射程（目测）有时能达到 3 mm。

These results, and others I have obtained in the course of the work, are very difficult to explain on the assumption that the radiation from beryllium is a quantum radiation, if energy and momentum are to be conserved in the collisions. The difficulties disappear, however, if it be assumed that the radiation consists of particles of mass 1 and charge 0, or neutrons. The capture of the α-particle by the Be^9 nucleus may be supposed to result in the formation of a C^{12} nucleus and the emission of the neutron. From the energy relations of this process the velocity of the neutron emitted in the forward direction may well be about 3×10^9 cm. per sec. The collisions of this neutron with the atoms through which it passes give rise to the recoil atoms, and the observed energies of the recoil atoms are in fair agreement with this view. Moreover, I have observed that the protons ejected from hydrogen by the radiation emitted in the opposite direction to that of the exciting α-particle appear to have a much smaller range than those ejected by the forward radiation. This again receives a simple explanation on the neutron hypothesis.

If it be supposed that the radiation consists of quanta, then the capture of the α-particle by the Be^9 nucleus will form a C^{13} nucleus. The mass defect of C^{13} is known with sufficient accuracy to show that the energy of the quantum emitted in this process cannot be greater than about 14×10^6 volts. It is difficult to make such a quantum responsible for the effects observed.

It is to be expected that many of the effects of a neutron in passing through matter should resemble those of a quantum of high energy, and it is not easy to reach the final decision between the two hypotheses. Up to the present, all the evidence is in favour of the neutron, while the quantum hypothesis can only be upheld if the conservation of energy and momentum be relinquished at some point.

(**129**, 312; 1932)

J. Chadwick: Cavendish Laboratory, Cambridge, Feb. 17.

如果碰撞过程中能量和动量都是守恒的，那么如果假定铍元素中释放出的这种射线是量子化的，就很难解释上述结果以及我在这项工作中得到的其他一些结果。然而，如果假设这种射线是由一种质量为 1，电荷为 0 的粒子（中子）组成的，那么这些困难将不复存在。我们可以设想，Be^9 原子核俘获 α 粒子后产生了一个 C^{12} 原子核并释放出中子。根据这一过程的能量关系，可以得出朝前方发射的中子的速度刚好约为 3×10^9 cm/s。这一中子穿过物体时与物体中的原子碰撞产生了反冲原子，通过实验对反冲原子能量进行观测得到的结果与这一观点基本吻合。此外，我还观测到，用与激发 α 粒子方向相反的这种射线轰击氢后发射出的质子的射程要比用正向射线轰击得到的质子的射程短得多。这一实验结果同样可以用中子假说来简单解释。

如果假设这种射线是由量子组成的，那么 Be^9 原子核俘获 α 粒子后将形成一个 C^{13} 原子核。对 C^{13} 质量亏损的精确测量结果表明，这一过程中释放出的量子的能量不可能大于 14×10^6 eV。因此，对于目前观测到的实验现象我们很难将其归因于这样一种量子。

可以预计，中子穿过物体时产生的很多效应都应该与高能量子产生的效应相似，因此对这两种假说的最终取舍并不容易。到目前为止，所有的证据都更倾向于中子假说，而量子假说只有在放弃能量和动量守恒原理的某些情形下才会得到人们的支持。

（曾红芳 翻译；刘纯 审稿）

Determinism

J. B. S. Haldane

Editor's Note

Arthur Eddington's essay on the decline of determinism provoked strong responses. The biologist J. B. S. Haldane here criticizes Eddington's remarks about human behaviour — specifically, that if the behaviour of atoms is indeterminate, then the human mind must also be. It was then considered that behaviour may be correlated between identical twins. Haldane points out that if one such twin breaks the law during a certain period, the chance that his twin will do so may be as high as 0.875. (These data were almost certainly flawed.) Eddington was effectively claiming that no amount of extra information could turn such a prediction into a certainty. But Haldane complains that this assertion is little more than an a priori pronouncement, unsupported by evidence.

IN his address on the decline of determinism,[1] Sir Arthur Eddington enunciates a very curious equation. "If the atom has indeterminacy, surely the human mind will have an equal indeterminacy; for we can scarcely accept a theory which makes out the mind to be more mechanistic than the atom." This statement will not bear too close an examination even from a non-quantitative point of view. Thus an attempt by myself to solve even a simple wave equation might lead to any of a large number of results; a similar attempt by Sir Arthur Eddington would lead to the correct solution with a high degree of probability. I do not think that this proves that his mind is more mechanistic than my own, whatever that may mean. Actually it is generally regarded as a compliment to describe a person as reliable, that is, to suggest that his conduct is predictable.

Fortunately, however, quantitative data exist which seem to show that, as regards moral behaviour, some minds are decidedly more determined than are some atoms as regards radiative behaviour. Consider a given man M_1, and the probability p that between times T_1 and T_2 he will commit an action such as to lead to his imprisonment for a breach of the law. If we have no further information regarding M_1, p is in most communities a small number, less than 0.01. If, however, M_1 has a monozygotic twin M_2 brought up in the same environment up to the age of 14, and M_2 is known to have been imprisoned for crime between the ages of 16 and 40, we can infer with a fairly high degree of probability that M_1, who has an identical nature and a similar nurture, has been or will be imprisoned between the same ages. Judging from Lange's[2] results, p is about 0.875 in south Germany when we have the above amount of information. If we increase the amount of information, for example, by excluding cases where M_2 has suffered from head injury, the value of p is raised still further. Now, if it could be shown that with sufficient information p became unity in a case of this kind, we should, I take it, have proved the determinacy of some kinds, at least, of moral choice. Actually the most that scientific method can

决定论

约翰·波顿·桑德森·霍尔丹

编者按

阿瑟·爱丁顿的那篇关于决定论在衰落的短文引起了强烈的社会反响。生物学家霍尔丹批驳了爱丁顿关于人类行为的观点，尤其是他认为：如果原子的运动是不可确定的，则人类的思想也不例外。随后霍尔丹列举了同卵双生双胞胎在行为上具有相关性的例子。他说如果其中的一个在某段时间内触犯了法律，那么另一个会触犯法律的可能性高达0.875。（这个数据肯定有问题。）爱丁顿指出这种说法缺少实际的证据。但霍尔丹辩解说爱丁顿的说法纯粹是由推理得到的结果，并没有得到实践上的验证。

阿瑟·爱丁顿爵士在他的关于决定论的衰落[1]的讲话中提到了一个非比寻常的对等关系，"如果原子具有不确定性，那么人的思维肯定也会具有同样的不确定性；因为把思维说成比原子更机械的理论是绝大多数人都不愿意接受的。"即便从非定量的角度上看，这种说法与事实也有一定的差距。比如：我在试图解一个很简单的波动方程时也许会得到众多解中的任意一个；而阿瑟·爱丁顿爵士在解同样的方程时会得到可靠性更高的解。我不会因此而认为他的思维比我的思维更机械，不管多么地像是这回事。事实上人们在说一个人可靠时通常是在称赞他，也就是说，他的行为具有可预见性。

然而，幸运的是，定量数据似乎表明，就道德行为而言，某些人的思想意识确实比一些原子的辐射行为更容易确定。对于某一个人 M_1，在 T_1 与 T_2 之间的任意时刻内，这个人付诸一个行动的概率为 p，比如说他会因违反法律而被监禁。如果我们不了解有关 M_1 的更多情况，则 p 对绝大多数人来说只是一个很小的数，小于0.01。然而，如果 M_1 有一个同卵孪生兄弟 M_2，他们在相同的环境下一直长到14岁，且已知 M_2 在16岁至40岁之间曾因犯罪被监禁过，那么我们就可以认为：M_1 的天性与 M_2 相同，后天培养与 M_2 相似，所以他很有可能在相同的年龄段内也曾被监禁过或将要被监禁。根据兰格[2]的研究结果，当我们已知以上信息时，在德国南部 p 约为0.875。如果我们能了解更多的情况，例如，知道 M_2 未曾有过头部损伤，则 p 的值还会更大一些。在这类情况下，如果我们了解的信息量足够多，可以把 p 的值取为1，我认为，这说明我们已经证明了某种决定论的存在，至少是在道德选择的层面上。实际上，科学研究的主要任务就是要证明 $p>1-\varepsilon$。如果阿瑟·爱丁顿

do is to prove $p>1-\varepsilon$. If Sir Arthur Eddington is correct, then no matter how complete our information, ε tends to a finite limit which is not very small. Clearly no amount of observation could prove it to be zero. But if it could be shown to be less than 0.01, we could neglect it to a first approximation in ethical theory, and if it proved to be less than 10^{-6} we might hazard the guess that the behaviour of human beings showed no more indeterminacy than that of other systems composed of about 2×10^{27} atoms.

I think that it is a legitimate extrapolation from the existing data that if we used all the available data in the above case, ε would be less than 0.05. It seems unfortunate that any attempt should be made to prejudge, on philosophical or emotional grounds, the magnitude of a quantity susceptible of scientific measurement. But from the heuristic point of view the deterministic theory has the advantage that it could be disproved, and would be if ε tended to a finite limit as the amount of available information increased indefinitely. On the other hand, indeterminism cannot be disproved unless its supporters state the value of ε, which they have so far carefully avoided. When the truth about human behaviour is discovered, it will probably appear that philosophers of all schools had failed to predict it as completely as they failed to predict Heisenberg's uncertainty principle. Human behaviour is a subject for scientific investigation rather than *a priori* pronouncements.

(**129**, 315-316; 1932)

J. B. S. Haldane: Royal Institution, Albemarle Street, London, W.

References:

1. *Nature*, 129, 240 (Feb. 13, 1932).

2. *Crime as Destiny* (1931).

爵士是正确的，则不管我们掌握的信息多么全面，ε 都将是一个比较大的有限数。显然再多的观测结果也不能证明 ε 为零。但如果能说明 ε 的值小于 0.01，则在伦理学理论的一级近似中我们就可以把它忽略掉，如果能证明 ε 小于 10^{-6}，我们也许可以大胆地说人类的行为不会比包含约 2×10^{27} 个原子的系统更难以预测。

我认为从现有数据进行的推理是合理的，如果我们利用上述事件中的所有可用信息，那么 ε 的值会小于 0.05。但遗憾的是：在哲学或情感的范畴内，对所有能够用科学方法测量的量进行预测都只是一种臆断。从发现的角度上看，决定论的优点是它可以被证伪，当可获得的信息量越来越不确定，ε 趋于一个有限的值时，我们就应该认为该命题不成立。从另一方面来说，非决定论是不能被证伪的，除非其支持者能够明确规定一个 ε 值，而这一点正是他们一直谨慎回避的。当人类行为的真正规律性被发现时，很可能所有学派的哲学家都没能完全预测出来，就像他们没有预测到海森堡测不准原理的出现一样。人类的行为是科学研究的一门学问，而不是先验的臆断。

（沈乃澂 翻译；李淼 审稿）

Determinism

Lewis F. Richardson

Editor's Note

The newly discovered principle of quantum indeterminacy inspired considerable philosophical discussion, and even a few attempts to apply it to human nature and the unpredictability of the individual human action. Here Lewis Fry Richardson writes to ridicule the latter proposition. Is it really necessary, he asks? Science is capable of making accurate predictions, as of the motion of a pendulum, for example, but only if no person interferes with it. The dynamics of something as large as the Moon can be treated with mathematical precision, precisely because no one can interfere with it. But when it comes to human action itself, we can never be sure that one person will not interfere with the workings of their own mind.

IS it really necessary to appeal to anything so *recherché* as Heisenberg's Principle of Indeterminacy in order to justify anything so familiar as personal freedom of choice? This question arises on reading Sir Arthur Eddington's interesting address in *Nature* of Feb. 13. Consider any one of the laws of physics commonly verified in the laboratory, say $T=2\pi\sqrt{l/g}$ for a simple pendulum. If, while one student is observing the pendulum, another student were to knock it about, the observations might misfit the formula. And so in general: the accepted laws of physics are verified only if no person interferes with the apparatus. We cannot interfere with the moon, because it is so massive and so far away: and that is part of the reason why the motion of the moon is almost deterministic; the "almost" referring to the extremely small Heisenbergian indeterminacy. But there is no great mass or great distance to prevent John Doe interfering with his own brain in the act of making his decision to buy a house from Richard Roe.

(**129**, 316; 1932)

Lewis F. Richardson: The Technical College, Paisley, Feb. 13.

决定论

刘易斯·理查森

编者按

新发现的量子测不准原理引发了哲学上的大讨论，有些人甚至用它来解释人类的本性以及个人行为的不可预知性。在这篇文章中，刘易斯·弗赖伊·理查森嘲讽了后者的观点。他认为没有必要用科学来预知人的行为。对于一个钟摆的运动情况，科学是可以准确预测的，但前提是在没有人干扰的情况下。月球等大型天体的动力学也可以通过数学方法精确预测，这也是因为没有人能干涉它运动的缘故。但是当论及人类的行为本身时，我们无法保证一个人不对自己的意识施加影响。

为了说明像个人选择自由这样熟悉的事也有必要求助于学究式的海森堡测不准原理吗？这个问题是我在阅读阿瑟·爱丁顿爵士于 2 月 13 日发表在《自然》杂志上的演说词时联想到的。对于任意一个在实验室里就可以证实的物理定律，比如一个单摆的周期，$T=2\pi\sqrt{l/g}$，如果当一个学生观察单摆的时候，另一个学生撞了一下这个摆，则观测结果可能就会与公式不符。所以可以这么说：大家普遍接受的物理定律只有在没有人干扰实验设备的情况下才能被证实。我们无法干扰月球，因为它的质量很大，而且离我们又很远：这就是为什么我们基本上可以预测月球运动的一部分原因；称“基本上”是因为确实存在极小的海森堡不确定性。但是再大的质量或再远的距离也不能阻止约翰·多伊考虑是否要从理查德·罗那里买一幢房子。

（沈乃澂 翻译；李森 审稿）

Disintegration of Lithium by Swift Protons

J. D. Cockcroft and E. T. S. Walton

Editor's Note

Using their recently developed proton accelerator, Cockcroft and Walton had begun experimenting on materials placed in the path of the proton beam. Here they report experiments with lithium. Scintillations on a mica window in the side of the vacuum tube, due to particle impacts, grew more frequent with increasing acceleration voltage. The physicists deduced that these particles were alpha particles, suggesting that lithium nuclei of atomic mass 7 in the target were absorbing protons and becoming unstable nuclei of atomic mass 8, which subsequently split apart into two alpha particles. This was the first observation of the artificial transmutation of an atomic nucleus.

IN a previous letter to this journal[1] we have described a method of producing a steady stream of swift protons of energies up to 600 kilovolts by the application of high potentials, and have described experiments to measure the range of travel of these protons outside the tube. We have employed the same method to examine the effect of the bombardment of a layer of lithium by a stream of these ions, the lithium being placed inside the tube at 45° to the beam. A mica window of stopping power of 2 cm. of air was sealed on to the side of the tube, and the existence of radiation from the lithium was investigated by the scintillation method outside the tube. The thickness of the mica window was much more than sufficient to prevent any scattered protons from escaping into the air even at the highest voltages used.

On applying an accelerating potential of the order of 125 kilovolts, a number of bright scintillations were at once observed, the numbers increasing rapidly with voltage up to the highest voltages used, namely, 400 kilovolts. At this point many hundreds of scintillations per minute were observed using a proton current of a few microamperes. No scintillations were observed when the proton stream was cut off or when the lithium was shielded from it by a metal screen. The range of the particles was measured by introducing mica screens in the path of the rays, and found to be about eight centimetres in air and not to vary appreciably with voltage.

To throw light on the nature of these particles, experiments were made with a Shimizu expansion chamber, when a number of tracks resembling those of α-particles were observed and of range agreeing closely with that determined by the scintillations. It is estimated that at 250 kilovolts, one particle is produced for approximately 10^9 protons.

The brightness of the scintillations and the density of the tracks observed in the expansion chamber suggest that the particles are normal α-particles. If this point of view turns out to be correct, it seems not unlikely that the lithium isotope of mass 7 occasionally

由快质子引起的锂衰变

约翰·考克饶夫，欧内斯特·瓦耳顿

编者按

考克饶夫和瓦耳顿利用他们最近研制的质子加速器开始对放在质子束路径中的物质进行实验。他们在此报告了用锂做实验的结果。粒子的碰撞会在真空管侧面的云母窗上发出闪烁，加速电压越大，发出闪烁的次数就越多。物理学家们推测这些粒子就是 α 粒子。他们指出靶中原子量为 7 的锂核因吸收质子而变成原子量为 8 的不稳定核，随后分裂成 2 个 α 粒子。这个实验使人类第一次观测到了一个原子核的人工裂变。

在先前写给《自然》杂志的一封快报 [1] 中，我们描述了一种通过加高压获得能量高达 600 千伏快质子稳定流的方法，同时还描述了测量这些质子在真空管外的射程的实验。现在我们把锂片放在真空管中，与粒子束成 45º 角。然后用同样的方法测量了快质子束轰击锂片产生的效应。在真空管的端口贴有一个阻止本领相当于 2 cm 厚空气的云母窗片。锂发出的辐射可以在真空管外用闪烁方法来测量。云母窗非常厚，它在所加电压达到最高值时足以阻挡散射的质子泄漏到空气中。

当加速电压达到 125 千伏时，我们立即看到了许多明亮的闪烁光，闪烁的次数随电压迅速增加，直到所使用的最高电压，即 400 千伏。在质子流为几个微安的情况下，我们在这时每分钟可以观测到数百次的闪烁。但当质子流被阻断或当锂片被一块金属屏蔽时，任何闪烁都看不到了。我们用在辐射路径上放置的云母片来测量粒子的射程，在空气中测量到的射程为 8 cm，这个值基本上不会随加速电压的变化而改变。

为了说明这类粒子的性质，我们用清水的膨胀室做实验，发现许多粒子的径迹类似于 α 粒子的径迹，而射程与用闪烁法测量的数值非常接近。我们估计在加速电压为 250 千伏时，大约每 10^9 个质子可以产生一个粒子。

在膨胀室中看到的闪烁亮度和径迹密度说明辐射粒子就是普通的 α 粒子。假如这个判断是正确的，那么质量数为 7 的锂同位素就可能俘获 1 个质子并在产生了质

captures a proton and the resulting nucleus of mass 8 breaks into two α-particles, each of mass four and each with an energy of about eight million electron volts. The evolution of energy on this view is about sixteen million electron volts per disintegration, agreeing approximately with that to be expected from the decrease of atomic mass involved in such a disintegration.

Experiments are in progress to determine the effect on other elements when bombarded by a stream of swift protons and other particles.

(**129**, 649; 1932)

J. D. Cockcroft and E. T. S. Walton: Cavendish Laboratory, Cambridge, April 16.

Reference:

1. *Nature*, 129, 242 (Feb. 13, 1932).

量数为 8 的原子核后又分裂成 2 个 α 粒子，每个 α 粒子的质量数为 4 且具有约为 8×10^6 eV 的动能。按照上述观点，每次分裂将产生约 1.6×10^7 eV 的能量，这与根据衰变过程中原子质量下降推算出的结果大致相等。

我们正在用实验测定快质子束和其他粒子束轰击别的元素产生的效应。

（沈乃澂 翻译；江丕栋 审稿）

The Cry of Tin

Editor's Note

Tin has been long known to release a "shriek" or "cry" when bent. Here Bruce Chalmers of University College London attempts to move beyond the vague notion that the sound is caused by the grinding of crystals against one another during deformation. He proposes that it stems from the phenomenon of twinning, in which two crystals share some of the same crystal lattice points, typically with one crystal being a mirror image of the other. Chalmers, who became a celebrated crystallographer, was then working with E. N. da C. Andrade, a pioneer in the study of the crystal structures of metals. Andrade endorses Chalmer's work here, saying that it may be important for understanding the general phenomenon of twinning.

IT is an observation of respectable antiquity that when a bar of tin is bent it emits a characteristic creaking, known as the "cry of tin". According to Mellor[1], who is one of the few authorities to refer to the subject, the cry "is supposed to be produced by the grinding of the crystals against one another during the bending of the metal", and I have been unable to find in the literature any more definite explanation of its origin. In the course of some experiments which I have been carrying out with Prof. E. N. da C. Andrade on single crystal wires I have made observations which, I think, make possible a rather more precise attribution of the sound.

I have obtained the cry with cadmium as well as with tin. When single crystal wires of these metals are stretched, the deformation takes place in two stages, a slip on the glide planes, which constitute in both cases a unique system, being succeeded by a mechanical twinning on a specified plane. The twinning does not take place until after a definite amount of glide extension has occurred, which enables the two phenomena to be studied separately.

With single crystal wires no sound is produced during the glide stage of extension, but with both metals twinning is accompanied by the characteristic creaking or tearing sound. The same sound also occurs when such wires are violently bent or twisted, a process which gives rise to the surface marking characteristic of twinning, although it does not allow the separation of twinning from other effects in the way that simple extension does.

The tin with which the phenomenon is normally observed is in a polycrystalline state. Although the cry occurs when polycrystalline cast rods of both tin and cadmium are bent, drawn wires of small diameter do not give it unless they are annealed, or, in other words, the production of the sound depends on the size of the crystallites being greater than a certain minimum—larger crystals, of course, being subjected to more severe strain when

锡叫

编者按

人们很早以前就知道金属锡在弯曲时会发出“尖叫声”或“哭泣声”。通常认为这种现象是由晶体在变形过程中的相互挤压造成的，而伦敦大学学院的布鲁斯·查默斯在这篇文章中试图给出更加清晰的解释。他认为叫声源于孪晶现象，即两个晶体共用晶格点阵上的某些点，尤其是当一个晶体作为另一个晶体的镜像存在时。当年查默斯与研究金属晶体结构的先驱者——安德雷德一起工作，后来他自己也成为了一位著名的晶体学专家。安德雷德对查默斯在这方面的研究表示认同，他说这对于认识孪晶中的普遍现象具有非常重大的意义。

这是一个相当古老的发现，当一根锡棒弯曲时会发出特有的吱吱嘎嘎声，即通常所称的“锡叫”。梅勒[1]是少数谈到过这一主题的作者之一，他认为这种声音“可能是金属在弯曲时由晶体彼此之间的相互摩擦而产生的”，但是我在文献中始终没能找到关于其起源的更为确定的解释。在我与安德雷德教授合作进行的关于单晶丝的实验过程中，曾经进行了一些我认为能够更精确地解释该声音的观测。

我发现镉也可以发出和锡一样的声音。在将这些金属的单晶丝拉伸时，变形过程分两个阶段，先是沿滑移面上滑移，这滑移存在于二个阶段中，随后是在某一特定晶面上发生机械孪生。孪生要在特定量的滑移形变出现后才会发生，这使得两种现象可以分别得到研究。

在滑移形变引起的延伸阶段，单晶丝没有产生声音，但是两种金属孪生时却伴随有特征的碾轧声或撕扯声。同样的声音在将这些金属丝强力弯曲或扭转时也出现了，这一过程产生了孪晶所特有的表面斑纹，尽管它还不足以将孪晶化与简单延伸过程中产生的其他效应相区分。

通常观测到这种效应时所用的锡处于多晶态。尽管锡和镉的多晶铸棒在弯曲时都会发出叫声，小直径的拉制金属丝却只有在退火后才能发出声音，或者换一种说法，这种声音的产生需要晶粒的尺寸大于某一特定下限——当然，较大的晶体在金属弯曲时要经受更为强烈的应变。有理由认为，在多晶态和单晶态中的发声都伴随

the metal is bent. It is reasonable to suppose that the cry is an accompaniment of twinning in the case of the polycrystalline state as well as in the single crystal state.

Preliminary measurements on cadmium indicate that whereas the heat evolved in the twinning is of the order of 0.1 calories per gram, less than one-tenth of this amount is produced during the whole extension accompanying gliding, although this extension is considerably greater than that due to the twinning. The measurements, which are being followed up with more accurate methods, give, at present, only rough approximations, but suffice to establish clearly this very much greater heating which accompanies twinning. This observation suggests that some of the mechanical energy that is supplied to the lattice to cause twinning is afterwards liberated as heat energy and, in particular cases, as sound energy, the cry of tin and cadmium being a manifestation of the latter.

It may be added that while a cry can be produced from zinc, which crystallises in the hexagonal system and twins readily, I have not been able to produce a cry with any metal crystallising in a cubic system, for which twinning does not take place.

Bruce Chalmers

* * *

The observations of Dr. Chalmers described in the above letter, which are being followed up, seem to me likely to prove of considerable importance for elucidating the problem of twinning. The generation of heat agrees with the view that, in twinning, the molecules, when sufficient energy is applied, slip from one equilibrium position to another, about which they then execute heavily damped vibrations, the energy of vibration dissipating itself in heat and probably in radiation of a frequency of the *Reststrahlen* order. The sound indicates that the twinning does not take place over the whole region of twinning simultaneously, for the sound frequency is much too low to be connected with the vibration of molecules or molecular units, but is propagated from layer to layer with a velocity or velocities of the order of sound velocity. It is possible that in the case of substances where sudden twinning is unaccompanied by audible sound, the sound exists, but is of too high a frequency to be heard.

E. N. da C. Andrade

(**129**, 650-651; 1932)

Bruce Chalmers: Physics Laboratory, University College, London, April 6.

Reference:

1. *Complete Treatise of Inorganic Chemistry*, 7, 296.

着孪生的发生。

对镉的初步观测结果，在孪生过程中所产生的热量只有 0.1 cal/g 的数量级，不足由滑移引起的整个延伸过程中所产生热量的 1/10，虽然，这种延伸明显大于孪生本身所导致的延伸。目前，以更为精确的方法进一步开展的观测只给出了粗略的近似值，但已足够清晰地确认这一伴随孪生过程而产生的较大热量。这一观测结果意味着，一些提供给晶格以产生孪晶的机械能后来以热能和（在某些特殊情况下）声能的形式释放出来，锡和镉的发声就是后一种情况的体现。

还可以补充的是，以六方晶系形式结晶并且容易形成孪晶的锌能够产生发声，但我一直无法用以立方晶系形式结晶的任何金属来产生声音，它们不产生孪晶。

布鲁斯·查默斯

* * *

上面这封信中所描述的查默斯博士即将进一步展开的观测，在我看来，很有可能会被证明对于孪晶问题的阐明具有相当的重要性。热的产生与下面的看法一致，即孪晶中的分子在外加足够能量时会从一个平衡位置滑移到另一个平衡位置，并在该位置发生强烈的阻尼振荡，振荡的能量以热的形式以及很可能以与**剩余射线**的频率相当的辐射形式消散。对声音的研究表明，孪晶化并不是在整个孪晶区域中同时发生的，因为声音的频率实在太低而无法与分子或分子单元的振动联系起来，只能以声速或具有声速水平的速度沿层间传播。有可能会有这样的情况，即突然形成的孪晶没有伴随着可听到的声音；声音是存在的，但是因其频率过高而无法被听到。

安德雷德

（王耀杨 翻译；沈宝莲 审稿）

The Neutron Hypothesis

D. Iwanenko

Editor's Note

Chadwick's hypothesis of the neutron rapidly drew attention from other physicists. Electrons were often expelled from the nucleus during radioactive decay, implying that they must reside somewhere within it. Here Russian physicist Dmitri Iwanenko suggested that electrons change their nature when inside the nucleus, losing their quantum-mechanical spin or magnetism. Inside the nucleus, they may be packed up inside neutrons or alpha particles (which were then still taken to be "elementary" particles along with protons and electrons). Indeed, beta decay involves the decomposition of a neutron into a proton and electron.

DR. J. Chadwick's explanation[1] of the mysterious beryllium radiation is very attractive to theoretical physicists. Is it not possible to admit that neutrons play also an important rôle in the building of nuclei, the nuclei electrons being *all* packed in α-particles or neutrons? The lack of a theory of nuclei makes, of course, this assumption rather uncertain, but perhaps it sounds not so improbable if we remember that the nuclei electrons profoundly change their properties when entering into the nuclei, and lose, so to say, their individuality, for example, their spin and magnetic moment.

The chief point of interest is how far the neutrons can be considered as elementary particles (something like protons or electrons). It is easy to calculate the number of α-particles, protons, and neutrons for a given nucleus, and form in this way an idea about the momentum of nucleus (assuming for the neutron a moment $\frac{1}{2}$). It is curious that beryllium nuclei do not possess free protons but only α-particles and neutrons.

(**129**, 798; 1932)

D. Iwanenko: Physico-Technical Institute, Leningrad, April 21.

Reference:

1. *Nature*, **129**, 312 (Feb. 27, 1932).

中子假说

德米特里·伊万年科

编者按

查德威克的中子假说很快引起了其他物理学家的注意。在发生放射性衰变时原子核中经常会放出电子，这意味着电子肯定位于原子核内部的某个位置上。俄罗斯物理学家德米特里·伊万年科认为当电子位于原子核内部时性质会发生变化，不再具有量子力学的自旋和磁性。在原子核的内部，电子可能堆积在中子或 α 粒子（那时人们还以为 α 粒子是与质子和电子并列的“基本”粒子）里。确实在 β 衰变中就包括一个中子分解成一个质子和一个电子的过程。

查德威克博士关于神秘铍辐射的解释 [1] 对于理论物理学家来说很有吸引力。难道不能认为中子在原子核的构成中也起着重要的作用，核电子都被包裹于 α 粒子或中子之中？因为缺乏关于原子核的理论，这一假设当然无法得到确认，但是如果我们记得核电子在进入核时其性质会发生根本性的变化，可以说失去了其诸如自旋和磁矩等个体特征，那么这个假设也许听起来并不是那么不可能。

我们需要关注的要点在于中子在多大程度上可以被看作是基本粒子（类似于质子或电子）。对于一种给定的核，很容易计算出其中 α 粒子、质子和中子的数目，并用这种方式形成一种关于核矩的概念（假定中子的核矩为 1/2）。奇怪的是，铍核并不具有自由质子而只有 α 粒子和中子。

（王耀杨 翻译；刘纯 审稿）

New Evidence for the Neutron

Irène Curie and F. Joliot

Editor's Note

The existence of the neutron here receives support from Irène Curie and Frédéric Joliot. Experimenting with particles emitted from lithium bombarded by alpha particles, they found a penetrating power less than that of gamma rays from radioactive decay. Moreover, these particles were absorbed more readily by paraffin than by lead, ejecting protons in the process, indicating that they were not electrons or light quanta (which would not do so because of their small or zero mass). The evidence points to the particles in question being different from previously known radiation. Curie and Joliot completed some of their experiments before Chadwick's, and might have discovered the neutron first had their interpreted them properly. Their early results stimulated Chadwick's own decisive experiments.

SEVERAL important communications dealing with the properties of rays emitted by atomic nuclei when bombarded with α-particles have recently appeared,[1] on which we should like to make a few comments.

It has been shown by F. Joliot[2] that the rays emitted by boron under the action of α-particles from polonium are much more penetrating than had originally been indicated. Their penetrating power, while superior to that of the most powerful γ-rays obtained from radio-active sources, is inferior to that of the rays obtained from beryllium bombarded by α-particles from polonium. This result does not agree with Webster's findings, but agrees with the fact that the protons ejected from boron are slower than those ejected from beryllium. Secondly, we have shown that the ejection of protons is a general phenomenon. By means of the Wilson chamber, we have photographed the paths of the helium nuclei ejected by beryllium rays, and from absorption measurements were able to conclude that other atoms are also ejected. Further, our experiments showed for the first time the important part played by the nuclei in the absorption of the rays emitted by beryllium under the influence of α-particles, a phenomenon which clearly marked them off from all previously known radiation.

J. Chadwick was led simultaneously to the same generalisation concerning the ejection of nuclei, and he put forward the view that the penetrating rays produced by the bombardment of beryllium by α-particles from polonium are neutrons. This interpretation is necessary if energy and momentum are conserved in the collision.

Recent experiments which we have carried out with M. Savel clearly show that the rays emitted by lithium have a penetrating power, in lead, less than that of the γ-rays of polonium (they are completely absorbed by 5 mm. of lead), and that they are much more readily absorbed, at equal surface mass, by paraffin than by lead. This proves that

关于中子的新证据

伊雷娜·居里，弗雷德里克·约里奥

编者按

中子的存在被伊雷娜·居里和弗雷德里克·约里奥的实验所证实。他们用 α 粒子轰击锂靶使之发射出粒子，然后发现该粒子流的穿透力低于放射性衰变中的 γ 射线。还发现石蜡比铅更容易吸收这些粒子。吸收过程中会放出质子说明它们不是电子或光量子（因为电子和光量子的质量很小或为零）。有证据表明该粒子也不同于以前所熟知的任何辐射。居里和约里奥的实验先于查德威克，如果他们能对实验结果作出正确的解释，那他们也许会首先发现中子。正是他们得出的前期结果引出了查德威克的那个具有决定意义的实验。

最近出现了几篇颇有影响力的论文，内容是关于受 α 粒子轰击的原子核所发射的射线的性质[1]，对此，我们想发表一些看法。

约里奥曾指出[2]，硼在从钋中发出的 α 粒子的作用下发射出的射线比原来对其穿透力的预计值大很多。它们的穿透力大于由放射源得到的最强的 γ 射线，但要低于从钋中发出的 α 粒子轰击铍得到的射线。这一结论与韦伯斯特的研究结果不相符，但是与硼发射的质子比铍发射的质子速度慢的现象一致。其次，我们曾说过发射出质子是普遍存在的现象。我们利用威尔逊云室拍摄了由铍辐射产生的氦核的路径，并通过吸收测量得知其他原子也会发射质子。第三，是我们的实验最先说明原子核在吸收被 α 粒子轰击的铍发射的射线中所起的关键作用，这一现象可以把它们与所有以前知道的辐射明确地区分开来。

查德威克与我们同时注意到了这一关于核发射的普遍现象，他还提出了以下观点，即用从钋中发出的 α 粒子轰击铍而产生的具有很强穿透力的射线就是中子。如果要保持能量和动量在碰撞中的守恒，就必须接受这一解释。

最近我们与萨韦尔共同完成的实验清晰地表明，由锂发射的射线在铅中具有穿透能力，但低于钋发射的 γ 射线的穿透能力（它们能被 5 mm 厚的铅完全吸收），当表面质量相同时，石蜡对它们的吸收要远远大于铅。这说明它们不可能具有电子或

they cannot be of an electronic or electromagnetic nature. Since for various reasons it is extremely improbable that we are dealing with hydrogen nuclei or α-particles (the energy of which would be enormous), these results prove—*independently of the ejection of nuclei and the laws of elastic collisions*—that the rays emitted by lithium under bombardment by α-particles from polonium are different from previously known radiation and are probably neutrons. The above reasoning does not apply to the rays ejected from beryllium, boron, or to those emitted by lithium when bombarded with the α-rays from the active residue of radium[3], because in such cases we do not have γ-rays of equivalent penetrating power, for comparison.

Our latest experiments, in collaboration with M. Savel, indicate that the protons ejected from beryllium form two groups. This suggests that there are also two groups of neutrons (each group not necessarily homogeneous); one group has a range of 28 cm. in air, and an energy of 4.5×10^6 electron volts; the other has a range of about 70 cm. and an energy of approximately 7.8×10^6 electron volts. We find it difficult to reconcile Chadwick's result of a *maximum range* of 40 cm. with the curves which we have obtained for the absorption of protons.

The mass of the neutron calculated by Chadwick[4] (based upon the experimentally estimated energy of the neutrons from boron), according to the reaction $B^{11} + \alpha = N^{14} + n$,[5] is about 1.006 (He = 4), and the atomic mass of Be^9, based on the energy of the fast neutrons (7.8×10^6 ev.), is 9.006. This suggests that the binding energy between the two α-particles and the neutron in the Be^9 nucleus is relatively weak. Further, we know that the rays emitted by beryllium are composed of neutrons and photons, and we may therefore suppose that they are emitted simultaneously according to the equation

$$Be^9 + \alpha = C^{12} + n + h\nu.$$

The photons of 2 to 4.5×10^6 ev. energy, which we have detected, would correspond to the group of neutrons of maximum energy 4.5×10^6 ev. (protons having a range of 28 cm.).

(**130**, 57; 1932)

Irène Curie and F. Joliot: Institut du Radium, Laboratoire Curie, 1, Rue Pierre-Curie, Paris (5e), June 25.

References:

1. Webster, H. C., Chadwick, J., Feather, N., and Dee, P. I., *Proc. Roy. Soc.*, A, **136**, 428, 692, 708 and 727 (1932).
2. Joliot, F., *C.R. Ac. Sci.*, **193**, 1415 (1931).
3. Broglie, M. de, and Leprince-Ringuet, L., *C.R. Ac. Sci.*, **194**, 1616 (1932).
4. Chadwick, J., *Proc. Roy. Soc.*, A, **136**, 702 (1932).
5. Curie, I., and Joliot, F., *C.R. Ac. Sci.*, **194**, 1229 (1932).

电磁学的特征。我们有许多证据证明在实验中出现的粒子绝对不可能是氢核或 α 粒子（它们的能量很大），这些结果表明：**与原子核发射粒子和弹性碰撞定律无关**，用由钋衰变得到的 α 粒子轰击锂发出的射线与以前所知的辐射都不相同，这类射线可能是中子。上述推论不适用于由铍和硼发射的射线，也不适用于用由镭的放射性剩余物产生的 α 射线轰击锂而发出的射线[3]，因为在这几种情况下，我们没有用具有相同穿透力的 γ 射线作为比照。

我们与萨韦尔合作完成的最新实验说明铍发射的质子可以分为两组。这意味着也存在两组中子（两组中子未必同质）；其中一组在空气中的射程为 28 cm，能量为 4.5×10^6 eV；另一组射程约为 70 cm，能量约为 7.8×10^6 eV。查德威克的结果是**最大射程**为 40 cm，而我们从质子吸收曲线得到的结论与他的结果并不一致。

查德威克根据反应式 $B^{11}+\alpha=N^{14}+n$[5] 计算出的中子质量[4]（基于实验测算出的由硼发射的中子能量）约为 1.006（He = 4），并从快中子的能量（7.8×10^6 eV）推算出 Be^9 的原子质量为 9.006。这说明 Be^9 核内两个 α 粒子与中子之间的结合能相当弱。此外，我们还知道铍发射的射线包括中子和光子，因而我们也许可以假定它们是按照下述反应式同时发出的：

$$Be^9+\alpha=C^{12}+n+h\nu$$

我们探测到光子的能量在 2×10^6 eV 到 4.5×10^6 eV 之间，它们应该对应于最大能量为 4.5×10^6 eV 的那一组中子（质子的射程为 28 cm）。

（沈乃澂 翻译；朱永生 审稿）

Mechanism of Superconductivity

J. Dorfman

Editor's Note

Superconductivity—the conduction of electrical current in cold metals without resistance—was observed in 1911, but lacked an explanation. Many suspected that some kind of coherent behaviour among the electrons must be responsible. Here Russian physicist J. Dorfman explores experimental evidence for analogies between superconductivity and magnetism. He studies the temperature dependence of the thermoelectric effect for lead (where heat creates electricity), and finds a sharp cusp near the superconducting transition temperature, leading him to suspect an analogy between the ferromagnetic and superconducting transitions. Both are now seen to belong to the same class of "critical" transitions, although magnetism per se is not involved in this kind of superconductivity.

IT was often assumed that the transition from normal conductivity to superconductivity may be connected with a kind of "spontaneous coupling" of the conduction electrons. Some authors were even inclined to identify this phenomenon with ferromagnetism. Although this extreme point of view seems to be very improbable, some analogies with ferromagnetism must surely appear if any kind of "spontaneous coupling" between electrons is responsible for superconductivity. For example, in this case the shape of the specific heat curve near the transition temperature must be analogous to that of ferromagnetic substances in the vicinity of the Curie point. W. Keesom and J. H. van den Ende,[1] and F. Simon and K. Mendelssohn[2] attempted to discover this anomaly of the specific heat in lead near the transition temperature (7.2° K), but they could not detect any trace of the effect. This result may be interpreted in two ways: either the hypothesis of the "spontaneous coupling" of the conduction electrons in superconductors is completely wrong, or the number of the electrons which are concerned in conductivity is here so small in comparison with the number of atoms that the specific heat anomaly of the conduction electrons cannot be detected with calorimetric methods.

As our measurements have shown, the specific heat anomaly of ferromagnetic bodies at the Curie point is so well pronounced in the thermoelectric effects (Thomson effect), that in spite of some difficulties concerning the sign of this effect, the order of magnitude of the specific heat anomaly can be computed from the purely thermoelectric constants in good agreement with calorimetric measurements. It is natural to try the same method in the domain of superconductivity. The recent investigations by J. Borelius, W. M. Keesom, C. H. Johansson, and J. O. Linde[3] of the thermoelectric force for lead and tin at the lowest temperatures permit us to compute the Thomson effect for these metals and to draw conclusions concerning the specific heat anomaly. Fig. 1 represents the Thomson coefficient for lead (as calculated from the experimental data) as a function of temperature. This curve is quite analogous to that of ferromagnetic substances, and it seems quite

超导电性机理

多尔夫曼

编者按

超导电性，即某些金属在低温下无阻的传导电流，是 1911 年被首次观察到的，但没有给出解释。有不少人猜测电子之间的某种相干行为应是产生超导性的原因，在下文中，俄罗斯物理学家多尔夫曼用实验揭示了超导性与磁性之间的相似性。他研究了铅的温度与热电效应（由热生电的现象）的关系，发现在超导转变温度附近存在一个尖点，这使他猜想在铁磁和超导转变之间存在着某种相似性。在今天看来，虽然磁性本身与超导性无关，但两者都属于同种类型的“临界”转变。

人们通常假设，从正常传导性到超导性的转变可能与导电电子的某种“自发耦合”有关。某些作者甚至倾向于认为超导现象等同于铁磁现象。虽然这个极端的观点似乎不存在成立的可能，但是，如果超导性真的是由电子间的某种“自发耦合”引起的，那么它就一定会表现出与铁磁性类似的行为。例如，超导体的比热曲线在转变温度附近的形状应该与铁磁体的比热曲线在居里点附近的形状相似。凯索姆与范登恩德[1]，以及西蒙与门德尔松[2]都试图去揭露铅的比热在超导转变温度（7.2 K）附近的这种反常行为，但是他们都未能探测到这种行为的任何迹象。该结果也许可以从两个角度来解释：要么是由于超导体中导电电子的“自发耦合”假说是完全错误的，要么就是由于在传导性中计入的电子的数目较之原子的数目而言太小，以至于无法用量热法探测出由传导电子引起的比热反常。

正如我们的测量结果所表明的那样，铁磁体在居里点的比热反常表现在热电效应（汤姆孙效应）上是非常显著的，以至于尽管有关此效应的正负号确定还存在一定的困难，但由纯热电常数计算出来的比热反常的数量级与量热法的测量值还是符合得很好。人们很自然地想到将这种方法用于超导性领域。最近，博雷柳斯、凯索姆、约翰松与林德[3]研究了在对最低温度下铅和锡的温差电势，这使我们能够计算这两种金属的汤姆孙效应，并可得到关于比热反常的结论。图 1 给出了铅的汤姆孙系数（由实验数据计算得出）与温度的函数关系。这条曲线和铁磁物质的曲线十分相似，且很可能代表了与传导性效应有关的电子的比热的普遍特征。不过，我

probable that it represents the general feature of the specific heat of the electrons concerned in the conductivity effects. It is not clear, however, why the temperature of the maximum of this specific heat curve (10.5° K) does not coincide with the transition point (7.2° K). Perhaps theory will be able to explain this discrepancy in the future.

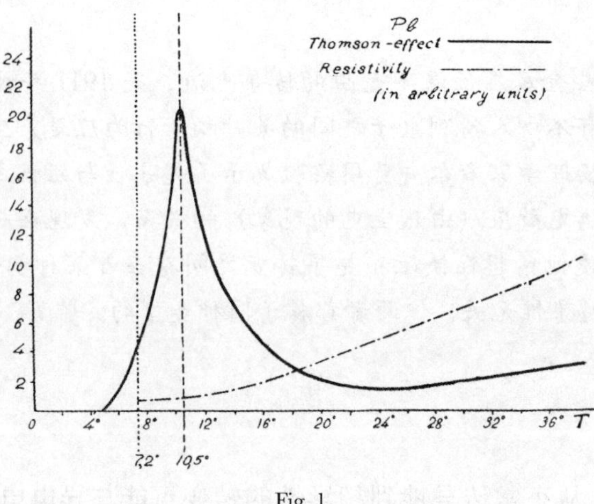

Fig. 1.

From these results two important quantities may be calculated: first, ΔC_ε (the height of the maximum of the specific heat curve), and secondly, ΔW_0 (the energy difference between the normal and the superconducting state at absolute zero), both for one electron.

	ΔC_ε cal./ degree.	ΔW_0 ergs.
Lead	8×10^{-25}	1.7×10^{-17}
Tin	$\sim 10^{-25}$	$\sim 0.6 \times 10^{-17}$

If the number of the electrons was equal to the number of atoms of lead, the specific heat anomaly could certainly be detected, its numerical value being of the same order of magnitude as the normal specific heat itself. The precision of the calorimetric measurements permits us to determine the upper limit of the number of the electrons involved in the conductivity effects of lead. Actually it seems that the number of the conduction electrons is less than 1/200 of the number of the atoms in this case.

It is well known that magnetic fields destroy the superconductivity, the threshold value of the field H increasing as the temperature is lowered. By extrapolating the experimental data the value of H_0 may be found corresponding to absolute zero. We assume that the threshold value of the field is given by the condition that the magnetic energy of the electron $|\mu H_0|$ (where μ is the spin moment) is equal to ΔW_0.

们还不清楚为什么这条比热曲线上最大值所相应的温度（10.5 K）与超导转变温度（7.2 K）不一致。也许未来的理论可以解释这一分歧。

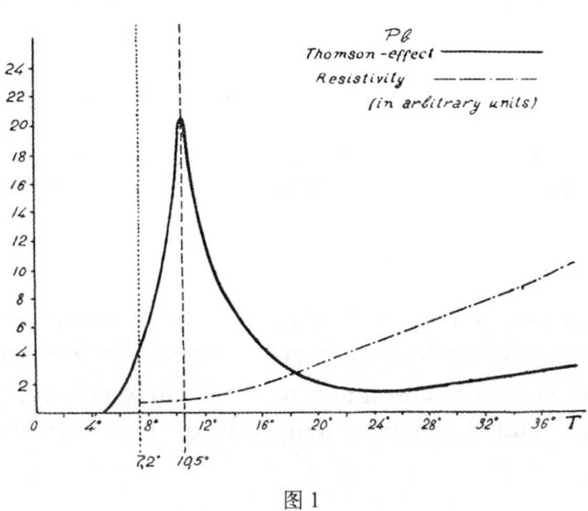

图 1

由这些结果可以计算出两个重要的物理量：首先是 ΔC_ε（比热曲线上最大值的高度），其次是 ΔW_0（绝对零度下正常态和超导态的能量差），两者都是针对一个电子。

	ΔC_ε 卡 / 度	ΔW_0 尔格
铅	8×10^{-25}	1.7×10^{-17}
锡	$\sim 10^{-25}$	$\sim 0.6 \times 10^{-17}$

倘若铅中的电子数目与原子的数目相等，则比热反常就一定能被探测到，因为它的数值与正常态的比热同量级。根据量热法的测量精度，我们可以确定铅中参与传导效应的电子数目的上限。实际上，导电电子的数目比此情况下原子数目的 1/200 还要少。

众所周知，磁场会破坏超导性，磁场 H 的阈值随着温度的降低而增大。从实验数据可以推出：临界场 H_0 的值可能与绝对零度有关。我们假设场的阈值通过电子的磁能 $|\mu H_0|$（μ 为自旋磁矩）等于 ΔW_0 来确定。

$$|\mu H_0|=\Delta W_0 \qquad (1)$$

This assumption means that the superconductivity must be destroyed when the energy of the external forces exceeds the energy of the "spontaneous coupling". Form (1) we may calculate H_0 for lead and tin, and compare them with the experimental results.

	H_0 expt. (gauss).	H_0 calc. (gauss).
Lead	2,000–2,500	2,000
Tin	560	~700

According to the recent experiments of McLennan and his co-workers,[4] superconductors cease to be superconducting for high frequency currents if the frequency v exceeds a certain threshold value. For tin at absolute zero, $v\sim10^9$ may be found by extrapolation of the experimental data obtained at higher temperatures. It is interesting to notice that by assuming

$$hv_0=\Delta W_0=|\mu H_0| \qquad (2)$$

(where h is Planck's constant), we obtain for the same metal $v_0 = 1\times10^9$.

The remarkable coincidence between the observed and the computed data seem to support the general trend of the assumptions developed in this note. It is interesting to notice that the frequency of the Larmor precession corresponding to H_0 is equal to v_0; thus the correlation between the two factors destroying the superconductivity may be found either on the lines of energetics or on the lines of the short time periods. Which of these interpretations corresponds to the real mechanism remains unsolved at this moment.

(**130**, 166-167; 1932)

J. Dorfman: Physical-Technical Institute, Sosnovka 2, Leningrad (21), U.S.S.R., May 23.

References:

1. *Comm. Leiden*, 230 *d* (1930).
2. *Z. Phys. Chem.*, B **16**, H. 1 (1932).
3. *Proc. Amsterdam Acad.*, **34**, No. 10 (1931).
4. *Proc. Roy. Soc.*, A, **136**, No. 829, 52 (1932).

$$|\mu H_0|=\Delta W_0 \qquad (1)$$

这个假设意味着：当外力的能量超过“自发耦合”的能量时，超导性一定会被破坏。我们可以由式（1）计算铅和锡的临界场 H_0，并把它们与实验结果作比较。

	H_0 的实验值（高斯）	H_0 的计算值（高斯）
铅	2,000~2,500	2,000
锡	560	~700

根据麦克伦南及其合作者近期的实验[4]，如果超导体中高频电流的频率 ν 超过了某一临界值，则超导现象就会消失。从较高温度下的实验数据可以推出，绝对零度下锡的 $\nu\sim10^9$。有趣的是，通过假设

$$h\nu_0=\Delta W_0=|\mu H_0| \qquad (2)$$

（h 是普朗克常数），对同一金属我们也可得到 $\nu_0 = 1\times10^9$。

实验值和计算值的明显致似乎支持了我们在文中提出的假设。有趣的是：对应于 H_0 的拉莫尔进动频率恰等于 ν_0；因此，两个破坏超导性的因素之间的关联既可以体现在能量的关系上，也可以体现在短周期的关系上。究竟这些解释中的哪一个对应真正的超导性机理，现在还没有答案。

（王静 翻译；陶宏杰 审稿）

Disintegration of Light Elements by Fast Protons

J. D. Cockcroft and E. T. S. Walton

Editor's Note

Here Cockcroft and Walton update their studies of how high-energy protons interact with the nuclei of light elements. Since their report of artificially induced disintegration of lithium, they had improved the sensitivity of their detecting apparatus and extended their study to the element boron. Here they record a huge number of alpha particles created by proton bombardment, suggesting a nuclear reaction in which the boron nucleus boron-11 becomes boron-12 on absorbing a proton, and subsequently disintegrates into three alpha particles.

SINCE the publication of our paper[1] on the disintegration of elements by fast protons, we have examined some of the light elements more carefully, using much thinner mica windows than we had previously employed on the high voltage tube. With the present arrangement, we can count particles which have passed through only 6 mm. air equivalent of absorber on their way from the target to the ionisation chamber.

In the case of lithium, we have found, in addition to the α-particle group of 8.4 cm. range, another group of particles of much shorter range. The number of these is about equal to that of the long range particles and their maximum range is about 2 cm. The ionisation produced by them indicates that they are α-particles. It will be of interest to examine whether any γ-rays are emitted corresponding to the difference of the energies of the α-particles in the two groups, but on account of the smallness of the effect to be expected, a sensitive method will be necessary.

In the case of boron, the number of particles observed increases rapidly as the total absorption between the target and the ionisation chamber is reduced. The maximum range of these particles is about 3 cm. and in our earlier experiments we determined the number of particles only after passing through the equivalent of 2.9 cm. of air, so that we were very nearly at the end of the range. Decreasing the absorber to 6 mm. of air gives an enormous increase in the number of particles. In this way about twenty-five times as many particles have been obtained from boron as from lithium under the same conditions. We estimate that there is roughly one particle emitted per two million incident protons at 500 kilovolts. The ionisation produced by the particles suggests that they are α-particles, and the energy of the main group would support the assumption that a proton enters the B^{11} nucleus and the resulting nucleus breaks up into three α-particles. There also seem to be present a small number of particles with ranges up to about 5 cm.

(**131**, 23; 1933)

J. D. Cockcroft and E. T. S. Walton: Cavendish Laboratory, Cambridge, Dec. 22.

Reference:

1. *Proc. Roy. Soc.*, A, 137, 229 (1932).

快质子引起的轻元素衰变

约翰·道格拉斯·考克饶夫，欧内斯特·瓦耳顿

编者按

考克饶夫和瓦耳顿改进了对于高能质子与轻元素原子核之间相互作用的研究。在报告了人工引发的锂衰变之后，他们提高了探测仪器的灵敏度，并把研究范围扩展到了元素硼。他们发现硼被高能质子轰击后会产生大量的α粒子，说明发生了这样的核反应：硼 11 核在吸收了一个质子之后生成硼 12，随后衰变成 3 个α粒子。

在发表了关于快质子引起元素衰变的文章[1]以后，我们对一些轻元素进行了更仔细的检查，我们所用的云母窗比以前在高压管上使用的要薄很多。采用目前的装置，我们能够对从靶出发穿越阻止能力仅相当于 6 mm 厚空气的吸收层到达电离室的粒子进行计数。

我们在检测锂元素的实验中发现，除了一组射程为 8.4 cm 的 α 粒子以外，还有另一组射程更短的粒子。这些粒子与长射程粒子的数量大致相等，它们的最大射程约为 2 cm。由它们产生的电离现象表明，它们也是 α 粒子。检验是否有与两组 α 粒子能量差相对应的 γ 射线产生对于我们来说是一件有趣的事，但考虑到该效应可能很微弱，所以需要一种很灵敏的方法。

在对硼元素进行实验时，如果我们减少靶与电离室之间的总吸收，则观测到的粒子数就会迅速增加。这些粒子的最大射程约为 3 cm，在早期的实验中，我们测定了穿越阻止能力相当于 2.9 cm 厚空气的粒子数，这非常接近于最大射程的终点。将吸收层厚度降至相当于 6 mm 空气时，得到的粒子数大幅增加。在相同条件下，由此方法从硼中得到的粒子约为从锂中得到的 25 倍。我们估计，当电压为 500 千伏时，大体上每 200 万个入射质子会激发出一个粒子。从这些粒子引起的电离来看它们就是 α 粒子，主要组 α 粒子的能量可以证明下面这个假设：一个质子进入 B^{11} 核，而生成的核又裂变为 3 个 α 粒子。似乎还存在少量射程高达 5 cm 的粒子。

（沈乃澂 翻译；江丕栋 审稿）

Energy of Cosmic Rays

E. Regener

Editor's Note

Physicists had obtained increasingly accurate measurements of the ionisation created in the upper atmosphere by cosmic rays. Using this data, Erich Regener here estimates the energy flux of the cosmic rays impinging upon the Earth. He estimates that 108 pairs of ions are created per second in each square cm. As impinging cosmic rays need an energy of 32 electron-volts to trigger ionisation, this led to an estimate for the total energy flux of 5.2 × 10^{-3} erg. cm.$^{-2}$ sec.$^{-1}$ As Regener notes, a body in thermal equilibrium under illumination from this flux would bear a temperature of 3.1 K, similar to Arthur Eddington's result for a body in equilibrium with the ordinary radiation coming from stars.

IN Nature for September 3, 1932, p. 364, I published the curve of the intensity of cosmic radiation in the high atmosphere, deduced from measurements made with a self-registering electrometer. It was possible by extrapolation to find the intensity I_∞ of radiation at its entrance in the atmosphere. The preliminary value given has now been corrected by the experimental determination of the factor which reduces the measurements with the ionisation chamber at 5 atmospheres to 1 atmosphere. Now the value I_∞ is found corresponding to a production of 333 pairs of ions cm.$^{-3}$ sec.$^{-1}$ in air at 0° and 760 mm. mercury pressure.

The graphical integration of the curve, giving the ionisation as a function of the height, makes it possible to calculate the total number of ions, produced by total absorption of cosmic rays by a column of air of 1 sq. cm. section. The high value of 1.02×10^8 pairs of ions is found. Some time ago, Millikan and Cameron[1] made a similar calculation, which gave a value of only 1.28×10^7 pairs of ions, due to an insufficient knowledge of the intensity in the high atmosphere. Taking the energy required to produce a pair of ions in air[2] as 32 electron-volts the flux S of energy coming to the earth from the cosmic rays is found to be 5.2×10^{-3} erg. cm.$^{-2}$ sec.$^{-1}$.

From an astrophysical point of view, the great energy of cosmic rays is remarkable. A body which absorbs all the cosmic rays would be heated by them. Equilibrium will be attained when the absorbed flux S of cosmic rays is equal to the heat radiation σT^4 of that body. T works out as 3.1° Kelvin. The value is equal to the temperature (3.18°) which Eddington[3] finds for a black body heated only by the heat and light radiation of stars. Eddington's calculation relates to a point in our local system of stars, but not in the neighbourhood of one of them. If at such a point the flux of energy of cosmic radiation is equal to that on the earth, the temperature of a black body, absorbing entirely the

宇宙射线的能量

埃里克·雷格纳

编者按

物理学家们一直在对宇宙射线在高层大气中产生的电离效应进行测定，所得的结果越来越精确。埃里克·雷格纳利用这些数据估算了宇宙射线在撞击地球时的能量通量。他预计每秒在每平方厘米的面积上宇宙射线将有 108 个离子对通过。因为只有能量大于 32eV 的宇宙射线才能产生电离效应，由此可以估算出宇宙射线的总能流为 5.2×10^{-3} erg · cm^{-2} · s^{-1}。雷格纳指出：一个处于热平衡状态中的物体在这么高能量射线流的照射下温度将升高 3.1 K。这个值近似于阿瑟·爱丁顿从多数来自恒星的辐射对平衡态天体的影响中得到的结果。

我在 1932 年 9 月 3 日《自然》杂志的第 364 页上公布了宇宙辐射在高层大气中的强度曲线，这条曲线是根据一台自动记录静电计的测量结果推演出来的。利用外推法可以得到宇宙射线在进入大气层时的辐射强度 I_∞。现在原始数据已经可以用确定的实验因子进行修正，这个因子可以把电离室在 5 个大气压下测量的结果折合成 1 个大气压下的结果。目前得到的 I_∞ 值相当于在 0℃，760 mm 汞柱空气中每秒每立方厘米产生 333 个离子对。

对曲线进行图解积分，发现电离度是随高度变化的函数，这样就可以计算出在横截面积为 1 平方厘米的空气柱完全吸收宇宙射线后产生的离子对总数。我们发现这个值高达 1.02×10^8 对离子。早些时候，密立根和卡梅伦[1]也进行过类似的计算，但由于他们对高层大气中的射线强度缺乏充分的认识，因而给出的数值仅为 1.28×10^7 对离子。如果在空气中产生一个离子对所需的能量[2]为 32 eV，那么宇宙射线带到地球上的能量通量 S 应该可以达到 5.2×10^{-3} erg · cm^{-2} · s^{-1}。

从天体物理学的角度来看，宇宙射线的能量如此巨大是很不寻常的。一个吸收了全部宇宙射线的天体将因此而被加热。当天体所吸收的宇宙射线的总通量 S 等于其热辐射 σT^4 时，该天体就会达到平衡，这时计算出的温度值 T 为 3.1 K，这与爱丁顿[3]发现的只被恒星发出的光和热所加热的黑体所具有的温度值（3.18 K）相等。爱丁顿的计算仅关系到我们所处的局部恒星系统中的一个点，而不涉及其中某一个恒星的周边区域。如果在这样一个点上，宇宙辐射的能流与在地球上获得的能流相等，那么根据 T^4 定律，一个完全吸收了这**两种**辐射后的黑体的温度只会升至 3.7 K。

two radiations, rises only to 3.7° Kelvin, according to the T^4 law. But at a point in space among the spiral nebulae, the ordinary radiation is very small and causes only a very small rise of temperature. Supposing that cosmic rays originate in such intergalactic space, they would produce an elevation of temperature corresponding to the flux of cosmic rays.

A more detailed report will be published shortly in the *Zeitschrift für Physik*.

(**131**, 130; 1933)

E. Regener: Physik. Inst. d. Techn. Hochschule, Stuttgart, Dec. 31.

References:

1. *Phys. Rev.*, **31**, 930 (1928).
2. Kulenkampff, H., *Phys. Z.*, **30**, 777 (1929).
3. *Internal Constitution of the Stars*, German edition, 468 (1928).

但对于处于旋涡状星云内的空间某一点来说，通常的辐射是非常小的，所以温度上升的幅度也微乎其微。假设宇宙射线起源于这样的星系际空间，它们将引发与宇宙射线通量相对应的升温现象。

更详细的结果将在近期的《物理学杂志》上发表。

（史春晖 翻译；马宇蒨 审稿）

Helium Liquefaction Plant at the Clarendon Laboratory, Oxford

F. A. Lindemann and T. C. Keeley

Editor's Note

In the early 1930s, liquid helium had become precious to physicists studying the properties of matter at very low temperatures. Here the physicist Frederick Lindemann and colleagues at the University of Oxford announced the development of a new means for producing large quantities of liquid helium. Their technique used the liquefaction of hydrogen under an abrupt change in pressure to cool helium, and could produce large volumes of the liquid in continuous operation with relatively cheap apparatus. The liquid helium lasted in their laboratory for about one and one-half hours. This method and subsequent developments would enable the Soviet physicist Pyotr Kapitza to discover the phenomenon of superfluidity in liquid helium in 1938.

THE main properties of liquid helium have been familiar to men of science for a great many years. The only object therefore in liquefying it is in order to cool other substances the characteristics of which it is desired to study in the neighbourhood of the absolute zero. It has long been known that the heat capacity of solids becomes extremely small at low temperatures. Thus the latent heat of evaporation of 20 mgm. of liquid helium is sufficient to cool 60 gm. of copper from the temperature to be attained with liquid hydrogen boiling under a reduced pressure to the boiling point of helium.

It is easy to design apparatus so that the substances the properties of which at low temperatures are under investigation, are cooled to the temperature of the surrounding liquid or solid helium and maintained at this temperature with a minimum of waste. It seemed preferable, therefore, to instal a small inexpensive apparatus requiring comparatively little liquid hydrogen, which can therefore be operated frequently or duplicated at comparatively small cost, rather than to indulge in a plant designed to produce liquid helium in large quantities. In any event, the financial resources available would have imposed this choice, even had the alternative procedure been considered desirable.

The apparatus which has been installed at Oxford is of a type developed by Prof. Simon and Dr. Mendelssohn in Berlin and Breslau. Two concentric cylinders capable of withstanding a pressure of some 150 atmospheres surround the space in which the substance under investigation is placed. Helium under a pressure of about 100 atmospheres is introduced into the space between the cylinders. The upper part of the annular space between the cylinders is separated from the lower, in which the helium is

牛津大学克拉伦登实验室的氦液化车间

弗雷德里克·林德曼，托马斯·基利

编者按

在 20 世纪 30 年代早期，液氦对于研究很低温度下物质性质的物理学家来说是十分珍贵的。在本文中，牛津大学的物理学家弗雷德里克·林德曼及其同事宣布研发出了一种能生产大量液氦的新方法。他们采用突然改变液态氢压力的方式来冷却氦，可以用这种较便宜的装置连续地得到大量的液氦。他们的实验室持续生产了 1.5 个小时的液氦。正是这一方法及接下来的一些发展使苏联物理学家彼得·卡皮查得以在 1938 年发现液氦中的超流性现象。

多年以来，液氦的主要性质就已为科学界的人士所熟知。所以人们把氦液化的唯一目的，是想用它来冷却其他物质以便研究它们在绝对零度附近的性质。长期以来人们就知道，固体的热容量在低温时会变得非常小。因此，蒸发 20 毫克液氦吸收的潜热足以使 60 克铜的温度从减压液态氢沸腾的温度降低到液氦沸点的温度。

为了研究低温下物质的性质，很容易设计一个实验装置把物质冷却到与其周围的液氦或固氦温度相同的温度，并以最小的损失使物质保持在这个温度。不过，我们更喜欢对液氢需求相对较少的小型廉价装置，而不是一个能大量生产液氦的工厂，因为选择小型设备可以降低经常运转或复制它的成本。无论如何，由于财力所限，即便是还有其他令人满意的方式可供选择，也只能选择这种装置。

安装在牛津大学的设备是由柏林的西蒙教授和布雷斯劳的门德尔松博士研制的。要研究的物质被装在由两个可承受 150 个大气压的同心圆桶围绕的空间中。在两个圆桶之间的空间注入压强约为 100 个大气压的氦。圆桶之间的环状区域被一金属板分隔成上下两部分，上部分形成了一个小的金属容器，并通过一根螺线型的细铜管与纯氢的供应源相连，氦在下半部分被压缩。整个装置用一根德银管保持在一

compressed, by a metal sheet, thus forming a small metal container which is joined by a spiral of thin copper tubing to a source of pure hydrogen. The whole is held in position on a German silver tube in the centre of a larger metal vessel containing hydrogen or helium gas at low pressure which can be evacuated by means of a mercury vapour pump. This outer vessel together with the copper spirals through which the hydrogen and helium are introduced is immersed in a Dewar flask containing liquid hydrogen.

When temperature equilibrium has been attained, hydrogen is introduced into the top vessel under a pressure of two or three atmospheres. Passing through the copper spirals, this liquefies owing to the excess pressure and runs down into the metal container over the double-walled helium cylinder. A tap to the mercury vapour pump is now turned on and a high vacuum produced in the metal box, so that the helium container with its superposed pot of liquid hydrogen is thermally insulated save for the necessary connecting tubes.

The yield of liquid helium is improved if the compressed helium is further cooled by boiling the hydrogen in the inner container under reduced pressure. If the helium is now allowed to expand, about half of it liquefies and the central space with the experimental substances it contains is cooled to the temperature of the surrounding helium. By evacuating the space above the liquid, that is, causing it to boil under reduced pressure, one can, of course, reduce the temperature to within one or two degrees of the absolute zero.

In the apparatus used at Oxford the helium lasts for about an hour and a half. If the experiment is not finished in this time, one can repeat the process in a few minutes at very small cost in liquid hydrogen. The helium expands into a rubber bag and is recompressed into a cylinder so that very little gas is lost. The temperature during the experiment can be observed on a large-scale manometer connected through a fine tube to a small vessel containing helium in the liquefaction space. The apparatus cost approximately £30. Since there is no need to recompress the helium rapidly, a small cheap compressor is sufficient.

The liquid hydrogen required is produced in a plant of the standard pattern designed in the Physico-Chemical Institute in Berlin which has been in use at Oxford for some years now without giving any trouble. Impurities in the hydrogen are condensed by a preliminary expansion and continuously removed by a slow stream of hydrogen. With a compressor capable of dealing with ten cubic metres of free gas an hour and an expenditure of approximately 1.4 litres of liquid nitrogen per litre of liquid hydrogen, this plant produces some $2\frac{1}{2}$ litres of liquid hydrogen per hour. Liquefier and compressor together cost approximately £350.

The liquid hydrogen is stored in pyrex Dewar flasks silvered and exhausted in the laboratory. As their efficiency equalled that claimed for the more complicated double vessels developed by Prof. Kapitza, they have been retained.

个更大的金属容器的中心，该容器中充有低压的氢气或氦气，可用汞蒸汽泵抽成真空。最外面的容器与输送氢和氦的铜螺线管一起浸没在一个含有液态氢的杜瓦瓶内。

当温度达到平衡时，压强为 2 或 3 个大气压的氢被注入到上部的容器中。氢气通过铜螺线管时由于所受压力过大而被液化，随后流入到在双壁氦圆桶上的金属容器中。这时打开连接汞蒸汽泵的阀门，金属容器内就会产生高真空，从而除了所需的连接管以外，氦容器与在它上面的液氢罐都与外界热绝缘。

如果通过减压使内层容器中的氢达到沸点以进一步冷却压缩状态下的氦，就可以提高液氦的产量。现在允许氦气膨胀，约一半的氦会被液化，放有实验物质的中间区域也会被冷却到与周围氦一样的温度。如果在液体上方抽真空，也就是让液体在减压的情况下沸腾，我们就可以把温度降低到距绝对零度一两度的范围之内。

牛津大学的设备能使液氦持续运转约 1.5 小时。如果在 1.5 个小时之内实验还没有做完，我们可在几分钟之内就开始重复上述过程，这只需消耗一些很便宜的液氢而已。氦气膨胀进入一个橡皮袋，然后被压缩到钢瓶内，所以丢失的气体很少。我们可以通过一个大量程的压力计观察实验时的温度，这个压力计通过一根细管子连接到液化区内的含氦气小容器上。这个设备的造价约为 30 英镑。因为不需要快速压缩氦气，所以使用一台小型廉价的压缩机就够了。

所需的液氢是用柏林物理化学研究所设计的标准设备生产的，目前这个设备已经在牛津大学无故障地运转了好几年了。氢气中的杂质先经过一次预膨胀使之凝聚，然后通过氢的低速流动继续去除杂质。使用的是每小时能处理 10 立方米自由气体的压缩机，每升液氢需要消耗约 1.4 升液氮，这个车间每小时可生产约 2.5 升的液氢。液化器和压缩机的总造价约为 350 英镑。

液氢被储存在镀银的派热克斯玻璃杜瓦瓶中并在实验室中耗尽掉。由于派热克斯玻璃杜瓦瓶与卡皮查教授研制的较复杂的双容器杜瓦具有相同的功效，所以我们一直在使用这些杜瓦瓶。

If low temperature work expands and a large number of experiments are in hand simultaneously, it may be necessary to consider the use of the continuous Linde process of liquefaction. In view of the cost of the gas and the precautions necessary for its recovery, its distribution involves considerable inconvenience, which for the time being are scarcely worth facing. The mere liquefaction, of course, offers no difficulties and there is little doubt that the Berlin type of apparatus, which is already in use in many laboratories, will prove as serviceable and efficient as the hydrogen liquefier, should it ever be necessary to change over to this system.

Finally, a word of thanks is due to Dr. Mendelssohn, who kindly brought the liquefier over from Breslau and placed all his knowledge and experience unreservedly at the disposal of the department. But for this, it would scarcely have been possible to obtain, without hitch or trouble, liquid helium within one week of the arrival of the apparatus in Oxford.

(**131**, 191-192; 1933)

如果低温工作扩展了，需要在低温下同时进行多项实验，恐怕就要考虑采用连续的林德液化过程了。考虑到气体很昂贵，以及必须仔细地回收气体，使用这一方法还有相当大的不便，所以目前还不太值得考虑。如果必须改为使用这种系统的话，仅从实现液化而言，柏林型装置当然是没有任何困难的，毫无疑问，它将被证明与氢液化器一样好用而高效，而且很多的实验室已在使用它。

最后，我要对门德尔松博士表示感谢，他慷慨地从布雷斯劳带来液化器并把自己所有的知识和经验毫无保留地贡献给这个系自由处理。否则，我们几乎不可能在设备到达牛津大学的一周之内，没有遇到任何麻烦就顺利地制备出液氦来。

（沈乃澂 翻译；陶宏杰 审稿）

Recent Researches on the Transmutation of the Elements*

E. Rutherford

Editor's Note

Ernest Rutherford here offers a snapshot of the rapidly advancing field of nuclear transmutation. It had been established that many atoms can be changed from one element to another by adding or subtracting any of the various particles believed to inhabit the nucleus. But so far no such reactions had been observed with heavy elements such as thallium, lead, bismuth or uranium. This might change, Rutherford says, as physicists around the world—notably Ernest Lawrence at the University of California—were developing accelerating devices of considerably higher energy. None, however, compared to the energies of cosmic rays, with which several experiments had recently found tentative evidence for a positively charged electron—later identified as the positron.

IT is now well established that the change of one atom into another can only be effected by the addition or subtraction of one of the constituent particles of the atomic nucleus, for example, an electron, proton, neutron or α-particle. Such a transformation was first accomplished in 1919 for the element nitrogen by bombarding it with swift α-particles from radioactive substances. About one α-particle in 100,000 comes so close to the nucleus that it enters and is captured by it. This violent disturbance results in the expulsion of a proton with high speed, and the formation of a new nucleus of mass 17. A number of light elements can be transformed by α-particle bombardment in a similar way, and in most cases a proton is ejected.

A new and strange type of transformation was discovered last year by Chadwick: when α-particles bombard the metal beryllium, uncharged particles of mass 1 called neutrons are expelled. These neutrons, which have remarkable powers of penetrating matter, are themselves very efficient agents for the transformation of atoms. Feather has shown that both nitrogen and oxygen are transformed by the capture of neutrons, with the expulsion of a fast α-particle. The types of transformations produced by the neutron are thus very different from those observed with the α-particle. The capture of an α-particle in general leads to the building up of a new nucleus three units heavier than before, while the capture of a neutron leads to the formation of a nucleus three units lighter.

During the past year, Cockcroft and Walton at Cambridge made the important discovery that comparatively low-speed protons are very effective in causing the transformation of a

* Substance of the Friday evening discourse delivered at the Royal Institution on March 10.

关于元素嬗变的最新研究*

欧内斯特·卢瑟福

编者按

欧内斯特·卢瑟福在这篇文章中发表了对快速发展的核嬗变领域的简评。已经确认许多原子都可以通过增加或减少核中的任意粒子使自己从一种元素转变成另一种元素。但迄今为止人们在像铊、铅、铋和铀这样的重元素中还没有观察到这种反应。卢瑟福说：这个结论可能会有所改变，因为全世界的物理学家，尤其是美国加州大学的欧内斯特·劳伦斯，正在研制能量更高的加速器。不过所得到的能量无法和宇宙射线的能量相比，最近一些与宇宙射线相关的实验显示可能存在一种带正电的电子——后来被人们确定为正电子。

目前我们完全可以认为：仅仅通过增加或减少原子核的一个组成粒子如电子、质子、中子或 α 粒子就可以使一种原子变为另一种原子。这种嬗变是 1919 年用放射性物质发射的快速 α 粒子轰击氮元素时首次完成的。有约十万分之一的 α 粒子非常接近原子核以至于进入其中而被其俘获。这种强力扰动除了导致放出高速的质子，还形成了质量为 17 的新原子核。用类似的方法以 α 粒子轰击可以嬗变一些轻元素，而且在大多数情况下，会放出一个质子。

去年，查德威克发现了一类新奇的嬗变：当 α 粒子轰击金属铍时，放射出质量为 1、被称为中子的不带电的粒子。中子对于物质有极强的穿透性，是产生原子嬗变非常有效的工具。费瑟指出，氮和氧都可以通过中子俘获发生嬗变，并放出一个快速 α 粒子。由中子引发的这类嬗变与那些由 α 粒子引发的嬗变有很大的差别。通常，α 粒子俘获生成的新原子核比之前的质量重三个原子质量单位，而中子俘获产生的原子核要比之前轻三个原子质量单位。

去年，剑桥大学的考克饶夫和瓦耳顿作出了重要的发现，相对低速的质子在引发许多元素的嬗变中都是非常有效的。质子在氢原子放电的过程中大量产生，随后

---

* 本文取自作者于 3 月 10 日在英国皇家研究院所作的周五晚间演讲。

number of elements. The protons are generated in large numbers by an electric discharge through hydrogen, and then speeded up by passing through an evacuated space to which a high potential of the order of 600,000 volts is applied. Under these conditions the protons acquire high speeds comparable with that of the α-particle from radium. When a stream of these swift protons corresponding to a micro-ampere falls on the element lithium, a large number of α-particles are emitted of energy comparable with that of the swiftest α-particle from radium. It seems that about one in 100 million of the protons enters a lithium nucleus of mass 7, and the resulting nucleus of mass 8 splits up into two α-particles, each of mass 4. Cockcroft and Walton have later found that the α-particles emitted consist of two groups differing widely in speed.

This transformation of lithium can be produced at surprisingly low voltages. With strong proton streams, the emission of α-particles can be observed for 30,000 volts; the number of particles increases rapidly with the voltage, and the variation has been examined by different observers over a very wide range, from 30,000 to 1.5 million volts.

Protons are also remarkably effective in disintegrating the light element boron, and again α-particles are emitted. It is possible in this case that the boron nucleus of mass 11, after capturing a proton, breaks up into three α-particles. The radiation observed is complex, and has not yet been analysed in detail. A number of other elements have been found to be transformed, apparently in all cases with the emission of α-particles.

In a special form of accelerating tube devised by Oliphant in the Cavendish Laboratory, a narrow, intense proton stream can be generated at voltages up to 200,000 volts. The protons, after being bent by a magnetic field, bombard a target of about one square centimetre in area. By special arrangements, it has been found possible to obtain in the detecting chamber at least a thousand times the number of particles observed by Cockcroft and Walton at the same voltage. By this method it is easy to observe the particles from very thin films of lithium and boron at comparatively low voltages, while the variation of number with voltage has been measured. For example, a number of α-particles are emitted from lithium with voltages so low as 30,000 volts. α-Particles from boron have been observed at 60,000 volts, but the number increases much more rapidly with voltage than in the case of lithium.

Special experiments have been made to test by this sensitive method whether the heavy elements thallium, lead, bismuth and uranium show any evidences of transformation for 200,000 volt protons, but no sign of emission of α-particles has been observed for these elements. At first, marked effects were observed, but these were ultimately traced to a minute contamination by boron, probably originating in the discharge tube. It seems not unlikely that the effect observed for uranium and lead in the original experiments of Cockcroft and Walton may have been due to an unsuspected contamination by a minute trace of the very active element boron.

在加有600,000伏高电压的真空中被加速。在这样的条件下，质子获得了与从镭中放射出的α粒子相当的高速度。当这些相应于微安级电流的快速质子束流轰击元素锂时，发射出大量的α粒子，其能量与从镭中放射出的最快速的α粒子的能量相当。大约有亿分之一的质子进入质量为7的锂原子核，产生质量为8的原子核，该原子核又分裂为两个质量为4的α粒子。考克饶夫和瓦耳顿后来发现，所发射的α粒子由速度相差很大的两群组成。

锂的嬗变可以在特别低的电压下产生。用强的质子流，可以观测到在30,000伏时放射出α粒子；粒子数目随着电压的加大而快速增加，在从30,000伏到1.5百万伏这个非常宽的范围内，不同的观测者都检测到了这种变化。

质子在蜕变轻元素硼时也是非常有效的，且会再次放出α粒子。在这种情况下，质量为11的硼原子核在俘获一个质子后破裂为三个α粒子是可能的。所观测到的辐射很复杂，详细的分析还未完成。许多其他的元素也已被发现可以发生嬗变，当然，在所有的情况下都会放出α粒子。

在由卡文迪什实验室的奥利芬特发明的特殊形式的加速管中，细而强的质子束流可以在高至200,000伏的电压下产生。经过磁场偏转的质子流轰击面积约为1平方厘米的靶。通过特殊的装置，在检测室中可以得到的粒子数比考克饶夫和瓦耳顿在相同电压下观测到的粒子数至少大几千倍。用这种方法很容易观测到在相当低电压下从非常薄的锂和硼薄膜中放射出的粒子，还测量了粒子数随电压的变化。例如，在电压低至30,000伏时从锂中可以放出许多α粒子。在电压为60,000伏时观测到了从硼放出的α粒子，但与锂的情况相比，α粒子的数量随电压增加得更快。

通过这种灵敏的方法人们已经进行了特殊的实验用以检验重元素铊、铅、铋和铀对200,000伏的质子是否显示任何嬗变的迹象，但并未观测到这些元素有发射α粒子的迹象。最初，观测到了明显的效应，但最终追查到这些效应是由硼的微量污染引起的，它可能来自于放电管。这似乎有可能就是考克饶夫和瓦耳顿在最初的铀和铅的实验中观测到的效应，它们可能是由微量非常活泼的元素硼的未知污染产生的。

During the last few years, much energy has been devoted throughout the world to developing methods of obtaining streams of very swift charged particles with which to bombard matter and effect its transmutation. In the apparatus of Cockcroft and Walton at Cambridge already referred to, a steady potential of 800,000 volts can be reached. A new and simple type of electrostatic generator has been designed by Van der Graaf and Atta in the Massachusetts Institute of Technology, with which they have obtained a steady potential of 1.5 million volts, and a larger apparatus is under construction with which they hope to obtain a potential of 15 million volts to apply to a large vacuum tube. Brasch and Lange have applied high momentary voltages to a discharge tube by using an impulse generator.

A new and ingenious method of multiple acceleration has been devised by Lawrence of the University of California with which he has already obtained protons of energy 1.5 million volts by using a potential less than 10,000 volts. The transformation of lithium has been examined at this high energy using a proton current of about a thousandth of a micro-ampere. It is hoped to develop this method so as to obtain protons of energy as high as 10 million volts or more.

Even if these new projects prove successful, the speeds of particles produced by their aid are much smaller than those observed for the very penetrating radiation in our atmosphere, where electrons and protons of energy from 200 million to 2,000 million volts are present. From the experiments of Anderson in Pasadena and Blackett and Occhialini in Cambridge, it seems certain that these very swift particles are very efficient in causing the transformation of nuclei, probably in novel ways. Strong evidence has been obtained of the production of a new type of positively charged particle which has a mass small compared with that of the proton. This may prove to be the positive electron, the counterpart of the well-known negative electron of light mass.

(**131**, 388-389; 1933)

在最近几年间，全世界都在投入大量精力用以发展获得快速带电粒子流的方法，以便利用这些带电粒子来轰击物质并使其发生嬗变。已经提到剑桥大学的考克饶夫和瓦耳顿的装置可以达到 800,000 伏的稳定电压。麻省理工学院的范德格拉夫和阿塔设计出了一台简单的新型静电加速器，他们用此得到了 1.5 百万伏的稳定电压。他们还在构建更大的装置，以期获得 15 百万伏的电压并将之应用到大的真空钢筒内。布拉什和兰格已将瞬时高电压通过脉冲发生器加到放电管上。

加州大学的劳伦斯发明了一种新的直接多次加速的巧妙方法，他用小于 10,000 伏的电压得到的质子能量为 1.5 百万伏。在使用约千分之一微安的质子流时，锂在这种高能下发生的嬗变得到了检验。这类方法被寄予厚望，因为它可能获得高达 10 百万伏甚至更高能量的质子。

即使这些新方案被证明是成功的，由它们产生的粒子速度仍远小于我们在大气层中观测到的穿透力很强的辐射，大气中电子和质子的能量在 200 百万伏到 2,000 百万伏之间。帕萨迪纳的安德森及剑桥大学的布莱克特和奥基亚利尼的实验似乎表明，这些非常快的粒子在引起原子核的嬗变中是很有效的，也许是新方式的嬗变。强有力的证据表明产生了一种比质子质量小且带正电的新型粒子。它可能被证明是正电子，即为人们所熟知的轻质量带负电荷的电子的反粒子。

（沈乃澂 翻译；张焕乔 审稿）

Light and Life*

N. Bohr

Editor's Note

This is an address given by Niels Bohr at the International Congress on Light Therapy in Copenhagen. Bohr had been invited to discuss the beneficial effects of light in curing diseases, but he spoke instead on the potential implications of the new quantum understanding of light for the science of living organisms. Despite the inadequacy of the wave picture for detailing the behaviour of light quanta, says Bohr, there could be no question of replacing it with a purely quantum-mechanical "particle" picture. He points out that any attempt to measure the trajectories of photons precisely inevitably destroys the phenomenon of wave interference. Such problems, he suggests, compel abandonment of a complete causal description of light phenomena.

In the second part of his address, Bohr argues that the implications of quantum theory stretch well beyond atomic physics. The assimilation of carbon by plants surely involved the quantum nature of interactions between light and matter, and the ability of the human eye to detect only a few photons suggests that its design probes the quantum limits of optics. Bohr suggests an analogy between uncertainty in the physical and biological worlds. Attempts to study organisms closely may ultimately interfere with their vital organs, just as physicists must disturb particles in order to study them. Perhaps life and consciousness may never be explained, but must be accepted a priori much as the quantum action is axiomatic in physics.

Part I

AS a physicist whose studies are limited to the properties of inanimate bodies, it is not without hesitation that I have accepted the kind invitation to address this assembly of scientific men met together to forward our knowledge of the beneficial effects of light in the cure of diseases. Unable as I am to contribute to this beautiful branch of science that is so important for the welfare of mankind, I could at most comment on the purely inorganic light phenomena which have exerted a special attraction for physicists throughout the ages, not least owing to the fact that light is our principal tool of observation. I have thought, however, that on this occasion it might perhaps be of interest, in connexion with such comments, to enter on the problem of what significance the results reached in the limited domain of physics may have for our views on the position of living organisms in the realm of natural science.

---

* Address delivered at the opening meeting of the International Congress on Light Therapy, Copenhagen, on August 15, 1932. The present article, conforming with the Danish version (*Naturens Verden*, **17**, 49), differs from that published in the Congress report only by some formal alterations.

光与生命*

尼尔斯·玻尔

编者按

这篇文章出自尼尔斯·玻尔在哥本哈根举行的国际光疗大会上的讲话。大会本来想邀请玻尔向大家介绍一下光对于治疗疾病的好处，但实际上玻尔讲的是光在新量子理论中的概念对生命科学的潜在意义。玻尔说：虽然描述光量子行为的波动理论还不够完善，但仍可能把光看作是量子力学中的一个“粒子”。他指出：任何想精确测量光子轨迹的尝试都必然会破坏波的干涉现象。他说这一问题使人们不得不放弃对光现象的纯因果描述。

玻尔在第二部分中指出，量子理论的拓展远远超过了原子物理的范畴。植物的碳同化作用肯定包含着光与物质之间发生相互作用的量子本质，人眼只能看到几个光子说明它的结构反映了光学上的量子极限。玻尔将物理学和生物学中的不确定性进行了类比。对生物体过于精密的研究可能会对它们的重要器官产生干扰，就像物理学家为了研究粒子必须干扰它们一样。也许生命和意识是永远不能被解释清楚的，只能把它们当成不言自明的公理，物理中的量子现象也是如此。

第一部分

作为一个研究领域仅限于非生命物质的物理学家，在接受盛情邀请来到这个科学界人士云集的大会介绍光对于治疗疾病的有利影响时，我还是有点犹豫的。因为我无法致力于这个对造福人类有重要贡献的精彩科学分支，我至多只能对纯粹的无机光学现象发表评述，光现象对古往今来的物理学工作者来说一直有着特殊的吸引力，一个重要的原因是光是我们的主要观测手段。然而，我认为，在这样的场合谈一谈物理学范畴内的研究成果对我们认识生物体在自然科学领域中的地位具有什么样的意义也许是一件有趣的事情。

* 这是 1932 年 8 月 15 日玻尔在于哥本哈根召开的国际光疗大会开幕式上发表的讲话。本文与丹麦版（《自然世界》，第 17 卷，第 49 页）是一致的，但与公开出版的会议报告不完全一样，区别在于把一些口语化的文字改成了比较正式的形式。

Notwithstanding the subtle character of the riddles of life, this problem has presented itself at every stage of science, since any scientific explanation necessarily must consist in reducing the description of more complex phenomena to that of simpler ones. At the moment, however, the unsuspected discovery of an essential limitation of the mechanical description of natural phenomena, revealed by the recent development of the atomic theory, has lent new interest to the old problem. This limitation was, in fact, first recognised through a thorough study of the interaction between light and material bodies, which disclosed features that cannot be brought into conformity with the demands hitherto made to a physical explanation. As I shall endeavour to show, the efforts of physicists to master this situation resemble in some way the attitude which biologists more or less intuitively have taken towards the aspects of life. Still, I wish to stress at once that it is only in this formal respect that light, which is perhaps the least complex of all physical phenomena, exhibits an analogy to life, the diversity of which is far beyond the grasp of scientific analysis.

From a physical point of view, light may be defined as the transmission of energy between material bodies at a distance. As is well known, such an energy transfer finds a simple explanation in the electromagnetic theory, which may be regarded as a direct extension of classical mechanics compromising between action at a distance and contact forces. According to this theory, light is described as coupled electric and magnetic oscillations which differ from the ordinary electromagnetic waves used in radio transmission only by their greater frequency of vibration and smaller wave-length. In fact, the practically rectilinear propagation of light, on which rests our location of bodies by direct vision or by suitable optical instruments, depends entirely on the smallness of the wave-length compared with the dimensions of the bodies concerned, and of the instruments.

The idea of the wave nature of light, however, not only forms the basis for our explanation of the colour phenomena, which in spectroscopy have yielded such important information of the inner constitution of matter, but is also of essential importance for every detailed analysis of optical phenomena. As a typical example, I need only mention the interference patterns which appear when light from one source can travel to a screen along two different paths. In such a case, we find that the effects which would be produced by the separate light beams are strengthened at those points on the screen where the phases of the two wave trains coincide, that is, where the electric and magnetic oscillations in the two beams have the same directions, while the effects are weakened and may even disappear at points where these oscillations have opposite directions, and where the two wave trains are said to be out of phase with one another. These interference patterns have made possible such a thorough test of the wave nature of the propagation of light, that this conception can no longer be considered as a hypothesis in the usual sense of this word, but may rather be regarded as an indispensable element in the description of the phenomena observed.

As is well known, the problem of the nature of light has, nevertheless, been subjected to renewed discussion in recent years, as a result of the discovery of a peculiar atomistic

尽管生物体中的秘密难以捉摸，但这个问题在科学发展的每一个阶段都会被提出来，因为科学上的探索就是要把对复杂现象的描述还原为简单的原理。然而现在，原子理论的最新进展表明，人们对自然现象的力学描述在本质上存在着不可抹杀的局限性，这一意想不到的发现重新燃起了大家对这个古老问题的兴趣。事实上，人们最先是在对光与物体之间相互作用的全面研究中认识到这一局限性的，该现象所揭示的特性无法与迄今为止已形成的物理学理论相统一。正如我将努力向大家说明的那样，物理学家们为解决这一问题所做出的努力，在某种程度上类似于生物学家对待生命现象时多多少少有点依据直觉的方式。但我马上要说明的是，光这种在物理学中可能是最不具有复杂性的现象，只是在这一点上与多样性远远超出科学分析范畴的生命现象相似。

从物理学的观点看，光可以被定义为相隔一定距离的物体之间的能量传播。大家都知道，利用电磁学理论就可以简单地解释这样的能量传递，这也许可以被看作是对包括超距作用和接触力在内的经典力学的直接拓展。根据这一理论，光可以被描述为耦合的电磁振荡，它们与无线电广播中使用的普通电磁波一样，只不过具有较高的频率和较短的波长。事实上，我们之所以可以通过直接观察或适当的光学仪器来判断物体的位置是因为光具有直线传播的特性，而这种特性完全依赖于光的波长比相关物体和观测工具的尺度要小很多。

然而，光具有波动性的思想不仅能为我们提供解释颜色现象的依据，在光谱学中颜色能提供有关物质内部结构的重要信息，而且它对于每一种光学现象的详细分析也是必不可少的。至于代表性的例子，我只需提及干涉条纹就可以了，当从一个光源发出的光沿着两条不同的路径投射到一个屏幕上时，就会出现干涉条纹。在这种情况下，我们发现在屏幕上两列波位相一致时，也就是说两束光的电磁振动方向相同时，分开的两个光束将在这些点上产生增强，而在电磁振荡方向相反也就是说两列波彼此异相时，光强将减弱甚至有可能消失。这些干涉条纹使人们能够精确地检测光传播的波动性，因而波动概念不再是一种通常意义上的假设，而有可能被认为是描述所观察到的现象中不可或缺的要素。

尽管如此，大家都知道，近几年人们又开始讨论光的本质这个问题了，这是因为人们发现在能量传输中有一个特殊的原子性质，这个特殊性质无法用电磁理论来

feature in the energy transmission which is quite unintelligible from the point of view of the electromagnetic theory. It has turned out, in fact, that all effects of light may be traced down to individual processes, in each of which a so-called light quantum is exchanged, the energy of which is equal to the product of the frequency of the electromagnetic oscillations and the universal quantum of action, or Planck's constant. The striking contrast between this atomicity of the light phenomena and the continuity of the energy transfer according to the electromagnetic theory places us before a dilemma of a character hitherto unknown in physics. For, in spite of the obvious insufficiency of the wave picture, there can be no question of replacing it by any other picture of light propagation depending on ordinary mechanical ideas.

Especially, it should be emphasised that the introduction of the concept of light quanta in no way means a return to the old idea of material particles with well-defined paths as the carriers of the light energy. In fact, it is characteristic of all the phenomena of light, in the description of which the wave picture plays an essential rôle, that any attempt to trace the paths of the individual light quanta would disturb the very phenomenon under investigation; just as an interference pattern would completely disappear, if, in order to make sure that the light energy travelled only along one of the two paths between the source and the screen, we should introduce a non-transparent body into one of the paths. The spatial continuity of light propagation, on one hand, and the atomicity of the light effects, on the other hand, must, therefore, be considered as complementary aspects of one reality, in the sense that each expresses an important feature of the phenomena of light, which, although irreconcilable from a mechanical point of view, can never be in direct contradiction, since a closer analysis of one or the other feature in mechanical terms would demand mutually exclusive experimental arrangements.

At the same time, this very situation forces us to renounce a complete causal description of the phenomena of light and to be content with probability calculations, based on the fact that the electromagnetic description of energy transfer by light remains valid in a statistical sense. Such calculations form a typical application of the so-called correspondence argument, which expresses our endeavour, by means of a suitably limited use of mechanical and electromagnetic concepts, to obtain a statistical description of the atomic phenomena that appears as a rational generalisation of the classical physical theories, in spite of the fact that the quantum of action from their point of view must be considered as an irrationality.

At first sight, this situation might appear very deplorable; but, as has often happened in the history of science, when new discoveries have revealed an essential limitation of ideas the universal applicability of which had never been disputed, we have been rewarded by getting a wider view and a greater power of correlating phenomena which before might even have appeared as contradictory. Thus, the strange limitation of classical mechanics, symbolised by the quantum of action, has given us a clue to an understanding of the peculiar stability of atoms which forms a basic assumption in the mechanical description of any natural phenomenon. The recognition that the indivisibility of atoms cannot be

解释。事实上，所有的光效应都可以被看作是一个单独的过程，在每一个这样的过程中，发生了所谓的一个光量子交换，光量子的能量等于电磁振荡频率与基本作用量子（或称普朗克常数）的乘积。这个光的原子性与电磁理论所要求的能量传输连续性之间的明显对立使我们陷入了一个有关迄今为止在物理学中仍不清楚的性质的困境。因为尽管波动理论还明显不够完善，但是毫无疑问，我们仍可以用在经典力学思想框架下的任何其他的光传播理论来取代它。

应该特别强调的是，引入光量子这个概念决不意味着对旧思想的回归，即认为光能的载体是有明确路径的实物粒子。实际上，任何尝试追踪单个光量子路径的行为都会干扰被研究的对象，这是所有光学现象共有的特征，波动性在描述光学现象时是不可缺少的；正如如果在光源和屏幕之间有两条路径，我们为了确保光只沿着其中的一条路径传播，就应该在另一条路径上引入不透光的物质，这样干涉条纹就会完全消失。因此，一方面是光传播的空间连续性，另一方面是光效应的原子性，它们应该被看作是一件事情相互补充的两个方面，从这个意义上说，每一个方面都代表了光现象的一个重要特征，尽管从力学角度上看它们是不可调和的，但也绝对不是完全对立的，因为要用力学方法进一步分析其中一个特征或另一个特征就需要排斥对方的实验装置。

然而，这种情况使我们不得不放弃对光现象的完整因果描述而安于概率计算，原因是用电磁理论描述光的能量传输在统计学意义上仍然有效。这种计算是所谓一致论的典型应用，它显示出我们在努力通过对力学和电磁学概念的有限运用获得对原子现象的统计学描述，这似乎是经典物理理论的一个理性概括，虽然经典物理学认为必须把作用量子看成是非理性的。

这种情形也许初看起来非常令人遗憾；但正如科学史上经常发生的，当我们从新发现中察觉到一个普遍适用性从未被人置疑过的思想存在本质上的局限时，我们就已经从相关现象中得到了更广的视野和更强的能力，而这种现象或许在以前就曾作为对立的情况出现过。因此，作用量子象征着经典力学出现了前所未有的局限性，这种局限性给我们提供了理解原子具有特殊稳定性的线索，而原子所具有特殊稳定性是用经典力学理论描述所有自然现象的基本假设。经典力学理论无法理解原子是

understood in mechanical terms has always characterised the atomic theory, to be sure; and this fact is not essentially altered, although the development of physics has replaced the indivisible atoms by the elementary electric particles, electrons and atomic nuclei, of which the atoms of the elements as well as the molecules of the chemical compounds are now supposed to consist.

However, it is not to the question of the intrinsic stability of these elementary particles that I am here referring, but to the problem of the required stability of the structures composed of them. As a matter of fact, the very possibility of a continuous transfer of energy, which marks both the classical mechanics and the electromagnetic theory, cannot be reconciled with an explanation of the characteristic properties of the elements and the compounds. Indeed, the classical theories do not even allow us to explain the existence of rigid bodies, on which all measurements made for the purpose of ordering phenomena in space and time ultimately rest. However, in connexion with the discovery of the quantum of action, we have learned that every change in the energy of an atom or a molecule must be considered as an individual process, in which the atom goes over from one of its so-called stationary states to another. Moreover, since just one light quantum appears or disappears in a transition process by which light is emitted or absorbed by an atom, we are able by means of spectroscopic observations to measure directly the energy of each of these stationary states. The information thus derived has been most instructively corroborated also by the study of the energy exchanges which take place in atomic collisions and in chemical reactions.

In recent years, a remarkable development of the atomic theory has taken place, which has given us such adequate methods of computing the energy values for the stationary states, and also the probabilities of the transition processes, that our account, on the lines of the correspondence argument, of the properties of atoms as regards completeness and self-consistency scarcely falls short of the explanation of astronomical observations offered by Newtonian mechanics. Although the rational treatment of the problems of atomic mechanics was possible only after the introduction of new symbolic artifices, the lesson taught us by the analysis of the phenomena of light is still of decisive importance for our estimation of this development. Thus, an unambiguous use of the concept of a stationary state is complementary to a mechanical analysis of intra-atomic motions; in a similar way the idea of light quanta is complementary to the electromagnetic theory of radiation. Indeed, any attempt to trace the detailed course of the transition process would involve an uncontrollable exchange of energy between the atom and the measuring instruments, which would completely disturb the very energy transfer we set out to investigate.

A causal description in the classical sense is possible only in such cases where the action involved is large compared with the quantum of action, and where, therefore, a subdivision of the phenomena is possible without disturbing them essentially. If this condition is not fulfilled, we cannot disregard the interaction between the measuring instruments and the object under investigation, and we must especially take into consideration that the various measurements required for a complete mechanical description may only be made

不可分割的，而毫无疑问通常突显了原子理论的特性；这一点至今没有本质上的更改，尽管后来的物理学已经用基本带电粒子替代了不可分割的原子，而且现在认为各种元素的原子以及化合物中的分子都是由电子和原子核这些基本粒子构成的。

然而，我在这里并不想谈及这些基本粒子的固有稳定性问题，而是要探究构成这些结构所必需的稳定性。事实上，以能量存在持续传递可能性为特征的经典力学和电磁理论是不能与对元素和化合物性质的解释达成一致的。的确，我们甚至不能用经典理论来解释刚体的存在，而为了测量空间和时间中的有序现象必须有赖于刚体。但是，由于作用量子的发现，我们了解到原子或分子能量的每一次改变都必须被看作是一个单独的过程，在这样的过程中原子从它的一个定态跃迁到了另一个定态。此外，因为在一个原子发射光或吸收光的跃迁过程中只会有一个光量子产生或消失，所以我们能够通过光谱观测直接测量每一个定态的能量。人们通过对原子碰撞和化学反应中能量交换的研究已经非常有效地确证了由上述方法得到的数据。

近些年来，原子理论取得了长足的进步，它足以为我们提供计算定态能值和跃迁概率的方法，因而我们根据一致论得到的有关原子性质的结论在完整性和自洽性方面几乎与用牛顿力学解释天文观测结果的水平不相上下。虽然只有在引入新的符号处理技巧之后才有可能合理地解决原子力学的困境，但在分析光学现象时的经验对于我们评估原子力学的发展仍然有非常重要的意义。因此，确立定态概念是对用力学法分析原子内部运动的补充；同样，引入光量子概念是对电磁辐射理论的补充。任何追踪跃迁详细过程的尝试都会使原子和测量仪器之间发生不可控的能量交换，会彻底干扰我们要研究的能量转移过程。

经典意义中的因果描述只有在相互作用远大于作用量子时才可能成立，因此，在基本不干扰它们的情况下是有可能对这些现象进行细分的。而在不满足上述条件的情况下，我们就不能忽略测量仪器与被研究物体之间的相互作用，为了得到完整的力学描述，我们需要考虑使用不同的测量方法，而这些测量只有通过相互排斥的实验装置才能得到。为了充分了解用力学分析原子现象的基本局限性，我们必须清

with mutually exclusive experimental arrangements. In order fully to understand this fundamental limitation of the mechanical analysis of atomic phenomena, one must realise clearly, further, that in a physical measurement it is never possible to take the interaction between object and measuring instruments directly into account. For the instruments cannot be included in the investigation while they are serving as means of observation. As the concept of general relativity expresses the essential dependence of physical phenomena on the frame of reference used for their co-ordination in space and time, so does the notion of complementarity serve to symbolise the fundamental limitation, met with in atomic physics, of our ingrained idea of phenomena as existing independently of the means by which they are observed.

(**131**, 421-423; 1933)

Part II

This revision of the foundations of mechanics, extending to the very question of what may be meant by a physical explanation, has not only been essential, however, for the elucidation of the situation in atomic theory, but has also created a new background for the discussion of the relation of physics to the problems of biology. This must certainly not be taken to mean that in actual atomic phenomena we meet with features which show a closer resemblance to the properties of living organisms than do ordinary physical effects. At first sight, the essentially statistical character of atomic mechanics might even seem difficult to reconcile with an explanation of the marvellously refined organisation, which every living being possesses, and which permits it to implant all the characteristics of its species into a minute germ cell.

We must not forget, however, that the regularities peculiar to atomic processes, which are foreign to causal mechanics and find their place only within the complementary mode of description, are at least as important for the account of the behaviour of living organisms as for the explanation of the specific properties of inorganic matter. Thus, in the carbon assimilation of plants, on which depends largely also the nourishment of animals, we are dealing with a phenomenon for the understanding of which the individuality of photochemical processes must undoubtedly be taken into consideration. Likewise, the peculiar stability of atomic structures is clearly exhibited in the characteristic properties of such highly complicated chemical compounds as chlorophyll or haemoglobin, which play fundamental rôles in plant assimilation and animal respiration.

However, analogies from chemical experience will not, of course, any more than the ancient comparison of life with fire, give a better explanation of living organisms than will the resemblance, often mentioned, between living organisms and such purely mechanical contrivances as clockworks. An understanding of the essential characteristics of living beings must be sought, no doubt, in their peculiar organisation, in which features that may be analysed by the usual mechanics are interwoven with typically atomistic traits in a

楚地、深刻地认识到，在物理测量中，绝不可能直接将物体和测量仪器的相互作用考虑在内。因为当仪器被用作观察手段的时候，就不能把它们包括在研究对象之中。正如广义相对论所表述的观点，物理学现象在本质上依赖于用于描述它们在空间和时间中坐标的参考系，因而原子物理学中的互补概念就用来表示基本的局限性，即我们根深蒂固的理念是现象不依赖于观测手段而独立存在。

（姜薇 翻译；张泽渤 审稿）

第二部分

人们对力学基本理论的修正引申出这样一个问题，即一个物理学解释可能会意味着什么，然而这不仅仅是解释原子理论中的现象所必需的，也为讨论物理学与生物学之间的相关性奠定了一个新的基础。当然，这绝不意味着我们在原子现象中发现的特征与普通的物理学效应相比更接近于生物体的性质。初看起来，原子力学所具有的基本统计学特点似乎难以与人们对所有生物体都具有的超精细组织的解释相匹配，这种超精细组织可以把本物种的所有特征植入一个微小的生殖细胞中。

然而，我们绝不能忘记：原子过程所特有的规律性不能用因果机制表示，只能借助互补模式描述，这种规律性至少在对生物体行为的描述和对无机物特性的解释上具有同等的重要性。因此，在植物的碳同化、同时也是动物养分的主要来源的生成过程中，我们正在研究的是一个现象，为了理解这个现象，必须确确实实地去考虑参与光化学反应的个体特性。另外，像在植物同化与动物呼吸作用中扮演重要角色的叶绿素和血红蛋白，它们所具有的复杂程度如此高的化合物的特性，也明显地表现出原子结构独特的稳定性。

然而，由化学知识类推解释生命体，诚然会比在古代将生命比作火更贴切，却不会强于人们常说的生命体与纯机械制造物如钟表发条之间的类比。毫无疑问，必须在特定的生物组织中探寻对生命本质特征的理解，在研究这些组织的特征时，将用普通力学进行分析并且以与无机物不同的方式交织着一些典型的原子特性。

manner having no counterpart in inorganic matter.

An instructive illustration of the refinement to which this organisation is developed has been obtained through the study of the construction and function of the eye, for which the simplicity of the phenomena of light has again been most helpful. I need not go into details here, but shall just recall how ophthalmology has revealed to us the ideal properties of the human eye as an optical instrument. Indeed, the dimensions of the interference patterns, which on account of the wave nature of light set the limit for the image formation in the eye, practically coincide with the size of such partitions of the retina which have separate nervous connexion with the brain. Moreover, since the absorption of a few light quanta, or perhaps of only a single quantum, on such a retinal partition is sufficient to produce a sight impression, the sensitiveness of the eye may even be said to have reached the limit imposed by the atomic character of the light effects. In both respects, the efficiency of the eye is the same as that of a good telescope or microscope, connected with a suitable amplifier so as to make the individual processes observable. It is true that it is possible by such instruments essentially to increase our powers of observation, but, owing to the very limits imposed by the properties of light, no instrument is imaginable which is more efficient for its purpose than the eye. Now, this ideal refinement of the eye, fully recognised only through the recent development of physics, suggests that other organs also, whether they serve for the reception of information from the surroundings or for the reaction to sense impressions, will exhibit a similar adaptation to their purpose, and that also in these cases the feature of individuality symbolised by the quantum of action, together with some amplifying mechanism, is of decisive importance. That it has not yet been possible to trace the limit in organs other than the eye, depends solely upon the simplicity of light as compared with other physical phenomena.

The recognition of the essential importance of fundamentally atomistic features in the functions of living organisms is by no means sufficient, however, for a comprehensive explanation of biological phenomena. The question at issue, therefore, is whether some fundamental traits are still missing in the analysis of natural phenomena, before we can reach an understanding of life on the basis of physical experience. Quite apart from the practically inexhaustible abundance of biological phenomena, an answer to this question can scarcely be given without an examination of what we may understand by a physical explanation, still more penetrating than that to which the discovery of the quantum of action has already forced us. On one hand, the wonderful features which are constantly revealed in physiological investigations and differ so strikingly from what is known of inorganic matter, have led many biologists to doubt that a real understanding of the nature of life is possible on a purely physical basis. On the other hand, this view, often known as vitalism, scarcely finds its proper expression in the old supposition that a peculiar vital force, quite unknown to physics, governs all organic life. I think we all agree with Newton that the real basis of science is the conviction that Nature under the same conditions will always exhibit the same regularities. Therefore, if we were able to push the analysis of the mechanism of living organisms as far as that of atomic phenomena, we should scarcely expect to find any features differing from the properties of inorganic matter.

通过研究眼睛的构造与功能，我们得到了有关这个已经发展得十分精细的组织的有用信息，也再一次地显示光现象的简单特性是非常有用的。我无需在这里详述细节，但想回顾一下眼科学是如何向我们揭示作为一种光学装置的人眼的理想特征的。实际上，由于光的波动性，干涉图像的尺寸规定了眼成像的限度，而干涉图像的尺寸与视网膜的分区相符，其中每个分区都有单独的神经与脑相连。此外，因为在这样的视网膜分区中，只要吸收几个光量子，或者也许只要吸收一个光量子就足以产生视感，有人甚至说眼睛的敏感程度已经达到了光效应的原子特性所设置的极限。在两个方面，眼睛的效能同那些连有合适放大器的高性能望远镜或显微镜一样，可以观测到单个过程。利用这样的仪器确实能从根本上提高我们的观察能力，但由于光的性质所设置的极限，在人们可以想象得出的仪器中，没有一种可以与眼睛的灵敏度相比。现在，眼睛这种只有借助物理学的最新发展才能被深刻认识的理想化结构也说明了其他的器官，不管它们的作用是接收来自周围环境的信息，还是对感觉印象做出反应，都同样会表现出与自身功能相适应的特点，在这类情况下，由作用量子所代表的个性特征以及一些放大机制起着关键性的作用。目前，我们尚无法知道除眼睛外其他器官的极限，而眼睛只取决于比其他物理现象简单得多的光现象。

然而，仅仅认识到运用原子论基本原理来说明生物体功能的必要性还不足以全面解释生物学现象。因此，我们要讨论的问题是，在我们能够用物理学观点解释生命现象之前，是否还缺少一些分析自然现象所需的基本特征。如果不去考证我们通过物理学解释而理解的内容，就很难得到这个问题的答案，更何况现实中生物现象的类别多得无法列举，这要比发现作用量子对我们的推动意义更深远。一方面，通过生理学研究不断揭示出的神奇特征与无机物的特征有着显著的差别，这使许多生物学者怀疑用纯物理学的方法是否可以真正理解生命的本质。另一方面，人们通常称为活力论的观点，不能正确地解释有一种特殊的生命力掌控着所有有机体的古老假设，在物理学中也不存在这样的力。我认为我们都赞同牛顿的观点，即科学的真正基础在于确信自然界在相同条件下会表现出同样的规律性。因此，如果我们能够把对生命体机理的分析推进到原子现象的层面，我们就不应该期待会发现任何有别于无机物的特点。

With this dilemma before us, we must keep in mind, however, that the conditions holding for biological and physical researches are not directly comparable, since the necessity of keeping the object of investigation alive imposes a restriction on the former, which finds no counterpart in the latter. Thus, we should doubtless kill an animal if we tried to carry the investigation of its organs so far that we could describe the rôle played by single atoms in vital functions. In every experiment on living organisms, there must remain an uncertainty as regards the physical conditions to which they are subjected, and the idea suggests itself that the minimal freedom we must allow the organism in this respect is just large enough to permit it, so to say, to hide its ultimate secrets from us. On this view, the existence of life must be considered as an elementary fact that cannot be explained, but must be taken as a starting point in biology, in a similar way as the quantum of action, which appears as an irrational element from the point of view of classical mechanical physics, taken together with the existence of the elementary particles, forms the foundation of atomic physics. The asserted impossibility of a physical or chemical explanation of the function peculiar to life would in this sense be analogous to the insufficiency of the mechanical analysis for the understanding of the stability of atoms.

In tracing this analogy further, however, we must not forget that the problems present essentially different aspects in physics and in biology. While in atomic physics we are primarily interested in the properties of matter in its simplest forms, the complexity of the material systems with which we are concerned in biology is of fundamental significance, since even the most primitive organisms contain a large number of atoms. It is true that the wide field of application of classical mechanics, including our account of the measuring instruments used in atomic physics, depends on the possibility of disregarding largely the complementarity, entailed by the quantum of action, in the description of bodies containing very many atoms. It is typical of biological researches, however, that the external conditions to which any separate atom is subjected can never be controlled in the same manner as in the fundamental experiments of atomic physics. In fact, we cannot even tell which atoms really belong to a living organism, since any vital function is accompanied by an exchange of material, whereby atoms are constantly taken up into and expelled from the organisation which constitutes the living being.

This fundamental difference between physical and biological investigations implies that no well-defined limit can be drawn for the applicability of physical ideas to the phenomena of life, which would correspond to the distinction between the field of causal mechanical description and the proper quantum phenomena in atomic mechanics. However, the limitation which this fact would seem to impose upon the analogy considered will depend essentially upon how we choose to use such words as physics and mechanics. On one hand, the question of the limitation of physics within biology would, of course, lose any meaning, if, in accordance with the original meaning of the word physics, we should understand by it any description of natural phenomena. On the other hand, such a term as atomic mechanics would be misleading, if, as in common language, we should apply the word mechanics only to denote an unambiguous causal description of the phenomena.

然而，当我们面临进退两难的困境时，我们必须记住，生物学和物理学的研究条件并不完全对应，因为前者要受到研究对象必须是活体的限制，而在后者中则不存在这样的限制。因此，如果我们试图对动物的器官展开研究以便描述单个原子在生命活动中的功能时，我们无疑会杀死一只动物。在对活体进行的每一次实验中，都肯定会存在因个体生理条件不同而导致的不确定性，这个观点本身说明，在这方面，即使我们允许生物体具有最低的自由度，也足以让生命体可以对我们隐藏其最根本的秘密。根据这个观点，我们应该把生命的存在看作是不能被解释的基本事实，而以此作为生物学研究的起点，这与作用量子类似，在经典机械物理学中，作用量子被看作是一种非理性的成分，而它与基本粒子一起构成了原子物理学的基础。从这个意义上说，宣称用物理或化学方法不可能解释生命体的功能就类似于用经典力学不能理解原子的稳定性。

然而，在进一步研究这个类比时，我们一定不要忘记物理学中的问题和生物学中的问题存在着本质上的不同。在原子物理学中，我们是从形式最简单的物质的性质入手的，而在生物学研究中具有根本性意义的是物质系统的复杂性，因为即便是最原始的生物也包含大量的原子。的确，经典力学的应用范围广泛，其中包括我们在研究原子物理时使用的测量仪器，要想在描述由大量原子组成的物体中应用经典力学就必须在很大程度上忽略伴随着作用量子的互补性。然而，在生物学研究中，人们通常不能像在原子物理学基础实验中那样控制单个原子所处的外部条件。实际上，我们甚至不知道哪些原子是真正属于生物体的，因为任何生命活动都伴随着物质的交换，原子不断地被构成这个生物体的组织吸入和排出。

物理学研究和生物学研究之间的这种本质不同暗示着人们无法明确地划分出在运用物理学原理解释生命现象时的界限，这也相当于因果力学描述的场和原子力学中固有量子现象之间的区别。然而，上述事实对这个类比的限制似乎从根本上取决于我们如何运用像物理学和力学这样的词。一方面，如果依照物理学这个词的原始意义，我们就应当把它理解为是对所有自然现象的描述，那么认为物理学在生物学中的运用受到限制就失去了意义。另一方面，如果在通用语言中我们用力学这个词只表示对现象的清晰的因果关系的描述，那么像原子力学这样的术语就会引起人们的误解。

I shall not here enter further into these purely logical points, but will only add that the essence of the analogy considered is the typical relation of complementarity existing between the subdivision required by a physical analysis and such characteristic biological phenomena as the self-preservation and the propagation of individuals. It is due to this situation, in fact, that the concept of purpose, which is foreign to mechanical analysis, finds a certain field of application in problems where regard must be taken of the nature of life. In this respect, the rôle which teleological arguments play in biology reminds one of the endeavours, formulated in the correspondence argument, to take the quantum of action into account in a rational manner in atomic physics.

In our discussion of the applicability of mechanical concepts in describing living organisms, we have considered these just as other material objects. I need scarcely emphasise, however, that this attitude, which is characteristic of physiological research, involves no disregard whatsoever of the psychological aspects of life. The recognition of the limitation of mechanical ideas in atomic physics would much rather seem suited to conciliate the apparently contrasting points of view which mark physiology and psychology. Indeed, the necessity of considering the interaction between the measuring instruments and the object under investigation in atomic mechanics corresponds closely to the peculiar difficulties, met with in psychological analyses, which arise from the fact that the mental content is invariably altered when the attention is concentrated on any single feature of it.

It will carry us too far from our subject to enlarge upon this analogy which, when due regard is taken to the special character of biological problems, offers a new starting point for an elucidation of the so-called psycho-physical parallelism. However, in this connexion, I should like to emphasise that the considerations referred to here differ entirely from all attempts at viewing new possibilities for a direct spiritual influence on material phenomena in the limitation set for the causal mode of description in the analysis of atomic phenomena. For example, when it has been suggested that the will might have as its field of activity the regulation of certain atomic processes within the organism, for which on the atomic theory only probability calculations may be set up, we are dealing with a view that is incompatible with the interpretation of the psycho-physical parallelism here indicated. Indeed, from our point of view, the feeling of the freedom of the will must be considered as a trait peculiar to conscious life, the material parallel of which must be sought in organic functions, which permit neither a causal mechanical description nor a physical investigation sufficiently thorough-going for a well-defined application of the statistical laws of atomic mechanics. Without entering into metaphysical speculations, I may perhaps add that any analysis of the very concept of an explanation would, naturally, begin and end with a renunciation as to explaining our own conscious activity.

In conclusion, I wish to emphasise that in none of my remarks have I intended to express any kind of scepticism as to the future development of physical and biological sciences. Such scepticism would, indeed, be far from the mind of a physicist at a time when the very recognition of the limited character of our most fundamental concepts has resulted

我不打算在这里进一步讨论这些纯逻辑学上的观点，但只想补充一点，即典型的互补性关系构成了这个类比的本质，这种关系存在于物理分析所需的细分与诸如自卫本能和个体繁殖这样的典型生命现象之间。事实上，正是由于这个原因，在力学分析中不曾有过的目的性概念，在必须考虑生命本质的领域中得到了应用。从这个方面来说，目的论在生物学中所扮演的角色使人回想起人们曾努力用对应的自变量将作用量子以一种理性模式引进到原子物理中。

在我们讨论用力学概念描述生物体时，我们对待它们和对待其他实物没有什么两样。然而，我无需强调这种生理学研究中特有的态度并没有任何对生命的心理问题的漠视。人们对力学概念在原子物理中所受限制的认识似乎更适合于理解生理学与心理学在观点上的明显对立。的确，在原子物理学中需要考虑测量仪器与研究对象之间的相互作用十分类似于在心理学分析中遇到的特定困难，该困难起源于当注意力集中在任何一个单一的功能上时，心理内容总在不断变化。

当适当考虑生物学的特性时，这个类比为阐明所谓的身心并行说提供了一个新的起点，这样说将使我们过于偏离我们要讨论的题目。然而，关于这一点，我想强调的是，这里谈到的观点完全不同于想借助在分析原子现象时因果描述受到的局限而极力推崇物质现象很可能会直接受到精神影响的看法。例如，当人们认为也许可以利用特定原子过程的规律来研究在有机体中的意志及其作用域时，原子理论只能为其建立概率的计算，我们正在谈论的这个观点与刚才提到的身心并行说水火不相容。确实，依据我们的观点，意志对自由的感受必须被看成是意识生活所特有的性质，人们必须在器官的功能中寻找与其对应的物质，既不允许力学因果描述也不能允许充分彻底地运用原子力学中的统计规律来进行物理学研究。除了形而上学的思索以外，我也许可以补充说，所有对一个解释的概念的分析都以放弃对我们自身意识行为的解释开始，也以放弃对意识行为的解释而结束。

最后，我想强调的是，我并不希望大家把我的话理解为是对未来物理学和生物学发展的怀疑。实际上，现在这个时代的物理学家绝不会抱有怀疑论的思想，因为科学已经在人们认识到最基本的概念存在局限性的基础上发生了深刻的变革。近期

in such far-reaching developments of our science. Neither has the necessary renunciation as regards an explanation of life itself been a hindrance to the wonderful advances which have been made in recent times in all branches of biology and have, not least, proved so beneficial in the art of medicine. Even if we cannot make a sharp distinction on a physical basis between health and disease, there is, in particular, no room for scepticism as regards the solution of the important problems which occupy this Congress, as long as one does not leave the highroad of progress, that has been followed with so great success ever since the pioneer work of Finsen, and which has as its distinguishing mark the most intimate combination of the study of the medical effects of light treatment with the investigation of its physical aspects.

(**131**, 457-459; 1933)

人们在生物学所有分支取得的惊人进展并没有受到放弃对生命问题本身的解释的影响，而且这些进展也相当有益于医术的发展。即使我们不能从物理学角度明确地区分健康与疾病，我们也没有理由怀疑在本次会议上所达成的对重要问题的解释，由于我们没有偏离科学发展的道路，因此从芬森的早期工作到现在已经取得了很大的进展，而此项工作明显的标志是光疗医学效应的研究和物理学研究的最紧密的结合。

（吴彦 翻译；张泽渤 审稿）

Nuclear Energy Levels

G. Gamow

Editor's Note

Although physicists still lacked any clear understanding of nuclear structure—the possibility that electrons inhabited the nucleus was still widely considered, for example—quantum theory was proving useful for interpreting nuclear spectroscopy. As George Gamow here points out, one could gainfully consider nuclear constituents to be bound by unknown forces inside a square-profiled well with infinitely high sides: a model tractable to quantum theory, which could predict the energies of states of different angular momentum. Gamow shows that spectroscopic data from an isotope called radium C' revealed transitions close to 11 of the 21 predicted by the theory. Further studies might indicate how the square-well model should be altered to make the theory more accurate.

IT is a plausible hypothesis that the forces acting on a particle inside the nucleus are comparatively weak in the internal region and increase rapidly to the boundary of the nucleus, the potential distribution being represented by a hole with more or less flat bottom and rather steep walls[1]. If we approximate this model by a rectangular hole with infinitely high walls, the energy levels of a moving particle will be determined by the roots of Bessel functions and can be easily calculated. For the real model, however, this theoretical level system will be deformed owing to the fact that the walls are neither quite steep nor infinitely high, producing compression of the upper part of the level system.

The best nucleus for testing this hypothesis is that of radium C', for which a lot of experimental evidence is available. For this nucleus we have the measurements by Rutherford[2] of long range α-particles (nine groups) giving us approximate positions of nuclear levels. The investigations of Ellis[3] give for a number of γ-lines (nine lines) their absolute intensities and, what is most important, the values of internal conversion coefficients enabling us, as has been shown by Taylor and Mott[4], to tell the dipole transitions from quadrupole transitions.

These data are sufficient to construct the level system of the radium C' nucleus, the main part of which is represented in Fig. 1, together with the theoretical one.

核能级

乔治·伽莫夫

编者按

虽然物理学家们对核结构还缺乏明确的认识，比如很多人仍然认为电子可能存在于原子核当中，但实践证明量子理论是可以用于解释核谱的。正如乔治·伽莫夫在文中所述：可以假设原子核中的各个成分被未知的力束缚在一口无限深的方井中，这样的假设是有益的：量子理论可以通过对这个模型的计算得到不同角动量状态的能量。伽莫夫指出：在该理论预测的 21 个跃迁中有 11 个接近于从镭 C' 同位素光谱中得到的数据。接下来的研究方向可能是：对方井模型进行怎样的调整才能使该理论更为精确。

我们似乎可以合理地假设作用于核内粒子上的力在核的内部区域相当弱，随着接近核的边界而迅速增大，势能的分布可以用一个近似平底深壁的孔来表示[1]。如果我们在模型中使用的是一个壁高为无限大的矩形孔，就可以由贝塞尔函数的根得到运动粒子的能级，计算方法非常简单。然而对于实际上的模型，由于壁并不是很陡，也不是无限高，因此理论能级系统会发生变化，能级系统的上部会被压缩。

镭 C' 的核最适于验证这个假说，因为我们有很多关于这种核的实验证据。卢瑟福通过对长程 α 粒子（9 组）的测量为我们提供了镭 C' 核能级的近似位置[2]。埃利斯[3]的研究给出了许多 γ 谱线（9 条谱线）的绝对强度以及内转换系数的值，后者为我们提供了非常重要的信息，正如泰勒和莫特曾经说过的那样[4]，内转换系数值是我们分辨四极跃迁和偶极跃迁的依据。

我们用这些数据足以构建镭 C' 核的能级系统，能级的主体部分示于图 1，图中还标出了理论值。

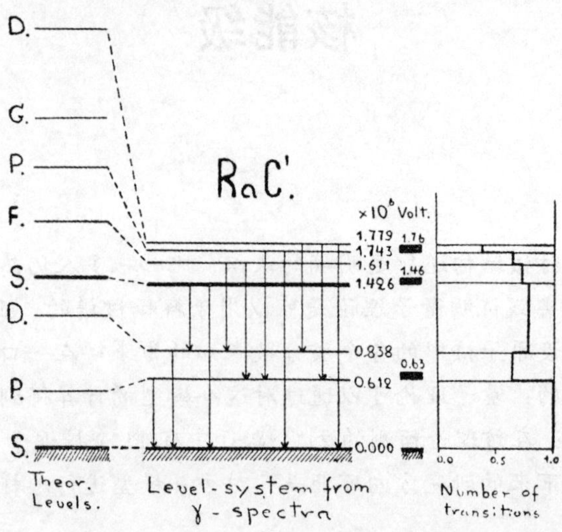

Fig. 1

We see immediately that not every level corresponds to a long range α-group; this is, however, to be expected, as the probability of α-disintegration from a level with large j is comparatively small, due to the additional barrier of centrifugal forces (for equal energies the probability for an α-particle escaping from *P*-, *D*-, *F*-, *G*-levels will be respectively 1.3, 4, 16 and 105 times less than for the *S*-level). The observed transitions are given in the accompanying table.

Constructed		Observed		Constructed		Observed	
$h\nu \times 10^{-6}$ volt	Δj	$h\nu \times 10^{-6}$ volt	Δj	$h\nu \times 10^{-6}$ volt	Δj	$h\nu \times 10^{-6}$ volt	Δj
0.588	2	0.589	..	1.131	0	1.130	0 ; 2
0.612	1	0.612	1	1.167	1	1.168	..
0.773	1	0.773	1	1.426	0→0	1.426	0→0
0.838	2	0.839	..	1.743	1	1.744	..
0.941	0	0.941	0 ; 2	1.779	2	1.778	0 ; 2
0.999	2	1.000	..				

From twenty-one mathematically possible transitions, eleven are actually found and, as can be seen from the table, have appropriate energies and obey the exclusion principle. From the remaining ten lines, two are not to be expected corresponding to $F \rightarrow S$-transitions, four fall in a spectral region not yet investigated and four are not observed, possibly due to comparatively small intensity. It is also of interest to construct an excitation diagram, building up the sums of the intensities for all lines crossing a given level interval. From this diagram we see that there must be a γ-line 0.226×10^6 volt with absolute intensity about 0.2 which at present is not known.

The similarity of theoretical and real level systems proves the correctness of our picture of

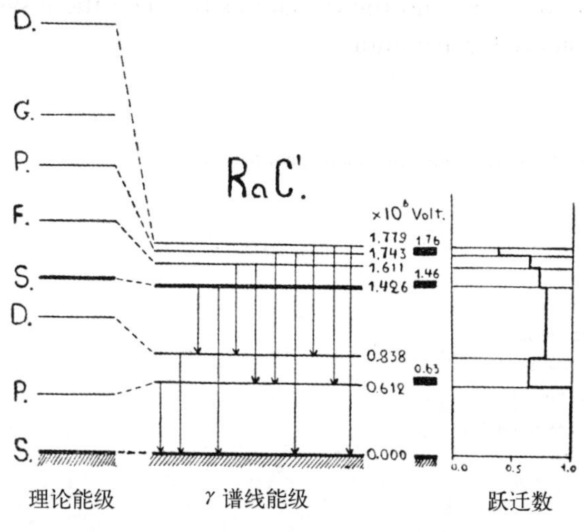

图 1

我们很快发现并不是每一个能级都对应于某个长程 α 粒子组；然而，这有可能是由于大 j 能级的 α 衰变概率很小的缘故，因为离心力会产生附加势垒（如果能量相同，一个 α 粒子从 S 能级逃逸的概率分别是从 P、D、F、G 能级逃逸的概率的 1.3、4、16 和 105 倍）。观测到的跃迁列于下表中。

理论值		观测值		理论值		观测值	
$h\nu\times10^{-6}$ V	Δj	$h\nu\times10^{-6}$ V	Δj	$h\nu\times10^{-6}$ V	Δj	$h\nu\times10^{-6}$ V	Δj
0.588	2	0.589	..	1.131	0	1.130	0 ; 2
0.612	1	0.612	1	1.167	1	1.168	..
0.773	1	0.773	1	1.426	0→0	1.426	0→0
0.838	2	0.839	..	1.743	1	1.744	..
0.941	0	0.941	0 ; 2	1.779	2	1.778	0 ; 2
0.999	2	1.000	..				

从数学的角度上看共有 21 种可能的跃迁，实际上只发现了其中的 11 种，如上表所示，能量值与理论预测相符并遵从不相容原理。至于剩下的 10 条谱线，其中 2 条 $F\rightarrow S$ 跃迁并不是预期应当有的谱线，另外 4 条处于研究范围之外的谱区，还有 4 条可能因强度太小而未被观测到。另一件有意思的事情是，我们可以通过构建跃迁图来计算穿过某个给定能级间隔的所有谱线强度之和。从这张图中我们发现了一条现在还未知的 γ 线，这条 0.226×10^{6} V 的 γ 线的绝对强度大约为 0.2。

理论与实际能级系统的相似性证实了我们对核内部势能的预测，而根据理论和

the potential inside the nucleus, and the deviations between these systems must permit us to calculate the real potential distribution.

(**131**, 433; 1933)

G. Gamow: Research Institute of Physics, University, Leningrad, Jan. 30.

References:

1. Gamow, *Proc. Roy. Soc.*, A, **126**, 632 (1930).
2. Rutherford, *Proc. Roy.* Soc., A, **131**, 684 (1931).
3. Ellis, *Proc. Roy. Soc.*, A, **129**, 180 (1930).
4. Taylor and Mott, *Proc. Roy. Soc.*, A, **138**, 665 (1932).

实测能级系统之间的偏差，我们一定能计算出实际的势能分布。

（沈乃澂 翻译；尚仁成 审稿）

New Evidence for the Positive Electron

J. Chadwick *et al.*

Editor's Note

Here James Chadwick, along with Patrick Blackett and Giuseppe Occhialini, confirms the existence of a positively charged anti-particle of the electron, initially reported by Carl Anderson from cloud-chamber experiments. Chadwick and colleagues placed pieces of polonium and beryllium just outside a cloud chamber, and near a lead target placed just inside the chamber. Alpha particles from the polonium excited neutrons or gamma rays from beryllium that in turn stimulated the emission of particles from lead. These left tracks in the cloud chamber with the properties expected for a positively charged electron. Further experiments strongly suggested that the unknown particles had the same mass as electrons, but a positive charge. They were later called positrons, the first known example of antimatter.

THE experiments of Anderson[1] and of Blackett and Occhialini[2] on the effects produced in an expansion chamber by the penetrating radiation strongly suggest the existence of positive electrons—particles of about the same mass as an electron but carrying a positive charge.

Some observations of the effects produced by the passage of neutrons through matter, and the experiments of Curie and Joliot[3] in which they observed retrograde electron tracks in an expansion chamber, led us to consider the possibility that positive electrons might be produced in the interaction of neutrons and matter, and we have recently obtained evidence which can be interpreted in this way.

A capsule containing a polonium source and a piece of beryllium was placed close to the wall of an expansion chamber. On the inside of the wall was fixed a target of lead about 2.5 cm. square and 2 mm. thick. This lead target was thus exposed to the action of the radiation, consisting of γ-rays and neutrons, emitted from the beryllium. Expansion photographs were taken by means of a stereoscopic pair of cameras. A magnetic field was applied during the expansion, its magnitude being usually about 800 gauss.

Most of the tracks recorded in the photographs were, from the sense of their curvature, clearly due to negative electrons, but many examples were found of tracks which had one end in or near the lead target and showed a curvature in the opposite sense. Either these were due to particles carrying a positive charge or they were due to negative electrons ejected in remote parts of the chamber and bent by the magnetic field so as to end on the lead target. Statistical examination of the results supports the view that the tracks began in the target and therefore carried a positive charge.

有关正电子的新证据

詹姆斯·查德威克等

编者按

在这篇文章中，詹姆斯·查德威克、帕特里克·布莱克特和朱塞佩·奥基亚利尼证实了一种带正电的粒子的存在，它是电子的反粒子，这种粒子是卡尔·安德森在云室实验中首先发现的。查德威克和同事把一小块钋和铍放在云室外面靠近云室内铅靶的地方。钋发射的α粒子从铍中激发出中子或γ射线，随后又使铅发射出粒子。这些粒子在云室中留下的径迹显示出了一个带正电的电子所具有的性质。更多的实验表明这种未知粒子的质量和电子相同，但带正电。后来它们被命名为正电子，是人们发现的第一种反物质。

安德森[1]以及布莱克特与奥基亚利尼[2]利用贯穿辐射在膨胀云室中产生的效应有力地证明了正电子的存在——一种与电子质量大致相等但带有正电荷的粒子。

人们在中子穿过物质时观察到的一些现象以及居里和约里奥[3]在膨胀云室中观察到反向电子径迹的实验使我们想到，正电子可能是在中子与物质发生相互作用时产生的，而且我们最近得到了可以用这种想法来解释的实验证据。

将装有钋放射源和铍金属片的密封容器放在靠近膨胀云室外壁的地方。在云室的内壁上固定一个面积为2.5 cm见方，厚2 mm的铅靶。这个铅靶会受到从铍中辐射出的一种由中子和γ射线组成的射线的照射。膨胀时的照片是利用立体像对照相机拍摄的。在膨胀过程中给云室外加一个磁场，强度通常为800高斯。

从弯曲方向来看，照片上记录的大多数径迹明显是由负电子产生的，但在实验中多次发现，有一些径迹的一端位于铅靶或铅靶附近并且向相反方向弯曲。之所以会出现这种现象，或许是因为产生这些径迹的粒子带正电荷，又或许是因为产生于云室远端的负电子在磁场作用下发生偏转，从而使它们的径迹最终在铅靶处结束。对该结果的统计分析支持了径迹始于铅靶的观点，所以产生该径迹的粒子是带正电的。

Strong evidence for this hypothesis was acquired by placing a metal plate across the expansion chamber so as to intercept the paths of the particles. Only a few good photographs have so far been obtained in which a positively curved track passes through the plate and remains in focus throughout its path, but these leave no doubt that the particles had their origin in or near the lead target and were therefore positively charged. In one case the track had a curvature on the target side of the plate, a sheet of copper 0.25 mm. thick, corresponding to a value of $H\rho$ of 12,700; on the other side the curvature gave a value $H\rho$ =10,000. This indicates that the particle travelled from the target through the copper plate, losing a certain amount of energy in the plate. The change in the value of $H\rho$ in passing through the copper is roughly the same as for a negative electron under similar conditions. The ionising power of the particle is also about the same as that for the negative electron. These observations are consistent with the assumption that the mass and magnitude of the charge of the positive particle are the same as for the negative electron.

The manner in which these positive electrons are produced is not yet clear, nor whether they arise from the action of the neutron emitted by the beryllium or from the action of the accompanying γ-radiation. It is hoped that further experiments now in progress will decide these questions.

Our thanks are due to Mr. Gilbert for his help in the experiments.

(**131**, 473; 1933)

J. Chadwick, P. M. S. Blackett, G. Occhialini: Cavendish Laboratory, Cambridge, March 27.

References:

1. Anderson, *Science*, 76, 238 (1932).
2. Blackett and Occhialini, *Proc. Roy. Soc.*, A, **139**, 699 (1933).
3. Curie and Joliot, *L'Existence du Neutron* (Hermann et Cie, Paris).

我们通过在膨胀云室中横放一块金属板以阻挡粒子路径的方法得到了支持这个假说的有力证据。虽然到目前为止，我们仅得到了几张效果较好的照片，照片上有一条明显弯曲的径迹穿过金属板，并且整条路径都比较清晰，但上述实验结果清清楚楚地说明这些粒子产生于铅靶或铅靶附近，因而带正电。在使用 0.25 mm 厚的铜板进行的一次实验中，我们在有铅靶的那一侧发现径迹出现弯曲，相应的 $H\rho$ 值为 12,700；而在铜板的另一侧，相应的值为 $H\rho$=10,000。这表明粒子从靶出发穿过铜板时，在铜板内损失了一部分能量。在穿过铜板前后粒子 $H\rho$ 值的变化与类似情况下负电子产生的结果大致相等。这种粒子的致电离能力也与负电子大致相同。以上实验结果都说明认为这种带正电的粒子与负电子质量相同且电荷数相同的假说是成立的。

我们现在还不清楚这些正电子是怎么产生的，也不知道它们是在铍发射的中子还是在随之发射的 γ 辐射的作用下产生的。希望正在进行的后续实验能够解决这一问题。

感谢吉尔伯特先生在实验中给予我们的帮助。

（王锋 翻译；李军刚 审稿）

Nature of Cosmic Rays*

A. H. Compton

Editor's Note

This address by Arthur Compton to a meeting of the American Physical Society describes recent studies of the geographical distribution of cosmic rays, and their variation with altitude. A marked increase in cosmic-ray intensity in temperate and polar latitudes, compared with tropical zones, could be explained if the primary cosmic rays were electrons arriving in two energy ranges, the higher-energy particles being largely unaffected by the Earth's magnetic field while those in the lower-band are strongly deflected and therefore unable to reach the tropics. The data argue against the particles being photons, but cannot distinguish between ordinary electrons, "positively charged electrons", or protons or alpha particles. Today we know that most primary cosmic rays are protons.

THERE are three kinds of experiments which seem to afford direct evidence regarding the nature of cosmic rays. These are: (1) the Bothe-Kolhörster double counter experiment, which compares the absorption of the particles traversing the counters with the absorption of cosmic rays; (2) measurements of the relative intensity of cosmic rays over different parts of the earth, designed to show any effect due to the earth's magnetic field; and (3) studies of the variation of cosmic ray intensity with altitude, which should follow different laws according as the rays are electrons or photons.

(1) The Bothe-Kolhörster experiment serves to measure the absorption in a block of gold or lead of the high-speed electrified particles that produce coincident impulses in two neighbouring counting chambers. It is found that this absorption is surprisingly small, about the same, in fact, as that of the cosmic rays themselves.

The simplest interpretation of this similarity in absorption is to suppose that the high-speed particles in question are the cosmic rays. There is, however, the alternative possibility that the primary cosmic rays are photons which eject high-speed electrons as recoil electrons when the photons are stopped, and that these recoil electrons are absorbed at about the same rate as the primary rays themselves. Theoretical calculations indicate that the recoil electrons should be absorbed five or ten times as rapidly as the photons which give rise to them. These calculations are somewhat uncertain because of extrapolation far beyond the wave-length region where the existing formulae have been tested. For this reason, the equal absorption coefficients of the cosmic rays and the high-speed particles

* Substance of an address presented at a symposium on cosmic rays, held by the American Physical Society at Atlantic City on December 30, 1932.

宇宙射线的性质*

阿瑟·康普顿

编者按

这是阿瑟·康普顿在美国物理学会的一次会议上就宇宙射线的地理分布以及它们的强度随高度变化的问题所作的演讲。与热带地区相比，宇宙射线的强度在温带和极区纬度上会有明显增强，这也许可以通过以下方式进行解释：如果初级宇宙射线是电子，它们以两个能量范围到达地球，那么高能粒子流几乎不会受到地球磁场的影响；而较低能段的粒子流会发生很大的偏转，因而根本无法到达赤道地区。这些数据表明构成宇宙射线的粒子不是光子，但不能区分是普通电子、“带正电的电子”、质子还是α粒子。今天我们知道大多数初级宇宙射线都是质子。

能够直接说明宇宙射线性质的实验有三类。它们是：(1) 博特–柯尔霍斯特的双计数器实验，可以用于比较穿过计数器的粒子的吸收和宇宙射线的吸收；(2) 测量宇宙射线在地球上不同地区的相对强度，目的是想了解地球磁场对其造成的影响；(3) 研究宇宙射线强度随海拔高度的变化，根据宇宙射线是电子或是光子而采用不同的定律。

(1) 博特–柯尔霍斯特实验用于测量金块或铅块对高速带电粒子的吸收，带电粒子可在两个相邻计数室中产生同步脉冲。实验发现，这类吸收出奇地小，实际上与宇宙射线自身的吸收大致相同。

对两者在吸收性质上相似的最简单的解释是假定宇宙射线就是高速运动的粒子。然而，还存在另外一种可能性，即认为初级宇宙射线是光子，当光子停止时会打出高速电子作为反冲电子，这些反冲电子以与初级宇宙射线本身大致相同的速率被吸收。理论计算表明，反冲电子被吸收的速度应该是产生它们的光子的吸收速度的5倍到10倍。这些计算结果不是很准确，因为外推远远超过了被现有公式检验过的波长范围。由于这个原因，虽然宇宙射线和高速粒子具有相同的吸收系数并不足以排除这些粒子就是宇宙射线激发的反冲电子。不过，这仍然难以想象，公式的误差会大

* 这是在专题讨论会上的一篇有关宇宙射线的演讲稿，该讨论会是由美国物理学会于1932年12月30日在大西洋城举办的。

does not necessarily rule out the possibility that the particles in question may be recoil electrons excited by the cosmic rays. It would, nevertheless, be surprising if the formulae were in error by so large a factor as five or ten, as would be implied if the coincidences are due to secondary electrons.

(2) If the cosmic rays consist of electrified particles coming into the earth's atmosphere from remote space, the earth's magnetic field should affect their geographical distribution. This effect has been investigated theoretically by Størmer, Epstein, and recently much more completely by Lemaître and Vallarta. It appears that for energies less than 10^9 electron volts, electrons approaching the earth can reach it only at latitudes north of about 60°. For energies greater than about 5×10^{10} volts the geographical distribution is not affected by the earth's magnetic field. For intermediate energies, there will be a difference in intensity with latitude according to the distribution of energy of the incoming electrons.

Experimental studies of the relative intensity of cosmic rays in different parts of the world have been made by J. Clay, who made several trips between Java and Holland, and found consistently a lower intensity near the equator; Millikan and Cameron, who found but slightly lower intensity in the lakes of Bolivia than in the mountain lakes of California, and no difference between Pasadena and Churchill close to the north magnetic pole; Bothe and Kolhörster, who carried a counting tube from Hamburg (53° N.) to Spitsbergen (81° N.) and back, and detected no variations in the cosmic rays larger than their rather large experimental error; Kennedy, who, under Grant's direction, carried similar apparatus from Adelaide, Australia, to Antarctica, and likewise found no measurable change; and Corlin, who on going from 50° N. to 70° N. in Scandinavia found some evidence of a maximum at about 55° N. The prevailing opinion regarding the significance of these measurements has thus been expressed by Hoffman in a recent summary: "The results so far have on the whole been negative. Most of the observers conclude that within the errors of experiment the intensity is constant and equal, and those authors who do find differences give their results with certain reservations."

During the past eighteen months, Prof. Bennett of the Massachusetts Institute of Technology, Prof. Stearns of the University of Denver, and I have organised a group of ten expeditions, with some sixty physicists, and we have attempted to make as extensive a study of the geographical distribution of cosmic rays as could be done in a limited period of time. Fig. 1 shows the position of the eighty-one stations in different parts of the world at which measurements have been made. These stations are about equally divided between the northern and southern hemispheres. They have extended from latitude 46° S. to latitude 78° N., and from sea level to about 20,000 ft. (Figs. 2 and 3). When these data are brought together, they show a marked difference in intensity between the cosmic rays at temperate and polar latitudes, as compared with the tropical latitudes. At sea level the difference in intensity is about 14 percent, at 2,000 metres about 22 percent, and at 4,400 metres about 33 percent. The change between the intensity for tropical as compared with that for temperate latitudes occurs rather sharply between geomagnetic latitudes 25° and 40°.

到 5~10 倍，以至于认为吸收系数的符合是由次级电子引起的。

（2）如果宇宙射线是从遥远太空进入地球大气的带电粒子，那么地球磁场应该会影响它们的地理分布。斯托默和爱泼斯坦曾对此进行过理论上的研究，最近勒梅特和瓦拉塔又大大完善了这一理论。他们的研究表明：对于能量小于 10^9 电子伏特的电子，它们在接近地球时只能到达纬度约 60°以北的地方；对于能量大于约 5×10^{10} 电子伏特的电子，其地理分布不受地球磁场的影响；如果能量介于两者之间，入射电子的强度将会按照其能量分布随纬度而异。

克莱在世界各地进行了宇宙射线相对强度的实验研究，他在爪哇与荷兰之间往返了好几次后发现赤道附近的强度总是比较低；密立根和卡梅伦也发现在玻利维亚湖附近的宇宙射线强度略低于加州山湖，而在帕萨迪纳与靠近北极的丘吉尔港之间的宇宙射线强度并无差别；博特和柯尔霍斯特带着一支计数管往返于汉堡（北纬 53°）和斯匹次卑尔根群岛（北纬 81°）之间，他们用误差较大的仪器没有检测到宇宙射线强度的变化；肯尼迪在格兰特的指导下，带了同样的装置从澳大利亚的阿得雷德到南极洲，同样没有测量到强度的变化；但科林在测量了斯堪的纳维亚半岛上从北纬 50° 到 70° 之间的宇宙射线强度后发现在大约北纬 55° 处有一个极大值。最近，霍夫曼总结了人们对这些测量结果的普遍看法：“迄今为止得到的结果从总体上说是负面的。大多数观测者认为，在实验误差范围内，宇宙射线的强度恒为一个常数，而那些发现强度确实存在差别的作者在公布结果时带有某种保留。”

在过去的 18 个月内，麻省理工学院的贝内特教授、丹佛大学的斯特恩斯教授和我组织了一组有 60 余位物理学家参加的 10 次考察，我们试图在限定的时间内对宇宙射线的地理分布进行尽可能大范围的研究。图 1 中标出了分布在世界各地的 81 个观察站的位置，在这些位置我们都进行了测量。这些观察站在南北半球上基本上是平均分布的。从南纬 46° 延伸到北纬 78°，从海平面一直到海拔约 20,000 英尺处（图 2 和图 3）。当把这些数据汇总在一起时，我们发现在温带和两极处的宇宙射线强度与赤道地区明显不同。在海平面上，强度差约为 14%；在海拔 2,000 米处，约为 22%；在海拔 4,400 米处，约为 33%。在地磁纬度 25° 和 40° 之间，赤道地区宇宙射线强度与温带地区宇宙射线强度之间的差别会格外明显。

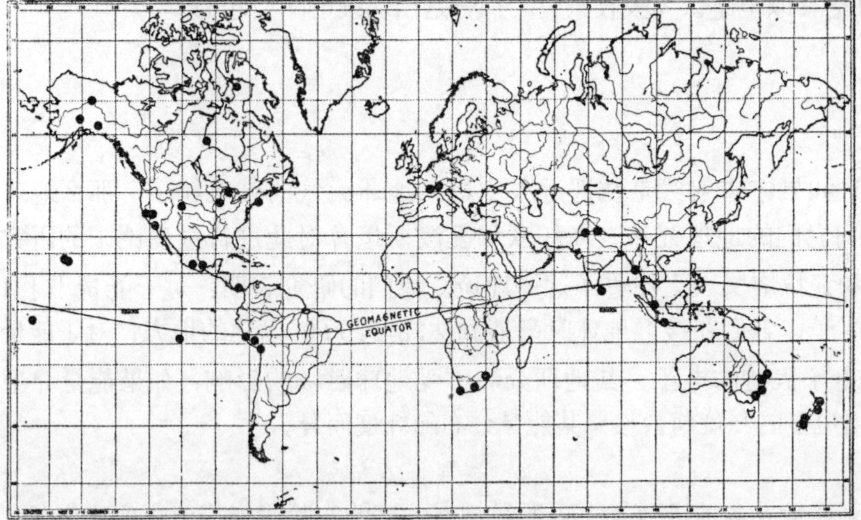

Fig. 1. Map showing major stations at which associated expeditions have made cosmic ray measurements during 1932.

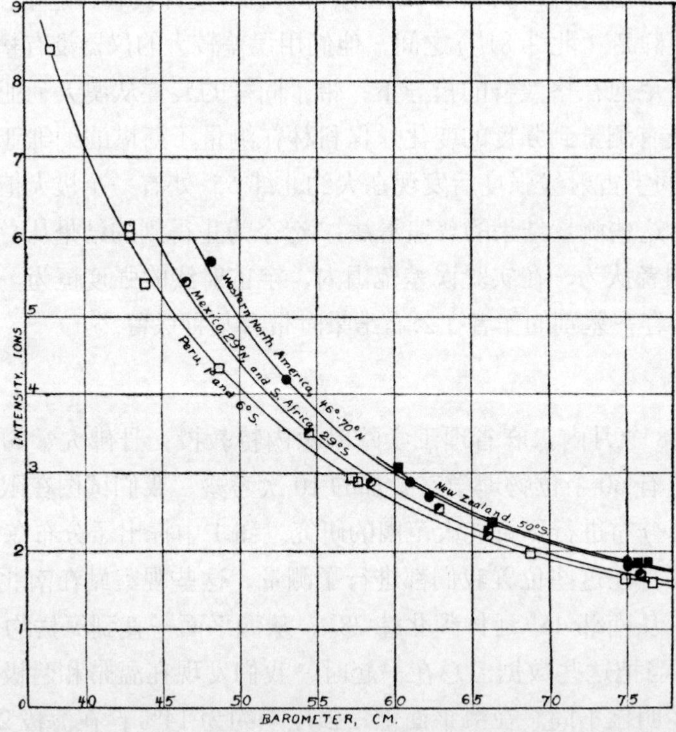

Fig. 2. Typical intensity–barometer curves, showing variation of intensity with altitude in different parts of the world. Circles, northern hemisphere; squares, southern hemisphere.

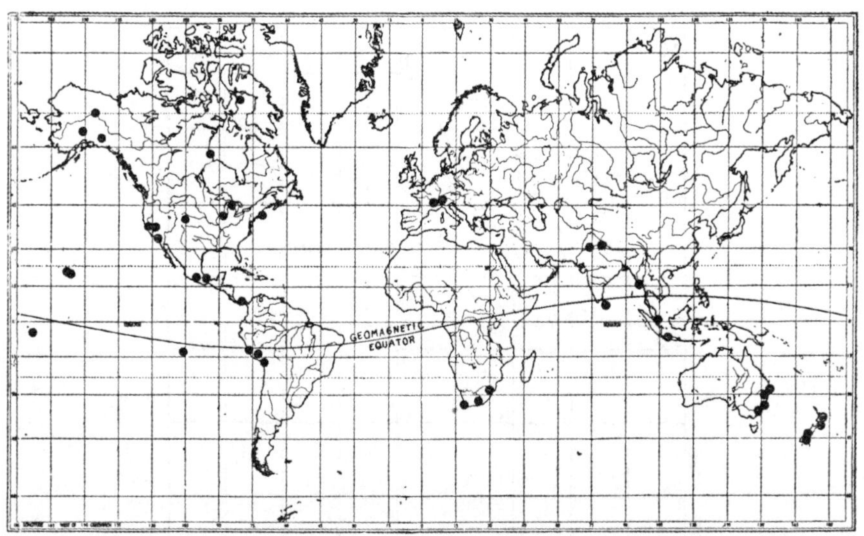

图 1. 考察队 1932 年在全球测量宇宙射线的主要站点分布图

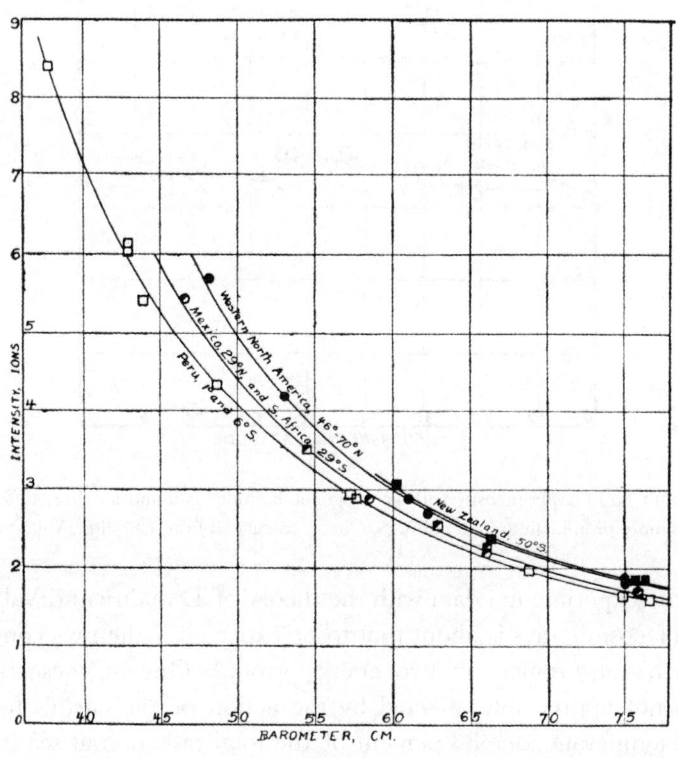

图 2. 典型的强度－气压曲线，这张图说明在世界上纬度不同的地区，宇宙射线的强度是不相同的。圆形，北半球；方块，南半球。

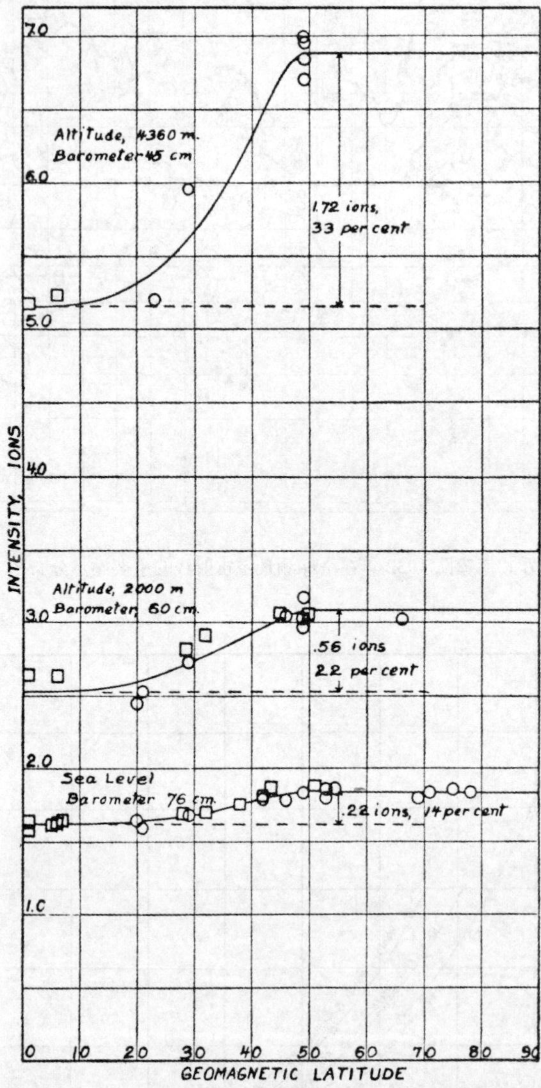

Fig. 3. Variation of cosmic ray intensity with geomagnetic latitude, as found for three different altitudes. The variation is more prominent at high altitudes. Curves calculated from Lemaître–Vallarta theory.

Comparison of the experimental data with the theory of Lemaître and Vallarta shows that the distribution of cosmic rays is about that to be expected if the rays consist of electrons entering the earth's atmosphere in two energy groups. One of these is of such great energy that it is not appreciably affected by the action of the earth's field. This group comprises in the temperate zone 88 percent of the total radiation at sea level, and might, so far as these experiments are concerned, be classified as photons. At 4,400 metres this component constitutes 75 percent of the total radiation. The second component is less penetrating and represents particles with an energy, if they are electrons, of about 7×10^9 electron volts. It is these particles which reach the earth at temperate latitudes but fail to reach it near the equator.

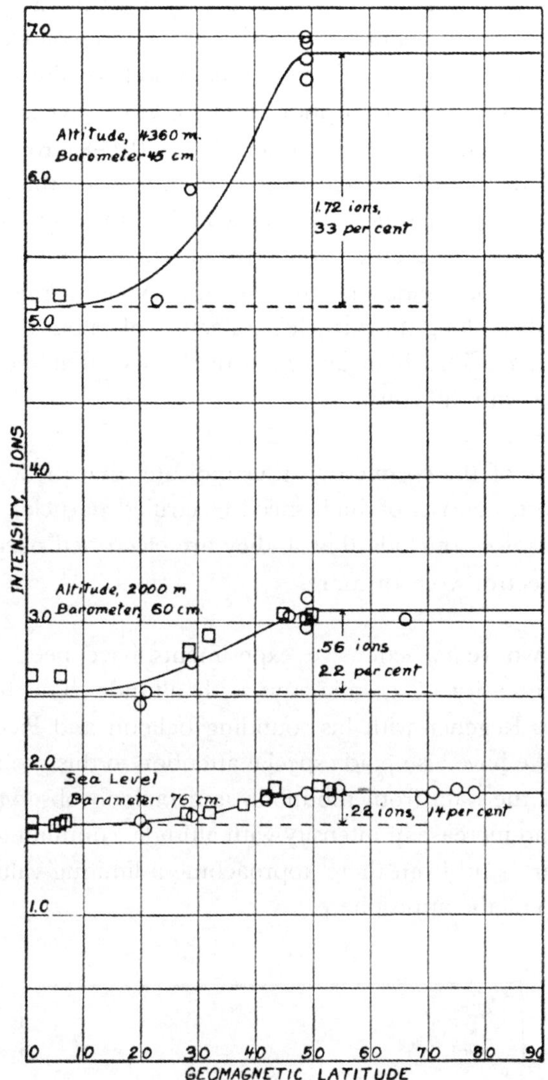

图 3. 宇宙射线强度随地磁纬度的变化，包括在海拔高度不同的三个地点测量得到的结果。海拔越高，变化越明显。曲线是根据勒梅特－瓦拉塔的理论计算出来的。

对这些实验数据与勒梅特和瓦拉塔理论的比较说明：如果进入地球大气层的宇宙射线由两个能量组的电子组成，则宇宙射线的分布就与人们预期的大体一致。其中的一组电子具有很高的能量，以至于基本不受地球磁场的影响。在温带地区的这组电子的能量占海平面处宇宙射线总辐射量的 88%，这组电子是迄今为止的实验所关注的可以看作是光子的那一类。在海拔 4,400 米处该成分占总辐射量的 75%。第二组成分穿透力稍差并表现为有相近能量的粒子，如果它们是电子，将具有约 7×10^9 电子伏特的能量。这些粒子可以进入地球的温带地区，但不能进入靠近赤道的地区。

It may be remarked that even these less-penetrating cosmic rays have energies which are much larger than those that could be accounted for as recoil electrons resulting from photons, if these photons were to constitute the main body of the cosmic rays. According to Millikan, measurements of the absorption of the cosmic rays at sea level indicate that their energy, if they are photons, is of the order of 2×10^8 electron volts. This is so much less (a factor of 35) than that of the electrons responsible for the difference in intensity between temperate and tropical zones, that we would seem to be safe in concluding that the particles reaching the earth are not of the recoil type. On the other hand, the energy of 7×10^9 electron volts demanded by the Lemaître–Vallarta theory for these particles would mean a range taken along the particle's trajectory of about three times the thickness of the earth's atmosphere. This is in accord with the fact that these rays penetrate the earth's atmosphere, but with difficulty.

This geographical study of the cosmic rays thus indicates that the less penetrating part of the cosmic rays, at least, consists of high-speed electrified particles. Regarding the more penetrating component, we conclude that if they are electrified particles, they must have an energy of 3×10^{10} electron volts or more.

(3) During the past two years, extensive experiments have been carried out studying the variation in the intensity of cosmic rays with altitude. The highest altitudes have been those reached by Regener with his sounding balloon and Piccard with his famous stratosphere balloon. We have also paid especial attention to this problem in our mountain experiments. In Fig. 4 the data from mountain and balloon observations are compared. These data show a rapid increase in intensity with altitude, continuing nearly exponentially to an altitude of 15 km., and from there approaching a limiting value as the apparatus is carried close to the top of the atmosphere.

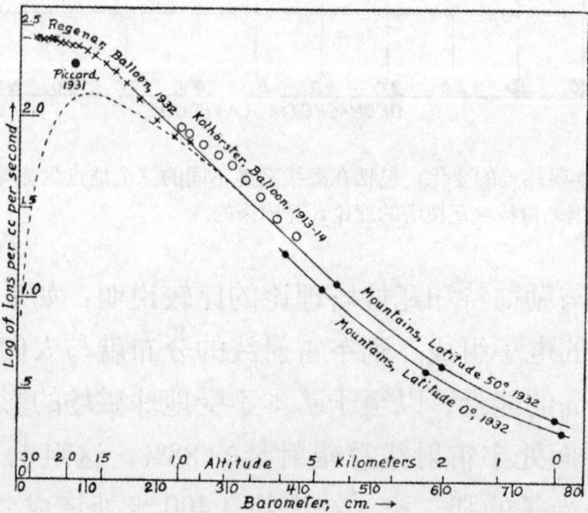

Fig. 4. Combined balloon and mountain data on intensity as function of altitude. Broken line, anticipated values for photons.

可以说，如果光子是组成宇宙射线的主体，那么即使是穿透性较差的那些宇宙射线所具有的能量也要远远高于由光子产生的反冲电子的能量。密立根认为，在海平面处对宇宙射线的吸收的测量表明：如果宇宙射线是光子，那么它们的能量就会达到 2×10^{8} 电子伏特数量级。这比使温带和热带地区宇宙射线强度产生差异的电子的能量小很多（相差 35 倍），以至于我们似乎可以肯定到达地球的粒子不是反冲电子。另一方面，勒梅特 – 瓦拉塔的理论要求粒子的能量应达到 7×10^{9} 电子伏特，这意味着粒子的射程约为地球大气层厚度的三倍。这一点符合这样的事实，即这些射线可以穿透地球大气，但是有一定的困难。

对宇宙射线地理分布的研究表明，宇宙射线中穿透力较差的部分起码会包含高速带电粒子。对于穿透性较强的成分，我们的结论是：如果它们是带电粒子，那么它们的能量就必须达到或超过 3×10^{10} 电子伏特。

（3）在过去的两年内，人们已经进行了大量的实验以研究宇宙射线强度随海拔高度的变化。目前以雷格纳的探空气球和皮卡德的著名平流层气球到达的高度最高。我们在山上作实验时也特别关注这个研究课题。在图 4 中，我们对山上的实验数据与气球观测的结果进行了比较。这些数据表明：随着高度的增加，宇宙射线的强度很快上升，持续呈指数型增长直至 15 km 高度，从那儿起，当测试仪器接近大气层顶部时，强度达到极限值。

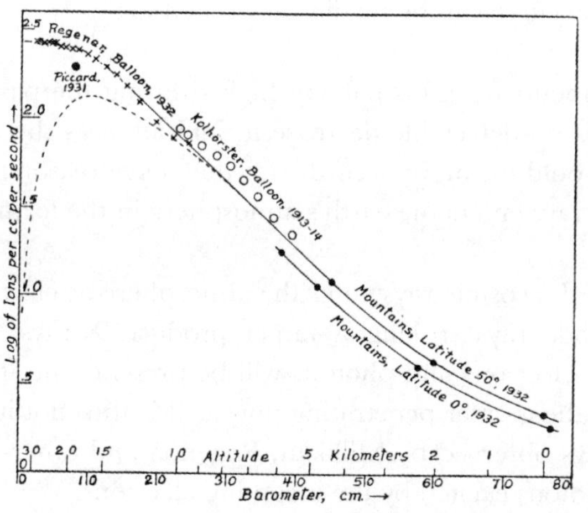

图 4. 综合气球实验和在山上所作的实验得到的宇宙射线强度随海拔高度的变化曲线。虚线是把宇宙射线看成光子后的预期值。

If we suppose that the cosmic rays enter the earth's atmosphere as photons, any secondary electrons that may have been associated with them in space will have been removed by the action of the earth's magnetic field. We should thus have a beam of pure photons entering the atmosphere. At the surface of the atmosphere these photons will produce very little ionisation, for the ionisation results directly from the secondary electrons that pass through the chamber. The secondary electrons, however, arise only at the absorption of the primary photons, and will not approach their normal intensity until the photons have traversed a thickness of air about equal to the range of the electrons. This means that at very high altitudes the ionisation due to a beam of photons entering the earth must be almost zero. The ionisation current should rather approach a maximum at a depth in the atmosphere at which the cosmic rays are somewhat less than half absorbed, and should then gradually diminish in intensity as sea level is approached. Our high mountain experiments confirm the recent balloon experiments as indicating that no such high altitude maximum exists. This would seem to rule out the possibility that the cosmic rays can be photons entering the earth's atmosphere from remote distances.

If we suppose, on the other hand, that the cosmic rays are electrons entering the atmosphere from above, we should expect very much the kind of intensity–altitude curve that the present experiments show. It is well known that the ionisation per unit path by high speed electrons remains almost constant over a wide range of energies. Thus, if a beam of such electrons enters the atmosphere, it will produce nearly uniform ionisation down to such depths that an appreciable number of the electrons are stopped by the air. If the initial electrons were all travelling downward, there would be a rather definite limit or range where there would be a rapid reduction in the ionisation by the cosmic rays. If, however, the initial electrons entered in all directions, some of them would be stopped even in the upper layers of the atmosphere. Thus, supposing that the cosmic rays consist of electrons entering the earth's atmosphere from outer space, the general characteristics of the intensity-altitude curve can be readily accounted for.

If Regener's measurements from his balloon flights during the past summer are reliable, it appears that there is no detectable decrease in ionisation as the top of the atmosphere is approached. This would mean, in accord with the above reasoning, that no appreciable portion of the cosmic rays enters the earth's atmosphere in the form of photons.

On the other hand, if the cosmic rays enter the atmosphere as electrons, they will produce photons just as cathode rays striking a target produce X-rays. Accordingly, at lower altitudes a mixture of electrons and photons will be present, and it may be expected that the photons will have the greater penetrating power. It is thus not impossible that the very penetrating cosmic rays observed by Millikan, Regener, and others at great depths under water may be such photons excited by the incoming electrons.

Although we have discussed the problem as if the electrified particles which seem to constitute the cosmic rays are electrons, it will be understood that the experiments that have been cited do not serve to distinguish between negatively and positively charged

假如我们把进入地球大气层的宇宙射线看作是光子，那么在太空中可能与这些光子相关联的次级电子都将因地球磁场的作用而偏离地球。因而我们应该有一束纯光子束进入大气层。在大气层表面，这些光子几乎不会发生电离，因为电离直接由穿过计数室的次级电子产生。然而，这些次级电子只有在吸收了初始光子的能量后才能产生，它们在光子穿越大气层的厚度小于电子的射程时是不会达到其正常强度的。这意味着在海拔很高的地方，由进入地球的光子束引起的电离应该接近于零。当到达大气层某一深度处的宇宙射线有将近一半的量被吸收时，电离电流将达到极大值，而在不断接近海平面的过程中，宇宙射线的强度应逐渐减小。我们在高山上所作的实验与最近的气球实验结果相同，都没有发现在高海拔处出现极大值。这似乎可以排除从遥远太空进入地球大气层的宇宙射线是光子的可能性。

另一方面，如果我们假设宇宙射线是从上方进入地球大气层的电子，我们应该能够很好地预期由现有实验得到的那类强度–高度曲线。众所周知，高速电子产生的每单位路径的电离在很宽的能量范围内几乎保持为一个常数。因此，如果一束这样的电子进入大气层，它们将产生几乎完全相同的电离直到这样一个深度，在这个深度上有数量相当可观的电子被空气阻止。如果初始的电子都是向下运动的，那么宇宙射线的电离发生快速下降的高度界线或范围将是非常明确的。然而，如果初始电子是从各个方向进入大气层的，则其中的某些电子会停止在大气层的上部。于是，假定宇宙射线是由外层空间进入地球大气层的电子所组成，实验的强度–高度曲线的总体特征就能很容易地得到解释。

如果雷格纳去年夏天用他的气球飞行实验得到的结果是可靠的，那么在接近大气层顶部时就不存在可探测的电离度下降。这意味着按照上述推理，在进入地球大气层的宇宙射线中，绝大部分都不是光子。

另一方面，如果宇宙射线以电子形式进入大气层，则它们也会生成光子，就像轰击靶子的阴极射线能产生 X 射线一样。于是，在海拔高度较低的地方将出现电子和光子的混合，可以预期光子将有更强的穿透能力。而密立根、雷格纳和其他研究者在水下很深处观测到的穿透力超强的宇宙射线并非不可能存在，它们也许就是被入射电子所激发的光子。

虽然我们推断组成宇宙射线的带电粒子可能是电子，但我们知道前面提到的那些实验都无法判断粒子带的是正电还是负电。这些实验结果同样可以用于说明宇宙

particles. The experiments are equally consistent with the view that the rays consist of protons or high speed α-particles. I find no way of reconciling the data, however, with the hypothesis that any considerable portion of the primary cosmic rays consist of photons.

(**131**, 713-715; 1933)

射线是由质子或高速 α 粒子组成的。然而，如果假设初级宇宙射线的主要成分是光子，我认为上述数据就没办法得到解释了。

（沈乃澂 翻译；马宇蒨 审稿）

A Possible Property of the Positive Electron

W. Elsasser

Editor's Note

The curious similarity of the neutron and proton, coupled with the discovery of the positron, suggested some intriguing possibilities for structures within the atomic nucleus. Some physicists had suggested that the proton might be a composite particle formed from a neutron and positron. Walter Elsasser suggests here that the hypothesis might also help explain recent discoveries about the proton's magnetic moment—a measure of its magnetic behaviour. He says the idea might account for why positrons seem to be found only in nuclei. Elsasser's suggestions were well motivated, but would soon be proven off the mark, especially by experiments showing the positron to have spin (a quantum property related to angular momentum) of 1/2, just like the electron.

THE detection of the positive electron (called positron) by Blackett and Occhialini[1] and by Anderson[2] makes it very probable that the positron has a great importance in the building up of nuclei. Anderson[2] suggests that the proton may consist of a neutron and a positron. In favour of this hypothesis we may mention the experiments of Stern (still unpublished), who found that the magnetic momentum of the proton is three times greater than it should be if the proton were to behave like an elementary particle in Dirac's theory. Following Heisenberg[3], both the proton and the neutron obey Fermi statistics and have a half integral spin momentum. This leads at once to the conclusion that, if the hypothesis of Anderson is true, the positron should obey Bose statistics and have an integral spin momentum (0 or 1). If this view should be confirmed by other experimental evidence we should understand better why the positrons can only be found in nuclei; for, since positrons have symmetrical wave functions, they can always be placed in the deepest energy levels. It seems to be an advantage of the proposed hypothesis, that contrary to Dirac's theory of "holes"[4] an essential asymmetry between positive and negative electricity is introduced into the laws describing the behaviour of elementary particles. Since the light-quanta also have whole number momenta, it seems that it may be a general rule, that symmetrical wave functions are combined with integral momenta and anti-symmetrical wave functions with half integral momenta.

(**131**, 764; 1933)

W. Elsasser: Physikalisches Institut, Technischen Hochschule, Zürich, April 25.

References:

1. *Proc. Roy. Soc.*, **139**, 699 (1933).
2. *Phys. Rev.* (March, 1933).
3. *Z. Phys.*, **77**, 1 (1932).
4. *Proc. Roy. Soc.*, **126**, 360 (1932).

正电子可能具有的性质

沃尔特·埃尔萨瑟

编者按

中子和质子的惊人相似以及正电子的发现使人们对原子核的内部结构浮想联翩。一些物理学家认为质子可能是由中子和正电子合成的。沃尔特·埃尔萨瑟在这篇文章中指出上述假说也许有助于解释最近的一项与质子磁矩有关的发现——一种对质子磁行为的测量。他说这个观点或许可以说明为什么正电子似乎只出现于原子核中。埃尔萨瑟的这种提法绝对不是空穴来风，但他的假设很快就被证明是错误的，实验证明正电子的自旋（一种与角动量相关的量子特性）为 1/2，和电子没有什么两样。

布莱克特、奥基亚利尼[1]和安德森[2]探测到了带正电的电子（被人们称为正电子），这很有可能意味着正电子是原子核的重要组成部分。安德森[2]指出，质子可能是由一个中子和一个正电子组成的。为了证明这个假说，我们要提到斯特恩的实验（尚未发表），他发现：根据狄拉克的理论，如果质子是基本粒子，那么它的磁矩应该是现在的 3 倍。按照海森堡[3]的说法，质子和中子都遵循费米统计规律并具有半整数自旋动量。由此我们马上可以得到这样的结论：如果安德森的假说成立，那么正电子应该遵守玻色统计规律并且具有整数自旋动量（0 或 1）。如果这个观点能够被其他实验现象所证实，那么我们就可以更好地理解为什么正电子只出现在原子核当中；因为正电子具有对称的波函数，所以它们可以一直处在最低的能级上。这似乎支持了上面提出的假说，但却与狄拉克的"空穴"理论[4]相反，狄拉克理论在描述基本粒子行为的物理规律中引入了正负电子之间重要的不对称性。由于光量子也具有整数自旋，看起来总的规律是：对称波函数与整数自旋动量并存，反对称波函数与半整数自旋动量并存。

（王静 翻译；李军刚 审稿）

Positrons and Atomic Nuclei

G. W. Todd

Editor's Note

Walter Elsasser had recently suggested that the proton might be a composite of neutron and positron, and that this hypothesis might help explain the positive charge of nuclei. Here George Todd criticizes that idea, and suggests another. He notes, however, that a particular transmutation of uranium can be understood in Elsasser's picture if the process involves the spontaneous creation of an electron and positron from nothing—a process that is now known to happen, although not here. Todd's proposed nuclear constitution leads to the conjecture that isotopes differ only by the number of neutrons in the nucleus, another prescient suggestion.

IN *Nature* of May 27, Dr. W. Elsasser offers evidence in favour of the suggestion that the proton consists of a neutron and a positron. Examining the question from a different point of view, I put forward the following as evidence against the suggestion.

If we allow that an atomic nucleus may contain α-particles, protons, neutrons, electrons and positrons, the number of possible structures for a nucleus of atomic mass P and atomic number Z increases rapidly with increase of P and Z, and for the heavy atoms it may run into hundreds. If we exclude the possibility of unattached electrons and positrons in the nucleus, then the structure becomes unique and is given by

$$\frac{1}{2}(Z-p)\ \alpha\text{-particles} + (P-2Z+p)\ \text{neutrons} + p\ \text{protons} \tag{1}$$

where $p = 0$ or 1, whichever value makes $\frac{1}{2}(Z-p)$ an integer.

It is also possible to get a unique structure by excluding the possibility of unattached electrons but allowing the possibility of positrons. The structure is then

$$\frac{1}{2}(Z-p')\ \alpha\text{-particles} + (P-2Z+2p')\ \text{neutrons} + p'\ \text{positrons} \tag{2}$$

where again $p' = 0$ or 1 as before. This is practically the suggestion which Elsasser supports.

Using these expressions, we can trace the changes which take place in the nuclei of radioactive elements during α- and β-ray transformations. The accompanying table shows a typical set of transformations.

正电子和原子核

乔治·托德

编者按

沃尔特·埃尔萨瑟最近指出质子可能是中子和正电子的合成物，而且这一假说很可能有助于解释原子核为什么带正电。在这篇文章中，乔治·托德反驳了这一观点，并提出了另外一个观点。不过托德指出：有一种特殊的铀嬗变也许可以用埃尔萨瑟的理论来解释，如果在这个过程中会凭空自发产生一个电子和正电子——现在我们知道这样的反应是存在的，虽然当时还不知道。由托德提出的核结构理论可以推出同位素之间的差别仅仅是原子核中的中子数不相同，这是托德的又一个真知灼见。

在5月27日的《自然》杂志上，埃尔萨瑟博士提出证据证明质子是由一个中子和一个正电子组成的。我从不同的角度出发用以下证据来反驳这一观点。

如果我们承认原子核有可能是由α粒子、质子、中子、电子和正电子组成的，那么对于一个原子质量为P、原子序数为Z的原子来说，它可能具有的结构个数将随着P和Z的增加急剧上升，而重原子可能具有的结构或许会达到数百种。如果我们排除原子核中存在独立正、负电子的可能性，那么就可以认为核结构只能有以下一种形式：

$$\frac{1}{2}(Z-p)\text{个}\ \alpha\ \text{粒子}+(P-2Z+p)\text{个中子}+p\text{个质子} \tag{1}$$

式中p取0或1，看哪一个值可以使$\frac{1}{2}(Z-p)$为整数。

如果只排除独立负电子，而允许独立正电子存在，也可以得到唯一的结构。此时原子核结构为：

$$\frac{1}{2}(Z-p')\text{个}\ \alpha\ \text{粒子}+(P-2Z+2p')\text{个中子}+p'\text{个正电子} \tag{2}$$

和前面一样，p'取0或1。事实上这就是埃尔萨瑟所支持的观点。

利用这些表达式，我们可以探索放射性元素的原子核在α和β射线衰变过程中发生的变化。下表中给出了一系列典型的衰变：

	Nucleus from (1)	Nucleus from (2)	Radiation
Ur I ↓	$46\alpha + 54n + 0p$	$46\alpha + 54n + 0p'$	} α
Ur X_1 ↓	45 54 0	45 54 0	} β
Ur X_2 ↓	45 53 1	45 54 1	} β
Ur II ↓	46 50 0	46 50 0	}
I_0	45 50 0	45 50 0	

The explanation of the α-ray changes is obvious. The β-ray changes in a radioactive series generally occur in pairs and the pair above shows changes identical with all other pairs of β-ray transformations. If the nuclear contents are expressed by (1), the changes take place in the following reasonable manner:—

$1n \rightarrow 1p + 1\beta$ the proton remaining in the nucleus,

and $3n + 1p \rightarrow 1\alpha + 1\beta$ the α- particle remaining in the nucleus.

On the other hand, if expression (2) gives the nuclear contents, then the changes which take place are

$0 \rightarrow 1p' + 1\beta$, a positron appearing in the nucleus,

and $4n + 1p' \rightarrow 1\alpha + 1\beta$, the α-particle remaining in the nucleus.

But where do the electron and positron come from in the first change, and how is the alteration in charge to be accounted for in the second change?

It is interesting to note that the expressions (1) and (2) give lower limits to the mass of an isotope. The minimum value from (1) is $P \geqslant 2Z - p$ and from (2) it is $P \geqslant 2Z - 2p'$. It will also be observed that isotopes only differ from each other in the number of neutrons in their nuclei.

(**132**, 65; 1933)

George W. Todd: Armstrong College, Newcastle-upon-Tyne, June 3.

	式（1）中的原子核	式（2）中的原子核	辐射
Ur I	$46\alpha + 54n + 0p$	$46\alpha + 54n + 0p'$	} α
↓ Ur X_1	45 54 0	45 54 0	} β
↓ Ur X_2	45 53 1	45 54 1	} β
↓ Ur II	46 50 0	46 50 0	}
↓ I_0	45 50 0	45 50 0	

关于产生 α 射线的转变的解释是显而易见的。而产生 β 射线的转变在放射系中一般成对出现，上表中出现的那对表现出与所有其他 β 射线转变对相同的变化。如果原子核的构成可以表示为式（1），那么发生转变的合理方式应该是这样的：

$1n \rightarrow 1p + 1\beta$，质子留在原子核中；$3n + 1p \rightarrow 1\alpha + 1\beta$，$\alpha$ 粒子留在原子核中。

另一方面，如果原子核的构成可以表示为式（2），那么发生的转变则是：

$0 \rightarrow 1p' + 1\beta$，一个正电子出现在原子核中；$4n + 1p' \rightarrow 1\alpha + 1\beta$，$\alpha$ 粒子留在原子核中。

然而，在第一种转变中，电子和正电子从何而来？在第二个转变中，又如何解释电荷变化？

有趣的是，我们注意到式（1）和式（2）给出了同位素质量的下限。由式（1）推出的最小值为 $P \geqslant 2Z - p$ 而由式（2）推出的最小值为 $P \geqslant 2Z - 2p'$。我们还发现，同位素之间的区别仅仅在于其原子核中的中子数不相同。

（王静 翻译；李军刚 审稿）

Magnetic Moment of the Proton

I. Estermann *et al.*

Editor's Note

Paul Dirac's theory of the electron predicted that it has a quantum-mechanical spin of 1/2 and a magnetic moment equal to 1 Bohr magneton. As the theory would seem to apply to any elementary particle of spin 1/2, physicists in the early 1930s expected the proton to behave similarly, having a magnetic moment smaller by the inverse ratio of their masses (1/1,840). But here Immanuel Estermann and colleagues announce a very surprising result: the magnetic moment of the proton is around 2.5 times larger than that. The result indicated that Dirac's theory failed in some important respects for the proton, and supplied the first experimental evidence that nuclear particles (nucleons) have internal structure.

THE spin of the electron has the value $\frac{1}{2} \cdot \frac{h}{2\pi}$, and its magnetic moment has the value $2\frac{e}{m_e c} \cdot \frac{1}{2} \cdot \frac{h}{2\pi}$, or 1 Bohr magneton. The spin of the proton has the same value, $\frac{1}{2} \cdot \frac{h}{2\pi}$, as that of the electron. Thus for the magnetic moment of the proton the value $2\frac{e}{m_p c} \cdot \frac{1}{2} \cdot \frac{h}{2\pi} = 1/1{,}840$ Bohr magneton $=$ 1 nuclear magneton is to be expected.

So far as we know, the only method at present available for the determination of this moment is the deflection of a beam of hydrogen molecules in an inhomogeneous magnetic field (Stern-Gerlach experiment). In the hydrogen molecule, the spins of the two electrons are anti-parallel and cancel out. Thus the magnetic moment of the molecule has two sources: (1) the rotation of the molecule as a whole, which is equivalent to the rotation of charged particles, and leads therefore to a magnetic moment as arising from a circular current; and (2) the magnetic moments of the two protons.

In the case of para-hydrogen, the spins of the two protons are anti-parallel, their magnetic moments cancel out, and only the rotational moment remains. At low temperatures (liquid air temperature), practically all the molecules are in the rotational quantum state 0 and therefore non-magnetic. This has been proved by experiment. At higher temperatures (for example, room temperature) a certain proportion of the molecules, which may be calculated from Boltzmann's law, are in higher rotational quantum states, mainly in the state 2. The deflection experiments with para-hydrogen at room temperature allow, therefore, the determination of the rotational moment, which has been found to be between 0.8 and 0.9 nuclear magnetons per unit quantum number.

质子的磁矩

伊曼纽尔·埃斯特曼等

编者按

保罗·狄拉克的电子理论预言：电子的自旋量子数为1/2，磁矩为1玻尔磁子。因为该理论似乎适用于所有自旋量子数为1/2的基本粒子，所以20世纪30年代早期的物理学家们认为质子也有类似的特征，它的磁矩大小与其质量成反比，是电子磁矩的1/1,840。但是，伊曼纽尔·埃斯特曼及其同事在这篇文章中报告了一个非常令人惊奇的结果：质子的磁矩大约是这一预计值的2.5倍。这一结果说明狄拉克的理论无法解释质子的一些重要特征，而且这也是证明核粒子（核子）具有内部结构的第一例实验证据。

电子自旋角动量为$\frac{1}{2}\cdot\frac{h}{2\pi}$，它的磁矩为$2\frac{e}{m_e c}\cdot\frac{1}{2}\cdot\frac{h}{2\pi}$，或者1玻尔磁子。质子的自旋角动量和电子具有相同的值$\frac{1}{2}\cdot\frac{h}{2\pi}$。由此，我们猜想质子磁矩为$2\frac{e}{m_p c}\cdot\frac{1}{2}\cdot\frac{h}{2\pi}$ =1/1,840玻尔磁子 =1核磁子。

据我们所知，目前能够确定这一磁矩的唯一有效方法就是氢分子束在非均匀磁场中偏转的实验（斯特恩–盖拉赫实验）。在氢分子中，两个电子的自旋是反平行的，所以互相抵消了。这样，分子磁矩就有两种来源：(1) 分子作为一个整体的旋转，这等价于带电粒子的旋转，因此可以产生由环形电流引发的磁矩；(2) 两个质子的磁矩。

在仲氢分子中，两个质子的自旋是反平行的，因此它们的磁矩相互抵消，只有转动力矩存在。在低温下（液化空气温度），几乎所有的分子都处于转动量子态为0的状态，所以是非磁性的。这一点已经被实验所证实。在较高的温度下（例如室温），可以用玻尔兹曼定律算出有一定比例的分子处于更高的转动量子态上，以2态为主。因此，室温下仲氢分子的偏转实验可以用来确定其转动力矩，我们已经得到的转动力矩为每单位量子数0.8到0.9核磁子之间。

In the case of ortho-hydrogen, the lowest rotational quantum state possible is the state 1. Therefore, even at the lowest temperatures, the rotational magnetic moment is superimposed on that due to the two protons with parallel spin. Since, however, the rotational moment is known from the experiments with pure para-hydrogen, the moment of the protons can be determined from deflection experiments with ortho-hydrogen, or with ordinary hydrogen consisting of 75 percent ortho- and 25 percent para-hydrogen. The value obtained is 5 nuclear magnetons for the two protons in the ortho-hydrogen molecule, that is, 2.5 (and not 1) nuclear magnetons for the proton.

This is a very striking result, but further experiments carried out with increased accuracy and over a wide range of experimental conditions (such as temperature, width of beam, etc.) have shown that it is correct within a limit of less than 10 percent.

A more detailed account of these experiments will appear in the *Zeitschrift für Physik*.

(**132**, 169-170; 1933)

I. Estermann, R. Frisch, O. Stern: Institut für physikalische Chemie, Hamburgischer Universität, June 19.

在正氢中，可能存在的最低转动量子态是 1 态。因此，即便是在最低的温度下，由于两个质子的自旋平行，转动磁矩也会叠加。然而，因为转动力矩已经由纯仲氢的实验测得，所以质子的磁矩可以从正氢或者从含 75% 正氢和 25% 仲氢的普通氢的偏转实验中得到。结果是，正氢分子中两个质子的磁矩是 5 核磁子，也就是说，质子的磁矩是 2.5（而不是 1）核磁子。

这是一个非常惊人的结果，但我们在后来的实验中提高了精度，并尝试了各种不同的条件（例如，不同温度，不同分子束宽度等），实验仍表明，这个结果的误差小于 10%。

我们将在《物理杂志》上发表有关这个实验的更多细节。

（王静 翻译；赵见高 审稿）

Interaction between Cosmic Rays and Matter

B. Rossi

Editor's Note

Italian physicist Bruno Rossi had recently shown that cosmic rays passing through matter often generate secondary particles. He had developed a sensitive technique to investigate the secondary "particle showers" by detecting the near-simultaneous passage of particles at three different detectors. Here he reports that showers seems to be more abundant in materials of higher atomic number, and that the particles penetrate lead for shorter distances than primary cosmic rays. Rossi's technique spurred the further development of "coincidence-detection" devices for high-energy and nuclear physics. Rossi himself later showed that cosmic rays near the Earth's surface have two components. The first component was ultimately identified as electrons and photons, the second with a new type of particle called the muon.

LAST year I showed by means of a coincidence method that a secondary corpuscular radiation is generated when cosmic rays pass through matter[1]. From the beautiful experiments of Blackett and Occhialini we know now that these secondary particles are produced in so-called "showers". This can also be shown by the method of coincidences; moreover, if a coincidence is observed between counters arranged in a triangle, we may conclude with a high degree of certainty that we have to do with a shower originating from a point in the neighbourhood of the counters[2]. The method of triple coincidences therefore offers a very useful means for investigating the frequency of occurrence of the showers.

Up to the present, the following results have been obtained:

(1) The showers occur more frequently in elements of high atomic number; the ratio of the numbers of coincidences caused by thin layers of lead, iron, aluminium of the same weight per cm.^{2} is 4:2:1 approximately[3].

(2) The number of showers emerging from a layer of lead, as a function of the thickness of this layer, increases at first, reaches a maximum at a thickness of about 20 gm./cm.^{2} and then decreases very rapidly; at 100 gm./cm.^{2}, for example, the frequency of the coincidences is less than one half of the maximum value. We conclude that the radiation which causes the showers has a mean range of a few centimetres in lead. It follows that this radiation cannot be identical with the primary cosmic rays.

(3) When the thickness of the layer is further increased, the frequency of the emerging showers decreases very slowly. The most probable hypothesis to explain this seems to be

宇宙射线与物质之间的相互作用

布鲁诺·罗西

编者按

意大利物理学家布鲁诺·罗西前不久指出，宇宙射线在穿过物质时经常会产生次级粒子。他还发展了一种很灵敏的技术，通过探测三个不同的探测器上几乎同时达到的粒子流来研究这些次级“粒子簇射”。在这篇文章中他指出，在较高原子数元素的物质中，簇射会更多，并且与初级宇宙射线相比，次级粒子在铅中的穿透距离更短。罗西的技术促进了高能物理和核物理学中所使用的“符合探测”装置的进一步发展。罗西本人后来发现，地球表面附近的宇宙射线中存在两种组分，其中一种组分最终被确定是电子和光子，而另一种组分中包含一种被称为 μ 介子的新型粒子。

去年，我用符合法证明了当宇宙射线穿过物质时会产生次级微粒辐射 [1]。现在，我们根据布莱克特和奥基亚利尼的精巧实验得知这些次级粒子是在所谓的“簇射”中产生的。这一点也可以用符合法来证明；此外，如果我们看到排列成三角形的计数器同时触发信号，那么就可以非常确定地得出以下结论：这种符合一定与这些计数器附近某一点上产生的簇射有关 [2]。因此，三重符合法对于研究簇射出现的频率是非常有效的。

到目前为止，我们已经得到了以下这些结论：

（1）簇射在高原子数元素的物质中出现得更频繁；由每平方厘米重量相同的铅、铁、铝薄层引发的符合次数比率大约为 4:2:1[3]。

（2）铅层中产生的簇射数是该层厚度的函数，随着厚度的增加，簇射数一开始增加，在厚度约为 20 克 / 平方厘米时达到最大值，然后迅速减少；例如，当厚度为 100 克 / 平方厘米时，符合出现的频率不到最大值的一半。我们的结论是：引发簇射的辐射在铅中的平均射程为几个厘米。由此可以肯定这种辐射不同于初级宇宙射线。

（3）当薄层的厚度继续增加时，簇射出现的频率减小得非常慢。对这一现象最有可能的解释是假设引起簇射的射线是在层中进一步产生的；因而这些射线被认为

to assume that a further production of the rays which cause the showers takes place in the layer; these rays are therefore to be regarded as a secondary radiation of the primary cosmic rays, the equilibrium value of which is roughly three to four times greater in air than in lead.

(4) The shower-producing rays are more readily absorbed by elements of higher atomic number. When 24.5 gm./cm.^{2} of lead is placed over the counters, 70±3 coincidences per hour are observed; this number was reduced to 36.7±1.4 by a further sheet of lead of 39 gm./cm.^{2} on top of the first one but only to 52.3±1.7 by a sheet of aluminium of the same weight per cm.^{2} and in the same position. From this and from (1) we conclude that the absorption of these secondary rays by an element and the number of showers which they produce depend in the same way on the atomic number. Thus it seems that the production of showers must be the main reason for their absorption. This is in agreement with the consideration that these rays should have an energy of at least some milliards of electron-volts, which could not be absorbed by a few centimetres of lead in the ordinary way.

(5) That the equilibrium value of the secondary radiation is lower in elements of high atomic number may be explained by their greater absorption, if we assume that the rate of production is roughly the same in all elements; which seems plausible from the experiments on the absorption of the primary rays.

(**132**, 173-174; 1933)

Bruno Rossi: Physical Institute, University of Padova, Italy, July 3.

References:

1. Rossi, B., *Phys. Z.*, **33**, 304 (1932).

2. Rossi, B., *Atti. R. Acad. Naz. Lincei* (in the press).

3. Rossi, B., *Z. Phys.*, **82**, 151 (1933).

是初级宇宙射线的次级辐射，它们在空气中的平衡常数是在铅中的 3~4 倍左右。

（4）产生簇射的射线更容易被较高原子数的元素所吸收。当把 24.5 克 / 平方厘米的铅置于计数器上时，每小时可以观察到 70±3 次符合；如果在原有铅层上再放置一层 39 克 / 平方厘米的铅，这个值就会减小到 36.7±1.4，但当在同一位置放置每平方厘米重量相同的铝层时，这个值只减小到了 52.3±1.7。根据这一现象和（1），我们可以得出这样的结论：元素对次级射线的吸收以及次级射线产生的簇射数以同样的方式取决于原子数的大小。由此看来簇射的产生一定是射线吸收的主要原因。这与认为这些射线至少具有数十亿电子伏特的能量，在通常情况下不可能被几厘米厚的铅层吸收的论点是一致的。

（5）如果我们假设簇射在所有元素中的产生率大致相同，那么就可以把原子数高的元素次级辐射平衡常数较小的原因归于原子数高的元素具有较强的吸收；从初级射线吸收实验的结果来看，这样的解释看上去是合理的。

（王静 翻译；刘纯 审稿）

Disintegration of Light Atomic Nuclei by the Capture of Fast Neutrons

W. D. Harkins *et al.*

Editor's Note

Observations of nuclear disintegration induced by neutrons provoked suspicions that such nuclear processes play a role in the Sun and other stars. Here William Draper Harkins and colleagues report experiments on the disintegration of nitrogen and several other nuclei by high-energy neutrons. They noticed that the incident neutrons must possess a certain threshold energy, which is converted to a mass increase in the product particles. For nitrogen this threshold is equivalent to the energy of particles in a gas with average temperature of about 10^{10} K. A small fraction of particles in a gas at typical solar temperatures would also have such velocities, suggesting that light atoms may undergo such disintegrations in stars.

ABOUT thirteen disintegrations of neon nuclei have been obtained in 3,200 pairs, and approximately 100 disintegrations of nitrogen nuclei in 7,600 pairs of photographs of a Wilson chamber through which neutrons were passing. The source of the neutrons consisted of beryllium powder intimately ground with a mixture of mesothorium and thorium-X. The neutron source used was on the average more powerful in the experiments with neon than with nitrogen. If all the factors are taken into account, it is found that with identical atomic concentrations of neon and of nitrogen in the chamber, the neon nuclei are disintegrated much less often than those of nitrogen.

The average energies of the neutrons which have been found to disintegrate light nuclei are, in millions of electron volts, 5.8 for nitrogen, 7.0 for oxygen, and 11.6 for neon. Here the value for oxygen is taken from the work of Feather. The mass data indicate that the energy needed to supply mass increases in just this order, and is respectively -1.4×10^6, 0 and $+2\times10^6$ electron volts, if the mass of the neutron is assumed to be that given by Chadwick, 1.0067, which is probably too high. Obviously the value assumed does not affect the differences between the energy values.

In a gas, ethylene, which consists of hydrogen and carbon, three disintegrations were obtained in 3,200 pairs of photographs. If carbon (12) is disintegrated by capture of the neutron the reaction is

$$C^{12} + n^1 \rightarrow C^{13} \rightarrow Be^9 + He^4$$
$$12.0036+1.0067\rightarrow 9.0155+4.00216$$

轻核俘获快中子产生的核衰变

威廉·德雷珀·哈金斯等

编者按

在观察到由中子引发的核衰变之后，人们猜测这样的反应可能会存在于太阳和其他恒星之中。在本文中，威廉·德雷珀·哈金斯和他的同事们报告了氮及其他原子核在高能中子诱发下产生的衰变。他们发现入射中子必定有一个能量阈值，这个能量转化为反应产物粒子的质量增加。对于氮来说，这一阈值等价于粒子处于气体平均温度约为 10^{10} K 时的能量。在太阳所处的温度下，气体中有一小部分粒子可以达到这样的速度，这说明恒星中的轻原子可能确实经历了这样的核衰变过程。

观察中子穿过威尔逊云室所得到的照片，在 3,200 对氖核照片中得到了大约 13 次核衰变，而在 7,600 对氮核照片中则大概得到了 100 次核衰变。中子源是由铍粉末与新钍和钍–X 的混合物充分研磨而成。平均而言，所使用的中子源在用氖进行实验时比用氮进行实验时强。如果考虑到所有的因素，我们就会发现当云室中的氖和氮具有相同的原子浓度时，氖核的衰变数通常会显著地少于氮核的衰变数。

我们已经知道能使轻核发生衰变的中子的平均能量，若以百万电子伏特计，对于氮是 5.8，对于氧是 7.0，对于氖则是 11.6。上述关于氧的数值是从费瑟的研究成果中得到的。质量数据表明：如果假定中子的质量等于查德威克给出的数值 1.0067——这个值有可能偏高，那么供给反应产物粒子的质量增加所需的能量也恰好按这个顺序增加，分别为 -1.4×10^6，0 和 $+2\times10^6$ 电子伏特。显然，假定的中子质量大小并不影响能量之间的差值。

在由氢和碳组成的乙烯气体中，我们从 3,200 对云室照片中得到了 3 次衰变。如果碳 (12) 通过捕获中子而发生了衰变，则其反应为：

$$C^{12} + n^1 \rightarrow C^{13} \rightarrow Be^9 + He^4$$
$$12.0036 + 1.0067 \rightarrow 9.0155 + 4.00216$$

or, if the mass assumed for the neutron is correct, $\Delta m = 0.0074$, which is equivalent to 6.9×10^6 electron volts.

This corresponds to a velocity of 3.6×10^9 cm. per sec., so only neutrons of a velocity higher than this should be effective in disintegrating carbon of mass 12. The smallness of the yield of disintegrations which we have obtained with carbon is thus to be expected, especially since probably less than one-fifth of the neutrons have velocities higher than this.

A remarkable relation which has been found to hold without exception is: in disintegrations by capture of a neutron the kinetic energy almost always decreases, is sometimes conserved, but in no case increases.

It has been pointed out previously by Harkins that the values for the energy which disappears suggest definite energy values for the γ-rays into which this energy is converted, but the accuracy of the work is not yet sufficient to prove that this is true.

It may be assumed that the neutrons in the stars are scattered by the atomic nuclei and thus take part in the temperature distribution of velocities of the atoms. If the neutrons of higher velocity are captured much more often than those of lower velocity, the distribution will be affected. Our experiments give no information concerning the capture of neutrons without disintegration, but only for those cases in which the capture is revealed by the accompanying disintegration.

It is of interest in this connexion to consider the minimum energy of the neutron which has been found to give a disintegration. The values, in millions of electron volts, are 1.9 for nitrogen, and 7.8 for neon. The corresponding maximum values are 16.0 and 14.5, the lower maximum for neon being due to the smallness of the number of disintegrations which have been obtained in this gas. An energy of 1.9×10^6 corresponds to a mean temperature of the order of 10^{10} degrees, but at 10^8 degrees a considerable number of neutrons should have this energy, and a moderate number even at 10^7, so it is not unreasonable to suppose that nitrogen nuclei are disintegrated by this process in the stars.

A part of this work was presented by Harkins on June 23 at a symposium on nuclear disintegration under the auspices of the Century of Progress Exposition, Chicago. Other papers were presented by Cockcroft, Lawrence and Tuve, and a general discussion of the theory was given by Bohr.

(**132**, 358; 1933)

William D. Harkins, David M. Gans and Henry W. Newson

也就是说，如果我们采用的中子质量是正确的，则 Δm=0.0074，相当于 6.9×10^6 电子伏特。

这对应于 3.6×10^9 cm/s 的速度，所以只有速度超过这个值的中子才能有效地使质量为 12 的碳发生核衰变。所以我们就可以理解为什么碳的衰变率比较低了，因为很可能只有不到五分之一的中子具有高于这个值的速度。

一个至今没有出现过例外情况的、值得注意的关系是：在俘获中子的核衰变过程中，动能几乎总是在减少，有时候会保持恒定，但绝不会增加。

哈金斯以前曾指出：反应中消失的能量值就是由该能量转变成的 γ 射线的确切能量值，但是目前该项实验的精确度还不足以证明这种说法是正确的。

我们也许可以假设恒星中的中子被原子核散射，因而参与了原子速度的温度分布。如果具有较高速度的中子比具有较低速度的中子更容易被俘获，那么上述分布情况就会受到影响。我们的实验没有给出与未引起核衰变的中子俘获相关的信息，而只涉及了那些伴随着核衰变的俘获现象。

与此相关，我们有兴趣了解目前已发现的能导致核衰变的中子所具有的最小能量。以百万电子伏特计，该值对于氮是 1.9，对于氖是 7.8。与两者对应的最大值分别为 16.0 和 14.5，氖的最大值偏低是因为在这种气体中得到的核衰变数较少。当平均温度的数量级达到 10^{10} 度时，对应的能量值为 1.9×10^6，不过在 10^8 度时就会有相当数量的中子具有这一能量，甚至在 10^7 度时就会有一些，所以假定在恒星中氮核可以通过这一过程实现核衰变没有什么不合理。

哈金斯于 6 月 23 日在芝加哥世纪进步博览会主办的关于核衰变的研讨会上介绍了该项研究中的一部分内容。考克饶夫、劳伦斯和图夫宣读了另外一些论文，对理论进行全面阐述的则是玻尔。

（王耀杨 翻译；朱永生 审稿）

Extremely Low Temperatures

W. J. de Haas

Editor's Note

Physicist Wander Johannes de Haas of the University of Leiden reports here on recent experiments reaching extremely low temperatures using the method of "adiabatic demagnetisation". In this method, a material magnetised by being held in a strong magnetic field can grow much colder if the field is abruptly removed. De Haas describes experiments with cerium fluoride. Preliminary tests were able to achieve a temperature of 0.27 K, and de Haas's team had more recently reduced this to 0.085 K. With the same apparatus, de Haas suggests, much lower temperatures should be possible. Given these extremes, he notes, current methods for measuring temperatures have reached their limits, and further progress will require new thermometers suited to very low temperatures.

IN 1926 Debye pointed out that temperature must decrease when a magnetised body is demagnetised adiabatically. Giauque made the same remark in 1927; and still earlier the same idea was expressed by Langevin for oxygen. Debye calculated the predicted effect for the case of gadolinium sulphate. His calculation was based upon the following considerations: a magnetisable body contains a great number of small elementary magnets. When such a body is magnetised these magnets are directed. The part of the entropy belonging to this order is decreased and, the process being supposed isentropic, the part of the entropy connected with the statistical movement must necessarily increase. When, on the contrary, the disorder of the elementary magnets is increased by demagnetisation, the part of the entropy connected with the magnetisation is increased and the part belonging to the statistical movement is decreased, so that the body is cooled down.

If we wish to obtain easily seen results, the following points require special attention: (1) the elementary magnets shall not exert a directing influence upon each other (no ferromagnetic body); (2) the elementary magnets shall have a moment as large as possible, subject to the restriction of (1); (3) at low temperatures the effect will be greatest, as here the part of the entropy belonging to the magnetisation becomes comparable with the other part, while at the same time the order strongly increases.

The condition to be fulfilled by the experimental arrangement in order to obtain extremely low temperatures is that the exchange of heat with the surroundings both by radiation and by connexion is cut down. In our experiments this condition is satisfied. The substance that is cooled down is at the same time used as a thermometer.

The experiments were made with cerous fluoride (a weakly paramagnetic salt of one of the rare earths) as suggested by Prof. Kramers, of Utrecht (see "Leipziger Vorträge",

极低温

万德·约翰尼斯·德哈斯

编者按

荷兰莱顿大学的物理学家万德·约翰尼斯·德哈斯在下文中描述了他最近利用“绝热退磁”法达到极低温的实验。在这一方法中，先使一种材料在强磁场中被磁化，然后突然去除磁场，则材料会变得更冷。德哈斯介绍了他用氟化铈所作的实验。最初实验，温度只能达到 0.27 K，近来德哈斯领导的团队已经把温度降低到 0.085 K。德哈斯认为，用这一装置还可以得到更低的温度。在给出这些极限值之后他注意到，在这么低的温度之下，当前的测温手段都已经达到了它们的极限，下一步的进展要求适用于很低温度的温度计。

1926 年德拜指出，被磁化了的物体在绝热去磁时其温度一定会降低。吉奥克在 1927 年也提出了同样的观点；在更早之前，朗之万就曾提到氧有这样的性质。德拜以硫酸钆为例，对预期效应进行了计算。他的计算基于以下几个原则：一种可磁化的物体中含有大量小的单元磁体。当该物体被磁化时，这些磁体会发生定向排列。属于这一有序的那部分熵会减少，而该过程被认为是一个等熵的过程，与统计运动相关的那部分熵就一定会增加。反过来，当单元磁体的无序度由于退磁而增加时，与磁化有关的那部分熵就会增加，而与统计运动相关的那部分熵则会减少，从而使物体的温度降低。

如果我们想得到显而易见的结果，就需要特别注意以下几点：(1) 单元磁体彼此之间不应该存在直接的相互作用（非铁磁体）；(2) 由于第 (1) 点的限制，单元磁体应具有尽可能大的磁矩；(3) 这种效应在低温时将达到最大，因为这时与磁化对应的那部分熵变得与另一部分熵大小差不多，与此同时，有序剧烈地增加。

为了获得极低的温度，在实验中必须保证完全切断与环境之间通过辐射或传导而进行的热交换。在我们的实验中这个条件是满足的。被冷却的物质同时也被用作温度计。

实验是用乌得勒支大学的克拉默斯教授建议的三氟化铈（一种稀土弱顺磁盐）（参见《莱比锡演讲》，1933 年）进行的。图 1 简单画出了我们所用的实验装置。棒

1933). Fig. 1 very diagrammatically represents the experimental arrangement. The rod A is fixed to a balance. It carries a small Dewar flask, which contains a small tube filled with cerous fluoride and suspended by a central carrier B. The whole apparatus is placed between the poles of the large electro-magnet of the Kamerlingh Onnes Laboratory in such a way that the salt occupies the spot of maximum $H.dH/dx$ (x=vertical co-ordinate). The lower extremity of the rod A together with the small Dewar flask is surrounded by liquid helium boiling at 1.26°K. The thermal insulation of the cerous fluoride is so good, that after 4–5 hours only a very small quantity of salt of exceedingly low heat capacity (the latter is inversely proportional to T^3) is cooled down to 1.26°K.

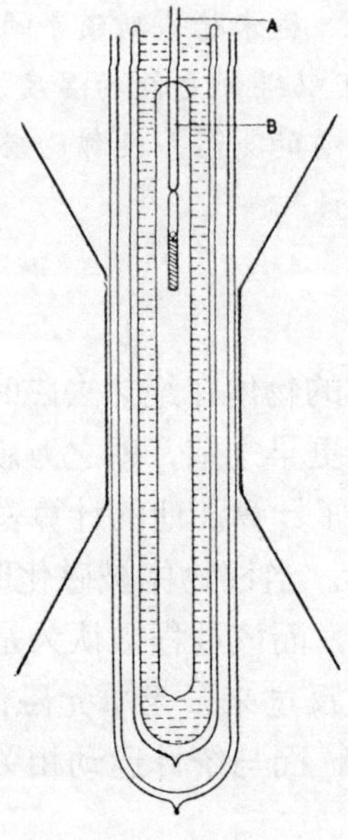

Fig. 1

We know that finally this temperature has been reached, because the salt is the indicator of its own temperature. This paramagnetic body is drawn into the field (31 k. gauss) by a force $K=M.dH/dx$, where M is the total moment of the body. As $M=\varphi(H, T)$ and as H remains constant, we easily see, by measuring K, when T has become constant.

As soon as this is the case the body is demagnetised by lowering the field to 2.7 k. gauss in the first experiments, to 1,000 or 500 gauss in later experiments.

In this weak field the force K is measured as a function of the time. In this way we obtain a curve as shown in Fig. 2.

A 固定在一平衡物上。它连着一个小的细长杜瓦瓶，杜瓦瓶中的小试管装有三氟化铈，小试管由中间导管 B 悬挂着。整个装置被置于卡默林·昂内斯实验室的大型电磁铁的两磁极之间，并确保顺磁盐样品处于 $H.dH/dx$ 最大值的位置（x 为垂直坐标）。棒 A 的最下端以及整个小杜瓦瓶都浸没在沸点为 1.26 K 的液氦中。三氟化铈的热绝缘非常好，以至于 4~5 个小时之后只有极少量的这种具有极低热容量（与 T^3 成反比）的盐被冷却到了 1.26 K。

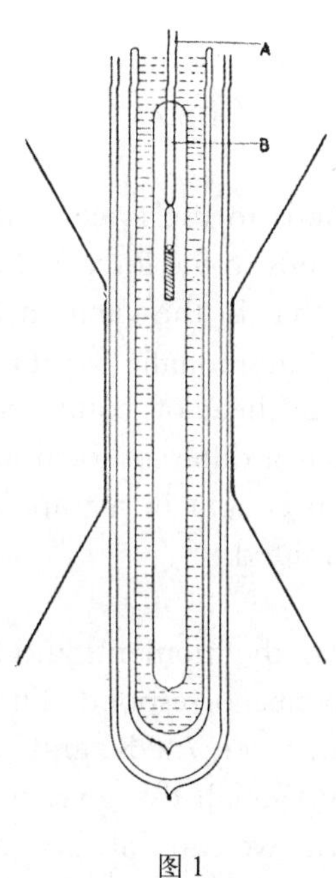

图 1

由于这种盐是其自身温度的指示器，我们知道最终被冷却到了 1.26 K。力 $K=M.dH/dx$ 将顺磁体吸入磁场中（强度为 31,000 高斯），其中 M 是顺磁体的总磁矩。由于 $M=\varphi(H, T)$，且 H 是保持不变的，我们很容易通过测量 K 而看出 T 是何时变为恒定的。

一旦满足了以上所说的情况，就通过降低磁场给顺磁体退磁，在头几次实验中把磁场降至 2,700 高斯，而在以后的实验中降到了 1,000 或 500 高斯。

在这种弱场中，测量了作为时间的函数的力 K。这样，我们得到了图 2 所示的曲线。

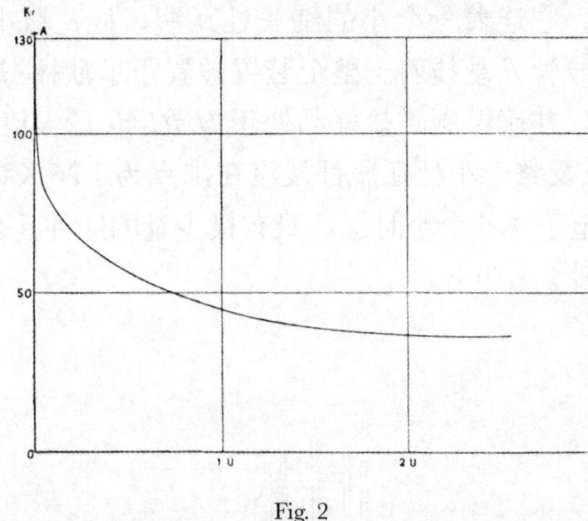

Fig. 2

The point A on the curve corresponds to the force at the lowest temperature, while the asymptote after a long time corresponds to the force at 1.26°K. The steep fall of the curve in the beginning is due to the fact that in the vacuum of the small Dewar flask a trace of helium (10^{-7} mm.) has been left intentionally for thermal contact. This helium gas is condensed under development of the heat of condensation. When the pressure of the helium gas is still lower, the curve has a different form and heating up with the time goes extremely slowly. The cold has been caught in a trap. The vacuum becomes practically absolute and heat is no longer transmitted.

The forces are directly proportional to the moments and the only thing we have to find out is to which temperatures these moments correspond. The connexion between temperature and moment has been determined between 7.2°K. and 1.3°K. The curve representing this connexion was extrapolated linearly, though this gives a stronger increase of the moment than the curve itself does. That is why we can only give an upper limit of the temperature reached.

The first experiments were made in March and April of this year with cerium fluoride and gave as the upper temperature limit 0.27°K. More recent experiments with dysprosium ethyl sulphate gave the upper limit 0.17°K. Finally, the experiments of July last made with cerium ethyl sulphate gave as the upper temperature limit 0.085°K.

It is possible that with the same experimental arrangement much lower temperatures can be reached. The choice of the right substance will decide further results. I am convinced that with the above-described arrangement the theoretical limit can be reached.

Up to the present, the lowest temperatures had been reached with the aid of liquid helium, boiling under very low pressures. The results of this method evidently depend upon the capacity of the pumps used and upon the perfection of the thermal insulation.

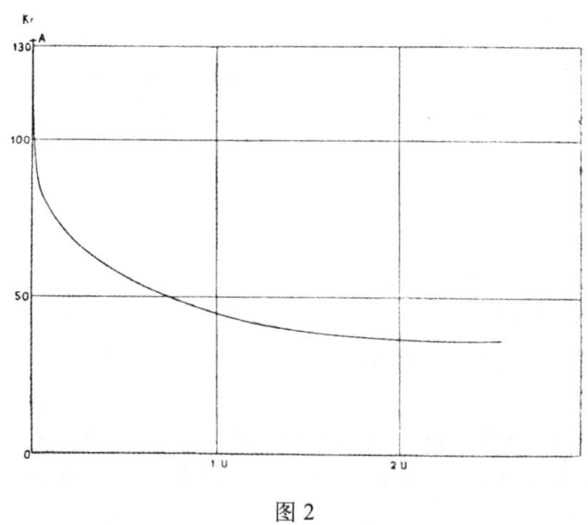

图 2

曲线上的 A 点对应于最低温度时力的值，而长时间后的渐近线趋近于温度为 1.26 K 时力的值。曲线在开始时急剧下降是因为在抽成真空的小杜瓦瓶中，为了保证热接触而有意留下了少量的氦（10^{-7} mm）。这些氦气在放出凝聚热的过程中被凝聚。当氦气的压强更低时，曲线具有不同的形式，随时间而升温的速度十分缓慢。低温被维持在阱里。这时的真空实际上变成绝对真空，而且不再有热传输了。

力正比于磁矩，因而我们唯一需要确定的是这些磁矩对应什么温度。在 7.2 K 到 1.3 K 之间，温度与磁矩的关系已经确定了。代表这一关系的曲线被线性外推，尽管这样做给出的磁矩随时间增加的幅度比实际曲线上增加的幅度还要大一些。这就是为什么我们只能给出所能达到的温度的上限的原因。

最初的实验是在今年 3 月和 4 月用氟化铈进行的，得到的温度上限为 0.27 K。较近期实验是用硫酸乙基镝进行的，给出的温度上限为 0.17 K。最后一次实验是在今年 7 月用硫酸乙基铈进行的，给出的温度上限为 0.085 K。

用同样的实验装置有可能达到更低的温度。正确地选用顺磁物质将决定未来的结果。我确信用上述装置可以达到理论的极限。

目前人们已经可以用在极低压强下沸腾的液氦达到最低的温度了。这一方法的最终结果很明显地取决于所用泵的排量，以及热绝缘的理想程度。

With this method Kamerlingh Onnes reached in October, 1921, a temperature of 0.82°K. Keesom worked with diffusion pumps of a capacity fifteen times higher than that of the pumps Kamerlingh Onnes had. In 1929 he reached a temperature of 0.71°K. It is, however, difficult to proceed much further in this way.

The conception of temperature is based upon the properties of the ideal gas. The temperature is determined with the aid of the helium thermometer (several corrections being applied). At the extremely low temperatures reached recently no gas thermometric measurement of the temperature is possible, so far as I can see.

A thermometric scale based upon another process must be fitted to the absolute temperature scale; only a magnetic scale can be used for this purpose.

Just as was done in gas thermometry, we shall have to find some substances, which within a considerable range of temperature show the same behaviour; a highly developed theory will then enable us to fix the temperature with the same accuracy by means of magnetic thermometry as by the use of gas thermometric methods.

A reservation must be made, however, for the case that at very low temperatures the substances used might become ferro-magnetic or might show a new kind of ferro-magnetism. In this case little might be said about the temperature.

Another great difficulty is this: if by means of the magnetic method we wish to cool down other substances, the question of the thermal contact becomes urgent. The radiation is negligible, and also the vapour pressure of the helium becomes so small that thermal contact by means of gaseous helium can scarcely be considered.

The experiments were made in collaboration with Dr. E. C. Wiersma, conservator of the Laboratory, to whom I express my warmest thanks for his help and for many suggestions. To Prof. H. A. Kramers of Utrecht I am indebted for his valuable theoretical assistance.

(**132**, 372-373; 1933)

1921 年 10 月，卡默林 · 昂内斯利用这种方法达到了 0.82 K 的低温。基桑使用的扩散泵排量是卡默林 · 昂内斯所用的 15 倍。他在 1929 年实现了 0.71 K 的低温。不过，这个方向的研究已经很难再有更大的进展了。

温度的概念是以理想气体的性质为基础的。我们是借助氦温度计（已进行过多次校正）来测定温度的。在我看来，在最近所达到的极低温度下，没有一种气体温度计的测温方法是可行的。

基于其他过程的温标必须拟合到绝对温标；为此目的，只有磁学标度是可用的。

正如在气体测温术所做的那样，我们必须找到某种在相当宽的温度范围内呈现同一行为的物质；然后，已经高度发展了的理论将使我们能够用磁测温术得到与气体测温法同样精确的温度。

但是我们必须有所保留，因为在极低的温度下，所使用的物质可能会变成铁磁性或者表现出新型的铁磁性。在这种情况下，就很难谈什么温度了。

另一个巨大的困难是：如果我们希望利用磁学方法来冷却其他物质，那么热接触就变成了非常紧迫的问题。辐射是可忽略的，而且氦的蒸汽压也变得非常小，以至于几乎不能考虑通过气态氦进行热接触。

实验是在实验室管理员威尔斯玛博士的协作下进行的，我要对他的帮助和诸多建议表示最真诚的谢意。我还要感谢乌得勒支的克拉默斯教授在理论方面所提供的有价值的帮助。

（王耀杨 翻译；陶宏杰 审稿）

Recent Developments in Television*

A. Church

Editor's Note

Companies in Europe and America were racing to develop the technology for television. As Archibald Church notes in this review, the scepticism widely expressed in 1926 after John Baird first demonstrated blurred and flickering television images was now being replaced by feverish excitement. In 1932, the Derby horse race had been televised and projected to a live audience in London, and the British Broadcasting Corporation had installed transmission equipment from Baird Television Ltd. Now engineers were developing improved displays based on cathode rays projected on fluorescent screens, and live broadcast was approaching feasibility. A finer screen resolution was needed, however and it was becoming necessary to allocate place in the broadcast spectrum for television signals.

ALL development of the art of television is recent. It is less than ten years since John Baird first obtained televised images of simple stationary objects such as a Maltese cross. He first demonstrated "real" television, the instantaneous reception of optical images of moving subjects, images of which had been transmitted by means of a variable electric current, on January 27, 1926. Most of the scientific workers and publicists present at that demonstration, while impressed by the achievement, were frankly sceptical of television ever achieving any position as a medium of entertainment or of its being put to other commercial uses. The received images were recognisable, but blurred and flickering, and to many scientific workers, a proof of the impossibility of advance in television by a mechanical system of transmission and reception. Other scientific observers, though less antipathetic to the mechanical system, were unconvinced that television broadcasting would ever be practicable owing to the wide range of frequencies which would have to be made available if images with detail comparing with that obtainable on a cinema screen were to be received. This was the more vital criticism of television, as it applied not only to the mechanical means by which Baird obtained his first results but also to any other means, for example, the utilisation of cathode rays, which might afterwards be enlisted in the service of television.

Only minor modifications had been made of the original apparatus used when Baird gave his first demonstration to members of the British Association at Leeds in 1927. In 1926 the subject to be televised was bathed in light from a battery of powerful electric lamps. Between the photoelectric cells and the illuminated subject was a scanning device, a disc in which thirty holes were punched at regular intervals on a spiral and making five revolutions per second. The subject was thus scanned by a rotating optical element strip by strip, each strip being presented in sequence to a light sensitive element,

* Paper read before Section A (Mathematical and Physical Sciences) of the British Association at Leicester on September 13.

电视技术的最新进展*

阿奇博尔德·丘奇

编者按

欧洲和美洲的公司在电视技术的研制方面正在展开竞争。正如阿奇博尔德·丘奇在这篇评论文章中所说的：当约翰·贝尔德在1926年首次向公众演示那些模糊和闪烁的电视图像时，人们普遍对此持怀疑态度，但现在这项技术正如火如荼地发展着。1932年，有人把德比马赛拍摄下来并实时传送给伦敦的观众；而且英国广播公司也已经安装了贝尔德电视有限公司的转播设备。现在工程师们正在想办法把阴极射线投射到荧光屏上以改善图像的质量，并使现场直播成为可能。这需要有较高的屏幕分辨率，而且还需要在广播频带中为电视信号安插一定的位置。

电视技术的所有发展都是最近涌现出来的。从约翰·贝尔德第一次获得简单静态物体，例如马尔他十字形的电视图像到现在还不到10年时间。在1926年1月27日，贝尔德第一次演示了“真正”的电视，这个装置可以即时接收到运动物体的光学图像，图像的传输是通过电流的变化实现的。很多科学工作者和媒体人士都出席了这次演示会，虽然他们被这项成就深深震撼，但仍坦言对电视是否能成为娱乐业的媒介之一或是否能被应用于其他商业用途持怀疑态度。尽管人们可以分辨出接收到的图像，但比较模糊而且忽隐忽现，所以许多科学工作者认为，电视的发展不可能通过机械式的传送和接收系统来实现。尽管其他的科学观察者对机械系统的反对不那么强烈，但也不相信可以这样来实现电视的转播，因为电视图像要是能与电影院屏幕上播放的图像一样清晰，就需要有很宽的频率范围。这个提法对电视的发展更为致命，因为它不仅否定了贝尔德在获得初步成功时所应用的机械手段，也否定了将来有可能应用于电视技术的其他手段，如阴极射线。

在贝尔德1927年首次在利兹向英国科学促进会成员展示电视的时候，这套设备仅仅做了很小的改动。在1926年时，人们用一组功率很高的电灯照射要被电视转播的物体。在光电管和被照射的物体之间有一个扫描设备，扫描设备是一个圆盘，在圆盘的周边沿螺线每隔一定间隔排列着30个小孔，圆盘每秒转5圈。一个旋转的光学元件一条一条地扫描要被电视转播的物体，每一条信息都会按顺序传递到一个光敏元件——光电管上。光电管中电流强度的变化会改变位于接收端的氖灯

* 这篇论文曾在英国科学促进会A分会（数学和物理学分会）于9月13日在莱斯特举行的会议上被宣读过。

the photoelectric cells. The varying strength of electric current transmitted by the photoelectric cells modified the light in a neon lamp at the receiving end, and this varying single light source was scanned in turn by a "Nipkow" disc synchronised with the disc at the transmitting end. The reconstituted image, two inches square, was seen by looking at the neon lamp through the scanning disc. Synchronism was obtained by the use of synchronous motors.

For the Leeds demonstration "noctovision" was used, the person televised being shielded from the direct glare of the lamps by a sheet of ebonite. In the meantime, however, Baird had made a further notable advance by his invention of the light spot method of scanning. To quote the text of his patent: "The scene or object to be transmitted is traversed by a spot of light, a light sensitive cell being so placed that light reflected back from the spot of light traversing the object falls on the cell." It is, in effect, an inversion of the flood-lighting method, and possesses the advantage that greatly increased signal strengths are obtained with considerable diminution in the intensity of the illumination to which the subject of transmission is exposed.

Abroad, the method was almost immediately applied by the Bell Telephone Company in the United States in carrying out a television transmission over a circuit between New York and Washington. The same year, 1927, Belin and Holweck achieved a measure of success in transmitting outlines and shadowgraphs using a cathode ray oscillograph (Fig. 1). The success of Baird had given an impetus to research in television in several countries as the patent records of England, the United States, Germany and France will testify.

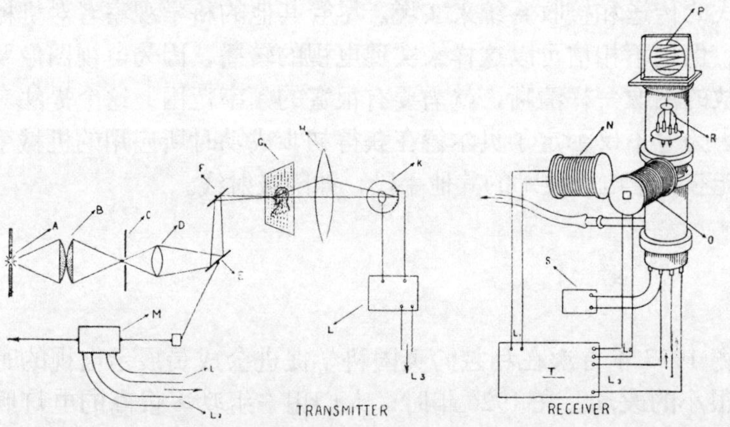

Fig. 1. Schematic diagram of Belin and Holweck's television apparatus.

On February 9, 1928, Baird achieved an ambition to be the first to televise across the Atlantic. The signals were picked up in the presence of Reuter's representative by an amateur operator at Hartsdale, a few miles from New York, the experimental receiver

的亮度，且这个不断变化的单光源被一个与发送端扫描圆盘同步的“尼普科夫”盘（译者注：以德国发明家保罗·尼普科夫命名，尼普科夫发明螺盘旋转扫描器，用光电管把图像的序列光点转变为电脉冲，实现了最原始的电视传输和显示。）依次扫描。重新组合而成的图像面积为 2 平方英寸，我们可以通过扫描圆盘观察氖灯看到这个图像。同步的产生是由同步电机实现的。

在利兹演示的时候，贝尔德使用了“红外线电视”，即为了避免被转播的人被灯光眩射，在人与灯之间隔了一层硬质橡胶。另一个显著的改进是贝尔德发明的光点扫描法。他在自己的专利中是这样写的：“用一束光扫过要传输的场景或物体，把光敏元件放置在这束照射到物体上又被反射回来的光能够经过的位置上。”这实际上是泛光照明法的一种倒置，它的优点是：可以大大增加所得信号的强度，因而可以适当减少照射到被电视转播的物体上的光亮度。

在国外，美国的贝尔电话公司很快就开始应用这种方法通过一个电路在纽约和华盛顿之间实现了电视传输。同一年，也就是 1927 年，贝林和霍尔威克使用阴极射线示波器（图 1）成功地传送了轮廓线和影像图。在贝氏成功的激励下，好几个国家都开始开展电视方面的研究，英、美、德、法的大量专利记录可以证明这一点。

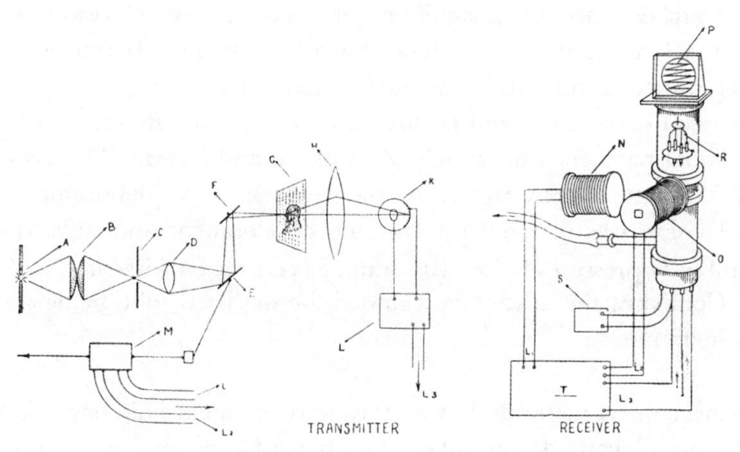

图 1. 贝林和霍尔威克所用电视设备的示意图

1928 年 2 月 9 日，贝尔德第一个实现了在大西洋两岸传送电视信号的目标。在距离纽约几英里远的哈茨戴尔镇，一个业余操作员当着路透社代理人的面接收了电视信号，他用实验接收设备在一个毛玻璃屏上显示出了大小约 3 平方英寸的图像。

showing an image about three inches square on a ground glass screen. This Baird followed almost immediately by a transmission from London to the s. s. *Berengaria* in mid-ocean. According to the chief staff engineer of the vessel, the "image varied from time to time in clarity, but movements could be clearly seen, and the image, when clear, was unmistakable". In these transmissions the wave-length used was 45 metres. The following year, using light spot transmission and cathode glow lamp with disc reconstruction, Baird demonstrated in engineering form at the British Association meeting in Cape Town, and the British Broadcasting Corporation agreed to provide facilities for a series of experimental television broadcasts by the Baird system on the London B.B.C. transmitter. At this time, the transmission of wording for instantaneous news broadcasts, telegram transmission in character, languages and other purposes was further developed by Baird and transmissions of this kind were featured in the experimental broadcasts.

In July 1930, the Baird Company gave its first public demonstration in a theatre, living artists and cinema films being transmitted from its studios in Long Acre, London, W.C.2, and reproduced on a multi-cellular lamp screen on the stage of the London Coliseum. The same year the youthful Baron von Ardenne in Germany commenced his researches on television, utilising the technique acquired in his development of cathode ray oscillograph tubes for the transmission and reception of television images, and within a year earned the distinction of being the first to demonstrate publicly cathode ray reception comparable with that produced by mechanical means. At first, von Ardenne received images transmitted by mechanical means, but later, by using a variable velocity constant intensity cathode stream instead of one of varying intensity and constant velocity, he was able to employ his cathode ray tubes for transmission and reception.

Meanwhile, researches into the possibilities of cathode ray television were engaging the attention of a large number of scientific workers in the laboratories of the Radio Corporation of America and its associated enterprises, independently by the Philco and other American companies, and by the Fernseh A. G. of Berlin, in which the Baird Company are equal partners with Bosch, Zeiss-Ikon, and Loewe. The last-named holds important von Ardenne patent rights. In the Fernseh A. G. laboratories, research in cathode ray television was directed towards the development and utilisation of "hard" tubes, that is, tubes at pressures below 10^{-5} mm., as contrasted with the "soft" tubes in use by the Loewe Company, the advantage claimed for the hard tube being its long life, an important consideration.

Proponents of mechanical methods, however, were by no means discouraged by the results obtained by the use of cathode ray tubes. The Baird Company, by using a mirror-drum instead of a Nipkow disc at the transmitting end, and at the receiving end using either a directly-modulated arc or a multiple Kerr cell in conjunction with a mirror-drum, was able to project fairly bright images on a screen about 6 ft. × 2 ft. in size, and demonstrated its results at the British Association centenary meeting in London, in the exhibition devoted to Mechanical Aids to Learning. This demonstration followed one in January 1931 in the Baird Laboratories at Long Acre of three-zone television, three 30-line mirror-drums

贝尔德很快又将伦敦的电视信号传送到航行在大洋中央的伯伦加莉亚号船上。该船的首席工程师说：“图像的清晰度不太稳定，但可以很清楚地看到运动变化，图片在清晰的时候是准确无误的。”转播使用的波长为 45 米。第二年，使用光点传输法和带有可实现图像重建的圆盘的阴极辉光灯，贝尔德在英国科学促进会于开普敦召开的会议上按照工程学的模式进行了演示，英国广播公司（BBC）同意为在伦敦 BBC 的发射机上使用贝氏系统进行一系列的实验性电视转播提供设备。同时，贝尔德还研制出实时播送新闻广播节目的语音以及用电报传输字符、语言和其他内容的方式，类似这样的传输构成了实验性转播中的主要特征。

1930 年 7 月，贝尔德公司在一家剧院进行了首场公开演示，艺术家的现场表演和电影胶片从位于伦敦 W.C.2 区朗埃克的演播室中被传送出去，并重现在伦敦剧院舞台上的一个多管灯屏幕上。同一年，年轻的冯阿登爵士在德国开始了他对电视的研究，他采用自己研制的、专用于传送和接收电视图像的阴极射线示波管，在不到一年的时间里，他就因第一个向公众展示能与机械式接收系统性能相媲美的阴极射线接收器而声名卓著。起初，冯阿登接收的是通过机械手段传送的图像，但是后来他利用一个速度可变、强度不变的阴极电子流来代替速度不变、强度变化的电流，这样就使他的阴极射线管既能传送图像又能接收图像了。

与此同时，对阴极射线电视的研究吸引了许多在美国无线电公司及其联营企业，主要是飞歌公司和其他美国公司，以及在德国柏林电视机股份公司工作的科学工作者们的注意，在柏林电视机股份公司中，贝尔德公司与博世公司、蔡司伊康公司以及洛伊公司的股权地位相当。洛伊公司是冯阿登专利的主要拥有者。柏林电视机股份公司的实验室在研制阴极射线电视时采用的是“硬”管，即压力低于 10^{-5} mm 的管子，与之相反洛伊公司使用的是“软”管，“硬”管所具有的优势是寿命长，而寿命是人们主要考虑的因素。

然而，机械方法的支持者丝毫没有因阴极射线管所取得的成功而感到沮丧。贝尔德公司在发送端用镜面鼓代替尼普科夫圆盘，在接收端或者用一个直接被调制的弧光灯或者用克尔盒与一个镜面鼓联用，这样就能把很亮的图像投射到一个大小约 6 英尺 ×2 英尺的屏幕上，并且贝尔德公司在英国科学促进会于伦敦召开的成立百周年纪念大会上的以“机械手段辅助学习”为主题的展览中演示了他们的成果。1931 年 1 月，在贝尔德位于朗埃克的三段式电视实验室中，他们又演示了一次他们

being used to obtain an extended image. Later in the same month the Gramophone Company gave a similar performance in London at the Exhibition of the Physical and Optical Society, at which cinema films were transmitted by the multi-channel process and reproduced by means of a Kerr cell and mirror drum apparatus on a translucent screen. In the same year the Derby was televised by the Baird process.

In 1932 five major events in the progress of television took place. Fernseh A. G., the company organised to develop the Baird processes in Germany, built and installed a complete transmission equipment for the Ente Italiano per le Audizione Radiofoniche in Rome; the Derby was televised and projected at the time of its occurrence upon the screen of a London cinema by the Baird Company; the British Broadcasting Corporation installed television transmission equipment designed by Baird Television Ltd., for regular transmissions from its London studio (Fig. 2); and the Baird Company designed and marketed a much improved home television receiver, the Nipkow disc and neon tube of the old type being replaced by a mirror-drum and Kerr cell combination for projecting the received image on a translucent screen: and Dr. Alexanderson, of the American General Electric Company, successfully transmitted and received television images over a light-beam, with apparatus and by methods similar to those demonstrated by the Marconi Company at Leicester at the recent meeting of the British Association.

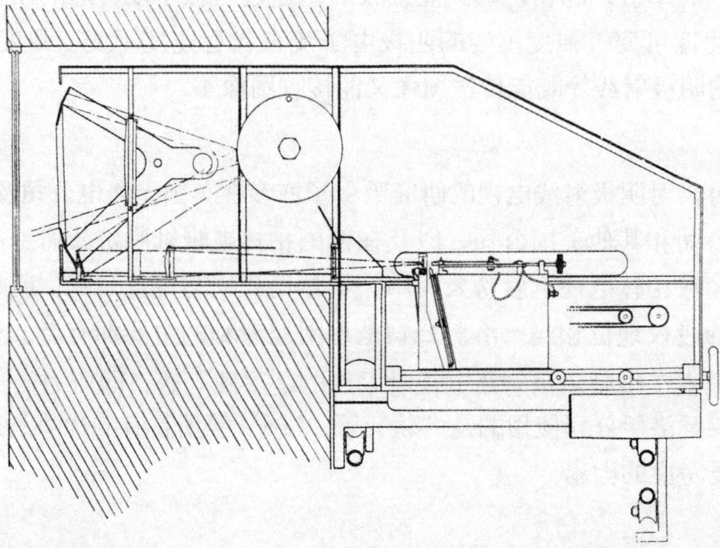

Fig. 2. A sectional scale drawing of the television transmitter as installed at Studio B. B., at Broadcasting House, London.

This year has been one of feverish activity on the part of all companies interested in the commercial exploitation of television, and of numerous independent research workers in various countries. Many interesting and ingenious modifications have been made in the cathode ray oscillographs. The Fernseh A. G. has made tubes with fluorescent ends with diameters up to 2 feet. Von Ardenne has devised a method of projecting the cathode ray

的成果，用3个30线的镜面鼓得到了持续时间较长的图像。在同一月的稍晚些时候，留声机有限公司在物理和光学学会于伦敦举办的展览上进行了类似的演示，他们用多通道方式传送电影胶片上的信息，并通过克尔盒和镜面鼓把重新生成的图像投射到一个半透明的屏幕上。同一年，贝氏方法被用于转播德比马赛。

1932年发生了5件对电视业发展影响重大的事情。德国人为了应用贝尔德的技术组建了柏林电视机股份公司，公司为广播电台设于罗马的意大利办事处安装了一整套电视转播设备；贝尔德公司通过电视转播把德比马赛的现场实时地传送到了伦敦电影院的大屏幕上；英国广播公司也安装了由贝尔德电视有限公司设计的电视转播设备，以用于其伦敦演播室的定期转播（图2）；贝尔德公司还设计了一种更先进的家用电视接收设备并投放到市场中，在这种新设备中，原来用于把接收图像投射到半透明屏幕上的尼普科夫圆盘和老式氖管被镜面鼓和克尔盒的组合体所取代；美国通用电气公司的亚历山德森博士成功地利用光束发送和接收到了电视图像，他所用的设备和方法与马可尼公司在英国科学促进会最近于莱斯特召开的会议上所作的演示类似。

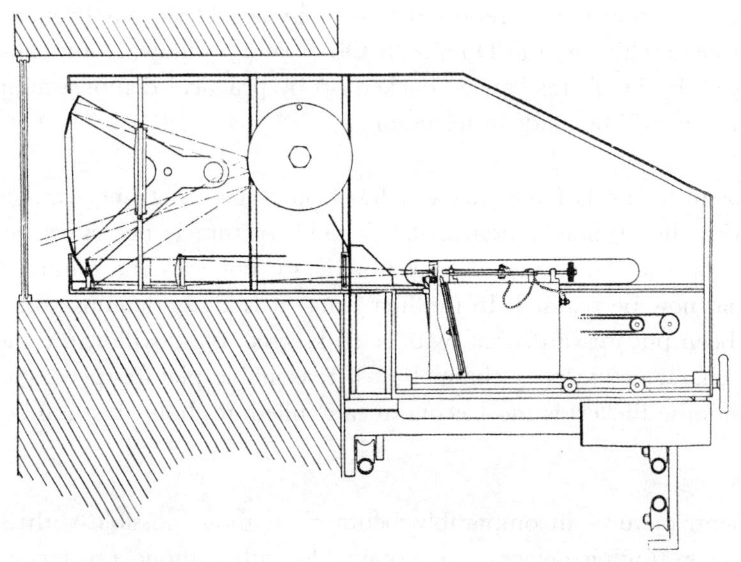

图2. 安装在英国广播公司伦敦广播大楼内演播室中的电视发射机的截面图

今年，所有对电视商业开发感兴趣的公司和来自不同国家的大批独立研究者都表现出了很高的研究热情。大家对阴极射线示波器进行了大量有趣和富有创造力的改进。柏林电视机股份公司制造出了末端发荧光的管子，其直径可达2英尺。冯阿登设计了一种可以把从管内一个板上发射出来的阴极射线束投射到外部屏幕上的方

beam from a plate within the tube on to an external screen. Von Mihaly has developed a mechanical system by which the modulated light of the receiver is swept by a small rotating mirror at the axis of a stationary drum across a number of mirrors fixed on its inside surface, for which he claims superiority over the revolving mirror-drum. The Fernseh A. G. laboratories have constructed beautifully accurate mirror-screws with 90 and 120 reflecting surfaces of stainless steel. Dr. Vladimir Zworykin, the American research engineer, has made sensational claims for what he terms his iconoscope, which has been described as consisting of two devices—a photoelectric mosaic on which a scene is focused by a lens system, and a cathode ray gun which fires at this mosaic screen a stream of electron projectiles. The signal plate on which the scene to be televised is focused may be about 4 in.×5 in. in size and on this surface are millions of small photocells, each consisting of a minute silver globule sensitised by caesium. These globules are deposited on an insulating plate, such as a thin sheet of mica, the back of which is made conductive by a metal coating. Within the same glass bulb as the mosaic screen is the electron gun, which throws a beam of electrons at the screen and is made to sweep across the screen horizontally and vertically by deflecting coils as in an ordinary cathode ray tube. Whenever an electron hits a photocell, it neutralises part of the charge on the associated condenser. This discharge current is picked up, amplified and transmitted to the receiving cathode ray beam which is moving across a fluorescent screen in synchronism with the scanning beam. The varying discharge currents modulate this receiving beam and hence the screen at the receiver. It is reported that a similar device has been invented by Dr. Francis Henroteau, chief of the Dominion Observatory, Ottawa, who calls his invention the "super-eye". If the claims can be backed up by practical demonstrations, a new and important advance will be made in television.

Again and again in the last two years, it has been urged that finer resolution than that obtainable with the 30-line standard, 2.4 : 1 ratio, picture is necessary before television will become popular. This may or may not be so, but the true nature of the present position should now be realised. In the first place, it is easily demonstrable (and it has in fact already been put forward) that, with a 10 kc./sec. band-width, the intelligence-time transmission characteristic of a channel (at a reasonable picture-speed such as 12.5 per second to minimise flicker) is most economically filled at about this number of lines and ratio.

At the moment, pictures incomparably better than those possible with 30 lines, using mechanical reconstituting devices, are obtainable with cathode ray receiving apparatus which has become available this year. Such pictures were first demonstrated publicly in Great Britain by the Baird Company at this year's meeting at Leicester of the British Association and by Loewe, Fernseh A. G. and others at the Berlin Radio Exhibition, comprise 120–240 scanning strips, and require side band widths of from 150 kc./sec. to 1,000 kc./sec. for their proper transmission. In view of the Geneva convention, under which absolutely no provision was made for the proper expansion and development of television in the broadcast band of wave-lengths, an entirely new radio technique will have to be developed. Local areas, served by ultra short wave radio transmitters, seem an ideal

法。冯米哈伊开发了一套机械系统，在该系统中接收器中的调制光被一个旋转的小镜子扫过，小镜子位于一个静止圆筒的轴线上，正对着固定在圆筒内表面上的许多镜子，他宣称这套系统优于旋转的镜面鼓。柏林电视机股份公司的实验室还造出了非常精确的镜面螺旋，在螺旋轴上装有 90 和 120 个不锈钢的反射面。美国研发工程师维拉蒂米尔·斯福罗金惊人地宣布他设计出一种他所谓的光电摄像管，该管由两部分组成，一个是透镜系统把场景聚焦在其上的嵌镶光电阴极；另一个是向嵌镶光电阴极发射电子束的阴极射线枪。要转播的场景被聚焦到大小约 4 英寸 × 5 英寸的信号板上，信号板表面有数以百万计的小光电管，每个光电管都包含一个被铯敏化的，易于感光的小银球。这些小球被置于一块绝缘板上，譬如薄的云母片上，在其背面涂有一层金属因而可以导电。电子枪与嵌镶光电阴极被放在同一个玻璃泡内，在与普通阴极射线管相同的偏转线圈的作用下，电子枪发射的电子束水平和垂直地扫过嵌镶光电阴极。一旦一个电子打在了光电管上，它就会中和对应聚光器上的部分电荷。这种放电电流被采集、放大并发送到接收端的阴极射线束上，这些阴极射线束随着扫描束同时移动到另一侧的荧光屏上。不断变化的放电电流使接收端的阴极射线束被调制，因而也调制了接收端的屏幕。据报道，渥太华自治领天文台的台长弗朗西斯·亨罗托博士也发明了一种类似的设备，他把自己的发明称作“超级眼”。如果这些设计都能付诸实践，那么在电视领域就将出现具有重大意义的革新。

人们在过去的两年中一直认为：为了使电视得到普及，就必须研制出图像清晰度高于已有的 30 线标准、画面比例为 2.4∶1 的电视。事实可能是这样，也可能不是这样，但我们应该清楚现在的真实情况。首先，很容易证明（实际上已经有人提出过）当带宽为 10 千赫时，一个频道的信息 – 时间传输特性（为了使闪烁最小化，应达到一个合理的图像速率，如每秒 12.5 次）采取这样的线数和比率最为经济。

目前，利用今年研制出的阴极射线接收设备所得到的图像的质量明显优于机械式接收设备重组而成的 30 线图像。贝尔德公司在英国科学促进会今年在莱斯特召开的会议上首次公开展示了这种质量的图像，洛伊公司、柏林电视机股份公司等其他公司在柏林无线电展览上也展示了同样的产品，包含 120~240 个扫描带，正常传输所需的边带带宽为 150 千赫到 1,000 千赫。因为日内瓦公约中没有任何一个条款在广播波段中为电视业的合理扩张和发展预留出一定的波长区，所以人们不得不想办法开发全新的无线电技术。在局部地区使用超短波无线电发射机似乎是一个理

solution. In practice, however, many difficulties arise, not the least of which is the shielding effect in populous areas of buildings, steel structures, trees, rises in ground contour, etc. Research in this direction is progressing; in fact, experimental short-wave transmissions of high quality pictures (that is, of 120-line definition or more) have already commenced in the London area, the Crystal Palace towers being utilised for this purpose. But it may be a year or two before an established service throughout the country is achieved. In the meantime, further problems arise in connexion with distortion in amplifiers against which the weapons provided by Oliver Heaviside, to whose classic researches on the underlying electrical principles of distortion in communication engineering too little credit is given, are powerless. In extending the band-pass of an amplifier, the "temperature effect" dealt with by L. B. Turner in his inaugural address to the Institution of Electrical Engineers becomes an important factor in determining the interference level of a system; all the more so because an increase in the number of scanning strips in a picture involves the diminution, usually according to some power-function, of light available to affect the light sensitive cells. Further carefully directed research in this direction has become imperative for the transmission of actual scenes, as opposed to film broadcasts. For the projection of television pictures to large audiences in cinema theatres and elsewhere, Fernseh A. G. has recently demonstrated an "intermediate-film" method (Fig. 3). In this the televised image is received on a cinema film which is then developed and passed through an ordinary cinema projector, the time interval between the reception and projection being about 6 seconds if an ordinary reel of film is used, about 20 seconds if an endless loop of film is used, the "base" being first emulsified and dried, then exposed to the receiving scanning device, developed, projected and de-emulsified. This method shows great promise but much further work remains to be done on it.

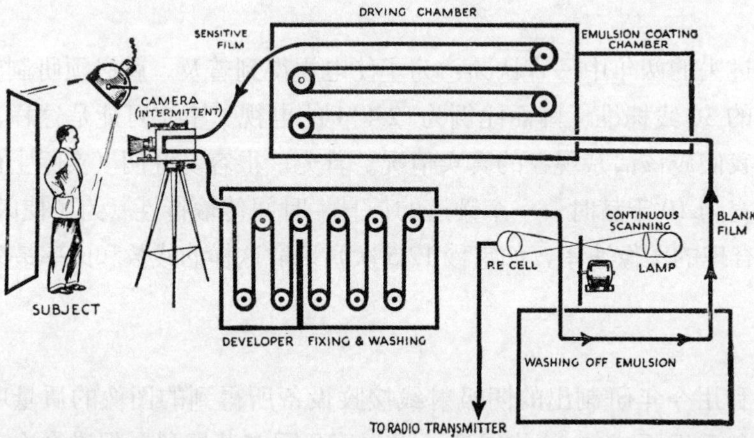

Fig. 3. Simplified diagram of the apparatus employed by Fernseh A. G. for the transmission of television by the intermediate film method.

(**132**, 502-505; 1933)

想的解决办法。然而，在实践中应用这种方法时遇到了很多困难，其中之一是许多地方因建筑物、钢结构、树木和地平面以上的高地等导致了屏蔽效应。目前人们正在进行这方面的研究；实际上在伦敦，人们已经开始试验用短波传输高质量的图像（即清晰度达到120线或更高的图像），为此目的用上了水晶宫的双塔。但要在全国范围内建立这样的服务恐怕还需要一两年。与此同时又出现了许多与放大器失真有关的问题，虽然奥利弗·赫维赛德对通信工程中的基础电学失真原理进行了可靠的研究，但他的研究成果在实践中的可信度太低，因而对解决这个问题无能为力。如果扩大一个放大器的带通，那么特纳在电气工程师学会的就职演说中提到的“温度效应”就会成为决定系统受干扰程度的一个重要因素；更因为在一幅图中扫描带数量的增加会使照射到光敏元件上的光以幂函数的形式迅速下降。与胶片播放不同，我们需要对实际场景的转播定向进行更加细致的研究。为了给电影院或其他场所的广大观众放映电视图像，柏林电视机股份公司最近展示了“中间胶片”的使用方法（图3）。在这个过程中，电视图像被传送到一个电影胶片上，然后被冲洗并通过一个普通的电影放映机播放出来，如果使用的是普通胶卷，则接收和放映过程大约需要6秒，如果使用循环胶卷，则所需的时间约为20秒，“片基”首先被乳化和干燥，接着进入接收端的扫描设备，然后冲洗、放映和去乳化。尽管这种方法具有广阔的前景，但仍有很多更进一步的工作需要做。

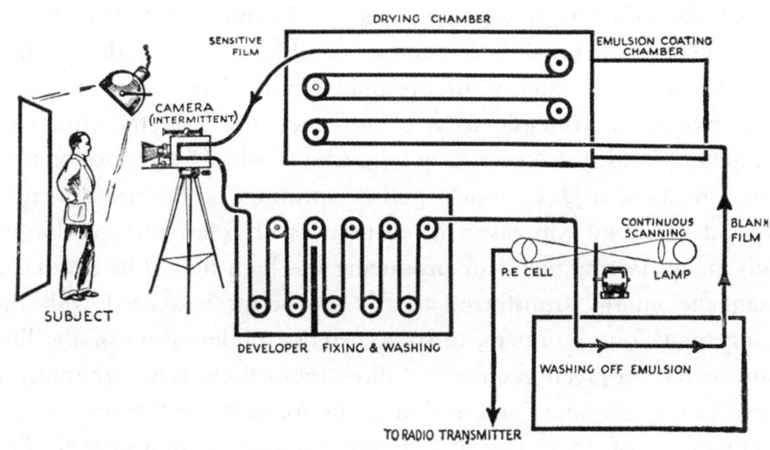

图3. 柏林电视机股份公司利用中间胶片法传送电视节目所用装置的示意图

（刘霞 翻译；赵见高 审稿）

Production of High Magnetic Fields at Low Temperatures

K. Mendelssohn

Editor's Note

Kurt Mendelssohn here reports an important experimental advance in the generation of powerful magnetic fields at low temperatures. Using superconducting alloys able to bear field strengths up to 22,000 gauss at a temperature of just 2 K, he and colleagues had experimented with a transformer-like arrangement of magnetically coupled coils. A magnetic field of 1,000 gauss created in a conducting primary coil was made to induce current in a secondary superconducting coil. With a dense coil solenoid as part of the second coil, the device could produce a field of 22,000 gauss in a small volume. This achievement, Mendelssohn points out, might be used to create extremely low temperatures by the method of adiabatic demagnetization.

THE use of a supra-conductor (therefore completely free from Joule heating) has been more than once suggested for the production of magnetic fields at low temperatures. The magnetic field obtainable by this means is limited by the magnetic threshold value at which supra-conductivity ceases. Still, considerable fields can be obtained by the use of alloys (investigated in Leiden[1]) the threshold value of which is 22,000 gauss at 2° K., a strength which is sufficient for many experiments.

The chief remaining difficulty lies in the heat conductivity of the leads to the supra-conducting coil. This problem of heat conduction through the leads can be eliminated by transferring the necessary energy for the magnetic field by induction. The suggested arrangement is similar in principle to a transformer, the primary circuit of which is normally conducting and the secondary circuit of which is supra-conducting. The primary circuit consists of a D. C. source and the primary of the transformer; the supra-conducting secondary circuit consists of a secondary with a few turns of large radius and a solenoid with many narrow turns for producing the high field. On closing the primary circuit the magnetic energy transferred to the secondary is shared with the solenoid. With this arrangement, one produces, to some extent, condensation of the lines of force. Calculation shows that for given geometrical dimensions there is an optimum ratio for the number of turns in the secondary coil to that in the solenoid. In this way, within the limits of the usual dimensions of an apparatus, it is easily possible with a primary field of about 1,000 gauss to obtain a field of 22,000 gauss in the space of a few cubic centimetres.

The method should be specially suitable whenever it is desired to produce fairly high magnetic fields at low temperatures in not too large a volume, as, for example, in the production of extremely low temperatures by the adiabatic demagnetisation of paramagnetic substances[2]. According to the experiments of Kürti and Simon[3], Giauque and MacDougall[4] and de Haas and his co-workers[5], it should be possible, with the above

低温下强磁场的产生

库尔特·门德尔松

编者按

在这篇文章中，库尔特·门德尔松报告了自己在制造低温强磁场的实验技术上取得的重大突破。在温度仅为 2 K 的情况下，利用可以承受高达 22,000 高斯强磁场的超导合金，他和同事们用一种类似于变压器的线圈排布方式的磁耦合线圈进行实验。他们首先在导电的初级线圈上施加 1,000 高斯的磁场以便在次级超导线圈上产生感应电流。次级线圈的一部分是缠得很密的螺线管，这台装置可以在一较小的空间内产生 22,000 高斯的磁场。门德尔松指出：这项技术的成功也许可以用于通过绝热退磁法产生极低温。

已经不止一次有人提出可以利用超导体（完全没有焦耳热）在低温下产生磁场。但用这种方式获得的磁场要受到超导电性消失所对应的磁场阈值的限制。尽管如此，利用温度为 2 K 时阈值可达 22,000 高斯的合金（在莱顿的研究结果 [1]）仍可以产生强度可观的磁场，这个强度对许多实验来说是足够了。

剩下的主要困难在于超导体线圈的引线存在热传导。我们可以通过磁场感应传递必要的能量来消除导线中的热传导问题。已提出的实验装置在原理上与变压器相类似，其主回路是正常传导的回路，而次级回路是超导回路。主回路包括直流源和变压器的初级回路；而次级超导回路则包含一个由几匝粗线圈绕成的次级回路和一个用来产生强磁场的由许多匝线圈密绕成的螺线管。当主回路接通时，传递到次级回路的磁能中的一部分分到螺线管上了。我们利用这台装置在一定程度上使磁力线变密了。计算表明，对于给定的几何尺寸，存在次级线圈匝数与螺线管匝数的一个最佳比值。这样，在通常的装置尺寸的极限内，我们很容易由 1,000 高斯的初级场在几个立方厘米的空间内获得高达 22,000 高斯的磁场。

这个方法应该特别适用于要求在低温下的一块不太大的体积内产生超强磁场的情况，例如利用顺磁物质的绝热退磁产生极低温的情况 [2]。根据库尔蒂和西蒙 [3]、吉奥克和麦克杜格尔 [4] 以及德哈斯及其同事们 [5] 的实验，用上述装置从 1 K 的初始温度得到 0.1 K 以下的温度应该是可以办到的。由于消除了沿着电流导线的热传导，

arrangement, to obtain temperatures below 0.1° K. from a starting point of 1° K. Since heat conductivity along the current leads is eliminated, and since heat capacities at helium temperatures are so minute, a few cubic centimetres of liquid helium should suffice to cool the whole arrangement.

Experiments with an apparatus embodying the above methods are being made in this laboratory.

(**132**, 602; 1933)

K. Mendelssohn: Clarendon Laboratory, Oxford. Sept. 28.

References:

1. de Haas, W. J., and Voogd, J., *Comm. Leiden*, No. 214b (1931).
2. Debye, P., *Ann. Phys.* (4) **81**, 1154 (1926); Giauque,W. F., *J. Amer. Chem. Soc.*, **49**, 1864 (1927).
3. Kürti, N., and Simon, F., *Naturwiss.*, **21**, 178 (1933).
4. Giauque, W. F., and MacDougall, D. P., *Phys. Rev.*, **44**, 235 (1933).
5. de Haas, W. J., *Nature*, **132**, 372 (Sept. 9, 1933).

而且在液氦温区热容量也很小，所以几立方厘米的液氦就足以冷却整台装置了。

我们正在实验室中利用根据上述方法设计的仪器进行实验。

（沈乃澂 翻译；赵见高 审稿）

New Results in Cosmic Ray Measurements*

E. Regener

Editor's Note

Erich Regener reports in more detail on his observations of the ionisation caused by cosmic rays in the upper atmosphere. Using electrometers born aloft in balloons, he and colleagues made a number of measurements at very high altitudes, where the air pressure fell below 50 mm Hg. In three trials, their devices gave consistent results; a fourth did not, but Regener noted that other researchers had observed a magnetic storm on this day which may have disturbed the measurements. In total, the data suggested that the cosmic radiation consisted of three distinct components of different penetrating power.

IN recent years I have endeavoured to explore the decay of the intensity of cosmic radiation over as wide a range as possible after its entrance into the earth's atmosphere. I believe that such an investigation is indispensable before a theory of the nature of cosmic radiation can be put forward. In *Nature*[1] and in the *Physikalische Zeitschrift*[2] I have already given some preliminary account of our measurements of the intensity of the cosmic radiation in the upper atmosphere. I propose to give here some further results, obtained in recent ascents with registering balloons, but first a few improvements of the apparatus which we have introduced must be described.

The balloon electrometer includes an electrometer system (a thin Wollaston wire, a quartz sling giving the directing power), the photographic objective, projecting the electrometer wire on the photographic plate. The wire is illuminated every four minutes from the side, so that it appears bright on a dark background on the photographic plate. There is also an aneroid barometer for the measurement of the air pressure and a bimetallic lamella for measuring the temperature. The movement of the aneroid, when the pressure decreases, limits by a pointer the image of the electrometer wire on the photographic plate. Since the measurement of the pressure is the most delicate part, especially when the pressure is low, we have added to the ordinary aneroid a second one, which only starts indicating when the pressure falls below one hundred millimetres. By observing the balloons in the air with two theodolites from a base of three to four kilometres, we have been able to prove that the measurements of the pressure with these two aneroids are fairly exact. The agreement with the height deduced from the pressure measurements was very good.

We have also employed another form of balloon electrometer. Our balloon electrometers hitherto constructed each had the ionisation chamber filled with air at a pressure of three or four to five atmospheres. The new electrometer has an *open* ionisation chamber, that is to say, a chamber communicating with the air outside through a tube containing

* Paper before Section A (Mathematical and Physical Sciences) of the British Association, delivered at Leicester on September 8.

宇宙射线测量中的新结果*

埃里克·雷格纳

编者按

埃里克·雷格纳详细地报告了他在研究高层大气中宇宙射线的电离作用时得到的观察结果。他和他的同事利用气球载静电计在气压低于50毫米汞柱的高海拔区做了多次实验。他们的装置在前三次实验中给出了一致的结果；但第四次的实验结果与前三次有偏离，不过雷格纳指出其他研究人员在第四次实验的那天观测到了磁暴，这也许是导致实验结果出现偏离的原因。所有这些数据都说明了宇宙辐射是由三种穿透力明显不同的成分组成的。

在最近几年里，我一直致力于探索宇宙辐射进入地球大气层后在尽可能广的范围内的强度衰减。我确信这项研究在我们提出一个有关宇宙辐射性质的理论之前是必不可少的。我已经在《自然》[1] 杂志和《物理杂志》[2] 上对我们测量高层大气中宇宙辐射强度的方法作了初步的说明。在这里我想给出一些更进一步的结论，是我们最近用探空气球的上升实验得到的，但首先我必须描述一下我们对这些仪器所做的改进。

探空气球静电计包括一个静电计系统（一根细的沃拉斯顿铂丝，一个提供下坠动力的石英吊坠）和一个将静电计的铂丝投影到摄影底片上的照相物镜。来自边缘的光每隔4分钟把铂丝照亮一次，所以铂丝在摄影底片的暗背景中出现一条亮线。还有一个用于测量气压的无液气压计和一个用于测量温度的双金属片。当压力下降时，无液气压计的运动通过一个指针来限制静电计的铂丝在摄影底片上的图像。因为气压测量是最精细的部分，尤其是当压力很低时，因此除这个普通的无液气压计外我们又添加了第二个无液气压计，新添的气压计只有在气压低于100毫米汞柱时才开始有显示。通过在3~4千米高的基地上用2个经纬仪观察空中的气球，我们已经能够验证用这两个气球上的无液气压计测量出来的压力是非常精确的，从压力测量中推导出来的高度与实际高度非常吻合。

我们也使用过另一种形式的气球静电计。至今为止，我们使用的每一只气球静电计都配有电离室，室内充有3或4到5个大气压的空气。而新的静电计有一个**开放**的电离室，也就是说，电离室是通过一根装有五氧化二磷的管子同外部大气相通

* 这篇论文曾在英国科学促进会A分会（数学和物理学分会）于9月8日在莱斯特举行的会议上被宣读过。

phosphorus pentoxide. The pressure in this chamber decreases as the balloon rises in the free atmosphere, and the ionisation chamber in this case must be larger in order to obtain adequate sensitivity. But such an arrangement is very convenient for measuring the absolute value of the ionisation due to the radiation, because it is much easier to obtain the saturation current at a low pressure. In the ordinary ionisation chamber, which is filled with gas at high pressure, it is well known that it is very difficult to obtain the saturation current.

Fig. 1 shows the results of the four best registrations of the cosmic radiation with the closed balloon electrometer. The minimum values of the air pressure on these four ascents are respectively:

August 12, 1932: 22 mm. mercury, 5.4 atmospheres pressure in the ionisation chamber.

January 3, 1933: 34 mm. mercury, 4.45 atmospheres.

March 9, 1933: 17.6 mm. mercury (this is the lowest pressure hitherto reached), 3.28 atmospheres.

March 29, 1933: 32 mm. mercury, 5.33 atmospheres.

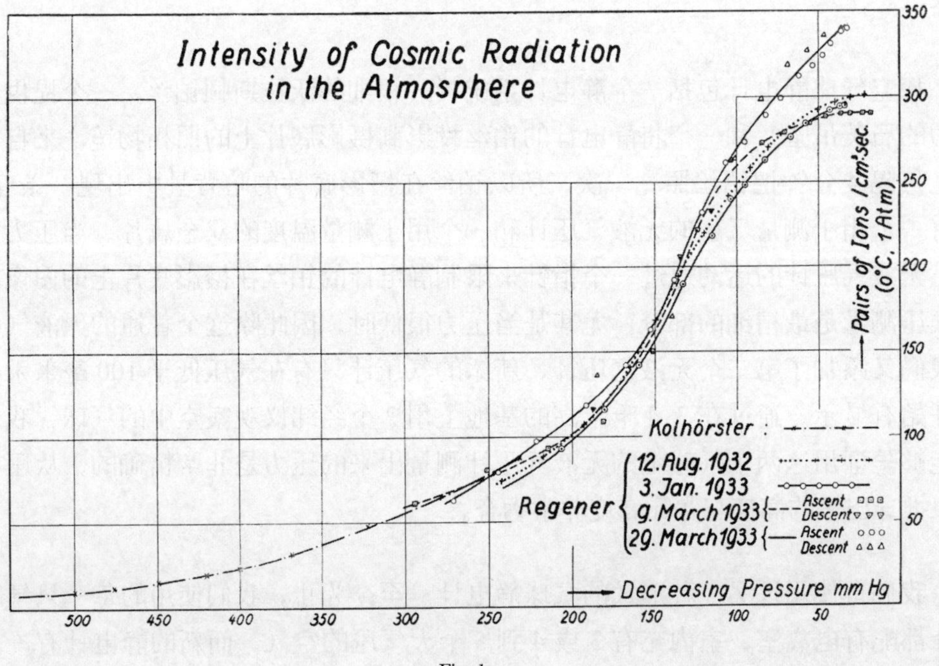

Fig. 1.

It is noteworthy that the first three ascents agree very well among themselves. Also the

的。因此，当气球在自由大气中上升时，电离室内的气压也会随着下降，而且此时的电离室必须做得更大一些，以获得足够的灵敏度。开放电离室的设计非常便于测量由辐射产生的电离的绝对值，因为在低压下更容易达到电流的饱和。大家都知道在充有高压气体的普通电离室中，电流很难达到饱和。

图 1 中显示的是我们用密闭气球静电计测量宇宙辐射的四次最佳实验结果。在这四次升空过程中达到的气压的最小值分别为：

1932 年 8 月 12 日，22 毫米汞柱，电离室内为 5.4 个大气压。

1933 年 1 月 3 日，34 毫米汞柱，室内 4.45 个大气压。

1933 年 3 月 9 日，17.6 毫米汞柱（这是迄今为止达到的最低气压值），室内 3.28 个大气压。

1933 年 3 月 29 日，32 毫米汞柱，5.33 个大气压。

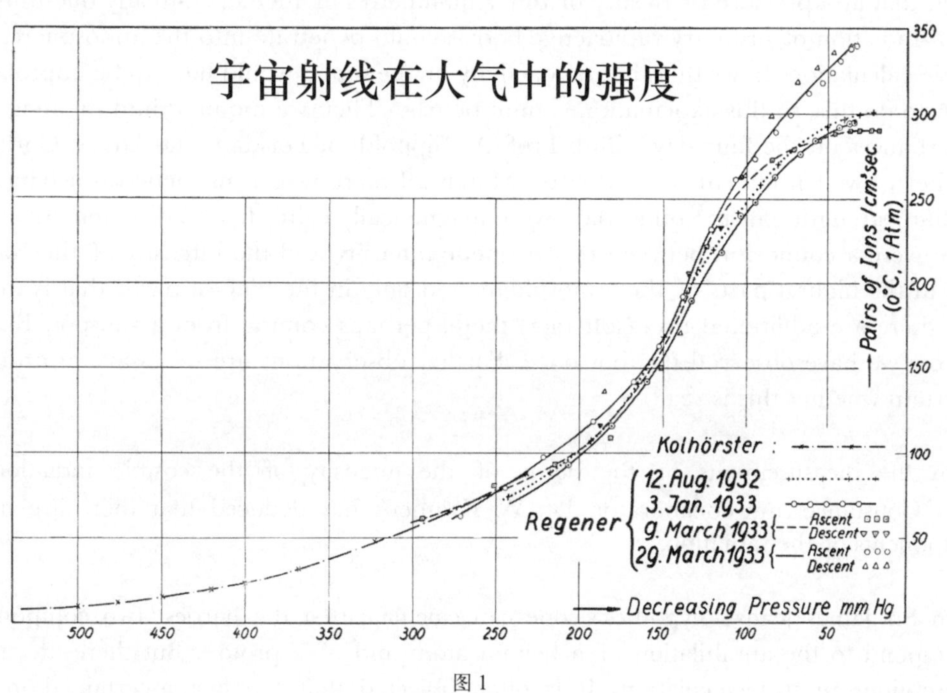

图 1

值得注意的是，前 3 次升空的结果相互之间很一致，第 4 次升空在气压为

fourth ascent agrees with the others very well at the medium heights at a pressure of 150 mm. mercury (that is, at a height of about twelve kilometres). But in the upper atmosphere, that is, at pressures of less than 100 mm., at heights greater than fifteen kilometres, and even more so at heights of twenty kilometres, the intensity begins to be much greater than on the other ascents, so that the maximum value is nearly fifteen percent greater than on the other ascents. This is probably *not* due to the inaccuracy of the measurements. It can be seen that the middle parts of the fourth curve agree very well with the others. Moreover the fourth registration, of March 29, 1933, is the best of all with the closed electrometers. It was also possible to obtain observations during the descent of the balloon (Fig. 1). These observations showed that the ordinary and the secondary aneroids worked very well.

The temperature during the hour in which the balloon was in the stratosphere varies comparatively little, from 6.5° to 11 °C. That is due to the "Cellophane" case, surrounding the electrometer like the glass of a forcing house and protecting it against the cold in the stratosphere.

We believe that the difference of the fourth curve from the others is real, and we have tried to find an explanation. We searched for the cause in the circumstances accompanying the four ascents. On the fourth ascent there was a new moon and we thought that perhaps radioactivity of the moon was the cause of the greater intensity on this day; for on the other ascents the moon was not in the sky. Incidentally, it should be noted that at a pressure of twenty or thirty millimetres of mercury, already one-third of the γ-radiation of ordinary radioactive bodies could penetrate into the atmosphere. But a little calculation shows that the radioactivity of the moon would have to be improbably great to do this, so this explanation cannot be true. Then we inquired into the magnetic disturbances on the four days. Both Prof. A. Nippoldt at Potsdam and Dr. A. Corlin in northern Sweden informed me that on March 29 there was a magnetic disturbance of medium strength, but the other days were magnetically calm. It would be remarkable if there were a connexion between the magnetic intensity and the intensity of the cosmic rays in the highest parts of the atmosphere, and only in the highest parts; that is to say, that there are additional rays (soft rays) there, perhaps coming from a sunspot. But up to now we have observed this but once. Further observations are necessary in order to ascertain whether this is real.

From the measurements of the decay of the intensity of the cosmic radiation in Lake Constance, my collaborator, Dr. W. Kramer[3], has deduced that there are many components of the radiation.

From Sir James Jeans's hypothesis, one can calculate that the hardest two components correspond to the annihilation of a helium atom and of a proton. But there are many assumptions in this calculation. It is often objected that the fact ascertained in our ionisation curve in the atmosphere, that the ionisation curve approaches a maximum value at the top of the atmosphere, is *not* in favour of the hypothesis, that the primary radiation is electromagnetic. Electromagnetic radiation coming from outside into the atmosphere

150 毫米汞柱左右的中高度区（即在高度约为 12 千米时）所得的结果与前 3 次也很一致。但是到了上层大气中，即在气压低于 100 毫米汞柱，高度大于 15 千米甚至达到 20 千米时，其强度开始超过其他三次实验，最大值比其他几次实验高出近 15%。这很可能不是由测量的不准确造成的。我们可以看到，第四个曲线的中部同其他曲线的中部非常吻合。而且，在 1933 年 3 月 29 日进行的第四次实验中我们采用了密闭的气压计，效果应该是最好的。在气球下降过程中，我们也可以记录观测数据（见图 1）。这些观测结果表明，普通无液气压计和第二个无液气压计的工作状态都非常好。

当气球在平流层飞行时，温度变化相对比较小——从 6.5℃到 11℃。那是因为“玻璃房”效应，静电计的周围被像温室的玻璃一样包裹着，使它能够抵御平流层的寒冷。

我们相信，第四次实验得到的曲线与前三次结果之间的差别是真实的，而且我们曾试过做出解释。我们在四次实验所处的环境中寻找原因。在第四次实验时，出现了一轮新月，我们认为，月面的放射性或许是造成那天强度更大的原因，因为在其他几次实验中，天上没有出现月亮。顺便说一下，我们应该知道，当气压低到 20 或 30 毫米汞柱时，一般的放射性物体发射的 γ 射线就已经有三分之一能够透过大气了。但在经过了简单的计算之后，我们发现月亮的辐射不可能造成这么大的影响，所以这个解释是不正确的。接着，我们又查询了这四天里磁场扰动的情况。德国波茨坦的尼波德教授和瑞典北部的科林博士告诉我，在 3 月 29 日出现过一次中等强度的磁扰，而在另外三天内没有发生磁扰。如果在大气层顶部的磁场强度和宇宙射线强度之间存在着关联，而且这种关联只发生在大气层顶部，这一现象是非常值得关注的；这意味着，那里还有其他射线（软射线）存在，它们也许来自于太阳黑子。但是因为这样的现象我们至今只观察到过一次，要证实这种关联是否存在有必要进行进一步的观测。

根据在康斯坦次湖进行的宇宙辐射强度衰变的测量结果，我的同事克雷默 [3] 博士曾经作出推断，即宇宙辐射是由多种成分组成的。

根据詹姆斯·金斯爵士的假设，我们能够计算出宇宙射线中最硬的两种成分与氦原子和质子的湮灭有关，但在计算过程中用到了许多假设。我们在整个大气层中得到的电离曲线证明，电离曲线在大气层顶部时会达到最大值；对于这一事实不支持初级宇宙射线为电磁辐射的假设，常常遭到众人的置疑。来自外太空的电磁辐射在进入地球大气层后会产生次级辐射，我们会在较低的大气层中，比如 20

of the earth will produce secondary radiation, and we shall find a maximum value of the intensity in the lower atmosphere, perhaps at twenty kilometres, and the intensity will diminish towards the top of the atmosphere.

It is easy to show, however, that the observed form of the ionisation curve agrees with the assumption that the radiation is electromagnetic. The observed curve is altered by the fact that the rays come from all directions. The rays coming from the side, that is, the rays which are already saturated with secondary radiation, because of the long distance they have travelled, are of greater account. My collaborator, Mr. B. Gross[4], has shown that it is possible to calculate from the curve observed for rays incident in all directions the corresponding curve of uni-directional rays. If the function for the rays from all directions be I_x, and ψ_x be the corresponding function for the intensity of rays coming from one direction only, then $\psi_x = I_x - I_x \frac{dI_x}{dx}$. The curve for rays entering the atmosphere vertically shows that their intensity diminishes towards the top of the atmosphere. Thus it agrees with the suggestion that at least a part of the radiation is electromagnetic. There is also a second maximum produced by the second component of the soft part of the radiation.

I would like to add a few words about the analysis of the radiation into its components. My collaborator, Dr. E. Lenz[5], has worked out a useful method for finding whether the radiation is monochromatic or contains more than one component. This method is independent of any assumption regarding the nature of the rays. Suppose that the intensity is a monotonic function of the absorption of monochromatic radiation. When the intensity I is multiplied by the thickness d of the absorbing layer, this gives a curve with a maximum at a certain value of d. If the radiation consists of two components of different penetrating power, then there are two maxima in the curve (let us say in the *deformed* curve), deduced from the original curve by multiplying the intensity by the thickness of the layer. When our experimental results are plotted in this way, the curve shows that the radiation in the atmosphere consists of two or three components of different penetrating power.

It is also possible in an experimental way to decide whether the decomposition of the radiation in components is real. When the intensity of radiation in the free atmosphere is observed with an open ionisation chamber, we do the same as we have just done in a mathematical way. The ordinary curve of the intensity of the radiation is a curve obtained with an ionisation chamber containing air at a pressure of one atmosphere. When we work with an open ionisation chamber, we find a value of the current in the chamber which is smaller in the same ratio as the pressure in the air, and thus in the chamber also, is smaller than the normal pressure. Thus we obtain directly a deformed ionisation curve, because the pressure is in proportion to the mass of the layer which is penetrated by the rays.

The two measurements made with such an ionisation chamber do not quite satisfy us yet. The first chamber employed was too small (volume only 22 litres) and therefore the sensitivity was inadequate. The second chamber had a volume of 105 litres and therefore the sensitivity was sufficient. A photographic record obtained on August 30 gave

千米处，发现射线强度达到最大值，而后在逐渐接近大气层顶部的过程中强度会有所下降。

然而，我们很容易证明，通过实验得到的电离曲线与宇宙射线是电磁辐射的假设并不矛盾。射线来自于各个方向的事实改变了观测到的曲线。来自于侧面的射线，也就是那些在穿越了很长距离之后次级辐射已经达到饱和的射线，是非常重要的。我的同事格罗斯[4]先生曾经指出，根据观察到的来自各个方向的辐射曲线可以计算出与单一方向对应的辐射曲线。假设来自各个方向的射线的函数是 I_x，只来自于一个方向的射线强度函数是 ψ_x，则 $\psi_x = I_x - I_x \frac{dI_x}{dx}$。我们从垂直射入大气层的辐射曲线中可以看到，它们的强度在到达大气层顶部时会下降。因此，这与至少有一部分射线是电磁辐射的观点相符。射线中软射线部分的次级成分还会产生第二个极大值。

对于分析射线的各种成分，我还想再多说几句。我的同事伦兹博士[5]已经找到了一种区分射线到底是单组分还是多组分的方法，该方法独立于任何有关宇宙射线性质的假说。假定强度是对单色辐射进行吸收的单调函数，则用强度 I 乘以吸收层的厚度 d 得到的曲线会在厚度 d 为某一特定值时达到一个最大值。如果射线包含两个穿透力不同的成分，则在原始曲线的强度值上乘以吸收层的厚度之后，曲线上就会出现两个最大值（让我们把这条曲线称作**变形后的**曲线）。当我们把自己的实验结果用这种方式处理时，从得到的曲线上可以看出，大气中的辐射包含两个或三个穿透力不同的成分。

我们也可以用实验方法来确认对辐射成分的分解是否是正确的。当我们用一个开放的电离室测量自由大气中的射线强度时，我们还用刚刚用过的数学方法来处理。普通的辐射强度曲线是用包含一个大气压空气的电离室得到的。当我们用一个开放电离室做试验时，我们会发现电离室中的电流值小于在同样比例的空气压力下——同时也是电离室中的压力下的电流值，也小于常压下的电流值。这样我们就直接得到了一个变形的电离曲线，因为压力与射线透过的物质层的质量成正比。

我们用这样的电离室做了两次实验，结果都不十分令人满意。第一次实验使用的电离室太小（体积仅为 22 升），因而灵敏度不够高；第二次实验所用的电离室体积为 105 升，这样灵敏度就足够高了。8 月 30 日得到的照相纪录给出了很好的结

good results, but unfortunately the temperature of the instrument went down very low, below −20 °C., and therefore the corrections needed were a little greater than usual. The apparatus had become too heavy (3.7 kgm.) for our balloons and we did not employ sufficient safeguards against the cold. But on working out the registrations, the curve (Fig. 2) is already better than those with the closed chamber, and agrees very well with the deformed curve calculated above. The second maximum is also noticeable as in the deformed curve, but this maximum is not very distinct and we shall try to ensure more favourable conditions so that as few corrections as possible are necessary. This part of the curve, I believe, is most important for the analysis and the explanation of the curve.

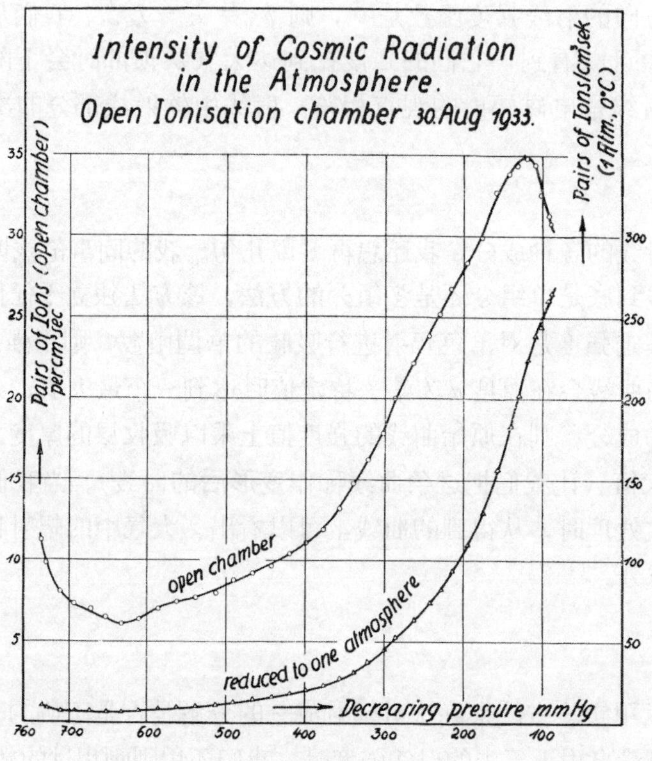

Fig. 2.

In general, the method of employing the open ionisation chamber is very convenient if one wishes to obtain the whole curve from sea-level to the top of the atmosphere, because the observed values with the open chamber vary only from 1 to 5, while the ionisation in the closed chamber varies from 1 to 150. Thus, in Fig. 2, the normal intensity curve is more accurate in the lower parts than with the closed chamber. The values for the normal curve are obtained from the values with the open chamber by multiplying them by p_0/p.

I offer my thanks to Mr. B. Auer for helping me in the measurements and to the

果，但不幸的是，这个装置的温度降到了 –20℃以下，因而所需的校正值略大于正常值。对于我们的气球来说，这个装置已经很重了（3.7 千克），所以我们没有安装抵御冷气的防护设施。但根据测量结果绘制的曲线（图 2）仍然会好于密闭电离室得到的结果，并且与用上述方法计算出来的变形曲线吻合得很好。在变形的曲线中也可以看到第二个最大值，但是，这个最大值并不是很明显，我们将设法确保更有利的实验条件，尽可能地避免各种修正。我认为曲线的这个部分对于我们分析和解释这条曲线至关重要。

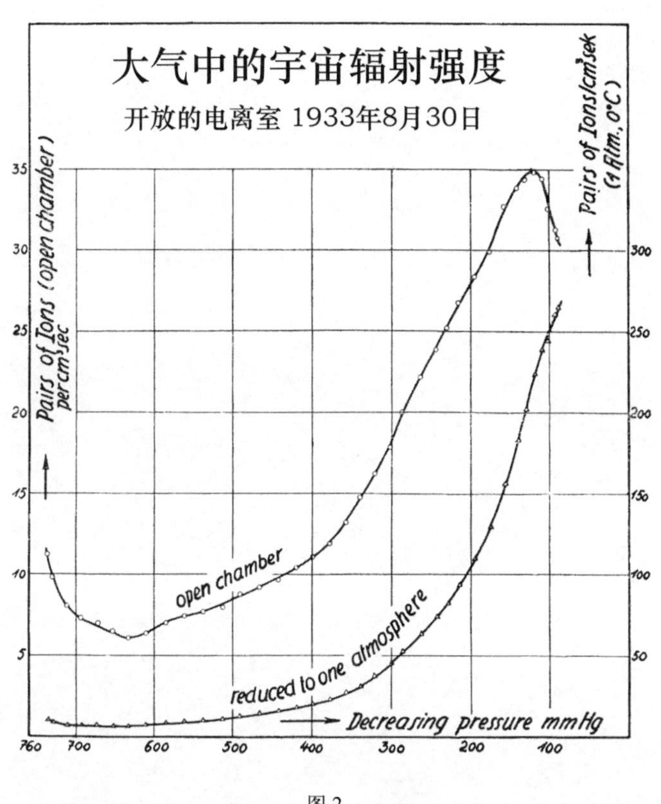

图 2.

一般而言，如果我们想得到一条从海平面到大气层顶部的完整曲线，用开放电离室进行测量是很方便的，因为开放电离室的观测值只在 1 到 5 之间变化，而封闭电离室的测量值则会从 1 一直变化到 150。因此，在图 2 中，开放电离室的标准强度曲线在较低部分会比密闭电离室的更精确，密闭电离室标准曲线上的值是由开放电离室的值乘以 p_0/p 得到的。

我要感谢奥尔先生在测量中给我的帮助，以及“德国科学临时学会”对这项研

Notgemeinschaft der Deutschen Wissenschaft for supporting my investigations.

(**132**, 696-698; 1933)

E. Regener: Technical High School, Stuttgart.

References:

1. Regener, E., *Nature*, **130**, 364 (1932).
2. Regener, E., *Phys. Z.*, **34**, 306 (1933).
3. Kramer, W., *Z. Phys.*, **85**, 411 (1933).
4. Gross, B., *Z. Phys.*, **83**, 214 (1933).
5. Lenz, E., *Z. Phys.*, **83**, 194 (1933).

究工作的支持。

（刘霞 翻译；马宇蒨 审稿）

Interaction of Hard γ-rays with Atomic Nuclei

Editor's Note

Two Chinese physicists from the National Tsing Hua University here describe experiments in which hard gamma rays kick electrons out of lead or aluminium. They suggested that the data imply some kind of nuclear disintegration triggered by the gamma rays. However, as Ernest Rutherford notes in a short addendum to the paper, the authors had sent him a letter along with their paper clearly indicating that they had not yet heard about recent results demonstrating the creation of electron–positron pairs from energetic gamma rays in the strong field in the vicinity of the atomic nucleus. This process of pair creation, he suggests, might well provide a more natural explanation for the experimental results.

IT is known that when a pencil of hard γ-rays of thorium-C″ passes through lead, in addition to the absorption by electrons of the shell, there exists a type of nuclear absorption, accompanied by the emission of characteristic radiations of frequencies different from the primary[1]. The intensity of such radiations has been estimated, and it has been found that the total energy of the characteristic radiations emitted is much smaller than the total energy of the primary radiation absorbed by the nuclei[2]. This would be expected, if we assume that a nuclear disintegration occurs in such a process, so that a part of the absorbed energy is spent. From this point of view, we have tried to detect electrons which might be ejected from the lead nuclei by the primary γ-quanta.

In our experiment, the γ-ray source was a radium–thorium preparation equivalent to 10 mgm. of radium. Two Geiger–Müller counters, one having an aluminium wall and the other a lead wall, were used. The counters had equal inner dimensions and approximately equal mass per square cm. of the wall (that is, 0.92 mm. thick for the aluminium counter and 0.22 mm. thick for the lead counter). Let N_{Al} and N_{Pb} be the number of electrons produced in equal time intervals by a given beam of γ-rays in the aluminium and lead counters respectively. The ratio N_{Pb}/N_{Al} as a function of the wave-length λ of the incident γ-radiation will at first decrease with decreasing λ, due to the diminishing photoelectric absorption of lead. As the wave-length further decreases, the ratio N_{Pb}/N_{Al} might, however, rise again, if the heavy lead nuclei begin to be disintegrated by γ-quanta of wave-length less than a certain value and the electrons ejected from the lead nuclei in the disintegration process add themselves to N_{Pb}.

By using a beam of γ-rays of thorium-C″ filtered through 2 cm. of lead and scattered by iron at different angles, we measured the ratios N_{Pb}/N_{Al} for γ-rays of different wave-lengths. The experimental result is shown in the accompanying table, where N_{Pb}/N_{Al} is multiplied by a constant k such that the value kN_{Pb}/N_{Al} is unity for the scattered radiation at 23°.

硬 γ 射线和原子核的反应

编者按

两位来自国立清华大学的中国物理学家在文中介绍了他们用硬 γ 射线将铅和铝中的电子轰击出来的实验。他们认为实验数据说明 γ 射线引发了某种形式的核衰变。然而，欧内斯特·卢瑟福在这篇论文后面的简短评论中指出：从作者随论文邮寄给他的一封信中可以明显看出，他们还不知道最近的一项研究成果，即在原子核附近的强场中，高能 γ 射线会产生电子－正电子对。卢瑟福认为用产生正负电子对来解释两位中国物理学家的实验结果也许会显得更合理一些。

众所周知，当一束由钍 C″ 放射出的硬 γ 射线穿过铅时，除了出现铅的外层电子对 γ 射线的吸收以外，还会出现铅原子核对 γ 射线的吸收，并伴随着发出和初级辐射具有不同频率的特征辐射 [1]。这种特征辐射的强度已经被估算过了，即放射出的特征辐射的总能量远远小于被原子核吸收的初级辐射的总能量 [2]。我们可以预见，如果假设原子核在这一过程中发生了衰变，那么原子核吸收的一部分能量就会在此过程中被消耗掉。从这种观点出发，我们测量了由初级 γ 粒子从铅原子核中打出来的电子。

我们在实验中采用的 γ 射线源是镭－钍合成的放射源，相当于 10 毫克的镭放射源。所使用的两个盖革－缪勒计数器，一个是铝壁，另一个是铅壁。两个计数器具有相同的内壁尺寸，并且每平方厘米壁的质量也大约相等（即 0.92 毫米厚的铝计数器和 0.22 毫米厚的铅计数器）。用 N_{Al} 和 N_{Pb} 分别表示在相同的时间间隔内，由给定的 γ 射线在铝计数器和铅计数器中产生的电子数目。比值 N_{Pb}/N_{Al} 是入射 γ 射线波长 λ 的函数，在开始阶段由于铅的光电吸收不断减弱，所以将会随着 λ 的减小而减小。然而，当波长 λ 进一步减小时，比值 N_{Pb}/N_{Al} 可能反而会变大。这是因为波长小于某个特定值的 γ 粒子可能会引起重铅核的衰变，在铅核衰变过程中放射出来的电子会增加 N_{Pb} 的值。

将一束由钍 C″ 放射出的 γ 射线用 2 厘米厚的铅板过滤，再经过铁散射到不同的方向，我们测量了对应不同波长 γ 射线的 N_{Pb}/N_{Al} 比值。下表中给出了相关的实验结果，其中 N_{Pb}/N_{Al} 要乘以一个常数 k，这样，我们把在辐射射线散射角度为 23° 时的 kN_{Pb}/N_{Al} 值作为单位 1。

	λ(x.u.)	kN_{Pb}/N_{Al}
Primary radiation	4.7	1.16±0.04
Scattered radiation at 23°	6.6	1.00
Scattered radiation at 46°	12.1	1.23±0.08

In the table, the ratio N_{Pb}/N_{Al} for λ = 6.6 x.u. is seen to be smaller than that for λ = 12.1 x.u., and for λ = 4.7 x.u. it again rises as was expected if electrons were ejected from the lead nuclei by the hard radiation. The difference of the two ratios for λ = 4.7 x.u. and 6.6 x.u. is about 16 percent. Now, the increase of the ratio N_{Pb}/N_{Al} for λ = 4.7 x.u. might also result from a difference in the scattering effect of the lead nuclei and aluminium nuclei towards the Compton recoil electrons produced in the counter walls by the incident γ-rays. If this were the case, the difference of the ratios for γ = 6.6 x.u. and 4.7 x.u. should be more pronounced by using counters of thicker walls, since the effect of scattering increases with thickness of the wall. But the same result, namely, a difference of about 16 percent between the two ratios, was obtained when the experiment was repeated with a lead counter with walls 0.3 mm. thick and an aluminium one with walls 1.2 mm. thick. Therefore the above result seems to support the view that the lead nuclei are disintegrated by the hard γ-rays.

The details of the experiment will be published elsewhere.

C. Y. Chao and T. T. Kung

* * *

It is obvious from a letter to me which accompanied the above communication that Prof. Chao and Mr. Kung have not yet heard of the recent work concerning the positive electron, and in particular of the creation of a pair of electrons, a negative and a positive, by the conversion of a γ-ray of high energy in the strong electric field of a nucleus. The experiments they describe provide valuable additional evidence of this phenomenon, and would doubtless have been interpreted by them in this way rather than as a nuclear disintegration. It is interesting to note that the magnitude of the effect is about the same as is found in other experiments.

Rutherford

(**132**, 709; 1933)

C. Y. Chao and T. T. Kung: Department of Physics, National Tsing Hua University, Peiping, China, Sept. 4.

References:

1. Chao, *Phys. Rev.*, 33, 1519 (1930); Gray and Tarrant, *Proc. Roy. Soc.*, A, 136, 662 (1932).
2. Gray and Tarrant, *Proc. Roy. Soc.*, A, **136**, 662 (1932). Chao, *Science Reports of National Tsing Hua University*, lst series, **1**, 159 (1932).

	λ（10^{-11} 厘米）	kN_{Pb}/N_{Al}
初级辐射	4.7	1.16 ± 0.04
辐射散射角为 23°	6.6	1.00
辐射散射角为 46°	12.1	1.23 ± 0.08

在上面的表格中，$\lambda=6.6\times10^{-11}$ 厘米时的 N_{Pb}/N_{Al} 值小于 $\lambda=12.1\times10^{-11}$ 厘米时的相应值，而当 $\lambda=4.7\times10^{-11}$ 厘米时，如我们所预料的那样，硬 γ 射线引发铅核放射出电子，从而使 N_{Pb}/N_{Al} 值增大。这个比值在 $\lambda=4.7\times10^{-11}$ 厘米时与 $\lambda=6.6\times10^{-11}$ 厘米时相差大约 16%。也许有人认为，当 $\lambda=4.7\times10^{-11}$ 厘米时比值 N_{Pb}/N_{Al} 的增加还有可能是由于铅核和铝核对入射 γ 射线在计数器管壁上产生的康普顿反冲电子的散射效应的差别导致的。如果真是这样的话，那么 $\lambda=6.6\times10^{-11}$ 厘米和 4.7×10^{-11} 厘米时的 N_{Al}/N_{Pb} 值之差在使用管壁较厚的计数器时应该更加显著，因为散射效应会随着管壁的增厚而增强。但是，当我们使用壁厚为 0.3 毫米的铝计数器和壁厚为 1.2 毫米的铝计数器重复上面的实验时，我们得到了相同的结果，也就是两个 N_{Pb}/N_{Al} 比值之间仍相差约 16%。这样看来，以上结果支持了硬 γ 射线引发铅核发生衰变的观点。

实验的具体细节将发表在其他地方。

赵忠尧，龚祖同

* * *

从与以上这篇通讯一块寄给我的信中可以明显看出，赵教授和龚先生还不知道最近关于正电子的研究工作，特别是关于在一个原子核的强电场中，高能 γ 射线可以转化产生一对电子，即一个正电子和一个负电子的情况。他们描述的实验又一次为这一现象提供了非常有价值的证据，毫无疑问，该实验现象应该被他们解释为高能 γ 射线产生了正负电子对，而不是原子核的衰变。值得注意的是，这一效应的强度与在其他实验中观测到的强度大致相同。

卢瑟福

（王静 翻译；尚仁成 审稿）

A Suggested Explanation of β-ray Activity

M. N. Saha and D. S. Kothari

Editor's Note

Radioactive beta decay had proven most baffling. Unlike alpha particles, beta particles were emitted not with a well-defined energy, but with a broad range of energies, and physicists had established that these energies were created in the nuclear process itself, not as the electron interacted with shell electrons. Here Indian physicists Meghnad Saha and Daulat Singh Kothari suggest that the mystery might be settled by the recently discovered process in which a gamma photon can create an electron and its antimatter partner, a positron. The electron so created might have any kinetic energy, depending on the photon's energy. This attempted explanation seemed promising, though physicists still knew nothing of the weak nuclear force or of neutrinos, which a proper explanation would require.

THE β-ray activity of radioactive bodies has until now proved to be a very baffling problem. The points at issue are summarised in Gamow's "Constitution of Atomic Nuclei", etc. (pp. 52–54), and in "Radiations from Radioactive Bodies" by Rutherford, Chadwick and Ellis (p. 385). They are also discussed at some length by Bohr in his Faraday lecture (1930).

Briefly speaking, the chief points under discussion are the following: the disintegration electrons (β-rays) from a radioactive body are not emitted with a single velocity as in the case of α-rays, but show a distribution of velocities over wide ranges, though the breaking-up of the atom is a unitary process, as is proved by the fact that the life-period is definite and there is one electron for each disintegrating atom. It has further been proved that the continuous distribution of velocities is a nuclear process, and not due to action of the surrounding shell of electrons.

It appears that the β-ray disintegration admits of a very simple explanation on the basis of the recent experiments by Anderson and Neddermeyer, and Curie and Joliot on the production of positrons by the impact of hard γ-rays with the nuclei of elements. These experiments have been interpreted by Blackett and Occhialini as indicating the conversion of a γ-ray quantum into an electron and a positron near the nucleus. Curie and Joliot have brought further evidence in favour of this view by showing that γ-rays of thorium C″ (energy 2.6×10^6 electron volts) are converted inside all matter into an electron (mass 9×10^{-28} gm., energy $m_0c^2 = 0.51\times10^6$ eV) and a positron (having the same mass and energy as the electron), the excess energy being distributed as the kinetic energy of the two particles, and the energy of the residual quantum. They have denoted this phenomenon by the term "materialisation of light quanta". They have further shown that a proton is a complex structure, being a compound of the neutron and a positron. As pointed out by Blackett and Occhialini, this explains the anomalous absorption of γ-ray quanta observed

β射线放射性的一种可能的解释

梅格纳德·萨哈，道拉特·科塔里

编者按

放射性 β 衰变是最令人困惑不解的。与 α 粒子不同，β 粒子的能量不是固定的，而是存在一个很宽的范围，物理学家们已经证实这些能量是核过程本身产生的，并非电子与壳层电子相互作用的结果。两位印度物理学家梅格纳德·萨哈和道拉特·辛格·科塔里在本文中提出：由于最近发现 γ 光子可以产生一个电子和它的反粒子——正电子，这个谜有可能会随之得到破解。在这个过程中产生的电子也许会具有一定的动能，这取决于光子的能量。这种尝试性的解释看似前景光明，但那时的物理学家对弱核力和中微子仍一无所知，而只有了解了弱核力和中微子，才能得到这个问题的正确解释。

放射性物质的 β 射线放射性一直以来都是一个非常令人困惑不解的问题。在伽莫夫的《原子核的结构》（第 52~54 页）及卢瑟福、查德威克和埃利斯的《放射性物质的辐射》（第 385 页）等著作中都总结了一些有争议的论点。玻尔在法拉第讲座（1930 年）中也对这些观点进行了详细的讨论。

简言之，争论的主要方面是：原子有确定的衰变寿命以及每个发生衰变的原子发射一个电子这些事实已经证明原子的衰变是一个单一的过程，但是由放射性物质放射出的衰变电子（β 射线）并没有像 α 射线那样具有单一的速度，而是出现了速度分布在一个很大范围内的情况。人们还进一步证明了速度的连续分布与核过程有关，而与核外电子壳层的作用无关。

最近，安德森、尼德迈耶以及居里、约里奥进行了硬 γ 射线轰击原子核产生正电子的实验，在此基础之上看似可以给出关于 β 衰变的一个非常简单的解释。布莱克特和奥基亚利尼对这些实验的解释是：一个 γ 粒子在原子核附近转化成了一个负电子和一个正电子。居里和约里奥为这个观点提供了进一步的证据，他们的实验表明钍 C″ 的 γ 射线（能量为 2.6×10^6 电子伏特）在所有的材料中都会转化为一个负电子（质量为 9×10^{-28} 克，能量为 $m_0c^2=0.51\times10^6$ 电子伏特）和一个正电子（质量和能量与负电子相同），多余的能量除了分配给正负两个电子作为动能外，还将分配给剩余的量子，他们把这种现象称为“光量子的物质化”。他们进一步指出，质子具有复杂的内部结构，包括一个中子和一个正电子。布莱克特和奥基亚利尼指出，格雷和塔兰特观察到的反常 γ 射线吸收也能用此进行解释，但根特纳发现，能量在 1.1 百万电

by Gray and Tarrant, which Gentner has found to commence with the γ-ray possessing the limiting energy 1.1 million electron volts.

The discovery, which is confirmed by so many workers, promises to be of great importance, as it establishes for the first time, on experimental grounds, the splitting up of a quantum into two charged particles of opposite sign. Many astrophysicists have postulated the probability of the annihilation of the proton and the electron with their mass energies converted into quanta, but the actual process, as revealed by these experiments, seems to be very different. For the quantum breaks up into charged particles possessing opposite charges, but having equal mass, and the positron being absorbed by the neutron forms the proton which is thus seen to be complex. The phenomenon is therefore not a "materialisation of the quantum" as Curie and Joliot suggest, for the neutron appears to be the fundamental mass-particle, but it consists in a splitting of the quantum into two fundamental opposite charges. We may call it "electro-division of the quantum".

Let us see how we can explain β-ray activity. If the "electro-division of a quantum" can be brought about by a nucleus when the quantum hits it from the outside, it is much more probable that the γ-rays produced within the nucleus itself should be completely split up into an electron and a positron. The electron will come out as a β-ray, but a positron will not be able to come out if the conversion takes place within the potential barrier. It will attach itself mainly to some one of the numerous neutrons which are present in the nucleus, and thus form a proton. The positive charge of the nucleus will therefore be increased by unity.

It is not difficult to explain the continuous distribution of β-ray energy. The γ-ray may suffer this "internal electro-division" anywhere within the nucleus, and hence the velocities imparted to the resulting electrons may vary within wide limits. The exact mathematical calculation can be carried out only when more data are forthcoming. The positron combining with the neutron will give rise to the softer γ-rays which are always present in a β-ray disintegration.

According to the above view, the β-ray emission is only a secondary process, the primary phenomenon which starts this chain of events being the generation of a primary γ-ray. We can now ask ourselves: How is this γ-ray generated? It must be due to the passage of an α-particle or proton from one barrier to another. Gamow, and also Condon and Gurney have postulated the existence of only one barrier in a radioactive nucleus for explaining the emission of α-rays, with definite velocity, but several lines of argument indicate that there may be more than one barrier present in the nucleus. When an α-particle crosses from one barrier to the other, the γ-ray responsible for the whole chain of events leading to the β-ray disintegration is emitted. The life-period is therefore determined by the time of leakage of an α-particle or proton from one barrier to another, and this explains why the life-periods of β-ray bodies are of the same order as those of α-ray bodies, and have a definite value.

(**132**, 747; 1933)

M. N. Saha and D. S. Kothari: Department of Physics, Allahabad University, Oct. 20.

子伏特以上的 γ 射线才能发生这种反常吸收过程。

这个发现已经被很多研究人员所确认，当然也必定有着非常重大的意义，因为这是第一次在实验基础上确立：一个量子可以劈裂成两个带有相反电荷的粒子。很多天体物理学家都认为一个质子和一个电子有可能会发生湮灭并且将其质能转化为量子，但是实际的过程，正如在实验中所看到的那样，是很难发生的。因为一个量子会分裂成一对带相反电荷但质量相等的粒子，并且其中的正电子会被中子吸收形成质子，由此可以认为质子具有复杂的结构。因此，这个现象并不是居里和约里奥所说的“量子的物质化”过程，因为中子看似一个有质量的基本粒子，但是它却存在于量子分裂成两个电量相反的电荷的过程中。我们可以把这个过程称为“量子的电分离”。

那么，让我们来看看 β 射线的放射性该如何解释。如果外部量子轰击原子核时会引发“量子的电分离”，那么在该原子核内部产生的 γ 射线更有可能会完全分裂成一个负电子和一个正电子。电子将以 β 射线的形式发射出来，但是如果这个分裂过程发生在势垒当中，那么正电子就不能被发射出来。正电子将和存在于原子核中的大量中子当中的一个相结合，形成一个质子。这样，原子核的正电量将因此增加一个单位。

而 β 射线能量的连续分布也不难解释。γ 射线可以在原子核中的任意位置发生“内部电分离”，这样，γ 射线转化产生的电子的速度就会在很宽的范围内变化。只有得到更多的实验数据，我们才能进行严格的数学计算。和中子结合的正电子将产生软 γ 射线，这种软 γ 射线经常出现在 β 衰变反应中。

根据上面的观点，β 射线的发射只是一个二级过程，初级 γ 射线的产生才是引发这个反应链的最初原因。现在我们可以问问自己：这种 γ 射线是如何产生的呢？一定是因为一个 α 粒子或者一个质子从一个势垒穿越到另一个势垒而产生的。伽莫夫以及康登和格尼曾经以放射性原子核中只存在一个势垒出发解释了具有确定速度的 α 射线的放射，但是一些争论表明，原子核中可能会有不止一个势垒。在一个 α 粒子穿越一个势垒来到另一个势垒的过程中产生了引发 β 衰变链的 γ 射线。所以 β 射线放射性的周期取决于 α 粒子或质子从一个势垒到另一个势垒所用的时间，这就解释了为什么 β 射线的寿命周期和 α 射线的寿命周期有相同的量级并且有确定量值。

（王静 翻译；尚仁成 审稿）

Latitude Effect of Cosmic Radiation

J. A. Prins

Editor's Note

Earlier experiments by Jacob Clay, Arthur Compton and others had suggested that the flux of cosmic rays has a minimum close to the equator. Here J. A. Prins reports results measured using an ionisation chamber carried by the S.S. *Springfontein* during a voyage from Holland to South Africa, which also showed a clear minimum near the equator. Unfortunately, he says, his apparatus had broken down, and he could not gather data from southern latitudes, which might have helped clarify an apparent difference noted by other researchers between the flux in the two hemispheres. But the results did suggest that most cosmic rays were charged particles, which would be influenced by the Earth's magnetic field and find penetration easiest near the poles.

IT was found for the first time by Clay[1] on voyages between Holland and Java that the intensity of cosmic radiation has a minimum in the neighbourhood of the magnetic equator. The extensive survey directed by Compton[2] confirmed the existence of this "latitude effect" and showed it to be more pronounced at higher altitude. More accurate results at sea-level are due to an investigation of Clay and Berlage[3]. As this again refers to the line from Holland to Java, I thought it would be worth while to perform analogous measurements on a trip from Holland to South Africa. During this investigation Hoerlin[4] published results he obtained on the line Peru–Strait of Magellan–Hamburg. These results and those of the other authors as given by Clay are represented in Fig. 1 by continuous curves, my own results by open circles. Clearly the latter lie somewhat closer to Clay's curve than to Hoerlin's.

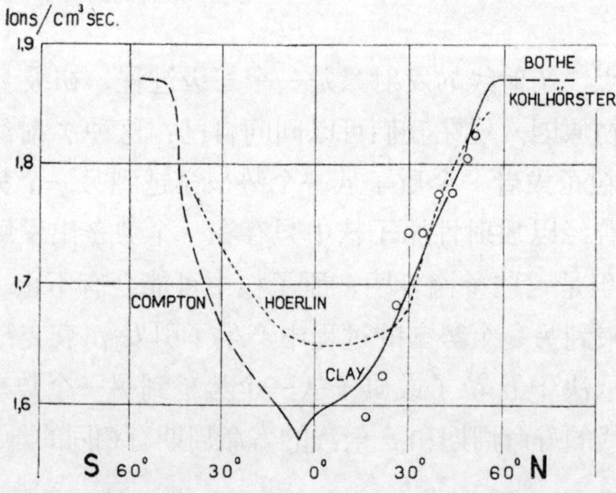

Fig. 1. Latitude effect of cosmic radiation. Circles indicate results of author. Vertical scale indicates number of pairs of ions in normal air.

宇宙射线的纬度效应

约翰·普林斯

编者按

雅各布·克莱、阿瑟·康普顿和其他人的早期实验曾经说明宇宙射线的通量在靠近赤道处有一个最小值。本文中，普林斯报告了他在从荷兰开往南非的斯普林方廷号远洋轮上利用装载于船上的电离室测量得到的结果，该结果同样说明在赤道附近会出现一个最小值。他说，不幸的是他的仪器坏了，所以没有能够收集到南纬的数据，这使他失去了一个证明其他研究人员所指出的宇宙射线在南北半球的通量存在明显不同的机会。但他的结果确实表明了大多数的宇宙射线都是带电粒子，它们会受到地球磁场的影响，而且在两极附近最易穿透。

克莱[1]在往返于荷兰和爪哇之间的旅行中首次发现宇宙射线的强度在地磁赤道附近存在一个最小值。康普顿[2]通过广泛的调查证实了“纬度效应”的存在，并发现在高纬度地区这种效应会更明显。克莱和贝尔拉赫[3]在海平面上进行测量得到了更为精确的结果。考虑到他们的调查也是在从荷兰至爪哇的航线上进行的，我认为有必要在从荷兰到南非的航线上进行类似的测量，以便于比较。就在我做这个实验的同时，霍尔林[4]发表了他在秘鲁–麦哲伦海峡–汉堡一线测量得到的结果。这些结果连同克莱给出的其他人的结果都画在图1中，用连续曲线代表；我的结果用空心圆圈表示。把我的结果分别与克莱和霍尔林的结果相比较后发现：我的结果与克莱的结果更接近。

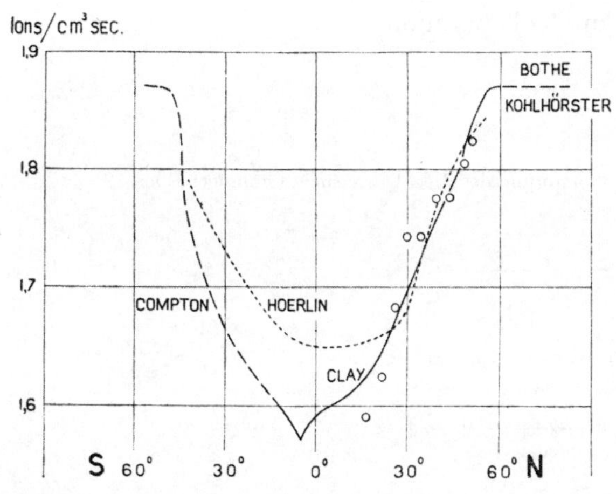

图1. 宇宙射线的纬度效应。圆圈代表作者的结果，纵坐标表示在标准大气中产生的离子对数量。

Unfortunately, my apparatus broke down in the tropics, so I have not been able to get evidence on the remarkable difference between the southern and northern hemispheres as indicated by Hoerlin's results. Though we may feel satisfied that an equatorial minimum of the same order of magnitude is found by all investigators (indicating that the cosmic radiation consists largely of a cosmic rain of charged particles) it would seem that an accurate repetition of this kind of measurement to obtain the exact shape of the curve is not superfluous.

Regarding my observations, the following particulars may be given. The ionisation chamber had a volume of 3 litres and contained argon at a pressure of 30 atm. It was shielded by 8 cm. of iron and was placed in a hut on board the S.S. *Springfontein* of the Holland Africa line, the deck over it being of negligible thickness. The wall of the ionisation chamber was brought to 120 v. and the ionisation current collected on an insulated rod connected to a Lindemann electrometer and to a small capacity (4 cm.). To start an observation, the earthing key of the rod was opened and a stop-watch set running at the same moment. The electrometer was kept at zero by gradually applying a potential to the capacity so as to compensate the charge due to the ionisation current. After some time (about 6 min.) the potential (about 3 v.) was read on a voltmeter. From this the number of ions produced in the chamber per $cm.^3$ per sec. may be deduced, assuming saturation. This number is called the "intensity" of cosmic radiation. A small correction for barometric pressure was applied to it (2.4 percent for 1 cm. mercury). In the graph in Fig. 1 these values (like those of Hoerlin) have been multiplied by such a factor (1/33.2) as to make the value at 50° coincide with the value given by Clay for normal air.

I wish to thank the Groninger Universiteitsfonds for a grant of money, Prof. Coster for allowing the apparatus to be made in the workshop of his laboratory, Prof. Clay for some kind advice and finally the directors of the Holland Africa line and the crew of the S.S. *Springfontein* for their kind collaboration.

(**132**, 781; 1933)

J. A. Prins: Natuurkundig Laboratorium der Rijks-Universiteit, Groningen, Oct. 19.

References:

1. Clay, J., *Proc. Amsterdam*, **30**, 1115 (1927); **31**, 1091 (1928).
2. Compton, A. H., *Phys. Rev.*, **43**, 387 (1933).
3. Clay, J., and Berlage, H. P., *Naturwiss.*, **20**, 687 (1932); Clay, J., *Naturwiss.*, **21**, 43 (1933).
4. Hoerlin, H., *Nature*, **132**, 61 (July 8, 1933).

不幸的是，进入热带地区以后我的仪器失效了，因此，我无法验证在霍尔林的结果中所指出的南北半球间的明显差异。虽然我们对所有调查者都发现在赤道附近宇宙射线存在同量级的最小值感到满意（这表明宇宙射线主要是由一系列宇宙中的带电粒子组成的），但精确地重复进行此类测量以得到该曲线的准确形状仍然是有必要的。

以下是我的观测过程。电离室的容积为 3 升，里面充有 30 个标准大气压的氩气。在用 8 cm 厚的铁板屏蔽后，电离室被放在往返于荷兰—非洲航线的斯普林方廷号远洋轮上的一个小屋里，铁板屏蔽上方的甲板厚度相对而言可忽略不计。电离室的壁上加有 120 伏的电压，在一根与一个林德曼静电计和一个小电容（4 cm）相连的绝缘棒上收集电离电流。在开始观测之前，先打开绝缘棒的接地开关，同时开始用跑表计时。为了保持静电计的示数为零，需要不断增加电容上的电压以补偿由于电离电流引起的电荷变化。过一段时间（大约 6 分钟）再读出电压表上的电位值（约为 3 伏）。假定这时已经达到了饱和状态，则由此可以推算出电离室中每秒每立方厘米产生的离子数，这个数被称为宇宙射线的“强度”。我对气压表的压力进行了小的修正（1 cm 汞柱修正了 2.4%）。为了使纬度为 50º 时的值与克莱在标准大气中得到的值一致，图 1 曲线上的这些值（和霍尔林的结果类似）都已经乘上了这样的一个因子（1/33.2）。

我要感谢格罗宁根大学基金会给予的资助，感谢科斯特教授让我在他的实验室中制作仪器，感谢克莱教授为我提供了一些好的建议，最后我还要感谢荷兰—非洲航线的负责人以及斯普林方廷号上全体船员的积极配合。

（刘霞 翻译；马宇蒨 审稿）

The Positive Electron

P. M. S. Blackett

Editor's Note

The discovery of the positive electron, although consistent with Dirac's theory of electrons, created problems for nuclear physicists eager for a plain account of how positive electrons are generated—most often by the interaction of energetic γ-rays with heavy particles or even atomic nuclei. Blackett's approach to the problem in this article is very much that of an experimentalist: how best to explain the fragmentary experimental data so far accumulated about positrons.

THE discovery of the positive electron arose from the study of cosmic radiation by the cloud method[1]. Amongst the tracks of the particles of very great energy, associated with cosmic radiation, were found some which differed from the tracks of negative electrons only by being curved by a magnetic field in the opposite direction. Terrestrial sources of positive electrons of lower energy are now also available, since it has been found that they are produced when hard gamma rays are absorbed by matter, and also in certain cases of nuclear transformation. The production of positive electrons in the laboratory is therefore an easy matter.

The charge and mass of a positive electron can be calculated from the ionisation it produces. For example, Anderson[2] has estimated that the difference between the ionisation due to fast positive and negative electrons with the same curvature in a magnetic field, is not as much as 20 percent. Since for *very fast* particles the ionisation depends on the square of the charge but scarcely at all on the mass, the charge on a positive electron cannot differ by as much as 10 percent from that on a negative electron. On the other hand, for *slow* particles with given charge, the ionisation varies as the mass, so the same equality of ionisation indicates that the masses must be within 20 percent. To obtain further information as to the properties of positive electrons, it is convenient to study in detail the simplest case known of their production; namely, that in which a beam of homogeneous gamma rays is absorbed by heavy elements.

The well-filtered gamma radiation from thorium-C″ is nearly homogeneous and has an energy of 2.62×10^6 volts. It has been found by Anderson and Neddermeyer[3], by Curie and Joliot[4] and by Meitner and Philipp[5] that when such rays fall on a heavy element, positive electrons are ejected.

Positive electrons are also produced when the radiation from beryllium, bombarded by alpha rays, is absorbed[6]. Though this radiation is complex, consisting of neutrons together with gamma rays of rather more than 5.0×10^6 volts energy, Curie and Joliot[6] have shown

正电子

帕特里克·布莱克特

编者按

正电子的发现虽然和狄拉克的电子理论相一致，但却给渴望清楚地解释正电子是如何产生的核物理学家们带来了一系列核物理上亟待解决的问题——通常情况下，正电子是在高能 γ 射线和重粒子甚至是和原子核的相互作用下产生的。布莱克特在这篇文章中对这个问题的处理非常符合一个实验科学家的身份：如何以最佳方式去解释当时断断续续积累起来的关于正电子的实验数据。

用云室法对宇宙射线的研究引发了正电子的发现[1]。在和宇宙辐射有关的高能粒子的轨迹中，发现了一些与负电子轨迹完全一致，只是在磁场作用下偏转方向与其相反的轨迹。现在在地球上，我们也能获得低能的正电子，因为已经有研究发现，在物质吸收硬 γ 射线以及在一些核反应发生时都可以产生正电子。因而，在实验室中获得正电子已经是一件很简单的事情了。

可以根据电离情况计算出正电子的电量和质量。比如，安德森[2]估计由于快正负电子在磁场中产生相同的偏转轨迹，因而它们之间的电离情况的差别不会大于20%。因为对**非常快**的粒子来说，电离情况依赖于粒子电量的平方，但是与粒子的质量几乎无关，而正负电子电量之间的差别不可能有10%那么大。另外，对于电量一定的**慢**粒子，电离情况随粒子质量的变化而变化，所以，相同的电离情况意味着正负电子之间的质量差别应该小于20%。为了得到与正电子性质有关的进一步的信息，我们可以详细地研究产生正电子的最简单的情况，也就是说，研究重元素吸收单色 γ 射线的情况。

由钍 C″ 放射出的 γ 射线经过严格的过滤后将基本达到单色的状态，并带有 2.62×10^6 伏特的能量。安德森和尼德迈耶[3]，居里和约里奥[4]，迈特纳和菲利普[5]都发现，当这样的射线照射重元素时，就会产生正电子。

铍受到 α 粒子轰击后的辐射产物被重元素吸收时也可以产生正电子[6]。虽然这种辐射产物很复杂，包括中子和能量大于 5.0×10^6 伏特的 γ 射线，但是居里和约里奥[6]已经通过吸收实验表明，正电子的确主要是由重元素吸收 γ 射线而产生的。

by absorption experiments that the positive electrons are certainly mainly due to the latter.

The following table, which is derived from the work of Curie and Joliot, Grinberg[7], and some unpublished results of Chadwick, Blackett and Occhialini, gives the numbers of positive electrons ejected in a forward direction from different elements by various radiations, the numbers being expressed as a fraction of the observed number of negative electrons. These percentages give only a rough indication of the frequency of production of positive electrons, since the actual angular distributions are not known, and since the effect of the particular experimental arrangement may be considerable.

Number of Positive Electrons produced when Gamma Rays are Absorbed.

Source	Energy of gamma ray	Absorber		
		U	Pb	Al
Ra	1.0 to 2.2×10^6 volts		3%	
ThC"	2.62×10^6 volts		10%	very small
Po+Be	5 to 6×10^6 volts	more than 40%	40%	5%

The ejected negative electrons comprise two groups, consisting of the photo-electrons with the whole energy of the quanta, that is with 2.62×10^6 volts, and the Compton electrons which have a maximum energy of 2.39×10^6 volts in a forward direction. The table shows that the number of positive electrons increases rapidly with the energy of the quanta and with the atomic number of the absorber.

If from these figures the effective area of a heavy atom for the production of a positive electron by a quantum of 5×10^6 volts is calculated, values are found which are rather larger than the area of cross-section of the nucleus. This fact makes it improbable that the production of the positive electrons is mainly a nuclear phenomenon.

This view is strengthened by consideration of the energies of the particles. The maximum energy of the positive electrons produced by a given radiation appears to be about the same for all absorbers. If the particles had a nuclear origin, a variation with the type of nucleus would be expected.

For the 5.0×10^6 and the 2.62×10^6 volt radiations, the maximum energies of the positive electrons are found to be about 4 and 1.6×10^6 volts respectively, that is, in each case about a million volts less than the energy of the quantum.

If the positive electrons are indeed produced outside the nucleus, many important conclusions follow:

(*a*) Since there is certainly no room, in atomic theory, for the permanent existence

综合居里、约里奥、格林伯格[7]的工作，以及查德威克、布莱克特、奥基亚利尼未发表的结果可以得到下表的数据。此表给出了不同元素在不同放射反应中前向发射的正电子数量与观测到的负电子数量的比值。因为我们并不知道正电子角分布的实际情况，而且可能还应该考虑某个特定的实验装置对实验结果的影响，因此，这些比值只是对正电子产生频率的粗略表示。

吸收γ射线所产生的正电子数量

射线源	γ射线能量	吸收物质		
		铀	铅	铝
镭	1.0×10^6~2.2×10^6 伏特		3%	
钍 C″	2.62×10^6 伏特		10%	极小
钋和铍	5×10^6~6×10^6 伏特	大于 40%	40%	5%

发射出的负电子包含两种成分，一种是带有量子全部能量的光电子，能量为 2.62×10^6 伏特；另一种是前向发射时最高能量为 2.39×10^6 伏特的康普顿电子。如表格所示，正电子的数量随着量子能量和吸收物质原子序数的增加而迅速增加。

如果从上面的数据出发，计算出重原子在吸收一个能量为 5×10^6 伏特的量子后产生正电子的有效面积，则得到的结果远远大于相应原子核的横截面积。这个结果表明，正电子的产生不可能主要是与原子核有关的现象。

对粒子能量的考虑会使以上的观点更令人信服。对于各种不同的吸收物质，在给定的辐射下产生的正电子的最高能量似乎都是大致相同的。如果产生的粒子是由γ射线与原子核发生反应所致，那么，不同类型的原子核所产生的粒子的能量应该是不一样的。

吸收了辐射能量为 5.0×10^6 伏特和 2.62×10^6 伏特的正电子，其最大能量分别约为 4×10^6 伏特和 1.6×10^6 伏特，也就是说，在各种情况下，正电子的能量都要比量子能量大约少 100 万伏特。

如果正电子确实是在原子核外产生的，那么就有以下几个重要结论：

(*a*) 根据原子理论，既然肯定没有正电子在原子核外长期存在的余地，那么来自

of positive electrons well outside a nucleus, then a positive electron that comes from there must be born there, and if born there, an equal negative electron must be born simultaneously in order to conserve electric charge. This is confirmed by the experimental observation that pairs of tracks do occur, which almost certainly are to be interpreted as due to the simultaneous ejection of a positive and a negative electron.

To produce such a pair of electrons with opposite charges requires an expenditure of energy $(m_1+m_2)c^2$. If both particles have the electronic mass, this energy amounts to 1.01×10^6 volts, so that in the case of the 2.62×10^6 volt radiation, no pair of positive and negative electrons can have more energy than 1.61×10^6 volts energy. Anderson has found this to be the case. Again, the maximum energy of a single positive electron producing an unpaired track should also be 1.61×10^6 volts. An experimental determination of this maximum energy is being made by Chadwick, Blackett and Occhialini, and their preliminary results* give the value of $1.58\pm0.07\times10^6$ volts, in excellent agreement with the theory.

(*b*) The positive electron must have a spin of $\frac{1}{2}$ and so obey the Fermi–Dirac statistics. For since energy is observed to be conserved during the birth process, it is to be expected that linear and angular momentum are also conserved. So if a quantum gives rise to a pair of particles, one of which has a spin of $\frac{1}{2}\frac{h}{2\pi}$, the other must have the same spin, since a quantum can only excite changes for which the angular momentum changes by 0 or 1. The argument is still valid even if possible changes in the nuclear spin are taken into account, for these must also be integral.

(*c*) A necessary consequence of the occurrence of the process whereby a quantum interacts with an atom to produce a pair of electrons of opposite sign, is the occurrence of the reverse process, in which a positive electron and a negative electron interact with each other and the field of an atom to produce a single quantum of radiation. Since the conditions for this occurrence cannot be rare, a positive electron cannot be expected to exist for more than a short time in matter at ordinary densities.

These conclusions as to the existence and the properties of positive electrons have been derived from the experimental data by the use of simple physical principles. That Dirac's theory of the electron predicts the existence of particles with just these properties, gives strong reason to believe in the essential correctness of his theory.

Dirac succeeded in formulating the wave equation for an electron moving in a potential field in such a way as to make it relativistically invariant. The solution of this new wave equation not only led, in the case of the hydrogen atom, to a complete explanation of

---

*The mass of the positive electron can be calculated from the equation

$$E_{max.} = hv - (m_1 + m_2)c^2$$

Using the values $hv = 2.62\times10^6$ volts, $E_{max.} = 1.58\pm0.07\times10^6$ volts.

We find $$m_2 = (1.04 \pm 0.14)m_1.$$

This calculation affords probably the most accurate estimate of the mass of a positive electron yet available.

那里的正电子必然是在那里产生的，而如果那里产生了正电子，考虑到电荷守恒，那里也必将同时产生一个带有相同电量的负电子。实验上确实观测到了成对的粒子轨迹，这显然可以解释成同时发射出正负电子对，实验有力地证明了以上的结论。

要产生这样一对正负电子需要消耗的能量为 $(m_1+m_2)c^2$。如果两个粒子的质量都与电子相同，那么这个能量就是 1.01×10^6 伏特，这样，在辐射能量为 2.62×10^6 伏特时，产生的正负电子对的能量就不可能超过 1.61×10^6 伏特。安德森证实了上述的情况。此外，没有形成成对轨迹的单个正电子的最大能量应该也是 1.61×10^6 伏特。查德威克、布莱克特和奥基亚利尼用实验测定了这个能量的最大值，他们得到的初步结果*是 $1.58\pm0.07\times10^6$ 伏特，和理论值非常吻合。

(*b*) 正电子应该具有 $\frac{1}{2}$ 的自旋，并遵循费米 – 狄拉克统计。由于观察到能量在正负电子产生阶段是守恒的，那么可以预见，在该阶段动量和角动量也是守恒的。因为一个量子只能激发角动量改变 0 或 1 的变化，所以如果一个量子产生一对粒子，其中一个带有自旋 $\frac{1}{2}\frac{h}{2\pi}$，另一个一定具有相同的自旋。即使把可能出现的把原子核自旋的变化也考虑进来，这个论点仍然成立，因为这些自旋的变化应该也是按整数变化的。

(*c*) 一个量子和一个原子相互作用产生一对电性相反的电子，这个过程的必然结果是会出现一个逆过程，在这个逆过程中，一个正电子和一个负电子相互作用并与一个原子场作用以产生一个量子的辐射。由于发生这种情况的条件并不罕见，因而可以预期，在一般密度的物质中，正电子是不可能长期存在的。

根据实验数据并运用简单的物理原理，我们得到了关于正电子的存在状态及其性质的结论。狄拉克的电子理论预言了具有这些性质的粒子的存在，因而有充分的理由相信他的理论是基本正确的。

狄拉克成功地构建了电子在势场中运动的波动方程，并保证了方程的相对论不变性。对于氢原子而言，这个新波动方程的解不仅完全解释了谱线的精细结构，还

---

* 正电子的质量可以通过以下方程

$$E_{\max}=h\nu-(m_1+m_2)c^2$$

并代入 $h\nu=2.62\times10^6$伏特，$E_{\max}=1.58\pm0.07\times10^6$伏特得到。

我们发现 $$m_2=(1.04\pm0.14)m_1$$

这个结果给出了迄今为止人们对正电子质量的最为精确的估计值。

the fine structure of the spectral lines, but also to a rational explanation of the spin and magnetic moment of the electron itself.

However, in addition to the solutions corresponding to the normal electronic levels found experimentally, were others which seemed to correspond to no observed facts. These solutions seemed to predict the existence of states in which the electrons possessed a negative kinetic energy, and therefore did not correspond to particles in any usual sense. These states could not be ignored, because transitions must theoretically occur between them and the normal states corresponding to positive kinetic energy. Dirac suggested that the difficulty might be avoided if it were supposed that all the negative energy states are normally occupied, and further, that the totality of electrons in such states produce no external field.

On this view, only an unoccupied state or "hole" would correspond to an observed particle. It followed from the theory that such unoccupied states should behave in an external field like particles with the same mass and spin as a negative electron but with a positive charge. The experimental discovery of the positive electron has therefore removed a very serious theoretical difficulty, and by so doing, has greatly extended the field of phenomena over which Dirac's theory may be applied.

Owing to analytical difficulties, the work of applying Dirac's theory to special cases has not progressed far, but Oppenheimer and Plesset[8] have calculated approximately the probability of the production of pairs of electrons of opposite charge when hard gamma rays are absorbed by matter. So far as these theoretical results go, they are in rough agreement with the experimental conclusions, both as regards the order of magnitude of the effect and its dependence on the energy of the quantum and the atomic number of the absorber.

The calculations give for the extra absorption by lead and tin of the 2.62×10^6 volt radiation, due to the production of positive electrons, the values of 25 percent and 15 percent of the absorption by the normal scattering and photoelectric processes. These figures are roughly those observed experimentally by Tarrant and Gray. So one may conclude that a large part of the anomalous absorption may be attributed to the production of positive electrons.

One would expect that the absorbed energy would be re-radiated in two ways. An ejected positive electron may disappear by the reverse process to that which produced it, that is, by reacting with a negative electron and a nucleus, to give a single quantum of a million volts energy (see (*c*) above). Or it can disappear, according to Dirac's theory, by another type of process, in which a positive electron reacts with a *free* or *lightly bound* negative electron so

给出了电子本身存在自旋和磁矩的合理解释。

然而，除了与实验中观测到的正常电子能级相对应的解以外，还有一些解不与任何可观测到的现象相对应。这些解似乎预见到了电子具有负动能的状态，因而无法对应于任何一般意义上的粒子。这些电子态不能被忽略，因为理论上存在这些电子态和对应于正动能的正常态之间的量子跃迁。狄拉克提出，如果假设正常情况下所有的负能态都被占满，并且这些态上的所有电子都不产生外场，那么这个理论上的困难就被排除了。

从这个观点来看，只有未被占据的负能电子态或“空穴”才能对应于实验上可观测到的粒子。根据理论进一步得出，这些未被占据的负能电子态在外场中表现得就像一些与负电子具有相同质量和自旋，但带有相反电荷的粒子。实验上发现了正电子进而解决了一个重大的理论难题，而在这个理论难题得到解决之后，狄拉克理论可以解释的现象也随之得到了极大的扩展。

由于分析上的困难，将狄拉克理论应用于某些特殊现象的工作并没有深入开展，但是奥本海默和普莱赛特 [8] 已经大致地计算了物质吸收硬 γ 射线时产生带相反电荷的电子对的概率。这些理论上的结果，无论在反应发生的量级上，还是在反应对量子能量和吸收物质原子序数的依赖情况上，都与实验结果大致吻合。

由理论计算得出：由于正电子的产生，铅和锡作为吸收物质额外多吸收了 2.62×10^6 伏特的辐射能量，这个能量是正常散射情况下所吸收能量的 25%，或是光电反应过程中所吸收能量的 15%。这些数据与塔兰特和格雷观测到的实验结果基本一致。因此我们可以得到这样的结论：大部分的反常吸收都有可能归因于正电子的产生。

我们可以预期，吸收的能量可以以两种方式再次辐射。一是发射出来的一个正电子可能会在与使其产生相逆的过程中消失，也就是说，和一个负电子以及原子核发生相互作用，产生一个能量为 100 万伏特量级的量子（见上面的结论 (*c*)）。二是按照狄拉克的理论，正电子也可以以另一种方式消失，即正电子和一个**自由的**或处于**弱束缚状态的**负电子发生相互作用后，两者都消失并放射出两个能量为 50 万伏特

that both disappear with the emission of two quanta of half a million volts energy.* It is remarkable that the re-emitted radiation is estimated by Gray and Tarrant to be composed mainly of just these two energies, of one half and one million volts. However, Fermi and Uhlenbeck[9] have found that the theoretical intensity of the hard component is far smaller than that observed.

This absorption of hard gamma rays by atoms, resulting in the production of pairs of oppositely charged electrons, may be thought of as a photoelectric absorption by the "virtual" electrons, that is, by electrons with negative kinetic energy, near the nucleus. According to Beck[10], these virtual electrons may be considered to have a binding energy of the order of 2 mc^2. Beck also shows that the number of these virtual electrons which are effective for the absorption are proportional to the square of the atomic number and that they amount to about one for each lead atom. The theory also indicates that the birth process takes place within a distance of $h/2\pi mc=3.85\times10^{-11}$ cm. of the nucleus, that is, well inside the K ring.

Curie and Joliot[11] have found that positive electrons are produced when *aluminium* and *boron* are bombarded by alpha particles, and that these positive electrons have a higher energy than the accompanying negative electrons. *Silver*, *lithium* and *paraffin*, however, give no positive electrons. Curie and Joliot suggest that the positive electrons originate in the disintegrating nucleus, but it seems possible that they may be produced mainly outside the nucleus by the internal conversion of a gamma ray emitted by the nucleus. To explain the effect in this way, the probability of internal conversion must be nearly unity.† The greater energy of the positives may be explained by the fact that a positive electron gains kinetic energy and a negative electron loses it, on escaping from the field of a nucleus. This resulting difference in kinetic energy will be the larger the nearer to the nucleus that the pair is born, and so should be larger in the case of such an internal conversion process, which depends on a spherical wave, than in the usual case of external absorption, which depends on a plane wave.

Though it was in association with cosmic radiation that positive electrons were first detected, the exact part they play in these complicated phenomena is not yet clear. But certain facts are established[12]. (*i*) Of the fast particles which produce the cosmic ray ionisation at sea-level, about half are positive and half negative electrons. Their energies range from a few million to nearly 10^{10} volts. (*ii*) The same ratio is found in the "showers". The showers appear therefore to represent the birth of multiple pairs of positive and negative electrons, as a result of one or more collision processes induced

---

* Dirac's calculation of this annihilation probability gives a positive electron a life of less than 10^{9} sec. in water, the life being inversely proportional to the density. If this predicted process is verified experimentally, it will be possible to assume the reverse process, the creation of a pair of electrons of opposite sign by the collision of two quanta of high energy. This latter process would then be the first case known of the "interference" of quanta; it is conceivable that this process has considerable cosmological importance.

† Oppenheimer and Plesset (*loc. cit.*) predict theoretically far smaller values.

的量子。* 值得注意的是，由塔兰特和格雷估计的再次辐射的能量主要就是由这两部分所组成，即 50 万伏特和 100 万伏特。但是，费米和乌伦贝克[9]发现，理论上硬成分的辐射能量的强度要远远小于实验上的观测值。

我们可以把原子吸收硬 γ 射线从而产生带有相反电荷的电子对的过程看作是一个通过“虚”电子来实现的光电吸收过程，这个“虚”电子即原子核附近具有负动能的电子。根据贝克[10]的理论研究，我们可以认为这些虚电子具有 $2mc^2$ 量级的结合能。贝克还表示，吸收过程中的有效虚电子数和原子序数的平方成正比，并且每个铅原子中大约有一个虚电子。理论结果还显示，电子对的产生过程发生在与原子核的距离小于 $h/2\pi mc = 3.85 \times 10^{-11}$ cm 的范围内，这个范围恰好在 K 壳层之内。

居里和约里奥[11]发现，当**铝**和**硼**被 α 粒子轰击时产生正电子，而这些正电子的能量高于伴随其产生的负电子的能量。然而**银**、**锂**、**石蜡**在 α 粒子轰击下没有产生正电子。居里和约里奥认为，正电子是在原子核的裂变过程中产生的，但是似乎是由原子核发射出的 γ 射线内转换而成，而这个过程有可能主要发生在原子核之外。为了用这种方式解释正电子的产生，γ 射线的内转化概率必须接近单位 1。† 正电子比负电子能量高可以解释为：为了从原子核势场中逃逸，正电子要获得能量，而负电子要损失概率。电子对产生的位置越靠近原子核则最终的动能差异也将越大，因而对于上述的内转换过程这个能量的差异也会大于通常的外部吸收过程，因为前者依赖于球面波，而后者依赖于平面波。

虽然正电子的最初发现和宇宙辐射有关，但目前还不清楚正电子在这个复杂现象中到底扮演什么样的角色。不过有以下几点是可以肯定的[12]：(*i*) 在海平面上发生宇宙射线电离的快粒子中，正负电子大约各占一半。它们的能量从几百万伏特到将近 10^{10} 伏特不等。(*ii*) 在“簇射”中也得到了相同的正负电子比。因此簇射似乎意味着由初级辐射引发的一次或多次碰撞过程导致了许多对正负电子的产生。狄拉

* 狄拉克计算了这种湮灭的概率，结果表明一个正电子在水中的寿命要小于 10^{-9} 秒，并且其寿命与密度成反比。如果这个预测的过程得到了实验上的证明，那么就可以假定逆过程是有可能存在的，即通过两个高能量子的碰撞产生一对符号相反的电子。后一个过程后来成为第一例已知的量子“干涉”；可想而知这个过程所具有的重大宇宙学意义。

† 奥本海默和普莱塞特（在上述引文中）预测的理论值要小得多。

by the primary radiation. Dirac's theory shows that the production of single pairs is of primary importance in the absorption of both gamma rays and particles of high energy[13], but has, as yet, given no hint of the cause of the multiple pairs forming the showers. (*iii*) It has been shown that the majority of the particles incident on the earth's atmosphere are positively charged[14].

Since protons are rarely observed at sea-level, it is probable that the positively charged incident particles are not protons but positive electrons. If this is so, the main part of the flux of cosmic radiation in inter-galactic space must be in the form of positive electrons; and since the total mass of this radiation has been estimated as possibly as large as 1/1,000 part of the mass of all the stars and nebulae[15], it appears that the positive electron, though rare, because ephemeral, on earth, is an important constituent of the universe as a whole.

(**132**, 917-919; 1933)

References:

1. Anderson, *Science*, **76**, 238 (1932). Blackett and Occhialini, *Proc. Roy. Soc.*, A. **139**, 699 (1933).
2. Anderson, *Phys. Rev.*, **43**, 491 (1933). *Phys. Rev.*, **44**, 406 (1933).
3. Anderson and Neddermeyer, *Phys, Rev.*, **43**, 1034 (1933).
4. Curie and Joliot, *C. R. Acad. Sci.*, **196**, 1581 (1933).
5. Meitner and Philipp, *Naturwissenschaften*, **24**, 468 (1933).
6. Chadwick, Blackett and Occhialini, *Nature*, **131**, 473 (1933). Meitner and Philipp, *Naturwissenschaften*, **15**, 286 (1933). Curie and Joliot, *C.R. Acad. Sci.*, **196**, 405 (1933).
7. Grinberg, *C. R. Acad. Sci.*, **197**, 318 (1933).
8. Oppenheimer and Plesset, *Phys. Rev.*, **44**, 53 (1933).
9. Fermi and Uhlenbeck, *Phys. Rev.*, **44**, 510 (1933).
10. Beck, *Zeit. Phys.*, **83**, 498 (1933).
11. Curie and Joliot, *C. R. Acad. Sci.*, **196**, 1885 (1933).
12. Anderson, *Phys, Rev.*, **41**, 405 (1932); Kunze, *Zeit. Phys.*, **80**, 559 (1933); Blackett and Occhialini, *loc. cit.*
13. Furry and Carlson, *Phys. Rev.*, **44**, 237 (1933).
14. Johnson, *Phys. Rev.*, **43**, 1059 (1933).
15. Lemaitre, *Nature*, **128**, 704 (1931).

克理论指出，在吸收 γ 射线和高能粒子的过程中产生了单个电子对，这具有重大的意义 [13]，但是到现在为止，这些还不能为形成簇射的多个电子对的产生机制提供线索。(*iii*) 事实表明，绝大多数入射到地球大气层中的粒子都带有正电荷 [14]。

由于在海平面上极少能探测到质子，那么带正电荷的入射粒子就很有可能是正电子而不是质子。如果事实的确如此，那么跨星系空间的宇宙射线流将主要由正电子构成；既然已经估算出这些辐射粒子的总质量大约是所有恒星和星云质量的 1/1,000[15]，那么可以认为，虽然正电子因为寿命短暂而在地球上很少见，但它仍然是整个宇宙的重要组成部分。

（王静 翻译；李军刚 审稿）

The Ether-drift Experiment and the Determination of the Absolute Motion of the Earth*

D. C. Miller

Editor's Note

The ether-drift experiment was that in which Michelson and Morley in 1892, using an optical interferometer they had constructed in Ohio in the United States, demonstrated that the supposed ether filling all of space did not exist. This paper uses similar equipment to work out what is called the "absolute" movement of the Solar System through space. The use of the term "absolute" pays scant attention to the theory of relativity, but the result does suggest how the Solar System is moving with respect to the surrounding galaxies.

THE ether-drift experiment first suggested by Maxwell in 1878 and made possible by Michelson's invention of the interferometer in 1881, though suitable for the detection of the general absolute motion of the Earth, was actually applied for detecting only the known orbital component of the Earth's motion. For the first time, in 1925 and 1926, I made observations at Mount Wilson of such extent and completeness that they were sufficient for the determination of the absolute motion of the Earth. These observations involved the making of about 200,000 single readings of the position of the interference fringes.

The ether-drift observable in the interferometer, as is well known, is a second order effect; and the observations correctly define the line in which the absolute motion takes place, but they do not determine whether the motion in this line is positive or negative in direction.

At the Kansas City meeting of the American Association for the Advancement of Science, in December, 1925, before the completion of the Mount Wilson observations, a report was made showing that the experiment gives evidence of a cosmic motion of the solar system, directed towards a northern apex; but the effects of the orbital motion were not found, though it seemed that the observations should have been quite sufficient for this purpose[1].

The studies of the proper motions and of the motions in the line of sight of the stars in our galaxy have shown that the solar system is moving, *with respect to our own cluster*, in the general direction of a northern apex in the constellation Hercules. This apex is near that

---

* Paper read before Section A (Mathematical and Physical Sciences) of the British Association meeting at Leicester on September 13, 1933.

以太漂移实验和地球绝对运动的确定*

米勒

编者按

迈克尔逊和莫雷于1892年用他们在美国俄亥俄州建造的光学干涉仪进行了以太漂移实验，实验表明原先假设充满整个宇宙的以太其实并不存在。这篇文章利用相似的仪器测量了太阳系在宇宙中的“绝对”运动。使用“绝对”这个术语表明作者没有考虑到相对论，但是实验结果的确表明了太阳系是如何相对于周围星系运动的。

1878年麦克斯韦首次提出了以太漂移实验，1881年迈克尔逊发明的干涉仪使该实验成为可能。虽然这个实验可以适用于探测地球总体的绝对运动，但实际上它仅在探测人们所熟知的地球运动的轨道分量时被应用过。1925 ~ 1926年，我在威尔逊山上进行了首次观测，这次观测无论从范围上还是完备性上都足以用来确定地球的绝对运动。这些观测数据包含大约200,000个单独的干涉条纹位置的读数。

众所周知，干涉仪中可观察到的以太漂移是一个二阶效应；通过观测可以准确地确定绝对运动的路线，但无法确定沿该路线运动的方向。

1925年12月，即威尔逊山的观测完成之前，美国科学促进会在堪萨斯城召开了一个会议，会议上的一个报告指出，实验证明，太阳系在宇宙中是朝着一个北向点运动的；尽管这个实验似乎应当足以观测到轨道运动的效应[1]，但是这一目的并未达到。

对银河系中恒星的自行和视向运动的研究表明，**相对于我们本星系团**，太阳系是大致朝着武仙座北向点的方向运动的。这个向点和前述报告中的以太漂移观测所示出的方向接近，这似乎是证明其正确性的有力证据。也许，这正是导致接下来的

* 本论文发表于1933年9月13日在莱切斯特举行的英国科学促进会会议A分会场（数学与物理科学）。

indicated by the ether-drift observations as just reported, and seemed to be confirmatory evidence of its correctness. Probably it was this that caused the continuation of the analysis of the problem, on the supposition that the absolute motion was to the northward in the indicated line. All possible combinations and adjustments failed to reconcile the computed effects of combined orbital and cosmic motions with the observed facts.

In the autumn of 1932, a re-analysis of the problem was made, based upon the alternative possibility that the motion of the solar system is in the cosmic line previously determined, but is in the opposite direction, being directed southward. This gives wholly consistent results, leading for the first time to a definite quantitative determination of the absolute motion of the solar system, and also to a positive detection of the effect of the motion of the Earth in its orbit.

The absolute motion of the Earth may be presumed to be the resultant of two independent component motions. One of these is the orbital motion around the Sun, which is known both as to magnitude and direction. For the purposes of this study, the velocity of the orbital motion is taken as 30 kilometres per second, and the direction changes continuously through the year, at all times being tangential to the orbit. The second component is the cosmical motion of the Sun and the solar system. Presumably this is constant in both direction and magnitude, but neither the direction nor magnitude is known; the determination of these quantities is the particular object of this experiment. The rotation of the Earth on its axis produces a velocity of less than four tenths of a kilometre per second in the latitude of observation and is negligible so far as the velocity of absolute motion is concerned; but this rotation has an important effect upon the apparent direction of the motion and is an essential factor in the solution of the problem. Since the orbital component is continually changing in direction, the general solution is difficult; but by observing the resultant motion when the Earth is in different parts of its orbit, a solution by trial is practicable. For this purpose it is necessary to determine the *variations* in the magnitude and in the direction of the ether-drift effect throughout a period of twenty-four hours and at three or more epochs of the year. The observations made at Mount Wilson correspond to the epochs April 1, August 1 and September 15, 1925, and February 8, 1926.

The point on the celestial sphere towards which the Earth is moving because of its absolute motion is called the apex of its motion. This point is defined by its right ascension and declination, as is a star, and the formulae of practical astronomy are directly applicable to its determination from the interferometer observations. The theoretical consideration of the determination of the apex of the motion of the Earth has been given in a paper by Prof. J. J. Nassau and Prof. P. M. Morse[2].

Table I gives the right ascensions and declinations of the apexes of the Earth's cosmical motion as obtained from the interferometer observations for the four epochs on the presumption of a southward motion, together with the right ascensions and declinations calculated upon the theory of an ether-drift.

研究都假设绝对运动是沿着所示路线向北的原因。然而在这种假定下，所有可能的组合和调整都不能使计算得到的轨道和宇宙运动的联合效应与实际的观测结果相符。

1932 年秋，人们基于另外一种可能性重新对这个问题进行了分析，即太阳系沿着之前确定的宇宙轨道运动，但方向相反，也就是向南运动。这一次得到了完全一致的结果，从而第一次定量地确定了太阳系的绝对运动，同时也首次明确地探测到了地球在其轨道内的运动效应。

地球的绝对运动可以被看成是两个独立运动分量的和。其中一个是绕太阳的轨道运动，其大小和方向都是已知的。在此项研究中，地球轨道运动的速度取为 30 千米 / 秒，其方向在一年中连续变化，并总是沿着轨道切线的方向。第二个分量是太阳和太阳系在宇宙中的运动。假设该运动的大小和方向都是恒定的，但均属未知；确定这些量正是这项实验的目的。在进行观测的纬度处，地球绕轴的自转将产生一个小于 0.4 千米 / 秒的速度，这相对于地球绝对运动速度而言可以忽略不计；但是地球自转对运动的视方向有着重要的影响，这也是求解此问题时的一个重要因素。由于轨道运动分量在方向上持续变化，所以要得到通解很困难；但是通过观测地球在轨道不同位置的合运动，从而得到实验解还是可行的。为此，必须在一年中选择三个或者更多个不同的时期，确定以太漂移效应的大小和方向在 24 小时内的**变化**。这项观测在威尔逊山上进行，其观测的时期分别是：1925 年 4 月 1 日、8 月 1 日、9 月 15 日以及 1926 年 2 月 8 日。

地球绝对运动所指向的天球上的那个点被称为地球运动的向点。天球就如同一颗恒星，这个点可由天球上的赤经和赤纬确定，通过干涉仪的观察结果，利用实测天文学的公式可以直接确定这个点的位置。纳索教授和莫尔斯教授的文章已经对确定地球运动的向点给出了理论思考 [2]。

表 1 给出了假定太阳系是在向南运动的情况下，利用干涉仪的观测结果得到的四个时期地球在宇宙中运动相应向点的赤经和赤纬，以及用以太漂移理论计算得到的相应值。

Table I. Location of resultant apexes

Epoch	α(Obs.)	α(Calc.)	δ(Obs.)	δ(Calc.)
Feb. 8	6^h 0^m	5^h 40^m	−77° 27′	−78° 25′
April 1	3 42	4 0	76 48	77 50
Aug. 1	3 57	4 10	64 47	63 30
Sept. 15	5 5	5 0	62 4	62 15

Apex of cosmic component $\alpha = 4^h\ 56^m$, $\delta = -70°\ 33'$

From these resultant apexes are determined four values for the apex of the cosmic component, which is the apex of the motion of the solar system as a whole. This apex has the right ascension $4^h\ 56^m$ and the declination 70° 33′ south.

Continuing the astronomical description, having found the elements of the "aberration orbit", these are used to compute the apparent places of the resultant apexes for the four epochs of observation. On the accompanying chart of the south circumpolar region of the celestial sphere (Fig. 1), the large star indicates the apex of the cosmic motion, and the four circles show the locations of the calculated apexes. These apexes necessarily lie on the closed curve representing the calculated aberration orbit, the centre of which is the apex of the cosmic component of the Earth's motion. This aberration orbit is the projection of the Earth's orbit on the celestial sphere, which in this case is approximately a circle. The observed apexes for the four epochs are represented by the small stars. The locations of the pole of the ecliptic and of the star Canopus are also shown. The close agreement between the calculated and observed apparent apexes would seem to be conclusive evidence of the validity of the solution of the ether-drift observations for the absolute motion of the Earth and also for the effect of the orbital motion of the Earth, which hitherto has not been demonstrated.

It may seem surprising that such close agreement between observed and calculated places can be obtained from observations of such minute effects, and effects which are reputed to be of such difficulty and uncertainty. Perhaps an explanation is the fact that the star representing the final result for the February epoch is, in effect,

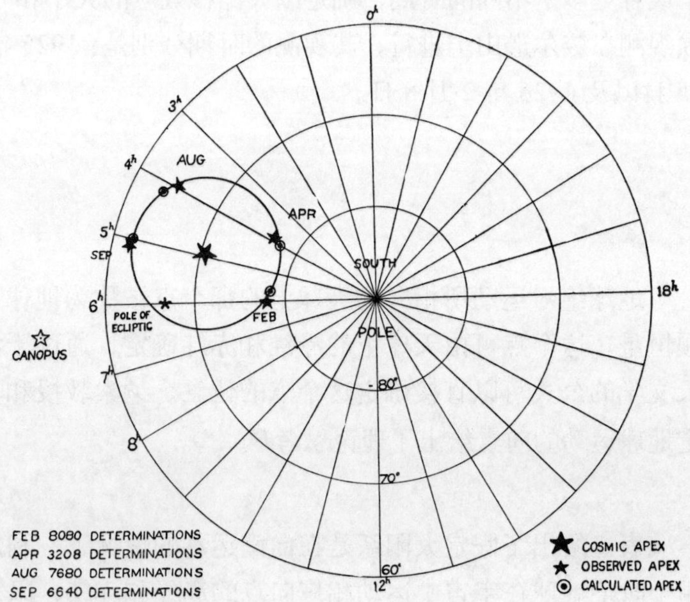

Fig. 1. Observed and calculated apexes of the absolute motion of the solar system.

表 1. 地球在宇宙中合运动的向点位置

时期	赤经（观测值）	赤经（计算值）	赤纬（观测值）	赤纬（计算值）
2 月 8 日	6 小时 0 分	5 小时 40 分	–77°　27′	–78°　25′
4 月 1 日	3　　42	4　　0	76　48	77　50
8 月 1 日	3　　57	4　　10	64　47	63　30
9 月 15 日	5　　5	5　　0	62　4	62　15

宇宙分量的向点 赤经 = 4 小时 56 分，赤纬 = –70° 33′

由上表中四个合运动的向点可以确定宇宙分量的向点，它也是太阳系作为一个整体运动的向点。这个向点的位置是赤经 4 小时 56 分，赤纬 –70° 33′。

让我们继续天文学的描述，在得出“光行差轨道”的根素后，即可以用其计算出四个观测时期合运动的向点的视位置。附图是天球南拱极区（图 1），图中大星号表示宇宙运动的向点，四个圆圈表示计算得到的向点位置。这些向点必须落在计算得到的光行差轨道的闭合曲线上，光行差轨道的中心即为地球运动宇宙分量的向点。此光行差轨道是地球轨道在天球上的投影，在这种情况下近似为圆。在四个时期观测到的向点在图中以小星号表示。图中还标出了黄极和老人星的位置。计算和观察得到的视向点的高度吻合似乎有力地证明了用以太漂移观测来求解地球绝对运动以及地球轨道运动效应的正确性，而对于后者迄今为止尚未得到实验证明。

令人惊奇的是，虽然以太漂移这种微弱的效应被公认为是难于观测且其结果会有很高的不确定性，但最后却得到了计算值和观察值如此高度一致的结果。也许，我们可以这样解释：这些观察结果实际上是由大量独立的位置测量平均得出的，二月

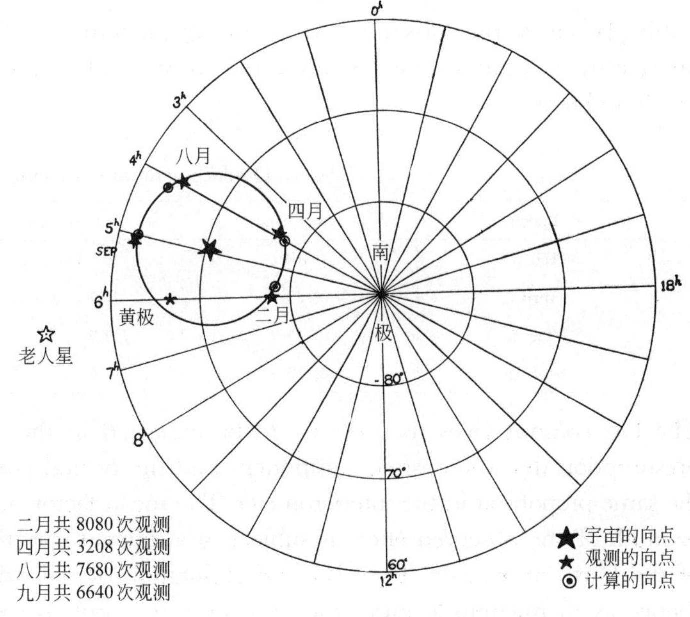

图 1. 太阳系绝对运动顶点的观测位置和计算位置

the average of 8,080 single determinations of its location; the star for the August epoch represents 7,680 single determinations, that for September, 6,640, and that for April, 3,208 determinations.

The location of the apex of the solar motion is in the southern constellation Dorado, the Sword-Fish, and is about 20° south of the star Canopus, the second brightest star in the heavens. It is in the midst of the famous Great Magellanic Cloud of stars. The apex is about 7° from the pole of the ecliptic and only 6° from the pole of the invariable plane of the solar system; thus the indicated motion of the solar system is almost perpendicular to the invariable plane. This suggests that the solar system might be thought of as a dynamic disc which is being pulled through a resisting medium and therefore sets itself perpendicular to the line of motion.

It is presumed that the Earth's motion in space is projected on to the plane of the interferometer, and the *direction* of this motion is determined by observing the variations produced in the projected component by the rotation of the Earth on its axis and by the revolution around the Sun. Both the magnitude and the direction of the observed effect vary in the manner and in the proportion required by an ether-drift, on the assumption of a stagnant ether which is undisturbed by the motion of the Earth through it. But the observed *magnitude* of the effect has always been less than was to be expected, indicating a reduced velocity of relative motion, as though the ether through which the interferometer is being carried by the Earth's motion were not absolutely at rest. The orbital velocity of the Earth being known, 30 kilometres per second, the cosmical velocity of the solar system, determined from the proportional variations in the observed effects, is found to be 208 kilometres per second.

Table II gives the observed periodic displacement of the fringe system as the interferometer rotates on its axis, and the corresponding velocity of relative motion of the Earth and ether.

Table II. Displacements and velocities

Epoch	Fringe Shift	Velocity (Obs.)	Velocity (Calc.)
Feb. 8	0.104 l	9.3 km./sec.	195.2 km./sec.
April 1	0.123	10.1	198.2
Aug. 1	0.152	11.2	211.5
Sept. 15	0.110	9.6	207.5

The last column gives the velocity to be expected in the stagnant ether theory on the presumption that the cosmic component and the orbital component are both reduced in the same proportion in the interferometer. The mean factor of reduction is $k = 0.0514$. The azimuth of the observed effect is subject to a diurnal variation, produced by the rotation of the Earth on its axis. The observed oscillations of the azimuth are in accordance with theory as to magnitude and time of occurrence, but for some unexplained reason, the axis of the oscillations is displaced from the meridian. In order to account for the results

有 8,080 次单次测定，八月有 7,680 次，九月有 6,640 次，而四月有 3,208 次。

太阳运动的向点位于南天的剑鱼座内，处在天空中第二亮的老人星以南约 20° 处，著名的大麦哲伦星云的中间。该向点偏离黄极 7° 左右，而仅偏离太阳系的不变平面的 6°，因此太阳系的运动几乎是垂直于其不变平面的。这表明，太阳系可以被看作是一个动力盘，这个圆盘被牵引着穿过某种阻尼介质，从而使其与运动路线垂直。

我们通常假设地球在宇宙中的运动被投影到干涉仪的平面上，通过观测由于地球的自转和绕太阳的公转所引起的这一投影分量的变化可以确定该运动的**方向**。观测到的效应的大小和方向均和假设以太静止时，即以太不被运动的地球所扰动时所预期的变化方式和比例相符。但是，观测到的效应的**值**总比预期的小，这表明相对运动速度的减小，这似乎表明载着干涉仪的地球所穿过的以太并不是绝对静止的。已知地球的轨道速度为 30 千米 / 秒，而根据观测到的效应成比例变化而确定的太阳系的宇宙速度为 208 千米 / 秒。

表 2 给出了干涉仪绕其轴旋转时观测到的条纹系统的周期性位移，以及相应的地球和以太的相对运动速度。

表 2. 位移和速度

时期	条纹位移	速度（观测值）	速度（计算值）
2 月 8 日	0.104 *l*	9.3 千米 / 秒	195.2 千米 / 秒
4 月 1 日	0.123	10.1	198.2
8 月 1 日	0.152	11.2	211.5
9 月 15 日	0.110	9.6	207.5

假设干涉仪所测得的宇宙分量和轨道分量都以相同的比例减小，最后一列给出的是在静止以太理论下的预期速度。平均减缩因子 k=0.0514。观测到的效应的方位角由于地球的绕轴自转存在周日变化。观测到的方位角振荡在大小和时间上与理论结果一致，但是由于一些未知原因，振荡轴偏离了子午线。为了解释这里给出的结果，似乎必须要接受修正后的洛伦兹 – 斐兹杰惹收缩理论，或者像斯托克斯所提出的那

here presented, it seems necessary to accept the reality of a modified Lorentz–FitzGerald contraction, or to postulate a viscous or dragged ether as proposed by Stokes.

The results here reported are, notwithstanding a common belief to the contrary, fully in accordance with the original observations of Michelson and Morley of 1887, and with those of Morley and Miller of 1904–5. The history of the ether-drift experiment and a description of the method of using the interferometer, together with a full account of the observations and their reduction, has been published elsewhere[3].

(**133**, 162-164; 1934)

Dayton C. Miller: Case School of Applied Science, Cleveland, Ohio.

References:

1. *Science*, **63**, 433; 1926. *Nature*, **116**, 49; 1925.
2. *Astrophys. J.*, March, 1927.
3. *Rev. Mod. Phys.*, **5**, 203, July, 1933.

样，假设以太有黏性或具有拖曳效应。

尽管这里给出的结果与普遍的观念相反，但却与1887年迈克尔逊、莫雷所进行的原始观测以及1904~1905年莫雷、米勒所进行的实验观测完全一致。另外，我们已经在其他刊物[3]上发表了相关的内容，包括以太漂移实验的发展、利用干涉仪进行实验的相关方法，以及进行观测及其归算过程的完整说明等。

（王静 孙惠南 翻译；邓祖淦 审稿）

Artificial Production of a New Kind of Radio-element

F. Joliot and I. Curie

Editor's Note

Scientists were still exploring the kinds of nuclear transmutations that could be induced by bombarding stable elements with particles such as alpha particles. Here Irène Curie and Frédéric Joliot report the formation of new, unstable isotopes of nitrogen, silicon and phosphorus made by alpha-irradiation of aluminium, boron and magnesium. The new "radio-elements" are evident from their decay over periods of several minutes, with emission of positrons. Uniquely, the two French scientists use chemical methods to separate the new isotopes and thereby identify their chemical nature. Short-lived positron-emitting isotopes, including the nitrogen-13 reported here, are now used in the medical imaging technique of positron emission tomography. (Note that the third paragraph seems to mistakenly mention "beryllium" in place of "boron".)

SOME months ago we discovered that certain light elements emit positrons under the action of α-particles[1]. Our latest experiments have shown a very striking fact: when an aluminium foil is irradiated on a polonium preparation, the emission of positrons does not cease immediately, when the active preparation is removed. The foil remains radioactive and the emission of radiation decays exponentially as for an ordinary radio-element. We observed the same phenomenon with boron and magnesium[2]. The half life period of the activity is 14 min. for boron, 2 min. 30 sec. for magnesium, 3 min. 15 sec. for aluminium.

We have observed no similar effect with hydrogen, lithium, beryllium, carbon, nitrogen, oxygen, fluorine, sodium, silicon, or phosphorus. Perhaps in some cases the life period is too short for easy observation.

The transmutation of beryllium, magnesium, and aluminium α-particles has given birth to new radio-elements emitting positrons. These radio-elements may be regarded as a known nucleus formed in a particular state of excitation; but it is much more probable that they are unknown isotopes which are always unstable.

For example, we propose for boron the following nuclear reaction:

$$_5B^{10} + {}_2He^4 = {}_7N^{13} + {}_0n^1$$

$_7N^{13}$ being the radioactive nucleus that disintegrates with emission of positrons, giving a stable nucleus $_6C^{13}$. In the case of aluminium and magnesium, the radioactive nuclei would be $_{15}P^{30}$ and $_{14}Si^{27}$ respectively.

一种新放射性元素的人工制造

约里奥，居里

编者按

科学家们仍旧在探寻粒子（比如α粒子）轰击稳定元素所导致的核嬗变的种类。这篇文章中伊雷娜·居里和弗雷德里克·约里奥报道了有关通过α粒子照射铝、硼和镁分别产生了氮、硅和磷的新的不稳定同位素的信息。从它们在数分钟内放出正电子的衰变来看，新“放射性元素”的存在是证据确凿的。特别的是，这两位法国科学家使用化学方法分离出了这些新的同位素，从而鉴定了它们的化学性质。短寿命的放射正电子的同位素，包括本文中所报道的氮–13，现在被用在正电子发射层析的医学成像技术中。（注意文中第三段似乎将“硼”错写成了“铍”。）

几个月之前，我们发现某些轻元素在α粒子的作用下会放出正电子[1]。我们最近的实验得到了一个令人十分惊喜的结果：以钋样品所产生的辐射照射铝箔，当移开放射性的钋样品时，正电子的发射不会立即停止。铝箔仍然保持着放射性，辐射就像普通的放射性元素一样以指数形式衰减。对于硼和镁，我们也观察到了同样的现象[2]。放射性活度的半衰期分别是：硼 14 分钟、镁 2 分 30 秒、铝 3 分 15 秒。

我们在氢、锂、铍、碳、氮、氧、氟、钠、硅和磷等元素中都没有观测到类似的效应。也许在某些情况下由于这一效应寿命过短，所以通过简易的观测无法观测到。

在α粒子作用下，铍、镁和铝元素嬗变产生了新的发射正电子的放射性元素。这些放射性元素可以视为是处于特定的激发态上的一个已知的核，不过它们是未知同位素的可能性更大，而通常这些同位素都是不稳定的。

例如，我们认为硼发生了下面的核反应：

$$^{10}_{5}\mathrm{B} + {}^{4}_{2}\mathrm{He} = {}^{13}_{7}\mathrm{N} + {}^{1}_{0}n$$

其中 $^{13}_{7}\mathrm{N}$ 是一个放射性核，蜕变时放出正电子，最终得到一个稳定的核 $^{13}_{6}\mathrm{C}$。对于铝和镁来说，所得到的放射性核分别是 $^{30}_{15}\mathrm{P}$ 和 $^{27}_{14}\mathrm{Si}$。

The positrons of aluminium seem to form a continuous spectrum similar to the β-ray spectrum. The maximum energy is about 3×10^6 e.v. As in the case of the continuous spectrum of β-rays, it will be perhaps necessary to admit the simultaneous emission of a neutrino (or of an antineutrino of Louis de Broglie) in order to satisfy the principle of the conservation of energy and of the conservation of the spin in the transmutation.

The transmutations that give birth to the new radio-elements are produced in the proportion of 10^{-7} or 10^{-6} of the number of α-particles, as for other transmutations. With a strong polonium preparation of 100 millicuries, one gets only about 100,000 atoms of the radioactive elements. Yet it is possible to determine their chemical properties, detecting their radiation with a counter or an ionisation chamber. Of course, the chemical reactions must be completed in a few minutes, before the activity has disappeared.

We have irradiated the compound boron nitride (BN). By heating boron nitride with caustic soda, gaseous ammonia is produced. The activity separates from the boron and is carried away with the ammonia. This agrees very well with the hypothesis that the radioactive nucleus is in this case an isotope of nitrogen.

When irradiated aluminium is dissolved in hydrochloric acid, the activity is carried away with the hydrogen in the gaseous state, and can be collected in a tube. The chemical reaction must be the formation of phosphine (PH_3) or silicon hydride (SiH_4). The precipitation of the activity with zirconium phosphate in acid solution seems to indicate that the radio-element is an isotope of phosphorus.

These experiments give the first chemical proof of artificial transmutation, and also the proof of the capture of the α-particle in these reactions[3].

We propose for the new radio-elements formed by transmutation of boron, magnesium and aluminium, the names *radionitrogen*, *radiosilicon*, *radiophsphorus*.

These elements and similar ones may possibly be formed in different nuclear reactions with other bombarding particles: protons, deutrons, neutrons. For example, ${}_7N^{13}$ could perhaps be formed by the capture of a deutron in ${}_6C^{12}$, followed by the emission of a neutron.

(**133**, 201-202; 1934)

F. Joliot and I. Curie: Institut du Radium, Paris.

References:

1. Irène Curie and F. Joliot, *J. Phys. et. Rad.*, 4, 494; 1933.
2. Irène Curie and F. Joliot, *C.R.*, **198**; 1934.
3. Irène Curie et F. Joliot, *C.R.*, meeting of Feb. 29, 1934.

铝放出的正电子似乎构成一个类似于β射线谱的连续谱，其最大能量约为 3×10^6 eV。就β射线的连续谱来说，为了在嬗变中满足能量守恒与自旋守恒的原则，允许同时放出一个中微子（或者一个路易斯·德布罗意的反中微子）似乎是必要的。

至于其他的嬗变情况，产生新放射性元素的嬗变正比于α粒子数的 10^{-7} 或 10^{-6} 。使用放射性强度达 100 毫居的钋样品，我们也只能得到大约 100,000 个放射性元素的原子。不过，使用计数器或者电离室来探测它们的辐射，就可以确定它们的化学性质。当然，前提是它们在放射性活度消失之前的几分钟内完成化学反应。

我们已经照射了化合物氮化硼（BN）。使用苛性钠溶液来加热氮化硼，可以产生气态的氨。实验结果表明，放射活性与硼元素分离并转移到氨中了。这个结果与此种情况中放射性核是氮的一个同位素这一假设吻合得很好。

将被照射的铝溶解于盐酸中，放射活性会被氢转移到气态，且可以被收集在试管中。该化学反应必定形成了磷化氢（PH_3）或者四氢化硅（SiH_4）。而在酸性溶液中，放射活性与磷酸锆一起沉淀下来，这似乎意味着放射性元素是磷的一个同位素。

这些实验给出了人工嬗变的第一个化学证据，同时也给出了在上述的反应中α粒子被俘获的证据 [3]。

我们提议将以硼、镁和铝嬗变形成的新的放射性元素命名为**放射性氮**、**放射性硅**和**放射性磷**。

在使用其他轰击粒子，如质子、氘、中子引起的一些核反应中，也有可能生成这些元素或者类似的其他元素。例如，$^{12}_{6}C$ 俘获一个氘后，释放出一个中子，就有可能会形成 $^{13}_{7}N$。

（王耀杨 翻译；张焕乔 审稿）

A Velocity-modulation Television System

Editor's Note

In the mid 1930s, many groups were developing technology for transmitting visual images over long distances. Here *Nature* reports on a recent advance. Most approaches produced an image by scanning a cathode ray rapidly over a fluorescent screen. Modulations of the ray intensity translate into lighter or darker parts of the image. But improved images could be obtained by modulating the velocity of the ray, rather than its amplitude, with the screen being naturally darker where the ray passes more quickly. The article notes that engineers had recently reported on advances in such a system, and demonstrated in laboratory tests that intensity modulation could be used in addition to improve the contrast of the images.

MANY of the investigators who are seeking at the present time to develop a practical system of television make use of the cathode ray oscillograph tube in one form or another, since the electron beam in such a tube provides an easily controlled means of scanning the picture to be transmitted. At the receiving end, the cathode ray tube is employed to build-up the received picture by varying the intensity of the beam in accordance with the light and dark portions of the picture. The ordinary type of cathode ray tube, however, gives only a small range of intensity control without the accompaniment of loss of focus of the spot on the fluorescent screen, and special electrode systems have to be arranged to obtain good intensity modulation in this manner. As an alternative to this method, the intensity of the cathode ray beam may be kept constant but its tranverse velocity may be varied as it moves over the picture, the beam being speeded up over the dark portions of the picture and slowed down over the light portions. The corresponding motion of the cathode ray beam at the receiving end thus gives varying illumination according to the speed of travel of the spot on the fluorescent screen, and with the aid of the phenomenon of persistence of vision, a true impression of the shades and contrasts in the picture received is obtained.

The conception of this velocity-modulation principle, or variable-speed cathode ray television, dates back to 1911, when it was described in a British patent by B. Rosing. Since that date the principle appears to have fallen into oblivion until it was revived in Germany by R. Thun in 1929. The first practical realisation of the method was achieved by M. von Ardenne in 1931 and reference was made to this work in *Nature* of October 7 last (p.573).

During the development of cathode ray oscillograph tubes for general scientific and technical purposes, the staff of Messrs. A. C. Cossor Ltd. realised the possibilities of the

一个基于速度调制的电视系统

编者按

20世纪30年代中期，许多研究组都在发展远距离传输可视图像的技术。《自然》的这篇文章介绍了当时的最新进展。大多数技术是通过使用阴极射线快速地扫描荧光屏来产生图像的。调制阴极射线的强度可以反映出图像的亮区和暗区。但是，相对于调制振幅而言，通过调制阴极射线的速度可以得到更加清晰的图像，当射线扫描速度较快时，扫过的区域会自然的较暗。这篇文章介绍了工程师们最近报道的这类系统的进展，同时也指出了在实验室测试中还可以用强度调制来增强图像的对比度。

目前，许多致力于开发实用电视系统的研究者们都在使用各种形式的阴极射线示波管，因为这种示波管中的电子束提供了一种很容易控制的方法来对将要传输的图片进行扫描。在接收端，阴极射线管根据原始图片的明暗区域相应地改变电子束的强度，以实现对接收图片的构建。然而，如果不考虑在荧光屏上聚焦点的发散作用，那么一般的阴极射线管能控制电子束的强度范围是很小的，要用这种方式，就必须使用特殊的电极系统才能获得良好的强度调制效果。作为这种方式的一个替代方案，可以保持阴极射线束的强度不变，而使其横向速度在扫过图片时发生变化：通过图片的暗区时，速度加快；通过明区时，速度变慢。于是，根据荧光屏上点的运动速度，接收端的阴极射线束的相应运动就会给出不同的照射强度，借助于视觉暂留现象，人们就能获得对于接收到的图片的明暗区域与对比度的真实现象。

这种速度调制或者变速的阴极射线电视的概念可以追溯到1911年，当时有一个名为罗辛的人在他的英国专利中提到过。之后这个理论概念就被人们遗忘了，直到1929年德国的图恩再次提起。这一理论的第一次实用化是在1931年由冯·阿登完成的，与此相关的参考文献发表在去年10月7日的《自然》上（第573页）。

在以一般科学和技术为目的的阴极射线示波管的发展过程中，科索尔有限公司的员工们认识到了上述电视系统的可行性，并且将之前在18个月中完成的大量深入

above system of television, and an account of the development work carried out during the past eighteen months was presented in a paper entitled "A Velocity-Modulation Television System", read before the Wireless Section of the Institution of Electrical Engineers by Messrs. L. H. Bedford and O. S. Puckle on February 7.

Consideration of the basic principles outlined above shows that it is impossible to realise a velocity-modulated picture from a uniformly scanned object; the scanning at the transmitter must also be of the variable-speed or velocity-modulated type, and must therefore be carried out by a cathode ray. It follows that a cathode ray oscillograph must serve as the source of light at the transmitting end, and, with oscillographs of the ordinary low-voltage type, the conditions of scanning-light economy will restrict the picture subject matter to cinematograph film material. This, however, is not considered to be a disadvantage of the method; many of the television systems being developed at the present time make use of a film as an intermediary, and processes are being devised in which the interval between the photography of the subject and the projection of the picture through the transmitter is reduced to the bare minimum.

The transmitting arrangements described by Messrs. Bedford and Puckle comprise the projection of light from the fluorescent screen of the oscillograph through the film picture on to a photoelectric cell. The output of the photo-cell amplifier operates, through a screen-grid valve and a thyratron, an electrical time-base circuit which supplies the potential difference to one pair of the deflecting plates of the oscillograph. The light from the cathode ray tube is thus swept in a straight line across the picture with a velocity which varies according to its transparency at different points. At the end of each scanning line, the discharge of the thyratron provides a "fly-back" action to the spot sufficiently rapid to be invisible. Simultaneously with this operation, a second valve and thyratron circuit provides a traversing time-base potential difference to the second pair of deflecting plates of the oscillograph tube. By this means the scanning line is traversed across the picture in successive steps.

From this description it will be realised that an image of the picture being transmitted is built up on the fluorescent screen of the cathode ray oscillograph, and this is found to be a useful feature of the system for monitoring purposes. Furthermore, for the reproduction of the image on the screen of another oscillograph tube at a distant receiving station, it is merely necessary to transmit to the second tube the voltages being applied to the two pairs of deflecting plates of the first tube. If these voltages are sent through two separate channels, the received picture is automatically synchronised with that at the transmitting end.

The authors of the paper referred to above have modified this arrangement to some extent, however, to enable all the intelligence to be sent along a single channel. Using a picture frequency of 25 per second with a detail corresponding to 120 or 160 scanning lines, the transmitted signals require a frequency band of the order of 240 kilocycles per second; and special amplifiers have been developed to give uniform amplification over this range. The size of the picture received depends upon the deflector voltages which

的研究工作都写在了一篇名为《一个基于速度调制的电视系统》的文章里，这篇文章于 2 月 7 日由该公司的贝德福德和帕克尔在电气工程师学会无线电分会上进行了宣读。

根据上面概括的基本原理可以看出，对物体进行匀速扫描是不可能得到速度调制的图片的；在发射端的扫描也一定是速度变化型或是速度可调型的，因此一定也要通过阴极射线来实现。在发射端，阴极射线示波器充当了光源的角色，当使用普通低压型的示波器时，这种扫描光较弱的情况会使扫描图片的材料受到限制，只能使用电影胶片。然而，并不能认为这是此方式的缺点，因为目前正在发展的许多电视系统都把胶片当成一种媒介，过程是这样设计的：将实际物体的相片和通过发射端的相片投影之间的距离减小到几乎为零。

贝德福德和帕克尔描述的传输设置是将来自示波器的荧光屏的光通过胶片图像投影到光电管上。光电管放大器的输出通过一个屏栅阀和一个闸流管来控制一个电子时基电路，它为示波器的一对偏转片施加电位差。于是，从阴极射线管出来的射线以一条直线扫过图片，根据图片不同位置透光率的不同，扫过的速度也不一样。在每一条扫描线的末端，闸流管的放电产生一个“回扫”的动作使之回到原点，而且速度极快以至于肉眼难以观察到。在进行这个操作的同时，第二个屏栅阀和闸流管电路也对示波管的第二对偏转片施加一个时基的电位差。通过这种方式，扫描线就一条一条地覆盖了整幅图片。

从上面的描述中可以看出，在阴极射线管的荧光屏上产生了待传输图片的图像，而且这被认为是一个对监视很有用的特点。此外，为了在更远处的接收端的另一个阴极射线管的荧光屏上再现图像，我们只需要将施加在第一个射线管的两对偏转片上的电压传送到第二个射线管中就可以了。如果这两路电压通过不同的通道传输，那么接收的图片就会自动地与发射端的图片进行同步。

但是，文章的作者们在谈及上面的问题时，对这种设置进行了某种程度的修改，使所有的信息都通过一个信道进行传输。如果使用的图片频率为每秒 25 张，每幅图 120 或 160 条扫描线，那么传输信号需要的频带的数量级就为每秒 240 千周；并且他们发展出了一些专用的放大器以保证在这个范围内对信号进行均匀放大。接收图

may be applied to the oscillograph electrodes, and it is anticipated that future design and manufacture will enable a suitable receiver tube with a 9-inch screen to be produced. Among the advantages of the method described above over that employing intensity-modulation are the increased picture brightness for a given receiving oscillograph and the concentration of detail in the light portions of the picture.

Although Messrs. Bedford and Puckle's experiments have so far been limited to transmission over wire lines, no particular difficulty is anticipated in applying the necessary signals to radio transmission, at least on the ultra-short wave-length of a few metres where such a large frequency band as 240 kc./sec. may be permitted. At the reading of the paper, a cinematograph film was shown illustrating typical pictures received in a laboratory test of the whole system. Among the features brought out in this demonstration was the fact that, when required to obtain a better contrast ratio in the received picture, intensity modulation may be superimposed with advantage upon the velocity-modulation signals, and means of achieving this very satisfactory combination are being investigated.

(**133**, 263; 1934)

片的大小取决于施加在示波器电极上的偏转电压的大小，而且可以预见到，未来的设计和制造技术将会生产出具有 9 英寸屏幕的接收管。上述方法相对于强度调节的优点包括：对于给定的接收示波管而言，接收到的图片的亮度会提高，图片亮区的清晰度也会提高。

虽然到现在为止，在贝德福德和帕克尔的实验中仍只限于用金属线进行传输，但是可以预见到利用无线电传输所需的信号时不会出现其他特殊的困难，至少在波长为几米的超短波长的范围内，频带达到每秒 240 千周是可以实现的。在读这篇文章的时候，我们已经在整个系统的实验室测试中使用电影胶片得到了典型的图片。从实验中呈现出来的特征可以看出：当接收的图片要求较高的对比度时，强度调制要优于速度调制，而将这两者完美结合还需要进一步探讨。

（刘东亮 翻译；赵见高 审稿）

The Positron*

C. D. Anderson

Editor's Note

In September 1932, Carl Anderson had announced the observation of the positron, a positively charged particle having the same mass as the electron. Here he reviews the discovery and what has been learned since that time. The definitive observation had been made possible, he notes, by having cosmic rays impinge upon a lead target and observing tracks in a cloud chamber held in a high magnetic field. In April 1933, other physicists had shown that a gamma ray in the strong field of a nucleus can create electron-positron pairs. Anderson reviews these experiments in some detail, and notes that the inverse process—the annihilation of an electron and a positron, producing gamma radiation—should also be important, although it had not yet been observed.

THE existence of free positive electrons or positrons was first reported by me in September 1932[1], from cosmic ray experiments carried out at the California Institute of Technology. In the original paper, all possible alternative interpretations of the effects there presented were discussed in detail, and it was shown that only by calling upon the existence of free positive electrons could those effects be logically interpreted.

As a part of Prof. R. A. Millikan's programme of cosmic ray research, in particular to make energy measurements of the cosmic ray particles by the use of a vertical cloud chamber in a very powerful horizontal magnetic field, photographs were first taken in August 1931 in such an apparatus involving the maintenance of a field of strength up to 20,000 gauss over a space measuring 17 cm. × 17 cm. × 3 cm. As reported in lectures in Paris and Cambridge, England, in November 1931 and published in March 1932 by Millikan and myself[2], this work brought to light for the first time the fact that nuclear effects are of primary importance in the absorption of cosmic rays, as demonstrated by the frequent occurrence of associated tracks or showers containing particles of positive charge as well as those of negative charge.

Through the insertion in May 1932 of a lead plate across the centre of the cloud chamber, it was possible to show definitely in several cases that the mass of these particles of positive charge could not possibly be as great as that of the proton. The direction of motion of the particles was given in two ways: first, by allowing them to pass through the lead plate and suffer a loss in energy, and secondly, by the observation in several instances of two or more tracks all originating at one small region in the material surrounding the chamber. For a given curvature of track, the specific ionisation showed that the mass was small compared with the proton mass, but even more definite evidence was gained from an observation of

* Address delivered at the Symposium on Nuclear Physics of the American Physical Society meeting in Boston, Mass., on December 27, 1933.

正电子*

安德森

编者按

1932 年 9 月，卡尔·安德森声明发现了正电子，一种与电子质量相同且带有一个正电荷的粒子。在这篇文章中他回顾了正电子的发现和从那时起所学到的知识。他指出，通过宇宙线轰击铅靶并在强磁场内的云室中观察产生的径迹，便可确认这一发现。1933 年 4 月，另外有物理学家发现，在核的强场中 γ 射线可产生电子 – 正电子对。安德森详细地回顾了这些实验，并指出相反的过程应该也是非常重要的，即电子和正电子的湮灭产生 γ 辐射，尽管这一现象当时还未被观测到。

1932 年 9 月，根据在加州理工学院进行的宇宙线实验，我本人首次报道了自由正电子或者正电子的存在 [1]。对于发生的那些效应，所有可能的解释都在这篇最初的文章中进行了详细讨论，该文还指出只有通过引入自由正电子的存在，才能从逻辑上解释那些效应。

作为密立根教授的宇宙线研究计划（具体来说就是利用在非常强的水平磁场内的垂直云室做宇宙线粒子能量的测量）的一部分，这种测量于 1931 年 8 月首次拍得照片，当时的装置情况是云室 17 cm × 17 cm × 3 cm，磁场强度保持在 20,000 高斯。正如 1931 年 11 月在巴黎和英国剑桥的讲座中所报道的，以及密立根和我本人 [2] 于 1932 年 3 月发表的结果一样，这项研究首次揭示了一个事实，即在宇宙线的吸收过程中核效应是最重要的。伴生的带正电荷和负电荷的粒子的径迹或簇射的频繁发生可以证明这一点。

1932 年 5 月将铅板插入云室中心的实验能够明确说明，在几种情况下，这些带正电荷的粒子的质量不可能像质子的质量那么大。这些粒子的运动方向可以由以下两个过程来确定：首先，允许它们通过铅板，并损失部分能量；随后，对同样来自云室周围物质中某个小区域的两条或多条径迹进行观测。对于一条给定曲率的径迹，比电离显示该粒子的质量小于质子质量，而从粒子射程的观测则获得了更加确定的

* 本演讲发表于 1933 年 12 月 27 日在美国马萨诸塞州波士顿举行的美国物理学会会议的核物理分会上。

the range of the particles. The observed ranges were several times, in some instances more than ten times, greater than the possible ranges of proton tracks of the same curvature.

These considerations were the basis of the report announcing the existence of the free positive electron or positron published in September 1932. Within the next five months a large number of confirmatory photographs revealing unambiguously the existence of positrons was taken, and a second report was published in March 1933[3] in which fifteen of these photographs were discussed. The specific ionisation exhibited by the positron tracks on these photographs showed that the magnitude of charge of the positron could not differ by as much as a factor of two from that of the free negative electron, and it was, therefore, concluded, unless one admits fractional values of the elementary unit of charge, that the free positive and negative electrons were exactly alike in magnitude of charge. This fact, together with the curvatures measured in the magnetic field of a positron before and after it penetrated a plate of lead, fixed its mass as not greater than twenty times that of the free negative electron.

Since then[4], an observation of a collision between a moving positron and a free negative electron in the gas of the chamber revealed, on the basis of the conservation laws, that its mass was equal to that of the free negative electron with an error of not more than 30 percent. More recent measurements[15] of the specific ionisation of the positives and negatives for both high and low speed particles, by actual ion-counts on the tracks in the magnetic field, showed the specific ionisation of the positives and the negatives to be equal to within 20 percent. This fixes the limits of difference between the positives and negatives with regard to their charges and masses at 10 percent and 20 percent respectively. Further details of the history of this discovery were presented at the American Association for the Advancement of Science meeting in Chicago in June 1933[4].

In March 1933 confirmatory evidence for the existence of positrons was presented by Blackett and Occhialini[5], based on similar experiments with a vertical cloud chamber operating in a magnetic field of 3,000 gauss and actuated by the responses of Geiger–Müller counters. In April 1933 Chadwick, Blackett and Occhialini[6], Curie and Joliot[7], and Meitner and Philipp[8] reported that the bombardment of beryllium by α-particles can produce radiation which results in the production of positrons, though in these experiments it was not possible definitely to identify the nature of the radiation producing the positrons. By absorption experiments, however, Curie and Joliot showed that the yield of positrons decreased approximately as was to be expected if the γ-ray rather than the neutron component of the radiation were responsible for their production.

The first experiments proving directly that a γ-ray photon impinging upon a nucleus gives rise to positrons were carried out at the Norman Bridge Laboratory, using the γ-rays from thorium C″, and reported in April 1933[9]. In this paper the fact that free electrons of both positive and negative sign are produced simultaneously by the impact of a single γ-ray photon, an observation of considerable theoretical import, was first presented. Preliminary results of energy measurements were given in June 1933 by Neddermeyer and myself[10].

证据。所观测到的该粒子的射程是相同曲率的质子径迹的可能射程的好几倍，在某些情况下甚至超过了 10 倍。

1932 年 9 月发表的宣称自由正电子或正电子存在的报道正是基于上述思考。在此后的五个月内又拍摄到了大量的验证性照片，这些照片明确地揭示了正电子的存在。1933 年 3 月刊登的第二篇报道 [3] 对其中的 15 张照片进行了探讨。这些照片上的正电子径迹所展现出的比电离显示，正电子的电荷大小与自由负电子的电荷大小相差不到两倍。因此除非我们允许电荷的基本单位为分数值，否则我们可以断定自由正电子和负电子的电荷大小是完全相同的。根据这个事实，连同在磁场中正电子穿透铅板前后测量到的曲率一起，就可以确定出其质量不会比自由负电子的质量高出 20 倍。

上述结论得出之后 [4]，对发生在云室气体中的一个运动正电子和一个自由负电子间的碰撞的观测表明，根据守恒定律，正电子的质量和自由负电子的质量是相等的，其误差不超过 30%。通过在磁场中的径迹上进行真实离子计数，得到的与高速和低速的正负电子的比电离有关的更多最新测量结果 [15] 表明，正电子和负电子的比电离差别不会超过 20%。这样就确定了对于电荷和质量，正电子和负电子的最大差值分别为 10% 和 20%。有关这个发现过程的更多细节发表于 1933 年 6 月在美国芝加哥举行的美国科学促进会的会议上 [4]。

1933 年 3 月，基于运行在 3,000 高斯磁场内的垂直云室中的类似实验，并利用盖革 – 米勒计数器的响应来启动云室，布莱克特和奥恰利尼 [5] 给出了正电子存在的确定性证据。1933 年 4 月，查德威克、布莱克特和奥恰利尼 [6]、居里和约里奥 [7] 以及迈特纳和菲利普 [8] 均指出，用 α 粒子对铍进行轰击可以产生辐射，从而导致正电子的产生，不过这些实验尚不能完全确定产生正电子的辐射的本质。然而通过吸收实验，居里和约里奥指出，如果正电子的产生是 γ 射线而不是辐射的中子组分造成的，那么产生的正电子的数量大体上会像预期中的那样减少。

首批直接证明 γ 射线光子撞击原子核会产生正电子的实验是在诺曼桥实验室进行的。该实验发表于 1933 年 4 月 [9]，所使用的 γ 射线来源于钍 C″。这篇文章首次报道了具有重要理论意义的观测，即用单个 γ 射线光子的碰撞会同时产生带有正电和负电的自由电子。1933 年 6 月，尼德迈耶和笔者 [10] 给出了能量测量的初步结果。

Curie and Joliot[11] in May 1933, and Meitner and Philipp[12] in June 1933, all of whom used γ-rays from thorium C″, also reported the detection of positrons from the same source. Curie and Joliot[13] have also shown that positrons are produced directly in the disintegration of aluminum and boron by α-particle bombardment. The positrons in the case of aluminum cannot here be produced by the internal conversion of a γ-ray photon unless the probability of such internal conversion is vastly greater than that to be expected on theoretical grounds[14]. Rather do these experiments indicate that an elementary positive charge is actually removed from the disintegrating nucleus and appears as a positron.

The foregoing furnishes in brief a historical survey of the early experimental work on positrons and their production.

A detailed study of the energy distribution and frequency of production of free positive and negative electron pairs by filtered thorium C″ γ-rays is of particular value because of the relative simplicity of these effects as compared with those appearing in the cosmic ray range of energies.

γ-ray Effects

A discussion will now be given of experimental evidence as it bears on the theory suggested by Blackett and Occhialini on the basis of the Dirac electron theory, which postulates the creation of a free positive-negative electron pair out of the absorption of a photon impinging upon a nucleus. The nucleus itself in this picture undergoes no disintegration, but plays merely the rôle of a catalytic agent. This discussion will be given in the light of (1) new statistical studies by Neddermeyer and myself on the thorium C″ γ-ray effects, and (2) new experiments on cosmic ray showers by Millikan, Neddermeyer, Pickering and myself.

The work of Curie and Joliot, and of Chadwick, Blackett and Occhialini on the radiation from thorium and that excited in beryllium by α-particle bombardment, together with our own work on the cosmic radiation[15], has shown that the absorption process which gives rise to positrons becomes increasingly important with high energy radiations and heavy absorbing materials. Further, we have made a statistical study based on a total of more than 2,500 tracks of single electrons, both positive and negative, and positive-negative pairs ejected from plates of lead, aluminum and carbon by γ-rays from radiothorium filtered through 2.5 cm. of lead (in some cases with unfiltered rays for comparison) to determine the frequency of occurrence of pairs and single positrons, and their energy distribution for absorbing materials of different atomic numbers. The ejection of the particles was observed from lead plates of 0.25 mm. thickness, aluminum plates of 0.5 mm. thickness and a graphite plate of 1.4 cm. thickness (used also for cosmic ray studies). The magnetic field was here adjusted to 825 gauss.

We will consider first of all the energies. Both the single positives and the pairs (the sum of the energies of the positive and negative components being taken) ejected from the

居里和约里奥[11]以及迈特纳和菲利普[12]分别于1933年5月和1933年6月，采用了钍C″的γ射线，并同样报道了从γ射线源可检测到正电子。居里和约里奥[13]还指出，经α粒子轰击后在铝和硼的蜕变中可直接产生正电子。就铝而言，这里的正电子不能由γ射线光子的内转换来产生，除非这种内转换的概率远大于理论预期值[14]。这些实验确实能够表明，基本正电荷实际上是从蜕变的原子核中转移出来的，并表现为一个正电子。

前面的论述简要地回顾了有关正电子及其产生的早期实验工作。

用过滤的钍C″的γ射线对自由正电子和负电子对的能量分布和产生频率进行详细研究具有特殊的价值，因为这些效应与在宇宙线能区出现的效应相比较为简单。

γ射线效应

现在将就支持布莱克特和奥恰利尼的理论的实验证据进行讨论，该理论以狄拉克的电子理论为基础，它假设光子撞击原子核的吸收过程会产生出一个自由的正负电子对。在此图像中，核本身不发生蜕变，而只是起催化剂的作用。讨论将依据以下两点：(1) 由尼德迈耶和笔者对钍C″的γ射线效应所进行的新的统计研究；(2) 由密立根、尼德迈耶、皮克林和笔者对宇宙线簇射所做的新实验。

居里和约里奥以及查德威克、布莱克特和奥恰利尼关于钍发出的辐射和用α粒子轰击铍而激发出的辐射的研究，以及我们关于宇宙辐射的研究[15]综合表明，产生正电子的吸收过程随着高能辐射和高吸收材料的发展而变得越来越重要。此外，我们基于总共超过2,500条单电子（包括正电子和负电子）和正负电子对的径迹进行了统计研究，这些单电子和电子对是由放射性钍通过2.5 cm厚的铅板过滤得到的γ射线照射铅板、铝板和石墨板而发射出来的（某些情况下会使用未过滤的射线以进行比较）。通过该研究可测定正负电子对和单个正电子产生的频率，以及它们对不同原子序数的吸收物质的能量分布。我们的观测中采用的是0.25 mm厚的铅板、0.5 mm厚的铝板和1.4 cm厚的石墨板（它们也同样被用于宇宙线的研究）。所使用的磁场为825高斯。

我们首先来考虑能量。从铅板上发射出来的单个正电子和电子对（取正负电子的能量之和）所显示的最大能量均约为1.6 *MV*（*MV*=兆电子伏），80%的单个正电子

lead plates showed a maximum energy of about 1.6 *MV* (*MV* = millions of electron-volts), 80 percent of the single positrons having an energy less than 0.8 *MV*. For the case of the unfiltered γ-rays, the positrons and the pairs, though occurring in relatively fewer numbers compared with those ejected by the filtered rays, showed also a maximum energy of 1.6 *MV*. Further, in the case of the positives and pairs ejected from the plates of aluminum, the maximum energy was about 1.6 *MV*.

The maximum energy of the single negative electrons in all cases was about 2.5 *MV*. Since the errors in the energy measurements may be as high as 15 percent, this is in good agreement with the highest energy to be expected for extra-nuclear electrons resulting from Compton encounters or photoelectric absorption of the 2.65 *MV* photons.

A maximum energy of 1.6 *MV* for the positives and the pairs, both from the lead and the aluminum, is in good accord with that to be expected on the Dirac picture if 1 *MV* is allowed for the energy required to create a pair of electrons. There occurred, however, one pair the total energy of which was 2.9 *MV*; it is conceivable, though not likely, that it may have been produced by cosmic rays, or again it may represent the rebound of an electron against the under surface of the lead plate.

Of equal importance with the distribution in energy is the distribution in number of single positive electrons and pairs as compared with the single negative electrons. Out of a total of 1,542 electrons ejected from the 0.25 mm. lead plate by γ-rays from radiothorium filtered through 2.5 cm. of lead, there were 1,387 single negatives, 96 single positives and 59 pairs. From an aluminum plate 0.5 mm. thick and ejected by the same radiation there were, out of a total of 943 electron tracks, 916 single negatives, 20 single positives and 7 pairs.

The negatives may be assumed to have arisen in general from Compton and photoelectric encounters with extra-nuclear electrons in the lead or aluminum. But the single positives and the pairs must all, of course, correspond to nuclear encounters. If we assume that on the average an equal number of positives and negatives results from nuclear impacts, we can calculate the ratio of the nuclear to extra-nuclear absorption. This amounts to about 20 percent for lead and about 50 percent for aluminum. These values are in reasonably good agreement with those obtained by Chao[16], Meitner[17] and Gray and Tarrant[18] by entirely different methods in the matter of the excess absorption shown by lead over that shown by aluminum and also in the general relation of nuclear to extra-nuclear absorption in both metals.

That the nuclear absorption in carbon is very small for the thorium C″ γ-rays is shown by the fact that, as compared with 415 negatives, there appeared only 2 pairs and 6 single positives.

On the whole, the energy relations of the positives and pairs, from both the aluminum and the lead, appear to be quite consistent with the pair-creation hypothesis, as are also the approximate values of the excess absorption in lead and aluminum calculated on this

的能量小于 0.8 *MV*。与使用过滤的射线的情况相比，在未使用过滤的 γ 射线的情况下，尽管发射出来的正电子和电子对的数量相对较少，但最大能量也显示为 1.6 *MV*。此外，在使用铝板的情况下，发射出的正电子和电子对的最大能量也约为 1.6 *MV*。

在所有情况下，单个负电子的最大能量都约为 2.5 *MV*。由于能量测量中的误差可能高达 15%，因此其与康普顿碰撞或 2.65 *MV* 光子的光电吸收后产生的核外电子的最高预期能量符合得很好。

如果 1 *MV* 是产生一个电子对所需要的能量，那么来自铅和铝的正电子和电子对的最大能量 1.6 *MV* 就能与狄拉克图像的预期很好地吻合。然而，产生一个电子对所需的总能量是 2.9 *MV*；虽然不太可能，但可以想到的是，它也许已经由宇宙线产生，再或许是它表示了一个电子对铅板下表面的反弹。

与能量分布同等重要的是，与单个负电子相比，单个正电子及电子对的数目分布。放射性钍通过 2.5 cm 厚的铅板过滤得到的 γ 射线照射 0.25 mm 厚的铅板，总共可发射出 1,542 个电子，其中有 1,387 个单个负电子、96 个单个正电子和 59 个电子对。用相同的 γ 射线照射 0.5 mm 厚的铝板，总共有 943 个电子径迹，其中 916 个单个负电子、20 个单个正电子和 7 个电子对。

通常可以认为负电子源于铅或铝中核外电子的康普顿碰撞和光电碰撞。但是单个正电子和电子对必然都对应于核的碰撞。如果我们假定在一般情况下由核碰撞产生的正负电子个数相等，那么我们便可以计算出核吸收与核外吸收之间的比值。对铅板而言，这个值约为 20%，而对铝板而言约为 50%。这些值与赵忠尧 [16]、迈特纳 [17] 及格雷和塔兰特 [18] 在关于铅的吸收远超过铝的吸收以及在两种金属中核吸收与核外吸收的普遍关系的研究中采用完全不同的方法得到的值相当吻合。

钍 C″ 的 γ 射线照射石墨板时，碳的核吸收是非常小的，这可以通过以下事实体现出来，即产生了 415 个负电子的同时，只出现了 2 个电子对和 6 个单个正电子。

总的来说，来自铝和铅的正电子和电子对的能量关系，似乎与有关电子对产生的假说非常一致，同样也与基于这个假说而计算得到的在铅和铝中的过剩吸收的近

assumption.

The ratio of the observed numbers of single positives compared with the pairs is also of great importance in this connexion. Whether a positive is always formed paired with a negative, or whether a positive not accompanied by a negative can in some cases be produced, is a question difficult to answer from the data so far obtained. An accurate calculation of the probability of removal of the negative, if a pair is generated, so that only the positive emerges from the plate, is not simple to make, depending as it does on energy loss and plural scattering in the plate, and on the initial space and energy distribution of the components of the pairs. But on the basis of very approximate considerations, it appears somewhat difficult to reconcile the appearance, for example, in the case of aluminum, of 20 single positives and only 7 pairs with the view that they are always formed in pairs. Experiments now planned in which the particles are ejected from very much thinner plates should decide this question.

One case should be cited in which two negatives and two positives were all observed to originate at one point in the lead plate. The possibility that this can represent two pairs accidentally associated in time and position is so remote that it is taken as evidence that *photons of energy even so low as those of the thorium C″ gamma-rays can occasionally give rise to showers such as are a common feature of the cosmic rays*[9].

Cosmic Ray Effects

Our recent stereoscopic photographs taken in a 17,000 gauss magnetic field show numerous showers of more than thirty electrons, some positives and some negatives, originating in lead plates placed across the chamber. In all the observed cases of shower production, it was clearly seen from the photographs that non-ionising particles produced the showers. Also photographs taken in a magnetic field of only 800 gauss showed many examples of single negatives, single positives, pairs and triplets, of energies of the order of only a million or two electron volts, ejected from plates of lead by the impact of non-ionising particles. These low energy ejections are in all respects identical with those produced by the thorium C″ γ-rays and are undoubtedly due to low energy photons. These electron effects cannot be ascribed to ordinary neutrons since a considerable study of neutrons in this very range of energies has shown that their absorption results in projected nuclei and not in electron projection or shower formation. The appearance of several such small electron showers on one photograph which contains evidences of showers which occurred above the chamber, brings to light a new fact, namely, that *in the absorption of the cosmic rays there are produced, in addition to the electron showers, in some instances, sprays of large numbers of secondary photons*. The evidences for this conclusion were presented at the November 1933 meeting of the National Academy of Sciences by Millikan, Neddermeyer, Pickering and myself[19], and a full discussion together with the photographs will appear shortly in the *Physical Review*. In one case, more than eighty low energy electron tracks simultaneously projected were photographed, their positions and orientations in the chamber showing that they must have arisen from nearly as many separate centres in

似值相一致。

就能量关系而言，观测到的单个正电子的数目与电子对数目之比也是非常重要的。一个正电子形成的同时是否总是伴随着一个负电子的形成，或者在某些情况下是否正电子产生的同时不伴随负电子的产生，这是一个难以通过目前所掌握的数据来回答的问题。当一个电子对产生时，板中只发出正电子，这时要对负电子移出概率进行精确计算并不容易，因为这受到多种因素的影响，其中包括发生这种情况时在板内的能量损失和多次散射，以及电子对组分的起始空间和能量分布。但是根据非常粗略的分析，结果似乎很难与实际情况相符，例如，对铝而言，有 20 个单个正电子，另外仅有 7 个电子对，而理论上来说它们应该总是成对出现的。在现在计划进行的实验中，粒子将从很薄的板上发射，这应当能对这个问题做出判定。

应当指出的一个例子是，实验中观测到两个负电子和两个正电子都在铅板的同一点产生。这表明在时间和位置上偶然出现关联的两个电子对的概率是极低的，因此这个现象可以证明，**即使能量像钍 C″ 的 γ 射线那样低的光子偶然也会产生簇射，如同宇宙线所具有的普遍特征一样** [9]。

宇宙线效应

我们最近在 17,000 高斯的磁场中拍摄的立体照片显示，大量多于 30 个电子的簇射（有些是正电子，有些是负电子）产生于穿过云室放置的铅板上。从照片中可以清楚地看到，在所有观测到的产生簇射的情况中，仅非电离粒子产生簇射。同样在仅为 800 高斯的磁场中拍摄的照片也显示了许多通过非电离粒子的碰撞，从铅板发射出单个负电子、单个正电子、电子对和三电子组的例子，它们的能量量级仅为 1 或 2 兆电子伏。这些低能的发射物在各个方面都与那些钍 C″ 的 γ 射线产生的发射物一致，并且无疑是由于低能光子产生的。这些电子效应不能归属于普通的中子，因为对这个能区的中子的大量研究已经表明，它们的吸收会导致核的发射，而不是电子的或簇射的形成。在一张照片中有几个这样的小型电子簇射的出现，这也是在云室之上出现簇射的证据，这一现象揭示了一个新的事实，即**在宇宙线的吸收中，除了电子簇射之外，在某些情况下，还会产生大量次级光子的喷射**。密立根、尼德迈耶、皮克林和笔者 [19] 于 1933 年 11 月在美国国家科学院的会议上给出了此结论的相关证据，结合这些照片所做的一个全面的讨论也将于近期发表在《物理学评论》上。在其中一个示例中，超过 80 个同时发射的低能电子的径迹被拍摄下来，它们在云室中的位置和方向表明，它们一定产生自包围云室的材料的许多分立中心，因此

the material surrounding the chamber, and must therefore be ascribed to such a spray of secondary photons.

That pair production or shower formation by a fast electron (positive or negative) is a relatively rare event is shown by the fact that more than a thousand fast electrons have been observed to traverse a 1 cm. lead plate, and only in one instance was a definite pair projected from the lead by a fast electron, while a large number of secondary negative electron tracks appeared as the result of close encounters with the extra-nuclear electrons in the lead plate. The immediate secondaries of fast electrons are therefore seen to consist largely of negative electrons and only in rare cases of positrons.

Because of the powerful magnetic field we are using, it is possible to deflect all but a very small number of the electrons projected in the showers by the photon impacts. In general, in a shower a pronounced asymmetry is noted in the numbers of positive as compared with negative electrons emerging from the lead plates, in one instance 7 positives and 15 negatives, and in a second case 15 positives and 10 negatives. These effects are only with some difficulty reconciled with the Dirac theory of the creation of pairs out of the incident photon. Rather might they indicate the existence of a nuclear reaction of a type in which the nucleus plays a more active rôle than merely that of a catalyst, as for example the ejection from it of positive and negative charges which then appear in the showers as free positive and negative electrons. The essential difference, however, between these two points of view may be merely that in one case the nucleus may change its charge, and in the other it does not do so.

To study nuclear absorption in a light element, more than four hundred successful photographs were taken in which a carbon plate of 1.4 cm. thickness replaced the lead plate. Many of these showed showers originating in a block of lead placed above the chamber, but in no instance was a secondary shower observed in the carbon plate. This indicates, in agreement with the thorium C″ data, the relatively small probability in comparison with lead of a carbon nucleus absorbing a photon by shower production.

A consequence of the pair-theory is that, in a suitably dense environment of negative electrons such as obtains in ordinary matter, a positron shall have a high probability of combining with a negative electron, resulting in the annihilation of both particles and the conversion of their proper and kinetic energies into radiation. The theory, though at present incomplete, states that the mean free path for annihilation is in general greater than the range of the positron, so that such annihilation should be evidenced by the appearance of quanta of about half a million electron-volts energy and a very small number of quanta of about one million electron-volts energy when positrons pass through matter[20]. The experiments by Gray and Tarrant[18] on the scattering of thorium C″ γ-rays showed the existence of secondary radiation of such energies, but some of the more recent experiments on the scattering of hard γ-rays fail to show a secondary radiation which can be attributed to the annihilation of positrons. Our cosmic ray photographs show that in the electron showers there are present large numbers of secondary photons, many of

这一定是由次级光子的喷射而产生的。

由快电子（正的或负的）产生电子对或形成簇射是相当罕见的事件，这可由以下事实说明，观测到多于 1,000 个快电子穿过 1 cm 厚的铅板，仅有一例确定是从铅板上由快电子射出的电子对，而大量的次级负电子径迹似乎均为在铅板中与核外电子近距离碰撞的结果。因此快速电子的即时次级发射似乎是由大量负电子构成的，而正电子仅是不多见的情况。

由于我们所用的磁场很强，因此除了由光子碰撞产生的簇射中发射出的极少量的电子没有发生偏转外，可能所有的电子都发生了偏转。通常在一个簇射中可注意到，从铅板发射出的正电子数目与负电子数目明显不对称。其中一个簇射中，有 7 个正电子和 15 个负电子，而在另一个簇射中，则有 15 个正电子和 10 个负电子。这些效应遭遇的唯一困难是，很难与入射光子产生电子对的狄拉克理论取得一致。很可能这些效应表明存在一类核反应，其中核的作用不仅是催化剂而是比之更加活跃，例如，正电荷和负电荷从核中发射出来，然后作为自由正电子和负电子出现在簇射中。然而，这两种观点本质上的区别可能只不过是，其中一个认为核是可以改变其电荷的，而另一种观点则认为其电荷不会改变。

为了研究在轻元素中的核吸收，于是将铅板替换为 1.4 cm 厚的碳板并成功地拍摄了 400 多张照片。许多照片显示出，位于云室上面的铅块中会产生簇射，但在所有情况下都没有在碳板上观测到次级簇射。这与钍 C″ 的数据得出的结论一致，即与铅相比，碳核经由簇射产生吸收光子的概率较小。

电子对理论的结论是，在负电子适当稠密的环境中（如同在普通物质中存在的那种），正电子与负电子有很高的概率相互结合，从而导致两种粒子的湮灭，而它们的固有能量和动能将转换为辐射。尽管这种理论目前还不完善，但它表明发生湮灭的平均自由程通常大于正电子的射程，因此，当正电子通过物质时，这类湮灭应当可以通过具有大约 0.5 兆电子伏能量的量子以及少量具有约 1 兆电子伏能量的量子的出现来证实[20]。格雷和塔兰特[18]关于钍 C″ 的 γ 射线的散射实验显示，存在具有这种能量的次级辐射，但在最近的一些硬 γ 射线的散射实验中却没有观测到次级辐射，这可以归因于正电子的湮灭。我们的宇宙线照片表明，在电子簇射中存在大量的次级光子，其中许多是在这个能区中，但还不能确定它们是否部分地产生于正电

which are in this range of energy, but it is not yet certain if they are produced in part by the annihilation of positrons. In two very recent papers, Joliot[21] and Thibaud[22] report the observation in experiments with artificially produced positrons of secondary photons of the energies to be expected if they arise from the annihilation of positrons. By control experiments with negative electrons, they showed that a beam of positrons impinging upon matter results in the production of a considerably greater quantity of photons than does an equal number of negative electrons.

(**133**, 313-316; 1934)

Carl D. Anderson: California Institute of Technology, Pasadena, Calif.

References:

1. Anderson, *Science*, **76**, 238; 1932.
2. Millikan and Anderson, *Phys. Rev.*, **40**, 325; 1932. See also Anderson, *Phys. Rev.*, **41**, 405; 1932: and Kunze, *Z. Phys.*, **80**, 559; 1933.
3. Anderson, *Phys. Rev.*, **43**, 491; 1933.
4. Millikan, *Science*, **78**, 153; 1933.
5. Blackett and Occhialini, *Proc. Roy. Soc.*, A, **139**, 699; 1933.
6. Chadwick, Blackett and Occhialini, *Nature*, **131**, 473, April 1, 1933.
7. Curie and Joliot, *C.R.*, **196**, 1105; 1933.
8. Meitner and Philipp, *Naturwiss.*, **21**, 286; 1933.
9. Anderson, A.A.A.S. meeting, April 28, 1933, and *Science*, 77, 432; 1933.
10. Anderson and Neddermeyer, *Phys. Rev.*, **43**, 1034; 1933.
11. Curie and Joliot, *C.R.*, **196**, 1581; 1933.
12. Meitner and Philipp, *Naturwiss.*, **24**, 468; 1933.
13. Curie and Joliot, *C.R.*, **197**, 237; 1933.
14. Oppenheimer and Plesset, *Phys. Rev.*, **44**, 53; 1933. Beck, *Z. Phys.*, **83**, 498; 1933.
15. Anderson, *Phys. Rev.*, **44**, 406; 1933.
16. Chao, *Proc. Nat. Acad. Sci.*, **16**, 431; 1930. *Phys. Rev.*, **36**, 1519; 1930.
17. Meitner and Hupfield, *Naturwiss.*, **19**, 775; 1931.
18. Gray and Tarrant, *Proc. Roy. Soc.*, A, **136**, 662; 1932.
19. Anderson, Millikan, Neddermeyer and Pickering, *Proc. Nat. Acad. Sci.* Autumn meeting Nov. 20, 1933. See also abstract by Anderson and Neddermeyer, A.A.A.S. meeting, Dec. 30, 1933.
20. Fermi and Uhlenbeck, *Phys. Rew.*, **44**, 510; 1933.
21. Joliot, *C.R.*, **197**, 1623; 1933.
22. Thibaud, *C.R.*, **197**, 1629; 1933.

子的湮灭。在最近的两篇文章中，约里奥[21]和蒂博[22]报道了如果次级光子是由于正电子的湮灭产生的，则观测利用预期能量的次级光子人工产生正电子的实验是可行的。通过用负电子控制实验，他们表示，与相同数量的负电子相比，一束正电子与物质碰撞时所产生的光子要多很多。

（沈乃澂 翻译；张焕乔 审稿）

Production of Induced Radioactivity by High Velocity Protons

J. D. Cockcroft *et al.*

Editor's Note

By the mid 1930s physicists were routinely producing new radio-isotopes by bombarding elements with alpha particles and protons. Here Cockcroft and colleagues describe the bombardment of graphite with high-energy protons, producing radioactivity that decreases with a half-life of about ten minutes. This suggested that the proton beam had created a significant amount of some radionuclide, presumably ^{13}N. Some of the emitted particles were positrons, although most were ordinary electrons with high energies that might be explained as due to "kicks" by gamma rays. These in turn might have been created by annihilation of positrons in the chamber walls. This was among the first observations of electron–positron annihilation, a process that was not well understood until the 1940s.

CURIE and Joliot[1] have reported that a number of new radioactive isotopes can be produced by the bombardment of various elements with α-particles, these isotopes emitting positive electrons. In particular, they showed that boron when bombarded by α-particles was transformed to the isotope N^{13}, radio-nitrogen, this isotope having a half life of 14 minutes. They suggested that the isotope might be produced by the bombardment of carbon with heavy hydrogen, the product, N^{14}, disintegrating with the emission of a neutron to radio-nitrogen.

We have bombarded a target of Acheson graphite with *protons* of 600 k.v. energy and have used a Geiger counter to search for any radiations produced after the bombardment ceased. After bombardment for 15 minutes with a current of about 10 microamperes of protons, the target was removed from the apparatus and placed against the Geiger counter. We then observed about 200 counts per minute, being about forty times the natural effect. The number of counts decayed exponentially with time, having a half life of 10.5±0.5 minutes.

We then carried out an experiment similar to that performed by Becquerel, in which the source was placed on one side of a 9 mm. thick lead plate with the counter on the opposite side, the whole being placed in a magnetic field, so that any electron emitted could only reach the counter by applying a field of appropriate sign and magnitude. We found that when the field was such that positive electrons could reach the counter, the number of counts increased by a factor of 3; when the field was in the reverse direction no definite increase was observed. We conclude, therefore, that the radiations consist in part at least of positive particles.

由高速质子产生的感生放射性

考克饶夫等

编者按

直到20世纪30年代中期，物理学家通常使用α粒子和质子轰击多种元素以产生新的放射性同位素。这篇文章中考克饶夫和他的合作者们描述了一种使用高能质子轰击石墨的实验，该实验产生了以半衰期约为10分钟的速率衰减的放射性。这表明质子束产生了某种数量可观的放射性核，有可能是^{13}N。发射出来的粒子中有一些是正电子，当然，大部分是高能的普通电子，普通电子可以解释为是由γ射线引起的"反冲"造成的。反之γ射线可能是由于正电子在云室壁上湮灭产生的。这是首次观察到的电子–正电子湮灭现象之一，而这一过程直到20世纪40年代才被人们很好地理解。

居里和约里奥[1]曾经报道，许多新的放射性同位素可以用α粒子轰击各种元素来获得，这些同位素会放射出正电子。他们特别指出，当用α粒子轰击硼时，硼会转变为氮的同位素^{13}N，即放射性氮，这种同位素的半衰期是14分钟。他们认为，有可能通过重氢轰击碳产生的^{14}N在放射出一个中子后衰变为这种放射性氮。

我们用600千电子伏能量的**质子**对艾奇逊人造石墨靶进行了轰击，并使用盖革计数器探测在轰击停止后产生的辐射。用强度约为10微安的质子流轰击15分钟后，将石墨靶从装置上移开并放置在正对着盖革计数器的地方。这时，我们观测到的计数约为每分钟200个，约为正常值的40倍。计数随时间呈指数衰减，其半衰期为10.5±0.5分钟。

然后，我们做了一个与贝克勒尔所做实验类似的实验，在这个实验中，放射源放在9 mm厚的铅板一侧，计数器放置在另一侧，整个装置放在磁场内，只有当磁场的方向和大小都合适时，发射出来的电子才能到达计数器。我们发现，当磁场能使正电子到达计数器时，计数增加到原来的3倍；当磁场反向时，未观测到计数的明显增加。因此我们认为，至少部分辐射是由带正电的粒子组成的。

We have also taken about 250 Wilson chamber photographs in a field of 2,000 gauss, placing the activated source against the outside of the chamber wall, which was about 3 mm. thick. Under these conditions, we observed only two electrons of positive curvature which could possibly have come from the source, these electrons having energies of the order of 500 k.v. We observed, on the other hand, 48 tracks of Compton electrons starting in the gas, having energies ranging from 100 k.v. to 500 k.v., suggesting the emission of γ-rays of energy between 500 k.v. and 1 million volts. These γ-rays may result from the annihilation of the positive electrons, presumably in the glass wall of the chamber. The deflection experiments, whilst not at present precise, tend to confirm that few of the positive electrons would have sufficient energy to penetrate the glass walls. Further experiments will, therefore, be carried out with the source inside the chamber.*

The observations suggest that the unstable isotope N^{13} is produced by the addition of a proton to C^{12}. The difference between the half life observed and that reported by Curie and Joliot may be due to the formation of N^{13} in a different excited state.

No marked increase in the number of counts was observed when a mixed beam of heavy hydrogen ions and protons was substituted for the proton beam.

We are very much indebted to Dr. K. T. Bainbridge, who supplied the Geiger counter with which the observations were made.

(**133**, 328; 1934)

J. D. Cockcroft, C. W. Gilbert and E. T. S. Walton: Cavendish Laboratory, Cambridge, Feb. 24.

Reference:

1. *Comptes rendus*, 198, 254; 1934.

* February 27. Experiments carried out with a counter having a mica window of small stopping power gave a great increase in the number of counts owing to the positive electrons now entering the counter. The absorption curve of the positive electrons is similar to that of negative electrons of 800 k.v. energy.

在 2,000 高斯的磁场中，我们将活化后的放射源放在约 3 mm 厚的云室壁的外侧，并拍摄了 250 张威尔逊云室的照片。在这种条件下，我们只观测到了两条正曲率的电子径迹，这些径迹可能是由放射源发射出的粒子形成的，而这些电子的能量在 500 千电子伏的数量级上。另一方面，我们观测到了从气体中发出来的 48 条康普顿电子的径迹，其能量分布在 100 千电子伏到 500 千电子伏的范围内，这表示 γ 射线的发射能量范围在 500 千电子伏与 1 兆电子伏之间。这些 γ 射线可能是由于正电子的湮灭而产生的，这一过程很可能发生在云室的玻璃壁内部。尽管偏转实验目前并不精确，但它仍然倾向于肯定少量的正电子将具有足够的能量穿透玻璃壁。因此，下一步的实验将把放射源置于云室内进行。*

观测到的这些现象表明，不稳定的同位素 ^{13}N 是由于 ^{12}C 增加一个质子而产生的。我们观测到的半衰期与居里和约里奥报道的数据不同，这可能是由于形成的 ^{13}N 处于不同的激发态。

当我们用重氢离子与质子的混合束代替质子束时，并未观测到计数的明显增多。

我们非常感激班布里奇博士，他提供了我们实验必需的盖革计数器。

（沈乃澂 翻译；夏海鸿 审稿）

* 2 月 27 日，在使用具有微小制动力的云母窗口计数器的实验中，正电子的进入使得计数器的数量极大地增加。正电子的吸收曲线与 800 千电子伏能量的负电子的吸收曲线相似。

Designation of the Positive Electron

H. Dingle

Editor's Note

As of 1934 there was no consensus about how to name the positive electron, discovered two years earlier. The popular (and ultimately the prevalent) choice seemed to be "positron", against which Herbert Dingle here argues. It is ugly and offends the literary sense with its hybrid character, he says, and moreover it suggests that the electron ought to be unhappily renamed the "negatron". Dingle suggests instead the poetic name "oreston", alluding to the sister and brother Elektra and Orestes of Greek myth. It never caught on.

I have been hoping that, following Lord Rutherford's proposal of a name for the heavy isotope of hydrogen, someone would suggest a more satisfactory word than "positron" for the positive electron. Since, however, no better qualified reformer has appeared, may I raise the question before it is too late? "Positron" is ugly; it offends literary purists by its hybrid character; and it not only bears no relation to the established name of the associated particle, the electron, but even suggests that that particle should be called the "negatron", which fortunately it is not.

In order to balance destructive by constructive criticism, I venture to propose the name "oreston" for the newcomer. The word is euphonious, pure Greek, and since, in one of the most beautiful of Greek stories, Orestes and Elektra were brother and sister, it implies an appropriate relation between the two particles. The name found favour among many physicists in Pasadena where Anderson first obtained evidence of the particle, when I mentioned it there last year. I do not propose, however, further to urge its claims, the purpose of this letter being mainly to cleanse the language of "positron", and only incidentally to nominate a substitute.

(**133**, 330; 1934)

Herbert Dingle: Imperial College, South Kensington, S.W.7, Feb. 12.

正电子的命名

丁格尔

编者按

截至 1934 年，物理学家对于两年前发现的带正电的电子的命名问题仍未达成一致。通常情况下（也是极为普遍的）会使用“positron”（正电子）一词，而这篇文章中赫伯特·丁格尔就此发表了反对意见。他认为这个单词是丑陋的，这种复合词的方式违背了文学的原则，此外，这还意味着“electron”（电子）应该不幸地被重命名为“negatron”。丁格尔建议代之以一个更加诗意的名字“oreston”，暗指希腊神话中的厄勒克特拉和俄瑞斯忒斯姐弟。然而这个名字并未流行起来。

在卢瑟福勋爵提议给氢的重同位素命名之后，我一直希望有人可以给带正电的电子起一个比“positron”更令人满意的名字。不过既然目前仍没有人给出更好的新命名，我是否可以在这里及时地提出这一问题？坦白而言，“positron”这个单词是丑陋的；它采用的复合词的形式在文学纯化论者看来是很反感的事情；它不仅与相关的粒子，即电子已确定的名字没有什么关系，甚至还容易让人以为电子 (electron) 应该被称为“negatron”，不过幸好大家没有这么做。

为了早日改变这一现状，在此我大胆建议为刚发现的带正电电子采用一个新的命名“oreston”。这个词非常悦耳并且是纯希腊语，此外还因为在最动听的希腊故事之一中，厄勒克特拉和俄瑞斯忒斯是亲姐弟，因此采用这一命名可以恰当地表达这两种粒子之间的关系。帕萨迪纳是安德森首次证实新粒子存在的地方，去年，当我在那里提议采用这个名字的时候，那里的物理学家都很赞成。然而，我并不会进一步要求大家非要接受这一命名。事实上，本文的主要目的是想消除“positron”这一命名，并顺便推荐我的命名。

（金世超 翻译；朱永生 审稿）

Transmutation Effects Observed with Heavy Hydrogen

M. L. Oliphant *et al.*

Editor's Note

While American physicists were calling the recently discovered isotope of hydrogen "deuterium", their British colleagues persisted with "diplogen". Here Mark Oliphant and Paul Harteck, working with Rutherford in Cambridge, describe experiments that seem to involve the reaction of two diplogen nuclei ("diplons"). They substituted diplogen for hydrogen in ammonium compounds and bombarded them with low-energy diplogen. The researchers note that if two diplogen nuclei combine to form a helium nucleus, its atomic mass would be slightly greater than the known atomic mass of helium. So the resulting nucleus would be highly unstable and would immediately decay, throwing out protons and neutrons. Identifying the putative decay products (helium-3 and hydrogen-3, or tritium) awaited further experimental refinement.

WE have been making some experiments in which diplons have been used to bombard preparations such as ammonium chloride (NH_4Cl), ammonium sulphate ($(NH_4)_2SO_4$) and orthophosphoric acid (H_3PO_4), in which the hydrogen has been displaced in large part by diplogen. When these D compounds are bombarded by an intense beam of protons, no large differences are observed between them and the ordinary hydrogen compounds. When, however, the ions of heavy hydrogen are used, there is an enormous emission of fast protons detectable even at energies of 20,000 volts. At 100,000 volts the effects are too large to be followed by our amplifier and oscillograph. The proton group has a definite range of 14.3 cm., corresponding to an energy of emission of 3 million volts. In addition to this, we have observed a short range group of singly charged particles of range about 1.6 cm., in number equal to that of the 14 cm. group. Other weak groups of particles are observed with the different preparations, but so far we have been unable to assign these definitely to primary reactions between diplons.

In addition to the two proton groups, a large number of neutrons has been observed. The maximum energy of these neutrons appears to be about 3 million volts. Rough estimates of the number of neutrons produced suggest that the reaction which produces them is less frequent than that which produces the protons.

While it is too early to draw definite conclusions, we are inclined to interpret the results in the following way. It seems to us suggestive that the diplon does not appear to be broken up by either α-particles or by proton bombardment for energies up to 300,000 volts. It therefore seems very unlikely that the diplon will break up merely in a much less energetic collision with another diplon. It seems more probable that the diplons unite to form a new

通过重氢观察到的嬗变效应

奥利芬特等

编者按

当美国物理家们将新发现的氢的同位素（氘）称为“deuterium”的时候，他们的英国同行则坚持使用“diplogen”这一叫法。这篇文章中，马克·奥利芬特和保罗·哈特克在剑桥同卢瑟福合作，描述了一个似乎包含两个氘原子核（“氘核”）反应的实验。他们用氘置换铵化合物中的氢，然后用低能的氘去轰击它们。研究者们指出，如果两个氘核结合在一起形成一个氦核，那么这个氦核的原子质量会比已知的氦的原子质量略大。所以该反应得到的原子核会是高度不稳定的，并且会立即开始衰变，放出质子和中子。对假定衰变产物（氦–3和氢–3，后者或称为氚）的鉴别有待进一步的实验改进。

我们进行了以下的实验，用氘核轰击氯化铵（NH_4Cl）、硫酸铵（$(NH_4)_2SO_4$）、正磷酸（H_3PO_4）等样品，使得化合物中的大部分氢被氘置换。当这些氘化合物受到强的质子束轰击时，这些化合物和普通氢化合物没有表现出太大区别。然而，当用重氢离子轰击时，即便是在20,000电子伏的能量下也能探测到大量快质子发射。在100,000电子伏时，这个效应则太强以致于我们的放大器和示波器不能对其进行跟踪。质子群有14.3厘米的固定射程，对应3兆电子伏发射能量。除此之外，我们还观察到了一个射程约1.6厘米的单电荷粒子群，数量和14厘米的质子群相同。在不同的样品中，还观测到了其他弱的粒子群，但是到目前为止，我们还不能确定这些反应就是氘核之间的初级反应。

除了这两个质子群，我们还观察到了大量的中子。这些中子中能量最大的约为3兆电子伏。对产生中子数目的粗略估算表明，产生中子的反应比产生质子的反应的频度要低。

现在得出确切的结论还为时过早，我们倾向于用以下方式来解释这些结果。当能量高达300,000电子伏时，无论用α粒子轰击还是用质子轰击，氘核都不会发生破裂，这对我们来说似乎是具有启发性的。因此，当氘核在更小的能量下与另一个氘核发生碰撞时，似乎就更不可能发生破裂了。更有可能的是，两个氘核结合而形

helium nucleus of mass 4.0272 and 2 charges. This nucleus apparently finds it difficult to get rid of its large surplus energy above that of an ordinary He nucleus of mass 4.0022, but breaks up into two components. One possibility is that it breaks up according to the reaction

$$D_1^2 + D_1^2 \rightarrow H_1^3 + H_1^1$$

The proton in this case has the range of 14 cm. while the range of 1.6 cm. observed agrees well with that to be expected from momentum relations for an H_1^3 particle. The mass of this new hydrogen isotope calculated from mass and energy changes is 3.0151.

Another possible reaction is

$$D_1^2 + D_1^2 \rightarrow He_2^3 + n_0^1$$

leading to the production of a helium isotope of mass 3 and a neutron. In a previous paper we suggested that a helium isotope of mass 3 is produced as a result of the transmutation of Li^6 under proton bombardment into two doubly charged particles. If this last reaction be correct, the mass of He_2^3 is 3.0165, and using this mass and Chadwick's mass for the neutron, the energy of the neutron comes out to be about 3 million volts. From momentum relations the recoiling He_2^3 particle should have a range of about 5 mm. Owing to many disturbing factors, it is difficult to observe and record particles of such short range, but experiments are in progress to test whether such a group can be detected. While the nuclei of H_1^3 and He_2^3 appear to be stable for the short time required for their detection, the question of their permanence requires further consideration.

(**133**, 413; 1934)

M. L. Oliphant, P. Harteck and Rutherford: Cavendish Laboratory, Cambridge, March 9.

成一个新的氦核，其质量为4.0272，电荷数为2。这个新的原子核的过剩能量比质量为4.0022的普通氦核的大，这一过剩能量明显是很难去除的，除非分裂为两部分。一种可能性是按照以下的反应发生分裂：

$$^{2}_{1}\mathrm{D} + {}^{2}_{1}\mathrm{D} \rightarrow {}^{3}_{1}\mathrm{H} + {}^{1}_{1}\mathrm{H}$$

在这种情况下质子的射程是14厘米，而我们所观察到的1.6厘米的射程与根据动量关系估算出的 $^{3}_{1}\mathrm{H}$ 粒子的射程吻合得很好。从质能关系计算得到这种新的氢同位素的质量为3.0151。

另一种可能发生的反应是：

$$^{2}_{1}\mathrm{D} + {}^{2}_{1}\mathrm{D} \rightarrow {}^{3}_{2}\mathrm{He} + {}^{1}_{0}n$$

其产生一个质量为3的氦同位素和一个中子。在之前的一篇文章中，我们指出，$^{6}\mathrm{Li}$ 被质子轰击后会转变为两个带双电荷的粒子，这样就产生了质量为3的氦同位素。如果最后的这个反应是正确的，那么 $^{3}_{2}\mathrm{He}$ 的质量应为3.0165，利用这个质量和查德威克给出的中子质量进行计算，得出中子的能量约为3兆电子伏。根据动量关系，反冲的 $^{3}_{2}\mathrm{He}$ 粒子的射程应该约为5毫米。由于存在很多干扰因素，观测并记录这样短射程的粒子是困难的，但是正在进行中的实验正试图去探测这样的粒子群。在探测所需的短时间内，$^{3}_{1}\mathrm{H}$ 和 $^{3}_{2}\mathrm{He}$ 原子核似乎是稳定的，它们的持久性问题还需要更进一步地探讨。

（王静 翻译；张焕乔 审稿）

Supraconductivity of Films of Tin

E. F. Burton

Editor's Note

Superconductivity had been discovered in 1911 by Heike Kamerlingh Onnes, who found that solid mercury lost its electrical resistance when cooled within a few degrees of absolute zero. Other materials such as tin and lead were soon also found to be "supraconductors", as they were then known, but there was no explanation for the phenomenon. Here E. F. Burton of the University of Toronto reported an important constraint on such a theory: a thin film of superconducting material such as tin loses its superconductivity when sandwiched between other materials that are not superconducting. Burton notes that the effect applies only to sufficiently thin superconducting films, less than a micrometre or so thick.

EXPERIMENTS on the relation of high frequency currents to the phenomenon of superconductivity led to work at Toronto with films of superconducting metals. The films (of tin) were produced by "tinning" the surface of fine wires which themselves were not superconducting: in the early experiments a coating of tin 2×10^{-4} cm. in thickness was "wiped" on constantan wire of 0.016 cm. diameter. In this way one obtains the equivalent of a thin cylinder of superconducting metal, and the resistance of the whole becomes zero below the transition temperature of the superconducting element used[1].

With the intention of studying further the effect of high frequency currents, samples of such coated wires were plated with other metals—for example, copper and nickel—which are not superconductors; an example of such a combination is constantan covered with tin and then plated with copper. The diameters of the wires forming the core were as follows: for constantan 0.056 cm., for copper 0.040 cm. and for nickel 0.045 cm.

Preliminary experiments were carried out on these samples to confirm their reaction with respect to direct currents—the ordinary superconductivity test—and it was found that thin films of tin cease to show superconductivity when these films are themselves plated over with a film of a non-superconducting metal, for example, copper or nickel. This surprising result shows itself only with thin films, but a number of repetitions of the experiments renders the results unmistakable. The accompanying table shows the nature of the phenomenon: so far, only the superconductor tin has been tested in this way.

锡膜的超导电性

伯顿

编者按

1911 年海克·卡默林·昂内斯发现了超导电性，他发现当温度低至绝对零度附近几度时固体汞的电阻会消失。随后另外一些材料诸如锡和铅也很快被发现是“超导体”，正如此后人们所知道的那样，但是当时并未能就这一现象给出解释。这篇文章中，多伦多大学的伯顿指出了超导理论的一个重要的限制条件：超导材料例如锡膜如果被夹在两层非超导的材料之间时，会失去它的超导电性。伯顿还指出这个效应只适用于足够薄的超导薄膜，厚度应小于或者等于 1 微米。

关于高频电流与超导电性现象之间关系的实验促成了在多伦多进行的对超导金属膜的研究。（锡）膜是在非超导材料制成的细线表面“镀锡”形成的：在早期的实验中，我们是将厚度为 2×10^{-4} 厘米的锡的镀层“包裹”在直径为 0.016 厘米的康铜丝上。按这种方法，我们得到了一个等效的超导金属的薄圆筒，而且当温度低于所用超导元素的转变温度时，整个薄圆筒的电阻会变为零 [1]。

为了进一步研究高频电流的效应，我们给这类镀丝样品镀上其他一些没有超导电性的金属，如铜和镍；例如，在镀了锡的康铜丝外再镀上铜，作为芯的几种金属丝的直径如下：康铜，0.056 厘米；铜，0.040 厘米；镍，0.045 厘米。

我们最初做的实验是为了证实这些样品对直流电的反应，即普通的超导电性测试，在实验中我们发现，当这些锡膜镀上非超导金属膜（例如铜或镍）时，它们就不再显示超导电性。虽然这个奇怪的结果只有在膜非常薄时才能得到，但多次的重复实验都倾向于表明这个结果正确无误。下表显示出这种现象的规律：到现在为止，我们只用这种方法对超导体锡做了试验。

No.	Sample	Thickness of Tin Film (cm. × 10^{-5})	Thickness of outer layer (cm. × 10^{-5})	Superconductive Action, direct current of 200 ma.
1a	Constantan-Tin	10	0	Transition point 3.69°K
1b	Constantan-Tin-Copper	10	100	Not superconducting at 2°K
2a	Copper-Tin	9	0	Transition point 3.58°K
2b	Copper-Tin-Copper	9	100	Not superconducting at 2°K
3a	Nickel-Tin	9	0	Transition point 3.42°K
3b	Nickel-Tin-Copper	9	100	Not superconducting at 2°K
4	Constantan-Tin	6.8	0	Transition point 3.49°K
5	Constantan-Tin	2	0	Transition point 2.48°K
6	Constantan-Tin-Copper	18	40	Not superconducting at 2°K
7	Constantan-Tin-Copper	4	20	Not superconducting at 2°K
8	Constantan-Tin-Nickel	15	30	Not superconducting at 2°K
9	Constantan-Tin(wiped)	90	0	Transition point 3.68°K
9a	Constantan-Tin-Copper	90	80	Transition point 3.44°K
10	Constantan-Tin (electro-plated)	200	0	Transition point 3.76°K
10a	Constantan-Tin-Copper	200	80	Transition point 3.73°K
11	Tin Wire	diameter	0.085	Transition point 3.77°K

It is seen that as the film of tin increases in thickness, a point is reached at which the superconducting property of the tin film is not lost by surface plating. This phenomenon will undoubtedly be of importance in framing a satisfactory theory of superconductivity—a consideration of utmost importance in dealing with metallic conduction. This work is being carried on by J. O. Wilhelm and A. D. Misener.

(**133**, 459; 1934)

E. F. Burton: McLennan Laboratory, University of Toronto, Feb. 17.

Reference:

1. E. F. Burton, "Superconductivity" (University of Toronto Press, and Oxford University Press), p. 70. J. C. McLennan, *Nature*, 130, 879, Dec. 10, 1932.

序号	样　品	锡薄膜的厚度（厘米 ×10^{-5}）	外层的厚度（厘米 ×10^{-5}）	超导作用，200 mA 的直流电流
1a	康铜 – 锡	10	0	转变点 3.69 K
1b	康铜 – 锡 – 铜	10	100	2 K 时无超导
2a	铜 – 锡	9	0	转变点 3.58 K
2b	铜 – 锡 – 铜	9	100	2 K 时无超导
3a	镍 – 锡	9	0	转变点 3.42 K
3b	镍 – 锡 – 铜	9	100	2 K 时无超导
4	康铜 – 锡	6.8	0	转变点 3.49 K
5	康铜 – 锡	2	0	转变点 2.48 K
6	康铜 – 锡 – 铜	18	40	2 K 时无超导
7	康铜 – 锡 – 铜	4	20	2 K 时无超导
8	康铜 – 锡 – 镍	15	30	2 K 时无超导
9	康铜 – 锡（焊接的）	90	0	转变点 3.68 K
9a	康铜 – 锡 – 铜	90	80	转变点 3.44 K
10	康铜 – 锡（电镀的）	200	0	转变点 3.76 K
10a	康铜 – 锡 – 铜	200	80	转变点 3.73 K
11	锡线	直径	0.085	转变点 3.77 K

我们发现，当锡膜厚度增加到一个点时，锡膜的超导性能不再因为表面镀层而丧失。这种现象对构造一个令人满意的超导理论无疑是极其重要的，这是处理金属电导时需要重点考虑的一个问题。威廉和麦色纳正在进行这项研究。

（沈乃澂 翻译；郑东宁 审稿）

Persistent Currents in Supraconductors

K. Mendelssohn and J. D. Babbitt

Editor's Note

Physicists in the early 1930s were struggling to understand superconductivity. Here Kurt Mendelssohn and J. D. Babbitt explored the surprising recent discovery by Walther Meissner and Robert Ochsenfeld that a superconductor expels magnetic flux from its interior. That finding confounded the expectation that persistent electrical currents in a superconductor would instead "capture" magnetic flux and sustain it even if the external magnetic field were removed. Yet the experiments reported here showed that the flux expulsion wasn't perfect: some magnetisation remains inside superconducting tin. This was the first evidence of persistent "supercurrents", although their value was only one-sixth of that predicted earlier.

UNTIL recently it was generally assumed that it was possible to predict, by the ordinary electromagnetic equations, the persistent current produced in a supraconductor cooled below the transition point in a constant external magnetic field after the field was switched off. Thus H. A. Lorentz[1] calculated the current induced in a supraconducting sphere, that is, the effective magnetic dipole when an external magnetic field is established.

According to results recently published by Meissner and Ochsenfeld[2], the matter is not so simple as might at first sight appear. Instead of the lines of force being "frozen in" as had been previously assumed would happen when a supraconductor was cooled below the transition point in a magnetic field, it appeared that the field increased in the neighbourhood of the supraconductor, which behaved as a body of zero permeability. If this were so, the flux of induction in the supraconductor should be zero and one might expect, in contradistinction to the old view, that no persistent current or effective induced dipole would be produced by switching off the external field.

The following experiments seem to show that although supraconductors do not conform to the older theory, neither do they behave as though they had zero permeability.

(1) A solid tin sphere of 1.5 cm. radius was cooled from 4.2°K. to 2.5°K. (the liquid helium was produced in a liquefaction apparatus utilising the expansion method of Simon) in a field of 70 gauss. When the field was switched off, the magnetic moment of the sphere was observed with a test coil. Its magnitude was about one sixth of that calculated according to the Lorentz equation.

The magnetic moment remained almost constant whilst the temperature of the sphere rose from 2.5° to 2.9°; with a further rise in temperature it decreased steadily, becoming

超导体中的持续电流

门德尔松，巴比特

编者按

20世纪30年代早期，物理学家们正在费尽心力地研究超导电性的产生机制。这篇文章中，库尔特·门德尔松和巴比特探讨了近期瓦尔特·迈斯纳和罗伯特·奥克森费尔德的惊人发现——超导体会将磁通量排除在外。在此之前人们猜想超导体中持续的电流应该能够“捕获”磁通量，并且即使外磁场被撤除，超导体中的磁通量也能够保持，但该发现否定了这一猜想。然而这篇文章中报道的实验显示磁通量的排除并不完全：超导态锡的内部仍保留有一定的磁化强度。这首次证明了持续的“超导电流”的存在，尽管它的大小只有之前预测值的1/6。

直至不久前，人们仍然普遍认为，根据一般的电磁方程就可以预测：在恒定外磁场中将超导体冷却至转变温度之下，关闭磁场时会产生持续的电流。据此，洛伦兹[1]计算出了超导球体中的感生电流，即在外磁场建立后的有效磁偶极。

按照迈斯纳和奥克森费尔德[2]最近公开发表的结果，事情并非如初看时那么简单。磁场中的超导体被冷却到转变点之下时，并不像我们以前假定的那样会将磁力线“冻结”，实验表明，超导体附近的磁场会增强，超导体表现得像一个完全的抗磁体。如果事实确实如此，那么超导体中的感应磁通量应为零，而且我们也可以推断出与旧观点相反的结论：关闭外磁场并不会在超导体中产生持续的电流或有效的感生偶极。

下列实验似乎表明，超导体虽然并不符合旧的理论，但也没有表现出磁导率为零的性质。

(1) 在70高斯的磁场中，将一个半径为1.5厘米的固体锡球的温度从4.2 K冷却到2.5 K（液氦是在一个液化装置中利用西蒙膨胀法产生的）。当关闭磁场时，我们用实验线圈测量固体球的磁矩。测量得出的磁矩大小约为根据洛伦兹方程计算所得值的1/6。

当球的温度从2.5 K上升到2.9 K时，磁矩几乎保持恒定；而在温度进一步上升时，磁矩稳步减少，且在3.7 K时变为零，这个临界温度就是锡的正常转变点。对磁

zero at 3.7°, the normal transition point of tin. Plotting the magnetic moment against the temperature, one obtains a curve of similar shape to that found for the magnetic threshold values.

(2) The same sphere was cooled to 2.5° without any external magnetic field, a field of 230 gauss (higher than the threshold value at this temperature) was switched on and immediately switched off. The magnetic moment thus produced in the sphere at 2.5° was 8 percent greater than that produced in the previous experiment using 70 gauss, but as the temperature rose it decreased and at 2.9° it reached the same value as the magnetic moment at this temperature in the previous experiment. From 2.9° to 4° the curve coincided with that found in experiment (1).

(3) Similar experiments to those described above were carried out with a hollow tin sphere of the same radius, the spherical space in the middle being equal in volume to one half the volume of the sphere. The magnetic moments produced in the hollow sphere were two to three times greater than those obtained with the solid sphere.

In all these experiments the magnetic field was produced by a cylindrical coil in the middle of which the sphere was placed, all iron being excluded. Although the field near the sphere was thus fairly homogeneous, we think it possible that the observed phenomena may be influenced by slight inhomogeneities of the external field. In a completely homogeneous field it would seem possible that the method of cooling might affect the results. In order to test this, we cooled the spheres from the poles and also from the equator. This did not seem to make any difference, the magnetic moment observed being of the same order of magnitude in either case.

As a result of these experiments, it seems certain that the effective permeability of substances when they become supraconducting decreases, as observed by Meissner and Ochsenfeld. On the other hand, it appears clear that under our experimental conditions the permeability does not vanish entirely, as might be expected in view of the almost infinite conductivity, or if it does vanish, it only does so in certain regions and not throughout the whole volume of the supraconductor.

In conclusion, we would like to express our thanks to Mr. T. C. Keeley for his advice and assistance in various phases of the work.

(**133**, 459; 1934)

K. Mendelssohn and J. D. Babbitt: Clarendon Laboratory, Oxford, Feb. 17.

References:

1. Comm. Leiden, Suppl., Nr. 50 b, 1924.
2. *Naturwiss*, 21, 787; 1933.

矩随温度变化的作图，我们得到一条与临界磁场随温度变化的曲线形状类似的曲线。

（2）同样的球体在无任何外磁场的情况下冷却到 2.5 K，打开并立即关闭 230 高斯的磁场（高于在此温度下的临界磁场）。结果在温度为 2.5 K 时产生的磁矩比上一实验中使用 70 高斯时产生的磁矩大 8%，而当温度上升时磁矩不断下降，在温度上升到 2.9 K 时，磁矩与在以前实验中温度为 2.9 K 时的磁矩值相同。在温度为 2.9 K~4 K 的范围内，得到的曲线与实验（1）中这个温度段的曲线重合。

（3）与上述实验类似，只是改用半径相同的空心锡球进行研究，中心的球型空心是整个球体积的一半。空心球产生的磁矩是实心球的 2~3 倍。

在所有这些实验中，磁场是由缠绕成圆筒状的线圈产生的，圆球置于线圈的中部，实验中没有使用任何铁质材料。尽管这样可以使圆球附近的磁场相当均匀，但我们还是认为，观测到的现象可能会受到外场细微的非均匀性的影响。而即便在一个完全均匀的磁场中，冷却方法也可能会影响到最终的结果。为了验证这一点，我们分别从圆球的两极和赤道位置对其进行冷却。然而这个做法并没有导致任何差异的产生，两种情况下观测到的磁矩的数量级是相同的。

根据这些实验的结果，我们似乎肯定，亦如迈斯纳和奥克森费尔德所观测的，当物质成为超导体时，它们的有效磁导率会减小。另一方面，很显然在我们当前的实验条件下，磁导率并不会如电导率趋于一个无限值状况所预期的那样完全消失，或者即便磁导率完全消失，那也仅是发生在部分区域内，而不会贯穿整个超导体。

最后，我们要感谢基利先生，他在我们工作的各个阶段都给予了建议和大量帮助。

（沈乃澂 翻译；郑东宁 审稿）

The Velocity of Light

M. E. J. Gheury de Bray

Editor's Note

There were many occasions in the early decades of the publication when *Nature* deliberately published items that were calculated to surprise and even challenge orthodox opinion. Thus the journal published in the 1930s three letters from a man called M. E. J. Gheury de Bray advocating the opinion that the velocity of light is not constant in time but that it had slowly decreased between 1924 and 1933. The possibility of a slow change in supposed "constants" of physics is now back in vogue among cosmologists, albeit so far without compelling evidence to support it.

IN 1927 there was published in these columns[1] a table of all the determinations of the velocity of light which I compiled from the original memoirs, together with a discussion, and I pointed out that except a pair of practically simultaneous values obtained in 1882 the final values (printed in heavy type) indicate a secular decrease of velocity. The last (and lowest) value given is 299,796 ± 4 km./sec. for 1926.

Since then, two determinations have been made: the first by Karolus and Mittelstaedt (1928) using a Kerr cell, to the terminals of which an alternating potential was applied, for interrupting periodically the luminous beam, instead of a toothed wheel[2]. A frequency can be obtained in this way, of the order of a million per second, which can be accurately calculated, thus permitting a very short base to be used (41.386 metre) without any loss of accuracy. The value found (mean of 755 measurements) was 299,778 ± 20 km./sec. The second recent determination is mentioned in *Nature* of February 3, p.169: it gives for the velocity of light in 1933 the value 299,774 ± 1 or 2 km./sec.

The determinations of this so-called constant made during the last ten years (the most accurate of the whole series) are therefore:

1924	..	..	299,802 ± 30 km./sec.
1926	..	..	299,796 ± 4 km./sec.
1928	..	..	299,778 ± 20 km./sec.
1933	..	..	299,774 ± 1 or 2 km./sec.

No physicist, looking at the above table, can but admit that the alleged constancy of the velocity of light is absolutely unsupported by observations. As a matter of fact, the above data, treated by Cauchy's method[3], give the linear law:

$$V_{\text{km./sec.}} = 299{,}900 - 4T_{(1900)\ \text{years}}.$$

光的速度

谷瑞·德布雷

编者按

在《自然》杂志发行的前几十年中，它对于发表那些旨在抨击甚至挑战传统观点的文章是非常慎重的，但事实上它仍然发表了很多这类文章。20 世纪 30 年代，《自然》就发表了一个名为谷瑞·德布雷的人的三篇通讯，他的观点是光速并不是永远恒定的，在 1924 年至 1933 年间光速随时间在缓慢地减小。天文学家中又掀起了这样一种风潮，即发现物理学中原以为“恒定”的事物存在缓慢变化的可能性，不过目前尚无确凿的证据证实这一观点。

本专栏在 1927 年[1]发表了我从原始文献中汇集的关于光速的全部测定结果的列表以及对此进行的讨论。我曾指出，除了一对几乎同时于 1882 年获得的数值之外，最终结果（黑体字印刷）显示出光速在不断地减少。表中所给出的最后一个（也是最小的一个）数值是 1926 年测得的 299,796±4 千米 / 秒。

在那以后，科学家又得到了两个测定结果：第一个来自卡罗卢斯和米特尔施泰特（1928），他们使用了克尔盒，并在其终端施加交变电压来代替齿轮，用以周期性地隔断光束[2]。用这种方式可以精确地计算并得到每秒百万数量级的隔断频率，从而可以使用一个极短的基线（41.386 米）而丝毫不损失精确性。由此而得到的光速数值（755 次测量结果的平均值）为 299,778±20 千米 / 秒。第二个近期测定结果是在 2 月 3 日出版的《自然》杂志第 169 页给出的，它测定于 1933 年，值为 299,774±1（或 2）千米 / 秒。

因此，可以将最近十年中对光速这一所谓常数的测定结果（整个序列中最为精确的）排列如下：

1924	..	..	299,802 ± 30 千米 / 秒
1926	..	..	299,796 ± 4 千米 / 秒
1928	..	..	299,778 ± 20 千米 / 秒
1933	..	..	299,774 ± 1（或 2）千米 / 秒

看过上面的表后，物理学家们不得不承认，所谓的光速恒定性显然没有得到观测结果的支持。事实上，经过柯西方法[3]处理后，上述光速测量值与测量年份满足线性关系：

$$V_{\text{千米 / 秒}} = 299{,}900 - 4T_{\text{20 世纪各年份}}$$

When I first pointed out this fact (in 1924) it was objected that the data available were inconclusive, because the probable errors of the observations were greater than the alleged rate of change. Sir Arthur Eddington has dealt the death blow to the theory of errors[4] and "this theory is the last surviving stronghold of those who would reject plain fact and common sense in favour of remote deductions from unverifiable guesses, having no merit other than mathematical tractability"[5]. Even "die-hards", however, may fruitfully meditate over the 2nd and the 4th values in the above table.

(**133**, 464; 1934)

M. E. J. Gheury de Bray: 40 Westmount Road, Eltham, S.E.9.

References:

1. *Nature*, **120**, 603, Oct. 22, 1927.
2. *Phys. Z.*, 698-702; 165-167; 1929.
3. *Engineer*, Sept. 13, 1912.
4. *Proc. Phys. Soc.*, 271-282; 1933.
5. Dr. N. R. Campbell, *loc. cit.*, 283.

我第一次指出这一事实时（1924年），有人反对说现有的数据是不确定的，因为观测中的或然误差大于所谓的变化速率。阿瑟·爱丁顿爵士曾对这一误差理论[4]施以致命一击："对于那些拒绝承认明确的事实与常识却喜欢从无法检验的猜测中得出间接推论的人来说，该理论是最后的生存堡垒，它除了数学便利性之外毫无优点可言"[5]。但是，即使是"老顽固们"也可能会对上表中第2和第4个值进行一番深刻的思考。

（王耀杨 翻译；张元仲 审稿）

Attempt to Detect a Neutral Particle of Small Mass

Editor's Note

In 1930, Wolfgang Pauli hypothesized that nuclear beta decay might involve a second particle being emitted from the nucleus in addition to the electron. He pointed out that the puzzling continuous variation in energy of beta-decay electrons could be understood if a second, electrically neutral particle carries away a varying amount of energy. Fermi named the proposed particle the neutrino, and here *Nature* reports on a failed attempt by Chadwick and colleagues to detect it. If such particles do exist, they found, they must have very small mass. Indeed they do, and because they interact so weakly with other particles, neutrinos are enormous difficult to detect. This was finally achieved in 1956. Physicists later discovered that neutrinos come in three flavours.

CHADWICK and Lea have recently published the negative result of an experiment designed to examine the possibility that the continuous β-ray spectrum is accompanied by the emission of penetrating neutral particles (*Proc. Camb. Phil. Soc.*, **30**, Part 1). The energies of these particles might be distributed in such a way that they combine with those of the β-particles to form a constant energy of disintegration, a low energy β-particle being associated with a high energy "neutrino". A strong source of radium D + E + F (radium E gives a well-marked continuous β-ray spectrum) was placed near a high-pressure ionisation chamber and an absorption curve was taken with lead screens. The radiation was all identified with the radium E and polonium γ-rays. If neutral particles are emitted, it is calculated that they cannot produce more than 1 ion pair in 150 kilometres path in air. A consideration of the possible nature of the particle shows that, if it exists, it must have small mass and zero magnetic moment.

(**133**, 466; 1934)

探测小质量中性粒子的尝试

编者按

1930 年，沃尔夫冈·泡利猜测原子核在 β 衰变过程中除了产生电子之外还会放出第二种粒子。他指出如果这第二种电中性的粒子携带了不同大小的能量，那么令人困扰的 β 衰变过程中电子能量的连续变化就可以得到解释了。费米将这一假想的粒子命名为中微子，《自然》在这篇文章中报道了查德威克和他的同事在探测中微子方面的一次失败的尝试。他们发现如果这一粒子确实存在，它们一定具有非常小的质量。事实正是如此，而且正因为中微子与其他粒子的相互作用非常微弱，所以中微子的探测异常困难。这一探测最终于 1956 年获得成功。之后物理学家又发现中微子有三种味。

最近，查德威克和利发表了他们实验得到的负面结果，设计该实验的目的是检验连续 β 射线谱伴随着发射出具有穿透性的中性粒子的可能性（《剑桥哲学学会会刊》，第 30 卷，第 1 部分）。上述中性粒子的能量可能是以这样的方式分布的：中性粒子能量与 β 粒子能量之和等于一个恒定的衰变能，由此把低能 β 粒子与高能“中微子”联系了起来。将一个包含镭 D、镭 E 和镭 F 的强放射源（镭 E 可以给出很清楚的连续 β 射线谱）置于高压电离室附近，并利用铅屏来获得吸收曲线。得到的谱线与镭 E 的射线谱及钋的 γ 射线谱是完全一致的。经过计算可得，即使有中性粒子发射，它们也不可能在空气中 150 km 的路径内产生出多于 1 个的离子对。对这种中性粒子可能的性质考虑得到的结果是，如果它们存在的话，其质量一定很小且磁矩为零。

（王耀杨 翻译；朱永生 审稿）

Developments of Television

Editor's Note

In a theatre in the West End of London on 20 March 1934, *Nature* reports here, shareholders of Baird Television Ltd. saw and heard the chairman address them from a studio at the Crystal Palace eight miles away. The feat relied on the a newly developed large cathode-ray oscilloscope, as well as high-quality photo-cells and amplifiers capable of processing signals over a wide frequency range. Transmissions from the Crystal Palace location could broadcast television signals to the whole of Greater London. The report notes that another demonstration is planned of a method reported earlier in *Nature* by which television images of topical events might be displayed on cinema screens and home receivers within only a few seconds of having happened.

AN application of science has enabled a chairman of a company to become a historic figure. At the annual general meeting of Baird Television, Ltd., held in a theatre in the west end of London on March 20, the shareholders heard and saw distinctly the chairman address them from a studio at the Crystal Palace, nearly eight miles distant. To the shareholders, and afterwards to representatives of the Press, the Baird Company arranged a programme of transmissions by radio from the Crystal Palace to enable the audience to see persons talking on various subjects, a cartoonist sketching at his easel, excerpts from popular films and "still" pictures. All these items were reproduced in the receiver with sufficient detail for an audience of more than a hundred persons to "look in", although the receiver was devised for use in the home rather than a theatre. The success of these demonstrations is attributed to the state of perfection of the large cathode ray oscillographs made exclusively for the Baird Co. by the research staff of a British industrial concern, the excellence of the photoelectric cells in use at the transmitting end, and the construction of amplifiers which are capable of dealing without phase distortion with a range of frequencies from 25 to 1,000,000 cycles per second. The subject matter to be televised is divided up into 180 lines (or strips) corresponding to 24 times the definition obtainable with the old 30-line apparatus. Vision is being transmitted from a dipole aerial on a wave-length of 6.0 metres, and sound on 6.25 metres.

Judging from the demonstrations given last week, the Baird Company's engineers have successfully overcome interference effects due to motors, lifts and other electro-magnetic disturbances met with at these short wave-lengths. A series of experiments have been carried out to ascertain the effective range of reception, as a result of which it is claimed that the Crystal Palace transmitting station can provide an ultra-short wave high definition television service for the whole of the Greater London area, which includes a population of about eight millions. Capt. A. G. D. West, who joined the board of the Baird Company last June to direct its technical development, is to be warmly congratulated on his

电视的发展

编者按

《自然》的这篇文章报道了1934年3月20日在伦敦西区一个剧院中，贝尔德电视有限公司的股东们观看并且听到董事会主席在8英里外的水晶宫演播室向他们进行的演讲。这一成就得益于最近研制出的大型阴极射线示波器、高质量的光电管和能够处理宽频率范围信号的放大器。从水晶宫输出的信号可以为整个大伦敦地区提供电视信号。这篇报道提到了《自然》先前报道的另一种计划用于电视方面的技术，该技术可以使时事新闻的电视图像在事件发生后的几秒内显示在电影院屏幕或者家用接收机上。

科学的一项应用使一家公司的董事长成为了历史性的人物。3月20日，贝尔德电视有限公司的年会在伦敦西区的一个剧院内召开，会上股东们清楚地听到并看见董事长在将近8英里外的水晶宫演播室向他们发表演讲。贝尔德公司给股东们，随后给出版社代表们安排了一个节目，节目通过无线电波从水晶宫传送信号，观众们能够看到谈论着不同话题的人们、一个正在画架上素描的漫画家以及一些从流行电影和“静止的”图画上摘选的内容。所有这些内容都被细致地再现于接收机上，供100多名观众“观看”，尽管接收机原本是为在家里而不是在剧院使用设计的。这些演示的成功主要归于以下几个原因：英国一家工业公司的研究人员专门为贝尔德公司制造的大型阴极射线示波器堪称完美，在发射端使用的光电管的性能非常优秀，新造的放大器能够无相位失真地处理每秒25周到每秒100万周频率范围内的信号。通过电视播放的内容被分解成180线（或带），这样其清晰度就是以前只能分解成30线的老设备的24倍。图像从一个偶极子天线处以波长6米的信号传出，而声音则以波长6.25米的信号传出。

从上周进行的演示来判断，贝尔德公司的工程师们已经成功克服了由汽车、电梯以及在这些短波长范围内的其他电磁干扰所产生的干涉效应。另外，工程师们也进行了一系列实验来确定有效的接收范围，据称，实验结果表明水晶宫的发射台能够为拥有约800万人口的整个大伦敦地区提供超短波的高清电视服务。去年6月加入贝尔德公司董事会主管公司技术发展的韦斯特上尉一定会因其成就受到热烈的祝贺，贝尔德公司也会因首次公开演示了高清电视播送的可能性而被人们赞颂。我

achievement; and the Company on the first public demonstration of the broadcasting possibilities of high-definition television. We understand that a demonstration will shortly be given of the intermediate film-method, described by Major A. G. Church in *Nature* of September 30, 1933, by means of which televised images of topical events will be thrown on screens in cinema theatres as well as on home-receivers within a few seconds of their occurrence. Another series of experiments on a new system of "scanning" invented by Mr. Baird is nearing completion. These experiments aim at securing sufficient illumination in a studio to enable "crowd" scenes to be televised directly with detailed fidelity.

(**133**, 488-489; 1934)

们可以预计，很快就会出现使用中间胶片法进行的演示。在 1933 年 9 月 30 日的《自然》上丘奇少校曾描述过该方法，通过这种方法，时事新闻的电视图像能够在事件发生后的几秒内显示在电影院屏幕或者家用接收机上。在贝尔德先生发明的新“扫描”系统上进行的另一系列实验也快要完成了。这些实验的目的是确保演播室有足够的照明度，以使“拥挤”的镜头能直接高保真地在电视上播放。

（刘霞 翻译；赵见高 审稿）

The "Neutrino"

H. Bethe and R. Peierls

Editor's Note

Hans Bethe and Rudolf Peierls here discuss the implications of Wolfgang Pauli's proposal that radioactive beta decay involves a hitherto unknown particle, the neutrino. If such a particle had small mass and spin 1/2, the continuous spread of energy of electrons emitted in beta decay would make sense, and energy and momentum conservation need not be abandoned for nuclear processes. They suggest that the neutrino is created at the moment of decay, and that a neutrino may initiate a reverse beta decay when impinging upon nuclei. However, by their estimates a neutrino should interact so weakly with other matter that observing the latter would be most unlikely. Indeed, they can see no way yet to observe a neutrino.

THE view has recently been put forward[1] that a neutral particle of about electronic mass, and spin $\frac{1}{2}\hbar$ (where $\hbar=h/2\pi$) exists, and that this "neutrino" is emitted together with an electron in β-decay. This assumption allows the conservation laws for energy and angular momentum to hold in nuclear physics[2]. Both the emitted electron and neutrino could be described either (*a*) as having existed before in the nucleus or (*b*) as being created at the time of emission. In a recent paper[3] Fermi has proposed a model of β-disintegration using (*b*) which seems to be confirmed by experiment.

According to (*a*), one should picture the neutron as being built up of a proton, an electron and a neutrino, while if one accepts (*b*), the rôles of neutron and proton would be symmetrical[4] and one would expect that positive electrons could also sometimes be created together with a neutrino in nuclear transformations. Therefore the experiments of Curie and Joliot[5] on an artificial positive β-decay give strong support to method (*b*), as one can scarcely assume the existence of positive electrons in the nucleus.

Why, then, have positive electrons never been found in the natural β-decay? This can be explained by the fact that radioactivity usually starts with α-emission and therefore leads to nuclei the charge of which is too small compared with their weight. The artificial β-emission was found for two unstable nuclei (most probably N^{13} and P^{30}) formed by capture of an α-particle and emission of a neutron, and therefore having too high a charge for their mass.

A consequence of assumption (*b*) is that two isobares differing by 1 in atomic number can only be stable if the difference of their masses is less than the mass of electron and neutrino together. For otherwise the heavier of the two elements would disintegrate with emission of a neutrino and either a positive or negative electron. There will be only a

“中微子”

贝特，佩尔斯

编者按

泡利认为放射性β衰变的过程中涉及了一种未知粒子——中微子，汉斯·贝特和鲁道夫·佩尔斯在这篇文章中对沃尔夫冈·泡利这一提议的含义进行了讨论。如果这种粒子质量很小且自旋为1/2，那么β衰变中放出的电子能谱的连续性就是可以理解的，而且核反应过程也因此而能满足能量和动量守恒。他们认为衰变发生时产生了中微子，并且中微子冲击原子核时会引发β衰变的逆过程。然而，根据他们的估算，中微子与其他物质的相互作用非常弱，因而难以观测。实际上，他们当时仍然没有找到任何办法来实现对中微子的观测。

最近有观点认为[1]，存在质量和电子差不多、自旋为 $\frac{1}{2}\hbar$ ($\hbar=h/2\pi$) 的中性粒子，而这种“中微子”是在发生β衰变时与电子一起发射出来的。以上这个假设使得能量和角动量守恒定律在核物理中也得以成立[2]。衰变过程中发射出来的电子和中微子均可被认为是 (*a*) 衰变发生前就存在于核中，或 (*b*) 衰变发生时创生出来的。在最近的一篇文章中[3]，费米利用假设 (*b*) 提出了一种β衰变的模型，而这一假设似乎得到了实验的证实。

按照 (*a*) 的解释，中子应该是由一个质子、一个电子和一个中微子组成的，而如果采纳 (*b*) 的解释，则中子和质子的作用就是对称的[4]，因此在核转变中也有可能同时发射出正电子和中微子。人们不太可能接受原子核中存在正电子的观点，因此，居里和约里奥[5]用人工正β衰变的实验有力地证明了 (*b*) 的解释是正确的。

那么，为什么从未在天然β衰变中发现过正电子呢？这可以用以下事实进行解释：放射现象通常先出现α发射，从而导致核的电荷与其质量相比很小。而人工β发射发现于两种不稳定核（最可能是 ^{13}N 和 ^{30}P）的形成过程中，即吸收一个α粒子并发射出一个中子，因此相对其质量而言，该不稳定的核具有很高的电荷。

由假设 (*b*) 可得到以下结论：原子序数相差1的两个同量异位素只有在它们的质量差小于电子和中微子的质量之和时才是稳定的。否则，两个元素中较重的一个将发生蜕变并发射出一个中微子和一个正电子（或负电子）。两个同量异位素的质量差如此之小的情况对应于质量亏损曲线上有限的区域，该区域大概位于中等原子量

limited region on the mass defect curve, probably at medium atomic weight, where such small differences are possible. In fact, neighbouring isobares have only been found with the mass numbers 87, 115, 121, 123, (187), (203), while isobares with atomic numbers differing by 2 are very frequent. In the first case, one of the two nuclei (Rb) is known to emit β-rays. In each of the last two cases one of the two isobares is stated to be exceedingly rare and its identification might be due to experimental error. The other three cases actually lie close together and have medium weight. A particular case of isobares are proton and neutron. Since all experimentally deduced values of the neutron mass lie between 1.0068 and 1.0078, they are certainly both stable even if the mass of the neutrino should be zero.

The possibility of creating neutrinos necessarily implies the existence of annihilation processes. The most interesting amongst them would be the following: a neutrino hits a nucleus and a positive or negative electron is created while the neutrino disappears and the charge of the nucleus changes by 1.

The cross section σ for such processes for a neutrino of given energy may be estimated from the lifetime t of β-radiating nuclei giving neutrinos of the same energy. (This estimate is in accord with Fermi's model but is more general.) Dimensionally, the connexion will be

$$\sigma = A/t$$

where A has the dimension cm.^{2} sec. The longest length and time which can possibly be involved are $\hbar/mc$ and $\hbar/mc^2$. Therefore

$$\sigma < \frac{\hbar^3}{m^3 c^4 t}$$

For an energy of 2.3×10^6 volts, t is 3 minutes and therefore $\sigma<10^{-44}$ cm.^{2} (corresponding to a penetrating power of 10^{16} km. in solid matter). It is therefore absolutely impossible to observe processes of this kind with the neutrinos created in nuclear transformations.

With increasing energy, σ increases (in Fermi's model[3] for large energies as $(E/mc^2)^2$) but even if one assumes a very steep increase, it seems highly improbable that, even for cosmic ray energies, σ becomes large enough to allow the process to be observed.

If, therefore, the neutrino has no interaction with other particles besides the processes of creation and annihilation mentioned—and it is not necessary to assume interaction in order to explain the function of the neutrino in nuclear transformations—one can conclude that there is no practically possible way of observing the neutrino.

(**133**, 532; 1934)

H. Bethe and R. Peierls: Physical Laboratory, University, Manchester, Feb. 20.

处。实际上，已发现的原子序数相差 1 的同量异位素的质量数只有 87、115、121、123、（187）和（203），而原子序数相差 2 的同量异位素是很常见的。对应上述第一个质量数，已知两种核之一（Rb）会发射 β 射线。而对应最后两个质量数，每个质量数的两种同量异位素中的任何一种都可以说相当稀少，而且很可能是实验误差导致大家以为存在着该同量异位素。其他三个质量数实际上很接近，且都具有中等的原子量。质子和中子是同量异位素的特殊情况。由于所有实验导出的中子质量值均在 1.0068 和 1.0078 之间，所以即使中微子的质量为零，中子和质子也肯定都是稳定的。

产生中微子的可能性必然意味着湮没过程的存在。这些现象中最令人感兴趣的是以下过程：一个中微子撞击一个核后，中微子消失并产生一个正电子或负电子，核的电荷数改变了 1。

给定能量的中微子发生以上过程时，其散射截面 σ 可以通过放射出相同能量中微子的 β 衰变核的寿命 t 来估算。（这个估计与费米模型的结果一致，但更具有普适性。）在量纲上，以上两个物理量的关系式可写成

$$\sigma = A/t$$

式中 A 的量纲为 $\mathrm{cm^2 \cdot s}$。该问题可能涉及的最长长度和时间分别是 $\hbar/mc$ 和 $\hbar/mc^2$。因此

$$\sigma < \frac{\hbar^3}{m^3 c^4 t}$$

当能量为 2.3×10^6 电子伏、t 为 3 分钟时，$\sigma<10^{-44}\ \mathrm{cm^2}$（相应于在固体物质中 10^{16} km 厚的穿透能力）。因此对于核转变时产生中微子的这类过程，我们绝对不可能观测到。

随着能量的增加，σ 也随之增大（在费米模型[3]中，当能量较大时，与 $(E/mc^2)^2$ 相当），但即使假定能量急剧增大，甚至是宇宙射线的能量，σ 也几乎不可能大到足以被我们观测到的程度。

因此，如果中微子除了上述的产生和湮没过程外与其他粒子没有相互作用——而我们也没有必要通过假设这种相互作用的存在来解释中微子在核转变中的作用——则我们可以断言没有实际可行的观测中微子的方法。

（沈乃澂 翻译；刘纯 审稿）

References:

1. W. Pauli, quoted repeatedly since 1931, to be published shortly in "Rapports du Septième Conseil Solvay, Brussels", 1933.
2. C. D. Ellis and N. F. Mott, *Proc. Roy. Soc.*, A, **141**, 502; 1933.
3. E. Fermi, *La Ricerca Scientifica*, **2**, No. 12; 1933.
4. This point of view was first put forward by I. Curie and F. Joliot at the Conseil Solvay, 1933.
5. I. Curie and F. Joliot, *Nature*, **133**, 201, Feb. 10, 1934.

Disintegration of the Diplon

P. I. Dee

Editor's Note

The recent experiments of Oliphant, Harteck and Rutherford with "diplons" (nuclei of the hydrogen isotope deuterium) suggested that they might react in pairs to form ^{3}He and ^{3}H. A colleague at the Cavendish Laboratory, Philip Ivor Dee, here reports partial confirmation from cloud-chamber experiments. The nuclear reaction of two deuterium nuclei is one of the most basic in nuclear physics, featuring in the fusion processes in the Sun.

IT has been shown by Oliphant, Harteck and Lord Rutherford in a recent letter[1] that the bombardment by high-velocity diplons of compounds containing diplogen gives rise to three groups of particles—two groups of equal numbers of singly-charged particles of ranges 14.3 cm. and 1.6 cm., together with neutrons of maximum energy of about three million volts. They suggest as possible explanations of these results the reactions:

$$ {}_1D^2 + {}_1D^2 \rightarrow {}_1H^1 + {}_1H^3 \qquad (1) $$

and

$$ {}_1D^2 + {}_1D^2 \rightarrow {}_2He^3 + {}_0n^1 \qquad (2) $$

an atom of ${}_1H^3$ of 1.6 cm. range and a proton of 14.3 cm. range satisfying the momentum relations in reaction (1). In this reaction it is to be expected that the proton and the isotope of hydrogen of mass 3 would recoil in opposite directions, except for a small correction due to the momentum of the captured diplon. The cloud track method is extremely suitable for an examination of this possibility, and I have recently taken expansion chamber photographs of the disintegration particles resulting from the bombardment of a target of "heavy" ammonium sulphate with diplons, to see if further information can be obtained.

The first set of experiments was made with a thin target contained in an evacuated tube at the centre of the chamber. Two opposite sides of the end of this tube were closed with mica windows of 6.3 mm. and 11.4 cm. stopping power respectively. The chamber was filled with a suitable mixture of helium and air to increase the lengths of the tracks of the short particles. Under these conditions, the particles of 14.3 cm. range emerging through the thick window and the particles of 1.6 cm. range emerging through the thin window end in the chamber and the usual reprojection permits precise determination as to whether the two tracks are co-planar and of the ranges. Owing to the fine structure of the grid supporting the thin window the efficiency of collection of pairs cannot be high; also the companion to a 14.3 cm. particle passing through the thin window would not be able to pass through the opposite thick window. In spite of these difficulties, opposite pairs of tracks of about 14.3 cm. and 1.6 cm. range are observed with far greater frequency than could be attributed to chance. The photograph reproduced as Fig. 1 is a fortunate

氘核的蜕变

迪伊

编者按

奥利芬特、哈特克和卢瑟福关于"氘核"（氢的同位素氘的原子核）的新近实验表明，它们可以成对发生反应生成 ^3He 和 ^3H。这篇文章中卡文迪什实验室的一位同事菲利普·艾弗·迪伊报道了来自云室实验的部分证据。两个氘核间的反应是核物理学中最基本的反应之一，它在太阳的核聚变过程中起着重要的作用。

在最近的一篇通讯[1]中，奥利芬特、哈特克和卢瑟福勋爵曾指出，用高速的氘核轰击含有重氢的化合物，可以产生三组粒子——其中两组射程分别为 14.3 厘米和 1.6 厘米的同等数目的单电荷粒子，以及一组最大能量约为 3 兆电子伏的中子。对于这些结果，他们提出下列反应作为可能的解释：

$$^2_1D + ^2_1D \rightarrow ^1_1H + ^3_1H \qquad (1)$$

和

$$^2_1D + ^2_1D \rightarrow ^3_2He + ^1_0n \qquad (2)$$

其中射程为 1.6 厘米的 ^3_1H 原子和射程为 14.3 厘米的质子就能满足反应式（1）中的动量关系。在这个反应中除了因俘获氘核的动量而做了一个小修正外，可以预期质子和质量数为 3 的氢同位素原子会向相反的方向反冲。采用云室径迹法来检验这一可能性是非常合适的。最近，为了考察能否获取更进一步的信息，我拍摄了一些用氘核轰击硫酸"重"铵靶而产生蜕变粒子的膨胀云室照片。

第一组实验把一个薄靶置于云室中心的真空管中。管的两端用具有阻断功能分别为 6.3 毫米和 11.4 厘米的云母窗封闭。云室充有适当比例的氦和空气的混合物以增加短径迹粒子的径迹长度。在这些条件下，射程为 14.3 厘米的粒子从真空管的厚窗中穿出，射程为 1.6 厘米的粒子则从另一端的薄窗中穿出并进入云室，而通常出现的二次投影可以精确测定粒子的射程以及其径迹是否共面。由于支撑薄窗的栅格的精细结构，粒子对的收集效率不可能高；另外，若射程为 14.3 厘米的粒子从薄窗中穿出，那么射程为 1.6 厘米的粒子就不能从对面的厚窗中穿出了。尽管存在这些困难，但我们观测到射程约为 14.3 厘米和 1.6 厘米的反向径迹对的频率远高于由于随机事件所导致的频率。图 1 这样的复印照片是一个幸运的例子，其中位于

example, the short track on the right being due to the new hydrogen isotope of mass 3. Detailed measurements of the lengths of the tracks and the angles between them are being made and will be published later.

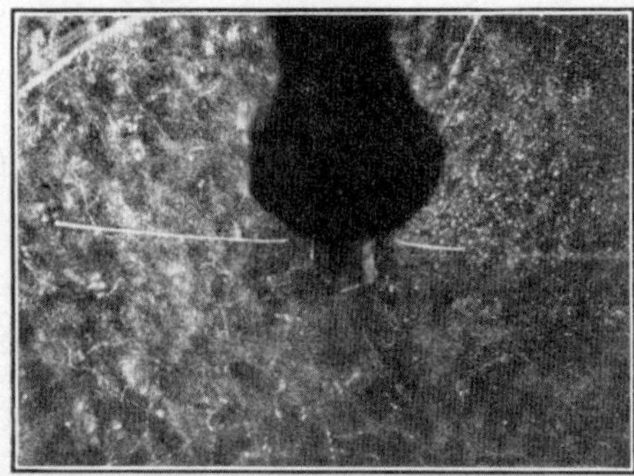

Fig. 1

To investigate the neutron emission, a second series of experiments has been made in which a target of the same material contained in a lead tube of 3 mm. wall thickness was bombarded in the same manner, the chamber being filled with a mixture of 50 percent helium in air. Under these conditions, thirty-one recoil tracks originating in the gas have been photographed. Assuming that these are due to impacts with neutrons, the latter appear to constitute an approximately homogeneous group of maximum energy of about 1.8 million volts. This energy appears to be in fair agreement with reaction (2) on substitution of the mass of ${}_2He^3$, which can be estimated from consideration of the energies of the short-range products resulting from the transformation of ${}_3Li^6$ by protons[2,3,4]. The ${}_2He^3$ group of reaction (2) with a possible range of about 5 mm. would not pass through the thinnest window used in these experiments, but special arrangements are being made to search for them in an expansion chamber.

These experiments are the first to be made with a new discharge tube constructed following a design due to Dr. Oliphant. I should like to acknowledge the much valuable advice which Dr. Oliphant has always so readily given me in the course of construction of this tube. I am also indebted to him for preparing the diplogen targets used in these experiments.

(**133**, 564; 1934)

P. I. Dee: Cavendish Laboratory, Cambridge.

References:

1. *Nature*, 133, 413, March 17, 1934.
2. *Proc. Roy. Soc.*, A, 141, 722; 1933.
3. *Nature*, 132, 818, Nov. 25, 1933.
4. *Nature*. 133, 377, March 10, 1934.

右方的短径迹是由质量数为 3 的氢的新同位素产生的。这些径迹的长度以及它们之间的夹角已被详细测量，稍后就会发表。

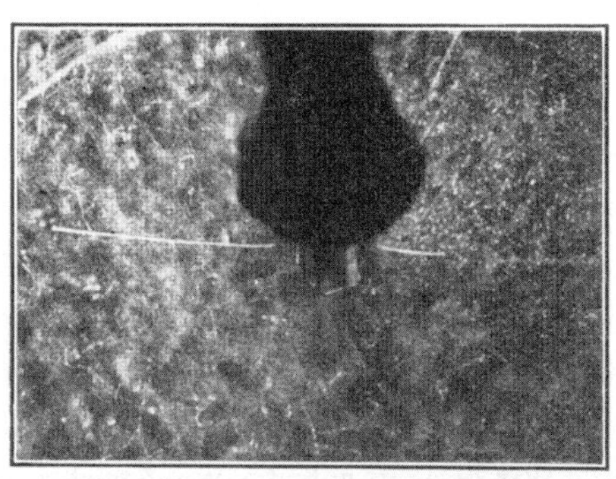

图 1

为了研究中子发射，我们进行了同系列的第二组实验，首先，云室中充入气体混合物为含氦量 50%的空气，用相同材料的靶置于 3 毫米厚的铅管中并以相同的方式进行轰击。在这样的条件下，我们拍摄到了 31 条在气体中产生的反冲径迹。假定这些径迹是由于中子碰撞产生的，那么后者似乎构成一个最大能量约为 1.8 兆电子伏的近似单能峰。该能量似乎与反应式（2）中以 ${}^{3}_{2}$He 的质量代入获得的能量值相当，这个质量可以根据由质子引起 ${}^{6}_{3}$Li 蜕变产生的短射程产物的能量来估算[2,3,4]。反应式（2）产生的 ${}^{3}_{2}$He 粒子群可能的射程约为 5 毫米，因此不能穿过这些实验中使用的最薄窗，但是我们正在制造一个特殊的实验装置以便能够在膨胀云室中找到它们。

我们的这些实验第一次使用了根据奥利芬特博士设计而制造出的新放电管。我还要感谢奥利芬特博士在我们制造放电管期间总是不厌其烦地给我们提供非常有建设性的建议。此外，我也很感激他在为上述实验准备氘靶时所提供的帮助。

（王耀杨 翻译；张焕乔 审稿）

Commercial Production of Heavy Water

Editor's Note

A plant of Imperial Chemical Industries, *Nature* here reports, has begun commercial production of heavy water, producing as much as 5 grams per day with a purity of 30% (pure heavy water was refined at a later date). The plant uses electrolytic method, inspired by Harold Urey and Edward Washburn's discovery that the concentration of deuterium increases in water undergoing electrolysis. The company believes it will be able to supply all heavy water required for commercial purposes in the near future. Production of heavy water became a major focus of nations in the Second World War, given its potential use in nuclear reactors.

THE recently discovered "heavy water", which has created so much interest in popular as well as scientific circles, is to be produced commercially in England. Plant has been developed at the Billingham works of Imperial Chemical Industries, Ltd., which is capable of producing a continuous supply of heavy water of approximately 30 percent purity at the rate of 5 gm. per day, while approximately pure "heavy water" will be produced at a somewhat later date. I.C.I. anticipate that they will be able to meet any commercial demand that may arise. Urey and Washburn, in the United States, discovered that the residual water in old electrolytic cells contained a larger proportion of heavy hydrogen than the normal. It was further found that by continued electrolysis, the concentration of the "heavy water" was enriched, ordinary light hydrogen being given off preferentially, and "heavy water" accumulating. This gave the key to a successful method of preparing "heavy water" in quantity, and the electrolytic method is the one in use at Billingham. Large-scale production of "heavy water" is only possible where exceptional resources of power and raw materials exist together. At Billingham, not only ordinary hydrogen in large quantities, but also residues in which "heavy water" has accumulated, are readily available. These resources, together with cheap power and convenient research facilities, make Billingham a logical centre for the large-scale production of the new compound. Since its discovery in the United States, its probable uses are becoming more evident, and it is eloquent testimony to the vitality of British chemical technique that in so short a space of time it should have been translated from a scientific curiosity to a marketable commodity.

(**133**, 604; 1934)

重水的商业化生产

编者按

《自然》杂志在这篇文章中报道了帝国化学工业公司的一个工厂开始了重水的商业化生产，他们每天生产 5 克纯度为 30% 的重水（后来才精炼得到纯的重水）。哈罗德·尤里和爱德华·沃什伯恩发现重氢在水中的浓度会随着电解的进行而升高，受此启发，这个工厂使用电解的方法生产重水。该公司相信在不远的将来他们有能力满足所有对商业用重水的需求。由于其在核反应堆中的潜在应用价值，重水的生产成了第二次世界大战期间各国关注的焦点。

最近发现的“重水”引起了公众及科学界极大的兴趣，并即将在英格兰进行商业化生产。帝国化学工业有限公司（简称 ICI）已经在比林罕厂区建立了厂房，该厂可以以每天 5 克的速率持续供给纯度大约为 30% 的重水，不久以后将会生产接近纯净的“重水”。ICI 估计他们将可以满足任何商业需求。美国的尤里和沃什伯恩发现，在旧电解池中剩余的水比一般的水含有更高比例的重氢。他们进一步发现，通过连续电解可以提高“重水”的浓度，普通的轻氢会被选择性地除去而使“重水”得到累积。这个发现为成功地进行大规模的“重水”制备提供了关键思路，而比林罕的工厂使用的正是电解的方法。只有在优越的动力和原材料同时满足的情况下才有可能大量生产“重水”。在比林罕，不仅普通氢原料充足，而且有其中累积了“重水”的残液便于利用。这些资源与廉价的动力以及方便的研究设备，使比林罕自然而然成为一个大量生产新化合物的中心。自“重水”在美国发现以来，其可能的应用价值日益明显，这也是英国化学工业技术具有生命力的有力明证，即：在非常短的时间内，重水将从科学珍品转变为具有市场价值的商品。

（沈乃澂 翻译；李芝芬 审稿）

The Neutrino

H. Bethe and R. Peierls

Editor's Note

Bethe and Peierls again write on the perplexing matter of the neutrino. It seems very unlikely, they note, that this particle would create any detectable ionisation after emerging from a nucleus. Nevertheless, its existence might still be demonstrated experimentally. One way, they suggest, would be to measure the recoil of a nucleus during beta decay. Though this energy in natural beta decays would be too small, it could be much bigger in artificial decays, for example of ^{13}N. If the neutrino hypothesis were correct, its existence would be evident from the missing energy or momentum in individual decays. Although this was not how the neutrino's existence would ultimately be demonstrated, the suggestion showed that doing so was not considered hopeless.

ALTHOUGH it seems very unlikely that neutrinos, after having been emitted in a nuclear process, give rise to any detectable ionisation[1], we would like to point out that it is not impossible in principle to decide experimentally whether they exist.

One possible experiment would be to check the energy balance for the artificial β-decay. Take, for example, the process

$$B^{10} + \alpha \rightarrow N^{13} + \text{neutron}$$
$$N^{13} \rightarrow C^{13} + e^{+} + \text{neutrino.}$$

One can safely assume that if the positive electron is emitted with the greatest possible energy, the kinetic energy of the neutrino will just be zero. The balance of energy in this case will therefore determine the mass of the neutrino. For this purpose one would have to know the mass defects of B^{10}, C^{13} and the neutron*, the kinetic energy of the α-particles and the neutrons and the upper limit of the spectrum of the emitted positive electrons.

A second way of deciding the question would be to observe the recoil of the nucleus in β-decay. With natural β-rays this is in practice impossible because the recoil energy is too small, but the nuclei involved in artificial β-decay are much lighter. The kinetic energy of recoil of a disintegrating N^{13} nucleus would be of the order of some hundreds of volts if there were no neutrinos. If the neutrino hypothesis is correct, there would be a defect of momentum which would be uniquely connected with the lack of observable energy in each individual process.

* The accuracy with which the mass of the neutron can be determined at present is, however, far from being sufficient for this purpose.

中微子

贝特，佩尔斯

编者按

贝特和佩尔斯又一次描述了中微子令人困扰的一面。他们注意到，这种粒子从原子核中出现之后，似乎不可能引起任何可观测的电离过程。然而，中微子的存在仍需要实验的验证。他们提出的一个方法是，测量原子核在β衰变过程中的反冲能。尽管天然β衰变中的核反冲能量很小，但在人工衰变过程中这个能量可以大得多，比如 ^{13}N 核的衰变。如果中微子假说是正确的，那么各种衰变过程中的能量和动量缺失就是它存在的证据。尽管并不能通过这一方法最终证明中微子的存在，但他们的方法表明验证这点也并不是毫无希望的。

虽然核转变过程中发射出来的中微子似乎不可能给出任何可观测的电离效应[1]，但我们要指出的是，原则上用实验来判断它们是否存在还是可能的。

一个可能的实验是检验人工β衰变前后能量是否守恒。例如下列实验过程：

$$^{10}B + \alpha \rightarrow {}^{13}N + \text{中子}$$
$$^{13}N \rightarrow {}^{13}C + e^{+} + \text{中微子}$$

我们完全可以假设，如果正电子是以最大可能的能量发射的，则中微子的动能将是零。这种情况下，根据能量守恒可以确定中微子的质量。为此，我们还必须知道 ^{10}B、^{13}C 和中子* 的质量亏损，α 粒子和中子的动能以及发射正电子谱的上限。

判定中微子是否存在的第二种方法是观测β衰变中核的反冲能。用天然β衰变实际上是不可能实现这一目的的，因为反冲能量太小了，但是人工β衰变中所涉及的核要轻得多。如果不存在中微子，^{13}N 核发生蜕变时的反冲动能将是几百电子伏的量级。而如果确实存在中微子，则每个独立过程中都将出现与可观测到的能量亏损唯一相关的动量亏损。

* 目前可以确定中子的质量，但远未达到所需的精确度。

In addition to the nuclear processes mentioned in our previous communication, it may also be expected that a nucleus catches one of its orbital electrons, decreases by one in atomic number, and emits a neutrino. (A corresponding process with increase in atomic number is not possible because of the absence of positive electrons.) This process further limits the possible mass differences between stable neighbouring isobares, and particularly between neutron and proton. If the hydrogen atom is to be stable, we must have (for the masses):

$$\text{Proton} + \text{electron} < \text{neutron} + \text{neutrino}.$$

The probability of such a process is less than that of a process involving emission only, the energy of the neutrino being the same. The reason is that the momentum of the electron, which enters in the third power, is about a hundred times smaller. But even for a surplus energy of 10^5 volts, the life-period of hydrogen would be only 10^{10} years, which seems incompatible with experimental facts. If therefore the neutrino is not heavier than the electron, the neutron must be at least as heavy as the proton.

(**133**, 689-690; 1934)

H. Bethe and R. Peierls: Physical Laboratory, University, Manchester, April 1.

Reference:

1. H. Bethe and R. Peierls, *Nature*, 133, 532, April 7, 1934.

除了我们以前文章中提到的核转变过程外，我们还认为核可能会捕获核外的一个轨道电子，从而原子序数减小 1 并发射出一个中微子。（与之对应的原子序数增加的过程是不可能的，因为不存在正电子。）这个过程进一步限制了稳定的相邻同量异位素之间，特别是质子与中子之间可能的质量差。如果氢原子是稳定的，则必须满足（对其质量而言）：

$$质子+电子<中子+中微子$$

如果中微子的能量是相同的，则发生以上过程的概率小于只有发射的 β 衰变的概率。理由是以三次方形式出现的电子动量小了百倍。但即使多赋予电子 10^5 电子伏的能量，氢的寿命也仅为 10^{10} 年，这似乎与实验事实相矛盾。因此，如果中微子并不比电子重的话，则中子至少应与质子同样重。

（沈乃澂 翻译；刘纯 审稿）

Liquefaction of Helium by an Adiabatic Method without Pre-cooling with Liquid Hydrogen

P. Kapitza

Editor's Note

Russian physicist Pyotr Kapitza reports on a new technique for liquefying hydrogen and helium, developed by himself and colleagues in Cambridge. These substances, so useful for modern physics research, were still being liquefied by methods dating from the beginning of the century. But his group had now developed a method based primarily on adiabatic expansion, in which no heat is lost or gained in the process. Their design eliminated the need for lubricants capable of working at such low temperatures, and could produce up to one litre per hour. The method would soon provide abundant supplies of liquid helium and hydrogen for low-temperature physics.

THE methods for the continuous liquefaction of hydrogen and helium at present in use are essentially the same as those originally used by Dewar and Kamerlingh Onnes when these gases were first liquefied. These methods are based on the use of the Joule–Thomson effect, combined with a regenerating heat exchange after the gas has been cooled below its conversion temperature by liquid air or hydrogen. Since these processes are essentially nonreversible, the efficiency of the method is very low: for example, Meissner[1] calculates that to produce liquid helium, one hundred times more power is required than if the process could be done reversibly. The advantages to be gained by using adiabatic expansion for the cooling of liquefying gases have long been realised, but owing to technical difficulties this method has only been used up to the present to liquefy small amounts of gas by a single expansion. Thus in 1895, Olszewski was the first to obtain a fog of liquid hydrogen drops by a sudden expansion of compressed hydrogen. Recently, Simon[2] has produced appreciable quantities of liquid helium also by a sudden expansion of highly compressed helium.

The technical difficulties in constructing an apparatus for continuous liquefaction by adiabatic expansion lie chiefly in the designing of a cooling expansion engine which will work at low temperatures. Two principal types of expansion engine can be considered. The first is a turbine, but this involves a number of technical difficulties which have not yet been overcome. The second type of machine is a reciprocating moving piston expansion engine; this also involves great difficulties, chiefly arising from the difficulty of finding a lubricant which will make the piston tight in the cylinder and retain its lubricating properties at the very low temperatures. Claude, however, managed to make such an expansion engine which would work at the temperature of liquid air by using the

无液氢预冷的绝热法氦液化

卡皮查

编者按

俄罗斯物理学家彼得·卡皮查报道了一种将氢和氦液化的新技术，这一技术是由他和他在剑桥的同事们共同研发的。这些对现代物理学研究非常有用的物质，仍在使用20世纪初的液化技术。卡皮查的研究小组发展出了一种主要基于绝热膨胀的技术，在绝热膨胀过程中，既没有热量失去也没有热量获得。他们的设计无须使用必须能够在同样的低温下工作的润滑剂，而且可以将产量提升至每小时1升。这种方法很快就为低温物理实验提供了充足的液氦和液氢。

目前所用的氢和氦连续液化方法，本质上与杜瓦和卡默林·昂内斯首次将这些气体液化时所用的方法相同。这些方法是基于焦耳–汤姆孙效应，并结合了再生式热交换，即利用气体被液态空气或液氢冷却到低于其转变温度后的部分返回气体实现的。由于这些过程本质上是不可逆的，因此这种方法的效率非常低：例如，迈斯纳[1]对这一过程的液氦产生效率进行了计算，发现它所需要的能量比可逆过程高出一百倍以上。人们很早以前就已了解到用绝热膨胀的方法对被液化气体进行冷却的优势，但由于技术上的困难，这类方法至今仅用于通过单次膨胀实现少量气体的液化。1895年，奥尔谢夫斯基通过将被压缩的氢突然膨胀的方法首次获得了液氢滴雾。最近，西蒙[2]同样采用将高度压缩的氦突然膨胀的方法，产生了数量可观的液氦。

制造通过绝热膨胀实现连续液化的设备，其技术难点主要在于对低温下工作的冷却膨胀机的设计。有两种主要的膨胀机类型可以考虑。第一类是涡轮机，但这涉及许多尚未攻克的技术难题。第二类机器是往复运动的活塞膨胀机；这也有很大的困难，主要是难以找到一种可以使活塞紧密贴近汽缸，同时在极低温下保持润滑的润滑剂。但克劳德通过采用被液化的气体作为润滑剂的方法，设法使这一类型的膨胀机可以在液态空气的温度下工作。不过，这个方法对于氢和氦的液化似乎并不

liquefied gas as the lubricant. This method, however, does not appear to be practicable for liquefying helium and hydrogen.

During the last year, in our laboratory we have been working on the development of a reciprocating expansion engine working on a different principle which does not require any lubrication of the piston at all, and which will work at any temperature. The main feature of the method is that the piston is loosely fitted in the cylinder with a definite clearance, and when the gas in introduced into the cylinder at high pressure, it is allowed to escape freely through the gap between the cylinder and the piston. The expansion engine is arranged in such a way that the piston moves very rapidly on the expanding stroke, and the expansion takes place in such a small fraction of a second that the amount of gas escaping through the gap is very small and does not appreciably affect the efficiency of the machine.

Fig. 1. Helium liquefaction apparatus at the Royal Society Mond Laboratory.

The principal difficulty in constructing such a machine was concerned with the valves in the expansion engine, which had to let in a considerable amount of gas in a small fraction of a second. Another difficulty was to find metals with the necessary mechanical properties for use at these low temperatures. All these difficulties have now been successfully overcome, and the liquefier is shown in the accompanying photograph (Fig. 1). The expansion engine is placed in the middle of the evacuated cylindrical copper casing, the dimensions of which are 75 cm. long and 25 cm. diameter. The casing also contains heat-exchanging spirals and a container of liquid air for the preliminary cooling of the

可行。

去年，我们实验室在研制一种基于不同原理工作的往复式膨胀机，它完全不需要对活塞进行任何润滑，并可以在任何温度下工作。这种方法的主要特征是，活塞宽松地装配于气缸中，之间留有一定的空隙，当气体在高压下被引入气缸时，允许其通过气缸与活塞之间的间隙自由地逸出。膨胀机被设计成在膨胀冲程内活塞运动得非常快，并且膨胀过程发生在远小于 1 秒的时间内，使得通过间隙逸出的气体量很少，不会明显地影响机器的效率。

图 1. 皇家学会蒙德实验室的氦液化装置

制造这种机器所遇到的主要困难是膨胀机中的阀门，它必须在远小于 1 秒的时间内让大量的气体进入。另一个困难是寻找在这样低的温度下具有必要力学性能的一些金属。现在所有这些难题均已成功得到解决，液化器如照片（图 1）中所示。膨胀机置于抽空的圆筒形铜制壳体的中间，其尺寸为长 75 cm，直径 25 cm。壳体内还包含热交换螺旋管，以及用于盛预冷氦的液空的容器。氦被压缩到 25 ~ 30 个大气压，并先被冷却到液态空气的温度，然后用膨胀机和再生式热交换螺

helium. Helium is compressed to 25–30 atmospheres and is first cooled to the temperature of liquid air and then cooled by the expansion engine and regenerating spiral to about 8°K. ; the final liquefaction is produced by making use of the Joule–Thomson effect. This combination proves to be the most efficient method of liquefaction. The liquid helium is drawn off from the bottom of the liquefier by means of a tap.

Following the preliminary cooling to the temperature of liquid nitrogen, the liquefier starts after 45 minutes to liquefy helium at a rate of 1 litre per hour, consuming about 3 litres of liquid air per litre of liquid helium. This output we hope will shortly be increased, but even now it compares very favourably with the original method of making liquid helium, in which, according to Meissner (*loc. cit.*), the consumption is 6 litres of liquid air plus 5 litres of liquid hydrogen per litre of liquid helium. It is also evidently a considerable advantage to be able to dispense with liquid hydrogen as a preliminary cooling agent. Theoretically it would be possible in our case also to dispense with liquid air, but the size of the liquefier would then be impracticably large. Using liquid hydrogen as a cooling agent, the output of the liquefier could be increased about six times.

The same liquefier has also been used for liquefying hydrogen, which was passed through a special circuit under a pressure of a few atmospheres.

A detailed description of the apparatus will shortly be published elsewhere.

(**133**, 708-709; 1934)

P. Kapitza: F.R.S., Royal Society Mond Laboratory, Cambridge.

References:

1. "Handbuch der Physik." Geiger and Scheel, vol. 11, p. 328.

2. *Z. Phys.*, **81**, 816; 1933.

旋管冷却到 8 K；最终的液化是利用焦耳－汤姆孙效应实现的。这种组合方式被证明是最有效的液化方法。液氦通过阀门从液化器的底部抽取出来。

在预冷到液氮温度之后，液化器在 45 分钟后开始以每小时 1 升的速率液化氦，每升液氦约消耗 3 升液态空气。我们期望产量很快会得到提高，不过即使是现在这样的水平，与原有的生产液氦的方法相比也是非常有优势的，按迈斯纳（在上引文中）的计算，原有方法每制备 1 升液氦要消耗 6 升液态空气加 5 升液氢。显然，新方法不使用液氢作预冷剂是一个非常大的优势。理论上讲，我们的方法中也可以不用液态空气，不过这样会使液化器的尺寸大得无法实现。如果用液氢作为冷却剂，液化器的产量大约可以增加为原来的六倍。

同样的液化器也已被用于氢的液化，其中氢气会在几个大气压下通过一个特殊管路。

有关装置的详细描述将于近期发表在其他文章中。

（沈乃澂 翻译；陶宏杰 审稿）

Mass of the Neutron

I. Curie and F. Joliot

Editor's Note

James Chadwick had estimated the neutron mass by analysing a reaction in which an alpha particle stimulates the transformation of boron into nitrogen with emission of a neutron. His calculations required the experimental values for the masses of several nuclei. Here wife-and-husband physicists Irène Curie and Frédéric Joliot report a more accurate and higher value for the neutron mass by examining other reactions, and without using any nuclear mass other than that of the proton. This higher value suggested that one might interpret nuclear beta decays as being triggered by the decay of a neutron into a proton within the nucleus, with the ejection of an electron and a neutrino, which today remains the modern view.

THE mass of the neutron has been calculated by Chadwick on the assumption that the neutrons of boron are emitted by the isotope ${}^{11}_{5}B$, according to the nuclear reaction

$${}^{11}_{5}B + {}^{4}_{2}He = {}^{14}_{7}N + {}^{1}_{0}n$$

Using the exact masses of ${}^{11}_{5}B$, ${}^{4}_{2}He$ and ${}^{14}_{7}N$ and the maximum energy of the neutron excited by the α-rays of polonium, one may calculate for the neutron a mass 1.0068 (taking ${}^{16}O = 16$).[1]

We have suggested[2] that the emission of the neutron of boron is due to the isotope ${}^{10}_{5}B$ and not to ${}^{11}_{5}B$. The nucleus ${}^{10}_{5}B$ can suffer two kinds of transmutation under the action of the α-particles of polonium, one with the emission of a proton, one with the emission of a neutron and a positive electron, according to the equations:

$${}^{10}_{5}B + {}^{4}_{2}He = {}^{13}_{6}C + {}^{1}_{1}H$$
$${}^{10}_{5}B + {}^{4}_{2}He = {}^{13}_{6}C + {}^{1}_{0}n + \overset{+}{\varepsilon}.$$

Our latest experiments on the creation of new radio-elements have confirmed our interpretation of the transmutation of boron. Similar reactions are observed with the nucleus ${}^{27}_{13}Al$ and with ${}^{24}_{12}Mg$. The reactions can be divided in two steps:

$${}^{10}_{5}B + {}^{4}_{2}He = {}^{13}_{7}N + {}^{1}_{0}n \qquad {}^{13}_{7}N = {}^{13}_{6}C + \overset{+}{\varepsilon}$$
$${}^{24}_{12}Mg + {}^{4}_{2}He = {}^{27}_{14}Si + {}^{1}_{0}n \qquad {}^{27}_{14}Si = {}^{27}_{13}Al + \overset{+}{\varepsilon}$$
$${}^{27}_{13}Al + {}^{4}_{2}He = {}^{30}_{15}P + {}^{1}_{0}n \qquad {}^{30}_{15}P = {}^{30}_{14}Si + \overset{+}{\varepsilon}$$

${}^{13}_{7}N$, ${}^{27}_{14}Si$, ${}^{30}_{15}P$ being unstable nuclei that disintegrate with the emission of positrons.

中子的质量

居里，约里奥

编者按

詹姆斯·查德威克通过分析α粒子引发的硼转变为氮并放出一个中子的反应，估算出了中子的质量。他的计算用到了几种原子核质量的实验值。在这篇文章中，物理学家弗雷德里克·约里奥和伊雷娜·居里夫妇通过对其他反应的研究给出了一个数值更为精确并且较大的中子质量值，并且他们的计算仅使用了质子的质量而没有用到其他任何原子核的质量。根据这一较大的中子质量数值，我们可以推测原子核的β衰变是由原子核中一个中子衰变为一个质子所引发的，同时放射出一个电子和一个中微子，这一观点一直沿用至今。

查德威克计算了中子的质量。在计算中，他假设硼放出的中子是由同位素 ${}^{11}_{5}B$ 放出的，对应于核反应：

$$^{11}_{5}B + {}^{4}_{2}He = {}^{14}_{7}N + {}^{1}_{0}n$$

利用 ${}^{11}_{5}B$、${}^{4}_{2}He$ 和 ${}^{14}_{7}N$ 的精确质量，以及钋放射出的α射线所激发出的中子最大能量，可以计算出中子的质量是 1.0068（取 ${}^{16}O = 16$）[1]。

但我们认为[2]，放射出中子的是同位素 ${}^{10}_{5}B$，而不是 ${}^{11}_{5}B$。${}^{10}_{5}B$ 的原子核在钋放射出的α粒子的作用下，可以发生两种嬗变，其中一种放射出一个质子，另一种放射出一个中子和一个正电子，核反应方程分别为：

$$^{10}_{5}B + {}^{4}_{2}He = {}^{13}_{6}C + {}^{1}_{1}H$$
$$^{10}_{5}B + {}^{4}_{2}He = {}^{13}_{6}C + {}^{1}_{0}n + \overset{+}{\varepsilon}$$

最近我们关于产生新的放射性元素的实验证实了我们对硼嬗变的解释。在 ${}^{27}_{13}Al$ 和 ${}^{27}_{12}Mg$ 中也观察到了类似的反应。反应可以分成两步：

$$^{10}_{5}B + {}^{4}_{2}He = {}^{13}_{7}N + {}^{1}_{0}n \qquad {}^{13}_{7}N = {}^{13}_{6}C + \overset{+}{\varepsilon}$$
$$^{24}_{12}Mg + {}^{4}_{2}He = {}^{27}_{14}Si + {}^{1}_{0}n \qquad {}^{27}_{14}Si = {}^{27}_{13}Al + \overset{+}{\varepsilon}$$
$$^{27}_{13}Al + {}^{4}_{2}He = {}^{30}_{15}P + {}^{1}_{0}n \qquad {}^{30}_{15}P = {}^{30}_{14}Si + \overset{+}{\varepsilon}$$

其中，${}^{13}_{7}N$、${}^{27}_{14}Si$、${}^{30}_{15}P$ 是不稳定的原子核，它们会发生衰变，同时放出正电子。

The complete reactions, with the masses and energy of all the particles are, for the two modes of transmutation of boron:

$$^{10}_{5}B + ^{4}_{2}He + W_\alpha = ^{13}_{6}C + ^{1}_{1}H + W_H + W_R,$$
$$^{10}_{5}B + ^{4}_{2}He + W_\alpha = ^{13}_{6}C + ^{1}_{0}n + \overset{+}{\varepsilon} + W_n + W_\varepsilon + W_R'$$

where W_α, W_H, W_n, W_ε, W_R, W_R' are the energies of the α-particle and the corresponding energies of the ejected particles and of the recoil atoms in the reactions. Subtracting the first of these equations from the second gives:

$$^{1}_{0}n = \text{mass of proton} - \text{mass of positron} + Q,$$

where $Q = W_H + W_R - W_n - W_R' - W_\varepsilon$.

One gets exactly the same equation using the transmutations of aluminium and magnesium.

Thus these equations enable us to calculate the mass of neutron without using the exact masses of any nucleus, except the proton.

According to our most recent measurements, the positrons emitted by the new radio-elements form a continuous spectrum of maximum energy 1.5×10^6 e.v. for $^{13}_{7}N$, 3×10^6 e.v. for $^{30}_{15}P$ and approximately 1.5×10^6 e.v. for $^{27}_{14}Si$. The emission of positrons is probably accompanied by the emission of neutrinos, but if the positrons have their maximum energy, the neutrinos will have a very small energy; the most recent hypotheses on the nature of this particle admits of a mass which is zero, or very small. So we need not take this particle into account in the calculations. The energy of the recoil atom in the disintegration with emission of a positron is negligible.

For the irradiation with the α-rays of polonium we have the following numerical values for the energies (expressed in 10^6 e.v.). One gets for the mass of neutron three values: 1.0098, 1.0092, 1.0089. These values agree approximately. Yet the first, deduced from boron, is the most precise. The energies of the neutrons of aluminium and magnesium and the energy of the positrons of magnesium are not well known.

	W_H	W_R	W_n	W_R'	W_ε	Q (10^6 e.v.)	Q in units of mass
B	8.05	0.23	3.3	0.59	1.5	+2.89	0.0031
Al	7.56	0.11	2	0.33	3.0	+2.34	0.0025
Mg	4.82*	0.21	1	0.48	1.5	+2.05	0.0022

*The maximum energy possible for the positrons does not correspond to a group effectively observed, but has been deduced by F. Perrin from the experiments of Bothe and Klarman, by the consideration of the energy balance relative to the groups of protons.

硼嬗变的两种模式的完整的反应方程式如下，其中包含了所有粒子的能量和质量：

$$^{10}_{5}B + ^{4}_{2}He + W_\alpha = ^{13}_{6}C + ^{1}_{1}H + W_H + W_R$$
$$^{10}_{5}B + ^{4}_{2}He + W_\alpha = ^{13}_{6}C + ^{1}_{0}n + \overset{+}{\varepsilon} + W_n + W_\varepsilon + W_R'$$

其中，W_α、W_H、W_n、W_ε、W_R、W_R' 分别是反应中 α 粒子的能量、放出粒子的相应能量，以及反冲原子的相应能量。用第二个反应方程式减去第一个反应方程式，可以得到：

$$^{1}_{0}n = 质子的质量 - 正电子的质量 + Q,$$

其中，$Q = W_H + W_R - W_n - W_R' - W_\varepsilon$。

利用铝和镁的嬗变反应方程式，也可以得到完全相同的方程。

这样，我们不必用到除质子以外的其他任何原子核的质量，就可以计算出中子的质量。

根据我们最近的测量，新形成的放射性元素放出的正电子的能量分布为连续谱，对 $^{13}_{7}N$、$^{30}_{15}P$ 和 $^{27}_{14}Si$ 而言，连续谱的最大能量分别是：1.5×10^6 eV、3×10^6 eV 和大约 1.5×10^6 eV。放出正电子时大多伴随着中微子的放出，但是如果正电子有其最大能量，那么中微子的能量将非常小。而且最近关于中微子性质的假说认为中微子的质量为 0，或者非常小。因此，在计算中我们不需要考虑中微子。在放出正电子的衰变过程中，反冲原子的能量也可以忽略不计。

我们得到钋放射的 α 射线的能量数值（单位是 10^6 eV），如下表所示。由此，我们分别得到三个中子质量：1.0098、1.0092、1.0089。这些值基本一致。但是第一个值最精确，即从硼原子推导出的中子质量。镁和铝中的中子能量以及镁中正电子的能量都还不是十分确定。

	W_H	W_R	W_n	W_R'	W_ε	Q（10^6 eV）	Q 以质量为单位
B	8.05	0.23	3.3	0.59	1.5	+2.89	0.0031
Al	7.56	0.11	2	0.33	3.0	+2.34	0.0025
Mg	4.82*	0.21	1	0.48	1.5	+2.05	0.0022

* 这个正电子能量的最大可能值与实际观测并不相符，但是通过考虑与质子组的能量平衡，佩兰已经从博特和卡拉曼的实验中推导出了这个最大值。

From considerations on the stability of the nucleus $^{9}_{4}$Be, the mass of the neutron should have a minimum value 1.0107. But an error of 0.001 in the determination of the mass of Be seems quite possible.

We may adopt for the mass of the neutron a value 1.010, in which the error probably does not exceed 0.0005.

With the mass 1.010 for the neutron, the maximum energy of the neutron ejected from beryllium by α-particles from polonium should be about 9×10^{6} e.v. The emission of slow neutrons when lithium is bombarded with α-particles from polonium, according to the reaction $^{7}_{3}\text{Li} + ^{4}_{2}\text{He} = ^{10}_{5}\text{B} + ^{1}_{0}n$, cannot be explained unless the mass adopted for $^{10}_{5}$B is too great, namely, by about 0.003.

If atomic nuclei contain only protons and neutrons, then the β-emission might be the consequence of the transformation of a neutron into a proton inside the nucleus, with the ejection of the negative electron and a neutrino, as has been suggested by several authors. The inverse processes would also be possible: transformation of a proton into a neutron with the ejection of a positron and a neutrino.

With the mass 1.010 for the neutron, the energy liberated in the transformation neutron → proton + $\bar{\varepsilon}$ is 2.1×10^{6} e.v.; the energy absorbed in the transformation proton → neutron + $\overset{+}{\varepsilon}$ is 3.1×10^{6} e.v.

(**133**, 721; 1934)

I. Curie and F. Joliot: Institut du Radium, Paris, 5.

References:

1. Chadwick, *Proc. Roy. Soc.*, 136, 692; 1932.
2. I. Curie and F. Joliot, *C.R.*, 197, 237; 1933.

考虑到 $^{9}_{4}Be$ 原子核的稳定性，中子的质量本应该有一个最小值 1.0107。但是在确定 Be 的质量时很可能存在 0.001 的误差。

我们可以取中子质量为 1.010，这个值的误差大概不会超过 0.0005。

取中子质量为 1.010 时，铍被钋放出的 α 粒子轰击所放出的中子的最大能量约为 9×10^6 eV。根据反应 $^{7}_{3}Li + ^{4}_{2}He = ^{10}_{5}B + ^{1}_{0}n$，除非认为我们采用的 $^{10}_{5}B$ 的质量太大，超过真实值 0.003，否则就无法解释当锂被钋放出的 α 粒子轰击时，会放射出慢中子的事实。

如一些作者所设想的那样，如果原子核只包含质子和中子，那么当核内部一个中子转变成一个质子，并放射出负电子和中微子的反应可能就是 β 发射产生的原因。相反的过程也有可能发生：一个质子转变成一个中子，并放出一个正电子和一个中微子。

取中子质量为 1.010 时，一个中子转变为一个质子和一个负电子的反应所释放的能量是 2.1×10^6 eV；一个质子转变成一个中子和一个正电子的反应所吸收的能量是 3.1×10^6 eV。

（王静 翻译；王乃彦 审稿）

The Explanation of Supraconductivity

J. Frenkel

Editor's Note

Superconductivity involves a transformation to a state of zero electrical resistance, apparently equivalent to an infinite conductivity. But here Russian physicist Yakov Frenkel suggests instead considering that the dielectric constant—a measure of a material's response to electric fields, due to polarisation—becomes infinite. While ordinary electrical current involves electrons moving freely through the material, a "polarisation current" might arise if the polarisation of atoms created by an applied electric field became infinite. Then, electrons could hop between atoms, creating coherent motion of one-dimensional chains of electrons sliding over the atomic lattice. Although Frenkel's contribution did not explain superconductivity, it did further an ultimately fruitful idea: that the phenomenon is somehow linked to coherent motion of charge carriers.

IT is customary to describe the supraconductive state of a metal by setting its specific electric conductivity σ equal to infinity. I wish to direct attention to another possibility, namely, that the supraconductive state can be described much more adequately by setting equal to infinity the *dielectric constant* ε of the substance, its conductivity σ remaining finite or even becoming equal to zero.

The actual meaning of the new definition can be seen from a comparison of the mechanism of ordinary electric conduction (σ finite) and ordinary polarisation (ε finite). In the former case the electrons called "free" move *independently*, the conduction current being constituted by a drift motion due to the action of an external electric field and superposed on the unperturbed random motion of the individual electrons. In the second case the electrons called "bound" are displaced by the electric field simultaneously in the same direction, the polarisation current being due to an orderly collective motion of all the electrons. Under normal conditions the displacement of the electrons with regard to the respective atoms remains small compared with the interatomic distances; this corresponds to a finite value of the dielectric constant. The assumption that the latter becomes infinite means that under the action of an infinitesimal field the electrons are displaced simultaneously over finite distances, each of them passing successively from an atom to the next one, like a chain gliding over a toothed track.

Such a collective motion of the "bound" electrons will constitute an electric current just as much as the individual motion of the free electrons, but a *polarisation* current rather than a *conduction* one. The electrostatic mutual action of the electrons moving collectively in a chain-like way will stabilise them against the perturbing action of the heat motion of the crystal lattice, which will result in the permanence of the polarisation current

超导电性的解释

弗伦克尔

编者按

超导电性涉及一个电阻突降为零的转变过程，而电阻为零似乎意味着一个无穷大的电导。但是在这篇文章中，俄国物理学家雅科夫·弗伦克尔则认为应该将介电常数视为无穷大，介电常数是表征材料由于极化引起的对电场响应程度的物理参数。虽然通常电流的形成就是电子自由通过材料的过程，但是如果由于电场导致的原子极化变为无穷大的话，也会产生“极化电流”。这样，电子可以在原子间跳跃，产生电子的一维链在原子晶格上滑移的相干运动。尽管弗伦克尔的这篇文章并没有解释清楚超导电性，但它确实促成了一个全新的想法：这一现象与载流子的相干运动存在某种联系。

在描述金属的超导状态时，我们习惯于将金属的电导率 σ 定义为无穷大。我希望能够引导大家关注另外一种可能性，而这有可能是对超导状态更加合适的描述，即设物质的**介电常数** ε 为无穷大，而电导率 σ 保持为有限值，甚至等于零。

这个新定义的真正含义可以从普通电传导（σ 为有限值）与普通电极化（ε 为有限值）机制的比较中看出。就前者而言，电子可被看成是在**独立地**“自由”运动，传导电流是由外电场作用引起的漂移运动和单个电子无干扰的随机运动叠加而形成的。而对后者而言，受“束缚”的电子在电场的作用下沿同一方向同时产生位移，极化电流是由所有电子有序的集体运动产生的。一般情况下，电子相对于各自所属原子的位移比原子间的距离要小；这对应于介电常数是有限值的情形。如果我们假设介电常数的值为无穷大，那就代表，在无穷小电场的作用下，所有电子会同时移动一段有限的距离，每个电子连续地从一个原子穿越至下一个原子，就像一根链子在锯齿形的轨道上滑动一般。

“束缚”电子的这类集体运动所形成的电流与自由电子的单个运动形成的电流大小相等，但它是**极化**电流，而不是**传导**电流。以类链方式集体运动的电子之间的静电相互作用会克服由晶格热运动引起的扰动，从而使电子的这种运动稳定下来，这

after the disappearance of the electric field by which it was started[1]. This permanence, which has been erroneously interpreted as corresponding to an infinite value of the specific conductivity, must be interpreted in reality as corresponding to an *infinite value of the dielectric constant.* Now, how is it possible to explain the occurrence of such an infinite value? This turns out to be a very simple matter, the appropriate mechanism having been considered already by Hertzfeld, who, however, failed to give it the correct interpretation. Consider a chain of equally spaced atoms with a polarisation coefficient α. This means that an isolated atom assumes under the action of an external field E an electric moment $p = \alpha E$. If the field E is acting in the direction of the chain, then in computing the polarisation of a certain atom we must add to it the field E' produced by all the other atoms in virtue of their induced electric moments. All these moments being the same, we get

$$E' = \frac{2p}{a^3} 2 \sum_{n=1}^{\infty} \frac{1}{n^3} = 4.52 \frac{p}{a^3};$$

and consequently

$$p = \alpha \left(E + 4.52 \frac{p}{a^3} \right),$$

whence

$$p = \frac{\alpha E}{1 - 4.52 \alpha / a^3} = \alpha' E. \tag{1}$$

We thus see that with a finite value of α for an isolated atom, an infinite value of the effective polarisation coefficient α' for the atom-chain is obtained if

$$4.52 \alpha \geqq a^3. \tag{2}$$

The sign > corresponding to a negative value of α' need not be distinguished from the sign =; in both cases the atom chain is characterised by the instability of the electron chain connected with it. This instability, which has been noticed previously by Hertzfeld, was interpreted by him as an indication of the fact that the electrons no longer remain bound, but become free "conduction" electrons. Thus the inequality (2) was considered as characteristic of the metallic state in general. I believe that it is characteristic not of the metallic state but of the supraconductive state, a supraconductor being rather a dielectric with freely movable electron chains (that is, with $\varepsilon = \infty$) than a metal.

According to a theory of the metallic state developed in a rather qualitative way by Slater[2] and recently greatly improved and generalised by Schubin[3], the normal conductivity of a metal is due to a *partial ionisation* of the atoms, a certain fraction s of all the atoms becoming positive ions, and an equal portion (to which the corresponding electrons are attached) negative ions. If these electrons are bound very weakly, they may be considered

使得在引发了该运动的电场消失后，极化电流还将一直持续[1]。这种电流的持续性曾经被错误地解释为是特定情况下电导率无穷大所致，而实际上应该用**介电常数无穷大**来进行解释。那么现在如何来解释介电常数无穷大的出现呢？事实上很简单，赫茨菲尔德已经想到了合适的机制，但是他并没有给出正确的解释。下面我们考虑由极化系数为 α 的等间距的原子组成的链。于是，在外场 E 的作用下，单个原子的电偶极矩就为 $p = \alpha E$。如果电场 E 沿链的方向，则在计算某一个原子的极化时，我们必须加上由于其他所有原子的感生电偶极矩所产生的电场 E'。假设所有这些电偶极矩都相等，我们得到

$$E' = \frac{2p}{a^3} 2\sum_{n=1}^{\infty} \frac{1}{n^3} = 4.52 \frac{p}{a^3}$$

因此

$$p = \alpha\left(E + 4.52 \frac{p}{a^3}\right)$$

所以

$$p = \frac{\alpha E}{1 - 4.52\alpha / a^3} = \alpha' E \tag{1}$$

由此可见，对于 α 是一个有限值的单个原子来说，如果

$$4.52\alpha \geqq a^3 \tag{2}$$

我们得到原子链的有效极化系数 α' 的值为无穷大。

(2) 式中符号“>”相当于 α' 是负值，与符号“=”没有区别；在这两种情况下，原子链可以用与其相连的电子链的不稳定性来表征。赫茨菲尔德以前就注意到了这种不稳定性，他还把这种现象解释为是电子不再受到束缚，而成为自由“传导”电子这一事实的一个标志。因此不等式 (2) 一般被认为是金属态的特征。我相信这并不是金属态的特征，而是超导态的特征，超导体是一种具有可自由运动电子链（即 ε = ∞）的电介质，而非金属本身。

按照由斯莱特[2]发展出的定性的金属态理论，以及最近由舒宾[3]对该理论的改进和推广，金属的正常导电性是由于原子的**部分电离**引起的，某一部分原子 s 变成

as "free" in the usual sense of the word. The conductivity of a metal is equal to the sum of the conductivities due to these free electrons or negative ions on one hand and the positive ions or "holes" on the other. The mechanism of electrical conduction consists in the *individual* jumping of an electron from a negative ion to one of the neutral atoms surrounding it (which is thus converted into a negative ion), or from a neutral atom to a positive ion, which thus becomes a neutral atom, its rôle being switched over to the "donor". We meet with the same type of electric conduction in electronic semi-conductors[4]. The chief distinction between a metal and a semi-conductor consists in the fact that in the former case $s > 0$ at the absolute zero of temperature (T) whereas in the latter case $s = 0$ at $T = 0$, increasing according to the Boltzmann equation ($s = ce^{-W/kT}$ where W is the ionisation energy) with the temperature.

The elements which are likely to become supraconductors form an intermediary group in the sense that at ordinary temperatures they are relatively poor conductors, like the ordinary semi-conductors; the dependence of their conductivity on the temperature is, however, of the same character as that of typical metals (negative temperature coefficient). This means that in the case of these intermediary elements or "half-metals", we have to do with substances which are characterised by a practically constant value of the ionisation fraction s. Their small conductivity can be explained either by a small value of s or by a small mobility of the individual electrons (which seems the more probable alternative in view of the correlation between supraconductivity and the Hall effect discovered by Kikoin and Lasareff). The fact that, in ordinary circumstances, that is, above the "transition temperature" T_c, these substances are not supraconductive, can be explained by the finite value of their dielectric constant as determined by the polarisability of ions stripped of the conduction electrons. The nature of the transition which takes place when the temperature T is decreased below T_c can thus be very simply interpreted by assuming that, at this temperature, s suddenly falls from a certain rather high value to *zero*, and that the polarisation coefficient α of the resulting *normal* atoms with their full complement of bound electrons satisfies the inequality (2)*. The very fact that the substance loses its conductivity (σ falling to zero along with s) thus transforms it from a metal into a dielectric with $\varepsilon = \infty$, that is, it becomes a supraconductor.

Both the necessity and the sharpness of the transition $s \to 0$ (that is, $\sigma \to 0$ and $\varepsilon \to \infty$) can be easily understood if we assume that the state $s = 0$ has a smaller energy than the state $s > 0$. It results from Slater's and especially from Schubin's calculations that the *lowest* energy level for polar (ionic) states *may* correspond to a *finite* value of s, whether this lowest level lies below or above the energy level corresponding to $s = 0$. It can further easily be seen that the distance between the successive levels in a band of levels corresponding to a given value of s is very small compared with kT, even for extremely low temperatures (of the order of a few degrees K.). If, further, the *total* width of the band was also small

* This inequality is probably satisfied for all metals, although not all of them are supraconductors, because for true metal s remains finite (and practically constant) down to the absolute zero of temperature, while for supraconductors it jumps to a finite value slightly above it .

正离子，而相等的另一部分（携带相应电子的部分）变成负离子。如果这些电子受到的束缚很微弱，通常可以称其是“自由的”。金属的电导率等于由这些自由电子或负离子产生的电导率加上正离子或“空穴”产生的电导率之和。电传导的机制主要是一个电子从一个负离子**单个地**跳到附近的一个中性原子上（于是这个中性原子就转换成一个负离子），或从一个中性原子跳到一个正离子上，于是正离子就成为中性原子，充当“施主”的角色。半导体[4]中具有相同类型的电传导机制。金属和半导体的主要区别在于：前者在温度（T）达绝对零度时，$s > 0$；而后者在 $T = 0$ 时，$s = 0$，且按照玻尔兹曼方程（$s = ce^{-W/kT}$，其中 W 是电离能）s 随着温度的上升而增大。

从某种意义上来说，能够成为超导体的元素形成了一个过渡状态的元素组，常温下，它们是传导性能相对比较差的导体，类似于普通的半导体；然而它们的电导率与温度的变化关系又和典型的金属（负温度系数）相同。当考虑这些过渡状态的元素或者“半金属”元素时，我们必然会遇到一些电离分数 s 实际为常数的物质。它们的弱导电性可以用很小的 s 值来解释，也可以用单个电子很小的迁移率来解释（考虑到基科因和拉萨雷夫共同发现的霍尔效应与超导电性之间的关系，这似乎是一个更合适的选择）。在通常情况下，也就是温度在“转变温度”T_c 以上时，这些物质不再是超导体，这一事实可以利用它们的介电常数为有限值来解释，这个有限值是由失去了传导电子的离子的极化率来决定的。对于当温度由 T 降至 T_c 之下时发生的相变的本质，我们可以简单地用以下两点假设进行解释：第一，在这个温度下，s 突然由某一相当高的值突变为**零**；第二，相应的带着所有束缚电子的**普通**原子的极化系数 α 满足不等式 (2)*。物质失去导电性（σ 与 s 变为零），于是由金属转变为 $\varepsilon = \infty$ 的介质，也就是变成了超导体。

如果我们假设 $s = 0$ 的态比 $s > 0$ 的态所具有的能量小，那么 $s \to 0$ 转变（即 $\sigma \to 0$ 和 $\varepsilon \to \infty$）所具有的必然性和陡峭特点就很好解释了。这样的结论可以从斯莱特尤其是从舒宾的计算中得到，这个计算显示：无论**最低**能级处于 $s = 0$ 对应能级之上或之下，极化（离子）态的最低能级都**可能**对应于**有限**的 s 值。从而我们很容易看出，即使在极低的温度下（几 K 量级），给定 s 值对应能带中连续能级的间距远小于 kT。如果能带的**总**带宽也小于 kT，那么就可以按式子 $k\lg g$ 来计算 $s > 0$

* 这个不等式几乎对所有金属来说都是成立的，尽管并不是所有的金属都是超导体，因为对于真正的金属而言，在温度下降到绝对零度的过程中，s 将保持一个有限值（实际上是常数），然而对超导体而言，它会跳跃成为比先前略大的一个有限值。

compared with kT, the entropy of the state $s > 0$ could be calculated as $k \lg g$, where g is the statistical weight of the whole band, that is, the number of ways in which the state s is realised. Taking all possible distributions of the ns electrons (negative ions) and ns positive holes (positive ions) between the n atoms, we get

$$g = \left[\frac{n!}{(ns)!\,(n - ns)!}\right]^2.$$

The transition $0 \rightarrow s$ is thus connected with an *increase* of entropy

$$\Delta\eta = 2k\,[n \lg n - ns \lg ns - (n - ns) \lg (n - ns)]. \tag{3}$$

In reality, the width of a band is of the order of 1 volt and therefore at least a thousand times larger than kT at the transition point. This will result in a much smaller entropy increase $\Delta\eta$.

So long, however, as $\Delta\eta > 0$ it follows that the state $s = 0$ must be stable at low temperatures and the state $s > 0$ at higher ones.

The transition temperature T_c as determined by the equality of the free energies of the two states is given by

$$T_c = \frac{\Delta\varepsilon}{\Delta\eta}\,(\Delta\varepsilon = \varepsilon_s - \varepsilon_0). \tag{4}$$

Taking $s = \frac{1}{2}$ (which is probably an exaggeration) and calculating $\Delta\eta$ with the help of (3), we get $\Delta\eta = 1.7k_n$. If $T = 4°$ (say) the transition energy $\Delta\varepsilon$ should be of the order of 14 small calories per gram atom. This value is greatly reduced if the width of the energy band under consideration is large compared with nT, its effective weight being accordingly small compared with g.

We thus see that the second condition for supraconductivity is expressed by the inequality $\varepsilon_s > \varepsilon_0$ at $T = 0$. But this is not all. Equation (1) is a good approximation so long as the chain-like displacement of the electrons x is small compared with the interatomic distance a. When x approaches $\frac{1}{2}a$, the electrons are pushed back by a force which varies more rapidly than the first power of x and can be overcome through the quantum mechanism of the tunnel effect. If a large number of electrons N are moving together in a chain-like way, they behave like a particle with an N-fold mass, the transition probability being correspondingly reduced. Now in his second theory of supraconductivity, Kronig[5] has shown that a chain or, as he puts it, a "linear lattice", of electrons, bound to each other in a quasi-elastic way, can be displaced through a periodic field of force (with a period a equal to the average spacing between the electrons) under the condition

$$h/b\sqrt{m} > a^2, \tag{5}$$

的态的熵，其中 g 是整个能带的统计权重，即得到态 s 所有方式的数目。考虑 n 个原子间 ns 个电子（负离子）和 ns 个正空穴（正离子）所有可能的分布，我们得到：

$$g = \left[\frac{n!}{(ns)!(n-ns)!}\right]^2$$

因此，与 $0 \rightarrow s$ 转变相应的熵**增加**是

$$\Delta\eta = 2k\left[n\lg n - ns\lg ns - (n-ns)\lg(n-ns)\right] \tag{3}$$

实际上，带宽的数量级是 1 电子伏，因此在温度处于转变点时，带宽至少是 kT 的一千倍。这将得到一个非常小的熵的增加量$\Delta\eta$。

然而，只要$\Delta\eta > 0$，$s = 0$ 的态在低温下一定是稳定的，$s > 0$ 的态在更高的温度下也一定是稳定的。

转变温度 T_c 由两个态的自由能相等决定，由下式给出：

$$T_c = \frac{\Delta\varepsilon}{\Delta\eta}(\Delta\varepsilon = \varepsilon_s - \varepsilon_0) \tag{4}$$

取 $s = \frac{1}{2}$（它可能是一个夸大值），根据(3)式计算$\Delta\eta$，我们得到$\Delta\eta = 1.7k_n$。假设 $T = 4$ K，则转变能 $\Delta\varepsilon$ 应是每克原子 14 卡的量级。如果在考虑的范围内能带的带宽比 nT 大，那么这个值会大幅度降低，其有效权重会相应地比 g 小。

于是，我们发现，超导电性的第二个条件应该表示为在 $T = 0$ 时满足不等式 $\varepsilon_s > \varepsilon_0$。但这还不是全部。只要电子的类链位移 x 小于原子间距 a，(1) 式就是一个很好的近似。当 x 约等于 $\frac{1}{2}a$ 时，电子则被一个力推回去，这个力比 x 的一次幂变化更快，并能被量子机制的隧道效应所克服。如果大量电子 N 以类链方式整体运动，它们就犹如具有 N 倍质量的一个粒子一样，转变概率将显著下降。克罗尼格[5]在他的第二个超导理论中已指出，电子链（他所说的“线形晶格”）以准弹性方式彼此束缚，在满足条件（5）式的情况下可以通过力的周期场（周期 a 等于电子之间的平均距离）。

$$h/b\sqrt{m} > a^2 \tag{5}$$

where h is Planck's constant, m the mass of an electron and b is the rigidity coefficient of the "electron lattice". Putting $b = \tau e / a^{3/2}$ where τ is a numerical coefficient of the order 1, Kronig finds that the condition (5) is fulfilled if a is of the order of less than a few Ångström units. This seems to show that a "linear lattice", that is, chain of electrons, is practically *always movable* with respect to the corresponding chain of atoms, provided the condition (2), which is much more restrictive, is fulfilled also. In fact, the latter condition seems to be the mathematical formulation of the possibility of treating the (bound) electrons as a kind of lattice. I do not believe in the reality of the three-dimensional lattices postulated by Kronig in his first paper. He has himself shown that such lattices, even if they exist, could not be moved through the ionic lattice. As a matter of fact, one-dimensional lattices or rather movable *chains* of bound electrons fully suffice for the explanation of supraconductivity. Such chains need not be movable in all directions. It is sufficient to assume that they should be movable in one particular crystallographic direction corresponding to the smallest spacing between the atoms, the dielectric constant being infinite for this direction and preserving a finite value for all the others.

In spite of its shortcomings, Kronig's theory is certainly the nearest approach to the correct explanation of supraconductivity published hitherto, the present theory differing from it more in form than in essence. The theory I advanced before, which was based on the supposed stabilisation of the free electrons (against heat motion) by their *electro-magnetic* action, was wholly erroneous in this particular respect. It was correct, however, in describing the motion of the electrons in the supraconductive state as an organised "collective" motion. This led to the result that a metal must possess when in this state an enormous diamagnetic susceptibility. This corollary subsists in the new theory and is corroborated by the fact recently discovered by Meissner that the magnetic permeability μ of a metal in the supraconducting state drops to zero. A supraconductor can thus be described as a body with $\mu = 0$ and $\varepsilon = \infty$, its electrical conductivity σ in the exact sense of the word being either finite or even zero.

A more complete account of the present theory will be published elsewhere.

(**133**, 730-731; 1934)

References:

1. The effects of heat motion of the crystal lattice on the individual electrons are mutually cancelled. Cf. R. Kronig, *Z. Phys.*, **80**, 203; 1933.

2. *Phys. Rev.*, **35**, 509; 1930.

3. In the press.

4. Cf. J. Frenkel, *Nature*, **132**, 312, Aug. 26, 1933.

5. *Z. Phys.*, **80**, 203; 933.

式中，h 是普朗克常数，m 是电子质量，b 是“电子晶格”的刚性系数。令 $b = \tau e/a^{3/2}$，其中 τ 是一个量级为 1 的系数，克罗尼格发现，如果 a 的量级小于几埃，则条件 (5) 式是满足的。这似乎表明，当限制性更强的条件 (2) 式也满足时，“线性晶格”即电子链相对于相应的原子链实际上**总是可运动的**。实际上，后一条件似乎是能否将（束缚）电子当作一种晶格来处理的可能性的数学描述。我并不认为，克罗尼格在他的第一篇文章中假设的三维晶格是真实存在的。他自己也表明，这类晶格即使存在也不能穿过离子晶格。事实上，一维晶格或可运动的束缚电子**链**完全可以解释超导电性。这类链并不需要在所有方向上都能运动。它们只要做如下的假设就足够了，即它们在一个特殊的晶体学方向上应该是可运动的，这个特殊方向对应于原子间的距离最小，而且介电常数在这个方向上为无穷大，而在其他方向上总保持一个有限值。

克罗尼格的理论虽然存在不足，但它确实是迄今发表的对超导电性最接近正确的解释，我的理论与其的差别更多是在形式上，而非本质上的。我以前提出的基于假设在**电磁**作用下自由电子（相对于热运动）的稳定性的理论在这个特定方面是完全错误的。但是，在将超导态中电子的运动描述为有组织的“集体”运动这点是正确的。这就要求金属在这种态下必须具有巨大的抗磁性的磁化率。这个推论在新理论中成立，而且被迈斯纳最近发现的事实所证实，迈斯纳的实验显示金属的磁导率 μ 在超导态中会下降到零。因此超导体可以被表述为 $\mu = 0$ 和 $\varepsilon = \infty$ 的物体，更确切地说，其电导率 σ 或者是一个有限值，或者就是零。

这一理论更完整的叙述将在其他地方发表。

（沈乃澂 翻译；郑东宁 审稿）

Modern Ideas on Nuclear Constitution

G. Gamow

Editor's Notes

George Gamow here reviews advances in the theory of nuclear structure. With the discovery of the neutron, physicists now considered stable nuclei to be composed of neutrons and protons. As Gamow points out, Heisenberg's theory of nuclear structure suggested that small nuclei should have roughly equal numbers of the two particles, and that an imbalance toward more neutrons should grow with increasing atomic mass, thereby counteracting the electrical repulsion between protons. Gamow considers the stability of nuclei to both beta and alpha decay, which could be understood only partially in current theory. The approximate proportionality of nuclear radius to the cube root of the atomic mass suggested that the density of the nucleus remains roughly constant.

WHEN the complexity of atomic nuclei was proved by the existence of spontaneous and artificial nuclear transformations, a very important question arose: From which of the elementary particles are the different nuclei built up? It seemed that this question could be simply answered as there were only two particles with pretensions to be elementary: the proton and the electron. The protons had to account for the main part of the nuclear mass and the electrons had to be introduced to reduce the positive charge to the observed value. For example, the nucleus of bismuth, with atomic weight 209 and atomic number 83, was considered to be constructed from 209 protons and $209 - 83 = 126$ electrons. It was also accepted as very probable that these elementary particles build up inside the nucleus certain complex units constructed from four protons and two electrons each (α-particles). All this construction was in good agreement with the experimental evidence, as electrons, protons and α-particles were really observed being emitted in nuclear transformations.

The theory treating the nuclei as constructed of α-particles, some protons and a certain number of electrons, was worked out by Gamow. Although this theory gave some interesting results as to the general shape of the mass-defect curve and the conditions of emission of α-particles, it met with serious difficulties. It was very difficult to understand, on the basis of the quantum theory of the electron, how the electron can exist in a space so small as that limited by the nuclear radius. It was also not clear why the nuclear electrons, behaving in quite a strange and obscure way, do not affect the processes of emission of the heavy nuclear particles, protons and α-particles.

About two years ago, it was shown by Chadwick that the experimental evidence forces us to recognise the existence of a new kind of particle, the so-called neutron, also with claims to be held to play an important rôle in nuclear structure. The discovery of neutrons

核组成的现代思想

伽莫夫

编者按

在这篇文章中，乔治·伽莫夫回顾了核结构理论的发展状况。随着中子的发现，现在物理学家认为稳定的原子核由质子和中子构成。正如伽莫夫指出的那样，海森堡的核结构理论认为小的原子核应该含有大体相等数量的两种粒子，但随着原子质量的增加平衡会被打破，中子的数目会变得更多，这会抵消质子间的静电斥力。伽莫夫认为当前的理论只能部分解释原子核相对α衰变和β衰变的稳定性。原子核半径与原子质量的立方根粗略成正比表明原子核的密度大致保持不变。

当自发和人工的核转变的存在证明了原子核的复杂性时，一个很重要的问题出现了：不同的原子核是由哪些基本粒子组成的？这个问题似乎可以简单地回答，因为自命是基本粒子的只有两类：质子和电子。原子核的质量主要是由质子贡献的，而电子的引入是为了减少正电荷以便符合实验的观测值。例如，原子量 209、原子序数 83 的铋原子核被认为是由 209 个质子和 209 − 83=126 个电子组成的。大家还认为这些基本粒子在原子核内构建某种由四个质子和两个电子组成的复合单元（α 粒子）也是很有可能的。所有这些结构都与实验结果吻合得很好，因为我们确实在核转变的实验中观测到了所放出的电子、质子和 α 粒子。

伽莫夫发展了一种理论，即原子核由 α 粒子、一些质子和一定数目的电子组成。虽然这个理论给出了一些有趣的结果，如原子核质量亏损曲线的一般形状和原子核发射 α 粒子的条件，但同时也遇到了严重的困难。根据电子的量子理论，我们难以理解电子如何能存在于原子核半径所限制的微小空间内。我们也不清楚原子核中电子的行为为何如此奇怪，且难以理解原子核的电子为何不会影响重核的粒子（质子和 α 粒子）的发射过程。

约两年前，查德威克指出，实验证据使我们不得不承认存在一类新的粒子，即所谓的中子，它们在原子核结构中扮演了重要角色。中子的发现大大简化了原子核内

considerably simplified the difficulties about electrons in nuclei. One could now suppose that the nuclei were completely constructed of neutrons and protons (for example, the nucleus of bismuth from 83 protons and 209 − 83=126 neutrons) which probably sometimes unite to form an α-particle (two neutrons and two protons). Thus the first of the above-mentioned difficulties was, so to say, hidden inside the neutron, while the second one was actually removed.

On the basis of these new ideas, Heisenberg succeeded in building up a general theory of nuclear structure, accounting for the main features of nuclear stability. The basis of his theory is certain assumptions about the forces acting between neutrons and protons. It seems most rational to accept the view that the interaction between particles of the same kind is only due to electric charges (that is, no forces between two neutrons and the usual Coulomb repulsion between two protons), while between two different particles (neutron and proton) strong exchange forces come into play. These last forces are probably of the same kind as the forces between atoms playing the main rôle in quantum chemistry, and may be considered as due to the exchange of charge between the two particles in question.

This hypothesis explains at once why the number of nuclear neutrons for heavy elements is considerably greater than the number of protons (that is, why the ratio of atomic weight to atomic number increases for heavier elements). In fact, if we neglect the coulomb forces, the most stable state of the nucleus will correspond to equal numbers of neutrons and protons, as in this case all the possibilities of binding (by attracting exchange forces), between protons and neutrons are saturated. The presence of the Coulomb repulsion between protons will, however, shift the optimum in the direction of a smaller number of protons and the position of lowest potential energy of our system will correspond to the larger proportion of neutrons. As the importance of the Coulomb forces increases with the nuclear charge, one can understand that an equal number of neutrons and protons is possible only for the lightest elements (first ten elements of the periodic system), while for heavier ones the number of neutrons predominates (126 neutrons and only 83 protons in bismuth).

Accepting the simplest form for the law of variation of the exchange forces with distance:—

$$I = a \cdot e^{-br} \tag{1}$$

and applying the quantum statistical method, Heisenberg calculated the behaviour of the nuclear model constructed from n_1 neutrons and n_2 protons. The result is that the particles are rather uniformly distributed inside a certain volume proportional to the total number of particles. This result fits very well with evidence otherwise obtained, that the density inside the nucleus is rather uniform and does not depend greatly on the atomic weight. The formula obtained for the total binding energy E of the nucleus as a function of n_1 and n_2 looks rather complicated and depends, of course, on the numerical values of the coefficients a and b in the expression (1) for the exchange force. Comparing this formula with experimental values of the mass defects of different nuclei, one can estimate the values of a and b. One finds thus: $a = 4.05 \times 10^{-5}$ erg; $b = 1.25 \times 10^{12}$ cm.$^{-1}$.

有关电子的困难。现在我们可以假设原子核完全是由质子和中子组成的（例如，铋元素的原子核是由 83 个质子和 209−83=126 个中子组成），而质子和中子有时也可能结合成一个 α 粒子（两个质子和两个中子）。因此，可以这么说，上面提到的第一个困难被隐藏在中子内部，而实际上第二个困难这时已经被排除。

根据这些新的思想，海森堡成功地建立了一个原子核结构的普遍理论，用来解释原子核稳定性的主要特征。他的理论基础是建立在某些关于质子和中子之间存在相互作用力的假设上的。看起来最合理且可接受的观点是，认为同类粒子之间的相互作用仅由电荷产生（即在两个中子之间不存在力的作用，而在两个质子之间通常有库仑排斥力），而在两类不同的粒子（中子和质子）之间有强的交换力起作用。这后一种力很可能与量子化学中起主要作用的原子间的作用力类同，可以认为是由相关的两个粒子之间电荷交换产生的。

以上的假设立即解释了为何重元素原子核中的中子数远大于其质子数（即为何原子量与原子序数之比对较重元素而言是逐渐增大的）。实际上，如果我们忽略库仑力，原子核最稳定的状态将相应于质子数与中子数相等的情况，在这种情况下，质子与中子之间所有可能的结合方式（通过吸引的交换力）趋于饱和。但是，由于质子之间存在库仑排斥现象使得原子核偏离最佳状态朝向较小的质子数，这时，系统的最低势能状态将对应于较大比例的中子数。由于库仑力的重要性随着核电荷的增大而增加，这时人们可以理解只有对那些最轻的元素（元素周期表中的前 10 个元素），同等数目的中子和质子才是可能的；而对于较重的元素，其中子数在数量上就会占优势（如铋元素由 126 个中子和 83 个质子组成）。

对交换力随距离变化的规律采用最简单的形式：

$$I = a \cdot e^{-br} \tag{1}$$

并应用量子统计方法，海森堡计算出了由 n_1 个中子和 n_2 个质子所构建的核模型的行为。结果是，在与粒子总数成正比的确定体积内，粒子的分布相当均匀。这项结果与从其他途径得到的结果是非常吻合的，即核内的密度分布相当均匀，且与原子量无太大关系。而以原子核总的结合能 E 作为 n_1 和 n_2 的函数得到的关系式似乎非常复杂，当然，E 与交换力表达式（1）中的系数 a 和 b 的数值也是有关的。将这个公式与不同原子核的质量亏损的实验值进行对照，我们可以估算出 a 和 b 的值，得到的结果是：$a = 4.05 \times 10^{-5}$ erg，$b = 1.25 \times 10^{12}$ cm^{-1}。

Very interesting consequences can also be obtained from Heisenberg's theory concerning the question of nuclear stability. It is easily understood that nuclei with a high positive electric charge must tend to emit positive particles. From the point of view of the energy balance, the most favourable case for such emission is the emission of an α-particle, as this removes from the nucleus a large amount of negative energy (the binding-energy of the α-particle itself), which is equivalent to the supply of an equal quantity of positive energy. The condition for the possibility of α-emission can be simply formulated if we consider it as a simultaneous subtraction of two neutrons and two protons from the nucleus in question. The work necessary for such subtraction is evidently

$$\frac{\delta E}{\delta n_1}\,\Delta n_1 + \frac{\delta E}{\delta n_2}\,\Delta n_2$$

or, as

$$\Delta n_1 = \Delta n_2 = -2,$$

$$-2\left(\frac{\delta E}{\delta n_1} + \frac{\delta E}{\delta n_2}\right).$$

To make a spontaneous α-disintegration possible, this quantity must be smaller than the above mentioned energy-supply due to the binding energy $\Delta M_\alpha c^2$ of the α-particle from neutrons and protons. (The difference appears as the kinetic energy of the emitted particle.) Thus the condition for α-decay will be:

$$-2\left(\frac{\delta E}{\delta n_1} + \frac{\delta E}{\delta n_2}\right) < \Delta M_\alpha c^2 \tag{2}$$

In Fig. 1, the ratio of the number of neutrons to the number of protons is plotted against the number of protons for all known isotopes. The α-stability curve as calculated from formula (2) is represented by a broken line (curve I) and one can see that it is situated too low. One notices, however, that the theoretical curve, apart from absolute values, gives a good idea of the general form of this stability limit. We may notice that the condition for the spontaneous emission of a proton:

$$-\frac{\delta E}{\delta n_2} < 0 \tag{2'}$$

will give us a stability limit located very far to the right of the α-stability curve, which means that spontaneous proton decay could only take place for very heavily charged nuclei (atomic number > 200). On the other hand, the condition for the emission of a neutron:

$$-\frac{\delta E}{\delta n_1} < 0 \tag{2''}$$

is never fulfilled, which can easily be understood if we remember that neutrons, having no charge, are not at all repelled by nuclei.

We must now turn our attention to the question of the emission of light particles. From the point of view of the neutron-proton model of the nucleus, we must accept the view that the process of ordinary β-emission is due to the transformation of a nuclear neutron into a proton with the liberation of negative charge in the form of an electron:

$$n \rightarrow p + \bar{e}.$$

从海森堡有关核稳定性问题的理论中，我们也能获得很有趣的结果。我们很容易理解具有高正电荷的原子核必然趋于发射出带正电的粒子。根据能量守恒的观点，对这类发射最有利的情况是发射一个 α 粒子，即从核中带走了大量的负能量（α 粒子自身的结合能），这相当于给原子核提供了同样多的正能量。如果把 α 粒子的发射认为是从相关的核内同时减去两个中子和两个质子，则 α 粒子发射的可能性条件可以被简单地表述出来。显然，这类过程扣除所需的能量为

$$\frac{\delta E}{\delta n_1}\Delta n_1+\frac{\delta E}{\delta n_2}\Delta n_2$$

或，当 $\Delta n_1 = \Delta n_2 = -2$时，

$$-2\left(\frac{\delta E}{\delta n_1}+\frac{\delta E}{\delta n_2}\right)$$

要使一个自发的 α 衰变成为可能，这个量必须小于上述由中子和质子形成 α 粒子时放出的结合能$\Delta M_\alpha c^2$。（两者的能量差将作为放出的粒子的动能。）因此，α 衰变的条件是：

$$-2\left(\frac{\delta E}{\delta n_1}+\frac{\delta E}{\delta n_2}\right)<\Delta M_\alpha c^2 \tag{2}$$

图 1 中画出了所有已知同位素的中子数和质子数的比值与其质子数的关系。根据公式 (2) 计算得到的 α 稳定性曲线用虚线（曲线 I）表示，从中可以看到，它所处的位置非常低。然而我们注意到，不考虑绝对值的大小，理论曲线还是让我们对 α 稳定性的界限有了很好的了解。我们可能会注意到，质子的自发发射条件为：

$$-\frac{\delta E}{\delta n_2}<0 \tag{2'}$$

这里给出的发射质子的稳定性界限远离 α 稳定性曲线的右边，这意味着只有很重的带电核(原子序数大于 200)才能发生自发的质子衰变。另一方面,对中子发射的条件：

$$-\frac{\delta E}{\delta n_1}<0 \tag{2''}$$

是永远不可能满足的，如果想到不带电荷的中子绝不被原子核排斥，我们将很容易理解这一点。

现在我们必须将注意力转回到轻粒子发射的问题上来。根据核的中子 – 质子模型，我们必须采纳以下观点，即通常的 β发射过程是由于核的中子以电子的形式释放出负电荷而转变为质子所引起的：

$$n\rightarrow p+\bar{e}$$

On the other hand, the recent discovery of the Joliots of elements emitting positive electrons suggests the possibility of the reverse process:

$$p \rightarrow n + \overset{+}{e}.$$

We can easily estimate the stability limits for such processes if we consider the emission of a nuclear electron as the subtraction from the nucleus of a neutron and simultaneous addition of a proton. The condition for the positive energy balance of such a transformation will evidently be:

$$-\frac{\delta E}{\delta n_1} + \frac{\delta E}{\delta n_2} < \Delta M_n \cdot c^2, \tag{3}$$

where ΔM_n is the mass defect of a neutron as constructed from a proton and an electron. In an exactly analogous way we obtain for the possibility of emission of a positive electron the condition:

$$-\frac{\delta E}{\delta n_2} + \frac{\delta E}{\delta n_1} < \Delta M_p \cdot c^2, \tag{4}$$

where ΔM_p is the mass defect of a proton as constructed from a neutron and a positive electron. From (3) and (4) we can conclude that the nucleus can be stable relative to electron emission only if

$$-\ \Delta M_p c^2 < -\frac{\delta E}{\delta n_1} + \frac{\delta E}{\delta n_2} < \Delta M_n c^2,$$

conditions which correspond in the stability diagram (Fig. 1) to a very narrow band (curves II and III)*, in contradiction with the experimental evidence.

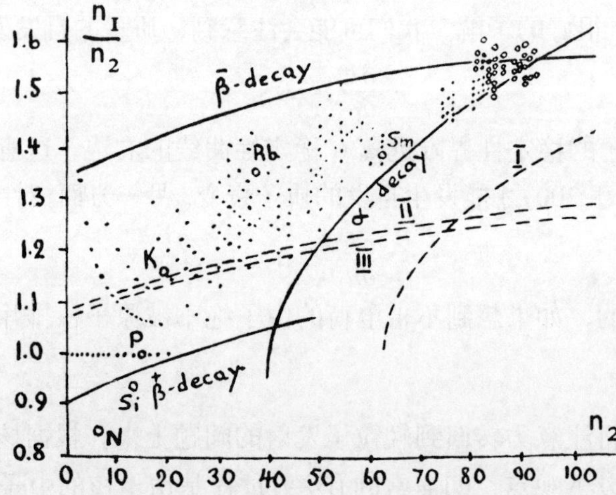

Fig. 1. A map of all known nuclei. Stable nuclei are indicated by full points, unstable nuclei by small circles.

* From the equations $n = p + \bar{e} + \Delta M_n c^2$ and $p = n + \overset{+}{e} + \Delta M_p c^2$, we obtain $\Delta M_n c^2 - (-\Delta M_p c^2) = \Delta M_n c^2 + \Delta M_p c^2 = \bar{e} + \overset{+}{e} = 2mc^2 = 1.6\times10^{-6}$ erg. This corresponds in Fig. 1 to a breadth of the stable region of about 0.016 units along the ordinate.

另一方面，约里奥夫妇最近发现了发射正电子的元素从而揭示了以上逆过程出现的可能性：

$$p \to n + \overset{+}{e}$$

只要我们认为核电子的发射相当于从原子核内拿走一个中子同时加入一个质子，则我们很容易估计出这个过程的稳定性极限。这种转变的正能量守恒条件显然将是：

$$-\frac{\delta E}{\delta n_1} + \frac{\delta E}{\delta n_2} < \Delta M_n \cdot c^2 \tag{3}$$

式中，ΔM_n 是一个质子和一个电子构成一个中子的质量亏损。通过严格类比，我们得到可能发射正电子的条件为：

$$-\frac{\delta E}{\delta n_2} + \frac{\delta E}{\delta n_1} < \Delta M_p \cdot c^2 \tag{4}$$

式中，ΔM_{p} 是一个中子和一个正电子构成一个质子的质量亏损。根据 (3) 式和 (4) 式，我们可以断定，仅在以下条件下，即

$$-\Delta M_p c^2 < -\frac{\delta E}{\delta n_1} + \frac{\delta E}{\delta n_2} < \Delta M_n c^2$$

核相对电子发射可能是稳定的，这个条件对应于稳定性图（图 1）中一条很窄的带（曲线 II 和曲线 III）*，这与实验结果相矛盾。

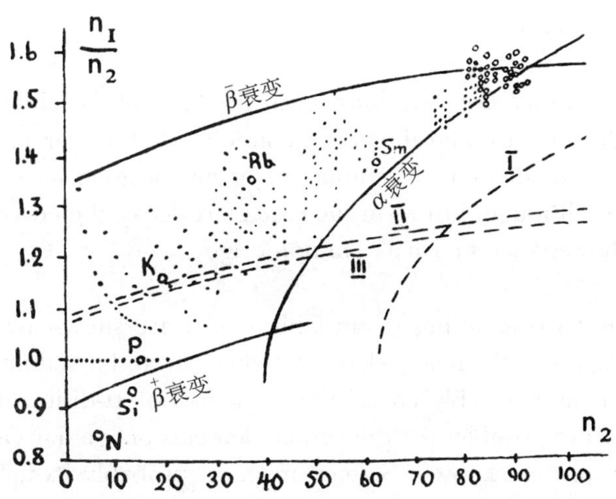

图 1. 所有已知原子核的图。稳定核用黑点表示，不稳定核用小圆圈表示。

* 从等式 $n = p + \bar{e} + \Delta M_n c^2$ 和 $p = n + \overset{+}{e} + \Delta M_p c^2$ 中，我们可得到 $\Delta M_n c^2 - (-\Delta M_p c^2) = \Delta M_n c^2 + \Delta M_p c^2 = \bar{e} + \overset{+}{e} = 2mc^2 = 1.6 \times 10^{-6}\mathrm{erg}$。在图 1 中由纵坐标上的一个约 0.016 单位稳定宽度范围所反映。

The stability region can, however, be made much broader if we consider more closely the energy conditions connected with electronic emission from nuclei. The point is that for a given total number of neutrons plus protons (that is, for given atomic weight) the nuclei are considerably more stable if the number of protons is even (even atomic number). The reason for this is that, with the increasing number of protons, each second one will lead to the formation of a new α-particle, and consequently correspond to larger liberation of energy. Thus if we plot the binding energy of isobaric nuclei against the atomic number (Fig. 2), the points corresponding to even-numbered elements will lie on a lower curve than those corresponding to the odd numbers. As can be seen from the diagram, this will have the result that for a series of elements extending some way both to the left and to the right side of the minimum, the emission of one electron (either negative or positive) will be energetically impossible. In such cases only the simultaneous emission of two electrons can be considered, but, as can be estimated from general theoretical considerations, such a double emission has extremely small probability. The possibility is not excluded that the natural β-activity of potassium and rubidium has its origin in such a double process, which would easily explain their extremely long period of life.

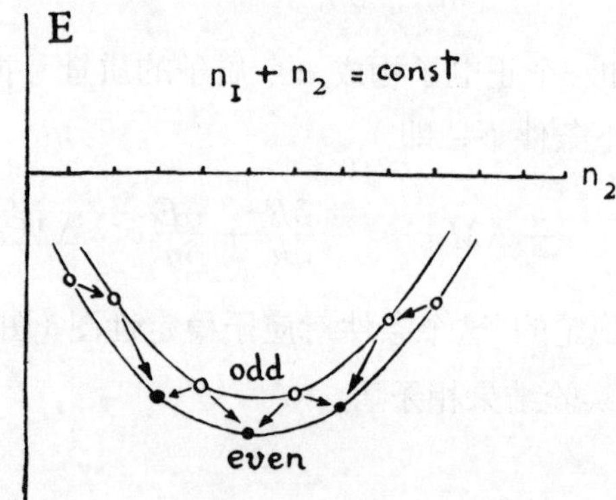

Fig. 2. Mass defect curves for typical isobaric nuclei. •, stable nuclei; ○, unstable nuclei.

According to these considerations we must push the limit of $\bar{\beta}$-stability upwards and the limit of $\overset{+}{\beta}$-stability downwards, and thus get a considerably broader stability region. It can be seen from Fig. 1 that theoretical limiting curves give a good idea of the form of the actual stability limits, although, just as in the case of α-decay, the curves go again too low. It seems that both discrepancies have a common origin.

In Fig. 1, the points corresponding to unstable nuclei are shown by small circles. One notices that in the region of the heavy elements, where α- and β-stability curves run rather close to one another (and possibly cross), the sequences of α-disintegrations followed by two β-disintegrations are possible. For the lighter elements only a few cases of spontaneous disintegration are at present known. Samarium (most probably its lightest isotope) emits α-particles of about 1.5 cm. range and has an average life of about 10^{12} years. The lightest isotopes of nitrogen, silicon and phosphorus (N^{13}_{7}, Si^{27}_{14}, P^{30}_{15}), unknown in Nature and produced artificially by the Joliots by α-bombardment of boron, magnesium and aluminium, give $\overset{+}{\beta}$-particles with an energy of 1–2 million volts and possess life-periods of several minutes.

然而，如果我们进一步考虑与原子核的电子发射有关的能量条件，以上讨论的稳定区域可以变得更宽。问题在于对一个给定的中子和质子的总数（即对于给定的原子量），如果质子数是偶数（偶原子序数），那么原子核是相当稳定的。理由是，随着质子数的增加,每两个质子将导致一个新的 α 粒子的形成,从而释放更多的能量。因此如果我们画出同量异位核的结合能相对其原子序数的关系图（图 2），则对应于偶数元素的点位于比奇数元素相应点更低的曲线上。如图所示，结果将是，对延伸到位于最小值两侧的一系列元素从能量角度上来看是不可能发射电子（不管是负电子或正电子）的。在这种情况下，只能认为两个电子同时发射。正如根据一般的理论估计的那样，这类成对电子发射的概率是很小的。但并不排除元素钾和铷的天然 β 放射性源于这类成对电子的发射过程，这也容易解释它们为何具有极长的寿命。

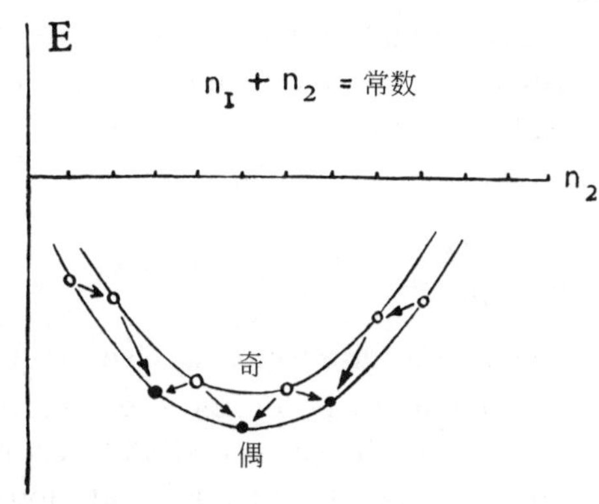

图 2. 典型的同量异位素核的质量亏损曲线。• 稳定核；○ 不稳定核。

根据以上的考虑，我们必须将 $\bar{\beta}$ 稳定性的极限向上推，而将 $\overset{+}{\beta}$ 稳定性的极限向下推，从而得到相当宽的稳定性区域。由图 1 可见，尽管与 α 衰变的情况一样，曲线的位置仍然很低，但是理论的极限曲线对实际的稳定性极限形状有一个很好的认识作用。似乎以上两类差异都出于相同的原因。

在图 1 中，对应的不稳定核用小圆圈表示。我们注意到，在重元素区域内 α 和 β 的稳定性曲线彼此很接近（可能有交叉），发生 α 衰变后紧接着两次 β 衰变的顺序是可能的。对于较轻的元素，目前我们只知道存在少数的自发衰变情况。如钐（极有可能是最轻的同位素）放出射程约为 1.5 cm 的 α 粒子，且平均寿命约为 10^{12} 年。而氮、硅和磷各自最轻的同位素（$^{13}_{7}N$，$^{27}_{14}Si$，$^{30}_{15}P$）在自然界中不存在，而是由约里奥夫妇利用 α 粒子分别轰击硼、镁和铝人工产生的，这些人工同位素放出能量为 1~2 兆电子伏，寿命约为几分钟的 $\overset{+}{\beta}$ 粒子。

The $\overline{\beta}$-emission from potassium and rubidium must be explained either as a double electron emission from their heavier isotopes (K^{41}_{19} and Rb^{87}_{37}) or as due to some unknown isotopes of chlorine and bromine resulting from a very short range α-emission of the above mentioned elements (probably from K^{40}_{19} and Rb^{86}_{37}). As these ranges in air, as calculated theoretically from the value of the corresponding decay constants, are 0.24 cm. and 0.63 cm. respectively, one can understand why the α-particles have not yet been detected. Thus we see that our general theoretical considerations fit rather nicely with the experimental evidence.

We now turn our attention to the details of the processes of emission of α-particles and electrons, and the connexion of the disintegration energy with the average period of life. The process of α-emission can be explained on the basis of the ordinary wave equation of Schrödinger as the velocities of the emitted α-particles are small compared with the velocity of the light. It was shown by Gamow, and independently by Gurney and Condon, that the long life of α-decaying bodies is due to the fact that the α-particle leaving a nucleus must cross a very high potential barrier, the transparency of which is extremely small and decreases very rapidly with the decrease of the energy liberated in the disintegration. Theory leads us to the following formula for the decay-constant λ as a function of the α-particle velocity v:

$$\lambda = \frac{4h}{mr_0^2} e^{-\frac{8\pi^2 e^2}{h}\frac{(Z-2)}{v} + \frac{16\pi e\sqrt{m}}{h}\sqrt{Z-2}\sqrt{r_0}} \tag{5}$$

where Z is the atomic number of the disintegrating element and r_0 the nuclear radius. Accepting r_0 for radioactive elements to be of the order of magnitude 10^{-12} cm., one obtains very good agreement between the calculated and measured values of λ and can explain theoretically the empirical relation between lgλ and v found by Geiger and Nuttall.

For complete agreement one must, however, accept the view that the nuclear radius r_0 changes from one element to another in such way that the density of the nucleus remains constant ($r_0 \sim \sqrt[3]{A}$). Formula (5) permits us also to estimate one of the values λ or v if the other is measured. Thus, for example, the range of the α-particles of radium C, estimated from this formula to be equal to 4 cm., is in good agreement with the value found later by Rutherford, and the period of life of the very short lived product radium C' given by this formula (10^{-3} sec.) fits well with the recent measurements of Jacobsen.

It is also interesting to notice that formula (5) may be successfully applied in the region of the lighter elements. According to (5) the period of life of samarium as estimated from the velocity of its α-particles must be about 10^{12} years, in good agreement with the observed value.

In the process of α-decay, it may often happen that the nucleus of the product of disintegration is constructed in an excited state, which corresponds to the emission of α-groups with slightly smaller energy (fine-structure of α-rays). The formula helps us to understand the relative intensities of such α-groups and also permits us to draw certain

钾和铷的 $\bar{\beta}$ 发射一定也可以解释成要么是它们较重的同位素（$^{41}_{19}K$ 和 $^{87}_{37}Rb$）发生了双电子发射，要么是源于上述元素（可能来自 $^{40}_{19}K$ 和 $^{86}_{37}Rb$）非常短射程的 α 粒子发射的一些氯和溴的未知同位素。根据相应的衰变常数进行理论计算，可以得到它们在空气中的射程分别为 0.24 cm 和 0.63 cm，这样我们就可以理解为何 α 粒子尚未被探测到。这里我们看到，通常的理论推算与实验结果较好地符合了。

我们现在将注意力转到 α 粒子和电子发射的详细过程以及衰变能量与平均寿命的关系上来。α 粒子发射过程基于普通的薛定谔波动方程可以解释为，发射的 α 粒子的速度是小于光速的。格尼、康登他们与伽莫夫都曾独立地指出，α 粒子衰变体的长寿命是由于从核中逃逸的 α 粒子必须穿过很高的势垒，而势垒的透射率极小，且随着衰变过程释放出能量的减小而迅速减小。理论导出以衰变常数 λ 作为 α 粒子速度 v 函数的公式：

$$\lambda=\frac{4h}{mr_0^2}e^{-\frac{8\pi^2e^2}{h}\frac{(Z-2)}{v}+\frac{16\pi e\sqrt{m}}{h}\sqrt{Z-2}\sqrt{r_0}} \tag{5}$$

式中，Z 是衰变元素的原子序数，r_0 是核半径。采用放射性元素的 r_0 是在 10^{-12} cm 的量级，这使得 λ 的计算值与测量值得到很好的符合，而且可以在理论上解释由盖革和努塔耳得到的 lgλ 和 v 之间的经验关系式。

然而，为了使理论和实验可以完全符合，我们必须采用以下观点，在核的密度保持恒定 $(r_0\sim\sqrt[3]{A})$ 的条件下，不同元素之间的核半径 r_0 是变化的。如果 λ 和 v 两个值中一个被测量，则通过式 (5) 我们可以估算出另一个值。例如，我们从式（5）估算得到镭 C 放出的 α 粒子的射程为 4 cm，这很好地符合了卢瑟福后来得到的数值，而由该公式得到很短存活产物镭 C' 的寿命（10^{-3} 秒）与最近雅克布森的测量值也符合得很好。

有趣的是我们还注意到，公式 (5) 也完全适用于较轻元素。根据（5）式，从钐放出的 α 粒子的速度可估算出钐的寿命约为 10^{12} 年，与观测值非常一致。

在 α 衰变过程中，衰变的产物的原子核通常总是处于激发态的，它对应于具有较小能量的 α 粒子群的发射（α 射线的精细结构）。这个公式有助于我们理解这类 α 粒子群的相对强度，也允许我们确认核的不同激发能级的量子数。另一方面，这

important conclusions about the quantum numbers of the different excited nuclear levels. On the other hand, it also explains the small number of so-called long-range α-groups corresponding to the disintegration of excited nuclei.

In contrast with the theory of α-decay, the understanding of the process of β-disintegration presents serious difficulties. First of all, the electrons emitted in β-decay possess a continuous distribution of energy, which seems to be in contradiction with the law of conservation of energy. It was pointed out by Bohr that the law of conservation of energy need not necessarily hold for processes involving nuclear electrons for which the modern quantum theory is not applicable. But, as was shown by Landau, the rejection of the conservation law for energy will be connécted with very serious difficulties in the general gravitational theory, according to which the mass present inside a certain closed surface is entirely defined by the gravitational field on this surface. It was proposed by Pauli that one might retain the energy conservation law by the introduction of a new kind of particle called a "neutrino". These neutrinos, having no electric charge and possessing very small (or even vanishing) mass, would be practically unobservable in all experiments and could easily take away the surplus energy of β-decay. The existence of such particles is, however, at present rather doubtful.

An attempt to construct a theory of β-disintegration on the basis of Dirac's relativistic wave equation, treating the emission of a nuclear electron in a similar way to the emission of light quanta by an atom, has recently been made by Fermi. In this theory, one accepts the view that the transformation of a nuclear neutron into a proton is connected with the creation of an electron and a neutrino, which, being born, leave the nucleus, dividing between them the energy liberated in this transformation. Accepting a definite value for the interaction energy giving rise to such transformations (of the order of magnitude of about 10^{-14} erg), Fermi obtains reasonable values for the decay constants of β-disintegrating elements and a good fit with the correlation curve between the decay constant and the maximum energy of the β-particles as found by Sargent.

An interesting consequence of this theory, which, however, is much more general and will hold for every theory treating electron emission as the result of the transformation of a neutron into a proton, is a definite exclusion rule for β-decay. According to this rule, β-transformations in which the original nuclei and those produced possess different spins are not all permitted, and can only happen with a rather reduced probability (about a hundred times less often than transformations in which the spin does not change). This explains at once the two different curves obtained by Sargent as due to permitted and not permitted transformations. It has been shown by Gamow that the application of the above mentioned exclusion rule for β-decay to the analysis of radioactive families gives very good results and permits us to give definite spin values to normal and excited states of radioactive nuclei.

(**133**, 744-747; 1934)

个公式也解释了少量的所谓长程 α 粒子群对应于激发态核的衰变。

与 α 衰变理论相比较，理解 β 衰变过程存在严重的困难。首先，β 衰变发射的电子具有连续的能量分布，这似乎与能量守恒定律相矛盾。玻尔曾指出，由于现代量子理论不适用于涉及核电子的过程，因此，能量守恒定律并不适用于该过程。但正如朗道所指出的，在广义引力理论中，某个闭合曲面中的质量完全是用该曲面上的重力场来定义的，违反能量守恒定律在广义引力理论中将会遇到非常严重的困难。泡利提出，我们可以通过引入称之为“中微子”的一类新粒子来保证能量守恒定律。这些中微子没有电荷，只有很小（甚至趋于零）的质量，实际上在所有的实验中都没有观测到它，但很可能就是它带走了 β 衰变多余的能量。但是，这类粒子存在与否目前尚存在争议。

最近，费米尝试在狄拉克相对论波动方程的基础上构建一套 β 衰变的理论，类比原子的光量子发射方式来处理原子核的电子发射。在该理论中，人们接受这种观点，即原子核的一个中子转变成一个质子，伴随产生了一个电子和一个中微子，它们一旦产生就会离开原子核，并将这个转变过程中释放的能量在它们之间分配。对导致这类转变的相互作用，能量采用某个确定值（约为 10^{-14} erg 的量级），费米由此得到了 β 衰变元素合理的衰变常数的值，并与萨金特得到的 β 粒子的衰变常数和最大能量的关系曲线吻合得很好。

然而，费米这套理论的一个有趣的结果是对 β 衰变给出了确定的不相容定则，这个结果是非常普遍的，适用于任何理论，即把电子发射认为是中子转变成质子的结果。按照这一定则，最初的原子核和产生具有不同自旋的那些原子核的 β 衰变过程并不都是允许的，且只能以较小的概率（约为自旋不变的转变概率的百分之一）发生。这立即解释了由萨金特获得的两条不同的曲线正是分别对应于允许转变和不允许转变。伽莫夫已指出，将上述的 β 衰变的不相容定则应用于分析放射族得到了很好的结果，并且允许我们给出放射性核正常态和激发态确定的自旋值。

（沈乃澂 翻译；张焕乔 审稿）

Radioactivity Induced by Neutron Bombardment

E. Fermi

Editor's Note

The impact of neutrons could induce radioactivity in many nuclei, as Enrico Fermi here reports. In experiments, he and colleagues had exposed a number of different elements to a source of neutrons, and found that in almost every case the exposed substances became radioactive. They found particularly strong effects with phosphorus, aluminium, silver, iodine and chromium, and measured half-lives varying from a few seconds to several hours. Fermi notes that it seemed likely nuclei were absorbing neutrons, which later transform into protons with ejection of an electron, thereby returning the nucleus to stability. Although very indirectly, these experiments hinted at the possibility of sustained nuclear reactions.

EXPERIMENTS have been carried out to ascertain whether neutron bombardment can produce an induced radioactivity, giving rise to unstable products which disintegrate with emission of β-particles. Preliminary results have been communicated in a letter to *La Ricerca Scientifica*, **5**, 282; 1934.

The source of neutrons is a sealed glass tube containing radium emanation and beryllium powder. The amount of radium emanation available varied in the different experiments from 30 to 630 millicuries. We are much indebted to Prof. G. C. Trabacchi, Laboratorio Fisico della Sanità Pubblica, for putting at our disposal such strong sources.

The elements, or in some cases compounds containing them, were used in the form of small cylinders. After irradiation with the source for a period which varied from a few minutes to several hours, they were put around a Geiger counter with walls of thin aluminium foil (about 0.2 mm. thickness) and the number of impulses per minute was registered.

So far, we have obtained an effect with the following elements:

Phosphorus—Strong effect. Half-period about 3 hours. The disintegration electrons could be photographed in the Wilson chamber. Chemical separation of the active product showed that the unstable element formed under the bombardment is probably silicon.

Iron—Period about 2 hours. As the result of chemical separation of the active product, this is probably manganese.

由中子轰击引发的放射性

费米

编者按

就像恩里科·费米在这篇文章中报道的一样，中子的碰撞会使许多原子核产生放射性。在实验中，他和其合作者们将许多种不同的元素暴露于中子源的辐射之下，发现几乎每一种实验中暴露的材料都会具有放射活性。并且他们在磷、铝、银、碘和铬中观察到了尤为强的效应，并测量得到它们的半衰期在数秒到几个小时之间不等。费米注意到似乎是原子核吸收了中子，并释放了一个电子，进而转变为质子，然后使原子核趋于稳定。这些实验尽管非常间接，但却隐含了持续核反应发生的可能性。

为查明物质在中子的轰击下是否会产生感生放射性的实验已经完成了，该实验产生不稳定的产物，而该产物将继续发生蜕变并放射出β粒子。初步的结果发表在1934年《科学研究》杂志第5卷第282页的通讯中。

中子源是一个密封的内含氡和铍粉末的玻璃管，在不同实验中，氡的用量从30到630毫居里不等。我们非常感谢公共卫生物理实验室的特拉巴基教授，他为我们提供了如此强的中子放射源。

把某些元素或含有这些元素的化合物压制成小的圆柱体形状，用放射源辐照其数分钟到几个小时不等，之后将它们放在以薄铝箔（厚约0.2 mm）为壁的盖革计数器周围，并记录每分钟的脉冲数。

迄今为止，我们已获得了下列元素在中子轰击下所产生的效应：

磷——强效应。半衰期约3小时。由磷物质蜕变放射出的电子可在威尔逊云室中被拍摄到。对放射性产物的化学分离情况显示，在中子的轰击下产生的不稳定元素可能是硅。

铁——半衰期约2小时。根据对其放射性产物进行化学分离的结果分析，产生的元素可能是锰。

Silicon—Very strong effect. Period about 3 minutes. Electrons photographed in the Wilson chamber.

Aluminium—Strong effect. Period about 12 minutes. Electrons photographed in the Wilson chamber.

Chlorine—Gives an effect with a period much longer than that of any element investigated at present.

Vanadium—Period about 5 minutes.

Copper—Effect rather small. Period about 6 minutes.

Arsenic—Period about two days.

Silver—Strong effect. Period about 2 minutes.

Tellurium—Period about 1 hour.

Iodine—Intense effect. Period about 30 minutes.

Chromium—Intense effect. Period about 6 minutes. Electrons photographed in the Wilson chamber.

Barium—Small effect. Period about 2 minutes.

Fluorine—Period about 10 seconds.

The following elements have also given indication of an effect: sodium, magnesium, titanium, zirconium, zinc, strontium, antimony, selenium and bromine. Some elements give indication of having two or more periods, which may be partly due to several isotopic constituents and partly to successive radioactive transformations. The experiments are being continued in order to verify these results and to extend the research to other elements.

The nuclear reaction which causes these phenomena may be different in different cases. The chemical separation effected in the cases of iron and phosphorus seems to indicate that, at least in these two cases, the neutron is absorbed and a proton emitted. The unstable product, by the emission of a β-particle, returns to the original element.

硅——非常强的效应。半衰期约 3 分钟。经辐照后产生的电子可在威尔逊云室中被拍摄到。

铝——强效应。半衰期约 12 分钟。经辐照后产生的电子可在威尔逊云室中被拍摄到。

氯——产生效应的周期比目前研究过的任何元素产生效应的周期都要长得多。

钒——半衰期约 5 分钟。

铜——效应很小。半衰期约 6 分钟。

砷——半衰期约 2 天。

银——强效应。半衰期约 2 分钟。

碲——半衰期约 1 小时。

碘——超强效应。半衰期约 30 分钟。

铬——超强效应。半衰期约 6 分钟。产生的电子可在威尔逊云室中被拍摄到。

钡——效应小。半衰期约 2 分钟。

氟——半衰期约 10 秒钟。

下列元素也显示出存在效应的迹象：钠、镁、钛、锆、锌、锶、锑、硒和溴。其中有些元素显示出两个或多个半衰期，这可能是由于某些元素存在多种同位素，也可能是由于某些元素发生了连续的放射性衰变。为了验证以上这些实验结果并将该研究拓展到其他元素，相关实验仍在继续进行中。

由上述实验现象所引起的核反应在不同情况下可能是不同的。铁和磷这两种元素的化学分离效应似乎表明，至少在中子轰击这两种元素时，原子核吸收了中子并放射出一个质子。在中子轰击下产生的不稳定产物通过放射出一个 β 粒子又转变为初始的元素。

The chemical separations have been carried out by Dr. O. D'Agostino. Dr. E. Amaldi and Dr. E. Segrè have collaborated in the physical research.

(**133**, 757; 1934)

Enrico Fermi: Physical Institute, Royal University, Rome, April 10.

化学分离的工作由达戈斯蒂诺博士完成。阿马尔迪博士和塞格雷博士合作进行了物理方面的研究。

（沈乃澂 翻译；夏海鸿 审稿）

Production of Large Quantities of Heavy Water

L. Tronstad

Editor's Note

Nature had recently reported on the establishment of a British plant for the manufacture of heavy water. Here Leif Tronstad of the Institute of Inorganic Chemistry in Norway tells of a similar project already undertaken by the Norwegians. Large quantities of heavy water were being produced from a facility run by Norsk Hydro near Rjukan in the Telemark region. In principle, this facility could produce up to 10 litres of heavy water in a single day, if this came to be demanded by increasing research activity in physics, chemistry and biology. The plant was sabotaged by the Norwegian resistance during the Second World War to prevent heavy water being used to develop a German atomic bomb.

FROM the discussion recently held in the Royal Society[1], and from several communications on heavy hydrogen published in *Nature*, it is obvious that larger quantities of heavy water are at present much needed for investigations in several branches of physics, chemistry and biology. To meet this demand, Imperial Chemical Industries, Ltd., is to undertake commercial production at Billingham[2]. It may also be of interest to report in this connexion, that various concentrates of the new water are now produced on a large scale in Norway by Norsk Hydro-Elektrisk Kvaelstofaktieselskab, Oslo. Large quantities of "1:300-water" can be obtained from the above company, and richer concentrates will be available at a later date.

This company at its works in Rjukan has one of the largest electrolytic hydrogen plants of the world, with a capacity of about 20,000 $m.^3$ per hour. Assuming the efficiency of separation by electrolysis so low as 10 percent[3], a quantity of about 10 litres of "pure" heavy water a day can be produced if the consumption requires.

In full agreement with other investigators, it has been found that the efficiency is only slightly affected by the conditions of the electrolysis[1,4]. However, certain difficulties arose using sulphuric acid with lead electrodes, due to the formation of porous lead on the cathodes and to the formation of fog. The efficiency of separation in both acid and alkaline solution agree fairly well with that found, for example, by Harteck[5]. Further details of the experimental results are to be published shortly in the *Zeitschrift für Elektrochemie*.

(**133**, 872; 1934)

Leif Tronstad: Institute of Inorganic Chemistry, Norwegian Technical High School, Trondhjem, Norway, May, 4.

重水的大量生产

特龙斯塔

编者按

《自然》最近对一个生产重水的英国工厂进行了报道。这篇文章中挪威无机化学研究所的莱夫·特龙斯塔指出，由挪威人承担的一个类似的项目已经启动了。大量的重水正在由泰勒马克行政区留坎附近挪威海德鲁公司运营的装置生产出来。原则上，这一装置每天可以生产多达 10 升的重水，以满足日益增长的物理学、化学和生物学研究的需求。这一工厂在第二次世界大战期间曾被挪威抵抗组织破坏，以防止重水被德国用来研制原子弹。

根据皇家学会近期举办的有关重氢的讨论会[1]以及在《自然》上发表的几篇关于重氢的通讯，我们可以明显地感觉到，目前在物理学、化学和生物学若干分支的科学研究中对重水有很大的需求。为了满足这种需求，帝国化学工业有限公司在比林罕进行了商业化生产[2]。目前挪威奥斯陆的挪威水电氮气有限公司（即现在的挪威海德鲁公司）大规模生产各种新水的浓缩物，对此进行报道也有可能引起人们的兴趣。从上述公司可以得到大量的“1:300 的水”，今后将可以获得浓度更高的浓缩物。

这家公司在留坎的制造厂拥有一个全球最大的电解氢工厂，其产量约为每小时 20,000 立方米。假定通过电解分离的效率只有 10%[3]，如果有消费需求，每天大约可以生产 10 升“纯净的”重水。

已经发现电解条件对其效率仅有轻微的影响[1,4]，这与其他研究者的结果完全一致。然而，由于会形成雾气以及在阴极上会形成多孔铅，在硫酸与铅电极一起使用的过程中出现了某些困难。正如哈特克[5]的发现，酸溶液和碱溶液两者的分离效率与上述发现非常一致。对实验结果更加详细的说明将于近期发表在《电化学杂志》上。

（沈乃澂 翻译；李芝芬 审稿）

References:

1. *Proc. Roy. Soc.*, A, **144**, 1; 1934.

2. *Nature*, **133**, 604, April 21, 1934.

3. Taylor, Eyring and Frost, *J. Chem. Phys.*, **1**, 823; (1933).

4. Compare, for example, Topley and Eyring, *Nature*, **133**, 292, Feb. 24, 1934. Bell and Wolfenden, *ibid.*, p.25.

5. Harteck, *Proc. Roy. Soc.*, loc. cit. and *Proc. Phys. Soc.*, **40**, 277; 1934.

Possible Production of Elements of Atomic Number Higher than 92

E. Fermi

Editor's Note

Physicists bombarding matter with charged particles had for the most part achieved induced radioactivity only with fairly light elements, because it was hard to reach the energies required to achieve an effect with heavier elements. But as Fermi here points out, this limitation did not apply to neutrons, which he had recently shown could induce radioactivity in a wide range of elements. Fermi describes new experiments with heavy elements, especially uranium. The resulting radiation showed half-lives from 10 seconds up to around 40 minutes. Experiments had ruled out the formation of isotopes of uranium, palladium, thorium, actinium, radium, bismuth and lead. The resulting radioactive isotope apparently had atomic number higher than 92, the maximum then known for any element.

UNTIL recently it was generally admitted that an atom resulting from artificial disintegration should normally correspond to a stable isotope. M. and Mme. Joliot first found evidence that it is not necessarily so; in some cases the product atom may be radioactive with a measurable mean life, and go over to a stable form only after emission of a positron.

The number of elements which can be activated either by the impact of an α-particle (Joliot) or a proton (Cockcroft, Gilbert, Walton) or a deuton (Crane, Lauritsen, Henderson, Livingston, Lawrence) is necessarily limited by the fact that only light elements can be disintegrated, owing to the Coulomb repulsion.

This limitation is not effective in the case of neutron bombardment. The high efficiency of these particles in producing disintegrations compensates fairly for the weakness of available neutron sources as compared with α-particle or proton sources. As a matter of fact, it has been shown[1] that a large number of elements (47 out of 68 examined until now) of any atomic weight could be activated, using neutron sources consisting of a small glass tube filled with beryllium powder and radon up to 800 millicuries. This source gives a yield of about one million neutrons per second.

All the elements activated by this method with intensity large enough for a magnetic analysis of the sign of the charge of the emitted particles were found to give out only negative electrons. This is theoretically understandable, as the absorption of the bombarding neutron produces an excess in the number of neutrons present inside the nucleus; a stable state is therefore reached generally through transformation of a neutron

原子序数大于92的元素的可能生成

费米

编者按

物理学家使用带电粒子对物质进行轰击，大部分情况下会产生感生放射性，而这仅仅是对于轻元素而言的，因为很难获得足以对重元素产生效应的能量。但费米在这篇文章中指出，使用中子进行轰击时不受此限制，他最近的研究表明中子可以使更大范围内的元素产生感生放射性。费米在本文中描述了关于重元素尤其是铀的新实验。得到的放射性元素的半衰期从 10 秒至 40 分钟左右不等。该实验排除了产生铀、镤、钍、锕、镭、铋和铅的同位素的可能。得到的放射性同位素的原子序数明显大于 92，这是当时已知的最大的原子序数。

人工嬗变得到的原子总是对应于一个稳定同位素，这一观点直到现在仍被广泛认可。但是约里奥夫妇首次通过实验结果证实事实并不完全是这样；在某些情况下，产生的原子具有放射性并具有可测量的平均寿命，且只有在放射出一个正电子之后它才达到稳定的状态。

由于库仑斥力的存在，只有轻元素可以发生嬗变，因此不论是用 α 粒子（约里奥）、质子（考克饶夫、吉尔伯特、瓦耳顿）还是用氘（克兰、劳里森、亨德森、利文斯通、劳伦斯）来轰击，可以被活化的元素的数目都必然要受到上述事实的限制。

但是在用中子轰击的情况下，这样的限制就失效了。中子引发蜕变的高效性完全弥补了中子源相对 α 粒子源或质子源较弱的缺点。事实上，不论原子量的大小，许多元素（截至目前试验过的 68 种元素中的 47 种）都可以被一个由铍粉和辐射剂量为 800 毫居里的氡填充的小玻璃管构成的中子源所活化[1]。这个中子源每秒释放出大约一百万个中子。

用这种方法活化的元素，其放射性强度强到足以通过磁学手段去分析放射出的粒子所携带的电荷符号，我们通过观察发现这些元素只能放出负电性的电子。这一点在理论上是可以理解的，由于轰击的中子被吸收而使核内的中子数增加；而只有当一个中子转变为一个质子时，体系才会达到稳定态，这个过程还会相应地放射出一个 β

into a proton, which is connected to the emission of a β-particle.

In several cases it was possible to carry out a chemical separation of the β-active element, following the usual technique of adding to the irradiated substance small amounts of the neighbouring elements. These elements are then separated by chemical analysis and separately checked for the β-activity with a Geiger–Müller counter. The activity always followed completely a certain element, with which the active element could thus be identified.

In three cases (aluminium, chlorine, cobalt) the active element formed by bombarding the element of atomic number Z has atomic number Z–2. In four cases (phosphorus, sulphur, iron, zinc) the atomic number of the active product is Z–1. In two cases (bromine, iodine) the active element is an isotope of the bombarded element.

This evidence seems to show that three main processes are possible: (*a*) capture of a neutron with instantaneous emission of an α-particle; (*b*) capture of the neutron with emission of a proton; (*c*) capture of the neutron with emission of a γ-quantum, to get rid of the surplus energy. From a theoretical point of view, the probability of processes (*a*) and (*b*) depends very largely on the energy of the emitted α- or H-particle; the more so the higher the atomic weight of the element. The probability of process (*c*) can be evaluated only very roughly in the present state of nuclear theory; nevertheless, it would appear to be smaller than the observed value by a factor 100 or 1,000.

It seemed worth while to direct particular attention to the heavy radioactive elements thorium and uranium, as the general instability of nuclei in this range of atomic weight might give rise to successive transformations. For this reason an investigation of these elements was undertaken by the writer in collaboration with F. Rasetti and O. D'Agostino.

Experiment showed that both elements, previously freed of ordinary active impurities, can be strongly activated by neutron bombardment. The initial induced activity corresponded in our experiments to about 1,000 impulses per minute in a Geiger counter made of aluminium foil of 0.2 mm. thickness. The curves of decay of these activities show that the phenomenon is rather complex. A rough survey of thorium activity showed in this element at least two periods.

Better investigated is the case of uranium; the existence of periods of about 10 sec., 40 sec., 13 min., plus at least two more periods from 40 minutes to one day is well established. The large uncertainty in the decay curves due to the statistical fluctuations makes it very difficult to establish whether these periods represent successive or alternative processes of disintegration.

Attempts have been made to identify chemically the β-active element with the period of 13 min. The general scheme of this research consisted in adding to the irradiated

粒子。

在某些情况下，利用通常采用的加入少量相邻元素的技术，可以对具有β放射性的元素进行化学分离。然后通过化学分析的手段将这些元素分离，并用盖革－米勒计数器分别检验它们的β放射性。这种放射性通常伴随着一个完全确定的元素，通过这个元素可以确定具有放射性的元素。

在三种情况下（铝、氯、钴），轰击原子序数为Z的元素会形成原子序数为Z–2的放射性元素。在四种情况下（磷、硫、铁、锌），放射性产物的原子序数为Z–1。在两种情况下（溴、碘），产生的放射性元素是被轰击元素的一种同位素。

上述证据似乎表明，可能有三种主要过程：(*a*) 俘获一个中子，即时放出一个α粒子；(*b*) 俘获一个中子，放出一个质子；(*c*) 俘获一个中子，放出一个γ量子，以带走过剩的能量。从理论上讲，过程（*a*）和（*b*）发生的概率主要依赖于放出的α粒子或H粒子的能量；元素的原子量越大，依赖程度越强。根据目前的核理论，我们只能对过程（*c*）发生的概率做出很粗略的估计；然而，这个估计值是观测值的1/1,000或1/100。

对于放射性重元素钍和铀给予特别的关注似乎是必要的，因为在这个原子量的范围内，原子核普遍的不稳定性会引发一系列的元素间的变换。基于这个原因，笔者与拉塞蒂和达戈斯蒂诺合作对这些元素进行了相关研究。

实验表明，在除去常见的放射性杂质之后，这两种元素都可以在中子轰击下被强烈活化。在我们实验中，使用盖革计数器对初始阶段感生放射性的强度的测量结果为每分钟约1,000个脉冲，这个计数器用0.2毫米厚的铝箔制得。这种放射性的衰减曲线表明这种现象是相当复杂的。对钍元素放射性的粗略观测表明，在这个元素的感生放射性的衰减过程中至少有两个周期。

铀得到了更好的研究；已经确定存在的周期包括10秒、40秒和13分钟，还包括介于40分钟到1天之间的至少两个以上的周期。统计涨落导致衰减曲线具有很大的不确定性，这使得人们难以确定这些周期是代表链式反应的衰变过程还是代表并行反应的衰变过程。

我们尝试对周期为13分钟的β放射性元素进行化学识别。这项研究的总体方案是向被辐照的物质（除去了衰变产物的硝酸铀的浓溶液）中加入一定量的普通β放

substance (uranium nitrate in concentrated solution, purified of its decay products) such an amount of an ordinary β-active element as to give some hundred impulses per minute on the counter. Should it be possible to prove that the induced activity, recognisable by its characteristic period, can be chemically separated from the added activity, it is reasonable to assume that the two activities are not due to isotopes.

The following reaction enables one to separate the 13 min.-product from most of the heaviest elements. The irradiated uranium solution is diluted in 50 percent nitric acid; a small amount of a manganese salt is added and then the manganese is precipitated as dioxide (MnO_2) from the boiling solution by addition of sodium chlorate. The manganese dioxide precipitate carries a large percentage of the activity.

This reaction proves at once that the 13 min.-activity is not isotopic with uranium. For testing the possibility that it might be due to an element 90 (thorium) or 91 (palladium), we repeated the reaction at least ten times, adding an amount of uranium X_1+X_2 corresponding to about 2,000 impulses per minute; also some cerium and lanthanum were added in order to sustain uranium X. In these conditions the manganese reaction carried only the 13 min.-activity; no trace of the 2,000 impulses of uranium X_1 (period 24 days) was found in the precipitate; and none of uranium X_2, although the operation had been performed in less than two minutes from the precipitation of the manganese dioxide, so that several hundreds of impulses of uranium X_2 (period 75 sec.) would have been easily recognisable.

Similar evidence was obtained for excluding atomic numbers 88 (radium) and 89 (actinium). For this, mesothorium-1 and -2 were used, adding barium and lanthanum; the evidence was completely negative, as in the former case. The eventual precipitation of uranium-X_1 and mesothorium-1, which do not emit β-rays penetrating enough to be detectable in our counters, would have been revealed by the subsequent formation respectively of uranium-X_2 and mesothorium-2.

Lastly, we added to the irradiated uranium solution some inactive lead and bismuth, and proved that the conditions of the manganese dioxide reaction could be regulated in such a way as to obtain the precipitation of manganese dioxide with the 13 min.-activity, without carrying down lead and bismuth.

In this way it appears that we have excluded the possibility that the 13 min.-activity is due to isotopes of uranium (92), palladium (91), thorium (90), actinium (89), radium (88), bismuth (83), lead (82). Its behaviour excludes also ekacaesium (87) and emanation (86).

This negative evidence about the identity of the 13 min.-activity from a large number of heavy elements suggests the possibility that the atomic number of the element may be greater than 92. If it were an element 93, it would be chemically homologous with manganese and rhenium. This hypothesis is supported to some extent also by the observed fact that the 13 min.-activity is carried down by a precipitate of rhenium sulphide

射性元素，使得它们在计数器中给出每分钟几百个脉冲的计数。通过其特征周期可以对感生放射性成分进行辨认，而如果感生放射性成分可以与加入的放射性成分进行化学分离的话，那么就可以合理地认定这两类放射性并不是源于多种同位素。

下面的反应能使我们能够将周期为 13 分钟的产物与大多数最重的元素分离开。将被辐照的铀溶液用 50% 的硝酸稀释；加入少量的锰盐，在沸腾的溶液中加入氯酸钠使锰以二氧化物（二氧化锰）的形式沉淀出来。二氧化锰沉淀物携带了大部分的放射性。

这个反应同时也证明了周期为 13 分钟的放射性物质不是铀的同位素。为了验证该放射性是否来源于 90 号元素（钍）或 91 号元素（镤），我们至少重复了 10 次这个反应，加上适量的铀–X_1 和铀–X_2，它们的放射性强度相当于每分钟约 2,000 个脉冲；为了稳定住铀–X 还加入了一些铈和镧。在这样的条件下，锰的反应只会携带周期为 13 分钟的放射性；尽管我们将测量过程控制在二氧化锰沉淀产生后 2 分钟之内完成，因为这样的条件下源自铀–X_2（周期 75 秒）的几百个脉冲应该是很容易检测到的，但我们在沉淀物中没有追踪到源自铀–X_1（周期 24 天）的 2,000 个脉冲，也没有发现铀–X_2 的脉冲。

利用类似的证据我们排除了原子序数为 88（镭）和 89（锕）的元素的可能性。在实验中，我们使用了新钍–1 和新钍–2，并加入了钡和镧；与前述情况类似，证据完全是否定性的。铀–X_1 和新钍–1 的最终沉淀物原被解释为分别生成了铀–X_2 和新钍–2，但实验中我们的计数器并未探测到 β 射线。

最后，我们向经辐照的铀溶液加入一些非放射性的铅和铋，这证明了二氧化锰沉淀反应的条件可以通过这种方法进行调整，从而得到具有周期为 13 分钟的放射性二氧化锰沉淀，且不含有铅和铋。

通过这样的方法，我们已排除了周期为 13 分钟的放射性是源于铀的同位素（92）、镤（91）、钍（90）、锕（89）、镭（88）、铋（83）和铅（82）的可能性，其行为也排除了钫（87）和氡（86）的可能性。

周期为 13 分钟的放射性可能来自一系列重元素的否定性证据，为我们指出了另一种可能性，即该放射性元素的原子序数可能大于 92。如果它是 93 号元素，那么它将与锰和铼同族。这个假设在某种程度上得到了以下观测事实的支持，即周期为

insoluble in hydrochloric acid. However, as several elements are easily precipitated in this form, this evidence cannot be considered as very strong.

The possibility of an atomic number 94 or 95 is not easy to distinguish from the former, as the chemical properties are probably rather similar. Valuable information on the processes involved could be gathered by an investigation of the possible emission of heavy particles. A careful search for such heavy particles has not yet been carried out, as they require for their observation that the active product should be in the form of a very thin layer. It seems therefore at present premature to form any definite hypothesis on the chain of disintegrations involved.

(**133**, 898-899; 1934)

E. Fermi: Royal University of Rome.

Reference:

1. E. Fermi, *Ricerca Scientifica*, 1, 5, 283; 6, 330. *Nature*, 133, 757, May 19, 1934. E. Amaldi, O. D'Agostino, E. Fermi, F. Rasetti, E. Segrè, *Ricerca Scientifica*, 8, 452; 1934.

13 分钟的放射性可以随不溶于盐酸的硫化铼沉淀而出。然而，由于还有其他一些元素也很容易以这种形式产生沉淀，因此，我们不能认定其为一种强有力的证据。

该元素原子序数为 94 或 95 的可能性很难与前述可能性相区分，因为它们的化学性质可能相当类似。此过程中包含的有价值的信息可以通过研究可能存在的重粒子发射行为而获得。不过对这类重粒子的仔细探索尚未展开，因为它们需要在薄层形式的放射性产物中进行观测。因此在目前情况下，我们还不能对所涉及的衰变链做出任何确定的假设。

（沈乃澂 翻译；朱永生 审稿）

The Factor $\frac{137}{136}$ in Quantum Theory

A. S. Eddington

Editor's Note

Arthur Eddington developed speculative theories to account for the values of the fine structure constant (which determines the strength of electrical forces) and the mass ratio of the proton to electron. In his theory these values were exactly 1/137 and 1847.6, in good agreement with experiments. But the measured ratio of electron charge to mass was slightly smaller than predicted, by a factor of more or less precisely 136/137. Here Eddington tries to explain how this extra factor might arise from the indistinguishability of electrons, which could cause overestimation of the electron mass by a factor of 137/136. Eddington's speculations did not stand the test of time, but reflected a preoccupation with finding simplicities among the known fundamental constants.

IT has been suggested by W. N. Bond[1] that, in some or all of the attempts to determine e/m experimentally, the quantity actually found is $\frac{136}{137}e/m$; for if the experimental results are corrected in accordance with this hypothesis, they are found to be in satisfactory accordance with my theoretical values of the fine-structure constant (137) and mass-ratio (1847.6). R. T. Birge[2] has confirmed this; and, quoting three important recent determinations of e/m, he has shown that the agreement is extremely close.

On theoretical grounds it seems likely that Bond's hypothesis is right. In my earliest paper on the subject[3], I gave the value of the fine-structure constant as 136, since I found the Coulomb energy of two elementary particles to be $1/136r$ in natural quantum units. This energy was $\frac{137}{136}$ times too large, because I had not allowed for the 137th degree of freedom arising from the indistinguishability of the particles. Bond's hypothesis implies that I am not the only victim of this mistake; current quantum theory in deriving from observational data the proper-energy or mass m of an electron has also obtained an energy $\frac{137}{136}$ times too large. If so, the cause is presumably the same, namely, neglect to take into account the degree of freedom due to indistinguishability.

There is nothing mystical in the effect of indistinguishability. It occasions, not an objective difference of behaviour, but a difference in what we can ascertain about the behaviour, and hence a difference of treatment. In the dynamics of two particles, we have to describe the change with time of the positions, momenta and spin-components (or of a probability distribution of them) of the particles which we call No.1 and No.2; and also we have to describe a growing uncertainty whether the particle, called No.1 at the time t, is the original No.1. If the probability that it is the original No.1 is $\cos^2\theta$ (so that the probability

量子理论中的137/136因子

爱丁顿

编者按

阿瑟·爱丁顿发展了一套推测性的理论，用来说明精细结构常数（其决定了电场力的强度）以及质子与电子的质量比的重要性。在爱丁顿的理论中，以上两个值与实验结果很好地吻合，分别严格等于 1/137 和 1847.6。但测量得到的电子荷质比略小于理论预期值，相差了大约 136/137 这个因子。本文中爱丁顿试图用电子的不可分辨性来解释这个多出来的因子，正是电子的不可分辨性导致估算得到的电子质量是实值的 137/136 倍。爱丁顿的假设虽然没有经得住时间的考验，但这反映出他当时专注于在已知的基本常数中寻求简单性。

邦德[1]提出，在某些或者说全部试图测定荷质比 e/m 值的实验中得到的结果都是真实的荷质比 e/m 值的 136/137；如果实验结果按此假设进行修正，则其与我从理论计算得到的精细结构常数（137）以及质子与电子的质量比（1847.6）的值是完全吻合的。伯奇[2]已经证实了这个结论；他引用了三次最近对荷质比 e/m 的重要测量，表明这些结果都是非常一致的。

在理论计算的基础上，邦德的假设看起来似乎是正确的。我在关于这方面研究的早期文章中[3]给出了精细结构常数的值为 136，因为我发现两个基本粒子的库仑能量可以用自然量子单位表示成 $1/136r$。事实上，我所得到的这个能量值相对而言远大于真实的结果，是真实值的 137/136 倍，原因是我没有考虑与粒子的不可分辨性相对应的第 137 个自由度。邦德的假设表明，我并不是这个错误的唯一受害者；目前的量子理论从观测数据得到电子的固有能量或质量 m 的结果表明，观测得到的结果也远大于真实值，是真实值的 137/136 倍。如果事实确实如此，则原因估计是相同的，即都没有考虑到与粒子的不可分辨性相对应的自由度。

粒子的不可分辨性本身并没有什么神秘之处。它不会使粒子的行为出现客观本质上的变化，而是引起了我们对粒子行为的主观推断的变化，进而使我们的处理方式有所不同。处理两个粒子的动力学行为时，我们必须给出这两个粒子的位置、动量和自旋分量随时间变化的情况（或者这些量的概率分布），我们把这两个粒子命名为 1 号粒子和 2 号粒子；另外，我们还必须考虑 t 时刻的 1 号粒子是否仍为原来的 1 号粒子这种不确定性的增长。如果 t 时刻的 1 号粒子是原来的 1 号粒子的概率是

that it is the original No. 2 is $\sin^2\theta$) the permutation variable θ will be a function of the time and have all the properties of a dynamical variable, giving therefore an extra degree of freedom of the system and having a momentum (energy of interchange) associated with it. When, however, the particles are distinguished without uncertainty, θ is constrained to be zero, and this degree of freedom is lost.

Thus for the treatment of two indistinguishable particles, we have to start with an a priori probability distributed over a closed domain of 137 dimensions, whereas for two distinguishable particles it is distributed over a closed domain of 136 dimensions. Naturally, the average values of characteristics of the distribution are slightly different in the two treatments. In particular, the energy tensor of the a priori probability distribution, which is identical with the metrical tensor $g_{\mu\nu}$ of macroscopic theory, is different. Hence *the two kinds of treatment are associated with different metrics of space-time*. It seems clear that a factor $\frac{137}{136}$ (neglected in current quantum theory) will be introduced by the change of metric when we equate the space occupied by the indistinguishable particles of quantum theory to the space occupied by the distinguishable parts of our measuring apparatus.

It may be asked: Why does this factor affect the mass of the electron but not that of the proton? The discrimination is, I think, not strictly between the proton and electron, but between the resultant mass $(M + m)$ which is nearly the mass of a proton, and the reduced mass of the relative motion $Mm/(M + m)$ which is nearly the mass of an electron; for it is in the relative motion that the question of distinguishing the two ends of the relation arises. It may also be asked why the factor $\frac{137}{136}$, which refers especially to a system of two particles, applies irrespective of the number of particles. The answer is that the metrical ideas of quantum theory are borrowed from those of relativity theory; and since the latter are based on the interval between two points, the former refer correspondingly to the wave function of two particles.

(**133**, 907; 1934)

A. S. Eddington: Observatory, Cambridge, June 5.

References:

1. W. N. Bond, *Nature*, 133, 327, March 3, 1934.
2. R.T. Birge, *Nature*, 133, 648, April 28, 1934.
3. A. S. Eddington, *Proc. Roy. Soc.*, A, 122, 358; 1929.

$\cos^2\theta$（因而是原来的 2 号粒子的概率便是 $\sin^2\theta$)，则置换变量 θ 将是时间的函数，并具有力学量的所有性质，因此系统多了一个与动量（能量交换）相关的自由度。然而，如果不考虑不确定性，即粒子可分辨时，θ 值就只能为零，系统就没有这类自由度了。

因此，在处理两个不可分辨的粒子时，我们在 137 维的闭域内利用先验的概率分布来处理问题；而对于两个可分辨的粒子的情况，则以 136 维的闭域内先验的概率分布为出发点。自然地，两类处理情况下的概率分布的平均特征值稍有差别。尤其是与宏观理论的度规张量 $g_{\mu\nu}$ 等价的先验概率分布的能量张量在这两种情况下是不同的。因此，**两类处理是与时空的两个不同度规相关的**。似乎很明显可见，当我们将量子理论中由不可分辨的粒子所占据的空间与由测量装置中可分辨的粒子所组成的空间视为等同时，由于两个空间度规的不同将引入 137/136 这个因子（在目前的量子理论中是忽略该因子的）。

可能有人会问：为何这个因子会影响电子的质量，而不影响质子的质量？我认为以上的两种质量并不是严格的质子和电子的质量，而是合成质量（$M+m$）与相对运动的约化质量 $Mm/(M+m)$，前者几乎是质子的质量，而后者几乎是电子的质量；正是由于考虑了相对运动，使得我们在辨别质量关系的两种极限时出现了问题。也许有人还会问，为何从两个粒子所组成的系统这一特殊情况中推出的因子 137/136 也适用于其他系统，而与组成系统的粒子数无关。原因是量子理论中度规的想法是借用相对论中度规的概念而来的；后者是基于两点之间的间隔，而前者与两个粒子的波函数有关。

（沈乃澂 翻译；李军刚 审稿）

Exchange Forces between Neutrons and Protons, and Fermi's Theory

Ig. Tamm

Editor's Note

Soviet physicist Igor Tamm, a pioneer in nuclear science and future Nobel laureate, here comments on Enrico Fermi's recent theory of beta radioactivity, which assumes that protons and neutrons can be transmuted into one another while emitting or absorbing electrons, positrons and neutrinos. As Tamm argues, Fermi's scheme suggests that the force binding nuclei might be an "exchange force" between nuclear particles, originating in the exchange of light particles in direct analogy to the photon-exchange process mediating the electromagnetic force. But a simple estimate of the magnitude of such a force yields a value that is far too small. Tamm's result implied that physicists were still seeking another fundamental force, which would later became known as the nuclear strong force.

FERMI[1] has recently developed a successful theory of β-radioactivity, based on the assumption that transmutations of a neutron into a proton and vice versa are possible and are accompanied by the birth or disappearance of an electron and a neutrino.

This theory implies the possibility of deducing the exchange forces between neutrons and protons, introduced more or less phenomenologically by Heisenberg. (This idea occurred also quite independently to my friend, D. Iwanenko, with whom I have since had the opportunity of discussing the question.) Consider two heavy particles *a* and *b*, *a* being in a neutron and *b* in a proton state. If *a* becomes a proton and *b* a neutron the energy remains unchanged. Now these two degenerate states of the system may be linked up by a two-step process: the emission of an electron and a neutrino by the neutron *a* which becomes a proton, and the ensuing re-absorption of these light particles by the proton *b* which becomes a neutron. The energy of the system will be in general not conserved in the intermediate state (compare the theory of dispersion). The emission and re-absorption of a positron and neutrino may also take place[2]. In this way the two degenerate states of the system considered are split into two energy states, differing by the sign of the exchange energy.

Since the rôle of the light particles (ψ-field) providing an interaction between heavy particles corresponds exactly to the rôle of the photons (electromagnetic field), providing an interaction between electrons, we may adapt for our purposes the methods used in quantum electrodynamics to deduce the expression for Coulomb forces.

Putting $\psi = \psi_0 + g\psi_1 + g^2\psi_2 + \ldots$, where g is the Fermi constant ($\sim 4 \times 10^{-50}$ erg. cm.[3]),

质子中子之间的交换力与费米理论

塔姆

编者按

苏联物理学家伊戈尔·塔姆是核科学界的先驱，并在之后获得了诺贝尔奖，这篇文章中他评论了恩里科·费米新近的β衰变理论，费米假定质子和中子可以相互转化，并在此过程中放出或者吸收电子、正电子和中微子。事实正如塔姆所质疑的那样，费米的理论方案表明束缚原子核的作用力可能是一种原子核粒子间的“交换作用”，这一想法源于对电磁相互作用中以光子交换过程为媒介的交换的直接类比，但是对于这种相互作用大小的粗略估算得到数值过小。塔姆的结果表明物理学家仍在寻找另一种基本相互作用，而这种相互作用后来被称为强相互作用。

最近，费米[1]成功地发展了一套关于β衰变的理论。这个理论基于这样的假设：中子转变为质子或质子转变为中子都是有可能的，并且在此过程中伴随着一个电子和一个中微子的产生或消失。

这一理论意味着推导出质子和中子之间的交换力是有可能的，交换力这一概念在某种程度上是由海森堡唯象地引入的。（这个概念也曾独立地出现在我的朋友伊万年科的脑海中，我曾有机会和他讨论过这个问题。）设有两个重粒子 a 和 b，a 处于中子态，b 处于质子态。如果 a 变成一个质子，而 b 变成一个中子，系统能量保持不变。这样，系统的这两个简并态就可以通过以下两个步骤联系起来：首先，中子 a 放出一个电子和一个中微子，变成一个质子；其次，质子 b 再吸收这些轻粒子，变成一个中子。整个系统的能量在中间态（与色散理论相比）并不守恒。正电子和中微子的发射和再吸收同样也有可能发生[2]，因此以这种方法所考虑的系统的两个简并态就劈裂成交换能符号不同的两个能态。

由于轻粒子（ψ 场）在重粒子相互作用中所扮演的角色与光子（电磁场）在电子相互作用中所扮演的角色非常类似，所以，我们可以采取量子电动力学中推导库仑力表达式的方法来研究我们的问题。

设 $\psi=\psi_0+g\psi_1+g^2\psi_2+\cdots$，其中 g 为费米常数（$\sim 4\times10^{-50}$ 尔格·立方厘米），

and using the theory of perturbations and retaining only that part of ψ which corresponds to the absence of light particles in the initial and final states, we obtain

$$\left(H_0 - i\hbar\frac{\partial}{\partial t}\right)\psi_2 \sim \left(K \mp \frac{1}{16\pi^3\hbar cr^5}I(r)\right)\psi_0,$$

where K is an infinite constant, r is the distance between a and b and $I(r)$ is a decreasing function of r, which is equal to 1 when $r \ll \hbar/mc$ (m is the mass of the electron). Neglecting K, one would obtain the same result if one introduced directly in the wave equation of the heavy particles an exchange energy $A(r)$:

$$A(r) = \pm\frac{g^2}{16\pi^3\hbar cr^5}I(r),$$

the sign of $A(r)$ depending on the symmetry of ψ in respect to a and b. Introducing the values of $\hbar$, c and g, we obtain

$$|A(r)| \ll 10^{-85} r^{-5} \text{ erg.}$$

Thus $A(r)$ is far too small to account for the known interaction of neutrons and protons at distances of the order of $r \sim 10^{-13}$ cm.

If the difference of masses of the neutron and of the proton is larger than the sum of the masses of an electron and a neutrino, the emission of light particles by a heavy particle may take place without violation of the conservation of energy. But again the corresponding value of the exchange energy may be shown to be far too small

$$|A(r)| < g\left(\frac{mc}{\hbar}\right)^3 \sim 10^{-18} \text{ erg.}$$

Our negative result indicates that either the Fermi theory needs substantial modification (no simple one seems to alter the results materially), or that the origin of the forces between neutrons and protons does not lie, as would appear from the original suggestion of Heisenberg, in their transmutations, considered in detail by Fermi.

(**133**, 981; 1934)

Ig. Tamm: Physical Research Institute, State University, Moscow.

References:

1. Fermi, *Z. Phys.*, **88**, 161; 1934.

2. Wick, *Rend. R. Nat. Acad. Lincei*, **19**, 319; 1934.

用微扰理论，且只保留初末态中与轻粒子的缺失有关的那部分 ψ，可以得到：

$$\left(H_0 - i\hbar\frac{\partial}{\partial t}\right)\psi_2 \sim \left(K \mp \frac{1}{16\pi^3 \hbar c r^5} I(r)\right)\psi_0,$$

其中，K 是一个无限大的常数，r 表示 a 和 b 之间的距离，$I(r)$ 是 r 的单调下降函数，当 $r \ll \hbar/mc$ 时，$I(r)$ 等于 1（m 是电子的质量）。不考虑常数 K，如果将交换能 $A(r)$ 直接引入重粒子的波动方程，我们将会得到相同的结果。$A(r)$ 的表达式为：

$$A(r) = \pm\frac{g^2}{16\pi^3 \hbar c r^5} I(r)$$

$A(r)$ 的符号取决于由 a 和 b 决定的 ψ 的对称性。将 $\hbar$、c 和 g 的值代入上式，我们得到：

$$|A(r)| \ll 10^{-85} r^{-5} \text{尔格}$$

因此，如果中子和质子的距离 $r \sim 10^{-13}$ 厘米量级，那么 $A(r)$ 的值就太小了，以致于无法解释中子和质子之间已知的相互作用。

如果中子和质子之间的质量差大于电子和中微子的质量和，那么重粒子发射轻粒子的过程就可能不违反能量守恒原理，但是，相应的交换能的值仍然太小：

$$|A(r)| < g\left(\frac{mc}{\hbar}\right)^3 \sim 10^{-18} \text{ 尔格}$$

我们得到的负面结果表明，不是费米理论需要根本的修正（因为使结果有实质上的改变绝非易事），就是中子和质子之间作用力的源头根本不像海森堡最初设想，且由费米细致考虑的那样存在于它们的相互转变之中。

（王静 翻译；王乃彦 审稿）

Interaction of Neutrons and Protons

D. Iwanenko

Editor's Notes

Russian physicist Dmitri Iwanenko here argues that Fermi's recent theory of beta radioactivity, while successful in some regards, makes some erroneous predictions. For decays observed in light nuclei, the theory gives a roughly correct link between the rate of decay and the maximum energy of the ejected particles. However, if one uses the theory to consider nuclear binding as an exchange of force-mediating particles similar to the electromagnetic force, one finds the force is only large enough at a distance between nucleons of about 10^{-17} m, which is far too small. Iwanenko concludes that Fermi's theory is thus only a very crude approximation.

AS electrons and positrons are expelled in some reactions from nuclei, we can try to treat these *light* particles like the photons emitted by atoms. Then the interaction of *heavy* particles (protons, neutrons) can be considered as taking place *via* light particles described by the equations of a ψ-field in the same manner as electromagnetic, for example, Coulomb, interaction takes place through an electromagnetic field, or photons.

The *first* order effects are the expulsion (or absorption) of an electron, which case was treated recently by Fermi, or of a positron. We may remark that the application of Fermi's formalism to positron disintegration of light nuclei (which we get by changing the sign of the charge number and taking for the latter the appropriate value) gives results which fit, though not very accurately, the observed relation between the half-period and the maximum energy of the disintegration particle[1]. Though there seems to be a quantitative disagreement between Fermi's theory (applied to positrons) and positron disintegration, on the other hand the calculated values for K and Rb support Fermi's assumption of the existence of quadripole transitions of heavy particles, giving too big values for the half periods in comparison with the usual dipole disintegrations. The exceptional position of K and Rb is in some way rather *anschaulich*. We may remark that the Sargent–Fermi rule, in contrast to the Geiger–Nuttall law, shows a less pronounced dependence on the charge number, so that for qualitative considerations even the wave functions of free particles can be used.

The *second* order effects give specially the probability of production of pairs, which is in the case of the ψ-field less effective than in the electromagnetic case, as the charge, e, is much bigger than Fermi's coefficient, g (the "charge" for the ψ-field). The most important second order effect is the subsequent production and annihilation of an electron and positron, in the field of proton and neutron, which leads to the appearance of an interaction *exchange* energy (Heisenberg's *Austausch*) between proton and neutron,

中子和质子的相互作用

伊万年科

编者按

俄国物理学家德米特里·伊万年科在这篇文章中指出费米新近关于β辐射的理论尽管在某些方面获得了成功，但它也给出了一些错误的预测。对于在轻原子核中观察到的衰变现象，虽然该理论给出的衰变速率和放出的粒子的最大能量之间的关系是大致正确的，但是，如果使用该理论时考虑束缚核的相互作用，从而认为其是类似于电磁相互作用的作用——媒介粒子的交换，人们会发现这种作用力只有在核子之间的距离大约为 10^{-17} 米时才足够强，而这个距离太短了。因此，伊万年科得出结论认为费米的理论只是一个非常粗略的近似理论。

由于电子和正电子都会在一些核反应中被原子核排斥出来，所以我们可以试着像处理原子所放射的光子一样处理这些**轻**粒子。于是，我们可以认为**重**粒子（质子、中子）之间的相互作用是通过 ψ 场方程所描述的轻粒子来实现的。这种相互作用与以电磁场或光子为介质的电磁相互作用（如库仑相互作用）具有相同的形式。

一阶效应是电子的释放（或吸收），费米最近研究过这个问题，当然正电子的情况类似。我们注意到，将费米理论应用到产生正电子的轻核衰变时（我们是通过改变电荷数的符号并且赋予后者适当的数值来实现这一过程的），得到的理论结果与实验观察到的衰变粒子的半衰期和其最大能量之间的关系[1]相符合，但不是精确吻合。虽然正电子的衰变和费米理论（应用于正电子）计算的结果似乎在数值上吻合得不是很好，但是从另一方面来看，K 和 Rb 的计算结果却支持费米关于存在重粒子四极裂变的假设，由其计算给出的四极裂变的半衰期要远大于通常的偶极裂变的半衰期。K 和 Rb 的特殊地位从某种程度上说可谓非常**直观**。我们注意到对照盖革 – 努塔耳定律，萨金特 – 费米规则反而很少依赖于电荷数，因此在做定性分析时，甚至可以用自由粒子的波函数。

二阶效应特别给出了产生粒子对的可能性。由于电荷 e 要远远大于费米系数 g（对应 ψ 场的“电荷”），所以相对于在电磁场中，在 ψ 场中产生粒子对的可能性要小。最重要的二阶效应是：在质子和中子场中，电子和正电子的产生和湮灭，导致了中子和质子之间相互作用的**交换**能（海森堡**交换**）的出现。这正如库仑相互作用可以看作是由两个电子之间光子的产生和吸收所导致的。不同于库仑作用中的比值

quite in the same way as Coulomb interaction can be conceived as arising from the birth and absorption of a photon in the case of two electrons. Instead of e^2/r one gets here an interaction of the order g^2/chr^5, which is easily verified dimensionally. The exact calculations were first carried out by Prof. Ig. Tamm, who also insisted on development of this method. With $g{\sim}10^{-50}$ (the computations were carried out by V. Mamasichlisov), which value is required by the empirical data on heavy radioactive bodies, we get an interaction energy of a million volts, no at a distance of 10^{-13} cm. but only at $r{\sim}10^{-15}$ cm., which is inadmissible. We may ask about the value of r, which would give a *self-interaction* energy of the order of the proper energy of a heavy particle. This value is of the order 10^{-16} cm., which is that of the classical radius of a proton.

The appearance of these small distances is very surprising and can be removed only by some quite new assumptions. Fermi's characteristic coefficient g appears to be connected also with distances of this order of magnitude.

(**133**, 981-982; 1934)

D. Iwanenko: Physical-Technical Institute, Leningrad.

Reference:

1. cf. D. Iwanenko, *C.R. Ac. Sci. U.S.S.R.*, Leningrad, 2, No. 9, 1934.

e^2/r，重粒子的交换能与 g^2/chr^5 成正比，这个值可以很容易地从数量级上进行验证。塔姆教授首先进行了准确计算，他仍在坚持发展这一方法。根据重放射性物质的经验数值的要求，取 $g \sim 10^{-50}$（由麻马斯克索夫计算得出），我们得到相互作用能为一兆电子伏，而这是在距离 $r \sim 10^{-15}$ cm 而不是距离 $r \sim 10^{-13}$ cm 时得到的，但是这个距离是不合理的。我们可能会对距离 r 值产生疑问，距离 r 应该给出一个对于重粒子来说数量级合理的**自作用**能量。这个值在 10^{-16} cm 量级，即质子的经典半径的量级。

出现这么小的间距实在是让人吃惊，只有引入某种新的假设才能解决这个问题。费米特征系数 g 也与这个间距的数量级大小有关。

（王静 翻译；王乃彦 审稿）

A "Nuclear Photo-effect": Disintegration of the Diplon by γ-rays

J. Chadwick and M. Goldhaber

Editor's Note

In 1934, some physicists still doubted if the neutron were a genuine subatomic particle, or a composite of the proton and the electron. Here James Chadwick and Maurice Goldhaber describe their experimental determination of a significantly more accurate neutron mass. They used photons to induce the disintegration of a "diplon"—a nucleus of heavy hydrogen—into a proton and neutron. As the masses of the atoms of hydrogen and heavy hydrogen were known very accurately, it was then easy to deduce a value for the mass of the neutron. This experiment suggested that the diplon is composed of a neutron and proton, and helped cement the status of the neutron as a fundamental particle.

BY analogy with the excitation and ionisation of atoms by light, one might expect that any complex nucleus should be excited or "ionised", that is, disintegrated, by γ-rays of suitable energy. Disintegration would be much easier to detect than excitation. The necessary condition to make disintegration possible is that the energy of the γ-ray must be greater than the binding energy of the emitted particle. The γ-rays of thorium C″ of $h\nu = 2.62 \times 10^6$ electron volts are the most energetic which are available in sufficient intensity, and therefore one might expect to produce disintegration with emission of a heavy particle, such as a neutron, proton, etc., only of those nuclei which have a small or negative mass defect; for example, D^2, Be^9, and the radioactive nuclei which emit α-particles. The emission of a positive or negative electron from a nucleus under the influence of γ-rays would be difficult to detect unless the resulting nucleus were radioactive.

Heavy hydrogen was chosen as the element first to be examined, because the diplon has a small mass defect and also because it is the simplest of all nuclear systems and its properties are as important in nuclear theory as the hydrogen atom is in atomic theory. The disintegration to be expected is

$$_1D^2 + h\nu \rightarrow {}_1H^1 + {}_0n^1 \qquad (1)$$

Since the momentum of the quantum is small and the masses of the proton and neutron are nearly the same, the available energy, $h\nu - W$, where W is the binding energy of the particles, will be divided nearly equally between the proton and the neutron.

The experiments were as follows. An ionisation chamber was filled with heavy hydrogen of about 95 percent purity, kindly lent by Dr. Oliphant. The chamber was connected to a linear amplifier and oscillograph in the usual way. When the heavy hydrogen was exposed to the γ-radiation from a source of radiothorium, a number of "kicks" was recorded by

一种“核的光效应”：γ射线引发的氘核蜕变

查德威克，戈德哈伯

编者按

1934年，仍有一些物理学家对于中子是真正的亚原子粒子抑或是电子和质子的结合体而感到疑惑。这篇文章中，詹姆斯·查德威克和莫里斯·戈德哈伯描述了他们对于中子质量更为精确的实验测定。他们用光子引发“氘核”（重氢的原子核）蜕变为一个质子和一个中子。由于氢原子和重氢原子的质量已经有了非常精确的数值，所以据此可以很容易推算出中子的质量。这一实验表明氘核由一个中子和一个质子构成，该实验也更加牢固地确立了中子是一种基本粒子的理念。

根据原子在光的照射下可以发生激发和电离，人们可以以此类推出，任何具有复杂结构的原子核都会被具有适当能量的γ射线激发或者“电离”（即蜕变）。蜕变要比激发容易探测得多。蜕变发生的必要条件是γ射线的能量大于所放出的粒子的结合能。钍C″放射出的γ射线的能量为$h\nu = 2.62 \times 10^6$电子伏，这是能够得到的且具有足够强度的射线中能量最高的，因此，人们期望核蜕变可以产生某种重粒子（例如中子、质子等），而只有在蜕变中质量亏损较小或者为负值的原子核才能产生这种蜕变；像^{2}D、^{9}Be以及放射性原子核则放射出α粒子。由γ射线引起的正电子或负电子发射很难被探测到，除非形成的原子核具有放射性。

首先选择重氢作为研究对象，因为氘核的质量亏损小，并具有最简单的核系统，而且氘核在核理论中的重要性就如同氢原子在原子理论中的一样。我们预期的蜕变反应是：

$$^2_1\mathrm{D} + h\nu \rightarrow {}^1_1\mathrm{H} + {}^1_0n \tag{1}$$

由于光量子动量非常小，而且质子和中子的质量几乎相等，所以可获取的能量$h\nu - W$（W是粒子的结合能）几乎会在质子和中子间平分。

实验过程如下：向电离腔中充满纯度为95%的重氢（重氢是奥利芬特博士慷慨赠予的），我们依照常规的方法将电离腔与一个线性放大器和示波器相连。当我们用放射性钍所放射的γ射线辐照重氢时，示波器就会记录到很多“突跳”。实验证明，

the oscillograph. Tests showed that these kicks must be attributed to protons resulting from the splitting of the diplon. When a radium source of equal γ-ray intensity was employed, very few kicks were observed. From this fact we deduce that the disintegration cannot be produced to any marked degree by γ-rays of energy less than 1.8×10^6 electron volts, for there is a strong line of this energy in the radium C spectrum.

If the nuclear process assumed in (1) is correct, a very reliable estimate of the mass of the neutron can be obtained, for the masses of the atoms of hydrogen and heavy hydrogen are known accurately. They are 1.0078 and 2.0136[1] respectively. Since the diplon is stable and can be disintegrated by a γ-ray of energy 2.62×10^6 electron volts (the strong γ-ray of thorium C″), the mass of the neutron must lie between 1.0058 and 1.0086; if the γ-ray of radium C of 1.8×10^6 electron volts is ineffective, the mass of the neutron must be greater than 1.0077. If the energy of the protons liberated in the disintegration (1) were measured, the mass of the neutron could be fixed very closely. A rough estimate of the energy of the protons was deduced from measurements of the size of the oscillograph kicks in the above experiments. The value obtained was about 250,000 volts. This leads to a binding energy for the diplon of 2.1×10^6 electron volts, and gives a value of 1.0081 for the neutron mass. This estimate of the proton energy is, however, very rough, and for the present we may take for the mass of the neutron the value 1.0080, with extreme errors of $\pm$ 0.0005.

Previous estimates of the mass of the neutron have been made from considerations of the energy changes in certain nuclear reactions, and values of 1.007 and 1.010 have been derived in this way[2,3]. These estimates, however, depend not only on assumptions concerning the nuclear processes, but also on certain mass-spectrograph measurements, some of which may be in error by about 0.001 mass units. It is of great importance to fix accurately the mass of the neutron and it is hoped to accomplish this by the new method given here.

Experiments are in preparation to observe the disintegration of the diplon in the expansion chamber. These experiments should confirm the nuclear process which has been assumed, and therewith the assumption that the diplon consists of a proton and a neutron. Both the energy of the protons and their angular distribution should also be obtained.

If, as our experiments suggest, the mass defect of the diplon is about 2×10^6 electron volts, it is at once evident why the diplon cannot be disintegrated by the impact of polonium α-particles[4]. When an α-particle collides with a nucleus of mass number M, only a fraction $M/(M+4)$ of the kinetic energy of the α-particle is available for disintegration, if momentum is to be conserved. In the case of the diplon, therefore, only one third of the kinetic energy of the α-particle is available, and this, for the polonium α-particle, is rather less than 1.8×10^6 electron volts. The more energetic particles of radium C′ should just be able to produce disintegration, and Dunning[5] has in fact observed a small effect when heavy water was enclosed in a radon tube.

这些突跳是氘核蜕变产生的质子引起的。如果我们用相同强度的镭源 γ 射线照射重氢，观察到的突跳就很少。由此我们推断，能量小于 1.8×10^6 电子伏的 γ 射线无法引发显著的蜕变，因为在镭 C 能谱的这个能量位置上有一条很强的线。

如果式（1）中假设的核反应过程是正确的，因为氢原子和重氢原子的质量都是精确已知的，其值分别是 1.0078 和 2.0136[1]，所以我们可以对中子质量进行可靠的估算。由于氘核是稳定的，而且可以由能量为 2.62×10^6 电子伏的 γ 射线（钍 C″ 放射的强 γ 射线）引发蜕变，所以中子质量一定在 1.0058 和 1.0086 之间；如果镭 C 发射的能量为 1.8×10^6 电子伏的 γ 射线不能引发蜕变，那么中子质量一定大于 1.0077。如果蜕变反应式（1）中释放的质子能量是可以测量的，那么中子质量就能较为精确地确定了。通过测量上述实验中示波器上突跳的大小可以对质子的能量进行粗略估算，由此得到的能量数值约为 250,000 电子伏。由这个值导出氘核的结合能为 2.1×10^6 电子伏，中子质量为 1.0081。不过，这种对质子能量的估算非常粗略，目前我们取中子质量为 1.0080，最大误差为 ±0.0005。

之前根据某些核反应的能量变化估算出来的[2,3]中子质量值为 1.007 和 1.010。然而，这些估算值不仅依赖于对于核过程的假设，还需依赖质谱测量，而有些质谱测量的误差会达到约 0.001 个质量单位。精确确定中子质量是十分重要的，我们希望这里给出的新方法能够成功解决这个问题。

对膨胀云室中氘核蜕变进行观测的实验正在筹备之中。这些实验将会证实前面假设的核反应过程，进而证实氘核是由一个质子和一个中子构成的。质子的能量以及它们的角分布也都将被测得。

正如我们的实验所表明的那样，如果氘核的质量亏损约为 2×10^6 电子伏，那么氘核不能在钋放射出的 α 粒子的轰击下发生蜕变[4]的原因就会立刻清楚了。当 α 粒子和一个质量数为 M 的原子核发生碰撞时，如果碰撞过程中动量是守恒的，那么在 α 粒子的动能中只有 $M/(M+4)$ 可以用于诱发蜕变。因此，在参与碰撞的原子核为氘核的情况中，α 粒子的动能只有 1/3 可用于蜕变，而对于钋放射出的 α 粒子，这个能量小于 1.8×10^6 电子伏。而镭 C′ 发射出的能量更高的粒子刚好可以引发蜕变，事实上，邓宁[5]已经在将重水封装于氡管里的实验中观察到了微弱的蜕变效应。

Our experiments give a value of about 10^{-28} sq. cm. for the cross-section for disintegration of a diplon by a γ-ray of 2.62×10^6 electron volts. In a paper to be published shortly, H. Bethe and R. Peierls have calculated this cross-section, assuming the interaction forces between a proton and a neutron which are given by the considerations developed by Heisenberg, Majorana and Wigner. They have obtained the transition probability in the usual quantum-mechanical way, and their result gives a value for the cross-section of the same order as the experimental value, but rather greater, if we take the mass of the neutron as 1.0080. If, however, we take the experimental value for the cross-section, the calculations lead to a neutron mass of 1.0085, which seems rather high. Thus the agreement of theory with experiment may be called satisfactory but not complete.

One further point may be mentioned. Some experiments of Lea[6] have shown that paraffin wax bombarded by neutrons emits a hard γ-radiation greater in intensity and in quantum energy than when carbon alone is bombarded. The explanation suggested was that, in the collisions of neutrons and protons, the particles sometimes combine to form a diplon, with the emission of a γ-ray. This process is the reverse of the one considered here. Now if we assume detailed balancing of all processes occurring in a thermodynamical equilibrium between diplons, protons, neutrons and radiation, we can calculate, without any special assumption about interaction forces, the relative probabilities of the reaction (1) and the reverse process. Using our experimental value for the cross-section for reaction (1), we can calculate the cross-section for the capture of neutrons by protons for the case when the neutrons have a kinetic energy $2(h\nu - W) = 1.0 \times 10^6$ electron volts in a co-ordinate system in which the proton is at rest before the collision. In this special case the cross-section σ_c for capture (into the ground state of the diplon—we neglect possible higher states) is much smaller than the cross-section σ_p for the "photo-effect". It is unlikely that σ_c will be very much greater for the faster neutrons concerned in Lea's experiments. It therefore seems very difficult to explain the observations of Lea as due to the capture of neutrons by protons, for this effect should be extremely small. A satisfactory explanation is not easy to find and further experiments seem desirable.

(**134**, 237-238; 1934)

References:

1. K. T. Bainbridge, *Phys. Rev.*, 44, 57; 1933.
2. J. Chadwick, *Proc. Roy. Soc.*, A, 142, 1; 1933.
3. I, Curie and F. Joliot, *Nature*, 133, 721, May 12, 1934.
4. Rutherford and A. E. Kempton, *Proc. Roy. Roc.*, A, 143, 724; 1934.
5. *Phys. Rev.*, 45, 586; 1934.
6. *Nature*, 133, 24, Jan. 6, 1934.

我们的实验给出了能量为 2.62×10^{6} 电子伏的 γ 射线所引发氘核蜕变时的碰撞截面为 10^{-28} 平方厘米。在近期即将发表的一篇论文中，贝特与佩尔斯假设中子和质子之间相互作用力符合海森堡、马约拉纳和维格纳提出的设想，从而计算出这个截面的数值。他们使用通常的量子力学方法计算了蜕变的概率，其理论计算给出的碰撞截面数值与实验值具有相同的数量级，但如果我们取中子质量为 1.0080，计算值就会相当大了。然而，如果我们设碰撞截面为实验值，那么计算得到的中子质量就是 1.0085，这个值似乎太大了。因此，我们只能说理论和实验基本相符，而并非完全一致。

另外还有一点需要提及。利[6]的一些实验表明，用中子轰击固体石蜡发射出一种硬 γ 射线，这种射线比单独轰击碳原子产生的射线的强度更强且量子能量更高。对此的解释是，中子和质子发生碰撞有时会结合成氘核，并释放出 γ 射线。这一过程是我们这里所研究的反应的逆过程。现在，如果我们假定氘核、中子、质子三者各自与射线之间的热力学平衡所涉及的所有过程都达到细致平衡，那么无需任何有关相互作用力的假定，我们就能计算出反应式（1）与其逆过程发生的相对概率。用我们在实验中测得的反应式（1）的截面数据，当中子具有 $2(h\nu-W)=1.0\times10^{6}$ 电子伏的动能，且质子在碰撞前静止时，可以计算出质子俘获中子的俘获截面。在这种特殊情况下，俘获截面 σ_c（在这里我们只考虑俘获后氘核处于基态的情况，忽略了可能存在的更高的激发态）要比“光效应”截面 σ_p 小很多。对于利的实验中相关的快中子，俘获截面 σ_c 不可能会非常大。因此，似乎很难将利在实验中的观测解释成为质子俘获了中子的结果，因为这个效应应该是非常微弱的。要想找到一个令人满意的解释并不是很容易，这似乎有待更进一步的实验。

（王静 翻译；王乃彦 审稿）

Quantum Mechanics and Physical Reality

N. Bohr

Editor's Note

In May 1935, Einstein, Podolsky and Rosen published a landmark paper arguing that quantum mechanics must be considered incomplete. They had described a thought experiment in which the values of some physical variables could be known before they were measured, implying that some aspect of physical reality must correspond to those variables, which quantum mechanics however failed to describe. Here Niels Bohr argues that their argument was flawed owing to an essential ambiguity in quantum mechanics, which demands that physical reality cannot be considered independently of the conditions that define an experiment. So began an argument over the completeness of quantum theory, and its apparent "non-local" character (the possibility of action at a distance), which continues to this day.

IN a recent article by A. Einstein, B. Podolsky and N. Rosen, which appeared in the *Physical Review* of May 15, and was reviewed in *Nature* of June 22, the question of the completeness of quantum mechanical description has been discussed on the basis of a "criterion of physical reality", which the authors formulate as follows: "If, without in any way disturbing a system, we can predict with certainty the value of a physical quantity, then there exists an element of physical reality corresponding to this physical quantity".

Since, as the authors show, it is always possible in quantum theory, just as in classical theory, to predict the value of any variable involved in the description of a mechanical system from measurements performed on other systems, which have only temporarily been in interaction with the system under investigation; and since in contrast to classical mechanics it is never possible in quantum mechanics to assign definite values to both of two conjugate variables, the authors conclude from their criterion that quantum mechanical description of physical reality is incomplete.

I should like to point out, however, that the named criterion contains an essential ambiguity when it is applied to problems of quantum mechanics. It is true that in the measurements under consideration any direct mechanical interaction of the system and the measuring agencies is excluded, but a closer examination reveals that the procedure of measurements has an essential influence on the conditions on which the very definition of the physical quantities in question rests. Since these conditions must be considered as an inherent element of any phenomenon to which the term "physical reality" can be unambiguously applied, the conclusion of the

量子力学和物理实在

玻尔

编者按

1935 年 5 月，爱因斯坦、波多尔斯基和罗森一起发表了一篇具有里程碑意义的论文，他们认为量子力学应该是不完备的。他们描述了这样一个理想实验，其中有些物理变量的值在测量前就是已知的，这意味着这些物理量对应于某些物理实在，但量子力学无法描述这些物理量。本文中尼尔斯·玻尔认为他们的理论有缺陷，因为在量子力学中有一个重要的地方未阐述清楚，即需要说清楚的是，不可能独立于实验所处的环境来考虑物理实在。由此引发了关于量子理论完备性和量子理论明显“非局域”特征（超距作用的可能性）的讨论，直到今天，这方面的讨论仍在继续。

最近，爱因斯坦、波多尔斯基和罗森在 5 月 15 日的《物理学评论》上发表了一篇文章，6 月 22 日的《自然》杂志对该文章进行了评论。基于“物理实在的判据”，他们在文章中讨论了量子力学描述的完备性问题并对该问题作了如下阐述：“假如系统没有受到任何形式的干扰，我们就可以确切地预言物理量的值，并且认为存在某个物理实在的要素与该物理量相对应。”

正如以上作者所述，像经典理论那样，在量子理论中通过对与所研究的系统有短暂相互作用的其他系统进行测量，可以预言与描述该力学系统有关的任何一个变量；但与经典力学不同的是，在量子力学中永远不可能同时给出两个共轭变量的确定值，上述文章的作者们根据其判据认为，物理实在的量子力学描述是不完备的。

然而，我想要指出的是，当把所谓的判据应用于量子力学的问题时将会出现基本定义模糊不清的问题。确实，前面已经假设在测量过程中不考虑系统与测量仪器之间任何直接的力学相互作用，但是进一步的研究表明，测量过程对现在我们所讨论的物理实在的确切定义有重要的影响。只有考虑测量过程对系统的影响并把这些影响看成是任何一个物理现象的内在属性，我们才可以明确地使用“物理实在”这个术语，因此，以上提到的那些作者们的结论似乎就不太合理了。不久，进一步完

above-mentioned authors would not appear to be justified. A fuller development of this argument will be given in an article to be published shortly in the *Physical Review*.

(**136**, 65; 1935)

N. Bohr: Institute of Theoretical Physics, Copenhagen, June 29.

整地论证这个问题的文章将发表在《物理学评论》上。

（沈乃澂 翻译；李军刚 审稿）

Isotopic Constitution of Uranium

A. J. Dempster

Editor's Note

Uranium has the highest atomic weight of any naturally occurring element, and exists predominantly in the form of uranium-238. Here, however, physicist Arthur Dempster of the University of Chicago reports that about 0.4% of the element is an unstable isotope uranium-235. This isotope, he suggests, is the origin of the actinium series of radioactive elements, as the elements protactinium-231, actinium-227, francium-223 and so on can be generated by radioactive alpha and beta decays. Dempster also notes that another isotope, uranium-234, appears to exist in traces of 0.008%. Uranium-235 would soon become very important as the only natural fissile isotope: one capable of sustaining a nuclear chain reaction driven by slow moving neutrons, thus making nuclear power and atomic bombs conceivable.

THE analysis of uranium rays from the volatile hexafluoride by Dr. Aston[1] has shown a single line at atomic weight 238. The element appeared to be simple to at least two or three percent, but its properties were not favourable for study in the gas discharge. As uranium is of great importance for the subject of radioactivity, the spark source described in *Nature* of April 6 (**135**, 542) was tried with uranium metal and gold as electrodes, and also with an electrode made by packing a nickel tube with pitchblende. It was found that an exposure of a few seconds was sufficient for the main component at 238 reported by Dr. Aston; but in addition on long exposures a faint companion of atomic weight 235 was also present. With two different uranium electrodes it was observed on eight photographs, and two photographs with the pitchblende electrode also showed the new component. The relative intensity could be only roughly estimated on account of the irregularity of the spark, but it appeared to be less than one percent of the intensity of the main component.

This faint isotope of uranium is of special interest as it is in all probability the parent of the actinium series of radioactive elements. In discussing Dr. Aston's analysis[2] of the isotopes in lead from radioactive minerals, Lord Rutherford[3] pointed out that the lead isotope of atomic weight 207 is probably the end product of the actinium series, so that the atomic weight of protoactinium would be 231, (207+6×4). This value has been verified by the recent chemical determination of the atomic weight by v. Grosse[4]. Protoactinium itself may be formed by α- and β-ray transformations from a hypothetical isotope of uranium, actino-uranium, with an atomic weight of 235 or 239[3]. The relative amount of actino-uranium at present on the earth would be 0.4 percent of the uranium according to a recalculation by Dr. v. Grosse[5,6]. The present observations thus support this theory, with the atomic weight of 235 for the isotope actino-uranium. A third isotope, uranium II, of

铀的同位素构成

登普斯特

编者按

自然界中存在的元素中，铀具有最大的原子量，且绝大部分以铀–238 的形式存在。而在这篇文章中芝加哥大学的物理学家阿瑟·登普斯特指出，约有 0.4% 的铀是不稳定的同位素铀–235。他认为这种铀的同位素是锕系列放射性元素的起源，因为镤–231、锕–227、钫–223 等锕系放射性元素可通过该同位素的 α 衰变和 β 衰变而获得。同时登普斯特还注意到了铀的另一个同位素——铀–234，它只占 0.008%。作为仅有的天然可裂变的同位素，铀–235 很快变得非常重要：它可以维持一个由慢中子驱动的链式核反应，这使得核能和原子弹成为可能。

阿斯顿博士[1]对挥发性六氟化铀中的铀离子束的分析显示出一条对应于原子量为 238 的单线。至少在 2% 或 3% 的精度上讲这种元素的同位素构成是简单的，但是它的性质却很不适于利用气体电离的方法进行研究。鉴于铀对于放射性研究的巨大重要性，我们使用了 4 月 6 日发表于《自然》第 135 卷第 542 页的论文中所描述的火花源的方法，我们以金属铀和金作为电极，还使用了填充沥青铀矿的镍管为电极。实验发现，对于阿斯顿博士所报道的原子量为 238 的主要成分来说，几秒钟的曝光时间就已足够；但是除此之外，在长时间曝光后，一个原子量为 235 的微弱伴线也出现了。采用两个不同的铀电极，我们在八张照片中观察到了这条伴线，同时在两张使用沥青铀矿电极的照片中也显示了这一新成分的存在。由于火花的不均匀性，对新成分的相对强度只能进行粗略地估算，不过它的强度似乎不到主要成分的 1%。

铀的这种少量的同位素引起了我们特别的关注，因为它很可能是锕系放射性元素的母体。在就阿斯顿博士对放射性矿物中铅同位素的分析结果[2]进行讨论时，卢瑟福勋爵[3]指出，原子量为 207 的铅同位素可能是锕系元素的最终产物，因此镤元素的原子量应为 231，即（207+6×4）。冯·格罗塞[4]采用新近的原子量化学测定法对这个值进行了确认。镤本身可以由一种假定的铀同位素（锕铀，原子量为 235 或 239）通过 α 衰变或 β 衰变而形成[3]。根据冯·格罗塞博士的重新计算[5,6]，目前地球上的锕铀的相对含量大约为铀的 0.4%。因此，就原子量为 235 的同位素锕铀而言，目前的观测结果是支持该理论的。第三种同位素——原子量为 234 的铀 II，理论上

atomic weight 234 amounts theoretically to only 0.008 percent of the uranium, and would be too faint for observation by the mass-spectrograph.

(**136**, 180; 1935)

A. J. Dempster: University of Chicago, July 12.

References:

1. Aston, *Nature*, **128**, 725; 1931.
2. Aston, *Nature*, **123**, 313; 1929.
3. Rutherford, *Nature*, **123**, 313; 1929.
4. A. v. Grosse, *Proc. Roy. Soc.*, **150**, 363; 1935.
5. A. v. Grosse, *Phys, Rev.*, **42**, 565; 1932.
6. A. v. Grosse, *J. Phys. Chem.*, **38**, 487; 1933.

只占铀总量的 0.008%，这个量对于质谱观测来说可能过少了。

（王耀杨 翻译；汪长征 审稿）

The Fundamental Paradox of the Quantum Theory

R. Peierls

Editor's Note

Rudolph Peierls here writes to counter a recent argument of George Temple, who had apparently demonstrated an inconsistency in quantum mechanics. Temple had assumed that one unique quantum-mechanical operator (a mathematical construct of the theory) must correspond to each of the classical variables of motion, such as positions and momenta. Peierls questions this assumption, noting that the order in which variables appear in the mathematical formalism matters in quantum theory, while it does not in classical theory. In this property of so-called non-commutation, in fact, lies the origin of Heisenberg's uncertainty principle.

THE question of the logical consistency of quantum mechanics has recently been discussed by Prof. G. Temple[1] and by Dr. H. Fröhlich and Dr. E. Guth[2]. Temple arrives at an apparent contradiction, starting from principles which he states to be an essential part of quantum theory. From these principles contradictory results follow, as Temple shows by a perfectly rigorous deduction. (The existence of this deduction seems to have been ignored by Fröhlich and Guth.)

The main assumption used by Temple is that to every function of the classical variables of motion (momentum, co-ordinate, etc.) there corresponds one unambiguously defined operator which may be taken as the representative of this function in quantum mechanics. Although this assumption can be found quite frequently in papers discussing the principles of quantum theory, and even in some text-books, I would like to emphasise that it is not at all necessary and—as one can see from Temple's argument—not even possible in quantum mechanics.

Quantum mechanics requires operators as representatives of physical variables for two purposes: to connect wave functions with experiments, actual or possible, and to calculate the time dependence of the wave function. In order to be able to apply quantum mechanics unambiguously to actual problems, one must, therefore, know (*a*) the operator representing the quantity measured by any given apparatus, and (*b*) the energy operator for any given physical system. Properly speaking, (*a*) is a special case of (*b*), for the properties of a measuring apparatus can always be analysed if its interaction energy with the object in question is known.

If, then, we had an apparatus built in such a way as to measure, say, qp^2q (p denoting the momentum, q the co-ordinate of a particle), this apparatus would be *different*

量子理论的基本佯谬

佩尔斯

编者按

鲁道夫·佩尔斯在本文中反驳了乔治·坦普尔最近的一个观点，当时坦普尔显然已经证明了量子力学中存在一个矛盾。坦普尔认为应该有唯一的量子力学算符（量子理论的数学表示）对应于每个经典的运动变量，如位置和动量。佩尔斯对坦普尔的这个结论表示质疑，他指出在量子理论中变量在数学表达式里的次序是很重要的，而在经典理论中是无关紧要的。事实上，海森堡测不准原理就是源于这种所谓的算符的不对易性。

最近，坦普尔教授[1]、弗勒利希博士和古思博士[2]都讨论了量子力学的逻辑一致性问题。坦普尔从他所陈述的量子理论的基本原理出发得出了一个明显的矛盾。这些矛盾的结果正是坦普尔从量子力学的基本原理出发通过严密精确的推导得到的。（弗勒利希博士和古思博士似乎并没有注意到坦普尔已做的相关推导。）

坦普尔的主要假设是：经典运动变量（如动量、坐标等）的每一个函数都有一个明确定义的算符与之相对应，这些算符可以被视为相应的函数在量子力学中的表示形式。虽然我们可以经常在讨论量子理论基本原理的文章中，甚至是相关的教科书中看到这个假设，但是，我想强调的是，从坦普尔的论证过程中可以看到，在量子力学中，这个假设并不必要，甚至是不合理的。

量子力学需要算符作为物理变量的表示形式有两个目的：一是将波函数与已经实现或可能实现的实验联系起来；二是计算波函数随时间的变化关系。为了能将量子力学明确地应用到实际的问题中，我们必须知道：(*a*) 任意给定装置所测量的量的对应算符，(*b*) 任意给定的物理系统对应的能量算符。严格地说，(*a*) 是 (*b*) 的特殊情况，因为如果测量装置与所考虑物体的相互作用能是已知的话，我们总可以对测量装置的性质进行分析。

譬如说，我们有这样一台用来测量 qp^2q（p 表示粒子的动量，q 表示粒子的坐标）这个量的测量装置，另外一台测量装置是用于测量 $\frac{1}{2}(p^2q^2 + q^2p^2)$ 的，虽然在经典力

from an apparatus measuring $\frac{1}{2}(p^2q^2 + q^2p^2)$, although in the limiting case of classical mechanics, where quantum effects are negligible, both would measure the same quantity.

Before applying such an apparatus (assuming, for the sake of argument, that it exists although it probably does not) one would have to make sure, either by experimental investigation or by applying quantum mechanics to its working mechanism, which operator actually is to be associated with it. In other words, an apparatus which is quite suitable for measuring a certain quantity in the classical limit may not satisfy our requirements if we rebuild it on a smaller scale, because then quantum effects will have to be taken into consideration.

That one never meets with any difficulty about the order of factors in the usual applications of quantum theory is due to the fact that only very simple operators occur in practice. The most typical of them is the energy of an electron in a field of force, $p^2/2m + V(q)$ (m = mass, V = potential energy). One generally assumes that the function in this exact form, without quantum corrections, has to be taken as the energy operator of a particle the energy of which would be given by the same expression in classical mechanics. This assumption seems very *natural*, but it cannot be *proved* on mere theoretical grounds. All one can say is that this expression is the *most plausible* amongst a variety of different expressions, which all become equal in the classical limit but differ by terms like, for example, $pV - Vp = (h/2\pi i)$ grad V (h = Planck's constant). It is for the experiment to show that the most plausible choice corresponds to reality and that such correction terms are absent.

That the absence of these correction terms cannot be inferred from mere theoretical considerations is proved by the fact that when relativity and spin corrections are taken into account, such terms actually *do occur*.

(**136**, 395; 1935)

R. Peierls: Physical Laboratory, University, Manchester, Aug. 5.

References:

1. *Nature*, 135, 957, June 8, 1935.
2. *Nature*, 136, 179, Aug. 3, 1935.

学的极限情况下，量子效应可以忽略不计，测量以上两个量的结果是一样的，但是以上两个测量装置本质上却是**不一样**的。

在应用这类测量装置之前（为了论证方便，虽然这个测量装置可能并不存在，但我们仍假定它存在）我们必须确定，不管是在实验研究中还是在应用量子力学的工作机制中，哪个算符与它们相联系。换句话说，一个在经典极限情况下适于测量某个量的装置，如果改装并放到小尺度空间中进行测量，则可能不符合测量要求，因为这时候我们需要考虑量子效应了。

我们在量子理论的日常应用中，并没在表达式中因子的次序上遇到问题，这是因为通常在实际情况中只会出现一些非常简单的算符。这些简单算符中最典型的是处在力场中电子的能量算符，$p^2/2m+V(q)$（m 表示电子的质量，V 表示势能）。我们通常认为，不考虑量子修正的话，这个严格表达式中的函数是粒子的能量算符，并且该能量算符给出的系统能量与经典力学中的表达式是一致的。这个假设似乎**顺理成章**，但是单凭理论推导并不能**证明**这个假设是对的。我们只能说这个表达式是众多不同的表达形式中**看似最合理**的一个，而这些不同的表达形式一般在经典极限下是等价的，除非有如 $pV - Vp = (h/2\pi i)\,\nabla V$（$h$ 表示普朗克常数）的项存在时，它们才不一样。实验表明，关于表达式最合理的选择应与现实情况相对应，这时这些修正项便消失了。

实际上，当我们考虑相对论和自旋修正的时候，这些量子效应的修正项**的确出现了**，这表明我们不能单纯地从理论推导的角度得出这些修正项不存在的结论。

（沈乃澂 翻译；李军刚 审稿）

Uncertainty Principle and the Zero-point Energy of the Harmonic Oscillator

R. A. Newing

Editor's Note

In 1913, Albert Einstein and Otto Stern suggested that measurements of the specific heat of hydrogen gas (a measure of how heat input changes its temperature) at low temperature could best be understood if the energy of the molecular vibrations could never be strictly zero, but had an irreducible residual energy. Theorists subsequently predicted such a residual, later known as the zero-point energy, from the equations of quantum mechanics. But the effect lacked intuitive justification. Here R. A. Newing shows that the zero-point energy can be seen as a consequence of Heisenberg's uncertainty principle, which forbids complete specification of position or momentum. Newing shows that the minimal possible value for the zero-point energy is consistent with that derived previously from quantum theory.

ACCORDING to quantum mechanics, an oscillator possesses a definite zero-point energy of vibration, and an attempt has been made to express this result *directly* in terms of some general principle. It has been found that the result may be deduced from the uncertainty principle, in view of the particular relation between position, momentum and energy in a simple harmonic field.

In a state of zero energy the vibrating particle would be at rest at the centre of the field, and its position and momentum would both be known accurately. But this would contradict the uncertainty principle, and the state is therefore not possible. The value of the minimum energy may be calculated from the uncertainty relation $\Delta p \Delta q \geqslant h/2\pi$. The linear harmonic oscillator is defined by the energy equation

$$W = \frac{1}{2}\mu\omega^2 q^2 + \frac{1}{2}p^2/\mu = \text{constant}.$$

If we interpret

$$\text{amplitude of } q = \Delta q = \text{uncertainty in position,}$$
$$\text{amplitude of } p = \Delta p = \text{uncertainty in momentum,}$$

then $$W = \frac{1}{2}\mu\omega^2(\Delta q)^2 = \frac{1}{2}(\Delta p)^2/\mu,$$

giving $$\mu\omega\Delta q = \pm\,\Delta p.$$

For real Δp the positive sign must be taken, and from the uncertainty relation, $(\Delta p)^2 \geqslant h\mu\omega/2\pi$,

测不准原理和谐振子的零点能

纽因

编者按

1913 年，阿尔伯特·爱因斯坦和奥托·斯特恩提出，如果在任何情况下，分子振动能量都不会完全等于零，而是存在着一个无法消除的残余能量，那么低温下氢气比热的测量（一项关于氢气温度随输入热量变化的测量）结果就能得到很好的解释了。紧接着，理论物理学家们从量子力学的方程中推导出了这一残余能量，其后来被称之为零点能。但是这一效应缺乏直观的合理性。这篇文章中，纽因认为零点能可以被看作是海森堡的测不准原理的推论，测不准原理认为不能同时精确的获知微观粒子的位置和动量。纽因的结果表明零点能的可能最小值与之前根据量子理论推导出的结果是一致的。

根据量子力学原理，一个振子具有确定的振动零点能，科学工作者也正在试图使用一些基本的原理来对这一结果进行**直接的**表述。在简单的谐波场中，考虑到位置、动量和能量之间的特殊关系，我们发现或许可以根据测不准原理推导出零点能。

处于零点能态的振动粒子如果会静止于场的中央，那么它的位置和动量都能准确地确定下来。但是这与测不准原理相矛盾，因此谐振子是不可能处于这种态的。我们可以根据测不准关系 $\Delta p\Delta q \geqslant h/2\pi$ 计算出能量的最小值。一维谐振子的能量是由下面的能量方程定义的

$$W=\frac{1}{2}\mu\omega^2q^2+\frac{1}{2}p^2/\mu=\text{常数}$$

如果我们这样理解：

$$q\text{ 的大小}=\Delta q=\text{位置不确定度}$$
$$p\text{ 的大小}=\Delta p=\text{动量不确定度}$$

于是有 $$W=\frac{1}{2}\mu\omega^2\ (\Delta q)^2=\frac{1}{2}\ (\Delta p)^2\ /\mu$$

得出 $$\mu\omega\Delta q=\pm\Delta p$$

对于实数Δp，必须取正号，根据测不准关系，我们推导出 $(\Delta p)^2 \geqslant h\mu\omega/2\pi$，因此可

and therefore $W \geqslant \frac{1}{2}h\omega/2\pi$. Taking the equality sign for the least value of the energy, it follows that the zero-point energy is $\frac{1}{2}h\omega/2\pi$.

(**136**, 395; 1935)

R. A. Newing: Department of Applied Mathematics, University, Liverpool, June 22.

得，$W \geqslant \frac{1}{2}h\omega/2\pi$。当求能量的最小值时取等号，于是得到零点能为 $\frac{1}{2}h\omega/2\pi$。

（沈乃澂 翻译；李淼 审稿）

The Slowing Down of Neutrons by Collisions with Protons

H. von Halban, Jr. and P. Preiswerk

Editor's Note

Fermi had recently shown that neutrons passing through substances containing hydrogen would be slowed by collisions with protons. Here Hans von Halban and Peter Preiswerk explore the process in detail. They note that neutrons of relatively low energy might well be slowed as they transfer energy to the molecules of the medium, in which case the slowing might be influenced by the molecular nature of the medium. So they passed neutrons through water, ethyl alcohol, benzene and liquid paraffin, measuring how the intensity of slow neutrons depended on distance travelled. The results showed that indeed the degree of slowing seemed likely to reflect differences in molecular motions. Slowing of neutrons was later to prove essential for controlled nuclear fission.

FERMI and others[1] showed that neutrons, passing through substances containing hydrogen, loose their energy by collisions with protons. It is of interest to discuss this process of slowing down somewhat further. So long as the energy of the neutron is higher than the energy with which the protons are bound in the molecules of the substance through which the neutrons pass, it seems evident that the latter give, on the average, half their energy to the proton at every collision. But when the neutrons are slowed down below this binding energy, they must excite rotation and oscillation of the hydrogen atom in the molecule in order to lose energy.

It is not certain whether the cross-section of protons for neutrons is a uniform function of the velocity of the neutrons, or if it shows discontinuities for energies comparable with the molecular bindings. In the latter case, it is possible that two substances, containing hydrogen held by different linkages, would show differences in slowing down the neutrons. We have carried out some experiments which indicate the existence of such differences.

Spheres with different radii (5–15 cm.) were alternately filled with water (0.11 gm. H/cm.^{3}), ethyl alcohol (0.10 gm. H/cm.^{3}), benzene (0.067 gm. H/cm.^{3}) and a liquid paraffin (0.14 gm. H/cm.^{3}). In the centre of the sphere a neutron source (radon + beryllium) was placed. The activation of a silver plate, which was fixed on the surface of the spheres and exposed for five minutes to irradiation, served as a measure of the intensity of slow neutrons.

Fig. 1 shows the number of slow neutrons per unit of the solid angle plotted against rd, where r is the radius of the sphere and d the quantity of hydrogen contained by 1 cm.^{3} of the liquid in question. The general aspect of these curves is already known. For small

与质子碰撞而导致的中子慢化

小哈尔班，普雷斯沃克

编者按

费米于近期指出中子穿过含氢物质时会与质子碰撞而减速。这篇文章中汉斯·冯·哈尔班和彼得·普雷斯沃克仔细探讨了这一过程。他们注意到能量相对较低的中子会由于将能量传递给介质分子而减速，在这种情况下，中子减速可能受到介质分子性质的影响。因此，他们使中子穿过水、酒精、苯和液体石蜡，以测量慢中子的强度是如何随穿透深度的变化而变化的。实验结果显示中子慢化的程度似乎反映出了介质分子运动的差别。后来，中子慢化被证明对于可控核裂变是非常关键的。

费米等人[1]指出，中子在通过含氢的物质时会因与质子发生碰撞而损失能量。进一步深入地讨论这一慢化过程是有趣的。只要中子的能量大于它要通过的物质分子中质子的束缚能，中子平均在每一次的碰撞过程中就会有一半能量传给了质子，这点似乎是很明显的。但当中子的能量减少到这种束缚能以下时，中子就一定会激发分子中氢原子的转动和振动，从而使中子的能量减少。

不能确定的是：中子对质子的碰撞截面是否为中子速度的单调函数，或者当中子能量与分子的束缚能可比时，碰撞截面是否呈现不连续性。在后一种情况下，两种含有不同结合形式氢的物质在中子慢化的过程中可能会表现出差异性。我们已经进行了一些实验，结果显示出了这种差异性的存在。

用水（氢含量为 0.11 g/cm^3）、乙醇（氢含量为 0.10 g/cm^3）、苯（氢含量为 0.067 g/cm^3）和液体石蜡（氢含量为 0.14 g/cm^3）分别装满半径不同（5 ~ 15 cm）的球。在球的中央放置着一个中子源（氡 + 铍），在球的表面安置一个银片并对其辐照 5 分钟，然后利用银片的放射性测量慢中子的强度。

图 1 显示了单位立体角中慢中子的数量随着 rd 的变化关系，其中 r 是球的半径，d 是 1 cm^3 液体的含氢量。这些曲线的一般意义是众所周知的。对小半径而言，随着

radii a rapid increase of the intensity with increasing radius is observed, due to the slowing down of neutrons by collisions with protons. After a certain point, an increase of the radius causes a reduction of the intensity. This clearly shows that not all neutrons which pass the surface of a sphere are reaching the next bigger sphere. The vanishing of slow neutrons must be ascribed to absorption.

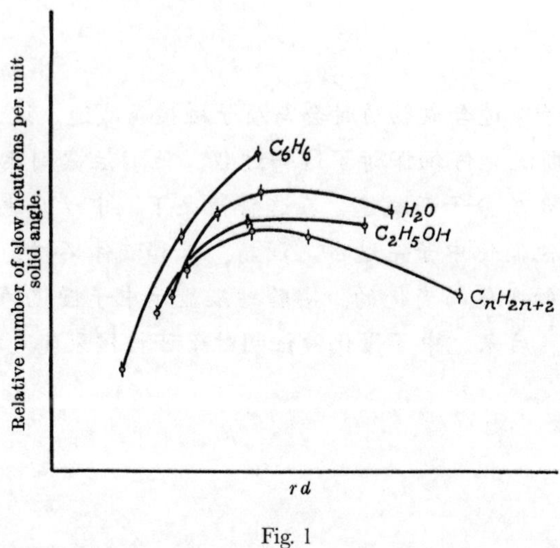

Fig. 1

The absorption of slow neutrons by paraffin and other substances containing hydrogen has been studied in detail by Bjerge and Westcott[2], who found that the number of slow neutrons is reduced to a half after diffusion through 1.6 cm. of water. By a different method we obtained the value of 2.5 cm. for paraffin in a preliminary experiment. A source of neutrons was placed in the centre of a paraffin wax cube of 14 cm. side. Five plates of paraffin wax, each 1 cm. thick, and finally a small silver plate were placed upon this cube. A screen of cadmium was interposed between the paraffin plates at different distances from the silver plate. The activity of the silver, obtained for equal times of irradiation, increased when the distance between the silver and the cadmium was increased. A curve was obtained which showed that influence of the absorption of the cadmium decreased to a half when the distance between the silver plate and the absorber was increased by 2.5 cm.

Fig. 1 shows that the maximum values of intensity are different for different liquids; these differences cannot be ascribed to the quantities of hydrogen contained by the liquids alone. Also, it is not possible to explain these results by absorption of slow neutrons by oxygen or carbon nuclei. A neutron has, for the same number of collisions with protons, passed twice the number of carbon atoms in benzene as in the liquid paraffin, and we see that the maximum value for the latter is much lower than for benzene, where the maximum seems to be just reached with the biggest sphere.

The differences in the influence of the four liquids examined, on the intensity of slow neutrons, cannot be ascribed to differences in the quantities of hydrogen, carbon or

半径的增大，观测到中子的强度显著增加，这是由于中子与质子碰撞而速度减慢。在某一个固定值之后，随着半径的增大，中子的强度反而减弱。这清楚地表明，并不是所有通过球表面的中子都会到达下一个更大的球面上。慢中子的消失必然是因为它被吸收了。

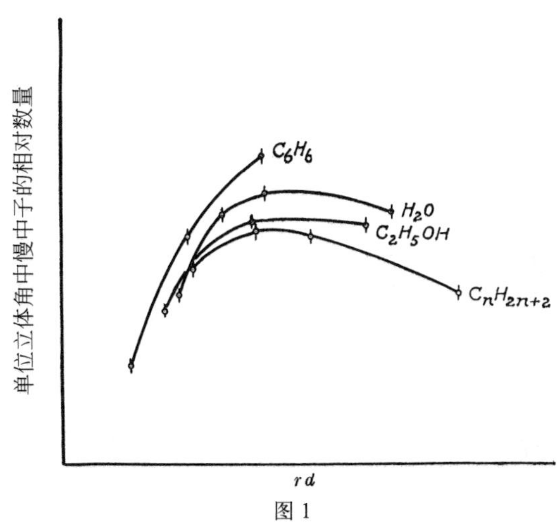

图 1

伯格和韦斯科特[2] 已经详细地研究了石蜡和其他含氢物质对慢中子的吸收，他们发现，其在通过厚 1.6 cm 的水之后，慢中子的数量减少了一半。在对石蜡进行的初步实验中，我们采用不同的方法得到的数值为 2.5 cm。把一个中子源放在边长为 14 cm 的立方体形石蜡的中心，在这个立方体石蜡上放置 5 个 1 cm 厚的石蜡板，最后放置一块小银板。在石蜡板之间插入一块镉片，并变换其位置以使其与银板的距离不同。在相等辐射时间内，当银板和镉板的距离增大时，银的放射性会增加。由实验得到的曲线表明，当银板与镉板之间的距离增大了 2.5 cm 时，镉吸收的影响减小到一半。

图 1 表明，对不同的液体，慢中子强度的最大值是不相同的；这些差别不能仅仅归因于液体中氢含量的不同。同样，也不能用氧核或碳核吸收慢中子来解释。在与质子碰撞次数相同的情况下，中子在苯中通过的碳原子数是在液体石蜡中通过碳原子数的两倍，而且我们还发现，液体石蜡中慢中子强度的最大值远远低于在苯中慢中子强度的最大值，而苯中的最大值似乎得用半径最大的球才刚能达到。

被测试的四种液体对于慢中子强度的影响力的差别，不能仅归因于是这些液体中氢、碳或氧含量的不同。这些液体的分子结构的不同也导致了它们之间的其他一

oxygen these liquids contain. Other differences between these liquids depend upon their molecular structure. Thus different probabilities for the slowing down of neutrons by excitation of rotation and oscillation of the hydrogen atoms in the different molecules may account for the discrepancies observed.

(**136**, 951-952; 1935)

Hans von Halban, Jr. and Peter Preiswerk: Institut du Radium, Laboratoire Curie, Paris, Nov. 5.

References:

1. Fermi and others, *La Ricerca Scientifica*, (v) 2, 1; 1934. (vi) 1, 1; 1935.
2. T. Bjerge and C. H. Westcott, *Proc. Roy. Soc.*, A, **150**, 709; 1935.

些差别。因此，或许我们可以这样解释观测到的差别，即不同分子中通过激发氢原子的转动和振动来减慢中子的概率不同。

（沈乃澂 翻译；王乃彦 审稿）

Viscosity of Liquid Helium below the λ-point

P. Kapitza

Editor's Note

Earlier experiments had noted strange behaviour in liquid helium at temperatures below about 2.18 K, where the liquid conducts heat with extraordinary efficiency. Here, physicist Pyotr Kapitza suggests that a dramatic decrease in the fluid's viscosity might explain the phenomenon, as it would make heat transport by convection much easier. Some recent experiments, he notes, had measured a decrease in the viscosity, but now his group, in a much more sensitive experiment, has found that the viscosity of the liquid below 2.18 K is at least 10,000 times smaller than that of any other known substance. Kapitza suggested this might be explained if liquid helium becomes a "superfluid" at low temperatures, perhaps even flowing with no viscosity at all.

THE abnormally high heat conductivity of helium II below the λ-point, as first observed by Keesom, suggested to me the possibility of an explanation in terms of convection currents. This explanation would require helium II to have an abnormally low viscosity; at present, the only viscosity measurements on liquid helium have been made in Toronto[1], and showed that there is a drop in viscosity below the λ-point by a factor of 3 compared with liquid helium at normal pressure, and by a factor of 8 compared with the value just above the λ-point. In these experiments, however, no check was made to ensure that the motion was laminar, and not turbulent.

The important fact that liquid helium has a specific density ρ of about 0.15, not very different from that of an ordinary fluid, while its viscosity μ is very small comparable to that of a gas, makes its kinematic viscosity $\nu = \mu/\rho$ extraordinary small. Consequently when the liquid is in motion in an ordinary viscosimeter, the Reynolds number may become very high, while in order to keep the motion laminar, especially in the method used in Toronto, namely, the damping of an oscillating cylinder, the Reynolds number must be kept very low. This requirement was not fulfilled in the Toronto experiments, and the deduced value of viscosity thus refers to turbulent motion, and consequently may be higher by any amount than the real value.

The very small kinematic viscosity of liquid helium II thus makes it difficult to measure the viscosity. In an attempt to get laminar motion the following method (shown diagrammatically in the accompanying illustration) was devised. The viscosity was measured by the pressure drop when the liquid flows through the gap between the disks 1 and 2; these disks were of glass and were optically flat, the gap between them being adjustable by mica distance pieces. The upper disk, 1, was 3 cm. in diameter with a central hole of 1.5 cm. diameter, over which a glass tube (3) was fixed. Lowering and raising this plunger in the liquid helium by means of the thread (4), the level of the liquid column in

液态氦在 λ 点以下的黏度

卡皮查

编者按

在以往的实验中曾记录到一种奇怪的现象，当温度低于约 2.18 K 时，液氦的热导率惊人的高。在这篇文章中，物理学家彼得 · 卡皮查认为这一物理现象是由于液体黏度的急剧减小所造成的，因为这样会使得对流传热更加容易。他注意到最近一些实验测量到黏度的减小，但现在他的研究组通过更为精确的实验发现，低于 2.18 K 时液氦的黏度至多是其他已知物质的测量值的 1/10,000。卡皮查认为如果液氦在这样的低温下变成了“超流体”，甚至流动时黏度为零，那么这个实验就能得到解释。

如同凯索姆首次观测到的那样，液氦 II 在 λ 点之下时表现出了异常高的热导率，这启发了我：也许可以用对流对它进行解释。如果这样解释的话就要求氦 II 的黏度非常小；到目前为止，仅有的对液氦黏度的测量是在多伦多进行的 [1]，测量结果显示，低于 λ 点时液氦的黏度是常压下的 1/3，是略微高于 λ 点时的 1/8。然而，在上述实验中，并没有完全确认液氦的运动形式就是层流，而不是湍流。

重要的是，由于液氦的比重 ρ 大约为 0.15，与普通流体没有很大差别，但其黏度 μ 与普通气体相比却非常小，这样就使得液氦的动力黏度 $\nu = \mu/\rho$ 格外的小。于是，当液氦在普通黏度计中流动时，雷诺数可能会很高，然而只有雷诺数保持很小的值时，才能使运动保持层流形式，对于在多伦多所使用的测量圆柱振子阻尼的方法尤其是这样。但是多伦多实验并不满足这一条件，因此推导出来的黏度值应该对应于湍流运动，而这比真实值大多少都有可能。

液氦 II 的动力黏度很小，这就使得测量其黏度比较困难。为尽量保持层流，我们设计出了如下实验方法（见附图）。利用液体流经圆盘 1 和 2 之间空隙时的压强差可以测得黏度；其中两个圆盘是用玻璃制成的，并且是光学级平整的，它们之间的空隙可以利用云母隔离片进行调节。上面的圆盘 1 直径为 3 厘米，中间有一个直径 1.5 厘米的洞，洞上固定着一个玻璃管（3）。利用线（4）使这个装置在液氦中上升和下降，就可以令管 3 中液柱的水平位置高于或低于外围杜瓦瓶中液面（5）。利用测高

the tube 3 could be set above or below the level (5) of the liquid in the surrounding Dewar flask. The amount of flow and the pressure were deduced from the difference of the two levels, which was measured by cathetometer.

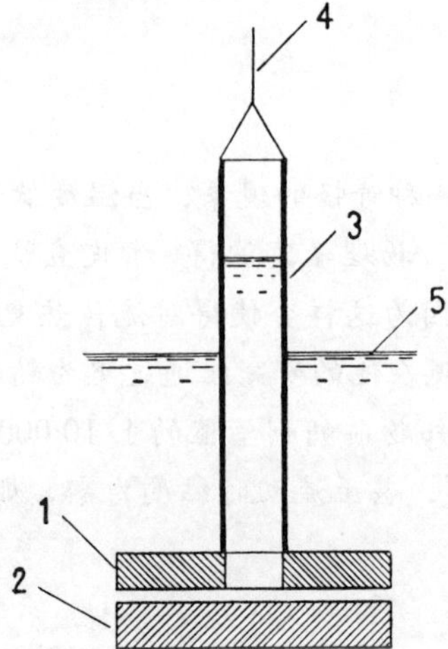

The results of the measurements were rather striking. When there were no distance pieces between the disks, and the plates 1 and 2 were brought into contact (by observation of optical fringes, their separation was estimated to be about half a micron), the flow of liquid above the λ-point could be only just detected over several minutes, while below the λ-point the liquid helium flowed quite easily, and the level in the tube 3 settled down in a few seconds. From the measurements we can conclude that the viscosity of helium II is at least 1,500 times smaller than that of helium I at normal pressure.

The experiments also showed that in the case of helium II, the pressure drop across the gap was proportional to the square of the velocity of flow, which means that the flow must have been turbulent. If, however, we calculate the viscosity, assuming the flow to have been laminar, we obtain a value of the order 10^{-9} C.G.S., which is evidently still only an upper limit to the true value. Using this estimate, the Reynolds number, even with such a small gap, comes out higher than 50,000, a value for which turbulence might indeed be expected.

We are making experiments in the hope of still further reducing the upper limit to the viscosity of liquid helium II, but the present upper limit (namely, 10^{-9} C.G.S.) is already very striking, since it is more than 10^4 times smaller than that of hydrogen gas (previously thought to be the fluid of least viscosity). The present limit is perhaps sufficient to suggest, by analogy with supraconductors, that the helium below the λ-point enters a special state which might be called a "superfluid".

As we have already mentioned, an abnormally low viscosity such as indicated by our experiments might indeed provide an explanation for the high thermal conductivity, and for the other anomalous properties observed by Allen, Peierls, and Uddin[2]. It is evidently possible that the turbulent motion, inevitably set up in the technical manipulation required

计读出两液面之间的差，我们就可以推导出流量和压力的大小。

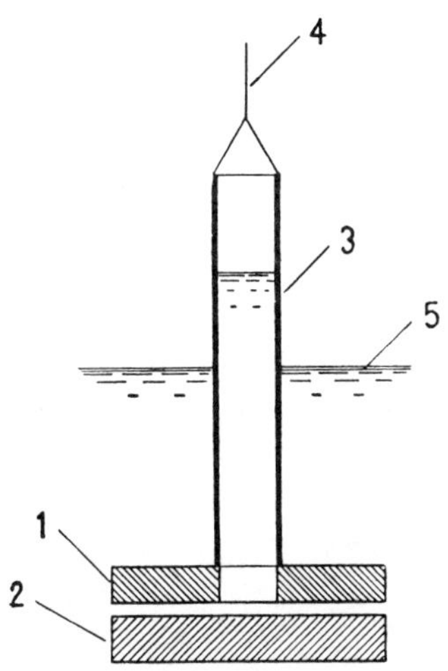

测量的结果相当明显。当两盘之间没有隔离片时，即令盘 1 和 2 接触（通过光学条纹的观测，估计其间距约为半微米），液氦流在高于 λ 点时需要经过几分钟才勉强能检测到，而在低于 λ 点时，液氦流动非常容易，管 3 中的液面在几秒钟内就稳定下来了。从这些测量我们可以推断，常压下氦 II 的黏度是氦 I 的 1,500 分之一。

实验还显示，氦 II 穿过空隙时的压强差正比于流动速度的平方，这表明这种流动一定是湍流。但是，如果我们假设它以层流的形式运动，并计算它的黏度，得到数量级为 10^{-9} 单位（厘米克秒制）的值，显然，这只是真实值的上限。按照这种估计，尽管间隙如此小，雷诺数仍然超过了 50,000 这个肯定会出现湍流的数值。

我们还在继续实验工作，期望进一步减小液氦 II 的黏度上限值，但是目前的上限（即 10^{-9} 单位（厘米克秒制））已经非常小了，因为它最多是氢气（过去认为是黏度最小的流体）黏度的 $1/10^4$。目前得到的极限也许足以表明，与超导体进行类比，氦在低于 λ 点时会进入一种特殊的状态，我们不妨称之为“超流体”。

如同我们之前提到的那样，我们的实验中显示的异常低的黏度，可能确实可以用来解释高热导率，以及由艾伦、佩尔斯和乌丁 [2] 所观测到的其他异常性质。在对液氦 II 进行所需的技术操作过程中，由于其具有较大的流动性，因此湍流的出现显

in working with the liquid helium **II**, might on account of the great fluidity, not die out, even in the small capillary tubes in which the thermal conductivity was measured; such turbulence would transport heat extremely efficiently by convection.

(**141**, 74; 1938)

P. Kapitza: Institute for Physical Problems, Academy of Sciences, Moscow, Dec.3.

References:

1. Burton, *Nature*, 135, 265 (1935); Wilhelm, Misener and Clark, *Proc. Roy. Soc.*, A, **151**, 342 (1935).
2. *Nature*, **140**, 62 (1937).

然是无法避免的，即便在用于测量热导率的小毛细管中；而这种湍流运动将通过对流效率极高地传输热。

（王耀杨 翻译；于渌 审稿）

Flow of Liquid Helium II

J. F. Allen and A. D. Misener

Editor's Note

In Cambridge, physicists John Allen and Don Misener had come to the same conclusion as Kapitza (see the previous paper). As they announce here, they had also attempted to measure the viscosity of liquid helium below the so-called "lambda point" of 2.18 K in experiments where the fluid flowed through narrow capillaries. But their results could only place an upper limit on the viscosity, comparable to Kapitza's value. They note that the flow they observed is very strange, being independent of the pressure difference applied. It is possible, they speculate, that the liquid is actually slipping over the surface. These results would later be explained by a general theory of liquid helium as a superfluid, a state that arises from quantum-mechanical behaviour.

A survey of the various properties of liquid helium II has prompted us to investigate its viscosity more carefully. One of us[1] had previously deduced an upper limit of 10^{-5} C.G.S. units for the viscosity of helium II by measuring the damping of an oscillating cylinder. We had reached the same conclusion as Kapitza in the letter above; namely, that due to the high Reynolds number involved, the measurements probably represent non-laminar flow.

The present data were obtained from observations on the flow of liquid helium II through long capillaries. Two capillaries were used; the first had a circular bore of radius 0.05 cm. and length 130 cm. and drained a reservoir of 5.0 cm. diameter; the second was a thermometer capillary 93.5 cm. long and of elliptical cross-section with semi-axes 0.001 cm. and 0.002 cm., which was attached to a reservoir of 0.1 cm. diameter. The measurements were made by raising or lowering the reservoir with attached capillary so that the level of liquid helium in the reservoir was a centimetre or so above or below that of the surrounding liquid helium bath. The rate of change of level in the reservoir was then determined from the cathetometer eye-piece scale and a stopwatch; measurements were made until the levels became coincident. The data showing velocities of flow through the capillary and the corresponding pressure difference at the ends of the capillary are given in the accompanying table and plotted on a logarithmic scale in the diagram.

液氦 II 的流动

艾伦，麦色纳

编者按

在剑桥，物理学家约翰·艾伦和冬·麦色纳得出了与卡皮查相同的结论（见前一篇文章）。正如他们在这篇文章中介绍的，他们利用液体流过毛细管的实验，也试图去测量低于“λ点”（2.18 K）时液氦的黏度。但是与卡皮查的测量值相比，他们的结果仅仅得出了黏度值的上限。他们注意到他们观察到的流动现象非常奇特，它与两端施加的压力差无关。他们推测液体实际上是掠过表面。后来液氦超流态的普遍理论成功解释了这些现象，这是一种源于量子力学行为的状态。

对液氦 II 各种性质的探索促使我们更细致地研究它的黏度。本文作者之一[1]曾经通过测定振动圆柱的阻尼推导出液氦 II 黏度的上限为 10^{-5} 单位（厘米克秒制）。我们得到的结论与卡皮查在前文中叙述的相同；也就是说，由于测量过程中雷诺数很高，这一测量可能反映的是非层流的流动形式。

本文的数据是通过观测液氦 II 在长毛细管中的流动获得的。其间用到了两种毛细管；第一种毛细管带有一个半径为 0.05 厘米、长为 130 厘米的圆形孔道，并与一个直径为 5 厘米的储液管相接；第二种毛细管是温度计毛细管，长 93.5 厘米，具有椭圆形的横截面，其半轴长分别为 0.001 厘米和 0.002 厘米，整个毛细管与直径为 0.1 厘米的储液管相连。测量方法是：将连接毛细管的储液管上下移动，以便于使储液管中的液氦平面位于周围液氦平面的上方或下方约 1 厘米处。利用测高计的目镜刻度和秒表便可确定储液管中液面的移动速率；当内外液面一致时即可停止测量。附表中数据显示了流经毛细管时的速度和相应的毛细管两端的压力差，对它们分别取对数绘于图中。

Capillary I		Capillary II			
T=1.07°K.		T=1.07°K.		T=2.17°K.	
Velocity (cm./sec.)	Pressure (dynes)	Velocity (cm./sec.)	Pressure (dynes)	Velocity (cm./sec.)	Pressure (dynes)
13.9	183.5	8.35	402	0.837	36.6
11.5	154.5	6.92	218	0.757	31.3
10.3	127.7	6.88	143	0.715	26.1
9.0	105.0	6.30	101	0.685	21.1
8.2	83.5	6.05	56	0.655	16.4
7.5	65.7	5.55	30	0.609	12.1
6.9	49.3	4.70	11.3	0.570	8.3
6.1	34.1	4.39	9.2	0.525	4.3
5.2^{5}	20.3	3.92	13.0	0.433	0.9
4.5^{5}	15.2	2.88	7.2		

The following facts are evident:

(*a*) The velocity of flow, q, changes only slightly for large changes in pressure head, p. For the smaller capillary, the relation is approximately $p \propto q^6$, but at the lowest velocities an even higher power seems indicated.

(*b*) The velocity of flow, for given pressure head and temperature, changes only slightly with a change of cross-section area of the order of 10^3.

(*c*) The velocity of flow, for given pressure head and given cross-section, changes by about a factor of 10 with a change of temperature from 1.07°K. to 2.17°K.

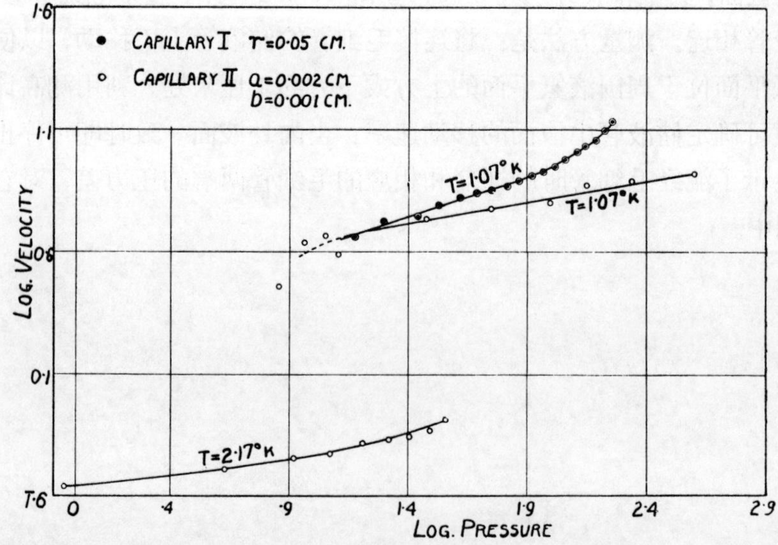

毛细管 I		毛细管 II			
T = 1.07 K		T = 1.07 K		T = 2.17 K	
速度 （厘米 / 秒）	压力 （达因）	速度 （厘米 / 秒）	压力 （达因）	速度 （厘米 / 秒）	压力 （达因）
13.9	183.5	8.35	402	0.837	36.6
11.5	154.5	6.92	218	0.757	31.3
10.3	127.7	6.88	143	0.715	26.1
9.0	105.0	6.30	101	0.685	21.1
8.2	83.5	6.05	56	0.655	16.4
7.5	65.7	5.55	30	0.609	12.1
6.9	49.3	4.70	11.3	0.570	8.3
6.1	34.1	4.39	9.2	0.525	4.3
5.2^5	20.3	3.92	13.0	0.433	0.9
4.5^5	15.2	2.88	7.2		

从上表明显可以看出：

（a）当压位差 p 大幅度改变时，流动速度 q 的变化不大。对于较细的毛细管，二者之间的关系近似为 $p \propto q^6$，但是在速度最低的时候，幂次似乎更高。

（b）对于给定的压位差和温度，当横截面面积以 10^3 数量级变化时，流动速度只发生微小的改变。

（c）对于给定的压位差和确定的横截面，在温度从 1.07 K 变化到 2.17 K 的过程中，流动速度变化了 10 倍左右。

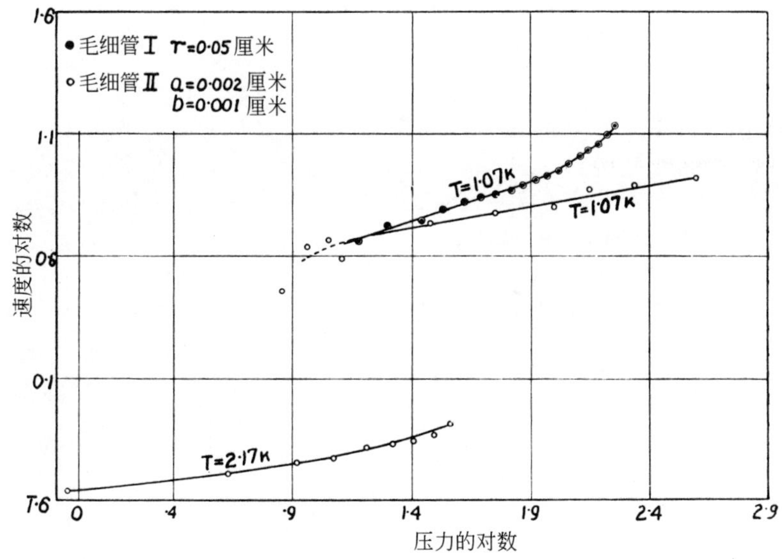

(*d*) With the larger capillary and slightly higher velocities of flow, the pressure-velocity relation is approximately $p \propto q^3$,with the power of q decreasing as the velocity is increased.

If, for the purpose of calculating a possible upper limit to the viscosity, we assume the formula for laminar flow, that is, $p \propto q$, we obtain the value $\eta = 4\times10^{-9}$ C.G.S. units. This agrees with the upper limit given by Kapitza who, using velocities of flow considerably higher than ours, has obtained the relation $p \propto q^2$ and an upper limit to the viscosity of $\eta = 10^{-9}$ C.G.S. units.

The observed type of flow, however, in which the velocity becomes almost independent of pressure, most certainly cannot be treated as laminar or even as ordinary turbulent flow. Consequently any known formula cannot, from our data, give a value of the "viscosity" which would have much meaning. It may be possible that the liquid helium II slips over the surface of the tube. In this case any flow method would be incapable of showing the "viscous drag" of the liquid.

With regard to the suggestion that the high thermal conductivity of helium II might be explained by turbulence, we have calculated that the flow velocity necessary to transport all the heat input over the observed temperature gradient in the Allen, Peierls and Uddin experiments[2] is about 10^4 cm./sec. On the other hand, the greatest flow velocity produced by manipulation and by the pressure difference along the thermal conduction capillary will not be likely to be greater than 50 cm./sec. It seems, therefore, that undamped turbulent motion cannot account for an appreciable part of the high thermal conductivity which has been observed for helium II.

(**141**, 75; 1938)

J. F. Allen and A. D. Misener: Royal Society Mond Laboratory, Cambridge, Dec, 22.

References:

1. Burton, E.F., *Nature*, 135, 265 (1935).
2. Allen, Peierls and Uddin, *Nature*, **140**, 62 (1937).

(d) 在较粗的毛细管中，对于略高的流动速度，压力－速度之间的关系近似为 $p \propto q^3$，其中 q 的幂次随速度增加而降低。

为了计算出黏度的可能上限，我们假定层流运动的计算公式即 $p \propto q$ 成立，那么就可以得到 $\eta = 4 \times 10^{-9}$ 单位(厘米克秒制)。这个结果与卡皮查给出的上限是一致的，因为他使用的流动速度比我们所用的高很多，所以得出的关系式为 $p \propto q^2$，他最后给出的黏度上限是 $\eta = 10^{-9}$ 单位（厘米克秒制)。

不过，当流动速度与压力无关时，我们所观测到的这种流动类型一定不属于层流，甚至不属于普通的湍流。因此，根据我们的测量数据任何已知的公式都不可能给出一个很有意义的“黏度”数值。液氦 II 可能掠过管壁，在这种情况下，任何流动方法都无法显示出液体的“黏滞阻力”。

有人认为用湍流也许能够解释氦 II 的高热导率，对此我们计算的结果是，只有流速大约为 10^4 厘米 / 秒左右时，才足以在艾伦、佩尔斯和乌丁的实验 [2] 中观测到的温度梯度条件下传输所有输入的热量。而另一方面，在实验操作下沿着热传导毛细管的压力差所能产生的最大流速可能不会超过 50 厘米 / 秒。由此看来，无衰减的湍流运动并不能解释在氦 II 中观测到的高热导率。

（王耀杨 翻译；于渌 审稿）

The λ-phenomenon of Liquid Helium and the Bose–Einstein Degeneracy

F. London

Editor's Note

In early 1938, *Nature* published the first reports of superfluidity in liquid helium, discovered independently by Pyotr Kapitza in Russia and by John Allen and Don Misener at Cambridge. Both reported that when cooled below 2.17 K, the viscosity of the liquid plunged apparently to zero. One proposed explanation was that the helium atoms adopted an ordered structure like that of diamond crystals. Here Fritz London proposes a radically different idea: that the phenomenon might be linked to the low-temperature quantum behaviour of particles of integer spin ("bosons"), whose collective statistical behaviour had been described by Bose and Einstein.

IN a recent paper[1] Fröhlich has tried to interpret the λ-phenomenon of liquid helium as an order–disorder transition between n holes and n helium atoms in a body-centred cubic lattice of $2n$ places. He remarks that a body-centred cubic lattice may be considered as consisting of two shifted diamond lattices, and he assumes that below the λ-point the helium atoms prefer the places of one of the two diamond lattices. The transition is treated on the lines of the Bragg–Williams–Bethe theory as a phase transition of second order in close analogy to the transition observed with β-brass. Jones and Allen in a recent communication to *Nature*[2] also referred to this idea. In both these papers, use is made of the fact, established by the present author, that with the absorbed abnormally great molecular volume of liquid helium (caused by the zero motion[3]) the diamond-configuration has the lowest potential energy among all regular lattice structures[4].

In this note, I should like first to show that the mechanism proposed by Fröhlich cannot be maintained and then to direct attention to an entirely different interpretation of this strange phenomenon.

(1) According to Fröhlich, a diamond lattice of He atoms, when partly formed, should offer, to any other He atom, a preference for being attached at those points which belong to the same diamond lattice, that is, the binding energy at a diamond point should be greater than anywhere else. It is, however, easy to see that just those points, which according to Fröhlich should become less favoured for low temperatures, have an appreciably greater binding energy. It is true these holes have four nearest neighbours at exactly the same distance (3.08 A.), as the lattice points of the diamond lattice have, but in addition they possess six second neighbours at the distance of 3.57 A. which the diamond lattice points do not possess, and these second neighbours contribute considerably to the binding energy just at the hole-places (about 50 percent to the potential energy).

液氦的 λ 现象和玻色–爱因斯坦简并

伦敦

编者按

1938 年初，《自然》刊载了两篇有关液氦超流动性的最早报告，那是俄国的彼得·卡皮查和剑桥的约翰·艾伦与冬·麦色纳各自独立发现的。两篇文章都提到当温度低于 2.17 K 时液氦的黏度似乎减小至零。有人认为这是由于氦原子排列成为类似金刚石晶体那样的有序结构。而这篇文章中弗里茨·伦敦提出了一个完全不同的看法：这种现象也许与某些具有整数自旋的粒子（“玻色子”）在低温下的量子力学行为有关，玻色和爱因斯坦曾经对玻色子的统计行为进行过描述。

弗勒利希最近的一篇文章中[1]试图将液氦的 λ 现象解释为在一个具有 $2n$ 个位置的体心立方晶格中，n 个空位和 n 个氦原子间的有序–无序相变。他认为，可以把一个体心立方晶格看作是由两套具有相对位移的金刚石晶格组合而成的，而且他假定在低于 λ 点时，氦原子更倾向于处在两套金刚石晶格中某一套的格点上。根据布拉格–威廉姆斯–贝特理论，这种转变被看作是一种二阶相变，非常类似于在 β 黄铜中发生的相变。琼斯和艾伦最近给《自然》的信[2]中也提到了这一观点。在这两篇文章中，都用到了本文作者确认的事实，即由于液氦占有的分子体积异常大（源于零点运动[3]），所以在所有规则结构的晶格中，金刚石结构具有最低的势能[4]。

在这篇短文中，我首先会说明弗勒利希提出的机制是不成立的，然后再引导大家关注对这种奇异现象的一种完全不同的解释。

（1）按照弗勒利希的观点，当 He 原子的金刚石晶格部分形成时，对于其他的 He 原子而言，属于同一套金刚石晶格上的格点应该更有优势，也就是说，金刚石晶格上的结合能应该大于其他点处的结合能。然而，我们很容易看出，按照弗勒利希的观点，在低温下对结合较为不利的那些位点反而具有比其他位点高得多的结合能。晶格中的空位在完全相同的距离（3.08 埃）处有 4 个最近邻，这与金刚石晶格的格点相同，但不同的是它们在距离 3.57 埃处还有 6 个次近邻，而且这些次近邻对空位的结合能贡献是很大的（约占势能的 50%）。因此，弗勒利希提出的这种合作现象是不会发生的。原子不仅易处于金刚石晶格的格点上，而且更易处于空位上，这意味

Therefore, actually no such co-operative phenomenon will appear as supposed by Fröhlich. The atoms would rather frequent the holes as much as the proper diamond lattice points, and this would signify that we have a body-centred lattice of $2n$ places for n atoms, every place being occupied with the probability $\frac{1}{2}$ only—even at the absolute zero. In this configuration every atom has on the average four nearest neighbours at a distance of 3.08 A, as in the diamond configuration, but in addition there are here on the average three second neighbours at the distance of 3.57 A. In the diamond lattice there are twelve second neighbours but at a distance of 5.04 A., where there is almost no Van der Waals field. It might be mentioned, by the way, that a face-centred lattice of $2n$ places for n atoms (on the average 6 first neighbours at a distance of 3.17 A.) has been found to have a still little lower energy than the configuration just discussed of the co-ordination number 4.

Complete numerical details cannot be given here; in any event it can be shown by such energetic discussions that a static spatial model of liquid He II of whatever regular configuration is certainly not possible. This has been previously suggested in consideration of the great zero point amplitude calculated for He^4. The determination of the most favourable co-ordination numbers of the first and second neighbours, however, maintains a good physical meaning: it may be considered as a rough Hartree calculus which yields the self-consistent field and the corresponding probability distribution of the atoms belonging to the minimum of energy.

(2) It seems, therefore, reasonable to imagine a model in which each He atom moves in a self-consistent periodic field formed by the other atoms. The different states of the atoms may be described by eigen functions of a similar type to the electronic eigen functions which appear in Bloch's theory of metals; and, as in Bloch's theory, the energy of the lowest states will roughly be represented by a quadratic function of the wave number K,

$$E = \frac{\hbar^2}{2m^*}K^2,$$

the effective mass m^* being of the order of magnitude of the mass of the atoms. But in the present case we are obliged to apply Bose–Einstein statistics instead of Fermi statistics.

(3) In his well-known papers, Einstein has already discussed a peculiar condensation phenomenon of the "Bose–Einstein" gas; but in the course of time the degeneracy of the Bose–Einstein gas has rather got the reputation of having only a purely imaginary existence. Thus it is perhaps not generally known that this condensation phenomenon actually represents a discontinuity of the derivative of the specific heat (phase transition of third order). In the accompanying figure the specific heat (C_v) of an *ideal* Bose–Einstein gas is represented as a function of T/T_0 where

$$T_0 = \frac{h^2}{2\pi m^* k}\left(\frac{n}{2{,}615}\right)^{2/3},$$

With m^* = the mass of a He atom and with the mol. volume $\frac{N_i}{n}$ = 27.6 cm.[3] one obtains $T_0 = 3.09°$. For $T \leqslant T_0$ the specific heat is given by

着对于 n 个原子来说，我们有 $2n$ 个位置的体心立方格子，每个位置被占据的概率即使在绝对零度下也仅为 1/2。每个原子在距离 3.08 埃处平均有 4 个最近邻，这与金刚石晶格结构相同，但除此之外，每个原子在距离 3.57 埃处平均还有 3 个次近邻。而在金刚石晶格中有 12 个次近邻，但距离是 5.04 埃，几乎没有范德瓦尔斯场。顺便提一下，对于 $2n$ 个位置，n 个原子的面心立方晶格（在距离 3.17 埃处平均有 6 个最近邻），其能量略低于刚才讨论过的配位数为 4 的结构。

在本文中我们不可能给出完整的数值计算过程；但是，通过对这种能量的讨论可以阐明，液体 He II 任何规则结构的静态空间分布模型肯定都是不可能的。这点在计算 He 巨大的零点振幅时就已被提出 [4]。然而，确定最近邻和次近邻的最佳配位数却有着清楚的物理意义，其可以看作是粗略的哈特里计算，由此我们可得到自洽场和相对应的处于能量极小值的原子的分布概率。

（2）因此似乎可以构造这样一个模型，其中每个 He 原子在其他原子形成的自洽周期场中运动。用本征函数来描述原子不同的态，类似于在布洛赫的金属理论中用电子本征函数描述电子态；而且，类似于布洛赫理论，最低态的能量可以粗略地表示为波矢 K 的二次函数，

$$E = \frac{\hbar^2}{2m^*} K^2$$

其中有效质量 m^* 与原子质量的数量级相当。不过我们现在必须用玻色–爱因斯坦统计代替费米统计。

（3）在一些著名的文章中，爱因斯坦已经讨论了“玻色–爱因斯坦”气体特有的一种凝聚现象；但是，随着时间的推移，人们开始认为玻色–爱因斯坦气体的简并性不过是一种假象。因此可能很多人并不知道这种凝聚现象实际上表现为比热的导数的不连续（三阶相变）。在附图中，**理想**玻色–爱因斯坦气体的比热（C_v）可表示为 T/T_0 的函数，

$$T_0 = \frac{h^2}{2\pi m^* k}\left(\frac{n}{2,615}\right)^{2/3}$$

式中取 m^* = He的原子质量，摩尔体积 $\frac{N_l}{n} = 27.6$ 立方厘米，可得 $T_0 = 3.09$ K。对于 $T \leqslant T_0$，所给出的比热为：

$$C_v = 1.92\ R(T/T_0)^{3/2}$$

and for $T \geqslant T_0$ by

$$C_v = \frac{3}{2}R\left[1 + 0.231\left(\frac{T_0}{T}\right)^{3/2} + 0.046\left(\frac{T_0}{T}\right)^3 + \cdots\right]$$

The entropy at the transition point T_0 amounts to 1.28 R independently of T_0.

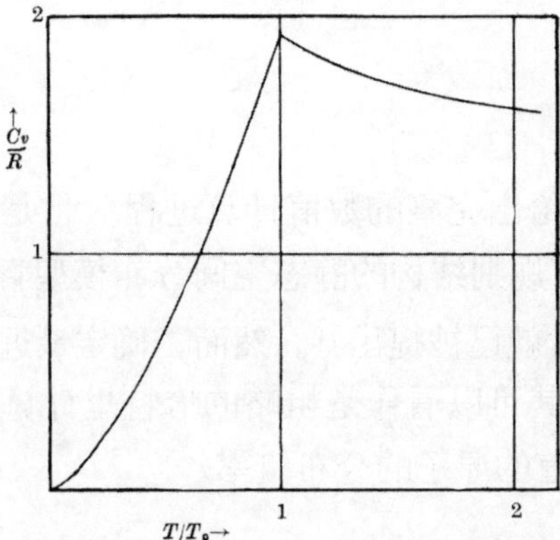

Specific heat of an ideal Bose-Einstein gas.

(4) Though actually the λ-point of helium resembles rather a phase transition of second order, it seems difficult not to imagine a connexion with the condensation phenomenon of the Bose–Einstein statistics. The experimental values of the temperature of the λ-point (2.19°) and of its entropy (~0.8 R) seem to be in favour of this conception. On the other hand, it is obvious that a model which is so far away from reality that it simplifies liquid helium to an ideal gas, cannot, for high temperatures, yield but the value $C_v = 3/2\ R$, and also for low temperatures the ideal Bose–Einstein gas must, of course, give too great a specific heat, since it does not account for the gradual "freezing in" of the Debye frequencies.

According to our conception the quantum states of liquid helium would have to correspond, so to speak, to both the states of the electrons and to the Debye vibrational states of the lattice in the theory of metals. It would, of course, be necessary to incorporate this feature into the theory before it can be expected to furnish quantitative insight into the properties of liquid helium.

The conception here proposed might also throw a light on the peculiar transport phenomena observed with He II (enormous conductivity of heat[5], extremely small viscosity[6] and also the strange fountain phenomenon recently discovered by Allen and Jones[2]).

A detailed discussion of these questions will be published in the *Journal de Physique*.

(**141**, 643-644; 1938)

$$C_v = 1.92\,R\,(T/T_0)^{3/2}$$

对于$T \geqslant T_0$，比热为：

$$C_v = \frac{3}{2}R\left[1 + 0.231\left(\frac{T_0}{T}\right)^{3/2} + 0.046\left(\frac{T_0}{T}\right)^{3} + \cdots\right]$$

在转变点 T_0，熵为 1.28R，与 T_0 无关。

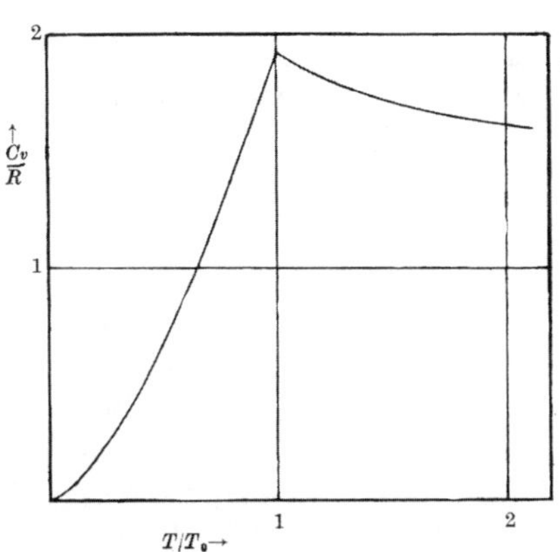

理想玻色–爱因斯坦气体的比热变化关系

（4）尽管氦的λ点实际上更类似于一种二阶相变，但还是很容易将其与玻色–爱因斯坦统计的凝聚现象联系起来。λ点处的温度（2.19 K）和熵（~0.8R）的实验值似乎支持了这一猜想。另一方面，我们可以很明显地看出，把液氦简化为一种理想气体是一个脱离实际的模型，这个模型在高温时只能给出 C_v = 3/2R，而在低温区，理想的玻色–爱因斯坦气体也必然会给出过大的比热值，因为此模型并没有考虑德拜频率的逐渐“冻结”。

按照我们的构想，可以说液氦的量子态既对应于金属理论中的电子态又对应于晶格的德拜振动态。当然，我们必须要把这个特点涵盖在理论中，才能有望对液氦的性质有定量的了解。

本文提出的构想或许对理解 He II 中特别的输运现象有一点启发（巨大的热导率[5]，极小的黏度[6]，以及最近艾伦和琼斯观察到的奇异的喷泉现象[2]）。

有关这些问题的详细讨论将发表在《物理学杂志》上。

（沈乃澂 翻译；于渌 审稿）

F. London: Institut Henri Poincaré, Paris, March 5.

References:

1. Fröhlich, H., *Physica*, 4, 639 (1937).

2. Allen, J. F., and Jones, H., *Nature*, **141**, 243 (1938).

3. Simon, F., *Nature*, **133**, 529 (1934).

4. London, F., *Proc. Roy. Soc.*, A, **153**, 576 (1936).

5. Rollin, *Physica*, **2**, 557 (1935); Keesom, W. H., and Keesom, H. P. *Physica*, **3**, 359 (1936); Allen, J. F. , Peierls, R., and Zaki Uddin, M., *Nature*, **140**, 62 (1937).

6. Burton, E. F., *Nature*, 135, 265 (1935); Kapitza, P., *Nature*, **141**, 74 (1938); Allen, J. F. and Misener, A. D., *Nature*, **141**, 75 (1938).

Mesotron (Intermediate Particle) as a Name for the New Particles of Intermediate Mass

C. D. Anderson and S. H. Neddermeyer

Editor's Note

In 1936, Carl Anderson and Seth Neddermeyer at the California Institute of Technology found evidence in cosmic-ray experiments for a particle having mass intermediate between those of the electron and proton. The particle carried one unit of charge, and penetrated matter much more strongly than the electron. Here the particle's discoverers propose a name for it. As its intermediate mass seemed the key property, they suggest the term "mesotron". This was later shortened to "meson", although the particle Anderson and Neddermeyer had discovered was actually what physicists today call the muon, a massive relative of the electron. The term meson later became used for intermediate-mass "hadrons" such as the pion or kaon, which are comprised of two quarks.

THE existence of particles intermediate in mass between protons and electrons has been shown in experiments on the cosmic radiation[1]. Since at present so little is known concerning the properties of these particles, for example, the exact value of the mass, the laws governing their production, their stability against disintegration, etc., it may yet be too early to assign to them a name. But inasmuch as several names have already been suggested, namely, dynatron, penetron, barytron, heavy electron, yukon and *x*-particle, it may be wise to consider the matter at this time.

The property which so far serves to distinguish the new particles from the other two types of particles which carry the same magnitude of electric charge, namely, the proton and the electron, seems to be the magnitude of their mass. Although from the experiments so far performed, it is not possible to say definitely whether the new particles exist with a unique mass only, or whether they occur with a range of masses, it does appear quite certain that the mass, whether unique or not, is greater than that of an electron and less than that of a proton. One must consider then three types of particles all carrying electric charges of equal magnitude: electrons, the new particles and protons. We should like to suggest therefore the word "mesotron" (intermediate particle) as a name for the new particles. It appears quite likely that the appropriateness of this name will not be lost, whatever new facts concerning these particles may be learned in the future.

(**142**, 878; 1938)

Carl D. Anderson and Seth H. Neddermeyer: California Institute of Technology, Pasadena, Sept. 30.

将介子（中间粒子）定为一种中间质量新粒子的名字

安德森，尼德迈耶

编者按

1936 年，美国加州理工学院的卡尔·安德森和塞思·尼德迈耶在宇宙射线实验中发现了一种质量介于电子和质子之间的粒子的径迹。这种粒子携带一个单位的电荷，穿透能力比电子强。本文中粒子的发现者提出为它命名。鉴于它的质量介于电子和质子之间是其重要性质，故建议选用单词“mesotron”，之后被缩写成“meson”，而安德森和尼德迈耶所发现的粒子实际上是今天物理学家所指的与电子有近亲关系的 μ 介子。介子（meson）一词后来用于表示中间质量的“强子”，例如由两个夸克组成的 π 介子和 K 介子。

在宇宙射线实验中发现了一种新粒子，其质量介于质子和电子之间 [1]。目前，人们对这种粒子的性质，例如粒子质量的精确值、产生的机制、对衰变的稳定性等了解甚少。因此现在给它们确定一个名字也许为时过早。但是，因为大家已经提出了几个候选的名字：（朝）代子、穿（透）子、重子、重电子、汤川子以及 x 粒子，所以也许现在来考虑这个问题是必要的。

迄今为止，只能通过质量的数值来区分这种新粒子与其他两种粒子——即携带相同数量电荷的质子和电子。虽然根据目前已做的实验尚无法肯定这种新粒子的质量是否有一个确定值，还是在一定的范围内变化，但可以肯定它们的质量，无论其数值是否唯一，都是大于电子而小于质子的，故可以认为三种类型的粒子，即电子、新粒子和质子，都带有相同数量的电荷。所以，我们建议用“mesotron”（中间粒子）作为这种新粒子的名称，这主要是因为我们认为这个名字的适用性不会因为将来发现了有关这个粒子的新性质而失效。

（胡雪兰 翻译；厉光烈 审稿）

Reference:

1. For historical summary see Wentzel, G., *Naturwiss.*, **26**, 273 (1938); and Bowen, Millikan and Neher, footnote, *Phys. Rev.*, **53**, 219 (1938).

New Broadcasting System

Editor's Note

An engineer in the United States, *Nature* reports here, has developed a new kind of radio broadcasting system. Radio systems at that time worked by amplitude modulation, which encodes the signal, such as voice, in modulations of the amplitude of the carrier wave. In contrast, the new system developed by Edwin Armstrong encodes information in modulations of the frequency of the carrier wave. That gave a sharp reduction in radio interference, although there was a considerable increase in the receiver's complexity. Although the basic principles of this scheme of frequency modulation, or FM radio, had long been known, Armstrong brought it to fruition with valves and circuits. The first transmitting station, in Alpine New Jersey, began operating soon after.

ACCORDING to a recent report by the New York Correspondent of *The Times*, a new type of wireless transmission and reception will be used in an experimental station now being erected at Columbia University by Major E. H. Armstrong, professor of electrical engineering in the University, and inventor of the now well-known supersonic-heterodyne receiver. The station will use the frequency-modulation system of transmission, as distinct from the amplitude-modulation system at present used by all broadcasting stations. In the former method, the frequency of the emitted carrier wave is varied by the speech and music modulation; whereas in the methods so far employed, the frequency of the carrier wave remains constant and its amplitude is varied by the applied audio-frequencies. The principles of frequency-modulation have been known since the earliest days of radio-telephony, but it has remained for Prof. Armstrong to demonstrate how these may be brought into practical use with modern valves and circuit arrangements. Among the advantages claimed for the new system are that it effects a considerable reduction of interference in radio reception, and that a much larger number of broadcasting channels will become available in any given wave-length band. Against these, however, is the serious disadvantage that special receivers are necessary for frequency-modulated transmissions, and this factor is likely to involve serious delay in the introduction of the new system into modern broadcasting technique. The first transmitting station on the new system will be at Alpine, New Jersey, opposite New York City, and it has a licence to broadcast on a frequency of 40 megacycles per second (wave-length 7.5 metres). It is also understood that suitable receivers are already being manufactured, so that the result of this practical experiment will be awaited with interest.

(**143**, 152; 1939)

新的广播系统

编者按

《自然》的这篇文章对一位美国工程师发展出的一套新的广播系统进行了报道。当时的广播系统是通过调幅，即对载波的振幅进行调制而将声音等信号进行编码的。与之相对应的是埃德温·阿姆斯特朗研制的新系统，该系统通过对载波的频率进行调制而编码信息。尽管增加了接收机的复杂程度，但这一技术大大降低了广播信号间的相互干扰。虽然人们早已经了解了调频（FM）广播的基本原理，但是阿姆斯特朗用电子管和线路将这一理论变成了现实。第一个采用这种技术的发射台不久将在新泽西州的阿尔派投入使用。

根据《泰晤士报》驻纽约记者最新的一则报道，一家实验电台将采用一套新型的无线发射和接收系统，该实验电台目前正由哥伦比亚大学电子工程系教授阿姆斯特朗少校在该校架设，阿姆斯特朗教授也是目前广为人知的超外差接收器的发明者。目前所有广播电台使用的都是调幅系统，与此不同的是，正在建设中的这座电台将使用调频发射系统。在调频系统中，发射载波的频率会随着语音和音乐的声调变化而变化，而在目前使用的调幅系统中，载波的频率是恒定不变的，其振幅则随着所用音频的变化而变化。在无线电话问世之初，人们就已经知道了调频的原理。但是在阿姆斯特朗教授之前，没有人知道如何利用现代的电子管和各种电路元件将其带向实际应用。新系统的优点是：它大大减少了无线电接收中存在的干扰，也大幅增加了在任意给定的波长范围内可用的广播频道。然而，这种系统也有严重的缺点，人们必须用特殊的接收机才能接收调频发射的无线电信号，这一缺点很可能会严重减缓现代广播系统中应用这种新系统的速度。使用这种新系统的第一家发射台将设在新泽西州的阿尔派（与纽约市隔河相望），该发射台已经获得了在每秒 40 兆周（即波长 7.5 米）的频率上进行广播的授权。据悉，合适的接收机正在生产中，因此，大家都在饶有兴趣地等待着这项应用性实验的结果。

（刘霞 翻译；赵见高 审稿）

Interpretation of Beta-disintegration Data

H. A. Bethe *et al.*

Editor's Note

Enrico Fermi had recently proposed a theory for radioactive beta decay. As Hans Bethe and colleagues point out here, the theory disagrees with experimental data by predicting, for example, too few electrons emerging at low energies. An alternative theory, while accounting for this feature, introduces other problems. Yet a resolution, the authors suggest, may lie in supposing that Fermi's theory is correct but that experiments so far have not observed simple radioactive decays but instead mixed together decay processes leading to different nuclear states. Evidence for this idea might be found by scrutinizing the gamma rays emitted during beta decays.

FERMI'S original theory of β-decay[1] made a definite prediction as to the energy distribution of the electrons emitted from a β-active element. It was found[2] that the experimental distribution curves did not agree in shape with this prediction in the sense that the number of electrons of low energy (relative to the upper limit of the spectrum) was considerably greater in the experimental than in the theoretical curves. A modified theory was then proposed by Konopinski and Uhlenbeck (K.U.)[3] which introduced in the distribution curve another factor proportional to the square of the momentum of the emitted neutrino. This theory appeared to agree with the facts.

Further experimental evidence, however, revealed a number of facts which did not fit in well with the modified theory: (1) The shape of the observed energy spectra did not fit the K.U. law near the upper limit, but seemed there to follow the original law of Fermi, although the latter did not fit with any other part of the curve[4]; if one determined the position of the upper limit by extrapolation from the K.U. law, one obtained values that were difficult to reconcile either with the observed spectrum or with the other data on the energy balance[5]. (2) The Sargent curves (decay constant against disintegration energy) seemed to agree better with the Fermi theory[4]. (3) The probability of capturing a *K*-electron, as compared to that of emitting a positron, was found to be much smaller than the K.U. theory would predict, but in reasonable agreement with the Fermi theory[6]. (4) An attempt to develop the K.U. theory into a mathematically consistent scheme showed that it was, at any rate, far more complicated than that of Fermi, and no proof has as yet been given that it can be consistently carried further than to the first order of approximation[7].

We therefore investigated the view that the original theory of Fermi correctly represents the elementary law; but that the observed spectra, in so far as they belong to "allowed" transitions, do not represent the effect of a single nuclear transformation, but rather a superposition of different spectra belonging to transitions to different levels of the final

对β衰变数据的解释

贝特等

编者按

恩里科·费米最近提出了放射性β衰变理论。正如汉斯·贝特和他的同事在这里所指出的，该理论的预言与实验结果不符，例如，预言认为几乎没有出现低能电子。另一种理论虽能解决这个难题，但却引入了其他问题。本文作者给出的解答是，不妨假定费米理论是正确的，只是迄今为止的实验还没有观测到简单的放射性衰变过程，而观测到的是混在一起的导致不同核状态的衰变过程。如果仔细检查β衰变中放出的γ射线，或许可以发现支持这一观点的证据。

费米最初的β衰变理论[1]明确地预言了从β放射性元素发射出的电子的能量分布。但是，人们发现[2]实验中能量分布曲线的形状与费米理论的预言不一致：在能量分布实验曲线中，低能电子（相对于能谱上限来说）的数目明显多于理论预言值。因此，科诺平斯基与乌伦贝克[3]对费米理论进行了修正（修正后的理论称为KU理论），他们在能量分布曲线中引入了一个与所放射的中微子的动量平方成正比的因子。修正后的理论似乎与实验结果相符。

然而，进一步的实验证据揭示了很多与修正理论不符的现象：(1) 所观测到的能谱形状在上限附近虽与KU理论不符，却与费米的原始理论一致，但费米理论并不能与曲线的其他部分相吻合[4]。如果通过KU理论推出上限位置，那么人们得到的值既难于与观察到的能谱吻合，也难于与基于能量平衡[5]的其他一些数据吻合。(2) 萨金特曲线（衰变系数依衰变能量变化的曲线）似乎与费米理论[4]符合得更好。(3) 和放射一个正电子的概率相比，俘获一个K电子的概率要远小于KU理论的预计，但是却在合理范围内与费米理论[6]一致。(4) 曾经有人尝试进一步发展KU理论，以求达到数学形式上的一致，然而这比对费米理论进行相同的尝试要复杂得多，并且到目前为止仍然没有证据可以支持比一阶近似更高的精确度[7]。

因此，我们认为，费米的原始理论正确地表示了基本法则。但是，因为观测到的能谱属于“允许”跃迁，所以并不表示单核转变的结果，而是向末态原子核不同能级跃迁形成的不同能谱的叠加。这样就可以清楚地看出，能量分布结果包含的（相

nucleus. It is then clear that the resulting energy distribution will contain rather more electrons of low energies, compared to the upper limit of the spectrum, than a single Fermi curve. If the nucleus is left in an excited state, it must eventually lose its energy by radiation, and the crucial test for the suggested point of view is the presence of γ-rays of suitable energy and intensity from all those radioactive bodies which have "allowed" transitions and the energy spectra of which are known to be different from the simple Fermi curve. The γ-rays might in some cases be absent because the excited state of the nucleus might be a metastable isomer; but this could not be true for all such elements. The restriction to allowed transitions is necessary because in "forbidden" transitions the shape, even of a single curve, is affected by more complicated factors[8].

Interpreting from this point of view the electron spectra of ^{12}B, ^{20}F, ^{17}F, ^{13}N, ^{15}O, as given by Fowler, Delsasso and Lauritsen[9], we have estimated the energy and intensity of the γ-radiation to be expected. The results are given in the following table.

Active element	Product nucleus	Energy of γ-ray in Mv.	Number of γ-quanta per disintegration
^{12}B	^{12}C	5	0.5
^{20}F	^{20}Ne	2	0.7
^{13}N	^{13}C	0.27	0.8
^{15}O	^{15}N	0.5	0.4
^{17}F	^{17}O	0.9	0.6

No great accuracy is claimed for these results since the curve-fitting is very sensitive to the high-energy ends of the electron spectra, which are not very accurately known, and also because we have assumed that only one excited level is involved, whereas there might be more.

The presence of a γ-ray accompanying the disintegration of ^{13}N was indeed reported by Richardson[10], who gave its energy as 0.3 Mv. A γ-radiation from ^{20}F was reported by Burcham and Smith[11], and measurements of the energy of this radiation, which are being made by Bower[12], indicate a value of about 2.2 Mv. Although γ-radiations of the predicted energies from the other elements of our table have not been reported, it is interesting to notice that an energy level in ^{17}O at 0.83 Mv. is known to exist, since it is excited in a number of other nuclear reactions[11,13], and similarly an excited state of ^{12}C at 4.3 Mv. is known[13]. We are indebted to Mr. P. I. Dee for this discussion of the experimental data.

Finally, we would like to point out that, although one cannot consider the above evidence as convincing confirmation of the point of view we suggest, it is certainly incompatible with the K.U. theory, since the existence of γ-rays shows that the observed curves must be superpositions of at least two simple spectra, whereas their shapes are not such as could be represented as sums of two K.U. curves with endpoints differing by the energy of the γ-rays.

(**143**, 200-201; 1939)

H. A. Bethe: Physics Department, Cornell University.

对能谱上限来说）低能电子比一条单峰费米曲线包含的更多。如果原子核处于激发态，它最终将通过辐射来释放能量，因此可以验证我们观点的关键实验现象包括：存在从所有具有“允许”跃迁的放射性物质中放射出的具有适当强度和能量的 γ 射线，以及与单峰费米曲线不同的能谱。在某些情况下，γ 射线也有可能不出现，因为原子核的激发态可能是一个亚稳的同质异能态，但是这种现象不可能对所有放射性元素都成立。对允许跃迁的限制是必要的，因为在“禁戒”跃迁中曲线的形状——即使是一条单峰曲线——都会受到更加复杂因素的影响[8]。

用以上观点来解释福勒、德尔萨索以及劳里森[9]给出的 ^{12}B、^{20}F、^{17}F、^{13}N 和 ^{15}O 电子能谱，我们估算了 γ 辐射的能量和强度的期望值。下表给出了计算结果。

放射性元素	生成原子核	γ 射线能量，单位兆电子伏	每次衰变辐射的 γ 量子数
^{12}B	^{12}C	5	0.5
^{20}F	^{20}Ne	2	0.7
^{13}N	^{13}C	0.27	0.8
^{15}O	^{15}N	0.5	0.4
^{17}F	^{17}O	0.9	0.6

我们没能给出非常精确的结果，因为曲线拟合对电子能谱的高能端非常敏感，而我们对此了解得又不够精确，并且我们在实验理论中假设只存在一个激发态，而事实上可能存在很多激发态。

理查森[10]曾报道过关于 ^{13}N 衰变伴随有 γ 射线出现的现象，并且给出 γ 射线的能量是 0.3 兆电子伏。伯彻姆和史密斯[11]报道过 ^{20}F 的 γ 辐射，其能量值由鲍尔[12]测得，约为 2.2 兆电子伏。虽然我们在表中列出的其他元素 γ 辐射的能量期望值还没有相关的实验报道，但有趣的是，^{17}O 的一个能级为 0.83 兆电子伏，这是已知存在的，因为它在其他很多核反应中是激发态[11,13]，类似地，^{12}C 一个能级为 4.3 兆电子伏的激发态也是已知的[13]。我们非常感谢迪伊先生与我们讨论了这些实验数据。

最后，需要指出的是，虽然不能将上述证据看作是对我们提出的观点的强有力证明，但是这些证据肯定与 KU 理论不符，因为 γ 射线的存在表明观察到的曲线一定至少是两条单峰能谱的叠加，然而它们的形状并不是两条由于 γ 射线能量造成其端点不同的 KU 曲线的叠加。

（王静 翻译；厉光烈 审稿）

F. Hoyle: Emmanuel College, Cambridge.
R. Peierls: University, Birmingham.

References:

1. Fermi, E., *Z. Phys.*, **88**, 161 (1934).
2. Kurie, F. N. D., Richardson, J. R., and Paxton, H. C., *Phys. Rev.*, **49**, 368(1936).
3. Konopinski, E. J., and Uhlenbeck, G. E., *Phys. Rev.*, **48**, 7 (1935).
4. Richardson, H. O. W., *Proc. Roy. Soc.*, A, **161**, 456 (1937).
5. Cockcroft, J. D., *Proc. Roy. Soc.*, A, **161**, 540 (1937).
6. Walke, H. (in the Press).
7. Fierz, M., *Helv. Phys. Acta*, **10**, 123, (1937).
8. Hoyle, F., *Proc. Roy. Soc.*, A, **166**, 249 (1938).
9. Fowler, W. A., Delsasso, L. A., and Lauritsen, C. C., *Phys. Rev.*, **49**, 569 (1936).
10. Richardson, J. R., *Phys. Rev.*, **53**, 610 (1938).
11. Burcham, W. E., and Smith, C. L., *Proc. Roy. Soc.*, A, **168**, 176 (1938).
12. Private communication.
13. Cockcroft, J. D., and Lewis, W. B., *Proc. Roy. Soc.*, A, **154**, 261 (1936).

Liquid Helium

J. F. Allen and H. Jones

Editor's Note

One of the sensations of the late 1930s was the discovery of the strange properties of liquid helium II—a form of liquid helium that appears when the ordinary material is cooled below 2.19K. The Royal Society Mond Laboratory was established in the early 1930s at the Cavendish Laboratory in Cambridge by Ernest Rutherford, who planned that it should be the workplace of Pyotr Kapitza. By 1939, however, when this article appeared, Kapitza had returned to Moscow, taking with him much of the electromagnetic equipment the Mond Laboratory had contained. One striking feature of the measurement described in this article is that the viscosity of liquid helium falls sharply as it is converted into helium II. The current explanation is that liquid helium II owes its distinctive properties to quantum mechanics and that the bulk of the liquid exists in the form of a "Bose–Einstein condensate"—essentially the liquid moves as an integrated whole.

THE properties of liquid helium can best be considered under two headings: (*a*) properties in thermal equilibrium, (*b*) transport effects. The equilibrium properties, which have been the subject of many careful investigations in Leyden and elsewhere, may be regarded as fairly well established. These include the determination of the specific heat for different temperatures, the variation of density with temperature at different constant pressures, and the relation of the saturated vapour pressure to the absolute temperature scale. The investigations of transport effects such as the flow of the liquid through tubes, heat conductivity and associated effects are still in an early stage of development, and no clear understanding of these phenomena has yet been reached. The subject was discussed at a meeting following the International Refrigeration Congress in July and also during the Cambridge meeting of the British Association. In this article we shall confine our attention largely to the interesting newly discovered transport effects, and give an account of recent experiments the results of which at present seem to find general acceptance.

Equilibrium Properties of Liquid Helium

Helium at atmospheric pressure liquefies at 4.22°K.; the critical temperature is 5.2°K. Generally speaking, the temperature range over which the properties of liquid helium have been measured extends only down to 1°K., since this is the lowest temperature conveniently reached by lowering the vapour pressure over the surface of the liquid. At 2.19°K. under its own vapour pressure, liquid helium undergoes a remarkable transformation. As the liquid is cooled through 2.19°K., the specific heat jumps suddenly from a value of 0.4 cal. per gm. per degree to more than 5 cal. per gm. per degree,

液 氦

艾伦，琼斯

编者按

20 世纪 30 年代末的轰动事件之一，就是发现了液氦 II（普通液氦冷却到 2.19 K 以下出现的一种形式的液氦）的奇特性质。20 世纪 30 年代初，欧内斯特·卢瑟福在剑桥卡文迪什实验室建立了英国皇家学会蒙德实验室，他希望彼得·卡皮查教授能留在这个实验室工作。不过，到 1939 年这篇文章问世时，卡皮查已经携带蒙德实验室中的很多电磁学设备返回了莫斯科。本文所描述的液氦的一个显著的特性是，液氦在转化为氦 II 时黏度会迅速下降。目前利用量子力学解释氦 II 的这种独特性质，即液体的整体是以“玻色－爱因斯坦凝聚”形态存在——从本质上讲，就是液体作为一个整体在移动。

液氦的性质最好可以从以下两个方面来考虑：(*a*) 热平衡性质，(*b*) 输运效应。经过莱顿和其他实验室大量仔细的研究，平衡性质已得到充分证实，其中包括不同温度下比热的确定、不同恒压下密度随温度变化的关系、以及饱和蒸气压与绝对温标的关系。而有关输运效应的研究尚处在初级阶段，诸如液体在管中的流动、热导率及相关效应等，人们对于这些现象还没有一个清晰的认识。在七月国际制冷学大会之后的一次会议，以及在英国科协的剑桥会议上，人们对液氦的输运效应进行了讨论。本文中，我们将主要讨论这个新发现的有趣的输运效应，并介绍一些目前看来结果已获得公认的最新实验。

液氦的平衡性质

常压下，氦在温度为 4.22 K 时液化；临界温度为 5.2 K。一般来说，能观测液氦性质的温度下限仅能达到 1 K，因为这是用降低液面蒸气压的方法所能顺利实现的最低温度。在温度为 2.19 K 以及自身蒸气压下，液氦发生了显著的转变。当液氦被冷却至温度为 2.19 K 时，比热从 0.4 Cal/(g · K) 突跃到 5 Cal/(g · K) 以上，之后以近于 T^5 快速下降。最近，西蒙曾指出，在通过铁铵钒绝热退磁产生的极低温度

thereafter falling rapidly, approximately as T^5. Simon has recently shown that at very low temperatures (0.02°–0.05°K.) produced by adiabatic demagnetization of iron ammonium alum, the specific heat of liquid helium varies as T^3.

The transformation at 2.19°K. has also a remarkable effect on the expansion coefficient of liquid helium. Above that temperature it is positive, while below it is negative, although there is no discontinuity in the value of the density itself at 2.19°K.

Phase transformations of the type that liquid helium undergoes at 2.19°K. are known in other branches of physics; for example, the Curie-point transformation of a ferromagnetic, the order-disorder transformation of certain alloys, and the transition between the superconductive and the normal state of a metal. The temperature of the transformation in liquid helium is known as the λ-point, a name introduced by Ehrenfest. The modification of the liquid below the λ-point is generally referred to as helium II, that above the λ-point as helium I.

Liquid helium at 1°K. can be solidified under an external pressure of 25 atmospheres. At higher temperatures, greater pressures are required to produce the solid. The properties of liquid helium I are such as would be expected of an ordinary liquid of very low boiling point. It is far otherwise with liquid helium II. In the first place, it is an immediate inference from the phase diagram that liquid helium II under its own vapour pressure remains liquid at absolute zero, for even in the neighbourhood of 1°K. the boundary line between the solid and liquid phases tends to become parallel with the temperature axis at a pressure of approximately 25 atmospheres. Since dp/dt (the slope of the boundary line) approaches zero as the temperature is lowered, and since the volume change, Δv, is not zero, it follows from Clapeyron's equation $dp/dt = \Delta s/\Delta v$, that Δs, the change in entropy on passing from liquid to solid, also approaches zero. There can thus exist the paradoxical situation of a liquid with zero entropy. On this account the possibility has been considered by several investigators that condensed helium at absolute zero has a space-ordered structure, and that the λ-point is of the nature of an order-disorder transformation.

An experimental investigation of this point has been carried out by Keesom and Taconis. They examined the reflection of X-rays in a column of liquid helium II at about 1.6°K., as well as in helium above the λ-point. No essential difference between the reflections in the two cases could be observed. Confirmation of the space-ordered theory, therefore, is still lacking, although it should be borne in mind that at 1.6°K. an appreciable degree of disorder would in any event be expected.

We may mention here that F. London, in a very interesting letter to *Nature*, has recalled a prediction by Einstein that a perfect Bose gas at sufficiently low temperatures should show a discontinuity in the temperature derivative of the specific heat. Applying Einstein's formula to liquid helium, London shows that the discontinuity would occur at 3.09°K. Although Einstein's discontinuity is only in the temperature derivative of the specific heat,

（0.02 K ~ 0.05 K）下，液氦的比热随 T^3 变化。

温度为 2.19 K 时的这一转变对液氦的膨胀系数也有明显的影响。在该温度以上，膨胀系数为正，而在该温度以下则为负，但液氦的密度本身在温度为 2.19 K 时并没有出现不连续性。

类似液氦在温度为 2.19 K 时发生相变的情况在物理学其他分支中也遇到过；如，在居里点的铁磁转变、某些合金的有序 – 无序转变、以及金属超导态与正常态的转变。液氦的相变温度被称为 λ 点，这是埃伦费斯特给它命的名。通常将低于 λ 点的液氦相称为氦 II，高于 λ 点的称为氦 I。

在温度为 1 K 下，外加 25 个大气压可使液氦固化。在较高温度时，则需要加更大的外压来使其固化。液氦 I 的性质正如预期所料，即与具有极低沸点的普通液体一样，而液氦 II 的性质则大不相同。首先，从相图上可以直接推断出，处于自身蒸汽压下的液氦II在绝对零度时仍保持液态，因为大约25个大气压时，甚至在1 K附近，固相与液相的边界线与温度轴趋于平行。由于温度降低时边界线的斜率 dp/dt 趋近于零，而体积的变化量 Δv 不是零，利用克拉珀龙方程 $dp/dt = \Delta s/\Delta v$ 可以得到，其从液相到固相的熵变 Δs 也趋近于零。因此，可能存在一种具有零熵的奇怪液体。由于这个原因，很多研究者认为，可能凝聚态氦在绝对零度时具有空间有序结构，而 λ 点则是一种有序 – 无序转变的本质体现。

凯索姆和塔康尼斯已对这点做了实验研究。他们考察了温度约为 1.6 K 时一段液氦 II 液柱和高于 λ 点时一段液氦 I 液柱的 X 射线反射。然而并没有观测到这两种情况的反射有什么本质不同。因此，关于空间有序理论的确证仍然不足，不过应该清楚的是，在温度 1.6 K 时，应该预期到总会有可观的无序度出现。

本文中我们可以提一下，伦敦在致《自然》的一封有趣的信中回忆到，爱因斯坦曾做过一个预言，即理想玻色气体在温度足够低时，其比热对温度的导数应该是不连续的。伦敦将爱因斯坦给出的公式应用于液氦，推断出这个不连续点应该出现在 3.09 K。尽管爱因斯坦提出的不连续性只是关于比热对温度的导数而非比热本身，

not in the specific heat itself, London suggests that there may be some connexion with the λ-point of liquid helium. The advantage of such a theory of the λ-point is that it appears not to necessitate a space-ordered structure for the liquid at low temperatures.

Transport Effects

Flow Phenomena. Experiments designed to measure the viscosity of liquid helium II have led to the most surprising and apparently contradictory results. The first attempt in this direction was made at Toronto by Burton and Misener, who measured the damping of an oscillating cylinder immersed in the liquid. The value for the viscosity was found to drop suddenly at the λ-point from approximately 10^{-4} C.G.S. units for helium I to 10^{-5} C.G.S. units for helium II. Recent and more precise measurements by the similar method of an oscillating disk, made by MacWood at Leyden, give values varying from 3×10^{-5} C.G.S. units for helium II just below the λ-point to 2×10^{-6} C.G.S. units at 1.1°K. Liquid helium II thus appears to be very much less effective in damping the motion of a body immersed in it than helium gas at room temperature.

Attempts to measure the viscosity of helium II by measuring the rate of flow through a tube were first made by Kapitza* and by Allen and Misener in the Royal Society Mond Laboratory at Cambridge. On account of the anticipated low viscosity, the latter used very narrow tubes to lessen the rate of flow, with the extraordinary result that the rate of flow was far in excess of what would occur in a liquid of viscosity 10^{-5} C.G.S. units. Moreover, the velocity of flow did not vary linearly with the pressure head. Kapitza found a velocity proportional to the square root of the pressure head, and interpreted this as evidence that the flow was turbulent. He gave a value of 10^{-9} C.G.S. units as an upper limit to the viscosity of helium II. Giauque in California has also observed the flow of helium II through an annular tube and has obtained a temperature variation of viscosity of from 10^{-6} to 10^{-8} C.G.S. units. Allen and Misener endeavoured to reduce the velocity of flow by using finer capillaries to obtain more nearly the condition for stream-line flow, and found that the rate of flow varied as a power of the pressure head of even less than one half. It was found that the dependence of velocity on the pressure head decreased with decreasing capillary size. For glass capillaries of 0.0015 cm. radius, it was found that the velocity varied as the 1/6 power of the pressure, whilst for capillaries of 5×10^{-5} cm. in radius, obtained by packing a metal tube with parallel wires and then drawing the tube through dies, the velocity became absolutely independent of the driving pressure. In the latter case, the velocity increased very rapidly with decreasing temperature and reached a value of 20 cm. per second at 1.1°K. The flow also appears to be non-classical in the case of the variation of length of the capillary, since a variation of length by a factor of 70 produced only a fourfold change in the velocity.

On the other hand, experiments made by Burton in Toronto showed that with relatively wide and short tubes and rapid flow, the velocity is linearly proportional to the pressure

* Actually Kapitza used the essentially similar method of radial flow between two parallel plates with small separation.

伦敦还是认为这与液氦的 λ 点可能会有某种关联。这样一种 λ 点理论的优势在于，在低温下似乎并不要求液氦必须具有空间有序结构。

输运效应

流动现象。为测定液氦 II 黏度而设计的实验得到了令人非常惊讶且明显矛盾的结果。多伦多大学的伯顿和麦色纳在这个研究方向上最先进行了尝试，他们测定了浸在液氦中的振动圆筒的阻尼。研究发现，黏度值在 λ 点突然下降，从氦 I 的大约 10^{-4} 单位（厘米克秒制）迅速变到氦 II 的 10^{-5} 单位（厘米克秒制）。最近，麦克伍德在莱顿用类似的方法用振动盘进行了更精确的测量，得到的结果是氦 II 的黏度值从稍低于 λ 点时的 3×10^{-5} 单位（厘米克秒制）变到了 1.1 K 时的 2×10^{-6} 单位（厘米克秒制）。由此看来，在阻止浸入其中的物体的运动方面，液氦 II 的能力比室温下的氦气还要差很多。

卡皮查 * 和英国剑桥皇家学会蒙德实验室的艾伦与麦色纳率先进行了另一种尝试，通过测量氦 II 流经管子的速率来测定其黏度。由于预先考虑到氦 II 的低黏度，后者使用了非常细的管子以降低流速。实验结果非常出人意料，氦 II 的流速大大超过了黏度为 10^{-5} 单位（厘米克秒制）的液体应有的流速。此外，流速并不随着压位差的变化而线性变化。卡皮查发现，流速正比于压位差的平方根，他认为这是湍流的证据。他给出了氦 II 黏度的上限数值为 10^{-9} 单位（厘米克秒制）。加利福尼亚的吉奥克观测了氦 II 通过环形管的流动，得到黏度随温度变化的范围是从 10^{-6} 单位到 10^{-8} 单位（厘米克秒制）。艾伦和麦色纳使用更细的毛细管来尽量降低氦 II 的流速，使之更接近于流线流的条件，研究发现流速随着压位差的不到 1/2 次幂而变化。而且，毛细管越细，速度对于压位差的依赖性越小。对于半径为 0.0015 厘米的玻璃毛细管来说，速度随压力的 1/6 次幂变化，而在使用半径为 5×10^{-5} 厘米的毛细管（用平行金属丝将一根金属管子填充，再用一系列拉丝模把管子拉出，即可制得）时，速度变得与驱动压完全无关了。在后一种情况下，速度随着温度降低而非常迅速地增加，并在温度为 1.1 K 时达到 20 厘米 / 秒。流速随毛细管长度的变化也表现为非经典的，因为长度变化了 70 倍，而速度只变化了 4 倍。

另一方面，多伦多大学的伯顿所进行的实验指出，在使用相对粗而短的管子和更快的流速时，速度正比于压位差，并且由此确定的黏度值与通过振动盘阻尼方

* 实际上，卡皮查使用一种本质上类似的方法，即两个近距离平行盘之间的径向流。

head, and that the viscosity so determined agrees with the value obtained by the damping of oscillating disks. The anomalous features appear, therefore, when the flow takes place through very long and fine capillaries (radii less than 10^{-3} cm.).

A little light is thrown on these curious results by the experiments of Mendelssohn and Daunt on the creep of liquid helium II in the form of mobile films over solid surfaces. If an open vessel be partially immersed in liquid helium II, it was observed that liquid gradually collected in the vessel until the liquid levels in vessel and bath were coincident. The rate of filling increased rapidly with lowering temperature. The mechanism was found to consist not in evaporation and recon-densation, but in the transfer of liquid by means of surface films. At all temperatures in helium II the rate of transfer of liquid by means of the surface films was found to be independent of the difference in level. Both the thickness and the velocity of propagation of a film have been measured. The thickness is of order 5×10^{-6} cm. and the velocity increases from zero at the λ-point to 20 cm. per second at 1°K., which gives a rate of transfer of about 10^{-4} cubic centimeters per second per centimeter width of film. If films are formed above the λ-point, that is, by helium I, they are not more than 10^{-7} cm. thick.

A rough synthesis of the experimental results on flow of helium II can be attempted as follows. In flow through a tube, two distinct but by no means separate processes are taking place*: (*a*) the normal flow of a fluid with a viscosity (of order 10^{-4} C.G.S. units) which increases with decreasing temperature, and (*b*) creep along the inside walls of the tube by means of a surface film (of thickness of the order of 5×10^{-6} cm.), the velocity of which increases rapidly with decreasing temperature. For wide tubes, effect (*a*) predominates and the flow approximates to that of a normal viscous fluid. As the size of the capillary decreases, effect (*b*) becomes more pronounced, whilst for capillaries less than 10^{-4} cm. in radius, effect (*a*) becomes negligible and the quantity of liquid flowing per second is directly proportional to the circumference of the capillary.

Heat Conduction. The first experiments in the transport of heat through liquid helium II were made by Rollin and Keesom; their observations showed that helium II was a most efficient agent for the transport of heat, being far more effective than copper at the same temperature. It was observed later by Allen, Peierls and Uddin, and shown more convincingly by Keesom and Saris, that the rate of heat transport was not proportional to the temperature gradient. It is thus impossible to measure a true thermal conductivity for the liquid. The rate of heat transport is greater the smaller the gradient, and reaches a value corresponding to a conductivity several thousand times as great as that of copper at room temperature with gradients of the order of 10^{-5} of a degree per cm. For a given gradient the "conductivity" increases rapidly below the λ-point to a maximum at 2.0°K. and then falls again. Simon and Pickard of Oxford have found that the anomalously high conductivity disappears at the temperature at which the specific heat becomes normal.

---

* To be published shortly by J. F. Allen and A. D. Misener.

法得到的数值是一致的。由此看来，当氦 II 流经很长且很细的毛细管（半径小于 10^{-3} 厘米）时，才会出现反常特性。

门德尔松和当特进行了液氦 II 以流动膜形式沿固体表面爬行的实验，这为理解上述异常的结果带来了一线希望。如果将一端开口的空容器的底部部分地浸入液氦 II 之中，就可以观察到液体沿着容器壁逐渐流于容器中，直到容器内外液面持平。液体流进容器的速度随着温度降低而迅速加快。目前已经知道，这一行为的机制并非蒸发与再凝结，而是液体以表面膜的形式迁移。研究还发现，在任何温度下，氦 II 以表面膜形式迁移的速度都与液面高度差无关。现在已经测定了膜的厚度与其传播速度。厚度约为 5×10^{-6} 厘米的数量级，而传播速度则由 λ 点的静止不动，增大到 1K 时的 20 厘米 / 秒，从而给出每厘米膜宽度的迁移速率大约为 10^{-4} 立方厘米 / 秒。如果膜是在 λ 点以上形成的，即由氦 I 形成，那么其厚度不会超过 10^{-7} 厘米。

下面试着对关于氦 II 流动的实验结果做一大致的综述。在液氦流经管子的过程中，有两个截然不同却不可分割的过程在发生 *：(*a*) 流体的正常流，黏度（为 10^{-4} 单位（厘米克秒制）的数量级）随温度降低而增加；(*b*) 流体以表面膜（其厚度为 5×10^{-6} 厘米的数量级）方式沿着管内壁爬行，流体的流动速度随着温度降低而快速增大。对于粗管子来说，以 (*a*) 效应为主，流动近似为普通黏性流体的流动。随着毛细管尺寸减小，(*b*) 效应变得越来越显著，对于半径小于 10^{-4} 厘米的毛细管来说，效应 (*a*) 变得可以忽略，流动速度随着温度降低而增大，而每秒液体的流量与毛细管的周长成正比。

热传导。罗林和凯索姆最早进行了液氦 II 的热输运实验；他们的实验观测表明，氦 II 是一种极为高效的热输运介质，比同温度下的铜还要高效很多。后来艾伦、佩尔斯和乌丁观测到，且经凯索姆和萨里斯令人信服地证明，热输运速率并不与温度梯度成比例。因此，不可能测得液体的真实热导率。热传递速率越大，温度梯度就越小且这个热传递速率相当于室温下，当温度梯度为 10^{-5} 厘米数量级时铜的热导率的几千倍。对于给定的温度梯度，“热导率”在 λ 点以下快速增大，在温度为 2.0 K 时达到最大值，此后再次减小。牛津大学的西蒙和皮卡德发现，反常高热导率出现在比热变为正常时的温度下。

* 艾伦和麦色纳将要发表的文章。

Fountain Effect. The fountain effect, which was discovered in the Mond Laboratory, shows in the most striking manner the fundamental difference between helium II and any other known liquid. In its simplest form, the effect may be described as follows. A tube is partly immersed in helium II; the lower end of the tube is a capillary; both ends are open and an arrangement is made to heat the liquid in the upper part. A steady flow of heat is thereby maintained down the capillary. Under these conditions, it is observed that the liquid inside the tube rises above the level outside, showing the existence of a pressure in the reverse direction to the heat flow. A more spectacular demonstration can be given by placing powder in the lower half of the immersed tube through which the heat flows. To produce the heat current in this case, it is sufficient to shine light on the powder. With this arrangement the liquid may be made to rise right out of the tube, and in fact a steady "fountain" several centimeters in height can easily be produced.

Quantitative measurements on the magnitude of the reaction pressure are very incomplete. Some data, however, are available from measurements* on the reaction to heat flowing in helium II through a tube filled with powder particles. The reaction pressure was found to attain a maximum value of approximately half an atmosphere for a gradient of 1° per cm. at 1.7°K. The value of the heat conductivity through the powder-filled tube was lower by a factor of a hundred than that through a smooth capillary of the same open cross-section. It seems, therefore, that the very large conductivity observed for helium II when the heat flows through smooth tubes is caused by violent convection currents which are set up by the reaction mechanism. This might be the reason for the apparent variation of conductivity with temperature gradient.

A complete understanding of the fountain effect must naturally await a satisfactory theory of the constitution of liquid helium II. However, a few interesting deductions can be made. In the first place, in the simple arrangement described above, the force holding the liquid above the bath level can only come from some form of interaction of the liquid with the walls of the tube or the attached heating wire. No other support is available, and the vapour pressure above the bath and inside the tube are sensibly the same. Secondly, this interaction must result in a downward thrust on the tube equal to the weight of liquid above the bath level. From the atomic point of view, this means that the interacting atoms are steadily transferring momentum in a downward direction to the tube, just as the thrust on the walls of a vessel containing gas implies that the atoms steadily lose momentum to the walls as they are reflected. This has the interesting consequence that the main heat transport in helium II at these temperatures (below 2.19°K.) cannot be due to the propagation of elastic waves as in ordinary liquids and solids, since elastic waves do not carry momentum.

Many and varied hypotheses have been made concerning the constitution of helium II. Michels, Bijl and de Boer and the present authors independently suggested that certain atoms which have more than the average energy in helium II have, as well, a larger than

---

* *Proc. Cam. Phil. Soc.* (in the Press).

喷泉效应。蒙德实验室发现了液氦 II 的喷泉效应，这一效应以非常惊人的方式表明了液氦 II 与其他任何一种已知液体的根本差别。这种喷泉效应最简单的形式可以描述如下：将一根管子部分浸入氦 II 中；管的下端为毛细管；两端都是开口的，并用一种装置给上半部分中的液体加热。由此保持着沿毛细管向下的热稳流。在此条件下，可以观测到管内液体上升到高于外部液面的位置，这表明在热流的相反方向存在着某种压力。关于喷泉效应更精彩的演示是，将一些粉末置于浸入管的下半部分，即有热流动的那一段。在这种条件下，为产生热流，只要将光照射于粉末之上就可以了。通过上述操作可以使液体直接上升到管口外，实际上很容易产生几厘米高的稳定“喷泉”。

对于反作用压力大小的定量测量还很不完善。不过，通过测定*氦 II 经过装满粉末颗粒的管子的热流的反作用力，已经得到一些数据。发现反作用压力，在温度为 1.7 K、温度梯度为每厘米 1 K 时反作用压力达到最大值，约为大气压的一半。与流经具有相同开口横截面的光滑毛细管相比，氦 II 流经装满粉末的管子的热导率数值要低一百倍。由此看来，热流经光滑管时，人们所观测到的氦 II 具有的极高热导率，是由反作用机制引起强烈对流所造成的。这可能就是热导率随温度梯度明显变化的原因。

当然，要完全理解喷泉效应，有待于一个令人满意的液氦 II 组成理论的出现。不过，现在已经可以对此作出一些有趣的推论。首先，在上面所描述的简单装置中，使管内液体保持在外部液池液面之上的力，只能来自于某种形式的相互作用，要么是液体与管壁之间，要么是液体与附着的加热导线之间。除此再没有其他支撑的可能，而且管内与液池上方的蒸气压也明显是相等的。其次，这种相互作用必然会在管壁上产生一个向下的推力，其大小等于高于液池液面之上的液体的重量。从原子层面来看，这意味着，参与相互作用的原子在向下的方向上向管壁稳定地传递着动量，就像盛有气体的容器器壁上所受的推力，即意味着原子在被反弹的过程中稳定地向器壁转移着动量。由此得到的有趣的结论是，在这些温度（低于 2.19 K）下，液氦 II 中的主要热输运不是像在普通液体或固体中那样归结为弹性波的传播，因为弹性波并不携带动量。

关于氦 II 的组合，目前已提出了各式各样的假说。米歇尔斯、比尔、德布尔以及本文作者各自独立地提出，氦 II 中某些具有高于平均能量的原子，在液体中也具

*《剑桥哲学学会会刊》(正在印刷中)。

average mean free path inside the liquid, and that heat flow represents a drift of these moving or "excited" atoms. Later, this idea of energetic particles moving through the unexcited or "condensed" atoms was developed by Tisza to include a theory of flow. This theory has not yet reached a quantitative stage, but has proved interesting and suggestive.

Note added in proof: Since this article was written, F. London has published (*Phys. Rev.*, 54, 947; 1938) an enlargement of the theory which he based on the consideration that helium at low temperatures exhibited Bose–Einstein condensation phenomena. His theoretical interpretation of the behaviour of liquid helium II appears to be quite in accordance with the experimental deductions given above, particularly with regard to the properties of flow of both heat and liquid.

(**143**, 227-230; 1939)

J. F. Allen: Royal Society Mond Laboratory.

H. Jones: Imperial College, London.

有较大的平均自由程，而热流则体现了这些运动的或“受激发的”原子的漂移。后来，这一具有活力的粒子在未激发的或“凝聚的”原子中运动的观点被蒂萨发展，并包含了流动理论。该理论尚未达到定量的程度，但已被证明是引人关注和有启发性的。

附加说明：自本文写就以后，伦敦考虑了氦在低温下出现玻色－爱因斯坦凝聚现象而发表了一篇文章对这一理论作了进一步的扩充（《物理学评论》，第 54 卷，第 947 页；1938 年）。他对于液氦 II 行为的理论解释，尤其是关于热和液体的流动性质，似乎与上面所给出的实验推论颇为吻合。

（王耀杨 翻译；陶宏杰 审稿）

Disintegration of Uranium by Neutrons: a New Type of Nuclear Reaction

L. Meitner and O. R. Frisch

Editor's Note

This is the first record in *Nature* referring to the stimulated disintegration of uranium nuclei by neutrons. Lise Meitner, an Austrian, had worked in Berlin with Otto Hahn until she was expelled in 1938 because of the German government's policy on people of Jewish origin. Otto Frisch, then working at Niels Bohr's institute in Copenhagen, was Meitner's nephew. The fission of uranium nuclei is of course the basis on which the first nuclear weapons were constructed.

ON bombarding uranium with neutrons, Fermi and collaborators[1] found that at least four radioactive substances were produced, to two of which atomic numbers larger than 92 were ascribed. Further investigations[2] demonstrated the existence of at least nine radioactive periods, six of which were assigned to elements beyond uranium, and nuclear isomerism had to be assumed in order to account for their chemical behaviour together with their genetic relations.

In making chemical assignments, it was always assumed that these radioactive bodies had atomic numbers near that of the element bombarded, since only particles with one or two charges were known to be emitted from nuclei. A body, for example, with similar properties to those of osmium was assumed to be eka-osmium ($Z = 94$) rather than osmium ($Z = 76$) or ruthenium ($Z = 44$).

Following up an observation of Curie and Savitch[3], Hahn and Strassmann[4] found that a group of at least three radioactive bodies, formed from uranium under neutron bombardment, were chemically similar to barium and, therefore, presumably isotopic with radium. Further investigation[5], however, showed that it was impossible to separate these bodies from barium (although mesothorium, an isotope of radium, was readily separated in the same experiment), so that Hahn and Strassmann were forced to conclude that *isotopes of barium* ($Z = 56$) *are formed as a consequence of the bombardment of uranium* ($Z = 92$) *with neutrons.*

At first sight, this result seems very hard to understand. The formation of elements much below uranium has been considered before, but was always rejected for physical reasons, so long as the chemical evidence was not entirely clear cut. The emission, within a short time, of a large number of charged particles may be regarded as excluded by the small penetrability of the "Coulomb barrier", indicated by Gamov's theory of alpha decay.

由中子引起的铀衰变：一类新型核反应

迈特纳，弗里施

编者按

这是《自然》中最早谈及铀核受中子激发而产生衰变的文章。奥地利学者莉泽·迈特纳曾在柏林与奥托·哈恩一起工作至1938年，后因德国政府的反犹政策而被驱逐。奥托·弗里施是迈特纳的侄子，当时在哥本哈根的尼尔斯·玻尔的研究机构中工作。当然，最早的核武器就是基于铀核裂变制造而成的。

费米及其合作者发现[1]，用中子轰击铀核以后，至少会产生出四种放射性物质，其中两种放射性物质的原子序数均大于92。通过进一步的研究[2]表明，事实上至少存在九种放射周期，其中有六种属于铀后面的元素，为了解释它们的化学行为及其衍生关系，必须假定存在核同质异能性。

人们在进行化学研究时，经常会假定这些放射性物质的原子序数与被轰击元素的原子序数相近，因为据目前所知，核受轰击后只会发射出带一个或两个电荷的粒子。例如，轰击铀核得到的类似于锇的化学性质的物质，被假定为类锇（$Z = 94$）而不是锇（$Z = 76$）或者钌（$Z = 44$）。

沿着居里和萨维奇[3]的观测结果继续探究下去，哈恩和施特拉斯曼[4]发现，中子轰击铀核时至少能形成三种放射性物质，它们的化学性质与钡类似，因而推测其为镭的同位素。然而进一步的研究显示[5]，这些放射性物质几乎不能与钡分离（然而在相同的实验中，新钍——一种镭的同位素——很容易与钡分离），因此哈恩和施特拉斯曼被迫得出这样的结论：**用中子轰击铀（$Z = 92$）核的结果是产生了钡（$Z = 56$）的同位素。**

乍看起来，这个实验结果似乎让人很难理解。因为人们以前轰击铀核时，也曾考虑过存在比铀原子序数小很多的元素形成，但是在化学证据尚未完全明晰之前，这种想法总是因为一些物理原因而被排除。伽莫夫的 α 衰变理论指出，由于穿透“库仑势垒”的可能性极小，所以在短时间内要发射出大量带电微粒是不可能的。

On the basis, however, of present ideas about the behaviour of heavy nuclei[6], an entirely different and essentially classical picture of these new disintegration processes suggests itself. On account of their close packing and strong energy exchange, the particles in a heavy nucleus would be expected to move in a collective way which has some resemblance to the movement of a liquid drop. If the movement is made sufficiently violet by adding energy, such a drop may divide itself into two smaller drops.

In the discussion of the energies involved in the deformation of nuclei, the concept of surface tension of nuclear matter has been used[7] and its value has been estimated from simple considerations regarding nuclear forces. It must be remembered, however, that the surface tension of a charged droplet is diminished by its charge, and a rough estimate shows that the surface tension of nuclei, decreasing with increasing nuclear charge, may become zero for atomic numbers of the order of 100.

It seems therefore possible that the uranium nucleus has only small stability of form, and may, after neutron capture, divide itself into two nuclei of roughly equal size (the precise ratio of sizes depending on finer structural features and perhaps partly on chance). These two nuclei will repel each other and should gain a total kinetic energy of *c*. 200 Mev., as calculated from nuclear radius and charge. This amount of energy may actually be expected to be available from the difference in packing fraction between uranium and the elements in the middle of the periodic system. The whole "fission" process can thus be described in an essentially classical way, without having to consider quantum-mechanical "tunnel effects", which would actually be extremely small, on account of the large masses involved.

After division, the high neutron/proton ratio of uranium will tend to readjust itself by beta decay to the lower value suitable for lighter elements. Probably each part will thus give rise to a chain of disintegrations. If one of the parts is an isotope of barium[5], the other will be krypton ($Z = 92-56$), which might decay through rubidium, strontium and yttrium to zirconium. Perhaps one or two of the supposed barium-lanthanum-cerium chains are then actually strontium-yttrium-zirconium chains.

It is possible[5], and seems to us rather probable, that the periods which have been ascribed to elements beyond uranium are also due to light elements. From the chemical evidence, the two short periods (10 sec. and 40 sec.) so far ascribed to ^{239}U might be masurium isotopes ($Z = 43$) decaying through ruthenium, rhodium, palladium and silver into cadmium.

In all these cases it might not be necessary to assume nuclear isomerism; but the different radioactive periods belonging to the same chemical element may then be attributed to different isotopes of this element, since varying proportions of neutrons may be given to the two parts of the uranium nucleus.

不过，以目前关于重核[6]行为的观点为基础，对于这种新型核衰变过程，我们想到了一种完全不同、本质上又很经典的假设。它的大致内容如下：由于紧密堆积和强烈的能量交换，预期重核中的微粒会以整体方式运动，有些类似于液滴的运动。如果外加的能量能使这种运动变得足够剧烈，这个“液滴”就可能会分裂为两个较小的部分。

在讨论核变形过程中所涉及的能量问题时，需要用到核物质表面张力的概念[7]，但人们只是在考虑核力存在的前提下估算过其数值。不过必须记住，带电微滴的表面张力因其所带的电荷而减小，并且核的表面张力随着核电荷的增加而减小，粗略的估计显示，当原子序数达到 100 时核的表面张力可能会减为零。

由此看来，铀核也许只具有较小的稳定性，因为其在俘获中子后可能会分裂为大小基本相同的两个核（两部分大小的精确比例取决于精细结构的特征，可能还有一部分偶然因素）。这两个核将彼此推斥，根据核半径与电荷进行计算，这两个核应该会获得约为 200 兆电子伏的总动能。似乎可以预期，这一能量值实际上可以利用铀与周期表中部元素的敛集率的差别计算得到。由于涉及的物质质量较大，所以相应的量子力学中的“隧道效应”产生的影响小到可以忽略不计，因此整个“裂变”过程可以使用经典的方式来描述。

具有高中子质子比的铀核分裂成两个新核以后，倾向于再进行 β 衰变，以使其调整到适合于较轻元素的较小比值。每个新核都可能会引起一条衰变反应链。如果其中一部分是钡[5]的同位素，另外一部分就将是氪（$Z = 92-56$）的同位素，氪可以经由铷、锶和钇衰变链一直衰变到锆。也许有一条或两条衰变链，我们假定其可能会发生钡－镧－铈衰变，而实际上发生的却是锶－钇－锆衰变。

有可能[5]，并且在我们看来非常有可能的是，那些曾被归结为铀以后元素的放射周期也是轻元素的。根据化学证据，到目前为止被归结为铀-239 的两个短的放射周期（12 秒和 40 秒）可能来源于锝的同位素（$Z = 43$），其经钌、铑、钯和银衰变链衰变到镉。

在上述所有情况中，都不必假定存在核同质异能性；至于属于相同元素的不同放射周期则可以归因于该元素具有不同的同位素，因为铀核分裂形成的两部分可以获得各种不同比例的中子。

By bombarding thorium with neutrons, activities are obtained which have been ascribed to radium and actinium isotopes[8]. Some of these periods are approximately equal to periods of barium and lanthanum isotopes[5] resulting from the bombardment of uranium. We should therefore like to suggest that these periods are due to a "fission" of thorium which is like that of uranium and results partly in the same products. Of course, it would be especially interesting if one could obtain one of these products from a light element, for example, by means of neutron capture.

It might be mentioned that the body with half-life 24 min.[2] which was chemically identified with uranium is probably really ^{239}U, and goes over into an eka-rhenium which appears inactive but may decay slowly, probably with emission of alpha particles. (From inspection of the natural radioactive elements, ^{239}U cannot be expected to give more than one or two beta decays; the long chain of observed decays has always puzzled us.) The formation of this body is a typical resonance process[9]; the compound state must have a life-time a million times longer than the time it would take the nucleus to divide itself. Perhaps this state corresponds to some highly symmetrical type of motion of nuclear matter which does not favour "fission" of the nucleus.

(**143**, 239-240; 1939)

Lise Meitner: Physical Institute, Academy of Sciences, Stockholm.
O. R. Frisch: Institute of Theoretical Physics, University, Copenhagen, Jan. 16.

References:

1. Fermi. E., Amaldi, F., d' Agostino, O., Rasetti, F., and Segrè, E. *Proc. Roy. Soc.*, A, **146**, 483 (1934).
2. See Meitner, L., Hahn, O., and Strassmann, F., *Z. Phys.*, **106**, 249 (1937).
3. Curie, I., and Savitch, P., *C.R.*, **206**, 906, 1643 (1938).
4. Hahn, O., and Strassmann, F., *Naturwiss.*, **26**, 756 (1938).
5. Hahn, O., and Strassmann, F., *Naturwiss.*, **27**, 11 (1939).
6. Bohr, N., *Nature*, **137**, 344, 351 (1936).
7. Bohr, N., and Kalckar, F., *Kgl. Danske Vid. Selskab, Math. Phys. Medd.*, **14**, Nr. 10 (1937).
8. See Meitner, L., Strassmann, F., and Hahn, O., *Z. Phys.*, **109**, 538 (1938).
9. Bethe, A. H., and Placzek, G., *Phys. Rev.*, **51**, 450 (1937).

用中子轰击钍元素，得到了曾被认为是镭和锕的同位素所具有的放射性[8]。其中某些元素的放射周期与轰击铀产生的钡和镧的同位素[5]具有的放射周期是近似相等的。因此，我们倾向于认为，上述放射周期应来源于钍的“裂变”，它类似于铀的裂变过程，并在一定程度上得到了相同的产物。当然，要是能从一种轻元素得到上述产物之一（例如，通过中子俘获的方式），那就更有趣了。

还要提一下，在化学上与铀相同且半衰期为 24 分钟[2]的那种元素，很可能就是真正的铀–239，它进而衰变为类铼。虽然类铼似乎没有放射性，但也可能是在缓慢地衰变，并有可能伴随着 α 粒子的发射。（根据对天然放射性元素已有的认识，铀–239 不可能发生超过一次或两次的 β 衰变；我们始终对观测到的衰变长链感到困惑。）这种物质的形成是典型的共振过程[9]；复合态所具有的寿命一定比核分裂所需要的时间长一百万倍。也许这种状态适合于某些不会发生“裂变”的核物质的高度对称性的运动方式。

（王耀杨 翻译；鲍重光 审稿）

Theory of Mesons and Nuclear Forces

C. Møller and L. Rosenfeld

Editor's Note

Mesons were so-called because they are intermediate in mass between the electron and the proton. In 1935, in a Japanese journal, Hideki Yukawa and colleagues proposed that mesons could account for the strong forces between nucleons (protons and neutrons) much as photons account for the electrical forces between charged particles of all kinds. This article by Rosenfeld and Møller, both protegés of Niels Bohr at Copenhagen, suggests how physical considerations require particular forms of the mathematical expressions (called wave functions) that agree better with experiments on the decay of light elements emitting β-particles. This prediction proved correct.

As was first pointed out by Yukawa, it is in principle possible to account for the short-range forces between nuclear particles by the assumption of virtual emission and absorption processes involving intermediary particles of integral spin, the so-called *mesons*[1], the mass of which is determined by the range of the forces. As has been shown by Kemmer[2], the simplest wave-equations for the mesons which satisfy, besides the claim of relativistic invariance, the condition of giving a positive definite expression for the energy, reduce to four types, characterized by different co-variance properties of the wave-functions, and each allowing the existence of neutral as well as positively and negatively charged mesons. Starting from such wave-equations, including the interaction of the meson field with the heavy nuclear constituents, the estimation of the resulting expressions for the nuclear forces has hitherto been carried out by using the ordinary perturbation method of quantum theory, and taking into consideration only the first non-vanishing approximation, in spite of the well-known lack of convergence of the method. It would thus seem desirable to discuss more closely the reliability of such results, and for this purpose a possible method of attack is suggested by an analogous situation in quantum electrodynamics, where a suitable canonical transformation allows us to separate, from the expression of the total energy of a system consisting of electrons and an electromagnetic field, a term depending only on the coordinates of the electrons and representing the Coulomb potential energy.

A similar method[3] is, actually, applicable to a system consisting of nuclear particles and a meson field. For such a system it is, in fact, possible to find canonical transformations effecting the separation of a "static" interaction between the nuclear particles, defined as the part of the interaction which is obtained when one neglects the time-variations of the variables characterizing the positions, spins and proton or neutron states of the heavy particles. This static interaction is in all cases exactly the same as that obtained as

介子和核力理论

默勒，罗森菲尔德

编者按

介子因其质量介于电子和质子之间而得名。1935 年，汤川秀树及其合作者在一本日本杂志中指出，介子可以用来解释核子（质子和中子）间的强力，这与光子可以用来解释所有带电粒子之间的电力类似。哥本哈根的尼尔斯·玻尔的两位高徒——罗森菲尔德和默勒在这篇文章中阐述了怎样用能与轻元素衰变放射出 β 粒子的实验吻合得更好的特定数学表达式（或称波函数）来满足物理学研究的需要。这个预言后来被证明是正确的。

正如汤川秀树最先指出的，如果假设核子在虚拟的发射和吸收过程中包括具有整数自旋的媒介粒子，即所谓的“介子”[1]，其质量取决于作用力的范围，那么原则上就可对核子之间的短程作用力作出解释。凯默[2]的研究表明：可以把最简单的介子波动方程——既满足相对论所要求的不变性，又能给出一个正定的能量表达式——归纳为四种类型，以波函数协变性的不同进行区分，并且每一种类型都允许中性、带正电或带负电的介子存在。从这些包含介子场与重核子之间相互作用的波动方程出发，人们只能通过采用通常的量子力学微扰理论（仅考虑一阶不为零的近似值）来估算核力的最终表达式，尽管这种方法具有众所周知的非收敛性。因此对该结果的可靠性进行进一步的讨论就显得十分必要，为此有人提出也许可以参照量子电动力学对类似情况的处理方法，即用一个适当的正则变换使我们可从由若干电子和一个电磁场组成的系统的总能量表达式中将只与电子坐标有关的项与库仑势能项分开。

其实，同样的方法[3]也适用于核子和介子场组成的体系。对于这样的一个体系，实际上有可能找到一些正则变换使核子之间的“静态”相互作用被分离出来，其中静态相互作用指的是相互作用中忽略了表征重粒子位置、自旋以及质子或中子状态的一些变量随时间变化的部分。这种静态相互作用在任何情况下都与微扰理论中的一级近似完全一致，而且，对于两个相距足以使静态相互作用远大于非静态相互作

a first approximation in the perturbation method, and there exists a lower limit, smaller but unfortunately not much smaller, than the range of the nuclear forces, to the mutual distances between two heavy particles for which the static interaction is more important than the additional non-static contributions arising from the terms, in the transformed Hamiltonian, which describe the remaining interactions between the heavy particles and the meson field.

Although no improvement can, of course, be obtained in this way as regards the self-energy difficulties, it would seem that consistent results can be derived from the transformed Hamiltonian by considering only the last-mentioned interactions as a perturbation and applying a method of treatment analogous to the correspondence methods used in electrodynamics. It is especially to be noted that if, following Yukawa, we also introduce an interaction between the meson field and electrons and neutrinos, the transformed Hamiltonian contains a term which represents a direct interaction between the heavy particles and the electrons and neutrinos, and which, when treated as a small perturbation, immediately gives the probabilities of β-disintegration processes. It is perhaps to be regarded as a satisfactory feature of the point of view just outlined that, contrary to previous treatments, where the nuclear forces came out in the same stage of the perturbation method as the probabilities of β-decay, account is here taken at the outset of at least the static part of the nuclear forces.

As regards the form of these static interactions, it is well known that the type of potential resulting from the four-vector meson field generally considered hitherto has the defect of including a term of dipole interaction energy which is so strongly singular for infinitely small mutual distances of the nuclear particles that it would not in general allow the existence of stationary states for a system of such particles. In order to remedy this defect, it seems necessary[2] to introduce besides the four-vector wave-function a further pseudoscalar wave-function for the meson field which has the property of giving rise to a static interaction of a form just capable of cancelling the dipole interaction term without affecting the others. The consideration of such a pseudoscalar meson field would also seem to be useful from the point of view of the theory of β-decay. While, for example, the four-vector theory yields[4] exactly the same form of the β-spectrum as Fermi's original theory, the introduction of a pseudoscalar wave-function in addition to the four-vector one gives rise to a modification of the energy distribution of the β-rays which seems to open a new possibility of a better adaptation to the experimental results.

A detailed account of our work will appear shortly in the *Proceedings of the Copenhagen Academy*.

(**143**, 241-242; 1939)

C. Møller and L. Rosenfeld: Institute for Theoretical Physics, Copenhagen, Jan. 6.

用项的重粒子，它们之间的距离存在一个下限，然而不幸的是，这个下限与核力力程相比虽小但没有小很多。所谓非静态相互作用项，指的是在变换后的哈密顿量中那些用来描述重粒子与介子场之间剩余作用的项。

尽管这种方法并没有解决自能的困难，但是我们在仅考虑最后提到的剩余相互作用项作为微扰并应用与电动力学中类似的处理方法时，可以从变换的哈密顿量中得到自洽的结果。特别值得注意的是，如果按照汤川秀树的理论，我们也在介子场、电子和中微子之间引入一种相互作用，则变换后的哈密顿量就含有描述重粒子和电子及中微子之间直接作用的一项，当我们把它视为小微扰的时候，就可以立即给出发生 β 衰变的概率。对于上述理论来说，一开始就考虑了核力中的静态作用部分，这或许可以被认为是一个令人满意的构想，但这与以前的处理方法相反，之前核力是与 β 衰变概率在微扰法中的同一步中出现。

迄今为止，人们普遍认为，这些静态相互作用的形式来自四矢量介子场的这种类型的势存在一个缺陷，就是包含一个偶极相互作用能项，当核子之间的距离趋于无穷小时，它如此奇异以至于在一般情况下不能允许这种粒子的系统存在定态。为了弥补这个缺陷，看来除了四矢量波函数以外，还有必要[2]把一个赝标量波函数引入介子场，这个波函数的作用是引出一个静态相互作用，其作用刚好能够抵消偶极相互作用项而又不影响其他项。从 β 衰变理论的角度来看，考虑这种赝标量介子场似乎也是有意义的。举例来说，由四矢量理论得到[4]的 β 光谱与早先费米理论中的 β 光谱形式完全相同，把赝标量波函数叠加到四矢量波函数上后，β 射线的能量分布将被修正，这为与实验结果更好地吻合提供了新的可能。

我们的详细工作报告很快将发表在《哥本哈根学会学报》上。

（胡雪兰 翻译；厉光烈 审稿）

References:

1. See Bhabha, *Nature* (in the Press).
2. Kemmer, *Proc. Roy. Soc.*, A, **166**, 127 (1938); *Proc. Camb. Phil. Soc.*, **34**, 354 (1938).
3. Independently of our work, essentially the same method has been proposed by Stückelberg (*Phys.Rev.*, **54**, 889; 1938), to whom we are very thankful for the kind communication of his manuscript.
4. Yukawa, Sakata, Kobayasi, Taketani, *Proc. Phys. Math. Soc. Japan*, **20**, 720 (1938).

Surface Transport in Liquid Helium II

J. G. Daunt and K. Mendelssohn

Editor's Note

Experimenters had recently discovered some strange behaviours in liquid helium—in particular, its seeming ability to flow with no viscosity. Here physicists John Daunt and Kurt Mendelssohn of the Clarendon Laboratory in Oxford report a further odd effect. They had placed a container, open at the top and holding liquid helium, into a bath of the same liquid. Applying heat inside the vessel, they had found that a thin film of liquid flowed up and over the vessel's lip, linking the two otherwise separate fluids. The authors suggest that the phenomenon may be linked to the recently noted "fountain effect", in which the heat from a weak light beam can expel a jet of liquid helium from a container.

IN our previous communications[1] on the "transfer" of liquid helium II by a surface film above the liquid level, we stressed the similarity of this "transfer" with the so-called transport phenomena in the free liquid, and suggested that the latter might be due to a process similar to the transfer above the liquid level. Later experiments on the formation of this film[2] seemed further to strengthen the conception that there existed a similar surface transport below the liquid level, and we have recently made two observations which seem to corroborate this hypothesis:

(*a*) A small Dewar vessel (see Fig. 1) containing a heating coil was suspended by a thread in a bath of liquid helium II. When no heat was supplied, the levels of the liquid both inside and outside the vessel adjusted themselves to the same height L_1, owing to the "transfer" through the film on the interconnecting glass surface. When, however, a current was passed through the heating coil, the level of the liquid inside the vessel *rose* above the outside level and took up an equilibrium position L_2. By increasing the connecting surface between the vessel and bath by a number of wires, differences between inside and outside levels up to 5 mm. could be obtained. This clearly shows that there exists a "transfer" of helium from a colder to a hotter place when a temperature gradient is imposed. On further increasing the heat supplied, however, the evaporation from the vessel became the predominant factor and the inside level fell below that of the bath.

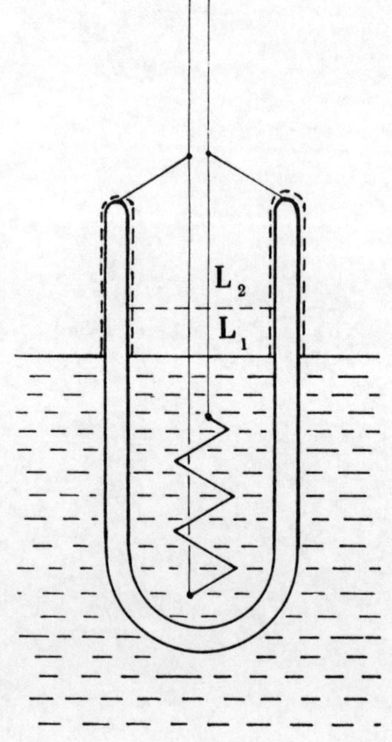

Fig. 1

液氦 II 中的表面传输

当特，门德尔松

编者按

最近的实验研究发现了液氦的一些奇特的行为，特别是它似乎具有完全无黏滞流动的能力。而在这篇文章中，牛津克拉伦登实验室的物理学家约翰·当特和库尔特·门德尔松报道了一个更为古怪的效应。他们将一个盛有液氦的上端开口的容器放置于液氦池中。在容器内加热时，他们发现有一薄层液膜向上流动，越过容器的边缘将两部分分开的液体连接在一起。本文作者认为这个现象可能与当时刚刚为人所知的“喷泉效应”有关，在“喷泉效应”中，微弱的光束产生的热量就可以使液氦从容器中喷射而出。

之前关于液氦 II 通过液面上的表面膜进行“迁移”的讨论中[1]，我们强调这种“迁移”与所谓自由液体中的传输现象非常相似，并指出后者可能是由一个类似于在液面上迁移的过程引起的。后来关于表面膜形成的一些实验[2]似乎进一步证实了这种说法，即在液面下存在着类似的表面传输，而且我们最近做的两项观察实验似乎也证实了这一假设：

(*a*) 用一根线系着一个装有加热线圈的小杜瓦瓶（如图 1）浸没在液氦 II 中。当不加热的时候，杜瓦瓶内外的液面通过自我调节达到相同的高度 L_1，这是由于液氦 II 以膜的形式在相互连接的玻璃表面“迁移”造成的。然而，当给加热线圈通以电流的时候，杜瓦瓶内的液面会**上升**到高于瓶外的液面并稳定在高度 L_2。如果用一些金属丝来增加杜瓦瓶与池中液氦的接触面积，瓶内外的液面高度差就可以上升到 5 毫米。这清楚地表明，当存在温度差的时候，氦会从温度较低的地方“迁移”到温度较高的地方。然而，当进一步加热时，瓶内的蒸发就成为主导因素，因此，杜瓦瓶内的液面会下降到低于池中的液面高度。

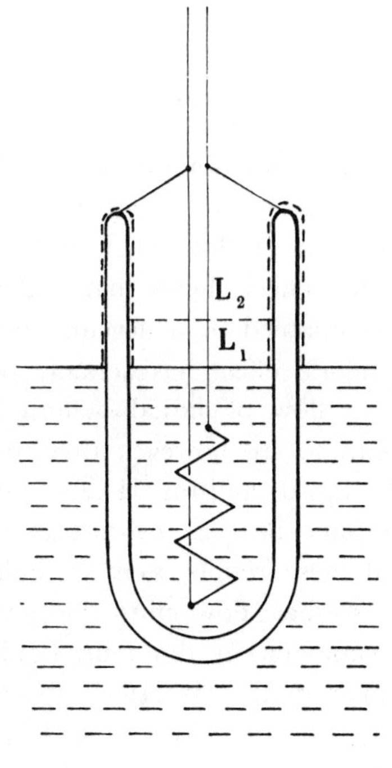

图 1

This effect is quite analogous to the "fountain phenomenon" in the bulk liquid, discovered by Allen and Jones[3]. However, in the present case the transfer of liquid must be carried out through the surface film above the liquid level, which shows that there exists a flow of helium against a temperature gradient, even if the two containers are not connected by free liquid. One may conclude therefore that the "fountain phenomenon" in the bulk liquid is probably also due to a surface transfer, though in this case along the surface below the liquid level. This hypothesis is further strengthened by the fact that the "fountain phenomenon" is more pronounced when tubes containing fine powder are used to connect the two volumes of liquid rather than a straight capillary, for which the available surface is comparatively small[4].

(*b*) A Dewar vessel (see Fig. 2) was closed at the top and had a hole at the lower end which was constricted by a plug, *P*, of fine emery powder. It contained a phosphor-bronze thermometer, *T*, and was suspended in a bath of liquid helium II. When the Dewar vessel was lifted out of the bath, the liquid ran rapidly out of the vessel through *P* and fell into the bath, and at the same time the temperature of the inside liquid was noticed to rise by about 0.01°. On lowering the vessel so that now liquid ran from the bath into the vessel, the liquid inside was cooled by a similar amount.

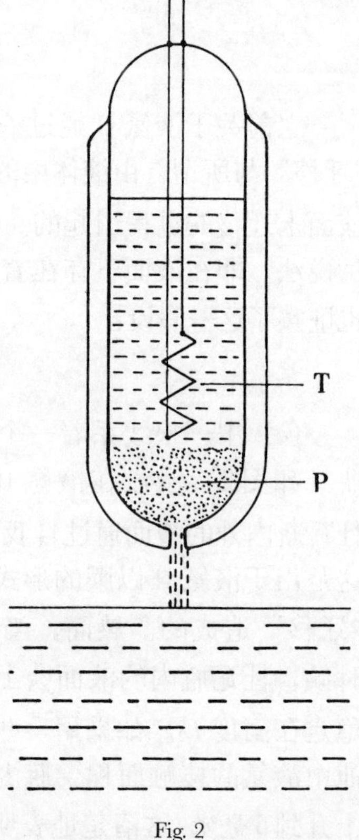

Fig. 2

This mechano–caloric effect is evidently the reverse of the "fountain phenomenon", for whereas the latter means that the setting up of a temperature difference results in a flow of liquid helium II, the mechano–caloric effect shows that a flow of liquid helium II is accompanied by a development of heat (or cold). Such a caloric effect has actually been postulated by Tisza[5] for a flow of liquid helium II through capillaries. It seems to us, however, that the anomalous phenomena of liquid helium II are not so much caused by capillary flow as by a transport along solid surfaces; and these results seem to indicate that the heat content of those atoms transported by surface flow must be lower than average. The hypothesis that the transport phenomena in the bulk liquid are due (at least primarily) to a surface transport similar to the "transfer" above the liquid level seems also to agree with observations by Allen and Misener[6] and H. London[7], as well as with theoretical considerations of F. London[8].

这一效应与艾伦和琼斯[3]在体相液体中发现的“喷泉现象”极为相似。然而，在我们的实验中，液体迁移必须通过液面上的表面膜才能实现，这说明：即使两个容器之间没有自由液体相连通，也会存在与温度梯度方向相反的液氦流。有人也许因此得出结论：瓶内液体中的“喷泉现象”可能也是由于表面迁移而产生的，尽管在这种情况下迁移处于液面之下。下面的事实则进一步证实了这一假设：当用装有细粉的管子连接两个容器中的液体时，产生的“喷泉现象”比用直毛细管时更加明显，因为毛细管可进行迁移的表面积相对较小[4]。

(*b*) 一个顶部封闭的杜瓦瓶（如图 2）底部有一个洞，洞口上方是由细金刚砂粉形成的填塞物 *P*。瓶中还有一个由磷青铜制成的温度计 *T*，它被悬挂在液氦 II 池中。当杜瓦瓶被升高到液氦 II 池面以上时，瓶内的液体很快地通过 *P* 流入液氦 II 池中，同时，瓶内液体的温度升高约 0.01 度。当降低杜瓦瓶使液体得以从液氦 II 池流入瓶中的时候，瓶内液体的温度也降低了大体相同的度数。

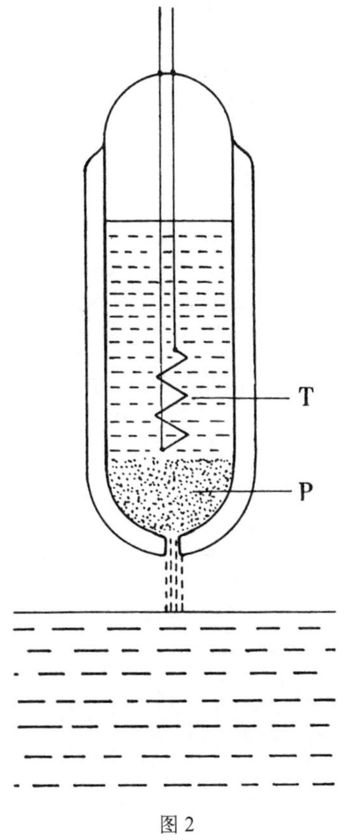

图 2

这一机械 – 热效应显然与“喷泉现象”正好相反，因为后者说明温度差导致了液氦 II 流的形成，而机械 – 热效应则显示液氦 II 流的形成伴随着温度的变化。实际上，蒂萨[5]根据通过毛细管的液氦 II 流曾提出过关于这一热效应的假设。不过，在我们看来，与其说液氦 II 流的异常现象是由毛细管流引起的，不如说是由固体表面传输造成的；这些结果似乎表明，那些在表面传输的原子的热容肯定低于平均值。体相液体中的传输现象是(至少主要是)由一种类似于液面上的“迁移”的表面传输而引起的，这一假设似乎与艾伦和麦色纳[6]以及伦敦[7]的观察结果相一致，同时也与伦敦[8]的理论研究相吻合。

A more detailed discussion of this tentative explanation with regard to these and other results will be given elsewhere.

(**143**, 719-720; 1939)

J. G. Daunt and K. Mendelssohn: Clarendon Laboratory, Oxford, March 21.

References:

1. Daunt and Mendelssohn, *Nature*, **141**, 911; and **142**, 475 (1938) and *Proc. Roy. Soc.*, in the Press.

2. To be published shortly.

3. Allen and Jones, *Nature*, **141**, 243 (1938).

4. Allen and Reekie, *Proc. Camb. Phil. Soc.*, **35**, 114 (1939).

5. Tisza, *Nature*, **141**, 913 (1938); *C.R.*, **207**, 1186 (1938).

6. Allen and Misener, *Nature*, **142**, 643 (1938); see also Allen and Jones, *Nature*, **143**, 227 (1939).

7. London, H., *Nature*, **142**, 612 (1938).

8. London, F., *Phys. Rev.*, **54**, 947 (1938).

有关上述以及其他一些研究成果的尝试性解释，我们将在别处进行更为详细的讨论。

（李世媛 翻译；于渌 审稿）